FARM
MANAGEMENT

FARM
MANAGEMENT

Michael D. Boehlje
Iowa State University

Vernon R. Eidman
University of Minnesota

John Wiley & Sons
New York   Chichester
Brisbane   Toronto   Singapore

Photo Credits

Chapter 1 Opener: Grant Heilman.
Chapter 2 Opener: Charlton Photos.
Chapter 3 Opener: Grant Heilman.
Chapter 4 Opener: Peter Menzel/Stock, Boston.
Chapter 5 & 6 Openers: Grant Heilman.
Chapter 7 Opener: Daniel S. Brody/Stock, Boston.
Chapter 8 Opener: David S. Strickler/The Picture Cube.
Chapter 9 Opener: Thomas Hovland/Grant Heilman.
Chapter 10, 11 & 12 Openers: Grant Heilman.
Chapter 13 Opener: Daniel S. Brody/Stock, Boston.
Chapter 14 Opener: Larry Rana/USDA.
Chapter 15 Opener: USDA.
Chapter 16 Opener: Courtesy Digital Equipment Corp.
Chapter 17 Opener: David Burnett/Contact Press Images.
Chapter 18 Opener: Barbara Alper/Stock, Boston.

Cover photo: Copyright Barrie Rokeach 1983.

Library of Congress Cataloging in Publication Data:

Boehlje, Michael.
    Farm management.

    Includes index.
    1. Farm management. I. Eidman, Vernon R., 1936–
II. Title.
S561.B557 1984   630'.68   83–23589
ISBN 0–471–04688–4
Printed in the United States of America
10 9 8 7 6 5 4 3 2 1

800 225–5945
One Wiley Dr.
Somerset, NJ
08875

PREFACE

The field of agriculture has witnessed many dramatic changes in the last two decades, and these changes have demanded a different approach to management of the farm business. No longer can the successful farmer have primarily a production and technology orientation; she or he must understand and skillfully apply management concepts in the areas of marketing and financial management as well. Management involves a broader set of activities than just developing plans and choosing acceptable production, marketing, and financial activities for the farm firm; it involves implementing those choices and plans once a decision is made, and then monitoring actual performance and comparing that performance to planned expectations. Farm management then is the application of planning, implementation, and control concepts to the activities of production, marketing, and finance. This integrated approach to farm management is the unique focus of this textbook.

Our objective in writing this book was to develop a comprehensive, integrated, and analytical—yet readable—discussion of the principles and procedures used in successful farm management. After a number of years of teaching experience using the traditional "production theory" approach, we felt that a broader framework was necessary—a management framework rather than the narrower focus of production theory. Our approach was not to reject or even deemphasize production theory, but to integrate it into a broader management framework—to recognize that farming is a unique business with a unique environment and set of problems, but a business nonetheless that can and should be managed through the application of business management principles and concepts.

The book is organized into four parts. The first part, "Introduction" (Chapters 1 and 2), introduces the basic concepts and principles of management, as applied to farm firms, and the accounting statements and information needed to evaluate the financial performance of a farm business. The second part, "Planning" (Chapters 3 through 11), focuses on the planning function of management. First, the concepts of production theory, enterprise budgets, and whole-farm budgeting are used to develop the production plan for the farm business. Then, the principles needed to develop the plan for buying inputs and selling products (the market plan) are discussed. Next, financial planning with a focus on investment analysis and tax management is reviewed. The final three chapters of this part develop the concepts needed to plan the farm business arrangement, to use linear programming procedures in farm planning, and to plan in an environment of risk and uncertainty.

The third part of the book, "Implementation" (Chapters 12 through 15), discusses the implementation function of management. The acquisition and management of the labor resource are discussed first, followed by a presentation of the concepts needed to efficiently acquire and manage the land resource. Acquiring and managing depreciable assets (buildings, equipment, machinery, and breeding stock) are then discussed, including the issue of replacement policies for depreciable items. Finally, this part is completed with a review of capital acquisition and management, with particular emphasis on debt utilization and the institutions that provide funds to farmers.

The final part of the book (Chapters 16 through 18) focuses on a management topic frequently underemphasized in farm management—"Control." The basic principles of control are developed, including systems concepts and the interface between the control function and the planning and implementation functions. These principles are then used to develop control systems for the production activities of the farm business. Finally, the control concepts are applied to the marketing and financing activities of the business.

The comprehensive, integrative nature of this book implies that it can be used for both introductory and advanced courses in farm management. The book is probably best adopted for a two-semester sequence at the junior or senior level; such a course sequence would enable one to focus on the planning function in the first semester and the implementation and control function in the second semester. An introductory farm management course with a traditional production-economics focus could utilize Chapters 1 through 6 on accounting information and production planning, and Chapters 12 through 15 on the acquisition and use of labor, land, depreciable assets, and capital. An advanced course building on this more traditional coverage might include a review of the first six chapters on production planning and concentrate on market and financial planning, the more advanced planning topics, and control, as presented in the last three chapters of the book.

As with any book, the final product is an amalgamation of concepts and ideas, many of which have been presented elsewhere. The originality of our presentation results from combining diverse ideas with our own experiences in a cohesive and understandable fashion. Students in farm management and finance courses at Oklahoma State University, the University of Minnesota, and Iowa State University have challenged our thinking and explanations over the years, encouraging corrections, refinements, and clarifications. Colleagues at these three institutions have also provided insightful criticisms and suggestions. In particular, we acknowledge and appreciate the comments and suggestions provided by Kenneth Thomas, Jerry Thompson, Earl Fuller, Jeffery Apland, and Joe Williams. And, finally, we thank our reviewers—Richard Perrin, North Carolina State University; George L. Casler, Cornell University; Myles J. Watts, Montana State University; and Joseph E. Williams, Oklahoma State University—who provided useful comments on various drafts of the manuscript. Obviously, the responsibility for any errors that remain is ours.

Michael D. Boehlje
Vernon R. Eidman

Contents

IMPLEMENTATION **495**

CHAPTER 12   LABOR ACQUISITION AND MANAGEMENT **496**

CHAPTER 13   LAND ACQUISITION AND MANAGEMENT **526**

CHAPTER 14   ACQUISITION AND MANAGEMENT OF DEPRECIABLE ASSETS **572**

INTRODUCTION

THE SCOPE OF FARM MANAGEMENT

CHAPTER CONTENTS

This book is concerned with the management of business firms producing primary agricultural products. U.S. and world agriculture is characterized by a wide range in the size and type of primary producing units. Some farms are owned by groups of individuals and hire most of the labor and management required. However, the great majority of farms in the United States are referred to as family farms.[1] Family farms serve as a residence

[1]Although the definition of a family farm is somewhat arbitrary, the essential characteristics relate to the contributions of labor and management. A family farm can be defined as a primary agricultural business in which the operator is the risk-taking manager and with his family performs most of the farm work and most of the managerial activities (Nikolitch, p. 1). Family farms account for 95 percent of all farms in the United States and about 60 percent of all farm products sold. As farms decrease in number and increase in size, the proportion of the total classified as family and other than family farms has remained approximately the same (Nikolitch, Schertz).

for the farm family as well as a place of business. In addition, many of the family's leisure-time activities revolve around the farm, making it difficult to separate the "business" and "way of life" dimensions of farming on the typical family farm. One meaningful way to make the distinction between these two aspects of farm life is to distinguish between the production (business) activities of farming and the consumption (way of life) activities.

In a broad sense, the production activities of farming involve those tasks required to produce the crop and livestock products and to generate the highest level of income in the most efficient manner. To perform these production activities most efficiently, the farmer must be acquainted with the basic functions of management; the principles and concepts of production economics; financing methods and investment analysis; the agronomic, biological, genetic, and engineering relationships that determine crop and livestock growth and production; the organization and operating procedures of the institutions that service or influence the farming operation; the legal regulations on farm business activities; and concepts of marketing and price determination. Farmers as business managers must incorporate price, input-output, and resource availability information along with appropriate analysis procedures to determine the most economically efficient production, marketing, and financial plan. Then they must carry out this plan in the most efficient manner and evaluate the consequences so that changes can be made to improve efficiency in the future.

Some farm decisions may not contribute to the efficient production of farm products. Although these decisions may appear to be irrational from a business or production point of view, they may be motivated by the desire of farmers to maximize their utility from the consumption as well as production activities of farming. These consumption activities may include cleaning up the farmstead and keeping the fence rows clipped, purchasing a bigger or newer tractor than is actually needed to accomplish the tillage operations, or not going into debt to purchase a farm. Such decisions are influenced by nonbusiness or consumption motivations as well as business or efficiency motivations.

This discussion of farm management will concentrate on the business activities involved in farming or ranching and the analysis of those activities with emphasis on efficiency. Such an approach is generally applicable to farms of various sizes and types in different areas of the United States and the world. The concepts of farming as a business and farming as a way of life are integrated where appropriate to make the discussion relevant to family as well as other than family farms. It should be noted that on most farms the purchase of consumption goods and services and the use of leisure time enter the operator's utility or preference function when he or she is viewed as a consumer. One of the extremely difficult tasks faced by the farmer is to develop a balance between the production (business) and consumption (way of life) activities. Thus, it is appropriate to consider some of the goals farm operators have and briefly discuss how the relative importance of these goals may vary from one farmer to another as well as how their relative importance may change for one individual over time.

GOALS OF THE FARMER

Economic analyses frequently assume that businesspeople operate as if they have only one goal—profit maximization. In fact, businesspeople—including farmers—possess numerous goals. Some of the more important personal goals from the standpoint of business decisions are listed below. The importance an individual attaches to one goal compared to another depends on that person's current financial situation, current and future financial needs, and set of values. Although a thorough examination of farm managers' needs, psychological motivations, and goals is beyond the scope of this discussion, a general understanding of these goals provides a useful starting point for the study of farm management.

Maximize Profit or Return

Obtaining the largest net return possible is frequently identified as the primary goal of most farmers. To achieve this goal, the farmer must choose that combination of crop and livestock enterprises where the marginal returns are equated to the marginal costs for all alternative enterprises. Economic concepts of the theory of the firm are particularly useful in guiding farm managers to accomplish this goal.

Both the measure of return to be used and the period of time over which income is to be maximized must be understood to evaluate the importance of this goal. Either annual net farm income, a measure of the return to the operator's land, labor, capital, and management, or an after-tax measure of net farm income is typically used. The concept of annual net farm income considers both cash and noncash sources of income and calculates returns to the operator's owned resources. (Typical sources of noncash income are inventory changes and the value of farm-produced products consumed in the home.) Annual net farm income is defined in detail in the discussion of farm accounting information in Chapter 2. That chapter also describes measures of returns which include not only annual net income, but also the change in the value of capital assets, particularly real estate. In recent years the return in farming has consisted of an increasingly larger proportion of capital appreciation and a smaller proportion of annual farm income. Thus, measures of returns that include both annual income and asset appreciation are important in many situations. Maximizing net income or return over a period of several years is usually more desirable than achieving a high income in one year at the expense of other years.

Numerous management specialists argue that farmers as well as other businesspeople really do not attempt to maximize income, but try to obtain at least a minimum income level. This behavior is frequently called satisfying behavior. An example of satisfying behavior is characterized by the statement, "If I can make $20,000 this year, I'll be happy." This behavior is exhibited by many farmers who, although they are capable and have an adequate resource base to grow, decide to maintain their current farm size rather than expand and increase their management responsibilities along with their income.

Contrary to common belief, satisfying behavior is not inconsistent with the basic principles of economic theory and maximization. Instead, a maximizing criterion is replaced with the objective of obtaining at least a minimum level of a particular goal. Microeconomic concepts concerning efficient factor use and product output are still applicable if a satisfying criterion is used.

Increase Net Worth

Over a period of years, the goal of increasing net worth is a major motivating objective for many farmers. This goal may not only influence the investment and production behavior of farmers as evidenced by how they operate and manage the farm firm, but it also may influence their consumption behavior. Thus, farmers may ask their families to sacrifice some of the luxuries of life, such as a new car every year and an expensive house, so that additional funds can be reinvested in the farm as a means of increasing productive capacity and farm net worth. The goal of increasing net worth may also be evidenced by the desire of a farmer to build a sizable farm operation which then can be transferred to a son or daughter who wants to farm. A further component of this motivation to increase net worth may be the desire to provide the children or grandchildren with an education and the financial support for a start in a business or profession. As we will discuss later, when investment analyses and the time value of money are considered, the maximization of net worth may not be consistent with the maximization of annual net income.

Control a Larger Business

Behavior exhibited by some farmers suggests that an important goal is to control a larger farm unit measured in terms of acres, number of dairy cows, or some other unit—regardless of the profitability of the larger operation. Some farmers feel that controlling a larger unit is a way of spreading their overhead and reducing average production costs. In many instances, it also is a means of obtaining recognition and prestige in the community.

Avoid Low Returns or Losses

Farming is a risky business, and farmers must frequently have a low aversion to risk. Like other businesspeople, however, most farmers will not consciously make business decisions that will endanger their chances of survival. Thus, risk reduction and the desire to maintain a viable farm business are important goals of many farm families. The desire to reduce risk so as to remain in the farming profession and to have a relatively stable income frequently results in business decisions emphasizing diversification, maintaining financial reserves, hedging, and other actions that may not maximize income but may reduce possible losses.

Reduce Borrowing Needs

Although most farmers recognize the productivity of and need for borrowed capital, the economic environment of the past, particularly the Depression years, has made many older farmers wary of using credit. The combination of high interest rates and low prices for many commodities during the early 1980s may have tempered credit use by many current-day farmers. Rather than expand their operation to increase its efficiency when cash is available, some farmers pay off loans rapidly to reduce their debt load. Although this may appear to be irrational behavior, if debt aversion is an important component of a particular farmer's utility function, this behavior cannot be unduly criticized.

Increase Family Living

This goal might embody the most specific acknowledgment of the consumption motivation of the farm family. In fact, it might be argued that the sole purpose of any attempt to increase income or net worth through production efficiency is really to use that income or wealth for the benefit of the family and for consumption purposes. It must also be recognized, however, that an increased level of current consumption by the family results in less income for reinvestment, a slower rate of expansion for the firm, and a lower level of income in the future which can be used for consumption goods. The trade-off between current consumption and reinvestment—which is a means of obtaining higher future consumption—provides evidence of the conflict between various goals of the farm family.

Increase Leisure Time

Leisure time is another component of the consumption function of the farm family. It can be a major goal for young farmers with families as well as for older farmers who want to enjoy the benefits of their many years of hard work and management. Many farmers are reluctant to express leisure time as a primary motivation for their activities, possibly because this goal is so alien to the Protestant work ethic which is predominant in rural areas. In defense of leisure time as an important goal of all farm families, it should be noted that social psychologists who have studied the management abilities of upper- and middle-level management personnel in industrial firms indicate that the ability to relax and the regenerative processes of leisure time can substantially increase the efficiency of performing the management functions. An appropriate amount of leisure time may, in fact, increase the efficiency and income of the farm operator rather than reduce it.

Have a Neat and Well-Kept Farmstead

Although keeping a neat farmstead may require resources that could be used in income-generation activities, the appearance of the farmstead is frequently an important social indication of a prosperous farm as well as a respected farmer. In addition, it is frequently argued that farmers who take pride in their possessions and attempt to keep them as attractive as possible also utilize the same diligence in making decisions to improve the efficiency and income of the farm operation. Thus, the desire to have a neat farmstead may not be inconsistent with that of economic efficiency. It may also be an important component of being accepted and respected in the community.

Provide Community Service

Many farmers devote themselves to their farm operation while others in their profession are actively involved in community affairs such as the school board, the church, and farm organizations. The desire to serve the community and to make it a "better place to live" may actually require sacrifices in farm income. Again this attitude should not be considered as evidence of irrational behavior, but only as an indication that the utility function of a farmer is multidimensional and that different farmers have different priorities.

As we will discuss shortly, the goals of the farmer are not static in nature. The relative importance of various goals in the farmer's objective function and decision process is influenced by changes in his/her wealth, family characteristics, and age. Thus, the farmer's goals are expected to change as the operator and the farm firm pass through different stages of the family-firm life cycle.

THE FAMILY-FIRM LIFE CYCLE

In the agricultural sector, which is characterized by the individual entrepreneur rather than a management team, the firm frequently exhibits a life cycle that parallels the life cycle of the farmer-entrepreneur. The farmer and the firm will pass through at least three stages during the operator's farming career.

The first stage is the *entry or establishment stage.* In this stage, the prospective farmer evaluates the opportunities in farming compared to other occupational alternatives and determines whether or not to enter the industry. This evaluation includes an analysis of the income potential in farming compared to nonfarm employment, the opportunities to accumulate net worth, the kind of life-style, the social and community involvements of farmers, the work and leisure activities associated with farming, the ability to handle the physical labor as well as the mental pressures of being a farmer-businessperson, and the opportunities to attain the necessary land and capital resources that are essential to have an economically viable farm unit. An individual who decides to accept the challenge of starting a farm business must then acquire the "critical mass" of capital resources and managerial ability which is necessary to establish a viable economic unit—a farm business that will generate a competitive income and have the capacity to grow.

Historically, a substantial number of new entrants moved into agriculture via the "agricultural ladder." Potential entrants began their career as hired laborers and through diligent work and wise spending accumulated sufficient funds to purchase a set of machinery. Subsequently, the new entrant first became a renter, then a part-owner of real estate, and finally, reached the pinnacle of success with full ownership of land as well as livestock and machinery. Although the process required family sacrifices, the resource requirements were sufficiently modest that this procedure could work successfully for a diligent worker. With the substitution of capital for labor, the rapid price increase in durable resources (particularly land and machinery), and the expanding capital requirements of the economically viable farm firm, the "agricultural ladder" may no longer be a viable source of new entrants. Not only is it virtually impossible to acquire sufficient capital resources through this historically successful procedure, but it also fails to provide the entrepreneurial training that is so important for successful new entrants into agriculture.

Substitutes for the "agricultural ladder" as a source of new entrants and as a means of acquiring the critical mass of resources necessary to meet family living requirements and have a reasonable chance to survive have not been well identified. Research indicates that most new entrants during the 1960s and 1970s used family help to establish their business. Although there are many alternative methods of using the financial backing and managerial experience of another family member to establish the farm business, three general approaches appear to be the most common. One approach is for an established farmer to take the new entrant (historically, a son, son-in-law, brother, or brother-in-law) into the business as a partner and expand the operation to an economic size for both families. A second approach is for the established farmer to provide the financial backing either through a loan or a guarantee (co-signature) of the new entrant's credit line to purchase a farm and equip it. A third approach is for the new entrant to be employed as a "working farm manager" to gain experience. The entrant can rent or buy land when it becomes available in the neighborhood and farm it using some or all of the established farmer's machinery. As the new entrant accumulates capital, the individual can "buy into" the current operation or move to a farm of his or her own.

As these three methods suggest, acquiring credit and a land base are prerequisite to becoming established in farming. Alternative strategies to acquire the critical mass of resources, such as renting, leasing, and installment sales contracts, are all a part of this entry stage.

The second stage in the family-firm life cycle can be identified as a *stage of growth and survival.*[2] During this stage, the farmer-entrepreneur attempts to expand the resource base by acquiring the services of additional inputs through purchase or lease. New techniques of production are evaluated as to their efficiency and profitability as well as their ability to increase production at reduced costs. Methods of improving labor productivity as the ratio of capital to labor increases are also adapted. Strategies to increase the volume of production through intensification with livestock or extensive expansion through acquiring a larger land base are also evaluated and implemented. The capital as well as labor requirements of the typical farm firm expand rapidly during this stage in the life cycle, resulting in continued utilization of debt as well as equity sources of funds. Thus, analysis of alternative sources of credit and evaluation of repayment ability are important issues to be faced during the growth and survival stage. In addition, a major consideration during this stage is maintaining a debt-equity structure that will guarantee survival during years of low income due to weather, disease, or low product prices. This consideration may require the maintenance of credit reserves, the acquisition of various types of production, income protection, and liability insurance policies, and the use of diversification strategies in production. In the later years of the growth and survival stage, emphasis may shift from expansion to consolidation of gains, reduction of costs, and stabilization of income.

The third stage in the family-firm life cycle is the *exit* or *disinvestment stage.* Two major processes are involved in this stage: the process of retirement and the intergeneration transfer of property. During retirement, the farmer attempts to reduce his/her management responsibilities while maintaining sufficient control of farm assets to generate adequate retirement income. Simultaneously, estate plans that implement lifetime or death transfers of farm property and the managerial responsibility associated with that property to the next generation are developed.

Unfortunately, few farmers plan for their retirement years. They give little consideration to such issues as the goals for the retirement period and how to accomplish these goals, retirement income needs, the sources of retirement income, the tax and social security problems of retirement, or the problems of renting or selling the farm. For some farmer-entrepre-

---

[2]An alternative term that has been used to identify this stage is *expansion and consolidation.* Regardless of the nomenclature used, the issues and objectives in this stage are the same.

neurs, successful retirement may include a contribution to society through holding public office or becoming active in civic affairs. Other retired farmers may be gainfully employed in the farm input supply or product merchandising sector of the agricultural economy.

Those farmers who desire to retire from the pressures of continued employment must evaluate numerous pre- and post-retirement strategies that may generate the income needed in the retirement years. Among these strategies are the investment of funds in insurance and/or annuity policies, mutual funds, bonds, and self-employed retirement income plans. After retirement, the income that is forthcoming from social security and various retirement funds, as well as rental income from the farm, and the sale of some farm assets will be major determinants of the consumption level and standard of living. Planning to meet retirement income needs is an important area of financial management for farmers in the exit stage.

Most farmers give as little consideration to the problems and process of transferring the firm and estate to future generations as they do to retirement. Substantial economic losses can occur if the proper strategy is not used to transfer a large estate from a retiring farmer to his/her heirs. These potential losses are attributable to estate, inheritance and gift taxes; liquidation losses and reduction in size economies; and legal and management fees incurred in the process of transferring property between generations. In addition, inadequate planning may result in family arguments and other noneconomic problems.

Recognition of estate planning as a legitimate component of farm management enables the farmer to consider the interrelationships between the creation of an estate and the transfer of that estate to future generations. Farmers must, therefore, not only determine the proper use of lifetime gifts, trust arrangements, wills, and other appropriate estate planning tools and techniques, but they must also evaluate the impact of investing in insurance, annuities, real estate, and other assets with respect to their implications for transfer of their estate to the heirs.

As suggested earlier, farmers possess multiple goals. In addition, the emphasis on various goals usually changes during the farmer's lifetime. For example, during the entry and establishment years, a farmer may place a high priority on income maximization and the opportunity for growth and expansion. Lower priorities usually are placed on the goals of leisure time, community involvement, and risk aversion during this initial stage. During the growth and survival stage, increased emphasis may be placed on risk aversion, and the farmer may want to spend more time with his or her family. He or she may also become involved in community activities to a greater extent. During the exit or disinvestment stage of the life cycle, income maximization typically becomes a low priority goal compared to security and risk aversion, leisure, and productive use of one's time and talents. Thus, it is not surprising to find many farmers following quite different investment, production, and marketing patterns because they are in different stages of the family-firm life cycle. An investigation of the changes in their goals and objectives from one stage in the life cycle to another typically explains the pattern of decisions.

One of the major issues that exists in agriculture, as well as any industry dominated by the sole proprietorship organization structure, is that of the efficiency of the firm over the family-firm life cycle. A related issue is the opportunity for new entrants to enter the industry. Although the family farm may be highly efficient during the prime of the farmer-entrepreneur's life, inefficiencies may exist during the entry and exit stages. At the entry stage, the limited size of many farm firms makes it diffucult, if not impossible, for the farmer to take advantage of the economies of size that may be available with larger units. During the exit stage, many farmers are attempting to reduce their commitment to farming, and income maximization is a lower priority goal for them. In addition, as indicated earlier, at the death of the farmer and the transfer of property to the heirs, a significant reduction in efficiency can occur because the new entrant frequently does not have the managerial experience to operate the new unit in the most efficient manner. The typical relationship between efficiency in the sole proprietorship and other business organizations, such as the partnership or corporation that provides for continuity of management, is shown in Figure 1.1. This figure indicates that increased efficiency can be obtained in the long run by coordinating the entry and exit processes. This coordi-

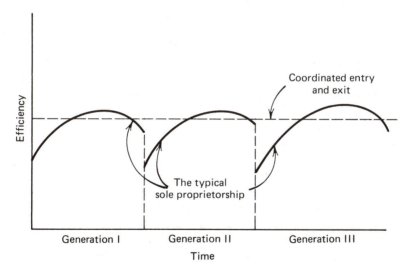

Figure 1.1.   Typical relationship between efficiency and operation of the business by succeeding generations. (*Source:* Presented by Dr. John Hopkin, "Getting In and Out of Agriculture," Seminar at Texas A and M University Research and Extension Center, Renner, Texas, February 19 and 20, 1974.)

nation can be obtained within the family farm structure through the use of multiowner business organizations, such as the corporation and partnership, through family arrangements that facilitate the son or son-in-law entering the farming business before the father has passed on, or through well-conceived and executed intergeneration transfer plans. This attention to coordination of the entry and exit processes in agriculture may not only provide opportunities for young farmers to enter the agricultural sector, but also generate a higher level of long-run efficiency than can be achieved by nonfarm businesses entering the industry.

MANAGEMENT

A cliché heard frequently in discussions among farmers and farm advisors is "the key to success in that enterprise is management" or "management makes the difference." Almost everyone recognizes that "management," whatever it is or however it is defined, is an important determinant of a firm's satisfactory accomplishment of its goals. Yet, when asked to define or describe "management," few people can articulate what the term means. Numerous definitions, all appropriately nebulous, are available. For example, some identify management as "the science or the art of combining ideas, facilities, processes, materials and people to produce and market a worthy product or service profitably" (Garoian, p. 19). This general definition applies to all types of firms, including farm businesses. Agricultural economists frequently define farm management simply as the allocation of limited resources to maximize the farm family's satisfaction.

Indeed, defining management is not an easy task. Many argue that to be understood it must be experienced. This does not imply that management principles, tools, concepts, functions, and responsibilities cannot be described and communicated through the spoken or written word. In fact, that is the primary purpose of this text. But the interrelationships between these management tools, principles, concepts, and functions can best be understood through application. This is one of the major reasons why most farm management courses utilize laboratory exercises and applications that enable the student to apply the concepts being discussed.

It is difficult (perhaps impossible) to develop a one-sentence definition of management that clearly describes the term and the activities it involves. Instead, the following discussion attempts to describe the tasks which the manager of a farm business must accomplish. This is done by first identifying the functions of management and then describing the fields or areas in which the farm manager must possess some knowledge and expertise.

The Functions of Management

A discussion of the functions of management provides a more complete indication of the activities encompassed by the term "management" than is possible with even the most detailed definition. Although experts may use

Figure 1.2 Major Activities of Each Function
of Management

Planning	Implementation	Control
1. Determine and clarify goals and objectives.	1. Acquire and maintain land and other real estate.	1. Develop a system to measure production, marketing, and financial performance.
2. Forecast prices and production.	2. Acquire, train, and supervise the work force.	2. Keep appropriate production, marketing, and financial records.
3. Establish the conditions and constraints within which the firm will operate.	3. Acquire and maintain machinery and equipment services.	3. Compare actual record results with standards established in planning.
4. Develop an overall plan for the long run, intermediate run, and the current year.	4. Acquire capital, credit, and purchased inputs required by the plan.	4. Identify corrective actions needed, if any.
5. Specify policies and procedures.	5. Schedule tasks to be completed.	
6. Establish standards of performance.	6. Communicate with employees, neighbors, landlords, bankers, and others as required to carry out the plan.	
7. Anticipate future problems and develop contingency plans.		
8. Modify plans in light of control results.		

different terms in their explanation of management functions, the concepts of planning, implementation, and control provide a useful and meaningful delineation of management functions. Some of the important activities of each management function are listed in Figure 1.2. These functional responsibilities of management are discussed in turn below.

Planning Planning is the most basic management function in that it provides the mode of operation to accomplish the firm's objectives. Essentially, planning involves selecting a particular strategy or course of action from among alternative courses of action with the objective of obtaining the greatest satisfaction of the firm's goals. Thus, planning is deciding in advance what should be done, how each task should be accomplished, when the task should be done, and who will be responsible for completing the task.

Planning is a prerequisite to satisfactory accomplishment of the implementation and control functions of management. A plan must be developed before the manager can determine what kind of resources will be required to accomplish the firm's goals. Unplanned action or activities cannot be controlled, for the essential concept of control is detecting and correcting deviations from a plan. Control without plans would, therefore, be meaningless, since plans furnish the standards and basic data needed for control.

The starting point for planning is to determine the goals and establish the constraints within which the firm will operate. Goals, such as those discussed earlier, provide the basis to judge the desirability of alternative plans. Data on the important resource availabilities, including land, labor, and capital, as well as the restrictions imposed by the social, political, and economic environment, are important in specifying the setting within which the firm must operate. Data on goals and restrictions provide the basis to proceed with the planning process.

Two basic tools used in the planning process are budgets and written policy and procedure statements. Essentially, a budget is a statement of expected results expressed in numerical terms. The development of expected annual cost and return budgets for specified enterprises, such as cattle or wheat, is a common activity of farm managers. These budgets, known as enterprise budgets, enable the farmer to determine the quantities of various inputs, such as feed and fertilizer needed, and the expected returns that will be generated for each unit of the enterprise. Using data from enterprise budgets, a farmer can then generate a whole-farm budget that summarizes the projected costs, returns, and net income along with the total resource requirements that result if the farm resources are used in the production of various combinations of crop and livestock enterprises. Financial budgets, such as a cash flow budget, provide information on the expected cash inflows and cash outflows that will occur during the forthcoming production period. This planning device can be useful in evaluating future capital and credit needs and in developing arrangements with lenders to satisfy credit requirements. Budgets of the financial and physical activities of the farm firm are, in fact, the fundamental planning tool used by farm managers.

In addition to budgets, policy and procedure statements can also be useful as planning tools. Policy and procedure statements are used to guide or channel thinking and short-run or operational decision-making. They can be used to assure that the decisions will be consistent with and contribute to objectives. Policy and procedure statements tend to predecide issues, thus avoiding repeated analysis of problems that arise regularly. Policy and procedure statements can also be useful in communicating plans and courses of action to employees as well as in providing guidelines for management in the planning and decision-making process.

Although policies and procedures can be communicated verbally or even through recurring management decisions that imply a specific policy or procedure, written policy statements are more easily interpreted by employees and also provide evidence of consistent standards for future management decisions. Policy and procedure statements may be one of the most underutilized planning tools available to farmers and farm managers, particularly those who hire labor or are involved in multiowner business organizations such as partnerships and corporations. Written statements of policies and procedures can be used in the selection, motivation, and compensation of employees and in the development of work schedules that can be used to implement plans for various enterprises. For example, management could use policy and procedure statements to communicate to employees the appropriate schedule for getting gestating sows ready for farrowing, placing them in farrowing stalls, giving iron shots to baby pigs, treating the pigs for diseases, creep feeding the pigs, and other relatively routine tasks. Similar policy and procedure statements would be useful in cow-calf operations and in many crop production enterprises.

Because planning involves predictions with respect to future events, forecasting is an essential component of the planning process. Forecasting involves assessing the future and making provisions for it in the plan. In the development of enterprise budgets, for example, the farm manager must forecast not only the expected price of the products being produced and the inputs being purchased off the farm, but also the physical efficiency of the production process and the amount of product that will be forthcoming during the production period. These forecasts must consider not only the output levels, but also demand conditions for the product being produced and for competitive products. Potential changes in government regulations and actions that might influence the market or the availability of production inputs must also be considered. Although forecasts are subject to error, effective planning of future events is impossible without forecast information.

Because of the uncertainty of future events and the resulting errors that may occur in forecasting, it is frequently desirable to develop contingency plans based on different sets of forecasts. This procedure of contingency planning enables the manager to adapt his or her decisions to changing conditions. Through the contingency planning approach, plans are developed using various forecasts of prices and production relationships. Then the plan associated with the "most likely" forecast is implemented. As the production period unfolds, however, changes in market or weather conditions that influence price and production expectations may change the "most likely" forecast of future events. If a contingency plan is available for this new "most likely" forecast, a change can be made from the original to the contingency plan before irreparable financial consequences have been incurred. This procedure of developing contingency plans is consistent with the adaptive decision-making process which is discussed in Chapters 16 to 18.

The material in the planning section of this text will relate the concept of planning to the unique problems faced by the farm manager. For example, the principles of production economics and budgeting will be presented as a means of developing enterprise plans as well as a necessary step in the development of plans for the entire farming operation. The use of hand budgeting and linear programming, as well as other computerized planning techniques, will be discussed as tools to assist in the planning process. The important role that price forecasting and different pricing alternatives (contracts, hedging, etc.) play in developing production and marketing plans will be reviewed. The use of financial planning tools and techniques, such as the cash flow budget and the present value method for analyzing investment alternatives, will be discussed as part of the planning process in farm management. The discussion of financial planning also will include considerations of leverage, the optimal debt-equity structure and the opportunity to use credit as a productive input in the firm as well as a source of liquidity. In addition, the tools and techniques for planning in an environment of risk and uncertainty will be reviewed. These concepts include probability and expected values, the development and evaluation of payoff matrices, the use of diversification strategies as a means of reducing risk, construction of flexible or nonspecialized production facilities, and the role of insurance in risk reduction. Finally, the legal and economic issues that must be considered in developing farm business arrangements will be discussed.

Implementation The second major function of the manager is to implement the plan that has been developed in the planning process. Essentially, implementation involves carrying out or putting into action the chosen plan. This function involves acquiring the personnel and other resources necessary to get the tasks done, organizing the work load to complete the tasks on schedule, and actually supervising and directing the accomplishment of the various tasks. Implementation is more than the physical labor associated with getting the job done. It also involves organizing and directing the physical activities whether they be performed by the farmers themselves or other employees.

Implementation of the farm plan requires the acquisition and coordination of the necessary land, labor, machinery, and capital resources. Acquisition of land through purchase alternatives, such as the mortgage or installment sale contract, must be evaluated. The alternative of leasing the land using a cash or crop share lease is also an important issue in implementing the production plan.

With respect to the labor resource, implementation requires the determination of whether full-time or part-time labor is required to complete the physical work. Obtaining information on prospective employees and choosing the right employee for the right job are also important components of the implementation process. Once labor has been hired, coordinating their activities to produce the crops and livestock in a timely fashion, providing necessary training and opportunities for personal development of the work force, and evaluating their performance and productivity are also necessary.

Capital acquisition and management involve the determination of alternative sources of debt and equity funds and the terms available from these different sources. Evaluation of credit terms and credit instruments with respect to their cost and other advantages and disadvantages is part of the implementation process. Negotiation of credit terms and the timing of loan repayments are also important components.

Acquisition of capital resources also includes negotiating for the purchase, physically taking possession, and having purchased inputs (such as feed, seed, fertilizer, herbicides, machinery, and livestock) available when needed. Deciding on the brand and dealer for each input, as well as negotiating the price and related terms, is an important part of the implementation function. These and other issues that must be resolved to carry out production, marketing, and financial plans will be discussed in the implementation section of this text.

Control The control function involves measuring performance and correcting deviations from expected behavior to assure the accomplishment of plans. Thus, control involves the traditional farm management activity of recordkeeping. Control is much broader, however, than simply keeping track of past performance through detailed historical records. The control function requires the farm manager to compare the actual outcome reported in the records with the projected budgets prepared during the planning process. Since the plans have been chosen by management to encompass the best means of accomplishing the firm's objectives, deviations from plans as evidenced by the control system provide a warning that current performance may not accomplish the specified goals. If the control system is properly designed, this deviation between planned and actual performance should provide the manager with some indication of what might be the causal problem. Consequently, the manager with an adequate control system can detect problems early in their development and make appropriate corrections to insure efficient satisfaction of the specified goals.

The basic control process involves three steps: (1) establishing standards, (2) measuring performance against these standards, and (3) correcting deviations from standards and plans. Standards are the criteria against which actual performance can be measured and are derived from the goals that have been specified by the manager. Standards may be physical in nature and express such performance criteria as pigs weaned per litter, average daily gains, yield per acre, or pounds of pork or beef per hundred pounds of feed fed. Alternatively, they may also be measured in financial terms such as rate of return on equity capital, production per dollar of expense, rate of capital turnover, debt to asset ratio, and production costs per acre. Standards for control purposes can and should be established for specific enterprises and even specific work activities as well as for the business as a whole. These standards normally should be specified as part of the planning function.

Although measurement of performance is an essential component of control, much of the performance data currently acquired by many farm firms is inadequate for control purposes. This inadequacy occurs for two reasons. First, many farmers measure performance only on an annual basis, and, consequently, an entire production period elapses before a serious attempt is made to compare performance with budgeted expectations. The primary concept of control is to correct deviations between planned and actual performance so that the objectives of the firm can be most efficiently accomplished. If these deviations are not detected until the end of the production period, few changes can be made that will improve performance. A control system must, therefore, provide timely information that will enable management to make appropriate adjustments early enough in the production process to have an impact on performance.

A second problem with many of the recordkeeping or control activities of farmers is that the performance variables being monitored provide little indication of potential problems. Although most farmers keep fairly accurate income and financial statements, these data provide only aggregate information on performance for the typical multienterprise farm firm. In many cases, a deviation between actual income and projected income cannot be traced to a particular enterprise or production activity. Even if the records are kept in sufficient detail to identify the enterprise that is generating less income than expected, most accounting data currently collected by farmers provide little, if any, indication of whether the deviation between actual and expected income occurred because of lower prices than expected or because of lower technical efficiency. Thus, many of the accounting systems currently used by farmers are inadequate for control purposes because they neither provide timely information nor indicate the causal forces that result in deviations between plans and actual performance. The development of records and accounting systems that provide timely data for control purposes is stressed in the accounting and control discussion in this book (Chapters 2 and 16 to 18).

The third step in the control process is that of actually correcting deviations from plans. In fact, the correction of deviations in performance is the point at which control interfaces with the other managerial functions. The manager may correct a deviation by modifying the goals, by redrawing the plans, or by making improvements in the implementation of the original plan. This overlap of the control function with the planning and implementation functions demonstrates the interrelationships of the manager's job. Consequently, management is an integrated process that requires the ability to plan and choose a particular course of action, to implement that action, and then to evaluate the results of that action compared to planned expectations through the control process so that improved performance can be obtained.

With respect to the control function, it is essential that the farm manager obtain detailed information on the cost and returns in each enterprise, labor and machinery utilization schedules, and appropriate production efficiency ratios to compare the actual technical efficiency to the expectations included in the production enterprise budgets. Likewise, financial information with respect to the income statement, the net worth statement, the cash flow statement, and the resulting financial and business performance ratios along with the tax record are important sources of control information which the farm manager must obtain to evaluate performance and make the appropriate adjustments. Finally, an evaluation of the sources of market information and a comparison of purchase and sale schedules with predicted marketing patterns are essential components of the farm manager's control function. All of these control records and their use in improving performance will be reviewed in the discussion on control of the farm business.

The Fields of Farm Management

While business management specialists customarily divide management along functional lines, agriculturists more typically emphasize fields or areas of expertise in their discussions of management. In reality, both the functions of management and areas of expertise must be interfaced. To adequately perform the planning, implementation, and control functions within the farm firm, the manager must have analytical expertise and access to data in the fields or areas of production, marketing, and finance.

Production   The most obvious area of responsibility for the farm manager is that of production phenomena. Plans must be made and implemented with respect to the production system to be used for each crop and livestock enterprise. This involves selecting the combination and timing of inputs for each product. Enterprise-specific decisions (such as what insecticide or herbicide will provide the desired control or whether a silage or high-concentrate ration should be fed to the cattle) are typical production decisions. Selecting the type and size of tractor needed to prepare the ground and plant the crop in a timely fashion, and deciding whether a confinement or open lot cattle feeding facility will provide the lowest cost of gain are other examples of production decisions.

The farm manager utilizes information on production efficiency and input-output relationships from numerous physical and biological sciences. Soil scientists can provide information on the response of crop yields to different fertilizer applications and the control effectiveness of alternative herbicides. Agronomists can suggest which variety is best suited to a particular soil type. Animal scientists can provide detailed information on the impact of the ration on butterfat content or milk production for dairy cows, or the gain that will result from different rations used in the feedlot. The effect of providing micronutrients and the use of various antibiotics in disease prevention and control can be provided by animal nutritionists. The impact of cross-breeding programs and preconditioning on the weaning weight of calves can also be provided by animal breeders.

Entomologists and plant pathologists can provide information with respect to the damage that might be done by various insect populations and diseases as well as the effectiveness of chemicals, resistant varieties, cultural practices, and other pest management strategies. Engineers can provide detailed blueprints to use in the construction of buildings and information on the best materials to use in that construction. They can also provide data on the fuel consumption of various size tractors and the acres per hour that can be plowed with various size power units or harvested with different size combines. These physical input-output or production relationships are a crucial component of the information required by the farm manager to make production decisions.

To determine the profitability of alternative production opportunities, the physical data must be combined with price and cost information as well as data on the availability of various land, labor, and capital resources. Thus, it is necessary to combine the information provided by the appropriate physical and biological sciences with price information and the appropriate decision-making procedures in making production decisions for the farm firm.

Marketing   The need for price and cost data to make adequate farm management decisions underscores the necessity for expertise in the second field of farm management—that of marketing. To maximize income or even to survive, farmers must not only produce the crop or livestock product efficiently, but they must also buy the inputs and sell the product at prices that result in a profit. The ability to analyze the market and to reflect changing market expectations in production schedules, input purchasing, and product-selling strategies is an essential component of profitable farm management. Basic decisions with respect to the scheduling or timing of production and sales require forecasts of future prices. Production scheduling decisions require the farmer to be acquainted with information on seasonal, cyclical, and trend movements in livestock and grain prices. Hence, the farmer must be aware of the supply and demand relationships for the particular product, the impact of consumer incomes and the availability of substitutes on product prices as suggested by income and cross-price elasticities of demand, and the expected response of other producers to current prices. The ability to gather and analyze these price expectation data is one of the basic marketing functions that must be performed by the farm manager.

Numerous other decisions require knowledge of market relationships and market phenomena. Such choices as which marketing channel to use, whether to sell cattle or hogs on a liveweight or grade and yield basis, and whether or not to sell grain at harvest or to dry and store it for sale at a later date are examples. The price premium paid for different grades of livestock is important market information that must be considered when deciding whether to feed dairy versus beef cattle, or yearlings compared to calves. The evaluation of the profitability of alternative hedging strategies or the potential for contracting part of the corn or soybean crop for future delivery also requires detailed analysis of market relationships and price expectations. The potential for contracting inputs, such as protein supplement, also involves basic market relationships and analysis. Evaluation of moisture discounts in grain production and the potential to sell specialty crops to local processing facilities are yet other marketing decisions that must be made by farmers. These examples illustrate the importance of accurate market information to the farm manager.

Finance In addition to the information on production efficiency and market and price relationships, data must also be available on resource availability for adequate farm management analysis. Except for the farmer's own labor and management resources, the acquisition of other productive inputs in farming such as land, machinery and equipment, and hired labor involves the outlay of money. Improving the labor and management skills through formal education also requires the use of money for tuition and other expenses. Thus, the field of finance and financial management is also an important area where the farm manager must have some expertise.

Basically, finance involves decisions with respect to the acquisition of funds and the use of those funds to acquire the services of various resources. For example, purchasing real estate with various combinations of equity and debt funds requires a financial management decision. Alternatively, purchasing cattle or hog protein or even hiring seasonal labor involves the commitment of working capital, which is also a financial management decision. Issues with respect to repayment capacity or the ability to generate sufficient cash to repay operating loans when they are due are included.

Whether capital should be committed to the purchase of real estate or whether the services of that real estate could more profitably be rented or leased is a financial management issue. Leasing machinery compared to purchase and the repayment schedule that will be required to amortize a loan on machinery are important finance questions. The choice among alternative sources of funds, including the appropriate combination of debt and equity, requires detailed financial analysis, as does the comparison among the terms and interest rates offered by alternative financial institutions. The management of working capital so as to take advantage of cash discounts on the purchase of feed and other inputs is another example of the importance of financial analysis in farm management. Financial management decisions also involve such questions as organizing a business to withstand expected risk, holding cash reserves for unexpected contingencies, acquiring insurance policies for protection against property damage, and developing estate plans. For adequate financial analysis, the farm manager must be acquainted with the concepts and procedures of cash flow to evaluate repayment capacity; understand the phenomenon of present value analysis and the basis for discounting in investment analyses; and have the ability to analyze financial statements, various tax management strategies, and alternative business organizations.

In summary, production decisions involve basic questions as to what to produce, how to produce it, and which combinations of inputs and outputs to use. In farm management these decisions must be integrated with basic marketing decisions as to where, when, and how to buy and sell inputs and products. Finally, the what, where, when, and how of production and marketing decisions must be integrated with the financial decisions of where will the funds be obtained, with what terms will they be acquired, how will they be repaid, and for what will they be used. Thus, the scope of farm management includes the three functions of management (planning, implementation, and control) and the fields of agricultural expertise (production, marketing, and finance) over the life cycle of the farm firm. The three-dimensional character of farm management is illustrated in Figure 1.3. In essence, farm management is concerned with planning, implementation, and control of the production, marketing, and financing dimensions of the farm firm throughout the entry, growth, and exit stages of the firm's life cycle.

The functions of management and areas of agricultural expertise are easily related to the systems concept of a business firm.[3] The farm business can be described as a system composed of many components or parts designed to accomplish an objective according to plan. The systems concept of the business emphasizes that the firm must have one or more objectives and that the system must be designed or planned. Then resources are acquired, the plan is put into effect, and the operation is monitored and controlled. The systems concept explicitly recognizes the interrelationship between the parts of the firm and stresses the importance of making management decisions for any one function or area considering the implications for the other components. For example, decisions on the production of an individual crop or livestock enterprise must be made

---

[3]See Richard A. Johnson, Fremont E. Kast, and James E. Rosenzweig, *The Theory and Management of Systems,* 2d ed. (New York: McGraw-Hill, 1967).

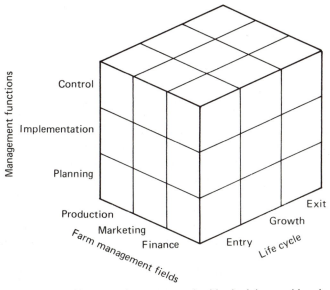

Figure 1.3. Farm management is concerned with decision-making in three functional areas and three fields of farm management over the life cycle of the farm firm.

considering the implications for other production enterprises, as well as marketing and finance, if the business is to achieve its objectives. Figure 1.3 also recognizes the interrelationship of functions and areas in making management decisions at any stage in the life cycle of the firm.

The systems concept of the farm business is implicit in the discussion throughout this book. However, the discussion does not stress systems terminology unless it is required to clearly present the topic. Thus, systems concepts are introduced more formally in the discussion of control procedures. Then they are used in developing control systems for the farm business.

The Decision-Making Process

Successful managers of farm businesses must be prepared to deal with decision-making in each of the areas outlined above. While the manager must deal with different circumstances and alternatives in each area, the management task is simplified somewhat in that the same decision-making process applies in all areas. Given the goals of the manager, the process can be described in five steps.

1. Define the problem or opportunity.

2. Identify alternative courses of action.

3. Gather information and analyze each of the alternative actions.

4.   Make the decision and take action.

5.   Accept the consequences and evaluate the outcome.

The first step is the identification or awareness that a problem or opportunity exists. In some cases the manager must resolve a difficulty such as how to market or dispose of a rain-damaged crop. In other situations the decision concerns how to best exploit an opportunity, such as greatly improved market prices. One basis to define problems and opportunities is to specify goals and develop detailed plans of how these goals are to be achieved. A keen awareness of changes in the social, economic, technical, and physical environment and an ability to project the impact of these changes on the operation of the business provide another basis to identify both problems and opportunities.

The second step is to formulate alternative courses of action that might be taken to solve the problem or exploit the opportunity. Identifying the cause or causes of a problem may suggest alternative solutions. There is seldom a problem or an opportunity for which reasonable alternatives do not exist. The more common problem is limiting the number of alternatives sufficiently so that each can be examined in detail.

Selecting an appropriate method to analyze the alternative solutions identified, collecting the appropriate data required for the analysis, and completing the analysis are all part of the third step. Selecting the type of analysis should normally precede data collection in order to avoid either collecting data that are not needed or failing to obtain information that is important in comparing the alternatives. When a large number of alternatives are identified, it may be helpful to use a simple screening process to reduce the list and to consider only the most promising alternatives in detail. The result of the analysis should estimate the effect of each alternative on the manager's more important goals.

The fourth step involves evaluating the estimated outcomes for each alternative and choosing the best alternative based on the goals of the operator. One alternatve may have the highest expected return but more risk and higher labor requirements than a second. A third alternative may be intermediate in profit, risk, and labor requirements. Which of these alternatives is selected depends on the emphasis or weight the manager gives each of these outcomes. Methodologies for considering the trade-offs between profit and risk proposed by decision theorists are discussed in Chapter 11. This step requires the intuition and judgment that are the "stock in trade" of experienced managers. After selecting the desired course of action, the manager must implement the alternative selected.

The fifth step is to accept the consequences, the unfavorable as well as the favorable outcomes of the decision. It is also necessary to monitor actual performance, as well as technology, laws, economic events, and the climate, and to reevaluate decisions when conditions change. As one reviews and learns, it may become apparent that a problem remains or that additional opportunities exist, leading to the repeated use of this decision-making process.

The five-step decision-making procedure can be used to deal with problems and opportunities in each of the three functions and each of the three fields of farm management. Whether a problem or opportunity exists in planning, implementation, or control, the manager can proceed by identifying alternative courses of action, analyzing each alternative, making a decision, and accepting the consequences for the decision. Thus, the procedure is used in each of the nine areas comprising farm management as shown in Figure 1.3.

Management—Art or Science?

The question is often asked whether management is a science or an art. Managing requires getting things done given the realities of the situation. Management, like all other professions (including medicine, law, music, landscape design, engineering, and athletics), is in part art. Like other professions, management is improved by making use of organized knowledge. This organized knowledge, whether crude or advanced, whether exact or inexact, to the extent it is well organized, clear, and pertinent comprises a science. Thus, management is part art, but the organized knowledge underlying it is science. The science and art of managing are complementary rather than mutually exclusive.

This book emphasizes the procedures that managers of farm businesses can use to obtain, analyze, and organize knowledge that is pertinent to business decisions on farms. As such it draws on concepts and tools from a wide range of disciplines and subject matter areas. It is worth mentioning some of these concepts and tools to emphasize the broad scope of disciplines from which we draw.

Accounting and Information Flows   Managers of farms rely on both internal and external flows of information to operate their business. Information on the past, present, and expected performance of the business and its operating environment are obtained using standard accounting procedures. Information on the capital, labor, and land markets, product price levels, government tax policies, environmental regulations, new technology, and weather forecasts must be obtained from external sources, such as literature and educational programs coordinated by the land grant system, the U.S. Department of Agriculture, the mass media, farm organizations, investment services, agribusiness firms, the national weather service, and others.

Economic Theory   Economic theory identifies key elements in the decision-making process and indicates the relationship between these elements.

Economic concepts indicate how certain problems can be analyzed and the relevant data for the analysis. Production economics principles including fixed versus variable costs, the principles indicating the most profitable level of inputs to use and combination of outputs to produce, and economies of size are integrated in the discussion of production planning. The appropriate supply and demand concepts provide the basis for analyzing resource acquisition decisions and the marketing of farm products.

Finance Principles Financial management principles and concepts, including considerations in the use of debt and equity capital to finance the business, analysis of investment alternatives, and risk management, are a key dimension of farm management. They are integrated with the economic and accounting concepts in the discussion.

Biological Sciences Data on input-output relationships and other relevant production information are obtained from the technical disciplines— soils, agronomy, horticulture, animal science, engineering, and others. These data provide the basis for developing the crop and livestock production systems as well as control systems for the production process.

Psychology and Sociology Managers pursue a variety of noneconomic goals rather than simple profit maximization. Principles of psychology and sociology are used to integrate these noneconomic goals into decision-making. They also provide a broader basis to discuss employer-employee relationships, land leases, and intergeneration transfer relationships.

Law and Political Science Law and political science provide much of the institutional setting within which the farm business must be operated. The conditions under which a partnership or corporation can be formed, the environmental restrictions that must be met in grain and livestock production, the grades and standards for products produced, income tax laws, and the estate transfer laws are examples of institutional restrictions that are important in the management of a farm business.

Mathematics and Statistics Budgets are commonly used as a way of recording the effect of all of the factors influencing the business during some period of time. Budgeting is an orderly method to assemble information that can be used by the decision-maker in choosing among alternatives. A wide variety of budgeting procedures are used in farm management. This book outlines the major types including enterprise budgets, partial budgets for whole-farm planning, cash flow budgets, and capital budgeting.

Many of the more advanced computational procedures used in management have been developed in applied mathematics and statistics. These include linear programming, statistical decision theory, and computer programming which are discussed in the following chapters.

Successful farm management requires the integration of concepts and tools from each of these disciplines. While this book emphasizes economic and business management principles and concepts, it also stresses the use of information from the other areas mentioned. An effort is made throughout to discuss the types of data needed and potential sources of such information, along with the analytical procedures to use in planning, implementation, and control.

SCOPE OF THE BOOK

This book is organized into an introductory section and separate sections for each of the three functions of management. The following chapter introduces the three major accounting statements that are basic to a discussion of farm management, although no attempt is made to discuss farm accounting procedures and methods in detail.

The planning function is discussed in Chapters 3 through 11. Production planning with budgeting is the topic of Chapters 3, 4, 5, and 6. Discussions of market planning (Chapter 7), financial planning (Chapter 8), and planning of farm business arrangements (Chapter 9) follow. The use of linear programming as a tool for production, market, and financial planning is discussed in Chapter 10. The planning section is completed with a discussion of planning within an environment of risk and uncertainty in Chapter 11.

The focus of the next section of the book is on acquiring the resources and organizing and directing the work load to carry out the plan—the process of implementation. The first three chapters in this section discuss the acquisition and management of labor and management services (Chapter 12), land (Chapter 13), and acquisition and management of depreciable assets (Chapter 14). Although the acquisition of each of these inputs involves capital, a thorough discussion of capital acquisition and management is reserved for Chapter 15.

The last section of the text emphasizes the control process. Chapter 16 discusses control concepts and procedures. The remaining two chapters deal with production control (Chapter 17) and marketing and financial control (Chapter 18). The three chapters in this section also emphasize the essential linkage between control and the successful completion of the planning and implementation functions.

Summary

Managers of farm businesses typically must consider multiple business and family goals in making decisions about how to operate the farm business. Some of these goals include maximizing profit or net return, increasing net worth, controlling a larger business, avoiding risk, reducing borrowing needs, increasing family living, having more leisure time, having a neat and well-kept farmstead, and providing service to the community. The decision-making principles and concepts discussed in this book recognize the farm operator's multiple goals, but focus primarily on the business and economic goals of the farm manager.

Because the individual entrepreneur dominates in the agricultural sector, the typical farm firm exhibits a life cycle that parallels the life cycle of the operator. This is an important concept, because a manager's goals or objectives may change over the life cycle of the individual and the firm. The first stage of the life cycle is that of entry or establishment where the overall objective is to become a successfully established farm operator. The second stage is growth and survival. During this stage the manager acquires additional resources, attempts to increase efficiency, and consolidates any gains to insure survival if difficult times are encountered. The third and final stage of the life cycle is that of exit or disinvestment. In this stage the entrepreneur attempts to phase out of the business and liquidates part of his or her assets, or transfers ownership to the succeeding generation of managers.

Good management is necessary to the efficient operation of any business, but the concept of management is nebulous and difficult to define. Insight into this concept can best be obtained by understanding the functions of management. These functions include: (1) planning which is the process of selecting a particular strategy or course of action from among the various alternatives with the objective of obtaining the greatest satisfaction of the firm's goals; (2) implementation which is the process of acquiring the resources and putting the chosen plan into action; and (3) control which involves evaluating performance to determine if it conforms to plans, and taking corrective action to improve performance when it is unacceptable. Higher goal levels can be achieved by managers who can complete these three functions efficiently in each of these areas: production, marketing, and finance. Production involves the process of efficiently converting inputs into products; marketing involves the activities of acquiring inputs and selling products; and finance includes decisions with respect to the acquisition of funds and the use of those funds within the farm business.

Management implies the making of decisions. Decision-making can best be accomplished by following a five-step procedure: (1) define the problem or opportunity, (2) identify alternative courses of action, (3) gather information and analyze each of the alternative actions, (4) make the decision and take action, and (5) accept the consequences and evaluate the outcome. To implement this decision-making process in the context of managing a farm, the manager must integrate knowledge from many areas including accounting and information systems, economic theory, finance principles, biological sciences and processes, psychology and sociology, law and political science, and mathematics and statistics. This book focuses on the procedures which managers of farm businesses can use to obtain, analyze, and integrate this knowledge in a manner useful for decision-making with respect to planning, implementation, and control of the production, marketing, and financing activities of the farm firm.

Questions and Problems

1. List several goals that you feel are important to farmers and discuss how they would be used in making decisions.
2. Identify the three stages of the family-firm life cycle, and discuss the major goals or objectives that might be dominant in each stage.
3. What relationship, if any, would you expect to exist between the economic efficiency of the firm and the family-firm life cycle?
4. What are the three functions of management? Briefly discuss each of these functions.
5. List some activities that are part of the planning function, some activities that are part of the implementation function, and some activities that are part of the control function.
6. What are the three key steps of the control process? Identify a particular decision that must be made in managing a farm and discuss how you might develop a control system or procedure for that decision.
7. What are the three fields of management? Discuss the activities that might be included in each of these fields of farm management.
8. Identify and discuss the steps of the decision-making process.
9. List characteristics of good and poor managers. Which of these characteristics can be taught in a course? How are the others obtained?
10. Select a crop or livestock production enterprise and list the data needed to develop a plan for the production of that enterprise. Then indicate the area(s) of knowledge providing each item of data.

Further Reading

Breimyer, Harold F. *Can the Family Farm Survive: The Problem and the Issues.* University of Missouri, Agr. Exp. Sta. Special Report 219, Columbia, 1978.

Garoian, Leon, and Arnold F. Haseley. "The Board of Directors in Agricultural Marketing Business." Oregon State University, May 1963.

Harmon, Wyatte L., Roy E. Hatch, Vernon R. Eidman, and P. L. Claypool. *An Evaluation of Factors Affecting the Hierarchy of Multiple Goals.* Oklahoma State University Agr. Exp. Sta. Tech. Bull. T-134, Stillwater, June 1972.

Johnson, Richard A., Fremont E. Kast, and James E. Rosenzweig. *The Theory and Management of Systems.* 2d ed., New York: McGraw-Hill, 1967.

Koontz, Harold, and Cyril O'Donnell. *Essentials of Management.* 2d ed., New York: McGraw-Hill, 1978.

Nikolitch, Radoje. "Family-Size Farms in U.S. Agriculture." U.S. Department of Agriculture, Econ. Res. Ser. Rep. ERS 499, February 1972.

Schertz, Lyle P. *Another Revolution in U.S. Farming?* U.S. Department of Agriculture, Agr. Econ. Rep. No. 441, pp. 13–41, December 1979.

Thierauf, Robert J., Robert C. Klekamp, and Daniel W. Geeding. *Management Principles and Practices.* New York: John Wiley and Sons, 1977.

2

FARM ACCOUNTING
INFORMATION AND
STATEMENTS

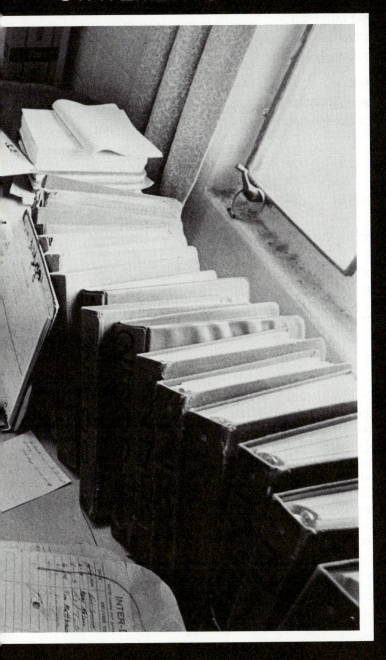

CHAPTER CONTENTS

Accurate and efficient production, marketing, and financial management decisions require extensive information. Some of this information can be acquired from farm records, while other data must be acquired from firms with which the manager deals and other public and private agencies. The purpose of this chapter is to provide an overview of farm records and to indicate the types of data managers can obtain from their accounting system. The need for farm accounting systems is discussed in the initial section. The following three sections provide an introduction to three financial statements—the income statement, the balance sheet, and the cash flow statement. These three accounting statements are particularly useful in planning, implementation, and control of the farm business.

Records and accounting can be tedious, complex, and time consuming. However, they can also be very rewarding when they provide the essential data for performance evaluation and assessment of progress that is important in managing the business. This chapter emphasizes the conceptual issues in developing and using accounts and financial statements for management purposes. Subsequent chapters further develop the use of these financial statements in carrying out the three functions of management. The mechanics of completing accounts and financial statements is beyond the scope of this discussion.[1]

THE NEED FOR FARM ACCOUNTING SYSTEMS

While many reasons can be given for keeping farm accounts, the use of accounts in managing a business can be summarized under four headings.

1. To provide data for use in forward planning.
2. To provide a history of performance for use in implementing the farm plan.
3. To aid in control of the operation.
4. To provide data necessary to file reports.

Data for Use in Forward Planning

Farm accounts can provide data on production levels of crop and livestock enterprises, the amounts of inputs used in producing these enterprises, the prices paid for inputs, and the cost and returns of individual enterprises. Records can also be summarized to indicate the cash receipts and cash expenses on a month-by-month basis. These data can be used in developing both short-run and long-run plans for the farm business. The data available in records are unique to the individual business. Farm planning typically requires that the data available from previous years be supplemented with additional data on expected prices, input requirements, and production levels for some systems of production that have not been used by the firm. Nevertheless, the data available from the past provide a starting point for the planning procedure.

A History of Performance for Use in Implementing the Farm Plan

Modern farming operations require relatively large amounts of capital for investment in both buildings and equipment as well as for operating purposes. The income statement (indicating the amount of money that has been made during the year) and the balance sheet (indicating the amount

---

[1]See Sydney C. James and Everett Stoneberg, *Farm Accounting and Business Analysis* (Ames: Iowa State Press, 1974), for a discussion of the procedures to follow in keeping farm accounts.

of equity that an operator has in his or her business) document past performance and are thus quite useful in securing financing for the farming
operation. Data from farm records can also be used to determine the
amount each individual in a partnership should receive, or the amount the
tenant and landlord with a livestock-share or crop-share leasing arrangement should be paid. Data on production levels and costs may also be
useful to illustrate an individual's "track record" to a prospective landlord.
These are a few of the ways that farm records can be used in implementing
the plan that has been developed.

Aid in Control of the Operation

Managers of farm businesses develop both short-run and long-run plans.
After the plans have been developed, they are concerned with implementing the plan, monitoring the actual outcome over time, and making adjustments in the plan as conditions warrant. In developing plans for various
enterprises, the operator sets physical and financial standards of performance. Accounts can be developed to record data on the physical and
financial performance measures that have been set for the individual enterprises. These data provide the operator with an opportunity to compare
the actual outcome with the performance standards that have been set and
to adjust the plan to changing conditions. The physical and financial standards for individual enterprises provide the basis for developing standards
for the whole-farm business. The standards set for the whole-farm business
are usually financial in nature. Managers prepare projected cash flows on a
month-by-month basis for the coming year and compare the actual cash
receipts and expenses to the projections that have been made. When a
significant difference between planned and actual cash receipts or cash
expenses occurs, it is immediately obvious to the operator. This gives operators an opportunity to adjust their cash inflows and outflows before a cash
flow problem develops. In the development of longer run plans, standards
are set on the rate of return on investment, the return per month of the
operator's labor, the rate of return on equity capital, and the amount of
change in net worth that should be achieved on a year-by-year basis. Farm
accounts can be used to indicate the extent to which farms are achieving
their goals and to identify the factors that are either allowing them to
achieve these goals or are preventing them from doing so.

Data Necessary to File Reports

Data on revenues and expenses from the business are required to calculate
the amount of state and federal income taxes that are to be paid, as well as
to file estate taxes and complete other reports requested by both private
and governmental agencies. Similar data are required to determine the
amounts of social security payments that must be paid on the operator's
income and wages paid to hired labor. Unfortunately, many farmers think

of this as the major reason for keeping farm records. While it is necessary to file some reports, the data required to complete these reports are typically much less detailed than is needed for efficient management of the business. The emphasis in this discussion is on the use of farm accounts for management purposes rather than completion of reports.

A question that frequently arises is, what accounts should a farmer keep? The answer to the question depends on the cost (the time, effort, and cash costs) of obtaining or keeping the records, and the value of the data obtained to the operator. Several types of records are needed on the firm's operation for use in making production, marketing, and financial decisions. It is sometimes argued that certain components must be included in a farm accounting system for it to be complete. However, the value of acquiring certain types of data varies from one firm to another. As a result, the components that make up the complete farm record system for one firm may be quite different from those for another. To be useful the account must provide data that the manager will consider in making decisions, and the data must be kept accurately and completely. In deciding on whether to keep a particular record, farm operators should use the following general rule: keep only those accounts that have an expected return that is greater than the cost of keeping the records. While it is often difficult to determine what the value of record data will be in the future, this rule provides a conceptual guideline for farm operators to use in selecting the components that they wish to include in their farm accounting system.

Accounts frequently included in accounting systems for farm businesses are the listing of receipts and expenses, the income statement, the balance sheet, the cash flow statement, the listing of accounts payable and receivable, and enterprise accounts. Three of these accounts—the income statement, the balance sheet, and the cash flow statement—are of primary interest as we begin to discuss farm planning. They are treated in the following three sections. A brief description of the other accounts mentioned is given at the close of the chapter.

THE INCOME STATEMENT

Most farmers are in business to make a profit. The purpose of the income statement is to determine the flow of income generated by the business over a period of time. The change in asset values is another source of return to farmers who own land and other assets, which will be discussed later. Farm businesses typically calculate the income statement annually to measure the profitability of the business over the previous business year. The annual nature of many crop and livestock enterprises makes measurement of profitability once each year meaningful. However, the income statement can be completed monthly, quarterly, semiannually, or for some other period if it is useful to do so for management purposes.

The flow of income can be measured using two different systems, the cash system and the accrual system. The cash system records income and expense items when cash changes hands or when checks are written or received. For example, if an input, such as fertilizer, were purchased in December but not paid for until January of the following year, the fertilizer expense would be included in the income statement for the following year rather than for the current year. Likewise, if livestock were sold before the end of the year but the check was not received until after January 1, the proceeds would be recorded as income in the following year. With the cash accounting system, inventory changes are ignored because they do not reflect cash transactions. Thus, an increase or decrease in the inventory of livestock or grain would have no impact on the income statement using the cash accounting system. So the cash method only recognizes cash transactions with the exception of depreciation which is included as an expense item. Because all inventory changes and deferred purchases or sales will eventually show up as cash over a period of years, the cash method accurately reflects the income of a business over time. However, it may not indicate the actual income produced during a particular accounting period since the amount of production maintained in inventory varies from year to year in the typical farm business.

With the alternative system of measuring income, the accrual accounting method, transactions that increase or decrease income are included in the statement when production occurs or the expense commitment is made, not when cash changes hands. For example, fertilizer purchased in December, even if delivery did not occur until January of the following year, would be included as an expense in this year's income statement. Similarly, the sale of livestock on December 31 would be included as an income item in this year's statement, even if the check were to be received in January of the following year.

The accrual accounting method also includes changes in inventory in the computation of net income. If the inventory of feed or livestock increases from the beginning to the end of the year, the accrual method recognizes this larger inventory as increased production and income during the current accounting year even though it has not been converted into cash. In contrast, a decrease in the value of inventory from the beginning to the end of the accounting period indicates that a larger volume of production was sold for cash during the year than was produced.

The accrual method attempts to reflect more accurately the actual production and expense commitments made during the accounting period. In the short run it provides a more accurate indication of the actual income generated by the physical and financial resources. It must be recognized, however, that there may be a significant difference between the accrual income and the cash flow of the typical farm business because of inventory changes, delayed sales, prepaid expenses, or other noncash adjustments that are part of the accrual accounting system.

Completing the Income Statement

Figure 2.1 illustrates a simple income statement for the farm business based on the accrual method of accounting. The first major category of items on the statement is farm operating receipts. This category includes all cash receipts from the sale of farm products. These include grain and forage crops, livestock (such as cattle, hogs, and poultry but excluding breeding animals which are considered capital assets), livestock products (such as milk, wool, and eggs), government payments (such as conservation payments or commodity program payments), payments received for custom work, and any other cash receipts from farming, including refunds on purchases and cash dividends from cooperative purchases. The total of these items equals gross farm operating receipts.

The next major category on the income statement is farm operating expenses. Operating expenses include the outlays for seed, fertilizer, chemicals, machine hire, storage, feed purchased, livestock purchased, breeding and veterinary expenses, livestock supplies, fuel and oil, utilities, machinery and other repairs, real estate and sales taxes, insurance, rents, trucking and marketing expenses, hired labor, interest paid, and miscellaneous expenses. Adding the farm operating expenses results in gross farm operating expenses. Subtracting gross farm operating expenses from gross farm operating receipts results in net cash operating income.

Adjustments must now be made to account for noncash transactions to compute net income on an accrual basis. The first adjustment is to account for the value of any products (such as meat, milk, and eggs) produced on the farm and consumed by the household without compensating the business in cash. The value of these products represents production by the business and should be added to net cash operating income to obtain a more correct accounting of accrual income. The expenses used to produce these products have normally been included in the appropriate farm operating expense categories. Thus, only the gross value of the products consumed needs to be listed on line 7.

The second adjustment is to account for inventory changes during the period. Inventory changes can be determined by comparing the beginning inventory and ending inventory obtained from the current and previous years' balance sheets. As noted in Figure 2.1, inventory items include feed and grain, farm supplies (feed, seed, fertilizer, and fuel), market livestock, accounts receivable, prepaid expenses and cash invested in growing crops, and accounts payable. Accounts payable include not only the production items charged at the feed, fertilizer, machinery, and other dealers, but also accrued rent, accrued interest, and accrued tax payments. If the ending

Figure 2.1 Income Statement

Period Covered: _____ *19___ to* _____ *19___*

Farm Operating Receipts			*Farm Operating Expenses*	
Livestock and livestock products			Seed	$_____
	Units		Fertilizer	_____
_____ (_____)	$_____		Chemicals and other crop supplies	_____
_____ (_____)	_____		Machine hire	_____
_____ (_____)	_____		Storage	_____
_____ (_____)	_____		Feed purchased	_____
Subtotal	$_____	(1)	Feeder livestock bought	_____
Crop sales			Breeding	_____
_____ (_____)	$_____		Veterinary	_____
_____ (_____)	_____		Livestock supplies	_____
_____ (_____)	_____		Fuel and oil	_____
_____ (_____)	_____		Utilities	_____
Subtotal	$_____	(2)	Machinery repairs	_____
Other operating receipts			Other repairs	_____
_____	$_____		Taxes, real estate, sales	_____
_____	_____		Insurance	_____
_____	_____		Rents	_____
Subtotal	$_____	(3)	Trucking and market	_____
Gross farm operating			Hired labor	_____
receipts (1) + (2) + (3)	$_____	(4)	Farm interest paid	_____
			Other	_____
			Gross farm operating expense	$_____ (5)
			Net cash operating income (4) − (5)	$_____ (6)

Value of Farm Products Produced on the Farm and Consumed by the Household $_____ (7)

Adjustment for Changes in Inventory

	Feed and Grain	Market Livestock	Accounts Receivable	Supplies and Prepaid Expenses	Accounts Payable	
Ending inventory	_____	_____	_____	_____	Begin _____	
Beginning inventory	_____	_____	_____	_____	End _____	
Net adjustment	_____	_____	_____	_____	_____	$_____ (8)

(continued)

Figure 2.1   (*Continued*)

Adjustments for Changes in Capital Items

	Breeding Livestock	Machinery and Equipment	Buildings and Improvements		
Ending inventory	___	___	___		
Plus sales	___	___	___		
Subtotal	___	___	___	(9)	
Beginning inventory	___	___	___		
Plus purchases	___	___	___		
Subtotal	___	___	___	(10)	
Net capital adjustment (9) − (10)	___	___	___	___	(11)

Farm profit or loss (6) + (7) + (8) + (11) (returns to unpaid labor, operator's labor, equity capital, and management)	___	(12)
Off-farm income	___	(13)
Total net income (12) + (13)	___	(14)
Proprietor withdrawal	___	(15)
Addition to retained earnings (14) − (15)	___	(16)

inventory exceeds the beginning inventory, a positive adjustment is required to reflect the fact that some of the products produced during this period remain in inventory and have not been sold. If the beginning inventory exceeds the ending inventory, a negative adjustment is required to indicate that, in addition to this year's production, some of the initial inventory items were sold and are already included in cash farm income. The net adjustments in accounts payable and receivable must also be included to achieve an accrual accounting of income produced during the period. The total net adjustment is calculated by algebraically summing the net adjustment for the five inventory categories.

The third adjustment required to compute net income on an accrual basis is to account for changes in the capital assets. These adjustments involve both cash transactions—purchases and sales of capital items (such as breeding stock, machinery, equipment, buildings, and improvements)—and depreciation which is a noncash transaction. The net capital adjustment for each category of capital assets is determined by adding capital sales to ending inventories and then subtracting the total of beginning inventories and capital purchases. This adjustment can be thought of as measuring the amount of depreciation that occurred during the period. The total net capital adjustment is calculated by algebraically adding the net capital adjustments for the three categories and entering the sum on line 11 of Figure 2.1.[2]

The farm profit or loss for the business is the sum of net cash operating income, the value of farm products produced on the farm and consumed by the household, total net adjustment for inventory, and the total net capital adjustment. The term "profit or loss" is used on the income statement in the accounting sense and does not reflect the economic definition of these terms. The economic interpretation of the value on line 12 of the income statement is the return to contributed resources. It is the accrual accounting measure of the income that has been generated with the unpaid labor, including the operator's labor, equity capital, and management resources available to the firm.

Off-farm income has become an increasingly important source of total income for U.S. farm families. Income from salaries and wages earned off the farm, income from nonfarm business enterprises, and earnings from nonfarm investments are examples of income that are to be listed on line 13.

Some farm businesses list the operator's salary or a family-living allowance as an operating expense, eliminating the need for line 15. However, many farm families withdraw money from the business in relatively small and more frequent sums as needed to pay family-living expenses. The total of these withdrawals for family-living expenses, nonfarm investments, and the proprietor's income tax and social security payments *plus* the value of farm products produced on the farm and consumed by the household (line 7) should be listed on line 15.

It is essential that separate family financial accounts be kept with either method of handling withdrawals for the business. Accurate personal accounts are necessary not only to complete the income statement, but also to complete the balance sheet and to file income tax returns. Certain personal expenditures, such as interest, medical expenses in excess of specified minimum levels, professional organization expenditures, and charitable contributions, are tax deductible when the individual itemizes deductions. Notice that the value of farm products produced on the farm and consumed by the household represents a proprietor withdrawal in kind and should be included in the total on line 15.

Subtracting proprietor withdrawals results in additions to retained earnings. If the proprietor has included all withdrawals for nonfarm investments on line 15, the additions to retained earnings indicate the

---

[2]The purpose of the income statement is to calculate the net returns to contributed resources from operating the business—not from owning assets. The farmer must be careful to calculate the beginning and ending values of assets in a manner that avoids including appreciation in machinery and buildings in the estimate of profit or loss. One way is to use values from the cost or basis column of the balance sheet as is illustrated in Figure 2.5. An alternative approach is to use the market value of these assets at the beginning of the year and use beginning year market value less a reasonable amount of depreciation as the end of the year value.

amount that has been added to the *proprietor's net worth* (or equity) *in the business assets.* It should be noted, however, that some of the money included under proprietor withdrawals might have been invested in nonfarm assets. Furthermore, the proprietor's nonfarm assets may have increased or decreased during the accounting period. For these reasons, the change in the individual's net worth discussed in the next section may be more or less than the addition to retained earnings.

Allocating Returns to Contributed Resources

The farm profit or loss (Figure 2.1, line 12) represents returns to the farm family's unpaid labor, operator's labor, equity capital, and management. In many cases, the unpaid labor includes not only the operator's labor, but also unpaid labor provided by other members of the family. Farm record systems commonly divide the farm profit or loss between capital, and labor and management. Any division of farm profit or loss into returns for the three inputs is somewhat arbitrary. The division is frequently made, however, to help evaluate if an operator is earning as large a return for capital and labor in farming as might be earned if the resources were used in other activities.

Farm profit or loss can be divided into returns to capital and returns to labor and management by specifying a charge for one resource and calculating the residual return for the other. Farm record systems typically subtract a charge for capital and calculate the residual return for the unpaid labor and management. A value can be assigned to the unpaid labor and management, leaving a residual return to equity capital. The steps in making these calculations are outlined in Figure 2.2. The average of the farm equity for the beginning and end of the period is typically used as the amount of equity capital. Unpaid labor is recorded separately for the operator and others since the salary or wage level might differ.

The interest rate and wage rates used in making the computations should be selected with care. The rate of interest to use is the rate that the farmer could earn on alternative investments having comparable risks. The value of unpaid labor should be based on the potential value that the individual could earn in the best alternative occupation. This represents the opportunity cost of unpaid labor and management resources in the farming operation.

The procedure outlined in Figure 2.2 provides a means of calculating residual returns to the operator's labor and management on both an annual and a monthly basis (lines 4 and 4a, respectively). Alternatively, the opportunity cost of labor can be subtracted (lines 3a and 3b) to obtain the total residual return to capital (line 5) and the residual return per dollar of equity capital (line 5a). Many farm management associations and recordkeeping services calculate returns to unpaid labor, management, and equity capital for farmers by area and type of farm. Such results are avail-

Figure 2.2 A Procedure to Allocate Returns
to Contributed Resources

	Calculating Residual Returns to:	
	Operator's Labor and Management	*Equity Capital*
(1) Farm profit or loss (returns to unpaid labor, operator's labor, equity capital, and management) (Figure 2.1, line (12))	_____	_____
(2) Interest on equity in the business $_____ × _____%	_____	XXX
(3) Opportunity cost of unpaid labor **a.** Operator's unpaid labor _____ months × $_____ per month	XXX	_____
b. Other unpaid labor _____ months × $_____ per month	_____	_____
(4) Residual return to operator's labor and management (1) − (2) − (3b)	_____	XXX
a. Operator's labor and management earnings per month ((4) ÷ _____ months operator's labor)	_____	XXX
(5) Residual return to equity capital in the business (1) − (3a) − (3b)	XXX	_____
a. Rate of return on equity capital ((5) ÷ $_____ equity)	XXX	_____

able through the Agricultural Extension Service and provide a basis to compare how well an individual farm is doing relative to similar farms in the area.

An Example
An example of a net income statement is presented in Figure 2.3 for Farmer A for year 1. Farmer A had cash receipts from milk, sales of young stock produced, slaughter hogs, and corn sales. Other cash operating receipts were government program payments, gas tax refunds, cooperative cash patronage refunds, and some miscellaneous cash receipts. Cash farm operating expenses are listed for the year. Notice that capital sales (of breeding stock, machinery, etc.) are not listed under operating receipts. Likewise, the purchase of capital items is not included under farm operating expenses. Farmer A's family consumed $427 of milk and pork produced on the farm during the year. Including the adjustment for changes in inventory and changes in capital items results in a farm profit of $41,194 for the year. The family had off-farm income of $8,493 and proprietor withdrawals (primarily for family-living expenses and payment of income tax) of $14,367, resulting in an addition to retained earnings of $35,320.

Figure 2.3 Income Statement: An Example

Period Covered: **January 1 Year 1 to December 31 Year 1**

Farm Operating Receipts				*Farm Operating Expenses*		
Livestock and livestock products				Seed	$ 2,200	
	Units			Fertilizer	10,764	
				Chemicals and other		
Milk	( )	$ 65,320		crop supplies	4,043	
Calves	( )	6,951		Machine hire	4,693	
Sl. Hogs	( )	7,460		Storage	2,377	
	( )			Feed purchased	6,031	
	Subtotal	$ 79,731	(1)	Feeder livestock bought	0	
Crop sales				Breeding	840	
Corn	( )	$ 22,644		Veterinary	1,228	
	( )			Livestock supplies	1,500	
	( )			Fuel and oil	4,356	
	( )			Utilities	2,066	
	Subtotal	$ 22,644	(2)	Machinery repairs	3,480	
Other operating receipts				Other repairs	4,068	
Program payments		$ 4,050		Taxes (real estate, sales)	2,942	
Gas tax refunds		673		Insurance	520	
Patronage refunds and				Rents	3,300	
miscellaneous		5,937		Trucking and market	0	
	Subtotal	$ 10,660	(3)	Hired labor	2,056	
Gross farm operating				Farm interest paid	19,433	
receipts (1) + (2) + (3)		$113,035	(4)	Other	2,000	
				Gross farm operating		
				expense	$ 77,897	(5)
				Net cash operating income		
				(4) − (5)	$ 35,138	(6)

Value of Farm Products Produced on the Farm and Consumed by the Household	$ 427	(7)

Adjustment for Changes in Inventory

	Feed and Grain	Market Livestock	Accounts Receivable	Supplies and Prepaid Expenses		Accounts Payable		
Ending inventory	78,810	600	5,572	3,000	Begin	13,547		
Beginning inventory	64,975	3,970	5,495	2,000	End	13,247		
Net adjust-ment	13,835	(3,370)	77	1,000		300	$ 11,842	(8)

(*continued*)

Figure 2.3 *(Continued)*

Adjustments for Changes in Capital Items

	Breeding Livestock	Machinery and Equipment	Buildings and Improvements			
Ending inventory	42,100	42,798	82,530			
Plus sales	3,488	0	0			
Subtotal	45,588	42,798	82,530	(9)		
Beginning inventory	31,525	36,398	80,949			
Plus purchases	300	19,344	8,613			
Subtotal	31,825	55,742	89,562	(10)		
Net capital adjustment (9) − (10)	13,763	(12,944)	(7,032)		(6,213)	(11)

Farm profit or loss (6) + (7) + (8) + (11) (returns to unpaid labor, operator's labor, equity capital, and management)	41,194	(12)
Off-farm income	8,493	(13)
Total net income (12) + (13)	49,687	(14)
Proprietor withdrawal	14,367	(15)
Addition to retained earnings (14) − (15)	35,320	(16)

In this example, Farmer A provided 12 months of labor, and the family provided the equivalent of one month for a total of 13 months of unpaid labor. The operator's net worth in the farm business was $291,445 on January 1 and $326,765 on December 31 for an average of $309,105. Farmer A estimates that a 6 percent rate of cash return is needed on equity in the business to provide a total return equal to the return that can be earned on alternative investments having comparable risks.[3] Furthermore, the opportunity cost for unpaid labor is $1,000 per month for the unpaid family labor and $1,500 per month for the operator.

The calculation of residual returns is illustrated in Figure 2.4. Subtracting a charge of 6 percent for the use of equity capital and $1,000 for the month of unpaid family labor results in residual returns to operator's labor and management of $21,648 or $1,804 per month. This is somewhat more than the opportunity cost of $1,500 per month the operator listed. Alternatively, it is not surprising that when the opportunity cost of $19,000

[3]Farmer A notes that approximately 80 percent of the business equity is the net worth in land and permanent buildings. These assets have increased an average of 5 percent per year over the last 20 years in Farmer A's county. The other assets will not appreciate in value. Thus, Farmer A estimates that approximately a 4 percent average rate of noncash return (0.80 × 0.05) will be earned on equity in the business. After accounting for the difference in taxes, Farmer A can earn a 10 percent rate of return on alternative investments having comparable risks. A 6 percent cash return on farm assets will, therefore, provide a comparable rate of return (cash and noncash) on farm assets.

Figure 2.4 An Example of Allocating Residual Returns to
Contributed Resources

		Calculating Residual Returns to:	
		Operator's Labor and Management	*Equity Capital*
(1)	Farm profit or loss (returns to unpaid labor, operator's labor, equity capital, and management) (Figure 2.1, line (12))	$41,194	$41,194
(2)	Interest on equity in the business $309,105 × 0.06	$18,546	XXX
(3)	Opportunity cost of unpaid labor		
	a. Operator's unpaid labor 12 months × $1,500 per month	XXX	$18,000
	b. Other unpaid labor 1 month × $1,000 per month	$1,000	$1,000
(4)	Residual return to operator's labor and management (1) − (2) − (3b)	$21,648	XXX
	a. Operator's labor and management earnings per month ((4) ÷ 12 months operator's labor)	$1,804	XXX
(5)	Residual return to equity capital in the business (1) − (3a) − (3b)	XXX	$22,194
	a. Rate of return on equity capital ((5) ÷ $309,105 equity)	XXX	7.18%

is allocated to unpaid labor and management, the residual rate of return on
equity capital (7.18 percent) is more than the 6 percent opportunity cost.
The operator may want to compare these returns with those obtained by
other family dairy farms in the area. In the event Farmer A's returns are
significantly below those of the high-return farms in the area, a comparison
of production and marketing efficiency with other farms may suggest areas
where Farmer A can improve his or her performance.

THE BALANCE SHEET

The balance sheet, also commonly known as the net worth or the financial
statement, provides a picture of the financial characteristics of the firm *at a
point in time*. The balance sheet systematically lists all assets and liabilities of
the business on a particular date. This provides a measure of the *stock* of
real and financial assets included in the firm, while the income statement
measures the *flow* of income generated by this stock of assets. Hence, the
balance sheet measures a stock at a point in time, whereas the income
statement measures a flow over a period of time.

The balance sheet provides data on two financial characteristics of the firm—solvency and liquidity. Solvency refers to the firm's ability to meet its financial obligations over a long period of time. Liquidity is a short-run characteristic; it refers to the firm's capacity to generate enough cash to meet its financial commitments as they fall due and to provide for unanticipated events. Lenders use the financial statement as the base document to analyze the ability of the farm operation to handle additional borrowed funds. Thus, it is important for the farm operator to understand how to prepare the balance sheet properly.

The structure of the balance sheet is derived from the basic accounting equation: Assets = Liabilities + Equity (or Net worth). An asset includes anything of value in possession of the operator and a claim on anything of value in the possession of others. Assets include such items as land, machinery and equipment, buildings, livestock, grain inventories, supplies, accounts receivable, and cash. Note that machinery leased or rented on a seasonal basis and land rented on a year-to-year basis for cash rent may be part of the capital managed by the farm operator, but such rented items are not included as assets on the farmer's financial statement or balance sheet. However, the longer term capital and operating leases for machinery and land represent a future flow of services which have a value and entail a commitment of future funds for lease payments (a liability). The value of these assets should be reflected on the balance sheet. Procedures to value these and other assets are discussed in the next section.

Someone has a claim on each asset of the farm business. This claim is based on the source of funds used to acquire the asset. The farm operator may have used earnings from previous years to acquire the machinery and equipment or breeding herd. Thus, the operator has a claim on those assets. In contrast, money may have been borrowed from the banker or Production Credit Association to purchase feeder cattle. Consequently, the lender has a claim on the feeder cattle in the amount of the loan used to purchase them. And the farmer may have used savings plus money borrowed from the Federal Land Bank or an insurance company to purchase land. Therefore, both the farmer and the lender have claims against the real estate. Thus, there are financial claims against all assets. Those claims held by the farm operator are referred to as equity, whereas claims held by individuals who are not part of the farm operation are known as debts or liabilities. Examples of liabilities are accounts payable, notes payable, and mortgages payable.

The basic accounting equation requires that the value of claims in the form of liabilities plus the value of claims in the form of equity must be equal to the value of assets. Because liabilities represent a higher priority claim than equity under the laws of business finance, the value of the equity claim or net worth is typically calculated as a residual value. Thus, assets are valued on the financial statement, the amount of debt obligations is subtracted, and any remaining value is imputed to equity or net worth.

In developing a balance sheet, it is necessary to distinguish between the balance sheet of the business and the balance sheet of the proprietor(s). It is partcularly important to make this distinction for businesses operated as partnerships and corporations. The distinction should also be made for a sole proprietorship. The distinction is important because farm operators typically have some assets and they may also have some liabilities that are not part of the farm business. Examples of assets commonly held by farm operators that are outside of the business are household belongings, stocks and bonds, the loan value of life insurance, recreational vehicles, and the house if it is separated from other farm real estate. Personal credit accounts and loans for consumer items are examples of liabilities that are not part of the business.

Agricultural lenders typically prefer personal as well as business balance sheets for the proprietor or proprietors of farm businesses. A summary of personal assets and liabilities as part of the farm balance sheet is the minimum requirement that is usually acceptable. Personal and business balance sheets enable lenders to review the total financial resources and obligations before agreeing to provide financing. The example balance sheets used here are arranged to provide totals for business assets and liabilities, while also including the personal assets and liabilities which lenders require. Farm businesses with multiple owners (partnerships and corporations) will want to maintain a separate balance sheet for the business and a personal balance sheet for each owner. A balance sheet listing combined personal assets and liabilities for two or more owners will not provide the data needed to calculate the liquidity and solvency of each owner. A personal balance sheet is required for each owner to provide this information.

Categories of Assets and Liabilities

The format of the balance sheet for a farm business is shown in Figure 2.5, with assets on the left side of the statement and liabilities on the right side. It is important for the farmer to understand the structure of the balance sheet and how it is completed. Incomplete information or improperly organized data can lead to incorrect analysis of the business's financial strength. Consequently, each section of the financial statement will be discussed in detail, with emphasis on classification of assets and liabilities. Then the discussion will turn to methods of valuing assets.

Business assets are divided into three categories—current, intermediate, and long-run. Personal assets are included in a fourth section of Figure 2.5 to separate them from the business assets. Current assets consist of cash and other property that will be converted into cash through the normal

Figure 2.5 The Balance Sheet

Name _____

Date _____

Assets

Current Business	Cost or Basis	Market Value
1. Cash and checking account	_____	_____
2. Farm notes and accounts receivable	_____	_____
3. Livestock held for sale	_____	_____
4. Crops held for sale and feed	_____	_____
5. Value in growing crops	_____	_____
6. Farm supplies	_____	_____
7. Prepaid expenses	_____	_____
8. Other	_____	_____
9. Total current assets	$_____	$_____

Intermediate Business

	Cost or Basis	Market Value
10. Machinery, equipment, and vehicles		
11. Breeding livestock		
12. Movable farm buildings		
13. Securities not readily marketed		
14. Other		
15. Total intermediate assets	$_____	$_____
16. Total current and intermediate assets	$_____	$_____

Liabilities

Current Business	Cost or Basis	Market Value
27. Accounts payable	_____	_____
28. Notes payable within 12 months	_____	_____
29. Principal payments on longer term debts due within 12 months	_____	_____
a. Real estate		
b. Other	_____	_____
30. Estimated accrued interest	_____	_____
31. Estimated accrued tax		
a. Property		
b. Income and social security		
c. Other	_____	_____
32. Accrued rent	_____	_____
33. Total current liabilities	$_____	$_____

Intermediate Business

	Cost or Basis	Market Value
34. Deferred principal owed		
35. Deferred accounts payable		
36. Deferred notes payable		
37. Contingent income tax liabilities	XXX	
38. Total intermediate	$_____	$_____
39. Total current and intermediate liabilities (33) + (38)	$_____	$_____

Long-Term Business

17. Farmland _____

18. Permanent buildings and improvements _____

19. Other $ _____

20. Total long-term assets $ _____

21. Total business assets (16 + 20) $ _____

Personal

22. Current
   a. Cash, checking account, savings _____
   b. Time certificates _____
   c. Readily marketable securities _____
   d. Other _____

23. Intermediate
   a. Retirement accounts _____
   b. Cash value of life insurance _____
   c. Nonfarm equipment _____
   d. Other _____

24. Long-term
   a. Contracts and notes receivable _____
   b. Nonfarm real estate _____
   c. Other _____

25. Total personal assets $ _____

26. Total assets (21 + 25) $ _____

40. Deferred principal on-farm real estate _____

41. Other _____

42. Contingent capital gains tax liability on real estate XXX

43. Total long-term liabilities $ _____

44. Total business liabilities (39 + 43) $ _____

45. Net worth (21 − 44) _____

46. Total business liabilities and net worth $ _____

Personal

47. Current
   a. Personal accounts payable _____
   b. Principal payments on personal longer term debts due within 12 months _____
   c. Other _____

48. Intermediate
   a. Life insurance loans _____
   b. Deferred principal payments on nonfarm intermediate accounts and loans _____
   c. Other _____

49. Long-term
   a. Deferred principal payments on nonfarm real estate _____
   b. Other _____

50. Total personal liabilities $ _____

51. Total liabilities (44 + 50) $ _____

52. Net worth (26 − 51) $ _____

53. Total liabilities and net worth (51 + 52) $ _____

operation of the farm during the course of the business year. Included here are all business assets held for sale or feeding, such as grain and forage inventories, as well as livestock that will be sold during the normal course of business, such as market hogs and feeder cattle. Farm supplies in inventory, such as veterinary and medical supplies, seed, fertilizer, and chemical inventories, are also included in current assets. Other business assets that can be quickly converted into cash without disrupting the normal operation of the business should also be included here. However, assets such as savings accounts or certificates of deposit may be personal assets and, if so, should be included in the personal category.

Intermediate assets include short-lived capital resources used in farm production that will not be sold or converted into cash during the coming year through the normal operation of the business. They are distinguished from long-term assets because they have a shorter useful life—usually less than 7 to 10 years. Intermediate assets include such items as machinery, equipment, vehicles, and breeding stock. Note that breeding livestock, machinery, and equipment are not included as current assets because they usually will not be sold during the current year "in the normal course of business." Although some intermediate assets such as breeding livestock and machinery could be sold to pay off debt obligations, such transactions involve selling off a productive part of the farm business.

Long-term assets are permanent and consist primarily of farmland and improvements. Many farm operators include land and improvements on the financial statement as a single entry, but separate values can be useful to indicate the amount and value of capital improvements made on a piece of property since its original purchase. The summation of current, intermediate, and long-term assets equals the total business assets of the farm operator.

Personal assets can also be subdivided into current, intermediate, and long-run categories. Current personal assets include personal checking and savings accounts, time certificates, and marketable stocks and bonds. Common personal intermediate assets include retirement accounts, the cash value of life insurance, personal vehicles, and household goods. Long-term personal assets include contracts and notes receivable and nonfarm real estate. Finally, the operator's total assets recorded on line 26 are the sum of both personal and business assets.

Liabilities or claims on assets are shown on the right side of the balance sheet as illustrated in Figure 2.5. The business liabilities are also categorized into three groups—current, intermediate, and long term.

Current liabilities include debt obligations that are payable on demand or within the operating year—normally 12 months. Examples of current liabilities include operating notes, cattle notes, and accounts payable. Accrued rent, taxes, and interest are also classified as current liabilities. In general, only that portion of the rent, property and real estate taxes, and interest *that have accrued to date* should be listed.[4] These accrued liabilities are included because they would have to be paid if the business were liquidated (with all assets sold and all liabilities paid) on the date the financial statement was prepared. They are listed as current liabilities because they must be paid from the sale of current assets. Current liabilities also include the principal payments on intermediate and long-term debts which are due within the next 12 months. Just like cattle loans, the funds that will be used to make the principal payments on intermediate and long-term debts must come from current assets. Including the proportion of these longer term obligations due this year as a current liability provides a better indication of these payments and the liquidity requirements of the farm business.

Intermediate liabilities typically have terms of one year up to 7 to 10 years. Examples of intermediate liabilities include nonreal estate loans for farm improvements, equipment purchases, and breeding stock. Long-term liabilities include obligations on long-term assets such as real estate mortgages and land contracts. Again, the amount of principal payment due this year on intermediate and long-term obligations is included as a current liability. Only the remainder of the loan principal (total principal less the principal due within the next 12 months) is listed as a deferred liability.

The treatment of leased property presents unique problems in the development of financial statements. In the past, a large majority of the leases for machinery and equipment, as well as farmland, were only of one-year duration. Financial leases for machinery and equipment are now commonly of three or more years' duration. An increasing proportion of land leases are also written for more than one year. Such leaseholds entail a liability that would have to be paid if the firm went out of business. They also entitle the firm to a flow of services that have a value and are an asset to the firm. Most lenders have been content to include a written notation at the bottom of the statement describing the leasing arrangement rather than reflect leasehold values directly in the financial statement. In many cases, leased machinery and equipment was ignored completely in the completion of the statement. But with recent innovations in financial leasing of machinery, equipment, and livestock facilities, lenders are becoming increasingly aware of the value and obligations associated with leasing arrangements and leaseholds. The accounting profession has established ac-

---

[4]When the balance sheet is completed at the end of the year and the income tax has not been paid, the estimated income tax should be listed as a current liability. However, the income tax liability is usually listed as a *personal liability* for sole proprietors, partnerships, and Subchapter S corporations, because income taxes are a personal rather than a business liability for these forms of business organization. Some individuals with significant amounts of off-farm income may want to make an exception to the rule, listing the portion of the income tax liability from farm earnings on line 31b and the tax liability for nonfarm earnings on 47c. Doing so may help them more accurately reflect the source of their total tax liability.

counting standards indicating that (1) the market value of capital items acquired using long-term financial leasing arrangements should be included as assets in the financial statement, and (2) the present value of the lease payments should be included on the liability side of the financial statement. This procedure recognizes that a leasehold has value, and the net benefit of leasing property may be positive or negative depending upon the financial and tax benefits of the leasing terms.

The procedure for handling financial leases suggested here is only a slight variation of that used by the accountant; the variation is recommended to simplify the computations. Assets that are acquired using long-term financial leasing agreements should be listed separately on the asset side of the balance sheet at fair market value in much the same fashion as those assets purchased with debt or equity funds. An offsetting entry should be placed on the liability side of the balance sheet to reflect the value of the lease payments for the leased asset. The most accurate entry would be the present value of the lease payments determined using the firm's cost of capital (as discussed in Chapter 8) as the rate of discount. A reasonable approximation is to include this year's lease obligation as a current liability, and the difference between the value of the leased asset and the current lease obligation as a deferred liability. If such a procedure is used, the liability associated with leased assets exactly offsets the value of those leased assets, and there is neither an addition to nor a subtraction from the firm's net worth resulting from the leasing arrangement. This method of handling leases explicitly recognizes the financial commitments that are part of a leasing agreement and the economic value associated with the leasehold. This appears to be a much preferred approach to that of ignoring leases and leased assets completely.

Net worth or equity in the business (line 45) is obtained by subtracting total business liabilities from total business assets. The sum of total business liabilities (line 44) and business net worth (line 45) must equal total business assets (line 21). This is a consequence of the basic accounting equation. As mentioned earlier, net worth is the owner's financial claim on assets; it indicates the amount the owner would receive if the business were sold and all financial obligations paid.

The personal liabilities also are divided into the current, intermediate, and long-term categories. After recording the personal liabilities, the operator's total liabilities and the operator's net worth or equity can be calculated (line 52).

Methods of Valuation for the Cost or Basis Column

The major problem in completing the balance sheet is placing a value on each asset.[5] Accounting convention holds to valuation of assets based on

---

[5]Completion of the balance sheet requires both a physical count of the assets and valuation of the assets. Some of the more difficult aspects of completing the physical count involve estimat-

costs even though some of the assets may have been purchased many years ago.[6] With cost-based accounting, capital assets are valued at the original cost or basis less the accumulated depreciation since purchase. Cost-based accounting may be used to compute beginning and ending values of capital assets included on the income statement. Completion of the balance sheet periodically (at the end of each accounting period), using cost-based valuation procedures for capital assets, also provides the data to analyze the financial progress that is made over time through reinvestment of earnings in the business. For these reasons, cost-based accounting procedures for capital assets are often used in completing the balance sheet.

Lenders, the farm operator, and others concerned with financial management of the farm business commonly express a need for a balance sheet based on current market values. The current market value of capital assets purchased in the past is frequently quite different from the cost or basis less depreciation value on the depreciation schedule kept for income tax purposes. Thus, cost-based accounting may not correctly reflect market values of capital assets and owner's equity. The legitimate reasons for using both cost-based accounting and current market valuation of capital assets often result in developing a balance sheet that is a combination of the two. This makes it difficult to interpret the balance sheet and use it for any purpose. A better alternative is to include two columns—one for cost or basis values of capital assets and a second for current market values.

An operator who completes the cost or basis valuation at the beginning and end of the year can calculate the change in net worth from reinvestment of farm income. By comparing the market value column at the beginning and end of the year, the operator can determine the combined change in equity from reinvestment in the business and change in the market value of assets. The difference between these two estimates of change in net worth represents the change in net worth due to change in asset values. Both estimates are of interest to farmers and lenders.

A true cost-based accounting system would generate a cost or basis value of assets produced, such as grain and livestock, through allocation of all direct and indirect costs incurred to the date of the statement. Farm record systems typically include data on the cost and depreciation taken for tax purposes on capital assets, but accounts that allocate all direct and indirect costs to the crop and livestock products produced are uncommon.

---

ing the quantity of purchased inputs (such as protein feed) and the amount of products (such as grain, silage, and hay) in bulk storage. Procedures to make these estimates are presented by other authors. For example, see *Midwest Farm Planning Manual,* 4th ed., Sydney C. James, editor (Ames: Iowa State University Press, 1979), pp. 246–47, and the *1981–82 Agricultural Engineers Yearbook,* American Society of Agricultural Engineers, St. Joseph, Michigan, June 1981.

[6]The application of the double column balance sheet to agriculture was introduced by Thomas L. Frey and Danny A. Klinefelter, *Coordinated Financial Statements for Agriculture,* Agr. Finance (Skokie, Ill., 1978).

Thus, a set of valuation procedures has been developed which are a modification of the procedures recommended by professional accounting associations for nonagricultural businesses.

Three methods of valuation—net selling price, cost, and original cost or basis less depreciation—are commonly used to value assets for the cost or basis column on the balance sheet. A brief description of each method and recommendations for their use are given below.

Net selling price refers to the market price of the asset less any selling costs that would be involved in disposing of the asset. This method is normally used to value crops that are held for either sale or feed and livestock on feed that will be sold within a one-year period.

The *cost* method is used to value purchased farm supplies such as fertilizer, feed, medication, and lubricants. For example, a farmer who purchased fuel for $1.00 per gallon before the end of the accounting period finds that the cost on the date of the inventory is $1.05 per gallon. This farmer should use the amount paid ($1.00) in valuing the inventory. If the manager purchases an input more than one time during the year, the inventory can be valued based on either the first-in first-out (FIFO) or last-in first-out (LIFO) method. Farmers traditionally use the FIFO method to keep inventory values more current.[7]

The *original cost or basis less depreciation* method (commonly referred to as the book value) is the value of the asset on the depreciation schedule that is kept for tax purposes. This method is normally used with a cost or basis accounting system for working assets such as machinery, purchased breeding stock, and buildings. The basis of the property is a way of measuring the investment in property for tax and other purposes. The basis of property purchased for money is usually its cost. This basis may be increased by improvements (such as the cost of overhauling a tractor or remodeling a building) and may be decreased by deductions for depreciation, depletion, and casualty losses (resulting from fire, storm, flood, and other types of involuntary conversions). This increased or decreased basis is called the adjusted basis. Perhaps the most common means of acquiring assets on farms is either purchasing for cash or trading in a similar item and paying some cash "to boot." The basis for cash purchases is typically the amount paid. The basis for when an item is traded in is usually calculated as the adjusted basis of the item traded in plus the additional cash paid to purchase the replacement item. For example, a machine that is purchased for $10,000 cash without an item traded in would have a basis of $10,000. If

---

[7]Farmers filing income tax on the accrual basis will probably want to use the LIFO method during periods of inflation and FIFO during periods of deflation to reduce the amount of income subject to taxation. Limitations on the use of these methods are given in the current edition of the *Farmer's Tax Guide,* Publication 225 (revised annually) U.S. Department of the Treasury, Internal Revenue Service, Washington, D.C.

the same item is purchased by trading in a similar machine with an adjusted basis (book value) of $1,000 and the farmer pays $8,000 cash in addition, the basis of the replacement machine is $9,000. The adjusted basis of this machine is calculated through time as the basis plus the value of improvements, minus the value of deductions for depreciation and the other items noted above. The reader is referred to a standard accounting text or the *Farmer's Tax Guide* for the exceptions and additions to these basic rules for calculating basis and adjusted basis.

The following guidelines are suggested to develop asset values for the cost or basis column of the balance sheet.

1. All assets that are to be sold within one year, such as feeder cattle, feeder lambs, hogs on feed, turkeys and broilers on feed, veal calves, and crop products being held either for feed or sale, should be valued at net selling price.

2. All supplies, including purchased feed, fertilizer, and other supplies, should be valued at cost.

3. Crops growing in the field are normally valued at the amount of cash costs that have been spent on the crops, unless, due to weather or other problems, it appears the value of the crops will not be as great as the cash costs.

4. Working capital assets, including machinery and breeding stock, should be valued at the adjusted basis listed in the depreciation schedule kept for income tax purposes.

5. Real estate improvements, such as farm buildings, fences, and tiling, should be valued at the adjusted basis listed on the depreciation schedule kept for income tax purposes.

6. Farmland is normally valued at its basis plus any adjustments on depreciable improvements included as part of the land.

Completion of the cost or basis valuation column emphasizes that assets valued at their adjusted basis may not reflect current market values. Furthermore, the cost or basis approach does not recognize that part of the return in agriculture, particularly with respect to real estate, occurs in noncash form as price appreciation.

Methods of Valuation for the Market Value Column

The purpose of the market value column is to more accurately reflect the asset values, liabilities, and owner's equity in current dollar values. Because the cost or basis valuation methods suggested are really a modified cost basis, identical values are used in both columns for all farm business assets except the machinery and equipment in the intermediate section and real estate in the long-term category. Personal assets that may have different market and cost values include readily marketable securities, nonfarm equipment, and nonfarm real estate.

Well-established markets should be used to estimate the current value of readily marketable securities using the net selling price method. But the market for many used capital assets, such as buildings, fence, grain bins, livestock equipment, concrete feeding floors, and other items either sold with the land or sold infrequently, is not well established. For this reason, three additional methods of valuing capital assets may be of interest. They are replacement cost less depreciation, replacement cost for equivalent function less depreciation, and income capitalization. Since two of these methods require the estimation of depreciation, we will discuss alternative methods of calculating economic depreciation and then explain the use of the three valuation methods.

Computing Depreciation   The concept of economic depreciation involves the prorating of the original cost of an asset over its useful life at a rate that reflects the approximate rate at which the asset's total service flow is consumed. Note that we suggest using this concept in estimating the value of an asset for the market value column only when well-established markets for used items do not exist. The initial part of the section discusses alternative depreciation methods and recommends the method that appears to most accurately reflect the loss in productive value for major groups of assets. These methods of calculating depreciation for tax purposes were replaced for assets purchased after 1980 by the accelerated cost recovery system. Current procedures for calculating the amount of an asset's value that can be deducted for income tax purposes are discussed in Chapter 8.

Two of the major decisions in calculating economic depreciation are to determine the length of time over which the asset should be depreciated (the length of useful life) and to ascertain which method of calculating depreciation should be used. The length of useful life should be based on the amount of use and obsolescence. Obsolescence provides an upper limit on the length of life. If the annual use is great enough to wear out the asset before it is obsolete, the length of useful life should be based on its wear-out life. For example, if the wear-out life of a machine is 2,250 hours and the machine is used 100 hours per year, this implies the machine can be used 22½ years before it will wear out. However, the farmer may plan to replace the machine after 10 years of use because of obsolescence. In this case, the farmer would use a life of 10 years rather than the wear-out life. If the individual uses the machine 300 hours per year, the wear-out life is 7 to 8 years. Thus, one can select the years of useful life based on wear-out life or obsolescence, whichever is shorter.

There are three common methods of figuring economic depreciation: the straight-line method, the declining balance method, and the sum-of-the-years'-digits method. The method of calculating each is explained briefly below.

The formula for calculating depreciation with the straight-line method is:

$$\text{Depreciation per year} = \frac{\text{Purchase cost} - \text{Salvage value}}{\text{Years of useful life}} \quad (2.1)$$

This is the easiest of the three methods to compute and the most commonly used method of calculating depreciation for farm assets. The example illustrated in Table 2.1 has a new cost of $10,000, a salvage value of $2,000, and a useful life of 10 years. Then:

$$\text{Depreciation per year} = \frac{\$10,000 - \$2,000}{10} = \$800$$

The depreciation shown in column 1 of Table 2.1 is the same in each of the 10 years, and the asset is depreciated to a remaining value (the salvage value) of $2,000 at the end of 10 years' life.

The declining balance method of depreciation uses a fixed rate of depreciation each year and applies the rate to the value of the asset at the beginning of the year. The formula is:

$$\text{Depreciation in a specific year} = \text{Rate} \times \begin{array}{l}\text{Undepreciated}\\\text{value at the}\\\text{start of the year}\end{array} \quad (2.2)$$

Notice in this method that the salvage value is not subtracted from the purchase cost in making the calculation. If we use the example above and the rate of 20 percent per year, the depreciation in the first year is 0.2 of $10,000 or $2,000—see column 2 of Table 2.1. The depreciation in the second year is 0.2 times $8,000 or $1,600. The calculation in succeeding years is made by determining the undepreciated value at the beginning of

Table 2.1 Annual Depreciation with Each of Three Methods for a Machine with a Purchase Cost of $10,000, a Salvage Value of $2,000, and a Useful Life of 10 Years

Year	Straight-Line Annual Depreciation (1)	Declining Balance Annual Depreciation (2)	Sum-of-the-Years'-Digits Annual Depreciation (3)
1	$-800.00	$2,000.00	$1,454.55
2	800.00	1,600.00	1,309.09
3	800.00	1,280.00	1,163.64
4	800.00	1,024.00	1,018.18
5	800.00	819.20	872.73
6	800.00	655.36	727.27
7	800.00	524.29	581.82
8	800.00	97.15	436.36
9	800.00	0.00	290.91
10	800.00	0.00	145.45
Total depreciation	$8,000.00	$8,000.00	$8,000.00

the year and multiplying by the same rate as was used in previous years. The calculation continues year by year until the undepreciated balance is reduced to the salvage value. The time required to depreciate the asset to its salvage value depends on the relative size of the salvage value compared to the purchase cost and the rate selected. In the example (Table 2.1), the asset is depreciated to salvage value during the eighth year.

The third method of calculating depreciation is the sum-of-the-years'-digits method. The amount of depreciation deducted in a given year is calculated using the following formula:

$$\text{Depreciation in a specific year} = \frac{\text{Number of years of useful life remaining}}{\text{Sum-of-the-Years'-Digits}} \times (\text{Purchase cost} - \text{Salvage value})$$

$$(2.3)$$

The sum-of-the-years'-digits is simply the sum of the numbers 1 through N, where N is the number of years of useful life. For instance, if the estimated life for the example asset is 10 years, then the fraction is:

$$\frac{\text{Number of years of useful life remaining}}{1 + 2 + 3 + 4 + 5 + 6 + 7 + 8 + 9 + 10} = \frac{n}{55}$$

The value of the fraction for the first year is 10/55, which is then multiplied times the purchase cost minus salvage value. Depreciation for the second year is 9/55 times purchase cost minus salvage value, and so on. An alternative to calculating the denominator of the fraction by summing the digits 1 through N is to use the formula $N(N + 1)/2$. If we calculate the depreciation for the $10,000 asset, the amount of depreciation in year 1 is 10/55 ($10,000 − $2,000) or $1,454.55 as shown in column 3 of Table 2.1. Depreciation for year 2 is equal to 9/55 ($10,000 − $2,000) or $1,309.09. In year 10, the depreciation is equal to 1/55 ($10,000 − $2,000) or $145.45.

In selecting the method of depreciation to be used in valuing assets for the market value column of the balance sheet, consideration should be given to how the asset is expected to decline in value over time. Most machinery and equipment items tend to depreciate more rapidly during the early years of their useful life. In addition, repairs tend to be greater in later years. For these types of assets, the declining balance method (with a rate 1.5 to 2.0 times the straight-line rate[8]) and the sum-of-the-years'-digits

---

[8]The "straight-line rate" referred to is obtained by dividing 100 percent by the number of years of useful life selected for the asset. For example, a machine with a useful life of 8 years has a "straight-line rate" of 100 ÷ 8 or 12.5 percent. Then 1.5 times the straight-line rate is 18.75 percent, while 2.0 times the straight-line rate is 25 percent. The example in Table 2.1 assumes a 10-year life and 2.0 times the straight-line rate for a rate of 20 percent or 0.20.

method do a better job of representing the remaining value of the asset than the straight-line method. The straight-line method tends to assign too little depreciation to early years and too much to later years. Because of the uncertain useful life, depreciation for breeding stock should be calculated with either the declining balance (with a rate of 1.5 to 2.0 times the straight-line rate) or sum-of-the-years'-digits method. On the other hand, many real assets, such as buildings and fences, decline very gradually in value over a period of time; the straight-line method does the best job of representing this gradually declining value.

The recommended method of calculating economic depreciation must be used with one of the following methods of valuing capital assets for the market value column of the balance sheet.

Replacement Cost Less Depreciation   Replacement cost less depreciation may be used during periods of inflation, instead of original cost or basis less depreciation used for the cost or basis column, to calculate the current value of long-lived assets. For example, assume that a pole-type building used for machinery storage was constructed 10 years ago at a new cost of $10,000. If the building is depreciated to a zero salvage value over 20 years on a straight-line basis, the current value will be equal to one-half the purchase cost, or $5,000. If it currently costs more than $10,000 to build the same structure, using replacement cost less depreciation will result in a somewhat higher current value. Suppose that the replacement cost is $18,000. In this case, the current value estimated with replacement cost less depreciation, assuming the 20-year life, a zero salvage value, and the straight-line method, is $9,000. This may be a useful method of valuing working assets when the purpose is to estimate the market value of owner's equity, but it *cannot be used* with a cost-based accounting system.

This method of estimating the value of the asset may result in over-valuation, either if the function of the asset has changed over its useful life or a technologically superior asset has been developed. In this event, the following method may be more appropriate than replacement cost less depreciation.

Replacement Cost for Equivalent Function Less Depreciation   Replacement cost for equivalent function, not necessarily equivalent asset, less depreciation considers the changing function of the asset and emphasizes the asset's effect on the income position of the business. It is particularly useful for buildings that have been converted from one use to another. Suppose a farmer constructed a concrete block building for swine feeding five years ago at a cost of $9,600. The farmer estimates that the building has a 20-year life. The current value of the building calculated with the original cost or basis less depreciation method, a zero salvage value, and a straight-line method of depreciation is $7,200. Now assume that the building is no longer used for swine feeding but has been converted to storage of machinery and supplies. The replacement cost for a similar amount of machinery storage is $5,200. Since the building is five years old, replacement cost for equivalent function is $5,200 minus $1,300, or $3,900. This is somewhat less than the value obtained using original cost or basis less depreciation. It is also reasonable to assume that it would be considerably less than the value obtained using replacement cost less depreciation, since the cost of building an equivalent structure would have also increased during the interim. Like the previous method, this approach may be a useful method of valuing working assets when the purpose is to estimate owner's equity on a market value basis, but it cannot be used with a cost-based accounting system.

Income Capitalization Income capitalization may be used to value assets that have a long life and contribute to the income of the business over the period. The approach calculates the current or present value of income to be received from the asset in the future. The current or present value equals the value of the income stream generated over its useful life. This method of valuation is commonly used in valuing land and capital investment decisions. A discussion of the procedures used to value assets using the income capitalization approach is given in Chapter 8.

In summary, most assets are assigned the same value in the market value column as they have in the cost or basis column on the balance sheet. The exceptions are noted below.

1. Working capital assets, including machinery and equipment, may be valued using the net selling price approach when the market is well established. When the market for the used item is not well established, replacement cost less depreciation can be used.

2. Real estate improvements, such as farm buildings, fences, and tiling, can be valued using either replacement cost less depreciation or replacement cost for equivalent function less depreciation.

3. Farmland is normally valued at its net selling price and/or with the income capitalization method.

An Example

An illustration of the balance sheet calculations for Farmer A is presented in Figure 2.6. Farmer A owns 259 acres and rents 74 acres. The 333 acres include 316.5 acres of tillable land that is used to produce corn, corn silage, and alfalfa hay. The major livestock enterprise on the farm is a dairy herd of approximately 40 cows and associated replacement heifer production. Farmer A also has a small farrow-to-finish swine enterprise.

Farmer A's assets are listed on the left side of Figure 2.6. Cash in the business checking account on January 1, Year 1, is listed on line 1. Many of

Figure 2.6 The Balance Sheet: An Example

Name *Farmer A*

Date *January 1, Year 1*

Assets			Liabilities		
Current Business	Cost or Basis	Market Value	**Current Business**	Cost or Basis	Market Value
1. Cash and checking account	1,672	1,672	27. Accounts payable	5,634	5,634
2. Farm notes and accounts receivable	5,495	5,495	28. Notes payable within 12 months	44,640	44,640
3. Livestock held for sale	3,970	3,970	29. Principal payments on longer term debts due within 12 months		
4. Crops held for sale and feed	63,975	63,975	a. Real estate	2,338	2,338
5. Value in growing crops	1,000	1,000	b. Other	12,132	12,132
6. Farm supplies	2,000	2,000	30. Estimated accrued interest	7,913	7,913
7. Prepaid expenses	0	0	31. Estimated accrued tax		
8. Other	0	0	a. Property	0	0
9. Total current assets	$ 78,112	$ 78,112	b. Income and social security	0	0
			c. Other	0	0
			32. Accrued rent	0	0
			33. Total current liabilities	$ 72,657	$ 72,657
Intermediate Business			**Intermediate Business**		
10. Machinery, equipment, and vehicles	36,398	47,420	34. Deferred principal owed	48,000	48,000
11. Breeding livestock	31,525	31,525	35. Deferred accounts payable	0	0
12. Movable farm buildings	0	0	36. Deferred notes payable	0	0
13. Securities not readily marketed	18,778	18,778	37. Contingent income tax liabilities	XXX	3,307
14. Other	0	0	38. Total intermediate	$ 48,000	$ 51,307
15. Total intermediate assets	$ 86,701	$ 97,723	39. Total current and intermediate liabilities (33) + (38)	$ 120,657	$ 123,964
16. Total current and intermediate assets	$ 164,813	$ 175,835			

(continued)

65

Figure 2.6 (Continued)

Assets

Long-Term Business

	Cost or Basis	Market Value
17. Farmland	128,500	310,800
18. Permanent buildings and improvements	80,949	100,800
19. Other	0	0
20. Total long-term assets	$ 209,449	$ 411,600
21. Total business assets (16 + 20)	$ 374,262	$ 587,435

Personal

22. Current		
a. Cash, checking account, savings	780	780

Liabilities

Long-Term Business

	Cost or Basis	Market Value
40. Deferred principal on-farm real estate	119,488	119,488
41. Other	0	0
42. Contingent capital gains tax liability on real estate	XXX	50,538
43. Total long-term liabilities	$ 119,488	$ 170,026
44. Total business liabilities (39 + 43)	240,145	293,990
45. Net worth (21 − 44)	134,117	293,445
46. Total business liabilities and net worth	$ 374,262	$ 587,435

Personal

47. Current		
a. Personal accounts payable	0	0

Assets		
b. Time certificates	0	0
c. Readily marketable securities	0	0
d. Other	0	0
23. Intermediate		
a. Retirement accounts	0	0
b. Cash value of life insurance	2,800	2,800
c. Nonfarm equipment	8,630	8,630
d. Other	0	0
24. Long-term		
a. Contracts and notes receivable	0	0
b. Nonfarm real estate	18,087	18,087
c. Other	0	0
25. Total personal assets	$ 30,297	$ 30,297
26. Total assets (21 + 25)	$ 404,559	$ 617,732

Liabilities		
b. Principal payments on personal longer term debts due within 12 months	0	0
c. Other	0	0
48. Intermediate		
a. Life insurance loans	0	0
b. Deferred principal payments on nonfarm intermediate accounts and loans	0	0
c. Other	0	0
49. Long-term		
a. Deferred principal payments on nonfarm real estate	0	0
b. Other	0	0
50. Total personal liabilities	$ 0	$ 0
51. Total liabilities (44 + 50)	$ 240,145	$ 293,990
52. Net worth (26 − 51)	$ 164,414	$ 323,742
53. Total liabilities and net worth (51 + 52)	$ 404,559	$ 617,732

the other entries for the current and intermediate assets are the beginning inventory values for inventory and capital assets listed on the income statement in Figure 2.3. Thus, most of the entries for business assets in the cost or basis column are self-explanatory. The accounts receivable represent the estimated payment expected for milk and one load of hogs that have been delivered. The book value for the depreciation schedule for tax purposes of machinery and equipment is listed on line 10. The raised breeding stock is valued at net selling price. The securities not readily marketable are composed of stock in several cooperatives with which Farmer A does business. These include stock which Farmer A purchased in the Federal Land Bank when a loan was obtained to buy the farm, and revolving stock received as a result of the business conducted with the input supply cooperative and the milk marketing cooperative. These securities will be converted to cash over a period of time, but they normally cannot be converted to cash on a short-term basis. The personal assets are listed on lines 22 through 24. The nonfarm equipment includes the personal share of the auto and household goods.

The only differences in asset valuation between the cost and the market value columns are for machinery and equipment, farmland, and permanent buildings and improvements. Estimates of market value were used for machinery (where available) and for the land. The replacement cost less depreciation method was used to estimate the value of other equipment and the permanent buildings.

Farmer A's liabilities under the cost or basis method include a real estate mortgage, an intermediate loan, notes payable, and accounts payable. The total amount of principal due, the amount of principal due in the next 12 months, and the interest due to date are summarized below. The principal due in the next 12 months is listed in the current liabilities section, with the deferred principal being listed on the appropriate lines of the intermediate and long-term liabilities' categories. Current liabilities should include the amount of interest due on the date for which the balance sheet is prepared. A more accurate statement of the claims others have on the

Liability	Principal Outstanding	Annual Interest Rate	Interest Paid Through	Interest Accrued to 1/1/Year 1	Principal Due in New Year	Deferred Principal
Real estate mortgage	$121,826	7%	3/31/Year 0	$6,396	$ 2,338	$119,488
Security agreement	60,132	8%	11/30/Year 0	401	12,132	48,000
Notes	44,640	10%	9/30/Year 0	1,116	44,640	0
Accounts	4,834	18%	12/31/Year 0	0	4,834	0

assets of the business is obtained by including interest that has accrued but has not been paid.[9]

The liabilities included in the market value column include two entries that are not listed under the cost column. They are the tax liabilities that would be incurred on the gain listed for machinery and equipment (line 10) and for real estate (lines 17 and 18). The value of machinery and equipment listed in the cost or basis column is assumed to be the adjusted basis for income tax purposes. The amount by which the market value exceeds the adjusted basis is the gain on machinery and equipment which is taxed as ordinary income. The estimated tax liability on machinery is estimated as 30 percent (Farmer A's marginal tax rate) of the gain. Because the increase in value of real estate held for more than one year is taxed as capital gain, only 40 percent of the capital gain is subject to tax. However, the additional income from sale of the farm would push Farmer A into a higher bracket for state and federal income taxes. The effect on tax brackets depends on how receipts from sale of the farm are spread over time and on Farmer A's other sources of income during that period. Given these considerations, Farmer A selects a 25-percent rate to estimate the tax on the real estate gain.[10]

Farmer A's total assets and total liabilities are shown on lines 26 and 51, respectively. Subtracting total business liabilities from total business assets results in net worth, or equity, of $134,117 on the cost basis and $293,445 on the market value basis. These represent two estimates of Farmer A's claim on the assets of the business. Including personal assets and liabilities increases Farmer A's net worth to $164,414 on the cost basis and $323,742 on the market value basis.

Measures of Solvency

The balance sheet provides information on the solvency of the firm or individual. Solvency is concerned with the relationship between the current market value of assets and the claims others have on the business. Various measures of solvency have been developed.

The amount of net worth in the business, $134,117 on the cost basis and $293,445 on the market value basis in the example, is an absolute measure of solvency. Lenders and others concerned with the vulnerability of the operation to changes in the valuation of assets may also want to calculate a relative measure of solvency. One relative measure is the debt/asset ratio.

---

[9]The reader attempting to contrast the entries on the income statement and the balance sheet should note that the entries on lines 27, 30, 31, and 32 are listed as accounts payable on the income statement. Thus, the beginning accounts payable in this example are $5,634 + $7,913 or $13,547.

[10]This is equivalent to a marginal rate of 62.5 percent on 40 percent of the gain.

$$\text{Debt/asset ratio} = \frac{\text{Total liabilities}}{\text{Total assets}} \tag{2.4}$$

In the example, the debt/asset ratio for Farmer A's business is $240,145/$374,262 = 0.64 on the cost basis and $293,990/$587,435 = 0.50 on the market value basis. This indicates that Farmer A has approximately $0.64 of debt for each dollar of assets on the cost or basis method of valuation and $0.50 of debt for each dollar of assets on the market value method. Hence, a large reduction would be required in the value of the assets (of 35 percent or more) to eliminate all of Farmer A's net worth. The absolute measure of solvency provides a measure of the dollars of net worth which the owner has, while the relative measure indicates how vulnerable the operation is to declining asset values. The lender may also want to calculate the debt/asset ratio based on both business and personal assets and liabilities. This is particularly important if the individual has large amounts of nonbusiness assets and/or liabilities. In this case, the debt/asset ratio is decreased to 0.60 and 0.48 for the cost and market valuation methods, respectively.

Perhaps the reason for using both an absolute and a relative measure of solvency can be clarified with a simple example. Assume Farmers X and Y each have a net worth of $200,000. However, Farmer X has assets of $2,000,000, while Farmer Y has assets of $220,000. The absolute measure of solvency is $200,000 in both cases. However, the debt/asset ratio is $1,800,000/$2,000,000 or 0.9 for Farmer X and $20,000/$220,000 or 0.09 for Farmer Y, indicating Farmer X is much more vulnerable to declining asset values.

Preparation of the balance sheet on an annual or more frequent schedule provides the basis to evaluate the direction and magnitude of changes in solvency over time. A comparison of the cost and market value columns also indicates if increases in net worth are being derived from the earnings of the business, an increasing market value of real estate or other assets, or a combination of the two.

Measures of Liquidity

Liquidity is concerned with the capacity to generate enough cash to meet financial obligations as they fall due. The financial structure of many farm businesses is composed of a large proportion of intermediate and long-term assets which typically earn a relatively low cash return. Consequently, the firm may have difficulty generating enough cash receipts to meet current financial obligations. Entries in the current assets and current liabilities sections can be used to indicate the liquidity position of the operation.

Current working capital provides an absolute measure of liquidity. It is calculated as:

Current working capital = Current assets − Current liabilities   **(2.5)**

For Farmer A, current working capital is $78,112 − $72,657 or $5,455. It is obvious that the amount of current working capital is very vulnerable to changes in the value of current assets. The current capital ratio is one relative measure of liquidity that indicates the vulnerability to change in asset values.

$$\text{Current capital ratio} = \frac{\text{Current assets}}{\text{Current liabilities}} \qquad \textbf{(2.6)}$$

The current capital ratio for Farmer A is $78,112/$72,657 = 1.08. This low current capital ratio suggests that Farmer A may have difficulty meeting current obligations from current business assets during the coming year. The current liabilities include a relatively large note, $44,640. Farmer A might investigate refinancing this note over a longer term, which would reduce the principal payments during the next year and shift some of the obligation to the intermediate or long term.

Lending agencies frequently establish minimum levels which they consider acceptable for the current capital ratio by type of farm and geographic area. They may require refinancing current liabilities when the current capital ratio is less than the prescribed standard. The firm may want to set higher standards reflecting the operator's attitude towards risk. We will return to the topic of selecting an appropriate level in our discussion of financial management and control.

Farmer A may want to prepare a projected monthly cash flow for the coming year to determine if expected cash inflows will be sufficient to pay cash requirements on a month-by-month basis. An introduction to the cash flow statement is included in the following section.

THE CASH FLOW STATEMENT

Effective financial control of the farm business requires thorough knowledge of the sources and uses of cash in the business. Some farm operations which have both a strong balance sheet and income statement find it difficult to generate cash when it is needed to meet cash commitments. The cash flow statement, also known as a sources and uses of funds or a flow of funds statement, summarizes all cash transactions affecting the business during a given period of time such as a month, quarter, or year. It provides a means of following movements of cash in the business.

The distinction between two types of cash flows—historic and projected—is important for this discussion. Historic cash flows record the actual sources and uses of cash for some previous period, such as the past year. A projected cash flow is an estimate of the sources and uses of cash for some future period, such as the coming year. The discussion in this chapter introduces the format and interpretation of the historic cash flow. Procedures to use in preparing projected cash flows will be discussed in Chapter 6.

Probably no record is as vital to the financial control system for the farm business as the historic cash flow statement. Although the income statement and balance sheet provide annual summaries of the financial condition and progress of the firm, they only give an indication of the performance of the firm once during the year. As was discussed earlier, the crucial concept of a control system is frequent monitoring of performance compared to expectations so that any deviations between plans and performance can be detected and adjustments made before disastrous financial consequences occur. The cash flow statement provides the mechanism for a continual monitoring of the firm's performance. Properly constructed, it also should provide an early warning system concerning the source of potential problems that might result in unacceptable performance.

Structure of the Statement

The cash flow statement includes all sources and uses of cash. The sources are:

- Beginning cash
- Cash receipts from crops and livestock
- Sale of capital items
- Nonfarm income
- Reduction in savings, stocks, and bonds
- New borrowings
- Outside equity in the form of gifts, sale of stock, etc.

The uses of cash are:

- Farm operating expenses
- Capital purchases
- Proprietor withdrawals, including taxes for individuals or cash dividends and taxes for corporations
- Principal and interest payments on farm debt
- Increases in savings, stocks, and bonds
- Ending cash

A useful format for the cash flow statement is shown in Figure 2.7. The statement includes four major parts—cash receipts, cash outflow, a

Figure 2.7 Example Cash Flow Statement Format

	Jan.–Mar.	*Apr.–Jun.*	*Jul.–Sept.*	*Oct.–Dec.*	*Annual*
Cash Receipts					
1. Grain and forage	___	___	___	___	___
2. Livestock and poultry	___	___	___	___	___
3. Custom work	___	___	___	___	___
4. Government payments	___	___	___	___	___
5. Capital sales					
• Breeding stock	___	___	___	___	___
• Machinery	___	___	___	___	___
6. Nonfarm income	___	___	___	___	___
7. Total cash receipts	___	___	___	___	___
Cash Outflow					
Operating expenses					
8. Seed	___	___	___	___	___
9. Fertilizer	___	___	___	___	___
10. Chemicals	___	___	___	___	___
11. Machine hire	___	___	___	___	___
12. Feed purchased	___	___	___	___	___
13. Feeder livestock purchased	___	___	___	___	___
14. Breeding, veterinary, and livestock supplies	___	___	___	___	___
15. Fuel and oil	___	___	___	___	___
16. Utilities	___	___	___	___	___
17. Repairs	___	___	___	___	___
18. Taxes, insurance, and rents	___	___	___	___	___
19. Hired labor	___	___	___	___	___
Other Outflows					
20. Capital purchases	___	___	___	___	___
21. Proprietor withdrawals including income tax	___	___	___	___	___
22. Intermediate loan payments					
• Principal	___	___	___	___	___
• Interest	___	___	___	___	___
23. Long-term loan payments					
• Principal	___	___	___	___	___
• Interest	___	___	___	___	___
24. Total cash outflow	___	___	___	___	___
Flow-of-Funds Summary					
25. Beginning cash balance	___	___	___	___	___
26. Cash receipts (line 7)	___	___	___	___	___
27. Cash outflow (line 24)	___	___	___	___	___
28. Cash difference	___	___	___	___	___
29. Borrowing this period	___	___	___	___	___
30. Payment on operating loan					
• Principal	___	___	___	___	___

(*continued*)

Figure 2.7 (*Continued*)

	Jan.–Mar.	Apr.–Jun.	Jul.–Sept.	Oct.–Dec.	Annual
31. • Interest	_____	_____	_____	_____	_____
32. Ending cash balance	_____	_____	_____	_____	_____

Loan Balances End of Period

	Balance End of Last Year					
33. Long term	_____	_____	_____	_____	_____	_____
34. Intermediate	_____	_____	_____	_____	_____	_____
35. Operating	_____	_____	_____	_____	_____	_____

flow-of-funds summary, and the loan balance summary. Each of the sources and uses of cash listed above is included in the first three sections. The cash receipts section includes operating sales, capital sales, miscellaneous farm income, and nonfarm income. The cash outflow section includes operating expenses, capital purchases, proprietor withdrawals for family living, payment of income taxes, and other expenditures and payments on intermediate and long-term loans. The third section uses the totals from the first two sections and summarizes the flow-of-funds for the period. Notice that the cash difference (the sum of the cash balance at the beginning of the period plus cash receipts minus cash outflows) can be either positive or negative for any period. If it is negative, money must be borrowed. The total money borrowed this period is the sum of new borrowing on current, intermediate, and long-term loans. It is appropriate to include all borrowing on this line because any capital purchases during the period have been included in total outflows. Finally, the principal and interest payments on operating loans are subtracted before obtaining the cash balance at the end of the period.

The fourth section is used to record the ending loan position. Notice that the cash receipts, cash outflows, and the flow-of-funds summary represent flows *during* the period. The loan balance indicates the loan position at the *end* of the period.

The headings across the top of Figure 2.7 reflect quarterly periods of the year. Monthly or bimonthly periods also are commonly used. The annual cash flow is recorded in the right-hand column. In addition to providing the annual sources and uses of funds, completion of this column provides a check on the periodic computations.

An Example

An example of a quarterly cash flow statement for Farmer A is given in Figure 2.8. Much of the data needed to complete the first two parts of the cash flow statement can be obtained from the income and expense records of the business. One simple way to obtain this type of data is to use the monthly bank statements that summarize the checking account transactions. Because many entries in the income statement are also based on the income and expense records, the similarities with the income statement entries are noted.

The cash receipts provide a quarterly breakdown of the crops, livestock, and miscellaneous receipts listed on the income statement (Figure 2.3). This section also includes the capital sales listed in the capital adjustments section of the income statement and the nonfarm cash income included on line 13 of Figure 2.3.

Lines 8–19 of the cash outflow section include each of the farm operating expenses on the income statement except interest. Interest on the intermediate and long-term loans is included under the appropriate quarter on lines 22 and 23. The interest on the operating capital, the remainder of the interest included in the income statement, is listed on line 31 of the flow-of-funds summary. Notice that this section includes the capital purchases listed on the income statement, as well as the cash part of proprietor withdrawals and principal payments on intermediate and long-term loans.

Turning to the flow-of-funds summary, notice that the cash difference was positive for the first quarter. Furthermore, enough cash was available on a day-by-day basis to meet cash commitments as they came due, and no new borrowing was needed (line 29). During the quarter, $4,000 was paid on the principal of the operating loan, thereby reducing the balance from $44,640 at the beginning of the quarter to $40,640 at the end of March. Interest paid on the operating loan during the quarter totaled $1,299. The cash balance at the end of March was $1,898. During the second quarter, the cash difference was negative. Furthermore, meeting cash commitments during the quarter required borrowing $15,000 of additional operating capital. A repayment of $5,000 was made on the operating loan during the quarter, making the ending balance $50,640. The operating loan was reduced to $40,640 by the end of the third quarter and increased through fourth-quarter borrowing to $54,640 at the end of the year. Also note that the long-term loan balance was reduced at the end of the first quarter and the intermediate loan balance was reduced during the fourth quarter as a result of the principal payments recorded on line 22 and 23 of the cash flow statement.

The annual cash flow is recorded in the final column of Figure 2.8. The entries in rows 1 through 24 are simply the sum of the cash receipts or cash outflow entries in the respective row for the four quarters. The beginning cash balance for the flow-of-funds summary is the value at the start of the year (January 1, Year 1). The entries in rows 26, 27, 29, 30, and 31 are the sum of the quarterly entries. The cash difference (line 28) and the ending cash balance (line 32) are calculated following the same procedure used for each of the quarterly calculations. Notice that the ending cash

Figure 2.8 Year 1 Cash Flow Statement for Farmer A

		Jan.–Mar.	Apr.–Jun.	Jul.–Sept.	Oct.–Dec.	Annual
Cash Receipts						
1.	Grain and forage	$ 0	$ 8,044	$ 14,600	0	$ 22,644
2.	Livestock and poultry	20,800	18,030	17,801	23,100	79,731
3.	Custom work					0
4.	Government payments	4,600	2,600	1,400	2,060	10,660
5.	Capital sales					
	• Breeding stock	790	898	890	910	3,488
	• Machinery					0
6.	Nonfarm income	2,800	987	2,693	2,013	8,493
7.	Total cash receipts	$28,990	$30,559	$37,384	$28,083	$125,016
Cash Outflow						
Operating expenses						
8.	Seed					
9.	Fertilizer	1,800	14,584	2,300	700	19,384
10.	Chemicals					
11.	Machine hire	0	390	1,303	3,000	4,693
12.	Feed purchased	1,901	730	1,650	1,750	6,031
13.	Feeder livestock purchased					0
14.	Breeding, veterinary, and livestock supplies	1,068	1,750	1,225	1,525	5,568
15.	Fuel and oil	750	1,806	775	1,025	4,356
16.	Utilities	480	360	420	806	2,066
17.	Repairs	530	2,018	3,450	1,550	7,548
18.	Taxes, insurance, and rents	970	2,331	1,015	2,446	6,762
19.	Hired labor	0	1,200	0	856	2,056
Other Outflows						
20.	Capital purchases	300	10,300	8,613	9,044	28,257
21.	Proprietor withdrawals including income tax	$ 4,800	$ 3,600	$ 2,800	$ 2,740	$ 13,940
22.	Intermediate loan payments					
	• Principal	0	0	0	12,132	12,132
	• Interest	0	0	0	4,810	4,810
23.	Long-term loan payments					
	• Principal	2,338	0	0	0	2,338
	• Interest	8,528	0	0	0	8,528
24.	Total cash outflow	$23,465	$39,069	$23,551	$42,384	$128,469
Flow-of-Funds Summary						
25.	Beginning cash balance	$1,672	$1,898	$1,770	$4,184	$1,672
26.	Cash receipts (line 7)	28,990	30,559	37,384	28,083	125,016
27.	Cash outflow (line 24)	23,465	39,069	23,551	42,384	128,469
28.	Cash difference	7,197	−6,612	15,603	−10,117	−1,781
29.	Borrowing this period	0	15,000	0	20,000	35,000

(*continued*)

Figure 2.8 *(Continued)*

		Jan.–Mar.	*Apr.–Jun.*	*Jul.–Sept.*	*Oct.–Dec.*	*Annual*
30.	Payment on operating loan					
	• Principal	4,000	5,000	10,000	6,000	25,000
31.	• Interest	1,299	1,618	1,419	1,759	6,095
32.	Ending cash balance	1,898	1,770	4,184	2,124	2,124

Loan Balances End of Period

		Balance End of Last Year				
33.	Long term	$121,826	$119,488	$119,488	$119,488	$119,488	—
34.	Intermediate	60,132	60,132	60,132	60,132	48,000	—
35.	Operating	44,640	40,640	50,640	40,640	54,640	—

balance in the annual column should be the same as the ending cash balance for the fourth quarter, providing a check on the calculation of the flow-of-funds summary.

Interpreting the Cash Flow Statement

Now the differences between the income statement and the cash flow statement can be summarized. The income statement includes inventory changes, changes in the value of capital items (depreciation), and unpaid operating expenses because these must be considered to measure profitability of the business on the accrual basis. Since none of these are cash items, they are excluded from the cash flow statement. The cash flow statement also includes principal payments because they are a use of cash; they are not considered in measuring profitability.

Consideration of all cash transactions is important in evaluating the liquidity of the firm and the ability to repay loans. The cash flow statement provides an interim or within-year indication of the actual performance of the business in terms of income generation and expense requirements. Comparing the quarterly totals to those for previous years will indicate whether the firm is generating more or less income during the first quarter or during the first half of the year. This information again can be useful in evaluating the performance of the firm during the year and in monitoring whether changes in plans or their implementation are required. For example, if accumulated cash income from the first of the year to the end of the second quarter is significantly less than in previous years, a farm manager may want to evaluate the production and/or marketing plan to determine whether adjustments should be made. Likewise, if accumulated expenditures during the first half of the year are running higher than in previous years, the manager should evaluate whether additional cost control measures should be implemented.

With the cash flow statement, potential problems can be detected during the year so that corrections can be made. If a farm manager relies solely on the annual income statement and net worth statement as the major components of the financial control system, problems that can have serious financial consequences are not detected until the end of the year. By year's end, losses may already be significant and control measures difficult to implement. The cash flow statement provides a more timely system for monitoring the financial condition of the firm, detecting potential problems, and suggesting alternative procedures that could be implemented to correct such problems. We will discuss the use of cash flow budgets for financial control in more detail in Chapter 18, after discussing the development of whole-farm plans.

OTHER ACCOUNTS FREQUENTLY INCLUDED IN FARM RECORD SYSTEMS

The three statements discussed—the income statement, the balance sheet, and the cash flow statement—are of primary concern in farm planning. Several additional accounts frequently included in farm accounting systems are a listing of receipts and expenses, accounts payable and receivable, and enterprise accounts. Each is discussed briefly below.

The *listing of receipts and expenses* is the record that comes to mind when farmers talk about an accounting system. It is used to record the daily transactions of receipts and expenses in a systematic manner. The receipts and expenses are normally divided into categories such as feed, seed, purchased livestock, medical expenses, fertilizer, herbicides, grain sales, sales of livestock on feed, and sales of capital assets. Farmers are encouraged to include the physical quantitites and the price or cost per unit as well as the total dollar amount received or paid. These data are needed to complete the income statement and the cash flow statement and to file federal income taxes.

A *listing of accounts payable and receivable* may be of use in managing the cash flow of businesses that maintain accounts with several input suppliers and/or sell products to one or more buyers on a regular basis. Many farm operators attempt to maintain this record mentally. As the number of individuals that may be charging input items on account increases, and as the number of products that are delivered with cash to be received at a later date increases, the need for an accounts payable and receivable record becomes more critical.

Enterprise accounts are frequently recommended for inclusion in a farm record system. An enterprise is defined as any portion of the farm business that can be separated from others by accounting procedures according to its receipts and expenses. Enterprises can be divided into three types.

1. Production enterprises, such as the dairy herd, corn, and alfalfa hay, are those enterprises that actually produce a marketable product.

2. Service enterprises, such as tractors, combines, the dairy barn, and specialized hog facilities, are those that provide services to each other and to the production enterprises but do not normally produce a marketable product.

3. Marketing enterprises, such as the purchasing and storage of inputs used by one or more production enterprises, and the marketing of products, can be defined as separate activities. This may be of interest to farmers who use cash contracts and the futures market to price inputs and/or products and want to analyze how much money has been made through their marketing efforts.

Relatively few farmers are willing to break down the total business into a complete list of production, service, and marketing enterprises. Many farmers using enterprise accounts keep the detailed information on a few major production enterprises and lump the rest of the business into a farm overhead enterprise. The decision of how many enterprises to include in the records can be made using the added cost-added return principle stated earlier.

Enterprise accounts include an inventory of assets for the enterprise, the listing of receipts and expenses, an income statement, and the cash flow. Thus, each enterprise is considered as a separate business within the firm. Each enterprise purchases inputs and sells products to other enterprises as well as off of the farm. For example, the dairy enterprise may purchase silage from the corn production enterprise and protein from an off-farm concern or marketing enterprise. The same dairy enterprise may sell milk to an off-farm concern, heifer calves to the replacement production enterprise, and cull cows and other cull animals to off-farm firms. The inputs purchased by the dairy enterprise are normally charged at current market prices. For example, hay would be charged to the dairy cows at the market price that could be received at the farm, even though the cost of production might be somewhat less. Such purchases and sales from one enterprise within the firm to another are referred to as noncash transfers.

To reiterate, the accounts a farmer should keep depend on the cost of obtaining or keeping the record and the expected value of the resulting record data. Many farmers keep only enough records for filing income taxes and making social security payments. This requires the kind of data that would normally be included in an inventory of depreciable assets and the listing of receipts and expenses. In addition, the operator of a sole proprietorship or partnership would need to keep data on deductible personal expenses. However, such records provide relatively few data that can be used for management decisions. Managers of family and commercial farms who understand the accounting system will normally find it profitable to include enterprise accounts for the major enterprises in the business, a balance sheet, a listing of receipts and expenses, an income statement, and a cash flow statement. In addition, payroll records and a listing of accounts payable and receivable may be advisable. A system of farm accounts including these parts provides much of the data that can be obtained from the current farming business for use in making management decisions. More detail on the use of the alternative accounts as part of a management control system will be provided in Chapters 16 through 18.

Summary

The manager of a farm business can use information obtained from farm accounts in four major ways: (1) to provide data for use in forward planning, (2) to provide a history of performance for use in implementing the farm plan, (3) to aid in control of the operation, and (4) to provide data necessary to file various reports. The three accounting statements described in this chapter are a basic part of any farm record system that provides data for these four uses.

Farm accounting systems should include the income statement, the balance sheet, and the cash flow statement. The income statement is used to determine the flow of receipts and expenses of the farm over a period of time. An income statement can be developed using the cash system or the accrual system of accounting; the cash system records transactions when cash changes hands; whereas the accrual system includes changes in inventory and recognizes that income is generated when production occurs and expenses are incurred at the date of commitment, not when cash changes hands. The purpose of completing an income statement is to obtain an accurate projection of farm profit or loss which the operator may want to allocate to the farm family's unpaid labor, equity capital, and management. The balance sheet summarizes the financial condition of the firm at a particular point in time and is based on the accounting equation that assets equal liabilities plus net worth.

The assets and liabilities must be evaluated consistently for the balance sheet to be meaningful. We advocate developing one set of calculations with capital assets evaluated on the cost or basis method, and a second set evaluating capital assets on a market value basis. The two sets of calculations allow the manager to estimate how much of the change in net worth is the result of reinvestment of earnings in the business and how much is the result of changes of asset values. By properly categorizing the assets and liabilities of the firm into current, intermediate, and long-term classes, the financial liquidity and solvency of the farm business can be evaluated. The solvency of the business can be measured by the debt-to-asset ratio or the amount of net worth in the business. Liquidity can be measured by the current capital ratio or the working capital position of the business.

The cash flow statement provides documentation of the sources and uses of cash in the farm business. This statement is crucial to the financial control concepts to be discussed later. In essence, the cash flow statement summarizes the amount and timing of cash that flows into the business from product sales and other sources, and the amount and timing of cash that goes out of the business for business expenditures, capital purchases, family living, and debt servicing. The cash flow statement is very useful in monitoring the financial performance of the business on a periodic (monthly or quarterly) basis.

Other accounts or records that may be useful in obtaining information about the efficiency and performance of the operation are the journal listing of receipts and expenses, the listing of accounts payable and receivable, enterprise accounts, inventories of depreciable and nondepreciable assets, and resource utilization records. These records and accounts provide information that can be used in developing plans, in implementing those plans, and in monitoring performance or control of the farm business.

Questions and Problems

1. Identify and discuss the four reasons for keeping farm accounts.
2. What information is needed to complete an income statement for a farm business?
3. What are the differences between cash and accrual accounting systems? Which system gives the more accurate information about the performance of the farm business?
4. Obtain an income statement for a farm business and analyze what it tells you about the performance of that business.
5. Discuss what is meant by the statement that any effort to allocate farm profit or loss to unpaid labor, equity capital, and management is arbitrary.
6. Identify the various methods of computing depreciation. What are the advantages and disadvantages of using each method?
7. What is a balance sheet? What does a balance sheet tell the manager about the financial performance and characteristics of the farm business?
8. Discuss the categorization of assets and liabilities into current, intermediate, and long-term classes. Why is this categorization important?
9. Identify the various methods of valuing assets. What are the advantages and disadvantages of each method?
10. Identify and discuss both an absolute and relative measure of solvency.

11. Identify and discuss both an absolute and relative measure of liquidity.
12. Obtain a financial statement for a farm operation and discuss the kind of information it provides about the efficiency, profitability, solvency, and liquidity of the business.
13. Describe a cash flow statement. How might a cash flow statement be used as part of a financial control system for a farm business?

Further Reading

Frey, Thomas L., and Danny A. Klinefelter. *Coordinated Financial Statements for Agriculture.* Agr. Finance, Skokie, Ill., 1978.

Helmkamp, John G., Leroy F. Imdieke, and Ralph E. Smith. *Principles of Accounting.* New York: John Wiley and Sons, 1983.

James, Sydney C., and Everett Stoneberg. *Farm Accounting and Business Analysis.* Ames: Iowa State Press, 1974.

Pyle, William W., and Merkit D. Larson, *Fundamental Accounting Principles.* 9th ed., Homewood, Ill.: Richard D. Irwin, 1981.

Walgenbach, Paul H., Norman E. Dittrich, and Ernest I. Hanson. *Principles of Accounting.* 2d ed., New York: Harcourt Brace Jovanovich, 1980.

PLANNING

DEVELOPING ENTERPRISE BUDGETS: ECONOMIC CONCEPTS

CHAPTER CONTENTS

Managers of farm businesses frequently must estimate costs and returns for future periods of time. In some cases, it may be necessary to estimate costs for one part of the business, such as the feed cost for the dairy herd and the ownership and operating cost of a new piece of machinery, or the manager may estimate the costs and returns for an enterprise, such as finishing a group of feeder pigs. In other cases, the manager may estimate the profitability of the total business for the coming year or, perhaps, several years. Familiarity with several economic concepts and computational procedures will aid managers in completing such analyses rapidly and correctly.

The next two chapters deal with the preparation of cost and return estimates or projections for individual enterprises. A projection of average annual costs and returns for an enterprise is commonly referred to as an enterprise budget or a gross margin calculation. The term "enterprise budget" is used in the discussion that follows. Notice the distinction between enterprise *accounts* and enterprise *budgets*. An enterprise account is a summary of costs and returns for some historic period such as the past year or an average of several years. It is obtained from the records of the business. An enterprise budget is a projection of costs and returns for some future period, such as the coming year.

While enterprise accounts may be useful in preparing enterprise budgets, they seldom provide all of the data needed. Even the operator who expects to produce the same enterprise using the identical system of production may expect different input prices, product prices, and/or production levels in the future than prevailed in the past. Furthermore, the change in prices expected may suggest that a different combination of variable inputs and/or timing of production would be more profitable in the future. In some cases, the operator may want to estimate the costs assuming a new set of buildings, machinery, or equipment is used which is more efficient but has higher overhead costs than the buildings and equipment used in previous years. In other cases, the operator may want to estimate costs and returns for an enterprise that has not been produced in the past. For these reasons even farm managers with good enterprise accounts should use the economic concepts and computational procedures discussed here to develop enterprise budgets. Managers without enterprise accounts must rely on the procedures discussed to an even greater extent.

This chapter discusses several economic concepts that are useful in making management decisions involving considerations of cost and the selection of inputs to produce a product or accomplish a task. The concepts will be illustrated in the context of developing enterprise budgets. Applications to other aspects of decision-making will be noted throughout the text. The concept of fixed versus variable costs indicates which costs are relevant for the decisions to be made. This concept is basic in making farm management decisions and is treated first in this section. The nature of the production relationships and two economic concepts to use in selecting the optimum level of variable inputs in production are discussed. They are the principle of diminishing marginal returns and the principle of input substitution. Both are explained and their application is illustrated. This chapter concludes with a discussion of cost relationships and the use of economies of size in constructing enterprise budgets. Computational procedures, examples of enterprise budgets, and their interpretation are discussed in Chapter 4.

FIXED VERSUS VARIABLE COSTS

Costs can be categorized in various ways, but division into two broad categories, fixed costs and variable costs, is appropriate for most economic analyses. Fixed costs are those that do not change with (are not a function of) the level of output. These costs remain the same whether or not output is produced. Depreciation on buildings, taxes on the farm, insurance, cash rent, and interest payments on loans are examples of costs that are usually fixed. If an individual is hired on an annual basis, the wage is also a fixed cost. Fixed costs remain the same in high-income as in low-income years. As will be discussed later, however, these costs are fixed only in the short run. In the long run they also become variable. On the other hand, variable costs change with the level of output. They are a function of the amount produced and do not occur unless the operator attempts to produce a product. Expenses for seed, fuel, pesticides, harvesting, drying, and labor hired on a daily or weekly basis are examples of variable costs. The money spent to purchase feeder cattle, feed, and medication, and to pay trucking and marketing expenses are other examples.

Variable costs include the money outlays for purchased inputs that are consumed in one production period. This category also includes the value of inputs produced on the farm that have a market value, such as home-grown feed and seed that are essentially consumed in the single production period. These inputs, such as shelled corn and forage fed to livestock, have an opportunity cost. If they were not used in production, they could be sold. Variable costs in crop production normally include seed, fertilizer, pesticides, fuel, lubrication, machinery repairs, expense for custom operations, hauling costs, and interest on operating inputs. Examples for livestock production include purchased feed, home-grown feed, veterinary and medical expenses, marketing and hauling, insurance on livestock, and interest on the money invested in both livestock on feed and operating inputs.

Fixed costs can be subdivided into cash and noncash costs. Fixed cash costs are those money outlays that would be borne by the business regardless of the size of the enterprise. Real estate taxes, personal property taxes, insurance on buildings and equipment, and maintenance on the farmstead are examples that normally are included in this category. Fixed noncash costs include costs that are borne over time either in cash or as an opportunity foregone, but that are not cash expenses within a short-run period such as one year. The depreciation and interest on buildings and equipment already on the farm are examples of fixed noncash costs. A land charge for the use of owned land (usually figured as a percentage return times the market value of the land) is another example of a noncash fixed cost.

The distinction between fixed and variable costs is important in decision-making. Only variable costs should be considered by the manager in deciding what to produce, how to produce, and how much to produce in the short run. Fixed costs will remain at the same level regardless of these decisions. Thus, neither fixed cash nor noncash costs should be considered in decision-making.

A distinction between variable cash and variable noncash costs can also be made. Intermediate products, such as grain and forage produced on the farm, are examples of variable noncash costs. The distinction between variable cash and noncash costs is important in preparing cash flows and certain types of financial analyses discussed later. Noncash inputs that are variable can be sold off of the farm, however. Thus, they have a short-run opportunity cost and must be considered as variable for decision-making.

The costs that are included in the variable category typically depend on the period of time considered. A farmer producing an annual crop, such as winter wheat, may consider the cost of using land (either the cash rent or interest and taxes on owned land) and the depreciation, interest, taxes, and insurance on the owned machinery required for wheat production as fixed costs. If the operator's labor has been committed to farming for the coming year, its cost may also be considered as fixed. Prior to the time tillage operations begin, however, all machinery operating expenses, seed, fertilizer, as well as other production and marketing expenses, are variable. If the crop is seeded during the fall, on January 1 the cost of tillage, any fertilizer applied up to that time, the cost of the seed, and other production expenses already incurred are fixed. Any decisions to use additional variable inputs, therefore, depend on the manager's estimate of the additional variable costs and expected returns at that time. Suppose that the operator topdresses the wheat crop with fertilizer and applies some insecticide during the spring. At harvest time all costs incurred up to that time are fixed. Thus, the decision to harvest or not to harvest the wheat crop depends on whether the expected returns from the crop are greater than the costs that are variable at that point in time—harvesting, hauling, and marketing costs. Even though a hail storm severely damages the wheat crop two weeks prior to harvest, the farmer should harvest the crop, providing expected returns exceed the variable costs that will be incurred in harvesting and selling the crop. Doing so will provide greater net returns (or smaller losses) than abandoning the crop.

In general, as the length of the planning period increases, the number of costs that are included in the variable category increases. A Corn Belt farmer estimating costs for feeding cattle during the coming year may treat the depreciation, insurance, interest, and taxes on feeding facilities as fixed costs. When the operator considers replacing the facilities, however, estimates of the long-run costs and returns from cattle feeding should consider the depreciation, interest, insurance, and taxes on these new facilities as variable rather than fixed costs. Of course, when the farmer commits resources to construction of the facilities, these costs then become fixed.

In developing enterprise budgets, it is important to distinguish between variable and fixed costs. The items included in the variable category depend on the period of time being considered. Farm management publications and materials typically include the variable costs under the heading operating costs. They also list fixed cash and fixed noncash costs under the heading ownership costs. Operating costs normally include all costs that are variable for a planning horizon of one production period. The ownership costs typically include depreciation, interest, insurance and taxes on buildings, equipment, and machinery. A charge for the use of the land and certain management costs may also be included under the ownership heading. Since the costs that are variable depend on the situation, the operator must be able to interpret the cost estimates and distinguish the variable from the fixed to use the information in decision-making. Later, when we discuss cash flows, it will also be important to distinguish between cash and noncash costs.

An example of an enterprise budget for soybeans is shown in Table 3.1. Gross returns are estimated to total $247.50 per acre based on an expected yield of 33 bushels per acre and an expected price of $7.50 per bushel. Assume that the estimates have been prepared for John Farmer's production conditions during the coming year. He already owns the machinery required and plans to use his own labor in producing the crop. His variable costs include all of the items under the operating costs heading in Table 3.1 except labor. In this case, his variable costs are $64.76 per acre. Net returns to Mr. Farmer's fixed resources are $247.50 minus $64.76 or $182.74 per acre. His labor is available to the farm and, like the machinery depreciation and housing, is a noncash fixed cost. Machinery insurance and land taxes are examples of fixed cash costs. Interest on the money invested in machinery and land may be either a fixed cash or a fixed

Table 3.1 Expected Costs and Returns per Acre of Soybeans

1.	Gross receipts	
	Soybeans—33 bushels @ $7.50	$247.50
2.	Operating costs	
	Preharvest machinery fuel, lube, and repair cost	9.58
	Seed	11.25
	Fertilizer	15.50
	Herbicides	13.20
	Crop insurance	6.19
	Combine operating cost	4.40
	Hauling	0.93
	Labor—1.93 hr @ $5.00	9.65
	Interest on operating capital	3.71
	Total operating cost	74.41
3.	Income above operating costs	173.09
4.	Ownership costs	
	Machinery depreciation, interest, insurance, and housing	32.88
	Interest on land investment	88.00
	Land taxes	12.00
	Total ownership costs	132.88
5.	Total costs shown	207.29
6.	Net returns above costs shown	40.21

noncash cost. Many farmers have loans on these assets for less than their value, thereby making these interest costs a combination of the two.

As the year progresses and the crop is produced, more of the costs are added to the fixed cash category. For instance, a few days prior to harvest the costs for preharvest machinery, fuel, lube and repairs, seed, fertilizer, herbicides, crop insurance, and (to a large extent) interest on operating capital have become fixed cash costs. Only the combine operating cost, hauling cost, and the interest on these two items of operating capital remain in the variable cash category. If yield and price expectations remain the same, net returns to fixed resources are approximately $247.50 minus $5.33 or $242.17.

If we consider a much longer run, an individual who has alternative employment and neither owns machinery nor controls the use of land could consider all of the costs listed as variable costs. In this situation, returns to fixed resources (the individual's management ability) are $40.21.

The level of variable inputs to include must be selected in developing an enterprise budget. Two economic principles, the principle of diminishing marginal returns and the principle of input substitution, provide a basis for choosing the level of variable inputs to use in producing a product.

PRODUCTION RELATIONSHIPS AND ECONOMIC CHOICE

Determining the amount of variable inputs to combine with one or more fixed resources is an important production decision. Selection of the appropriate amount of variable inputs to use depends largely on (1) the technical relationship between the inputs and the output, and (2) the prices of the inputs and the price of the output. The technical relationships are discussed first, followed by an explanation of the profit-maximizing conditions.

Nature of the Production Function

The technical relationship between the amount of variable inputs used and the output produced is referred to as the factor-product relationship or the production function. It is also commonly referred to as the input-output relationship in farm management, while biological scientists frequently refer to it as the response curve. The relationship is defined per unit of time. It relates the amount of yield or product that can be produced for alternative combinations of inputs within a specified time interval, such as one year. If X_i represents the amount of the ith input and Y represents the amount of the product produced, then the production function can be written as:

$$Y = f(X_1|X_2, \ldots, X_n) \tag{3.1}$$

This relationship indicates that the amount of product Y is a function of the amount of the variable input X_1 and the level of the fixed inputs X_2, through X_n. The number of factors that are fixed and variable depends on the length of the planning horizon as noted in the first section of this chapter. The initial part of this discussion considers only one variable input and its effect on production. The analysis is extended to include two variable inputs later in the chapter.

The relationship between the amount of a single variable input and the output of a single product can take one of three general forms: constant productivity, diminishing productivity, and increasing productivity of the variable input. Constant productivity exists when each unit of variable input added to the fixed factor(s) increases output by the same amount. If the application of each pound of nitrogen on an acre of corn increases yield one bushel, the nitrogen has constant productivity. A production function with constant productivity is illustrated in Figure 3.1. Plotting the production function with output Y on the vertical axis and the variable input X_1 on the horizontal axis results in a straight line. Constant returns are indicated by the dotted triangles which have three units of additional output for each unit of variable input.

While productivity is generally not constant when adding increasing amounts of variable input to one acre of a crop or one animal, the concept is applicable to whole-farm planning where the fixed factors, such as livestock facilities and land, are readily divisible. An operator with a limited supply of land, livestock facilities, capital, or labor can apply the same amount of variable input to each acre or animal produced and obtain the same increase in output for each unit of variable input. If the use of one

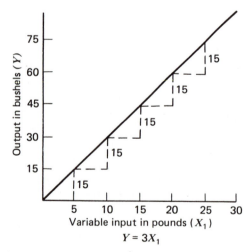

Figure 3.1. Physical production function indicating constant physical productivity.

pound of herbicide per acre increases soybean yield four bushels and the farmer has 200 acres of soybeans, then a linear response exists between the use of the herbicide and soybean production up to 200 pounds of herbicide. If the farmer can produce 33 bushels of soybeans using one acre of land, two hours of labor, and $60.00 of operating capital, then this combination of labor and capital can be used on other acres of comparable quality and achieve the same increase in soybean production. This concept is used later in discussing whole-farm planning with budgeting and linear programming.

Diminishing productivity of a variable factor exists when each additional unit of the variable input adds less to total output than the previous unit. For example, a farmer considering alternative levels of fertilizer in small-grain production would expect to find that as increasing amounts of fertilizer were added to one acre, the increase in the yield would become smaller and smaller. Diminishing productivity is illustrated in Figure 3.2. The amount added to total yield by the first five-pound unit of input is 13.75. The second through the sixth units add 11.25, 8.75, 6.25, 3.75, and 1.25 to total product, respectively. Adding a seventh unit of the variable input would make the output decline somewhat. The total physical product relationship curves toward the horizontal axis; it is concave from below, indicating diminishing physical productivity. More examples of relationships that exhibit diminishing physical productivity are presented later in this section.

Increasing productivity of a single factor exists when each additional unit of the variable input adds more to total product than the previous one. Increasing productivity is illustrated in Figure 3.3. The first unit of five

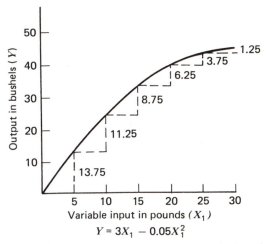

Figure 3.2.   Physical production function indicating diminishing physical productivity.

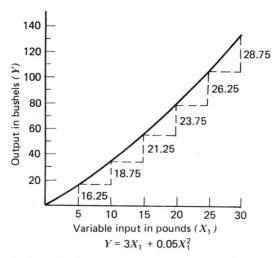

$$Y = 3X_1 + 0.05X_1^2$$

Figure 3.3. Physical production function indicating increasing physical productivity.

pounds increases output 16.25 bushels, while units 2 through 6 increase output by 18.75, 21.25, 23.75, 26.75, and 28.75, respectively. Thus, each unit of variable input adds more to output than the previous unit.

Although a production function may show increasing productivity for one variable input with all other factors held fixed, ordinarily increasing productivity only represents one segment of the production function. Irrigation water applied to produce a grain crop in a low rainfall area may exhibit increasing physical returns. A minimum amount of water is required for germination and vegetative growth of the crop. Properly applying and timing this minimum amount allow the crop to remain alive until the reproductive stage, but produce little, if any, yield. Adding somewhat more water can be expected to result in an increasing amount of yield per acre inch of water applied. After this range of increasing physical productivity, adding a larger and larger amount of irrigation water can be expected to result in diminishing productivity to water applied. Hence, the increasing productivity represents only one range within a full production function. Data on small farms short on capital may also suggest that increasing productivity exists for capital. Because capital is only one input in the production function, increasing capital to higher and higher levels would eventually result in diminishing productivity. Again, increasing productivity represents only part of the overall production function.

A production function exhibiting both increasing and decreasing marginal productivity is illustrated with some hypothetical small-grain yield data in Table 3.2. These data illustrate the response of yield per acre to alternative seeding rates. The fixed factors (X_2, ..., X_n) include the land area (one acre), the tillage practices, the fertilizer that has been applied, the

Table 3.2 Per Acre Small-Grain Yields
for Alternative Seeding Rates

Units of Seed per Acre[a] X_1	Total Physical Product (yield per acre) Y	Marginal Physical Product (additional yield per unit of seed) MPP_{X_1}	Average Physical Product (yield per unit of seed) APP_{X_1}
0	0		–
		7.6	
1	7.6		7.6
		8.6	
2	16.2		8.1
		9.2	
3	25.4		8.5
		9.3	
4	34.7		8.7
		9.1	
5	43.8		8.8
		8.3	
6	52.1		8.7
		7.2	
7	59.3		8.5
		5.7	
8	65.0		8.1
		3.7	
9	68.7		7.6
		1.3	
10	70.0		7.0
		−2.5	
11	67.5		6.1

[a]A unit of seed in this example is 10 pounds.

pesticides that have been used, and the management ability of the individual. The variable factor (X_1) is the amount of seed being applied per acre. The figures in Table 3.2 indicate that when one unit of seed (10 pounds) is seeded per acre, the yield per acre is 7.6 bushels. When two units of seed are applied, the yield increases to 16.2 bushels per acre, and the yield is 25.4 bushels for a seeding rate of three units. Each unit of seed added through the tenth unit increases the yield per acre. Applying the eleventh unit of seed reduces the yield 2.5 bushels per acre. The data listed in Table 3.2 are shown graphically in Figure 3.4a. The figure indicates that increasing productivity exists from zero to between four and five units of the variable input. This is followed by a range of diminishing physical productivity.

Marginal and Average Physical Product

The discussion to this point has dealt with one technical relationship, the production function. Two other technical relationships—the marginal physical product and the average physical product—can be derived from the production function and are important in selecting the amount of a variable input to use.

The marginal physical product of the ith input is equal to the change in the amount of product produced over the change in the amount of the ith input that is used. Indicating the amount of the added variable input by ΔX_i and the added output by ΔY, MPP_{X_i} equals ΔY divided by ΔX_i or

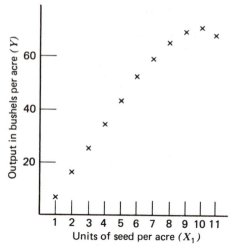

Figure 3.4a. Yield level for discrete levels of the variable input seed in small-grain production (data from Table 3.2).

$$MPP_{X_i} = \frac{\Delta Y}{\Delta X_i} \qquad (3.2)$$

The third column of Table 3.2 indicates the marginal physical product or the additional yield resulting from each unit of seed applied. The marginal physical product of the first unit of seed is 7.6 bushels. That is, using one unit of seed increases the total product from zero to 7.6 bushels per

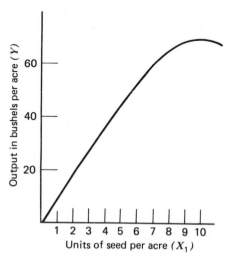

Figure 3.4b. The production function for small-grain yield data of Table 3.2.

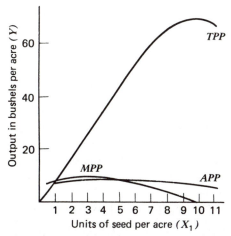

Figure 3.4c.   Total physical product, average physical product, and marginal physical product for small-grain yield data (Table 3.2).

acre. Applying the second unit of seed increases the yield by 8.6 bushels; so the marginal physical product of the second unit of seed is 8.6 bushels. Notice that the marginal physical product increases as the third and fourth units of seed are applied. When the fifth and succeeding units of seed are applied, however, the amount added to yield per acre declines. In this example, increasing marginal physical productivity is exhibited for zero through four units of the variable input, and diminishing marginal returns are exhibited for the fifth through the eleventh units.

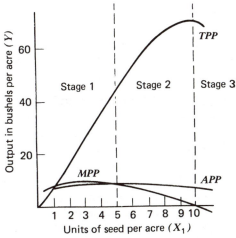

Figure 3.4d.   Three stages of production for small-grain yield data (Table 3.2).

The average physical product (APP_{X_i}) is equal to the average output per unit of variable input and is calculated as total product divided by the amount of variable input used. In equation form:

$$APP_{X_i} = \frac{Y}{X_i}$$

(3.3)

The average physical product is given in the fourth column of Table 3.2. For instance, when two units of seed are used per acre, total product is 16.2 bushels. The average physical product is equal to 16.2 divided by two or 8.1 bushels per unit of variable input. In this example, the average physical product increases through the fifth unit of seed and declines from the sixth through the eleventh unit.

The three relationships, the total physical product, the marginal physical product, and the average physical product, are plotted in Figure 3.4 for the small-grain data shown in Table 3.2. The yield data listed in the second column of Table 3.2 are plotted in Figure 3.4a. Only the discrete points of the production function for the alternative units of seed per acre are plotted. For many variable inputs, such as seed, the amount of variable input is divisible into very small units. Thus, it is reasonable to approximate the relationship that exists between the amount of output and the amount of variable input with a continuous relationship. This relationship or the production function is shown in Figure 3.4b. The marginal physical product relationship can also be plotted on the same diagram. When working with discrete data of the type listed in Table 3.2, it is useful to plot the marginal physical product at the midpoint of the units of variable input. The marginal physical product for one unit of seed is, therefore, plotted at 0.5 units of seed in Figure 3.4c, while the marginal physical product for two units of seed is plotted at the midpoint between one and two units. Notice that the marginal physical product increases through four units of seed and then declines to zero, at the point of maximum total product. It becomes negative as total product declines. The average physical product relationship is also shown in Figure 3.4c. The average physical product relationship is simply plotted at the units of variable input that are being used.

The Three Stages of Production

The three relationships—total physical product, marginal physical product, and average physical product—can be used to define three stages of production. These three stages of production are shown in Figure 3.4d. Stage 1 is defined as the area in which marginal physical product is greater than average physical product. Notice that in Stage 1 marginal physical product increases, reaches a maximum, and then declines. Average physical product increases throughout the stage and reaches a maximum at the boundary between Stage 1 and Stage 2. Total physical product is increasing throughout Stage 1. Also notice that marginal physical product is equal to average physical product at the boundary between Stage 1 and Stage 2.

Stage 2 is also characterized by increasing total physical product. Total physical product increases through Stage 2 and reaches a maximum at the boundary between Stage 2 and Stage 3. Within this stage marginal physical product is declining. It is less than average physical product throughout Stage 2 and reaches zero at the boundary between Stage 2 and Stage 3. Average physical product is declining and positive through Stage 2.

Stage 3 is characterized by declining total physical product. In this stage, marginal physical product is declining and negative. Of course, average physical product continues to decline in Stage 3.[1]

The three stages of production provide useful information concerning the rational range of production. It would be irrational for the manager to operate in either Stage 1 or Stage 3 regardless of the level of input and product prices. It is obvious that one would not want to operate within Stage 3. Applying additional units of the variable input and forcing production into Stage 3 reduce the amount of total product produced. If it is assumed that the price of the variable input and the product are constant and positive, the manager would make more money by leaving some of the variable input unused.

Perhaps it is less obvious why the operator would not want to operate within Stage 1. Again, assume that the price of the seed and the price of the output are positive and constant. Now let us limit the amount of seed to 10 units and the amount of land to 10 acres. Notice that the seed available is quite small relative to the 50 to 100 units required to seed all 10 acres and operate in Stage 2. We can illustrate why the manager would not want to operate within Stage 1 by considering alternative ways of allocating the 10 units of seed to the 10 acres of land. (Any other combination that limits the amount of seed relative to the amount of land available could also be used for purposes of illustration.) Alternative ways of allocating the 10 units of

---

[1] Some precision can be added to the definition of the marginal physical product curve and the relationship of the marginal physical product to the other physical product curves using calculus. If we use the production function in equation 3.1, we have

$$TPP = Y = f(X_1 | X_2, \ldots, X_n)$$

The average physical product for X_1 can be written as:

$$APP_{X_1} = \frac{Y}{X_1} = \frac{f(X_1 | X_2, \ldots, X_n)}{X_1}$$

The marginal physical product of X_1 is the rate of change of its total physical product with respect to changes in X_1, which is given by the partial derivative of the production function with respect to X_1:

$$MPP_{X_1} = \frac{\delta Y}{\delta X_1} = \frac{\delta f(X_1 | X_2, \ldots, X_n)}{\delta X_1}$$

The first partial derivative provides an estimate of marginal productivity at a point on the production function, which is more precise than calculating the average marginal productivity over a range in input level X_1 as we did in equation 3.2. This permits more precise determination of the input level at which $MPP_{X_1} = APP_{X_1}$, the boundary between Stages 1 and 2.

Table 3.3 Alternative Ways of Allocating 10 Units of
Small-Grain Seed on 10 Acres and the Resulting Production
(based on the production function in Figure 3.2)

Seeding Rate per Acre	Acres of Land Seeded	Yield per Acre (bu)	Total Production Using 10 Units of Seed (bu)
1	10	7.6	76.0
2	5	16.2	81.0
3	3⅓	25.4	84.7
4	2½	34.7	86.8
5	2	43.8	87.6
6	1⅔	52.1	87.1
7	1³⁄₇	59.3	84.6
8	1¼	65.0	81.2
9	1⅑	68.7	76.3
10	1	70.0	70.0

seed on the 10 acres of land are given in Table 3.3. If one unit of seed is
used per acre, 10 acres of land can be seeded with the 10 units of seed. The
yield in this case is 7.6 bushels per acre, and total production on the 10
acres is 76 bushels. If, on the other hand, 2 units of seed are seeded per
acre, only 5 acres of land can be seeded to small grain. The yield increases
to 16.2 bushels per acre, however, and the total production increases to 81
bushels. Increasing the seeding rate to 5 units of seed per acre maximizes
the amount of production from the 10 units of seed. If we seed 5 units of
seed per acre, only 2 acres of land can be seeded. The yield per acre is 43.8
bushels, resulting in 87.6 bushels of production from the 10 units of seed.
Seeding at higher rates results in higher yields per acre, but less total production with the 10 units of seed. If the price of the input and the price of the
product are constant and positive, it will be more profitable to operate at
the level of variable input where marginal physical product is at least equal
to average physical product (even if some of the fixed input is idle) than at
any amount of variable input to the left of this value. This indicates that
Stage 1, like Stage 3, is an irrational area of production. Production should
always occur in Stage 2, but the decision-maker must consider the price of
the input and the price of the product to determine exactly the profit-
maximizing level of the variable input to use in production.

Biological production responses to alternative levels of variable inputs
can be expected to follow the principle of diminishing marginal physical
productivity. The principle can be stated as follows. When one or more
factors are held fixed in amount, the amount added to total product by
using an increasing amount of a second factor must eventually decline.
Notice in this statement that a variable factor is being combined with one or
more fixed factors. Also notice that the statement permits a segment of
increasing marginal physical productivity preceding a range of diminishing

marginal physical productivity. (The data in Table 3.2 exhibit increasing marginal physical productivity in the range of 1 to 4 units of seed and diminishing marginal physical productivity from 5 through 11 units). The statement of the principle permits a segment of increasing returns but does not require it. Typical applications of the theoretical model in production agriculture may *not* exhibit both increasing and decreasing marginal physical productivity and, hence, may not have all three stages of production. Frequently, some output is achieved without the manager applying any of the variable input. In applications involving the response of a crop to either fertilizer or irrigation water, some of the variable input is typically available naturally without the manager applying it. Thus, the portion of the production function being observed does not start from a zero level of variable input, but rather with some amount of the variable input already available. In many of these cases, the manager does not observe Stage 1, but finds declining marginal physical productivity as the variable input is increased and marginal physical product less than average physical product throughout the entire range being observed.

Selecting the Profit-Maximizing Level of One Variable Input

The manager must consider the price of the input and the price of the product in determining the profit-maximizing amount of the variable input to use. If we define the marginal factor cost (MFC_{X_i}) as the cost of the last unit of the variable input, and the value of the marginal product (VMP_{X_i}) as the price of the product times the marginal physical product ($P_Y \cdot MPP_{X_i}$), profits are maximized where the value of the marginal product is just greater than or equal to the marginal factor cost.[2] Some students of farm management confuse *marginal factor cost* with *marginal cost*. Notice that

---

[2]The decision rule can be derived as follows. Given the objective of profit maximization (π), the production function in equation 3.1, and given (fixed) input and product prices P_{X_1} and P_Y, respectively, we can write the profit function to be maximized as:

$$\pi = P_Y \cdot Y - P_{X_1} \cdot X_1 - \text{Fixed costs (FC)}$$

The production function can be substituted for Y making profit a function of X_1. To maximize profit, we differentiate with respect to X_1. If we assume constant prices for both X_1 and Y we obtain:

$$\frac{\delta\pi}{\delta X_1} = \frac{\delta Y}{\delta X_1} \cdot P_Y - P_{X_1} \geq 0$$

In this relationship $\frac{\delta Y}{\delta X_1} \cdot P_Y$ is the value of the marginal physical product and P_{X_1} is the marginal factor cost. Thus, we can write

$$VMP_{X_1} \geq MFC_{X_1}$$

It should also be noted that profit maximization with this rule requires that the level of X_1 selected is in its range of declining marginal physical productivity. The existence of declining marginal physical productivity for the examples used in this chapter can be determined by inspecting the data.

Table 3.4   Comparison of Value of the Marginal Product with Marginal Factor Cost to Decide Most Profitable Output Level and Amount of Variable Resource to Use with a Fixed Resource

Input or Quantity of Nitrogen Fertilizer to Use (lb per acre)	Total Physical Product (yield per acre)	Marginal Physical Product (added yield)	Marginal Factor Cost (added cost of fertilizer with N =)		Value of Marginal Product (price × marginal physical product)	
			$0.20	$0.40	Price = $2.50/bu	Price = $4.00/bu
0	90					
40	110	20	$8.00	$16.00	$50.00	$80.00
80	126	16	8.00	16.00	40.00	64.00
100	133	7	4.00	8.00	17.50	28.00
120	139	6	4.00	8.00	15.00	24.00
140	144	5	4.00	8.00	12.50	20.00
160	149	5	4.00	8.00	12.50	20.00
180	153	4	4.00	8.00	10.00	16.00
200	156	3	4.00	8.00	7.50	12.00
220	158	2	4.00	8.00	5.00	8.00
240	159	1	4.00	8.00	2.50	4.00
260	159.5	0.5	4.00	8.00	1.25	2.00
280	160	0.5	4.00	8.00	1.25	2.00
300	159	−1	4.00	8.00	−2.50	−4.00

Source: Based on data for Waseca County in Minnesota for the years 1970–73 reported by W. E. Fenster, C. J. Overdahl, G. W. Randall, and R. P. Schoper, *Effect of Nitrogen Fertilizer on Corn Yield and Soil Nitrates,* Miscellaneous Report 153–1978, Agricultural Experiment Station, University of Minnesota, St. Paul, 1978, p. 6.

marginal factor cost is the cost of the last unit of the variable input used. In the example in Table 3.2, the marginal factor cost would be the cost of one unit (10 pounds) of seed. In contrast, the *marginal cost* is the additional cost of producing another unit of output.

Calculating the marginal factor cost, the value of the marginal product, and finding the profit-maximizing level of the variable input to use in production are illustrated in Table 3.4. These data indicate the response of corn to alternative levels of nitrogen fertilizer on soils at Lucan, Minnesota. The yield data indicate that an average yield of 90 bushels per acre was achieved when no nitrogen fertilizer was applied to the soil. Applying 40 pounds of *N* per acre resulted in an expected yield of 110 bushels per acre—an increase of 20 bushels over the 0 application rate. Applying 80 pounds of *N* per acre resulted in a yield of 126 bushels, 16 bushels greater than the yield achieved with 40 pounds of *N* per acre. Notice that the application of heavier rates of nitrogen increases yield up to an application rate of 180 pounds of N per acre. The marginal physical product is shown in the third column of Table 3.4. It is calculated as discussed earlier.

The marginal factor cost is shown for two prices of nitrogen fertilizer. In the fourth column, the cost of nitrogen is calculated at $0.20 per pound of N. Thus, the marginal factor cost of the first 40-pound increment is $8.00. The marginal factor cost of the second 40-pound increment is also $8.00, whereas the remaining increments are 20 pounds each, resulting in a marginal factor cost of $4.00. In the fifth column, the price of nitrogen is assumed to be 40 cents per pound. Thus, the marginal factor cost is exactly double the amount shown in column four in each case. The value of the marginal product is listed in the sixth column assuming a corn price of $2.50 per bushel. The value of the marginal product is equal to the price times the marginal physical product. Thus, the value of the marginal product for the first 40 pounds of nitrogen is equal to $2.50 times 20 bushels or $50. When a price of $4.00 per bushel is assumed, the value of the marginal product of the first 40 pounds of N is $4.00 times 20 bushels or $80 per acre.

By comparing the marginal factor cost and the value of the marginal product for the appropriate prices of nitrogen and corn, the manager can determine the profit-maximizing level of nitrogen to use. The manager would want to add additional nitrogen as long as the value of the marginal product exceeded the marginal factor cost. If we assume that the price of nitrogen is $0.20 per pound and the price of corn is $2.50 per bushel, the marginal factor cost at 220 pounds of nitrogen is $4.00 and the value of the marginal product of the last 20 pounds of nitrogen is $5.00 per acre ($2.50 times 2 bushels). This is the last 20-pound increment of nitrogen that has a VMP_{X_1} greater than or equal to the MFC_{X_1}. Thus, the profit-maximizing level of nitrogen to use is 220 pounds. This results in an expected yield of 158 bushels per acre. On the other hand, if the price of nitrogen is $0.20 per pound of N and the price of corn is expected to be $4.00 per bushel, the manager would apply 240 pounds of nitrogen because the additional 20 pounds of nitrogen would result in a value of the marginal product of $4.00, which is exactly equal to the marginal factor cost. The most profitable level of nitrogen to apply is shown for four combinations of nitrogen price and corn price in Table 3.5.

Two additional considerations should be noted in applying the rule to choose the optimal level of variable input to use in production. The price of the product used in the computation should be the net price—net of additional cost associated with handling and marketing the additional production. For instance, in the corn example the price of corn used should be net of additional harvesting, hauling, and drying costs. If the farmer assumes that the market price of corn will be $2.80 per bushel and that each additional bushel produced will result in an *additional* combining cost of $0.07 per bushel, an additional hauling cost of $0.05 per bushel, and a drying

Table 3.5 Most Profitable Fertilizer Levels
for Alternative Nitrogen and Corn Prices Shown in Table 3.4,
Waseca County, Minnesota

Price		Profit-Maximizing Amount	Expected Yield
Per Pound N	Per Bushel Corn	of N per Acre (lb)	per Acre (bu)
$0.20	$2.50	220	158
0.40	2.50	180	153
0.20	4.00	240	159
0.40	4.00	220	158

and storage cost of $0.18, then the price used in the computation should be $2.80 less the $0.30 of additional costs or $2.50 per bushel. This consideration is particularly important when applying the principle to products that have a large proportion of their cost associated with harvesting and marketing of the product.

The second consideration is that input costs should include all charges including financing costs, and that using increasing amounts of variable inputs may result in additional costs not included in the price of the input itself. For instance, applying nitrogen requires a machinery operation. The cost of this machinery operation could be included as part of the marginal factor cost of the first increment of N. If heavier application rates require more machinery time per acre fertilized, this additional cost should also be included as part of the marginal factor cost for the higher application rates.

In many cases, managers have much fewer response data than indicated in Table 3.4, but the principle of diminishing marginal returns may still be quite useful in selecting the level of the variable input to use. Many experiments include only three or four rates of an input. Consider the data on yield response of grain sorghum to alternative nitrogen fertilization rates in Table 3.6. Only four levels of nitrogen fertilizer—0, 100, 200, and 300 pounds per acre—were considered in the trial. The marginal factor cost of the fertilizer is $25. The value of the marginal product exceeds the marginal factor cost for the first and second units, but not for the third. A direct application of the procedure to select the profit-maximizing level of N suggests that 200 pounds should be applied. It may be profitable to fertilize at a slightly higher level, however. The marginal physical product is calculated over a 100 pound increment of fertilizer. Without further information on the shape of the marginal physical product curve, it is reasonable to assign the average value of marginal products of $41.80 and $13.00 to 150 and 250 pounds of N, respectively. If we assume a linear value of the marginal product curve, the marginal value at 200 pounds would be ($41.80 + $13.00) ÷ 2 or $27.40. If the marginal physical prod-

Table 3.6 Three-Year Average Irrigated Grain Sorghum
Yield on Richfield Clay Loam Soil at Goodwell, Oklahoma

Lb of N per Acre	Yield of Grain Sorghum (lb per acre)	Marginal Physical Product	Marginal Factor Cost with Nitrogen at $0.25 per lb	Value of Marginal Product with Grain Sorghum at $0.05 per lb
0	3,761			
100	6,910	3,149	$25.00	$157.45
200	7,749	836	25.00	41.80
300	8,007	260	25.00	13.00

Source: Based on data by H. Eugene Reeves and B. B. Tucker, "Grain Sorghum Nitrogen Fertility," *Panhandle Research Station Report 1975,* Agricultural Experiment Station Research Report P–722, Oklahoma State University, Stillwater, 1975, pp. 61–62.

uct curve has the typical shape and decreases at an increasing rate, a relatively large proportion of the 260 pounds of marginal physical product probably could be obtained by applying 220 to 230 pounds of nitrogen per acre. Thus, it may be profitable to apply somewhat more than the 200 pounds of *N.* This illustrates that the principle of diminishing marginal returns provides a way of analyzing even the limited data that are available for such decisions.

Selecting the Profit-Maximizing Level of Variable Input with Nonconstant Product Prices

The principle of diminishing marginal returns is also applicable to livestock production. In feeding livestock, it is common to observe that increasing amounts of feed are required per pound of gain as an animal grows and matures. Consider the data in Table 3.7 on the amount of feed required for 10 pounds of gain for hogs of different weights. The data assume that 13 percent protein rations are being fed throughout the weight interval shown. Notice that the expected amount of feed required for an additional 10 pounds of gain increases from 37.2 pounds for hogs weighing 180 pounds to 51 pounds of feed for hogs weighing 290 pounds. In this example, the marginal physical product is 10 pounds in each case, and the amount of input required (pounds of feed per 10 pounds of gain) changes. Using a feed price of $0.09 per pound and multiplying by the amount of feed required per 10 pounds of gain result in a marginal factor cost of feed of $3.35 for an increase in weight from 180 to 190 pounds. The marginal factor cost increases as the amount of feed required per additional 10 pounds of gain increases.

If the price received for the animal does not change as the weight of the animal increases, the computation of the profit-maximizing feeding

Table 3.7 Selecting the Most Profitable Selling Weight for
Swine Using the Principle of Diminishing Returns

Weight of Pig (lb)	Days Required for 10 lb of Gain	Additional Feed for 10 lb of Gain	MFC of Feed @ $0.09/lb	Calculating the Marginal Value Product			Market Price per cwt
				Change in Value of the MPP	+Value of the Current Weight	Marginal =Value Product	
180							$49.00
190	5.0	37.2	$3.35	$4.90	$ 0	$4.90	49.00
200	4.9	38.4	3.46	4.90	0	4.90	49.00
210	4.9	39.6	3.56	4.90	0	4.90	49.00
220	4.8	42.0	3.78	4.90	0	4.90	49.00
230	4.8	42.5	3.82	4.90	0	4.90	49.00
240	4.8	44.0	3.96	4.85	−1.15	3.70	48.50
250	4.7	45.5	4.10	4.80	−1.20	3.60	48.00
260	4.7	46.0	4.14	4.80	0	4.80	48.00
270	4.5	47.5	4.28	4.70	−2.60	2.10	47.00
280	4.5	48.0	4.32	4.70	0	4.70	47.00
290	4.5	49.5	4.46	4.60	−2.80	1.80	46.00
300	4.5	51.0	4.59	4.60	0	4.60	46.00

Source: Unpublished data for "very good" feed efficiency developed by Dr. Jerry Hawton,
Department of Animal Science, and Dr. Paul Hasbargen, Department of Agricultural and
Applied Economics, University of Minnesota, St. Paul. The data are based on an average feed
efficiency of 3.52 pounds of feed per pound of gain for pigs from 40 to 230 pounds.

level would be identical to that used in the fertilizer examples above. Typically, however, the market price of the animal does change as the animal
increases in weight. Assume that the market price is $49 per 100 pounds
for hogs weighing from 180 through 230 pounds. Also assume that the
price decreases to $48.50 when hogs reach 240 pounds and $48 per 100
weight at 250 pounds. A further decrease to $47 per 100 pounds occurs at
270 pounds, and another decrease to $46 per 100 pounds occurs at 290
pounds. In this case, the decision-maker is faced with a changing product
price and must calculate a marginal value product that considers both the
additional amount of pork produced and the change in price as the animal
increases in weight.[3]

The marginal value product or the change in value of the hog can be
divided into two parts. In each case, the additional feed results in 10 more
pounds of pork to sell. Initially, an additional 10 pounds at $0.49 per

[3]The term "value of the marginal product" is used for situations with constant product prices,
while the term "marginal value product" is used when the product price is not constant.

pound has a value of $4.90. Since there is no change in the price of pork as the weight of the hog increases from 180 to 230 pounds, the marginal value product is $4.90 for each 10-pound increment. However, as the decision-maker considers adding another 10 pounds to a 230-pound hog, the additional weight only brings 48½ cents per pound or $4.85. In addition, the price of the initial 230 pounds will drop $0.005 per pound, decreasing the value of the 230 pounds by $1.15. The net effect of feeding the hog from 230 to 240 pounds is a marginal value product of $4.85 minus $1.15 or $3.70. The figures in Table 3.7 indicate that the producer should feed the hog to 230 pounds, but not beyond that weight because the next 10 pounds would result in a lower marginal value product than the marginal factor cost, and thus lower net returns per head.

There may also be additional costs that vary as an animal is fed to heavier weights. In the hog example, heavier weights may result in additional costs for use of the facilities, labor, interest on the money invested in the animal, and insurance costs on the animal that also need to be included as part of the marginal factor cost. If these costs are related to the amount of time required for production, they can simply be included as part of the marginal factor cost.

There are many examples in agriculture where changing the amount of the variable input will change the quality of the product, and hence the price or value of that product. In some cases, adding additional quantities of fertilizer per acre will also improve the protein level or the quality of the product and thus the value of that product in feeding livestock, or it will increase the price that can be received on the market. When these relationships can be established, they can be included in the manner illustrated with the data in Table 3.7. In these situations, additional units of variable inputs would increase the price and thus add to the value of the product rather than result in a decrease in price as illustrated in Table 3.7.

Selecting the Profit-Maximizing Level of Two Variable Inputs

Frequently, managers are concerned with the proper combination of two variable inputs to use in production. In some situations, each input can be analyzed independently if there is no interaction between them. In many cases, however, an interaction does exist between the inputs. This is particularly true with regard to fertilizer. Consider the data in Table 3.8 on the response of corn to alternative levels of nitrogen and phosphate fertilizer. The numbers in the table indicate expected yield of corn per acre for alternative combinations of nitrogen and phosphate fertilizer. For instance, the expected yield with 100 pounds of nitrogen and 60 pounds of phosphate per acre is 102 bushels per acre. Other numbers in the table can be interpreted in the same way.

The yield information in this table can be analyzed in a manner exact-

Table 3.8 Yield of Corn per Acre for Alternative Levels of Nitrogen (N) and Phosphate (P_2O_5) Fertilizer

Pounds of P_2O_5 per Acre	*Pounds of N per Acre*				
	60	*80*	*100*	*120*	*140*
			bushel		
20	76	90	95	97	96
40	78	92	100	104	105
60	78	93	102	105	108
80	77	92	103	106	109

ly analogous to that used to analyze the data in Table 3.4.[4] However, marginal factor cost and value of the marginal product are calculated mentally instead of formally written for each comparison. Assume that the price of nitrogen is $0.20 per pound and that the price of corn, net of the appropriate harvesting and marketing cost, is $2.50 per bushel. The profit-maximizing combination of nitrogen and phosphate fertilizer can be deter-

[4]The decision rule derived in footnote 2 can be extended to more than one variable input in a direct manner. Assume we have r inputs that are variable where $r < n$. The production function can be written as:

$$TPP = Y = f(X_1, X_2, \ldots, X_r | X_{r+1}, \ldots, X_n)$$

The profit function to be maximized can be written as

$$\pi = P_Y \cdot Y - P_{X_1} \cdot X_1 - P_{X_2} \cdot X_2 - \cdots - P_{X_r} \cdot X_r - \text{Fixed costs}$$

To maximize profits we differentiate with respect to each of the r variable inputs. Assuming constant input and product prices, the partial derivatives can be written as:

$$\frac{\delta \pi}{\delta X_1} = \frac{\delta Y}{\delta X_1} \cdot P_Y - P_{X_1} \geq O$$

$$\frac{\delta \pi}{\delta X_2} = \frac{\delta Y}{\delta X_2} \cdot P_Y - P_{X_2} \geq O$$

$$\cdot$$
$$\cdot$$
$$\cdot$$

$$\frac{\delta \pi}{\delta X_r} = \frac{\delta Y}{\delta X_r} \cdot P_Y - P_{X_r} \geq O$$

The r equations must be solved simultaneously to determine the profit-maximizing level of the variable inputs. Changing the amount of one input may affect the marginal productivity of the other inputs. The procedure described in the text provides a way of simultaneously equating VMP and MFC for the two variable inputs with discrete data, albeit in a somewhat crude fashion.

To ensure profit maximization, the marginal productivity of the variable inputs must be declining. This can be determined by inspection for the examples used in the text. Readers are referred to a standard graduate-level economic theory text, such as Hal R. Varian, *Microeconomic Analysis* (New York: W. W. Norton and Company, 1978), Chapter 1, for a more complete discussion of the second order conditions required for profit maximization.

mined by first holding the level of one input constant and calculating the marginal physical product, the marginal factor cost, and the value of marginal product as one moves across a row or down a column for the other factor. Once the profit-maximizing level of the first input for a given level of the second has been found, the next step is to hold the first factor constant and vary the second. The manager should alternate in this way until the profit-maximizing combination of the two has been located. For example, using the data of Table 3.8, first hold the amount of phosphate constant at 20 pounds per acre and consider the marginal physical product, the value of the marginal product, and the marginal factor cost resulting from increasing amounts of nitrogen per acre. Increasing the application from 60 pounds of nitrogen to 80 pounds results in an increase in the yield from 76 to 90 bushels per acre. The marginal physical product is 14 bushels per acre. The value of the marginal product is $2.50 × 14 bushels, or $35.00 per acre, while the marginal factor cost is $0.20 × 20 pounds, or $4.00 per acre.

Then compare an application rate of 100 pounds of nitrogen with that of 80 pounds of nitrogen. Again the value of the marginal product ($12.50 per acre) exceeds the marginal factor cost of $4.00 per acre. An additional 20 pounds of nitrogen (from 100 to 120) has a value of marginal product of $5.00 compared to a marginal factor cost of $4.00 per acre. However, the value of the marginal product of the next 20 pounds of N (an application rate of 140 pounds of N per acre) is negative when 20 pounds of phosphate is applied per acre. So the nitrogen level of 120 pounds seems best at this point. The next step is to hold the amount of nitrogen constant at 120 pounds and to add increasing amounts of phosphate per acre. Moving down the 120-pound nitrogen column, we see that the marginal physical product of an additional 20 pounds of phosphate is 7 bushels. The value of the marginal product is $17.50 compared to a marginal factor cost of $0.25 × 20 pounds or $5.00. The marginal physical product of increasing the phosphate application another 20 pounds from 40 to 60 pounds is 1 bushel. The value of the marginal product in this case is less than the marginal factor cost.

At this point the manager should again check the possibility of increasing the amount of nitrogen applied per acre. Increasing the rate from 120 to 140 pounds of N per acre, holding the phosphate application at 40 pounds, results in an increased yield or a marginal physical product of 1 bushel. The value of the marginal product is $2.50, which is less than the marginal factor cost of the additional 20 pounds of nitrogen. Before concluding that the profit-maximizing combination of nitrogen and phosphate has been found, the manager should check the possibility of adding both an additional unit of 20 pounds of nitrogen and 20 pounds of phosphate. That is, the value of the marginal product at 40 pounds of phosphate and 120 pounds of nitrogen should be compared with 60 pounds of phosphate and 140 pounds of nitrogen. Making this calculation results in a marginal physical product of 108 minus 104 or 4 bushels per acre. The value of the marginal product is $10.00 per acre. The marginal factor cost of this change is equal to $0.20 × 20 pounds of nitrogen plus $0.25 × 20 pounds of phosphate, or $9.00 per acre. Thus, it is profitable to increase the fertilizer levels to 140 pounds of nitrogen and 60 pounds of phosphate per acre. The manager should further check to see if it is profitable to add an additional 20 pounds of phosphate. In this case, the marginal physical product is only 1 bushel, and the value of the marginal product is less than the marginal factor cost. Therefore, the profit-maximizing combination with the prices given is 140 pounds of nitrogen and 60 pounds of phosphate per acre. The expected yield is 108 bushels per acre.

The manager could have held nitrogen at 60 pounds per acre and added increasing amounts of phosphate initially and followed the same procedure as described above. An individual following this path would have arrived at the same combination of nitrogen and phosphate per acre.

The Principle of Input Substitution

Managers frequently face the question of how to combine two or more inputs in production. In some cases, the problem is to select the level of the two variable inputs and the level of output that maximizes profits. This type of problem was analyzed above. In other situations, however, the inputs can be substituted for each other or used in varying proportions to produce a *given level of output*. With a given level of output and price, the manager's problem is to find the cost-minimizing combination of variable inputs because doing so also maximizes profits for that level of output.

There are many examples of input substitution in agricultural production. Various levels of hay and grain may be fed to dairy cows. A cow may be fed a low level of grain and a relatively high level of hay. Increasing the level of grain fed will cause the cow to consume less roughage. Thus, grain may be substituted for hay in the ration. Feeder cattle can also be fed on various combinations of concentrates and roughage to produce the same amount of gain. In this case, it may take more time to achieve the same weight of finished animal if the higher roughage ration is fed. Nevertheless, various combinations can be used to produce the specified amount of output. There are also many examples of input substitution in crop production. Hay can be harvested in a variety of ways. On the one hand, it can be cut with a mower, raked, and baled, and the bales loaded by hand and placed in the barn. Another alternative is to cut the hay with a swather, bale the hay with a baler that automatically throws the bales onto a wagon, unload the bales from the wagon, and stack them in the barn using an elevator. A third alternative is to windrow the hay, bale it with a large round baler, and move the bales mechanically from the field to storage. The first alternative requires more labor but less capital per ton of hay harvested. The last alternative represents a higher capital cost per ton of hay harvested but less labor. In this situation, the alternative combinations of inputs are represented by discrete alternatives rather than a continuous substitution relationship between labor and capital. Many other examples of labor-capital substitution can be found in feed handling and manure handling for livestock.

The Factor-Factor Relationship

The functional relationship expressing all combinations of two variable inputs that can be used to produce the same level of output is known as an isoquant. For two variable inputs X_1 and X_2, a given level of output Y^O, and fixed amounts of other factors X_3 through X_n, the functional relationship can be written as:

$$X_1 = g(X_2 | Y^O, X_3, ..., X_n) \qquad (3.4)$$

If the product Y can be produced with any combination of X_1 and X_2, the isoquant is a continuous relationship. In many situations, such as the labor-capital alternatives mentioned above, the isoquant is composed of a number of discrete points or combinations of the two inputs. The examples given assume continuous relationships, but the procedures discussed can be used to analyze discrete alternatives as well.

An isoquant is illustrated in Figure 3.5. The relationship shows quantities of X_1 and X_2 that can be used to produce 10 units of product Y. Combinations of 3.6 units of X_1 and 1 unit of X_2, 4 units of X_1 and 1.5 units of X_2, or any other combination on the isoquant, can be used to produce the same amount of product. Factors X_1 and X_2 substitute for or replace each other. For instance, the change from point a to point b involves substituting 1 unit of X_2 for 0.93 units of X_1. Moving along the isoquant to the right results in smaller and smaller quantities of X_2 being replaced by 1 unit of X_1.

The marginal rate of substitution is used to measure the amount of one input that substitutes for another. The marginal rate of substitution is defined as the decrease in the amount of one input as the amount of the second input is increased by one unit. The marginal rate of substitution of X_2 for X_1 can be denoted as $\Delta X_1 / \Delta X_2$ where Δ refers to "change in." Thus, $\Delta X_1 / \Delta X_2$ is read as the change in X_1 over the change in X_2. The quantity is always negative because the slope of the isoquant is negative. Since it is negative in all rational areas, the sign is frequently not shown. It should be understood, however, that the value is negative.

The marginal rate of substitution can be calculated as the average rate of change between two distinct combinations of inputs.[5] In Figure 3.5 the

---

[5] The marginal rate of input substitution can be calculated at a point by differentiating the function for the isoquant, equation 3.4, with respect to X_2. The result is dX_1/dX_2. Again, this is negative given the shape of the isoquant.

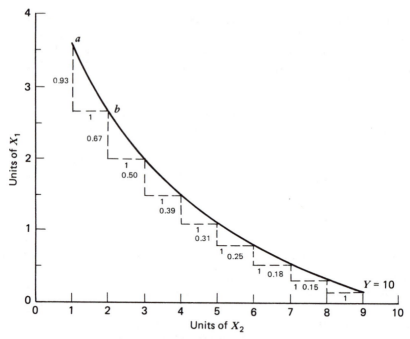

Figure 3.5. An isoquant illustrating diminishing rates of substitution.

average marginal rate of substitution is 0.93/1 between the first and second combinations. Increasing the amount of X_2 by one-unit increments replaces increasingly smaller amounts of X_1 as one moves along the isoquant to the right. The marginal rate of substitution decreases from 0.93/1 to 0.15/1. This is an example of two inputs with a diminishing marginal rate of substitution. Many of the factor-factor relationships in agriculture involve a diminishing marginal rate of substitution. Ration problems where forage is substituted for concentrate are typically of this type. In contrast, the isoquant in Figure 3.6 is a straight line and has a constant rate of substitution. Each unit of X_2 added replaces 0.4 units of X_1, for a 0.4/1 substitution ratio. This is one of the limiting cases for an isoquant—the two inputs are perfect substitutes. Examples include two types of corn with different protein levels that substitute for each other at a constant rate in livestock feeding, and substituting a larger machine that does more work per hour for a smaller machine. The second limiting case is technical complements—the case where inputs do not substitute for one another. In this case, the isoquant forms a right angle with the lines parallel to the axes. An example is the requirement to have one operator for one tractor. The combination of two machines, such as a tractor and plow, to perform an operation is another example. The shape of the isoquant for a situation of technical complements is illustrated in Figure 3.7.

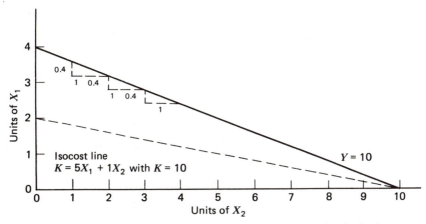

Figure 3.6.   An isoquant illustrating constant rates of substitution.

Selecting the Least Cost Combination

After the isoquant has been obtained, the next step is to define an isocost (equal cost) line and find the least cost combination of the two variable inputs that can be used to produce the specified level of output. Let P_1 and P_2 be the prices of X_1 and X_2, respectively. In addition, let K represent the level of cost. The isocost relation can be written as:

$$K = P_{X_1}X_1 + P_{X_2}X_2 \tag{3.5}$$

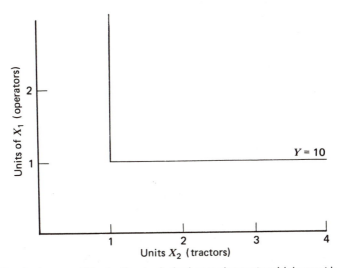

Figure 3.7.   An isoquant illustrating technical complements which must be combined in fixed proportions.

This relation indicates all combinations of X_1 and X_2 that can be purchased with a fixed amount of money K. This relationship can be plotted on the isoquant map. To do so, rearrange equation 3.5 and express X_1 as a function of X_2. Doing this results in equation 3.6.

$$X_1 = \frac{K}{P_{X_1}} - \frac{P_{X_2}}{P_{X_1}} X_2 \tag{3.6}$$

For instance, if $P_{X_1} = \$2$, $P_{X_2} = \$1$ and $K = \$2$, equation (3.6) can be written as:

$$X_1 = \frac{2}{2} - \frac{1}{2}X_2 = 1 - 0.5X_2 \tag{3.7}$$

and plotted on Figure 3.8 as a dashed isocost line with $K = 2$. This isocost line shows all combinations of X_1 and X_2 that can be purchased for $2.00. Since the price of the two inputs is assumed constant regardless of the amount of input purchased, the isocost relationship is a straight line. By changing the value of K to higher and higher levels, a family of isocost lines can be developed. The objective is to find the isocost line that is tangent to the isoquant. This is indicated by an isocost line having $K = 7$ in Figure 3.8. The least cost combination of inputs needed to produce 10 units of output is 1.75 units of X_1 and 3.5 units of X_2. Any other combination of X_1 and X_2 would be on a higher isocost line and, thus, have a greater cost.

Figure 3.8 indicates that the isocost line is tangent to the isoquant at the least cost combination of the two variable inputs. The slope of the isocost line in equation 3.6 is $-P_{X_2}/P_{X_1}$. The slope of the isoquant is $-\Delta X_1/\Delta X_2$. These two slopes are equal at the point of tangency. Thus, the cost-minimizing combination of two variable inputs used in the production of a single product is achieved when the ratio of factor prices is inversely equal to the marginal rate of substitution. The condition can be expressed as:

$$\frac{\Delta X_1}{\Delta X_2} = \frac{P_{X_2}}{P_{X_1}} \quad \text{or} \quad \frac{\Delta X_2}{\Delta X_1} = \frac{P_{X_1}}{P_{X_2}} \tag{3.8}$$

Again the reader is reminded that both ratios are negative. Both negative signs must either be included or omitted.[6]

The cost-minimizing principle is normally illustrated with an isoquant having a diminishing marginal rate of substitution as illustrated in Figure

---

[6]The slope of the isoquant at a point, dX_1/dX_2, defined in footnote 5, can be substituted for $\Delta X_1/\Delta X_2$ to obtain a more precise solution. The resulting equilibrium condition is $dX_1/dX_2 = P_{X_2}/P_{X_1}$. As indicated above, both ratios are negative.

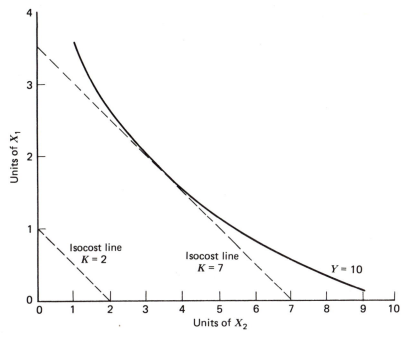

Figure 3.8. Tangency of isocost line and isoquant showing least cost combination of two variable inputs.

3.8. However, the same general concept applies to isoquants for perfect substitutes and perfect complements. The difference is that a tangency cannot be obtained. Instead, the lowest isocost line that touches the isoquant indicates the cost-minimizing combination of variable inputs. In the case of perfect substitutes, this will always result in all of one input being used, as illustrated in Figure 3.6. In the example, 10 units of X_2 and 0 units of X_1 are the least cost combination. An isocost line has not been added to the technical complements example of Figure 3.7. The reader can verify that a wide range of price ratios would result in combining the inputs in a fixed proportion.

An Application to Swine Feeding

The application of input substitution principles can be illustrated with data on the alternative combinations of soybean meal and corn used to feed hogs from 50 to 215 pounds (Tables 3.9 and 3.10). For instance, the data in Table 3.9 indicate that 16.7 pounds of soybean meal and 124.8 pounds of corn would produce 50 pounds of gain on a pig weighing 50 pounds at the start of the feeding period. Likewise, any of the other eleven combinations of soybean meal and corn listed in Table 3.9 would also produce 50 pounds of pork on a pig starting at 50 pounds. The total cost of each combination

Table 3.9 Combinations of Corn and Soybean Meal
Required per Pig for the 50- to 100-Pound Feeding Period

| Approximate Percentage Protein | Average Daily Gain | Pounds Soybean Meal per Head | Pounds of Corn per Head | Change in Feed Requirements | | Substitution Ratio |
				Increase in SBM	Decrease in Corn	ΔLbs Corn / ΔLbs SBM
13.5	1.26	16.7	124.8			
				1.2	4.2	3.50
14.0	1.27	17.9	120.6			
				1.2	4.0	3.33
14.5	1.29	19.1	116.6			
				1.3	3.7	2.85
15.0	1.31	20.4	112.9			
				1.3	3.5	2.69
15.5	1.33	21.7	109.4			
				1.4	3.3	2.36
16.0	1.35	23.1	106.1			
				1.4	3.0	2.14
16.5	1.38	24.5	103.1			
				1.4	2.9	2.07
17.0	1.41	25.9	100.2			
				1.5	2.8	1.87
17.5	1.44	27.4	97.4			
				1.5	2.6	1.73
18.0	1.48	28.9	94.8			
				1.6	2.5	1.56
18.5	1.50	30.5	92.3			
				1.6	2.2	1.38
19.0	1.53	32.1	90.1			

Source: Douglas G. Tiffany, *Economic Ration Formulation for Growing and Finishing Swine*, unpublished Plan B Project Paper, Department of Agricultural and Applied Economics, University of Minnesota, St. Paul, April 1977.

Table 3.10 Combinations of Corn and Soybean Meal
Required per Pig for the 100- to 215-Pound Feeding Period

| Approximate Percentage Protein | Average Daily Gain | Pounds Soybean Meal per Head | Pounds of Corn per Head | Change in Feed Requirements | | Substitution Ratio |
				Increase in SBM	Decrease in Corn	ΔLbs Corn / ΔLbs SBM
10.0	1.27	12.7	432.4			
				3.3	22.6	6.85
10.5	1.36	16.0	409.8			
				3.8	19.8	5.21
11.0	1.44	19.8	390.0			
				4.3	17.3	4.02
11.5	1.51	24.1	372.7			
				4.8	15.1	3.15
12.0	1.57	28.9	357.6			
				5.0	13.0	2.60
12.5	1.62	33.9	344.6			
				5.2	11.2	2.15
13.0	1.65	39.1	333.4			
				5.3	9.5	1.79
13.5	1.68	44.4	323.9			
				5.0	7.9	1.58
14.0	1.69	49.4	316.0			
				4.6	6.5	1.41
14.5	1.69	54.0	309.5			
				4.0	5.0	1.25
15.0	1.69	58.0	304.5			
				3.2	3.8	1.19
15.5	1.67	61.2	300.7			
				2.2	2.4	1.09
16.0	1.64	63.4	298.0			

Source: Douglas G. Tiffany, *Economic Ration Formulation for Growing and Finishing Swine*, unpublished Plan B Project Paper, Department of Agricultural and Applied Economics, University of Minnesota, St. Paul, April 1977.

of soybean meal and corn could be calculated to find the cost-minimizing ration. However, it is easier and quicker to use the following procedure which uses the principle noted earlier. First, define the substitution ratio, which is equal to the change in units of the replaced input over the change in the units of the added input. Letting X_1 represent the replaced input and X_2 the added input, this is $\Delta X_1/\Delta X_2$. Second, define the price ratio as equal to the cost per unit of the added input over the cost per unit of the replaced input. This is equal to P_{X_2}/P_{X_1}. As illustrated above, the least cost input combination occurs where the substitution ratio is equal to the price ratio, that is, where equation 3.8 holds. In applying the principle to discrete data, such as the alternative combinations of corn and soybean meal in Table 3.9, an equality may not be found. In these cases, select the two adjacent substitution ratios, of which one is greater and the other is smaller than the price ratio. The combination of variable inputs between these substitution ratios is the cost-minimizing solution.

The increase in the pounds of soybean meal in the ration, indicated in the fifth column of Table 3.9, ranges from 1.2 to 1.6 pounds. The decrease in the pounds of corn, indicated in the sixth column, diminishes from 4.2 to 2.2 pounds. The substitution ratio, defined as the change in the pounds of corn over the change in the pounds of soybean meal, ranges from 3.50/1 to 1.38/1. If we assume a price of soybean meal of $10.00 per 100 pounds and a corn price of $0.05 per pound ($2.80 per bushel), the price ratio is 0.10 over 0.05 or 2.0/1. This price ratio falls between the substitution ratios of 2.07 and 1.87. Thus, a combination of 25.9 pounds of soybean meal and 100.2 pounds of corn would be selected as the cost-minimizing combination of corn and soybean meal to feed.

Table 3.10 lists thirteen combinations of corn and soybean meal that can be used to feed hogs from 100 to 215 pounds. Notice that the substitution ratio (soybean meal for corn) ranges from 6.85 to 1.09. If we assume a price ratio of 2/1 ($0.10 per pound of soybean meal and $0.05 per pound of corn), the least cost ration is 39.1 pounds of soybean meal and 333.4 pounds of corn. Again, this ration is selected because the price ratio is between the substitution ratios of 2.15 and 1.79.

It should be remembered that the above rule is a cost-minimizing rule; it is not necessarily a profit-maximizing rule. It may be more profitable to produce at a higher output level, that is, on a higher isoquant, than on the one selected. Furthermore, there may be other inputs which in reality are varying at the same time that X_1 and X_2 are varying. For instance, in the hog feeding example analyzed in Tables 3.9 and 3.10, it takes more time to produce the gain with some rations than others as indicated by the average daily gain listed in column 2. For some farmers, the labor cost and the overhead cost of facilities may be fixed per head. However, the interest on the investment and the risk of disease outbreak are certainly not fixed.

The market price of the hogs may also change over time. This could result in either more advantageous or less advantageous prices for longer feeding periods. Furthermore, the quality of the product may vary for alternative combinations of X_1 and X_2. In cattle feeding, for example, the combination of roughage and grain may affect the quality of the animal produced, which in turn affects the price and value of the output. Consequently, a decision-maker using the principle of input substitution must consider these additional factors in deciding if the cost-minimizing solution is also a profit-maximizing one.

COST RELATIONSHIPS AND ECONOMIC CHOICE

Data made available for decision-making are frequently presented in terms of total, average, and marginal costs. The total, average, and marginal physical product relationships we have been discussing can be related to these cost concepts in a direct manner. This section discusses the relationship between the product curves and the cost curves. First, we will discuss short-run cost relationships where one or more than one input is held fixed in amount. Second, the rule to select the most profitable level of production based on cost curves will be developed. Finally, we will consider the use of long-run cost relationships in farm planning.

Short-Run Cost Relationships

The shape of the classical production function or total physical product curve (*TPP*) with one variable input is shown in Figure 3.9. We will consider the case of only one variable input (X_1) so that the relationship between the product and cost curves can be studied. All other inputs ($X_2, ..., X_n$) are considered fixed in the analysis, a very short-run situation. The relationship for total physical product is given by equation 3.1. The relationships for the corresponding marginal physical product and average physical product are given by equations 3.2 and 3.3.

The total cost curves for the production of various levels of output Y are shown in Figure 3.10. Cost curves are graphed with the dollars of cost on the vertical axis and the amount of output on the horizontal axis. The relationships for total variable cost, total fixed cost, and total cost are given by equations 3.9, 3.10, and 3.11, respectively.

$$TVC = P_{X_1}X_1 \tag{3.9}$$

$$TFC = \sum_{i=2}^{n} P_{X_i}X_i \tag{3.10}$$

$$TC = TVC + TFC = \sum_{i=1}^{n} P_{X_i}X_i \tag{3.11}$$

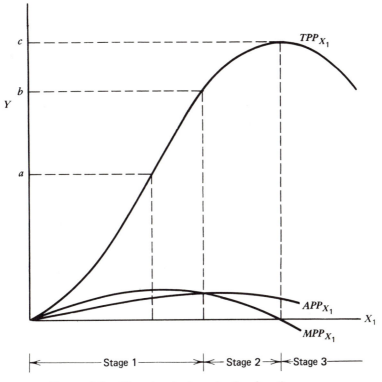

Figure 3.9. The classical production function.

Notice the corresponding relationship between the production function and the total variable cost curve. Output increases at an increasing rate until output level a is achieved. The total variable cost curve increases at a decreasing rate within the same range of output. Within Stage 2 (output level b to output level c), total output is increasing at a decreasing rate and total variable costs are increasing at an increasing rate. The total cost curves become vertical at output level c, the boundary between Stages 2 and 3. The vertical curve reflects the fact that costs continue to increase while the addition to output is zero. What if we plot the total cost curves for Stage 3 of production (not shown in Figure 3.10)? Total variable costs and total costs would continue to increase as the output level declined, resulting in total cost and total variable cost curves that bend back to the left as they increase. The shape reflects the irrationality of producing in Stage 3. Higher cost levels would result for production of a given amount of output in Stage 3 than in Stage 2.

The average and marginal cost curves for the classical production function are shown in Figure 3.11. The relationships for the average fixed

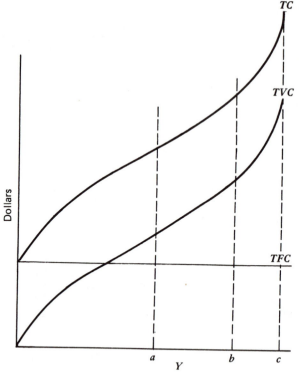

Figure 3.10. Total cost functions

$$AFC = \frac{TFC}{Y} \tag{3.12}$$

$$AVC = \frac{TVC}{Y} = \frac{P_{X_1}X_1}{Y} = \frac{P_{X_1}}{APP_{X_1}} \tag{3.13}$$

$$ATC = AFC + AVC \tag{3.14}$$

$$MC = \frac{\Delta TC}{\Delta Y} = \frac{P_{X_1}\Delta X_1}{\Delta Y} = \frac{P_{X_1}}{MPP_{X_1}} \tag{3.15}$$

costs, average variable costs, average total costs, and marginal cost are given by equations 3.12 through 3.15, respectively.

Notice that with one variable input the minimum average variable cost occurs at the output level having the maximum average physical product, the boundary between Stages 1 and 2. This is indicated by output level *b* in Figures 3.9 through 3.11. The marginal cost is the change in total cost per

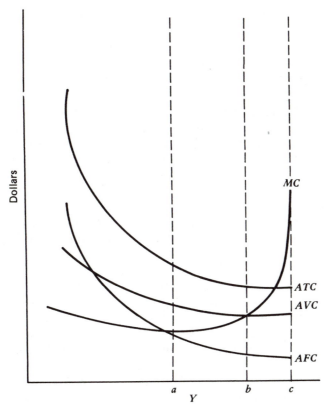

Figure 3.11. Average and marginal cost functions.

unit of output. Since the only input level being changed is X_1, the marginal cost is the change in the cost of X_1 per unit of Y. With constant input prices, this is P_{X_1} times the reciprocal of the marginal physical product as shown in equation 3.15. Thus, marginal cost is a minimum where the MPP_{X_1} is a maximum—output level a in Stage 1 of production. Marginal cost increases as MPP_{X_1} declines. It is equal to average variable cost at the boundary between Stages 1 and 2, where $APP_{X_1} = MPP_{X_1}$. The reader can confirm this conclusion by using equations 3.13 and 3.15. Marginal cost continues to increase and becomes vertical at the maximum TPP, where $MPP_{X_1} = 0$. The marginal cost ceases to have meaning for higher levels of variable input use where the $MPP_{X_1} < 0$.

The producer operating within Stage 2 will increase net returns (or reduce losses) by using higher levels of the variable input (and producing more output) as long as the $MC < P_Y$. The simple logic is that each additional unit of output produced adds more to gross returns than to cost when $MC < P_Y$.[7]

Managers are frequently interested in cost minimization, but the reader should notice that various cost-minimizing rules are unlikely to lead to profitable output levels. For example, a rule to minimize average variable costs would result in selecting the input level b at the boundary between Stages 1 and 2. If P_Y is greater than MC at this level, profit can be increased by operating at a higher output level. In the event P_Y is less than the minimum AVC, the manager will minimize losses (i.e., reduce them to TFC) by ceasing production. Attempting to minimize MC would also result in producing within Stage 1, a less profitable alternative than operating in Stage 2 where $P_Y = MC$, or ceasing production if P_Y is less than the minimum AVC.

The cost calculations are illustrated in Table 3.11 using the data from Table 3.4 on the response of corn to nitrogen fertilizer. In this example, nitrogen is the only variable input. It has a price of $0.20 per pound. All other inputs cost $200 per acre and are considered fixed. The values in Table 3.11 are calculated using equations 3.9 through 3.15. In the earlier example, we found that with corn priced at $2.50 per bushel the most profitable production level was 158 bushels using 220 pounds of N per acre. Comparing a corn price of $2.50 with the marginal cost per bushel in Table 3.11 also indicates that fertilizing at more than 220 pounds of N per acre will result in a marginal cost that exceeds the price of corn. If only these discrete levels of nitrogen application are considered, the marginal cost calculation indicates that the same level of N would be applied for any corn price between $2 and $4 per bushel.

The number of variable inputs increases as the length of the planning horizon increases, providing more opportunities to adjust the combination of inputs in a cost-minimizing manner. For example, a farmer could probably reduce the average total cost of corn production for low N application rates in Table 3.11 by using the rules developed in the previous section to select the optimum seeding rate, herbicides, and level of other nutrients. Over a period of several years, the operator also may be able to adjust the machinery and equipment complement and achieve certain efficiencies that reduce the cost of producing the product.

The effect of shifting costs from the fixed to the variable category

---

[7] This can be derived more formally by differentiating the profit function introduced in footnote 2 with respect to Y:

$$\pi = P_Y \cdot Y - P_{X_1} \cdot X_1 - FC$$

Assuming prices are constant:
$$\frac{\delta \pi}{\delta Y} = P_Y - P_{X_1} \cdot \frac{\delta X_1}{\delta Y} \geq 0$$

Thus:
$$P_Y \geq P_{X_1} \cdot \frac{\delta X_1}{\delta Y}$$

or
$$P_Y \geq MC$$

Table 3.11 Production Costs Derived
from the Production Function

X_I (lb)	Y (bu)	TFC ($/acre)	TVC ($/acre)	TC ($/acre)	AFC ($/bu)	AVC ($/bu)	ATC ($/bu)	MC ($/bu)
0	90	$200	$ 0	$200	$2.222	$0	$2.222	
40	110	200	8	208	1.818	0.073	1.891	$0.40
80	126	200	16	216	1.587	0.127	1.714	0.50
100	133	200	20	220	1.504	0.150	1.654	0.571
120	139	200	24	224	1.439	0.173	1.612	0.667
140	144	200	28	228	1.389	0.194	1.583	0.80
160	149	200	32	232	1.342	0.215	1.557	0.80
180	153	200	36	236	1.307	0.235	1.542	1.00
200	156	200	40	240	1.282	0.256	1.538	1.33
220	158	200	44	244	1.266	0.278	1.544	2.00
240	159	200	48	248	1.258	0.302	1.560	4.00
260	159.5	200	52	252	1.254	0.326	1.580	8.00
280	160	200	56	256	1.250	0.350	1.600	8.00
300	159	200	60	260	1.258	0.377	1.635	—

lowers the average fixed cost curve and raises the average variable cost curve, providing input prices are constant. If the identical level of inputs is used, equation 3.11 indicates that the average total cost curve will be unaffected. However, if a cost-reducing adjustment can be made for some output levels, the average total cost curve will be reduced for those output levels. For this reason, we expect longer run average total cost curves (those considering more inputs variable) to be at the same level or below the shorter run cost curves. If the short-run cost curve uses the most efficient combination of inputs over a range of output, the long-run and short-run curves will be identical for that output range. However, the long-run curve will lie below the short-run curve in those output ranges where more efficient input combinations can be achieved over the longer period.

Economies of Size

The shape of the long-run average cost curve by size of operation is important in agriculture. Producers selecting a system of production for an enterprise want to identify the most efficient combination of fixed and variable inputs by size of enterprise. They are also interested in how the average total cost varies over alternative sizes of enterprise, to indicate whether they will be at a competitive advantage or disadvantage relative to producers with other sizes of operation.

Economies of size in agriculture indicate that the average total cost per unit of output for an enterprise initially declines as size increases and then reaches a relatively constant level. Average total cost decreases initially because of both technical efficiencies and pecuniary or pricing economies.

Technical efficiencies refer to the more efficient use of labor, machinery, equipment, and other inputs. For example, expanding the size of a dairy enterprise with a given housing system would result in lower buildings and equipment investment, more efficient labor use due to more specialization of labor, and more efficiency in manure handling per unit of milk produced. Similarly, expanding a corn enterprise should result in more efficient use of equipment and labor. Pecuniary economies can exist either because of lower prices for purchased inputs or higher prices for the product produced. Farmers with larger operations are typically able to purchase inputs in large quantities at somewhat lower prices and to market their products more efficiently, resulting in lower marketing costs per unit of output.

To develop accurate enterprise budgets, the probable size of the enterprise and the business must be considered. This is important in estimating per acre machinery and equipment costs as well as labor requirements for crop production. In livestock production, size considerations are particularly important in developing building and equipment expense estimates as well as in determining labor requirements.

Estimates of capital investment, labor requirements, and feed and nonfeed costs per head for an open-lot shelter cattle feeding system used on Iowa farms illustrate the economies of size considerations (Table 3.12). Capital investment decreases from $257 per head for a lot of 100 head capacity to $133 per head for a lot of 1,000 head capacity. Much of the reduction is due to lower investment costs per head for feed storage, feed handling, and waste storage. Feed costs are essentially the same for all sizes of lot, but some of the nonfeed costs per head decline as the lot size increases. Nonfeed costs include the costs for labor, fixed costs on buildings, facilities and equipment, interest on the feed and livestock inventory, and direct cash costs such as veterinary expense and power expense. The nonfeed costs decline primarily because the investment in facilities per

Table 3.12 Labor, Capital Investment, Feed and Nonfeed Costs for Open-Lot Shelter Yearlings on Iowa Farms

	Lot Capacity in Head			
Item	*100*	*300*	*600*	*1000*
Capital investment per head	$257	$182	$146	$133
Hours labor per head	4.4	2.4	2.0	1.8
Feed costs per cwt gain	32.05	32.05	32.05	32.05
Nonfeed costs per cwt gain	24.22	19.46	17.67	16.93
Total	$ 56.27	$ 51.51	$ 49.72	$ 48.98

Source: James McGrann, John Ball, Michael D. Boehlje, and Gene Rouse, *Evaluation of Feedlot Systems*, Cooperative Extension Service, Iowa State University, Ames, July 1979.

head is lower. Economies of labor use from 4.4 to 1.8 hours per head result from increased specialization of labor in the larger lots and somewhat more mechanization.

A final point to note is that there tend to be relatively large economies of size at small sizes of the enterprise, but relatively little reduction in the average cost per unit of output as the size of the enterprise increases. The information in Table 3.12 also illustrates this point. Capital investments decline from $257 per head for a lot of 100 head capacity to $146 per head for a lot of 600 head capacity. Increasing the size to 1,000 head only reduces the investment cost an additional $13.00 per head. The decrease in nonfeed costs and labor requirements is also much greater between 100 and 600 head capacities than between 600 and 1,000 head. Thus, the individual preparing an enterprise budget must be aware of the effect of economies of size on the estimates being prepared.

Summary

Production costs can be classified as fixed or variable. The distinction between fixed and variable costs is basic to all economic decision-making in managing the farm business. Fixed costs do not change with the level of output. Fixed costs for a farmer who is planning for next year typically include depreciation, taxes, insurance, and interest payments. Variable costs change with the level of output, and in planning for the next year, they include such items as fuel, seed, fertilizer, feed, and veterinary expenses.

As the length of the planning period increases, more costs are included in the variable category. In the long run, virtually all inputs can be changed and thus become variable costs. Only variable costs should be considered by the manager in making production decisions. The difference between cash and noncash costs is also an important consideration, particularly in determining the cash flow consequences of specific enterprise decisions.

The basic technical relationship underlying an enterprise budget is the production function. A production function expresses the relationship between the use of inputs and the products produced; it indicates the technical efficiency in converting inputs into products. Production functions can exhibit increasing, constant, or decreasing marginal productivity. The marginal physical product is measured as the additional output that results from an additional unit of input. Average product or productivity is defined as the total amount of output divided by the total amount of input utilized.

The production relationship can be divided into three important stages using these concepts of average and marginal product: Stage 1 where average product is increasing and marginal product exceeds average product; Stage 2 where average and marginal product are decreasing and marginal product is greater than zero; and Stage 3 where both average and marginal product are decreasing and marginal product is less than zero. The rational range of production is Stage 2; the optimal level of production within Stage 2 is determined by equating the value of the marginal product with the marginal factor cost. The value of the marginal product is determined as the price of the output times the marginal physical product, and the marginal factor cost is the cost of the last unit of the variable input.

When more than one input is used in the production process, the principle of input substitution must be used to determine the combination of inputs that will minimize costs for a given production level. This principle is based on the factor-factor relationship which indicates the alternative combinations of inputs to produce a given level of product. The substitution relationship is measured by the marginal rate of substitution, and the optimal or least cost combination of inputs is determined by equating this marginal rate of substitution with the ratio of input prices.

Economies of size suggest that the average total cost per unit of output declines (at least initially) in agriculture as size increases. Thus, the probable size of the enterprise must be considered in developing enterprise budgets to accurately reflect the technical efficiency for the size of enterprise being planned.

Questions and Problems

1. Why develop enterprise budgets? How are these budgets used in making management decisions?
2. Define fixed and variable costs. Why is this distinction between fixed and variable costs important?
3. What is a production function? What are the reasons for increasing, constant, and decreasing productivity?
4. Define average and marginal physical product. How are average and marginal physical product calculated?
5. Collect data from an agronomist or animal scientist on the productivity of various crop or livestock products (for example, yield response to fertilizer use or livestock gain as a function of protein or concentrate consumption), and determine the total product, average physical product, and marginal physical product relationships.
6. Define the three stages of production. Why is Stage 2 considered to be the rational stage of production?
7. What adjustments should a producer make who is producing in Stage 1? What if the individual is producing in Stage 3?
8. Discuss the economic concepts that can be used to choose the profit-maximizing level of one variable input to use in the production of a product.
9. Apply the profit-maximizing principle to the data collected earlier in Question 5.

10. How is the single product-single factor profit-maximizing principle modified to handle nonconstant product prices?

11. Discuss the economic concepts used to select the least cost combination of inputs to use in the production of a given level of one product.

12. What is the principle of input substitution? Define the marginal rate of substitution.

13. Describe how the shape of the average variable cost and the marginal cost curves compares to the shape of the average physical product and the marginal physical product curves.

14. Why are size economies important in developing enterprise budgets?

15. Collect data from an agronomist or animal scientist on the use of various fertilizer nutrients in crop production or protein and concentrates in livestock production and apply the appropriate principles to choose the optimal combination of inputs for the production process.

16. Define the input substitution relationship of perfect complements. Can you think of any examples of perfect complements? Define perfect substitutes. Can you think of real world examples of perfect substitutes?

Further Reading

Dillon, John L. *The Analysis of Response in Crop and Livestock Production.* 2d ed., New York: Pergamon Press, 1977, Chapter 1.

Doll, John P., and Frank Orazem. *Production Economics: Theory with Applications.* 2d ed., New York: Wiley, 1984, Chapters 2, 3, 4, and 7.

Ferguson, C. E., and J. P. Gould. *Microeconomic Theory.* 4th ed., Homewood, Ill.: Richard D. Irwin, 1975, Chapters 5, 6, and 7.

Ferguson, C. E., and S. Charles Maurice. *Economic Analysis.* Rev. Ed., Homewood, Ill.: Richard D. Irwin, 1974, Chapters 6, 7, and 8.

Heady, E. O., and John L. Dillon. *Agricultural Production Functions.* Ames: Iowa State Press, 1961, Chapter 2.

Hirshleifer, Jack. *Price Theory and Applications.* 2d ed., Englewood Cliffs, N.J.: Prentice-Hall, 1980, Chapter 9.

Stanton, B. F. "Perspective on Farm Size." *Amer. J. Agr. Econ.* 60 (1978):727–37.

4

ENTERPRISE BUDGETS: COMPUTATIONAL PROCEDURES

CHAPTER CONTENTS

The enterprise budget is a convenient means of summarizing enterprise cost and return projections for use in many management decisions. These decisions include selecting the system to use in producing a crop or livestock enterprise, developing a leasing arrangement, choosing the appropriate investment, and selecting the whole-farm plan. For ·example, a comparison of an enterprise budget for nonirrigated production with one for irrigated production of a crop on the same land area would show the differences in yield, seeding rate, fertilizer application, pesticide application, harvesting cost, irrigation cost, monthly labor and capital requirements, and the difference in expected net returns. Similarly, a farmer might use two enterprise budgets to compare two systems of producing the same species of livestock. The cost and return for a crop can be analyzed to estimate the amount of cash rent or the proportion of the crop that can be paid for land rent. Investment decisions in land, buildings, and equipment typically require data on cost and return for the crop and/or livestock enterprise(s) to be produced with the potential investment. These data can be summarized efficiently with enterprise budgets and used with the capital budgeting procedures discussed in Chapter 8 to complete the analysis. Furthermore, enterprise budgets are an efficient way to summarize the basic data needed in planning the whole-farm business, whether the analysis is completed with the budgeting procedures described in Chapter 6 or with the linear programming procedures explained in Chapter 10. Thus, the procedures explained in this chapter to estimate the individual components of costs and returns, as well as the enterprise budget format, are used regularly in analyzing alternatives available to the farm manager.

The preparation of enterprise budgets requires a working knowledge of both the economic concepts discussed in the previous chapter and the computational procedures discussed here. The enterprise budget is based on the system of production that identifies the specific output(s) to be produced (such as corn for grain, 450-pound choice feeder calves, or sweet corn for processing), the sequence of operations, the approximate time the operations are to be performed, and the inputs required for the production process. The economic principles discussed in the previous chapter should be used in combination with an understanding of the underlying biological growth and development process in selecting the level and timing of input use. For example, an understanding of the timing and level of nutrient uptake by a crop or the change in nutritional requirements by livestock over the animal's life is basic to the appropriate application of the economic principles we have studied. Likewise, knowledge of the life cycle of major weeds, insects, and diseases, and the impact these pests may have on crop yields is important in selecting the most economic pesticides to include in the production system. A more detailed discussion of specifying the production system is presented in Chapter 17 where it is used as the basis for development of the enterprise control system. The discussion here presumes that the system has been specified and concentrates on the computational issues and procedures involved in preparing the enterprise budget.

The first two sections of this chapter discuss methods of estimating the costs of durable inputs (machinery, equipment, and buildings) used in crop and livestock production. The discussion deals with the issues of estimating the costs during periods of rising prices for new and used capital assets as well as during periods of stable prices. The estimation of seasonal input requirements and interest on the operating capital used is discussed in the third section. The final section presents examples of enterprise budgets, discusses their interpretation, and indicates how they can be used.

ESTIMATING COSTS OF MACHINERY SERVICES

Machinery and equipment expense represents a major category of costs in crop production. To prepare accurate enterprise budgets, individuals need a procedure to estimate machinery costs and to determine how these costs vary for alternative systems of production. Managers are frequently faced with the problem of estimating costs and returns for two or more methods of producing a given crop. For example, a farmer may be interested in comparing the costs of corn production using a 100 horsepower tractor with the costs of using a 150 horsepower tractor for tillage operations. Alternatively, the operator may want to compare the costs of corn production for minimum and conventional tillage. One cannot simply use average machinery costs from farm records or from other studies in many cases because the amount of use per year, fuel prices, auxiliary equipment required, speed of operation, initial investment cost, operating conditions, and other factors may be different from the average of those in the published sources. A consistent method of estimating machinery and equipment costs is needed for use in preparing enterprise budgets for both crop and livestock alternatives. It is also important for farm managers to be able to estimate machinery costs (1) to compare the costs of purchasing new versus used machinery, (2) to make decisions on equipment purchases, (3) to use in comparing ownership with leasing and custom hire, and (4) to decide when to replace machines. This section presents procedures one can use in estimating annual machinery costs.

Many items of machinery are used in more than one enterprise. A combine on a Corn Belt farm may be used to harvest corn, soybeans, and one or more small-grain crops. In these cases, the manager may also be interested in allocating the annual machinery expense to each enterprise. A method of making these allocations is also illustrated below.

Costs To Be Estimated

The annual costs of providing machinery services can be divided into two categories. The fixed or ownership costs are those costs that do not vary with the amount of machine use per year. For an owned machine, these costs include capital recovery (or depreciation plus interest on the money invested in the machine), housing, insurance for fire, theft, storm and vandalism, and personal property taxes. The fixed costs for leased equipment depend on the terms of the leasing arrangement. Typically, costs that do not vary with the amount of use per year include an annual lease payment, a housing cost, and insurance (if that is not provided as part of the lease). Fixed costs for custom hire are typically zero. The second category includes those costs that vary with the amount of use and are referred to as variable or operating costs. The operating costs for an owned machine include fuel, lubrication, repairs and maintenance, and labor. The operating costs for a leased machine are frequently the same as for an owned machine. The leasing company sometimes provides the repairs, and they are not part of the farmer's operating costs. Variable costs of custom hire are the custom charge per acre. The sum of annual ownership and annual operating costs equals the annual costs of providing machinery services.

In distinguishing between fixed and variable costs, the discussion in Chapter 3 emphasized that only variable costs are relevant for decision-making, and the costs that are variable depend on the length of the planning horizon. Now we can apply these concepts to select the machinery costs to be estimated. Enterprise budgets are typically developed for one of two lengths of run: (1) the long-run which is usually considered to be a period of five years or more; and (2) the short-run or the next production period.

- The individual developing enterprise budgets for use in *long-run* planning should consider *both* the *ownership* and *operating* costs as variable. Over a period of five or more years, the machinery complement could be changed and some machines would normally be replaced. Thus, the ownership as well as the operating costs should be estimated and included in the enterprise budget.

- The farmer planning for the *short-run* or next year typically has the set of equipment and machinery required. The variable costs in this case are *fuel, lubricants,* and *repairs.* The hours of labor required are also estimated for use in planning the total farm. When the individual also intends to project a cash flow, it may be necessary to estimate the fixed cash costs—insurance and personal property taxes.

We will now develop methods to estimate costs for long-run budgets. Those interested only in developing short-run budgets may want to skip ahead to the discussion of operating costs and then read only the discussion of estimating taxes and insurance in the ownership cost section for use in projecting cash flows.

Historic Versus Current Replacement Costs

Relatively rapid changes in the purchase price and other machinery costs have occurred during recent years. Many farmers trading or selling machinery during periods of rapid price increases have received approximately as many dollars for an item of equipment as they paid for it several years earlier, although it should be recognized that in most cases the dollars received did not have the same purchasing power as those spent. Thus, it is important to discuss the adjustments that should be made in the calculation procedures when the decision-maker either estimates costs for machines purchased in the past or expects significant price changes in the future.

The discussion here is in terms of price changes rather than inflation. Inflation is defined as an increase in the general price level. If some prices in the economy are increasing while others are decreasing, the weighted average or the general price level may not change and no inflation occurs. However, the individuals and industries affected must deal with these price changes. The price system is the major mechanism to allocate changing supplies and demands for resources (such as land, energy, labor, capital, steel, and other resources) and products in the economy. As inputs become scarce relative to their demand or the demand for a product increases relative to its supply, the price level increases to equate supply and demand. If other prices in the economy decline at the same time, however, the general price level may remain unchanged. It is, therefore, important to deal with changing price levels for capital assets regardless of whether the general price level is rising, constant, or declining. As we note in our discussion of the interest charge, the change in the general price level is of primary importance in estimating the opportunity cost of using the capital invested in the assets.

The enterprise budget was defined earlier as an estimate of the average annual cost and returns for the enterprise. It is important to state all costs and returns included in the enterprise budget in dollars of the same purchasing power. This can be accomplished by stating all prices for outputs, operating inputs, and capital inputs in current year prices. The current year market prices of the product, seed, feed, and fertilizer, the interest rate for operating capital, and other annual inputs are routinely used in preparing such estimates.

Price of Capital Assets   To complete the enterprise budget on a current basis, we must also include the current market cost of the services provided by capital assets—machinery, equipment, buildings, and land. The current market value is most easily approximated by calculating ownership and operating costs of capital assets based on *current year market or replacement prices*. As discussed in Chapter 2, some capital assets, such as major items of farm machinery, have reasonably well-defined markets. In these cases, the *current market value* for the used machine can be assumed to reflect its value in dollars of current purchasing power. Other assets do not have well-defined markets, and the replacement cost of a similar asset (or one of the asset valuation methods described in Chapter 2 for the market value column of the balance sheet) should be used. Current market prices for used capital assets and current replacement prices for new capital assets represent the economic cost of purchasing these assets in current year dollar values, once we take into account any inflation or deflation, plus other changes in the price level of these assets that have occurred since they were purchased. During periods of moderate changes in the general price level, we recommend using the beginning of the year asset values from the market value column of the balance sheet (Figure 2.6) in making these calculations.

Unless all costs are expressed in current dollars, the profitability of the enterprise is incorrectly stated. During periods of increasing capital asset prices, the ownership and operating costs that are calculated, based on the original cost of the asset some years earlier, will result in understating the machinery and building costs and overstating the profitability of the enterprise(s) using these assets. Of course, the opposite situation would tend to occur either during a period of declining capital asset prices, or for a capital asset whose price has declined because of technological obsolescence.

Selecting the Interest Rate   After selecting the current market value or replacement cost as the basis to calculate the cost of the service, the next concern is the appropriate charge for the use of the invested capital. The nominal interest rates observed for borrowed funds, as well as the return received for money invested in the business, can be divide into two parts—the real rate and the inflation premium. The real rate is the return for the use of the capital, which can be thought of as being composed of the rate of return for a risk-free asset plus a risk premium. The inflation premium recognizes that the purchasing power of the money invested is expected to decline over the period. Basing the machine cost calculation on the current market or replacement price of the asset takes account of any change in the value of money up to the current time. The market price of capital assets can be expected to adjust with the change in the general price level. Thus, the real rate of interest or return on the money represents the cost of using the capital in current dollar terms. Using the nominal rate of interest with current replacement costs overstates the economic cost in current dollars by the amount of the inflation rate. For this reason the real rate is recommended.

The general rate of inflation for this purpose should be measured with either the producer price index (PPI) or the implicit gross national product (GNP) deflator. The annual average values for these two indexes are shown in Table 4.1 for 1968–82. The PPI is a *fixed* weight price index that measures price changes for a group of nonlabor inputs used in various production, refining, and distribution processes. The implicit GNP deflator is a *variable* weight measure of domestic inflation constructed from the price changes for the major components of GNP—consumption, investment, government expenditures, and net exports. The weights for sub-

Table 4.1 Price Deflators and the Interest Rates on Outstanding Farm Debt

Year	GNP Price Deflator[a] 1972 = 100	Percentage Change from Previous Year	Producer Price Index[a] Finished Goods 1972 = 100	Percentage Change from Previous Year	Federal Land Banks[b] Average Rate on New Loans[c]	Production Credit Associations[b] Average Cost of Loans[d]
1968	82.6	4.6	87.7	2.9	6.8	7.3
1969	86.7	5.0	90.9	3.6	7.8	7.8
1970	91.4	5.4	94.1	3.5	8.7	9.0
1971	96.0	5.0	97.0	3.1	7.9	7.3
1972	100.0	4.2	100.0	3.1	7.4	7.0
1973	105.9	5.9	109.1	9.1	7.5	8.1
1974	116.0	9.5	125.9	15.4	8.1	9.4
1975	127.2	9.7	139.4	10.7	8.7	8.9
1976	133.7	5.1	145.6	4.4	8.7	8.2
1977	141.7	6.0	155.0	6.5	8.4	7.9
1978	152.9	7.9	167.2	7.9	8.4	8.8
1979	165.5	8.2	185.8	11.1	9.2	10.7
1980	178.6	7.9	210.8	13.5	10.4	12.9
1981	195.5	9.5	230.2	9.2	11.2	14.5
1982	207.2	6.0	239.4	4.0	12.3	14.6
1968–82 average	–	6.7	–	7.2	8.8	9.5

[a]*Sources:* Council of Economic Advisors, *Economic Indicators*, Washington, D.C.: U.S. Government Printing Office, February 1983, and *Statistical Abstracts of the United States*, Washington, D.C.: U.S. Department of Commerce, 1980, p. 479.

[b]*Sources:* *Agricultural Statistics 1981*, Washington, D.C.: U.S. Department of Agriculture, p. 481, and Emanual Melichar and Paul T. Balides. *Agricultural Finance Datebook*, Washington, D.C.: Board of Governors of the Federal Reserve System, February 1983, p. 32.

[c]Contract rates on loans through the Federal Land Bank Associations.

[d]The rate represents interest (less patronage refunds), service fees, cost of record searches, and filing and releasing mortgages, etc., paid by borrowers as a percentge of average loans outstanding during the year.

components and components vary from one period to the next, reflecting current spending patterns in the economy. The implicit GNP deflator is usually considered a more accurate measure of the change in the general price level because it reflects both the changes in prices and the changes in spending patterns. An important effect of the variable weight index is to moderate the percentage change in the index from one year to the next as illustrated by the data in Table 4.1. The annual change in the implicit GNP deflator and the PPI from 1968 to 1982 averaged 6.7 and 7.2 percent, respectively. However, either the PPI or the implicit GNP deflator can be used here.[1]

A comparison of the percentage change in the GNP deflator and the PPI with the average rates of interest shown in Table 4.1 indicates that the real rate of interest fluctuates a great deal from one year to the next. It may be negative in some years, such as 1975 when the percentage change in either index exceeded the nominal interest rate, and quite large in others, such as 1982. During the fifteen-year period shown, the real rate of interest averaged approximately 2.0 percent for Federal Land Bank loans and 2.6 percent for Production Credit Association loans using the GNP deflator as the measure of the increase in the general price level. The real rates are 1.5 percent and 2.1 percent using the PPI.[2] It should be noted that the real rates for 1968–1982 are lower than the 3 to 5 percent which many economists agree is more typical for these sources of borrowed funds.

The task at hand is to select the appropriate real rate of interest to use in the machinery cost calculations. The first step is to estimate the expected increase in the general price level for the long-run planning period. This should abstract from the year-to-year fluctuations noted above and may be based on estimates of future changes in the general price level prepared by the federal government and private forecasters. Then the farmer must also estimate the appropriate nominal rate for the business. The appropriate nominal rate is the opportunity cost for the use of capital in the business, that is, the nominal rate of return in its best alternative use. An operator using equity capital should estimate the opportunity cost and consider this the nominal rate. If a farmer can borrow all of the capital that can profitably be used in the business, the nominal opportunity cost can be approximated by the market rate of interest. Most farmers are limited in the total

---

[1]The important distinction to note is that either the PPI or the implicit GNP deflator is preferred to the consumer price index (CPI). The CPI reflects price changes for a fixed combination of consumption goods. This index is considered inappropriate because it is based on consumption goods and because the combination of consumption goods actually purchased from period to period varies. The CPI also includes a large weight on home mortgage interest rates which probably overstates the importance of that expenditure in the purchases of most farm families. For these and other reasons, the CPI is considered a less useful measure of the change in the price level than either the PPI or the implicit GNP deflator. See Paul T. Prentice and Lyle P. Schertz, *Inflation: A Food and Agricultural Perspective,* U.S. Department of Agriculture, Economic Research Service, Agricultural Economic Report No. 463, February 1981, pp. 4–7 for a discussion of the three deflators.

[2]The real rate is commonly calculated as the nominal rate minus the rate of inflation. However, the relationship between the real rate (i) and the inflation rate (f) is multiplicative rather than additive. The nominal amount of \$1 at the end of one period is given by $(1 + r) \cdot \$1$ where r is the nominal rate, which is equal to $(1 + i)(1 + f) \$1 = (1 + i + f + if) \$1$. In simply subtracting f from r to estimate i, we ignore the product "if." Given this multiplicative relationship the real rates quoted were calculated as $(1 + r)/(1 + f) = (1 + i)$.

amount of capital they can borrow, however, and the opportunity cost is higher than the interest rate paid. In this situation, the appropriate real rate is also higher than would be calculated using the market interest rate.

The computational procedures for machine costs are described using current replacement costs and real interest rates. Then the estimates based on historic purchase prices and nominal interest rates are also shown in the worksheet on pages 153–155 (Table 4.7). The results of the two methods are contrasted at the end of the section.

The equations estimate *average* ownership and operating costs. Thus, the average annual depreciation, interest, insurance, housing, taxes, and the average repair cost per hour of life are calculated. Different estimates of machinery costs for the first, second, third, and subsequent years could be prepared, but such a procedure is unnecessarily complex for many decisions. The exceptions are annual cash flow projections and the machinery replacement decision. Methods to use in making the calculations of annual cash costs are mentioned at the close of this section, while the more complete specification on a year-by-year basis is discussed in Chapter 14 in conjunction with a discussion of the machinery replacement decision.

Two machines are used to illustrate the computational procedures. Costs are calculated for a tractor having 120 power-take-off horsepower (PTOHP), as well as a 6-16 inch moldboard plow. The original list price of the five-year-old tractor and plow was $25,440 and $4,650, respectively. The buyer was able to obtain a discount, but then had to pay sales taxes and some minor delivery charges, making the purchase price (the amount actually paid five years ago) $23,900 for the tractor and $4,370 for the plow. The current list prices for the comparable tractor and plow are $37,400 and $6,830, respectively. The purchase price would be $35,100 for the tractor and $6,420 for the plow at the current time. The tractor has been used 600 hours and the plow 150 hours annually. The same amount of annual use is expected for the remainder of their 10-year useful life. A nominal interest rate of 12 percent, a general rate of inflation of 7 percent (and a real rate of 5 percent), and a diesel fuel price of $1.25 per gallon are to be used in the cost computations.

Ownership Costs

Depreciation and Interest Depreciation is the reduction in the market value of a machine due to wear, obsolescence, and age. Notice that this is the economic concept of depreciation, not necessarily the amount of depreciation to take for tax purposes. The amount of wear may cause a machine to be somewhat more or less valuable than similar machines of the same age. The development of a new and more efficient machine may make another machine obsolete, causing a sharp decline in its remaining value. Farmers customarily trade machines before they are worn out, making wear a minor consideration in calculating depreciation. The effect of obsolescence is difficult to predict. Hence, future depreciation for a machine is normally estimated based on the age of the machine.

Two decisions must be made before average annual depreciation can be calculated: the years of useful life and a salvage value at the end of the useful life must be estimated. The useful life of the machine is the period of years the owner plans to use the machine. It is usually less than the wear-out life since, as noted above, farmers normally trade machines before they are worn out. Estimates of the wear-out life in hours for various machines are given in Table 4.2. For example, the wear-out life of a tractor is 12,000 hours of use. The tractor in our example is used 600 hours per year for 10 years, or a total of 6,000 hours. The plow in the example is to be used 150 hours per year, or 1,500 hours in 10 years. This is also less than the 2,500 hours of wear-out life listed in Table 4.2. Therefore, wear-out should not reduce the useful life specified by the operator.

The salvage value is an estimate of the remaining market value of the machine at the end of the useful life. This is junk value if the machine is to be used until it is worn out, or zero if the machine is simply to be retired behind the barn.

Estimates of the remaining farm value as a percentage of list price for four categories of machinery are given in Table 4.3. They were developed from published reports of used equipment values and are estimates of the average "as-is" value of machines in average mechanical condition at the farm. These figures must be used as multipliers of the replacement list price of the machine, regardless of whether it was purchased for a lower (or higher) cost. The salvage values are stated in terms of current market values. Thus, the difference between current replacement cost and salvage value is the amount of depreciation to be taken in current dollars.

Once the length of useful life has been selected and the salvage value estimated, average annual depreciation can be calculated using equation 4.1:

$$\text{Depreciation per year} = \frac{\text{Purchase price} - \text{Salvage value}}{\text{Years of useful life}} \qquad \textbf{(4.1)}$$

For the example tractor, the current replacement list price is $37,400 and the useful life is 10 years. As shown in Table 4.3, a remaining farm value of 29.5 percent or $11,033 is predicted at the end of 10 years. The example plow has a current replacement list price of $6,830 and a useful life of 10 years. In Table 4.3, we find the plow is in remaining value group 4 and has a remaining farm value of 17.7 percent, or $1,209 is predicted at the end of 10 years. The salvage values are entered on line 1 of the first and second columns of the worksheet (Table 4.7, page 154). Then using equation 4.1

Table 4.2 Estimated Wear-Out Life, Total Lifetime Repair Cost as a Percentage of List Price, and Repair Cost Category

Machine	Estimated Wear-Out Life (hrs)	Total Lifetime Repair Cost as Percentage of List Price	Repair Cost Category (see Table 4.5)
Stationary power unit	12,000	120	2
Tractor, 2-wheel drive	12,000	120	2
Tractor, 4-wheel drive	12,000	100	1
Tractor, crawler	12,000	100	1
Combine PTO	2,000	100	7
Combine, self-propelled	2,000	60	3
Cotton picker, mounted	2,000	80	5
Cotton picker, self-propelled	2,000	60	3
Swather, self-propelled	2,500	100	8
Wagon and box	5,000	100	–
Corn head	2,000	100	7
Corn picker	2,000	80	5
Cotton stripper	2,000	60	3
Cutter, rotary or stalk	2,000	60	3
Fertilizer equipment, dry and liquid	1,200	120	9
Floats and scrapers	2,500	60	4
Harvester, flail	2,000	80	5
Harvester, potato	2,500	80	6
Harvester, sugar beet	2,500	80	6
Hay conditioner	2,500	100	8
Land plane	2,500	60	4
Loader, ensilage	2,000	100	7
Loader, front end	2,500	60	4
Manure spreader	2,500	60	4
Mower	2,000	120	11
Rake, side delivery	2,500	100	8
Seeding equipment, planters	1,200	100	10
Sprayer, mounted	1,200	100	10
Tillage tools (plows, cultivators, harrows, etc.)	2,500	120	12
Truck, farm	2,000	80	5
Truck, feed	2,500	60	4
Truck, pickup	2,000	60	3
Wagon, feed	2,500	100	8
Baler, PTO	2,500	80	6
Baler, with engine	2,500	60	4
Blower, ensilage	2,000	80	5
Harvester, forage, pull-type	2,000	80	5
Harvester, forage, self-propelled	2,000	60	3
Sprayer, self-propeller	2,000	80	5

Source: American Society of Agricultural Engineers. Data ASAE D230.2, Agricultural Machinery Management Data, revised February 1971, *1976 Agricultural Engineers Yearbook*, St. Joseph, Michigan, p. 329. Reprinted by permission.

Table 4.3   Remaining Farm Value, Expressed
as a Percentage of List Price, for Various Categories of Farm
Equipment at Selected Years of Age

| | *Equipment Categories[a]* | | | |
| | *(1)* | *(2)* | *(3)* | *(4)* |
Age of Equipment (year)	*Wheel Tractors, Stationary Power Units*	*Combines, Cotton Pickers, SP Windrowers*	*Balers, Blowers, Forage Harvesters, SP Sprayers*	*All Other Field Machines*
1	62.6	56.6	49.6	53.1
2	57.6	50.1	43.9	47.0
3	53.0	44.4	38.8	41.6
4	48.7	39.3	34.4	36.8
5	44.8	34.7	30.4	32.6
6	41.2	30.8	26.9	28.8
7	37.9	27.2	23.8	25.5
8	34.9	24.1	21.1	22.6
9	32.1	21.3	18.6	20.0
10	29.5	18.9	16.5	17.7
11	27.2	16.7	14.6	15.7
12	25.0	14.8	12.9	13.9
13	23.0	13.1	11.4	12.3
14	21.2	11.6	10.1	10.8
15	19.5	10.2	9.0	9.6

Source: Equations are adapted from American Society of Agricultural Engineers. Data ASAE D230.3, Agricultural Machinery Management Data, revised September 1977, *1981–82 Agricultural Engineers Yearbook,* Section 6, p. 236. Reprinted by permission.

[a]The remaining farm values (RV) as a percentage of list price are calculated from the following equations:

$$\text{RV Category } 1 = 68 \ (0.92)^n$$
$$\text{RV Category } 2 = 64 \ (0.885)^n$$
$$\text{RV Category } 3 = 56 \ (0.885)^n$$
$$\text{RV Category } 4 = 60 \ (0.885)^n$$

where n is the number of years.

$$\text{Tractor depreciation per year} = \frac{\$35,100 - \$11,033}{10} = \$2,407$$

$$\text{Plow depreciation per year} = \frac{\$6,420 - \$1,209}{10} = \$521$$

The interest cost is a charge for the use of the capital invested in the machine. This should be calculated as the real cost of using capital in current dollars. The real interest rate (the nominal rate minus the general rate of inflation) is multiplied by the average investment to calculate the interest charge. The average investment in the machine over its life is approximated by dividing the sum of the purchase cost and salvage value by 2. The annual interest charge is calculated using equation 4.2

$$\text{Interest per year} = \frac{\text{Purchase price} + \text{Salvage value}}{2} \times \text{Interest rate} \tag{4.2}$$

For the example tractor and plow with a real interest rate of 5 percent:

$$\text{Interest per year on tractor} = \frac{\$35,100 + \$11,033}{2} \times 0.05 = \$1,153$$

$$\text{Interest per year on plow} = \frac{\$6,420 + \$1,209}{2} \times 0.05 = \$191$$

These calculations are entered in the first and second columns of the worksheet (Table 4.7, page 154) on lines 3 and 4.

Capital Recovery as a Substitute for Depreciation and Interest   Depreciation and interest combined represent the amount of money set aside to pay the loss in value (depreciation) of the asset and interest on its remaining value. The methods discussed in equations 4.1 and 4.2 for machinery, referred to as the traditional methods, do not provide a large enough sum annually to cover both depreciation of the capital and to pay interest on the unrecovered amount at the specified interest rate in the designated number of years. Thus, an alternative but more complex computational procedure, the capital recovery method, may be used to more accurately reflect these costs.

This method is based on the use of the capital recovery factor—the amount of money required at the end of each year to pay interest on the unrecovered capital at the designated rate and recover the investment within the specified number of years. The capital recovery factors in Appendix Table IV are on a dollar basis. For example, the capital recovery factor for four years and 9 percent interest is 0.3087. That is, the annual cost of interest and recovery of the investment over four years at 9 percent interest is $0.3087 per dollar invested. Other values in Appendix Table IV are interpreted in a similar manner.

To illustrate the improved accuracy of the capital recovery method compared to straight-line depreciation and interest on the average investment, assume an asset costing $100 is purchased. The asset is to be depreciated to a zero value in four years. The interest rate is 9 percent. Using equations 4.1 and 4.2:

$$\text{Depreciation per year} = \frac{100 - 0}{4} = \$25.00$$

$$\text{Interest per year} = \frac{100 + 0}{2} \times 0.09 = \$4.50$$

Table 4.4 An Example of Capital Recovery

| Year | Unrecovered Capital at Start of Year | Amount at End of Year | | |
		Total Payment	Interest	Amount Applied to Unrecovered Capital
1	$100.00	$30.87	$9.00	$21.87
2	78.13	30.87	7.03	23.84
3	54.29	30.87	4.89	25.98
4	28.31	30.87	2.55	28.32

Thus, an annual charge of $29.50 is included in ownership costs, but a somewhat larger payment ($100 × 0.3087) or $30.87 per year, based on capital recovery factors in Appendix Table IV, is required to recover the capital and pay interest on the unrecovered portion as illustrated in Table 4.4.

The entries in the first line indicate that $100.00 is invested. At the end of the first year $9.00 interest has accrued on the $100.00. A payment of $30.87 is made, of which $9.00 is used for interest and the remaining $21.87 is applied to recovery of the capital investment. The unrecovered capital during the second year is $100.00 minus $21.87 or $78.13. Interest on this amount at 9 percent for one year is $7.03, allowing $23.84 of the payment at the end of year 2 to be used to recover part of the capital invested. Note that the payment at the end of the fourth year is just adequate (actually $0.01 more than needed due to rounding) to pay the interest and recover the remaining capital investment. Clearly, a smaller annual payment would have been inadequate both to pay interest at 9 percent each year on the unrecovered capital and to recover the $100.00 within the four-year period. The difference in amount calculated by the two methods is $30.87 minus $29.50 or $1.37 per year in the example. Furthermore, using straight-line depreciation and interest on the average investment consistently underestimates the actual cost. The amount of "error" increases as the size of the investment increases and as the interest rate increases.

The capital recovery method of calculating the annual charge for depreciation and interest is given in equation 4.3.

$$\text{Annual Capital Recovery Charge} =$$

$$\left[\left(\begin{matrix} \text{Purchase} \\ \text{Price} \end{matrix} - \begin{matrix} \text{Salvage} \\ \text{Value} \end{matrix} \right) \times \left(\begin{matrix} \text{Capital} \\ \text{Recovery} \\ \text{Factor} \end{matrix} \right) \right] + \left[\left(\begin{matrix} \text{Salvage} \\ \text{Value} \end{matrix} \right) \times \left(\begin{matrix} \text{Interest} \\ \text{Rate} \end{matrix} \right) \right]$$

(4.3)

This equation allows for the possibility of a positive salvage value, making it more complex than the simple calculation for the example above. The computation includes calculating the amount of capital required per year to recover the difference between the purchase price and salvage value and to pay interest on the uncovered part at the designated interest rate. Notice that the interest on the salvage value is added to the amount required for interest and recovery of the depreciated value. When the salvage value is zero, as in the $100.00 example above, the last term is zero and drops out of the equation.

The capital recovery procedure is applied below to calculate depreciation and interest costs per year for the tractor and plow analyzed in the machinery costs section of the chapter. The capital recovery factor for 5 percent interest and 10 years of life is 0.1295 (Appendix Table IV).

Annual capital recovery charge for the tractor =
[($35,100 − $11,033) × 0.1295] + [$11,033 × 0.05] = $3,668
Annual capital recovery charge for the plow =
[($6,420 − $1,209) × 0.1295] + [$1,209 × 0.05] = $735

The values are entered in columns three and four of the worksheet (Table 4.7, page 154) on line 5. The value calculated for the tractor with this method is $108.00 greater than the $3,560 calculated using equations 4.1 and 4.2. The value for the plow is $23 greater than the $712 calculated using the traditional method.

The major advantage of the capital recovery approach is that it more accurately estimates the amount of cost involved. The major disadvantage is that the user must have access to the appropriate capital recovery factor(s) to use it. Thus, both approaches will continue to be used in the foreseeable future.

Taxes, Insurance, and Housing (TIH)   These three costs are usually much smaller than depreciation and interest, but they should be included. Personal property taxes are assessed on farm machinery in some states but not in others. The amount of personal property taxes can be determined by multiplying the assessed value times the tax rate. Some states have an assessed value that is equal to 100 percent of market value, while in others it is much less. To place the tax rate on a market value basis, multiply the proportion of market value times the tax rate in dollars per $1,000. For example, if machinery is normally assessed at 40 percent of the fair market value and the tax rate is $5.00 per $1,000 of assessed value, the rate on a market value basis is 0.40 × $5.00 or $2.00 per $1,000.

Insurance for losses resulting from storm, fire, and vandalism is normally carried to replace the machines in the event of damage or loss from these sources. If insurance is not carried, the risk is assumed by the rest of the business. Insurance rates vary by area. They range from $6 to $10 per $1,000 of valuation in many areas of the United States.

Housing provided for machinery varies from farm to farm in an area as well as across the country. Providing housing, tools, and maintenance equipment generally results in less deterioration from weather and better maintenance. This should result in less loss of field time, lower repair costs, and a higher trade-in value. The cost of providing housing and maintenance facilities is part of the cost of providing machinery services on the farm and should be included in machinery costs. The cost of providing housing and maintenance facilities on many farms ranges from 0.5 to 3.0 percent of average investment annually. A rate of 2 percent is used in the example calculation below.

To simplify the calculation of taxes, insurance, and housing, they can be lumped together and multiplied by the average investment (equation 4.4).

$$\text{TIH per year} = \frac{\text{Purchase price} + \text{Salvage value}}{2} \times \text{Rate} \qquad \textbf{(4.4)}$$

Assume for the tractor and plow example that personal property taxes are 0.2 percent, insurance is 0.6 percent, and housing is 2.0 percent. These three ownership costs would be:

$$\text{Tractor TIH per year} = \frac{\$35,100 + \$11,033}{2} \times 0.028 = \$646$$

$$\text{Plow TIH per year} = \frac{\$6,420 + \$1,209}{2} \times 0.028 = \$107$$

The results of these calculations are entered on line 6 of the worksheet (Table 4.7, page 154). They are the same for either the traditional or the capital recovery methods.

Total Ownership Costs   Total annual ownership costs for a machine are the sum of either depreciation and interest or the annual capital recovery, plus taxes, insurance, and housing costs. For the example, these total $4,206 for the tractor and $819 for the plow using the depreciation and interest formulas. The totals are somewhat higher: $4,314 for the tractor and $842 for the plow in this example using the capital recovery method. Ownership costs per hour of operation are calculated using equation 4.5.

$$\text{Ownership cost per hour} = \frac{\text{Total annual ownership cost}}{\text{Hours of use per year}} \qquad \textbf{(4.5)}$$

Assuming the tractor is used 600 hours and the plow 150 hours annually:

$$\text{Tractor ownership cost per hour} = \frac{\$4,314}{600} = \$7.19$$

$$\text{Plow ownership cost per hour} = \frac{\$842}{150} = \$5.61$$

The total annual ownership cost of a machine remains the same for a wide range in hours of use, but the ownership cost per hour varies with the hours of use.[3] If the tractor in the example is used only 200 hours per year:

$$\text{Tractor ownership cost per hour} = \frac{\$4,314}{200} = \$21.57$$

In general, the more the equipment is used the lower the ownership cost per hour.

Operating Costs

Fuel   Fuel costs per hour are estimated based on fuel consumption per hour of operation. Fuel consumption per hour can be estimated in one of the following ways. First, if the tractor or other machine is currently in use on the farm, the manager can estimate future fuel consumption based on past performance. Second, average fuel consumption by make and model of tractor is available from tractor test data.[4] Third, average fuel consumption can be estimated using equation 4.6, 4.7, or 4.8:

$$\begin{aligned}\text{Gasoline requirement in gallons per hour} =\\ 0.06 \times \text{Maximum PTOHP}\end{aligned} \qquad \textbf{(4.6)}$$

$$\begin{aligned}\text{Diesel requirement in gallons per hour} =\\ 0.044 \times \text{Maximum PTOHP}\end{aligned} \qquad \textbf{(4.7)}$$

$$\begin{aligned}\text{LP requirement in gallons per hour} =\\ 0.072 \times \text{Maximum PTOHP}\end{aligned} \qquad \textbf{(4.8)}$$

The fuel cost per hour can be estimated using equation 4.9:

[3]Either the useful life or the salvage value should be reduced in those situations having an amount of use over the useful life (amount of use per year times years of useful life) that approximates or exceeds the machine's wear-out life. Notice that reducing the useful life only affects depreciation or capital recovery; the other costs remain the same. Reducing the salvage value and holding the useful life the same increase the depreciation or capital recovery per year, but reduces the average investment. Hence, the other annual ownership costs are reduced. If the machine is used relatively few hours per year, the remaining value may be higher than predicted by tabulated values, although this result is highly dependent on the obsolescence of the type and size of machine.

[4]Average annual fuel consumption for a specific make and model of tractor can be approximated from the Nebraska Tractor Test Data, available from the Department of Agricultural Engineering, University of Nebraska, Lincoln, Nebraska.

Fuel cost per hour = Fuel cost per gallon ×
Fuel requirement in gallons per hour   **(4.9)**

If we assume a maximum power take-off horsepower of 120 and a diesel price of $1.25 per gallon for the example, fuel consumption is calculated using equation 4.6 and the fuel cost per hour using equation 4.9:

Diesel requirement in gallons per hour =
0.044 × 120 = 5.28
Fuel cost per hour = $1.25 × 5.28 = $6.60

This figure is entered on line 10 of the worksheet (Table 4.7, page 154).

Lubrication   Surveys indicate that lubrication costs average about 15 percent of fuel costs.[5] After calculating fuel cost per hour, lubrication cost per hour can be calculated using equation 4.10:

Lubrication cost per hour =
0.15 × Fuel cost per hour   **(4.10)**

In the example:

Lubrication cost per hour = 0.15 × $6.60 = $0.99

This is entered on line 11 of the worksheet (Table 4.7, page 154). Notice that lubrication costs are calculated only for machines that have an engine. Lubrication costs are included as part of the repair and maintenance costs for other machines.

Repairs and Maintenance   Repairs occur because of routine deterioration and wear, accidental breakage, carelessness, and operator neglect. Repair costs for a particular machine vary widely from one geographic region to another because of soil type, terrain, and climate. Within a local area repair costs also vary due to soil type and terrain, as well as operator maintenance policies and method of operation.

Past repair expenses are the best data for predicting repair costs. Records of repair expenses were used to derive the data in Table 4.5. The data in each column indicate the total accumulated repair costs as a percentage of the original list price of the machine for alternative amounts of accumulated use.

Notice the relationship implied by the data in each column. The

[5]American Society of Agricultural Engineers, *1981–82 Agricultural Engineers Yearbook*, ASAE Data: ASAE D230.3, Agricultural Machinery Management Data, Section 6, p. 237.

Table 4.5 Total Accumulated Repairs as a Percentage of List Price by Hours of Use[a]

Repair cost category (Table 4.2)	1	2	3	4	5	6	7	8	9	10	11	12
Estimated wear-out life in hours	12,000	12,000	2,000	2,500	2,000	2,500	2,000	2,500	1,200	1,200	2,000	2,500
Total repairs as percentage of list price	100	120	60	60	80	80	100	100	120	100	120	120
Accumulated hours of use												
100	0.1	0.1	0.9	0.7	1.2	0.9	1.5	1.1	3.7	3.1	2.4	1.8
250	0.3	0.4	3.3	2.4	4.4	3.2	5.5	4.0	13.4	11.2	8.0	6.0
500	0.9	1.0	8.7	6.4	11.5	8.4	14.4	10.5	35.4	29.4	19.8	14.8
1,000	2.4	2.9	23.0	16.8	30.4	22.2	38.0	27.8	93.4	77.7	48.7	36.4
1,200	3.2	3.8	29.6	21.7	39.2	28.7	49.1	35.9	120.0	100.0	61.7	46.2
1,500	4.4	5.3	40.5	29.6	53.6	39.2	67.1	49.1	—	—	82.4	61.7
2,000	6.8	8.2	60.0	44.3	80.0	58.6	100.0	73.4	—	—	120.0	89.7
2,500	9.5	11.1	—	60.0	—	80.0	—	100.0	—	—	—	120.0
3,000	12.5	15.0	—	—	—	—	—	—	—	—	—	—
4,000	19.3	23.1	—	—	—	—	—	—	—	—	—	—
5,000	26.9	32.3	—	—	—	—	—	—	—	—	—	—
6,000	35.4	42.4	—	—	—	—	—	—	—	—	—	—
8,000	54.4	65.3	—	—	—	—	—	—	—	—	—	—
10,000	76.1	91.3	—	—	—	—	—	—	—	—	—	—
12,000	100.0	120.0	—	—	—	—	—	—	—	—	—	—

Source: Equations are adapted from American Society of Agricultural Engineers Data ASAE D230.2, Agricultural Machinery Management Data. revised February 1971, *1976 Agricultural Engineering Yearbook*, p. 325. Reprinted by permission.

ᵃThe accumulated repair and maintenance costs at any point in a machine's life are estimated from an equation of the form:

$$TAR\% = aX^b$$

where

$TAR\%$ = total accumulated repairs as a percentage of initial list price.

a = coefficient shown below.

X = 100 times the ratio of the accumulated hours of use to the wear-out life given in Table 4.2.

b = an exponent shown in the equation below.

Estimated Wear-Out Life	Total Repairs as Percent of List	Equation
12,000	100	$TAR\% = 0.100(X)^{1.5}$
12,000	120	$TAR\% = 0.120(X)^{1.5}$
2,000 or 2,500	60	$TAR\% = 0.096(X)^{1.4}$
2,000 or 2,500	80	$TAR\% = 0.127(X)^{1.4}$
2,000 or 2,500	100	$TAR\% = 0.159(X)^{1.4}$
1,200	120	$TAR\% = 0.191(X)^{1.4}$
2,000 or 2,500	120	$TAR\% = 0.301(X)^{1.3}$

For example, the accumulated repairs and maintenance costs for a two-wheel drive tractor with 4,000 hours of use is calculated as $TAR\% = 0.120 (100 \cdot 4000/12,000)^{1.5} = 23.059$ or 23.1 percent. This value is shown in the above table for repair cost category 2.

amount of repairs per hour increases as the total amount of use increases. Because of the difference in the wear-out life and mechanical nature of various machines, the machines have been divided into groups and a separate repair cost relationship has been estimated for each group. The repair cost category for various machines is listed in Table 4.2. By selecting the correct repair cost category and accumulated repair cost factor from Table 4.5, repair costs per hour can be calculated using equation 4.11:

$$\text{Repair and maintenance cost per hour} = \frac{\text{List price} \times \text{total accumulated repairs as a percentage of list price}}{\text{Hours use over machine's useful life}}$$

(4.11)

The example two-wheel drive tractor and plow are in repair cost categories 2 and 12, respectively. Accumulated repairs for 6,000 hours (600 hours per year times 10 years) of tractor use and 1,500 hours (150 hours per year times 10 years) of plow use are 42.4 and 61.7 percent of the list price, respectively. Using equation 4.11:

$$\text{Tractor repair costs per hour} = \frac{\$37,400 \times .424}{6,000} = \frac{\$15,858}{6,000} = \$2.64$$

$$\text{Plow repair costs per hour} = \frac{\$6,830 \times .617}{1,500} = \frac{\$4,214}{1,500} = \$2.81$$

These results are entered in the worksheet (Table 4.7, page 154) on lines 9 and 12.

Labor   Different size machines require different amounts of labor to perform a given task, such as plowing or harvesting. It is, therefore, important to calculate the amount of time required. It may also be important to include the value of the labor in machinery cost analyses.

Because of the time required to lubricate and service machines, as well as time delays in getting to and from the field, the actual man-hours of labor usually exceed actual field time by 10 to 20 percent. Consequently, hourly labor costs can be estimated by multiplying the labor wage rate times 110 to 120 percent of the machine hours used for a particular operation. For the example, labor is estimated as 120 percent of $5.00 per hour. The value is entered on line 14 of the worksheet on page 154. The additional 20 percent is also included in the hours of labor required for the enterprise for use in whole-farm planning. The hours of machine time required is multiplied by 1.2, and the result is entered on line 28 of the worksheet on page 155.

Total Operating Costs Operating costs for machinery include fuel, lube, repair, and labor costs. The fuel, lubrication, and repair costs, which are variable costs, are summed and listed on line 13 of the worksheet (Table 4.7, page 154). They total $10.23 and $2.81 per hour for the example tractor and plow, respectively. Labor, which may be variable, fixed, or a combination of the two, is listed separately. We will also include separate lines for these two categories in the enterprise budgets.

Machine Costs per Acre

It is necessary to calculate the cost of machinery operations per acre to develop enterprise budgets. This can be accomplished by calculating the acres that can be covered per hour and the hours required per acre, and then multiplying the hours required per acre times the cost per hour to estimate machinery cost per acre. Consider this procedure one step at a time.

First, the number of acres a machine can be expected to cover in one hour of operation under normal field conditions depends on the machine's width, ground speed, and field efficiency. The acres covered per hour can be calculated using equation 4.12, where S is speed of the machine while it is in motion in miles per hour, W is width of the machine in feet, and E is field efficiency expressed as a decimal. Typical speed and efficiency factors are given in Table 4.6.

$$\text{Acres per hour} = \frac{S \times W \times E}{8.25}^{6} \qquad \textbf{(4.12)}$$

The width of the example plow is 6 times 16 inches or 8 feet. Assume the tractor pulls the plow at a speed of 4.5 miles per hour and with an efficiency of 80 percent. If we use equation 4.12, acres covered per hour are calculated as:

$$\text{Acres per hour} = \frac{4.5 \times 8.0 \times 0.80}{8.25} = 3.49$$

This value is entered on line 17 of the worksheet (Table 4.7, page 155).

---

[6]The formula is derived as follows:

$$\text{Acres per hour} = \frac{\text{Sq ft covered in 1 hour}}{\text{Sq ft in one acre}} = \frac{S \times F \times W \times E}{43,560}$$

where F is 5,280, the number of feet in one mile. Dividing the numerator and denominator by 5,280 results in equation 4.12.

Table 4.6 Machinery Performance Data

Machine	Speed or Performance Rate MPH	Typical Range for Field Efficiency
Tillage implements	3.0–6.0	0.70–0.90
Rotary tiller	1.0–4.5	0.70–0.90
Rotary hoe	5.6–11.0	0.70–0.90
Row cultivators	1.5–4.0	0.70–0.90
Fertilizer and chemical application	3.0–5.0	0.60–0.75
Row crop planting with fertilizer and herbicides	3.0–6.0	0.50–0.85
Grain drill	2.5–6.0	0.65–0.85
Cutterbar mower	5.0–6.0	0.80–0.90
Mower-conditioner	2.0–4.2	0.88–0.90
Side delivery rake	4.0–6.0	0.85–0.90
Balers	2.5–5.0	0.70–0.90
Forage harvesters	1.5–4.0	0.50–0.75
Windrower, small grain	5.0–7.0	0.75–0.85
Combines	2.0–3.5	0.65–0.80
Corn pickers	2.0–3.5	0.60–0.75
Cotton pickers	1.5–3.0	0.60–0.75
Cotton strippers	2.0–3.5	0.60–0.75
Sugarbeet harvester	2.5–4.0	0.60–0.80

Source: American Society of Agricultural Engineers Data ASAE D230.3, Agricultural Machinery Management Data, revised September 1977, *1981–82 Agricultural Engineers Yearbook,* St. Joseph, Michigan, Section 5, p. 236. Reprinted by permission.

Second, the hours required to cover one acre can be calculated as the reciprocal of acres per hour. The procedure is given in equation 4.13.

$$\text{Hours per acre} = \frac{1}{\text{Acres per hour}} \qquad (4.13)$$

For the example situation, hours per acre equals 1/3.49 or 0.29. This value is entered on the worksheet on line 18 (Table 4.7, page 155).

Third, average machine costs can be calculated for an individual acre by multiplying the appropriate ownership and operating costs per hour by the number of hours required to perform the operation on one acre (equation 4.14).

$$\text{Machine cost per acre} = \text{Machine cost per hour} \times \text{Hours per acre} \qquad (4.14)$$

For example, ownership costs per hour for the example tractor and plow using the capital recovery method and current replacement cost are ($7.19 + $5.61) or $12.80 per hour. They are ($12.80 × .29) or $3.71 per

Table 4.7 Worksheet for Estimating Farm Machinery Costs

Part I: Estimating Costs per Hour

Information		Example Figure	Example Figure
(a)	Machine	Tractor	Plow
(b)	Initial list price	$25,440	$4,650
(c)	Purchase price	$23,900	$4,370
(d)	Accumulated hours	3,000	750
(e)	Annual use, hours	600	150
(f)	Useful life, years	10	10
(g)	Nominal interest rate, percentage	12	12
(h)	Engine or PTO horsepower	120	—
(j)	Fuel type	Diesel	—
(k)	Fuel price/gallon	$1.25	—
(m)	Labor rate, dollar/hour	$5.00	—
(n)	Taxes, insurance, and housing, percentage of average investment	2.8	2.8
(p)	Current list price of replacement machine	$37,400	$6,830
(q)	Current purchase price of replacement machine	$35,100	$6,420
(r)	General inflation rate, percentage	7	7

(continued)

153

Table 4.7 *(Continued)*

	Based on Current Replacement Costs				Based on Historical Costs	
	Traditional Method		Capital Recovery		Traditional Method	
	Tractor	Plow	Tractor	Plow	Tractor	Plow
Estimating Ownership Costs						
(1) Salvage value = (from Table 4.3) × (p)	$11,033	$1,209	$11,033	$1,209	$ 7,505	$ 823
(2) Average investment = $\frac{(q)+(1)}{2}$	23,066	3,814	23,066	3,814	15,702	2,596
(3) Depreciation = $\frac{(q)-(1)}{(f)}$	2,407	521	XXX	XXX	1,640	355
(4) Interest = $(g-r)$ × (2)	1,153	191	XXX	XXX	1,884[a]	312[a]
(5) Capital recovery = $[(q)-(1)][CRF$ from Appendix Table IV] + $[(1)\times(g-r)]$	XXX	XXX	3,668	735	XXX	XXX
(6) Taxes, insurance, and housing = (n) × (2)	646	107	646	107	440	73
(7) Total ownership cost per year = (3) + (4) + (5) + (6)	4,206	819	4,314	842	3,964	740
Estimating Operating Costs						
(8) Total accumulated hours = $(e)\times(f)$	6,000	1,500	6,000	1,500	6,000	1,500
(9) Total accumulated repairs = (from Table 4.5) $\times$ (p)	15,858	4,214	15,858	4,214	10,787	2,869
(10) Fuel cost/hour = (value from equation 4.6, 4.7, or 4.8) $\times$ (k)	6.60	—	6.60	—	6.60	—
(11) Lubrication cost/hour = $0.15 \times$ (10)	0.99	—	0.99	—	0.99	—
(12) Average repair cost/hour = $\frac{(9)}{(8)}$	2.64	2.81	2.64	2.81	1.80	1.91
(13) Total fuel, lubrication, and repair cost/hour = (10) + (11) + (12)	10.23	2.81	10.23	2.81	9.39	1.91
(14) Labor cost/hour = $1.2 \times (m)$	6.00	—	6.00	—	6.00	—
(15) Ownership cost/hour = $\frac{(7)}{(e)}$	7.01	5.46	7.19	5.61	6.60	4.93
(16) Total cost/hour = (13) + (14) + (15)	23.24	8.27	23.42	8.42	21.99	6.84

Table 4.7 (Continued)

Part II: Estimating Costs per Acre

Information		Based on Current Replacement Costs		Based on Historical Costs
		Traditional Method	Capital Recovery	Traditional Method
(s)	Operation	Plowng	Plowing	Plowing
(t)	When performed	November	November	November
(u)	Tractor used	120 PTOHP	120 PTOHP	120 PTOHP
(v)	Implement used	6-16" Plow	6-16" plow	6-16" plow
(w)	Speed	4.5	4.5	4.5
(y)	Width	8.0	8.0	8.0
(z)	Efficiency	0.80	0.80	0.80
(17)	Acres per hour = $\dfrac{(w) \times (y) \times (z)}{8.25}$	3.49	3.49	3.49
(18)	Hours per acre = 1/(17)	0.29	0.29	0.29
Ownership Costs				
(19)	Tractor ownership cost/hour (15)	$ 7.01	$ 7.19	$ 6.60
(20)	Implement ownership cost/hour (15)	5.46	5.61	4.93
(21)	Total ownership cost/hour = (19) + (20)	12.47	12.80	11.53
(22)	Ownership cost/acre = (18) × (21)	3.62	3.71	3.43
Operating Costs				
(23)	Tractor fuel, lubrication, and repair cost/hour (13)	$10.23	$10.23	$ 9.39
(24)	Implement fuel, lubrication and repair cost/hour (13)	2.81	2.81	1.91
(25)	Total fuel, lubrication, and repair cost/hour = (23) + (24)	13.04	13.04	11.30
(26)	Total fuel, lubrication, and repair cost/acre = (18) × (25)	3.78	3.78	3.28
(27)	Labor cost/acre = (18) × (24)	1.74	1.74	1.74
(28)	Hours labor/acre = (18) × 1.2	0.35	0.35	0.35

[a]These figures are based on the nominal rather than the real rate of interest, because the nominal rate has traditionally been used with machinery cost computations based on historic costs. We recommend using real interest rates and current replacement costs to more accurately represent the current economic cost of machine services.

acre. The entries for ownership costs are made on lines 19 through 22 of the worksheet (Table 4.7, page 155).

The fuel, lubrication, and repair costs are summarized on lines 23 through 26 of the worksheet. For the example situation fuel, lube, and repairs are ($10.23) + ($2.81) or $13.04 per hour and ($13.04 × 0.29) or $3.78 per acre. Both labor cost per acre and hours of labor per acre are estimated using the same procedure; they are entered on lines 27 and 28 of the worksheet (Table 4.7, page 155).

Machinery costs can be estimated for one acre of crop production using the machinery cost worksheet (Table 4.7) by specifying the machine operations to be performed, estimating the machinery costs per acre for each operation, and summing the costs per acre for all operations. A summary of the worksheet calculations for corn production is given in Table 4.8. The manager has listed the machinery operations he or she expects to perform, the machines to be used, and the month during which each operation is to be performed. The manager has estimated the ownership and operating costs per hour for each machine using Part I of the worksheet (Table 4.7). The hours, ownership costs, fuel, lube and repair costs, and labor costs have been estimated per acre using Part II of the worksheet. The hours of labor per acre are summed to provide an estimate of the labor required for machinery operation per acre of corn production. An estimate of per acre machinery ownership and fuel, lube, and repair costs is obtained by summing the costs for each operation. The calculations indicate that corn production on the farm will require 2.30 hours of direct labor for machinery operation. Estimated machinery ownership costs are expected to total $54.37 per acre, while the fuel, lube, and repair costs per acre sum to $44.72.

Comparison of Cost Estimation Methods

The most commonly used approach for estimating machinery ownership and operating costs bases the estimates on the historic list price and the nominal interest rate. This approach is a rather natural outgrowth of cost-based accounting concepts for income taxes, profit or loss statements, and cost-based balance sheets. As we noted earlier, historic costs do not necessarily reflect current economic costs for the service provided by the assets. Our purpose in this section is to contrast the estimates that result from the methods using current replacement costs and the real interest rate with analogous estimates based on the historic list price and the nominal interest rate.

The machinery ownership and operating costs for the example tractor and plow are shown for each of the estimation methods in the worksheet, Table 4.7. As the example illustrates, higher depreciation results from using current replacement prices, but the amount of interest from the two methods depends on the relation between the average investment calcu-

Table 4.8 Summary of Estimated Machinery Costs for Corn Production with Conventional Tillage

Worksheet Line (Table 4.7)

								Alternative Machinery Operations					
(s) Operation	Discing	Fertilizer	Apply herbicide	Discing	Harrow	Plant	Cultivate	Harvest	Haul grain	Dry grain[a]	Shred stalks	Plow	Total per acre
(t) When performed	April	May	May	May	May	May	June	October	October	October	October	November	
(u) Tractor used	120 HP	60 HP	60 HP	120 HP	120 HP	60 HP	60 HP	Combine	60 HP	–	60 HP	120 HP	
(v) Implement used	Disc	Fertilizer spreader	Sprayer	Disc	Harrow	Planter	Cultivator	Combine	Wagon	Dryer	Shredder	6-16"	
(18) Hours per acre	0.11	0.09	0.07	0.11	0.03	0.13	0.13	0.37	0.32	0.62	0.23	0.29	–
(22) Ownership cost per acre	$4.30	$0.82	$0.75	$4.30	$1.47	$4.87	$3.63	$15.56	$3.88	$7.84	$3.24	$3.71	$54.37
(26) Total fuel, lube, and repairs/acre	$1.77	$0.92	$0.31	$1.77	$0.39	$1.69	$1.56	$11.61	$2.43	$16.80	$1.69	$3.78	$44.72
(28) Hours labor per acre[b]	0.13	0.11	0.08	0.13	0.04	0.16	0.16	0.44	0.38	0.04	0.28	0.35	2.30

[a]The costs are based on drying corn from 23 percent to 15 percent moisture.
[b]Labor requirements are 120 percent of the time required for field operations.

Table 4.9 Estimating Monthly Labor Requirements to Finish Two Groups of 480 Feeder Pigs Annually

Task	Jan.	Feb.	Mar.	Apr.	May	Jun.	Jul.	Aug.	Sept.	Oct.	Nov.	Dec.
Purchasing and obtaining pigs			25							25		
Routine feed preparation and care	72	58	23	72	72	72	58			23	72	72
Marketing finished hogs		14					14					
Cleaning facilities		24					24					
Total	72	96	48	72	72	72	96	0	0	48	72	72

157

lated with current replacement costs, the average investment calculated with historic costs, the real interest rate, and the nominal interest rate. Thus, a larger dollar value for depreciation and interest could result from either method. The current replacement cost and the average investment are higher for the current replacement cost method during periods of increasing capital asset prices. Taxes, insurance, housing, repairs, and maintenance will, therefore, be higher when they are based on current replacement costs. The procedures used to estimate fuel, lubrication, and labor costs will provide the same estimates for either approach provided the quantity of fuel consumed is the same.

As these comments suggest, one cannot indicate which method will result in the largest or smallest cost estimate. But basing machinery ownership and operating costs on current replacement costs (for capital assets providing the service efficiently) and real interest rates is the most accurate method of estimating the current economic cost of providing the service. We recommend using this approach with the capital recovery method.

Some of the machinery costs are cash items and must frequently be projected over time for preparation of cash flows. The cash costs are taxes, insurance, fuel, lubrication, and repairs. These costs can be projected for future years by estimating the nominal percentage change in annual price of the item, r, and multiplying the estimate based on current replacement costs above by $(1 + r)^n$, where n is the number of years into the future. If the price of repairs were expected to increase 8 percent per year for the near future, the cost of repairs estimated for the current year would be multiplied by (1.08), $(1.08)^2$, $(1.08)^n$ for the next year, the following year, and for the nth year in the future, respectively. We will deal with this matter in projecting cash flows in later chapters.

ESTIMATING COSTS OF BUILDING SERVICES

Building services are important for the convenient, efficient, and profitable operation of farm businesses. They contribute to the income of the business both directly and indirectly in several ways. First, the farmstead in general and certain buildings in particular provide a base of operations for the business. In many cases, part of the operator's house serves as the office of the firm's manager. In other cases, part of another building is used. The office serves as a recordkeeping center, a communications center, and the place where many decisions are made. Although the contribution is indirect, it is important to the profitability of the business. Second, buildings provide storage for inputs used by the business. Storage of inputs preserves their quality until they are used in the production process. The storage of machines, seed, chemicals, fertilizer, and feed contributes to the income of the business indirectly. The storage of products from harvest until they are sold represents a third way buildings contribute to business income. Storage may increase the efficiency of harvesting nonperishable crops by avoiding the time required to haul the product to an elevator or processing plant. Storage may also enable the manager to sell the product at a higher price. Fourth, buildings are an integral part of many livestock production systems. Facilities are often designed to provide the space, sanitation, temperature, humidity, and other environmental conditions required to produce healthy animals (including hogs, broilers, and slaughter cattle) and quality livestock products (such as milk and eggs). Managers who want to build or remodel facilities must typically decide how much to invest to achieve more efficient production and reduced labor requirements. The principle of input substitution can be used as a guide in selecting which labor-capital combination to use in production of a product. Finally, buildings can be used to improve the quality and/or process products produced. Grain drying, egg grading and storage, and milk storage are examples.

Estimating Annual Costs

Like other capital assets, building cost estimates should be based on the current replacement cost of a structure that efficiently provides the same quantity and quality of service. This rule is relatively straightforward for situations in which the existing building would be replaced with essentially the same type of structure. It is more difficult to apply it to structures that have been built for one purpose and have since been converted to another use. However, the asset valuation method outlined in Chapter 2, referred to as replacement cost for equivalent function and not necessarily equivalent asset, applies in these cases.

The annual costs of providing building services are frequently referred to as the "DIRTI" costs. The five letters in this acronym refer to depreciation, interest, repairs, taxes, and insurance. The formulas for depreciation, interest, real estate taxes, and insurance are similar to those used in calculating machinery costs. Like machinery, the real interest rate should be used in calculating the charge for the use of capital. Building repair and maintenance costs are affected by the type of construction and climate, as well as by the intended use and frequency of use. Because farm buildings have not been a standard manufactured item, repair and maintenance standards analogous to those used for farm machinery have not been developed. Repair and maintenance costs are normally calculated as 1.0 to 3.5 percent of original cost. A rate of 2 percent is suggested for normal use, with somewhat higher rates for buildings used continuously. Thus, equations 4.15 through 4.18 can be used to estimate average annual building costs.

$$\text{Depreciation per year} = \frac{\text{Purchase price} - \text{Salvage value}}{\text{Years of useful life}} \qquad \textbf{(4.15)}$$

$$\text{Interest per year} = \frac{\text{Purchase price} + \text{Salvage value}}{2} \times \text{Interest rate} \qquad \textbf{(4.16)}$$

$$\text{Repairs and maintenance per year} = \text{Purchase price} \times \text{Rate per year} \tag{4.17}$$

$$\frac{\text{Real estate taxes and}}{\text{insurance costs per year}} = \frac{\text{Purchase price} + \text{Salvage value}}{2} \times \text{Rate} \tag{4.18}$$

The salvage value is normally considered zero for buildings. If a structure has a remaining value at the end of its useful life, it normally does not exceed the cost of removing the building. Working with a zero salvage value makes it possible to simplify the computations, by calculating the percentage of new cost which each of the five annual costs represents. The percentage for depreciation is 100 percent divided by the years of useful life. Interest, real estate taxes, and insurance are based on the average investment. With a zero salvage value, however, the average investment is exactly one-half of the original cost. Thus, the rate for these costs can be halved and multiplied by the purchase price. The rate for repairs and maintenance costs is already based on purchase price and does not require an adjustment.

Consider the following data on a building for finishing pigs.

Replacement cost of the building	$60,000
Salvage value	0
Years of useful life	20
Real estate taxes	1.4% of value
Insurance rate	0.6% of value
Nominal interest rate	12%
General inflation rate	7%
Repairs and maintenance	2.0%
Capacity	500 head

In this example the building has a 20-year life. Hence, the percentage of purchase price for depreciation is 100 divided by 20 or 5 percent. Halving the stated rates results in 2.5 percent for real interest and 1 percent for real estate taxes and insurance. The resulting values for the example are listed below.

Cost Item	Percentage of Replacement Cost
Depreciation	5.0
Interest	2.5
Repairs and maintenance	2.0
Taxes and insurance	1.0
Total	10.5

Then

$$\text{Annual building cost} = \$60,000 \times .105 = \$6,300$$

Again the capital recovery procedure discussed earlier can be substituted for depreciation and interest. The capital recovery factor for a 20-year life and a 5 percent interest rate (Appendix Table IV) is 0.0802. Because the salvage value is zero, the last term of equation 4.3 is zero and the capital recovery factor can be combined with those for the other costs as shown below.

Cost Item	Percentage of Replacement Cost
Capital recovery	8.02
Repairs and maintenance	2.00
Taxes and insurance	1.00
Total	11.02

Then

$$\text{Annual building cost} = \$60,000 \times 0.1102 = \$6,612$$

Allocating Building Costs

Of the five building costs only repair cost is variable, and even a portion of this item depends more on deterioration due to weather and other environmental factors than to the amount of use. Thus, once the structures are in place the cost of providing annual building services is largely fixed in nature. As in the case of machinery services, any method to allocate fixed costs to individual units is arbitrary. A method commonly used to allocate such costs to one unit of an enterprise (acre, head, hundredweight) is based on the proportion of the annual building service used by that unit. For instance, assume that the swine feeding structure analyzed above holds 500 head of pigs at one time. The building costs are $6,612 divided by 500 = $13.22 per pig space. Allocating these costs per pig depends on the number of batches fed per year and the effect of increasing use on the repairs and maintenance. This example assumes that repairs increase 0.5 percent or $300 annually for each additional batch. Thus, total annual building costs increase to $6,912 for 2 batches and $7,212 when the building is used three times per year. The building costs per pig would be:

Number of Batches per Year	Building Cost per Pig
1	$12.60
2	$6.91
3	$4.81

Consider a second example of computing building costs. A building 46 feet by 72 feet is to be used for machinery storage. The manager wants to estimate the annual building costs and to allocate these costs to the items of machinery stored in the building.

Current purchase cost of 46 ft × 72 ft building	$16,560
Salvage value	0
Years life	25
Taxes	1.4% of value
Nominal interest rate	12%
General inflation rate	7%
Insurance	0.6% of value
Repairs and maintenance	1.5% of value

Cost Item	*Percentage of Replacement Cost*
Capital recovery (Appendix Table IV)	7.1
Repairs and maintenance	1.5
Taxes and insurance	1.0
Total	9.6

The annual building costs are $16,560 × 0.096 or $1,590. These costs can be allocated to individual machines based on either their value or the square footage they occupy. If we base the estimate of housing costs on the square footage of space used, the average annual building cost per square foot for this example is:

$$\frac{\text{Annual building cost}}{\text{Sq ft in building}} = \frac{\$1,590}{46 \times 72} = \frac{\$1,590}{3312} = \$0.48$$

The annual building cost for an individual machine can then be computed by multiplying the cost per square foot by the square footage occupied by the machine. If a 100 HP tractor stored in the above building requires 127 square feet, then 127 × $.48 = $61 is the annual housing cost for that tractor. If a 60 HP tractor stored in the above building requires 104 square feet, then 104 × $.48 = $50 is the annual housing cost for that tractor. This procedure of estimating annual housing costs for machinery can be used in place of the standard percentage based on the machine's purchase price discussed in the previous section of this chapter.

It is not suggested that farmers should routinely take time to allocate housing costs to individual units of livestock and machines. These costs are fixed for existing buildings and should not be considered in decision-making. However, in those cases where such costs are to be allocated, the above procedures provide a reasonable way of estimating the average ownership costs per unit. For example, part or all of the building may be rented, and the above procedure could be used to estimate a rental charge. The above procedures may also be used in allocating ownership costs for a new building. Ownership costs for a new building are variable until the manager has committed the funds to build the structure.

ESTIMATING SEASONAL INPUT REQUIREMENTS

Although enterprise budgets are simply estimates of average annual costs and returns per unit of the enterprise, it is typically useful to estimate the seasonal requirements for certain inputs as a basis for whole-farm planning. The importance of estimating the seasonal requirements for an input depends on the seasonal variation in the amount of input required and the likelihood that the availability of that input may limit the farm organization. The variability in requirements of labor, operating capital, pasture, building space, and certain machinery services is frequently great enough to warrant making seasonal estimates. Estimates of seasonal labor and operating capital requirements are usually necessary to develop a detailed farm plan that does not require more labor than is available and more capital than can be obtained. Estimates of seasonal cash operating expenses are also useful as a basis to project the seasonal cash flow for the coming year(s). The amount of grazing required by (or assumed to be available for) livestock is often seasonal. In these situations, it is desirable to estimate the requirements by month of the year. The enterprise sometimes requires building services (such as space in a farrowing facility or grain storage) and/or certain machinery services (such as planting or harvesting) which may limit the size of enterprise that can be accommodated with the building and machinery services available. Seasonal requirements for such inputs should also be estimated.

The seasonal input requirements provide a basis for estimating the total amount of an input required by the farm plan. For instance, the amount of labor required by 200 acres of corn in May can be estimated by multiplying the hours required per acre during May by 200. Then the total for the farm organization in May is obtained by summing the labor required by all enterprises included in the organization.

The number of "seasons" used depends on the production process involved. Requirements are often estimated monthly for labor, capital, pasture, and other inputs. However, the length of the season selected should depend on the nature of the production process—not tradition. Therefore, if certain inputs must be applied or certain operations performed within a two-week period, it may be necessary to define the two-week period as one season. Perhaps longer periods can be defined for other periods of the year.

Seasonal Labor Requirements

Estimates of seasonal labor requirements can be developed by listing the tasks to be performed by season and the amount of time required to perform each task. In crop production the machinery operations can be listed by the month performed. The amount of machine time required per acre provides an estimate of the amount of direct labor required for the operation (assuming only one person is required to perform the machinery operation). Monthly hours of labor for machinery operations can be obtained by summing the entries on line 28 of the worksheet (Table 4.7, page 154) for estimating machinery costs. For example, the labor required for the machine operations listed in Table 4.8 total 0.13 hours in April, 0.52 hours in May, 0.16 hours in June, 1.14 hours in October, and 0.35 hours in November. Labor required for other operations (such as preparation of machinery, supervising hired or family labor, monitoring the progress of the crop, and hand operations such as gleaning corn fields) should be added to obtain the total for the enterprise.

Seasonal labor requirements for livestock may be somewhat more difficult to estimate because engineering relationships analogous to the field time required for machinery operations are unavailable. However, the general procedure of specifying the tasks to be performed and estimating the amount of time required for each of the tasks to be performed can be used. Using this approach allows an individual to make estimates for the system of production and working conditions on a particular farm. The procedure is illustrated for finishing feeder pigs in Table 4.9. Notice that the estimates in Table 4.9 are total requirements for two groups of 480 pigs per group. Hours of labor per pig can be estimated by dividing hours per month by number of pigs. The farm management extension specialists in most states within the United States and their counterparts in other areas of the world also maintain current estimates of seasonal labor requirements for alternative enterprises and systems of production. These estimates can be used if an individual is unwilling to specify the tasks and the amount of time required per task.

Seasonal Cash Operating Capital

Seasonal cash expenses can be estimated by specifying when the operator pays for each cash input used in production. Thus, the individual's plan for the purchase of inputs must be known in order to specify the seasonal cash expenses. Cash operating expenses are specified by month for the corn example in Table 4.10. The example assumes that the operator plows the land in November of the previous year (the money for this item is spent almost one year before the crop is harvested); that the operator purchases the seed, fertilizer, and chemicals in March, even though the inputs are not used until the next month; and that the machinery operating expenses (fuel, lubrication, and repairs) occur in the month the operation is performed. A cash expense for the crop insurance premium is included during June. Estimated monthly cash operating costs are listed near the bottom of Table 4.10. They are $3.78 for November, $78.01 for March, $1.77 for April, $5.08 for May, $10.56 for June, $32.53 for October, and zero for

Table 4.10 Summary of Monthly Cash Expenses per Acre for the Corn Production Example

Cash Expenses	Jan.	Feb.	Mar.	Apr.	May	Jun.	Jul.	Aug.	Sept.	Oct.	Nov.	Dec.	Annual Total
Plowing											3.78		
Apply fertilizer					0.92								
Disc				1.77	1.77								
Apply herbicide					0.31								
Harrow					0.39								
Planting					1.69								
Seed			15.12										
Nitrogen			23.74										
Phosphate			15.00										
Potash			4.95										
Herbicide-insecticide			19.20										
Cultivation						1.56							
Crop insurance						9.00							
Combine										11.61			
Hauling										2.43			
Drying										16.80			
Stalk shredding										1.69			
Estimated monthly cash operating costs	0	0	$78.01	$1.77	$5.08	$10.56	0	0	0	$32.53	$3.78	0	$131.73
Interest factor	0.10	0.09	0.08	0.07	0.06	0.05	0.04	0.03	0.02	0.01	0.12	0.11	
Interest charge	0	0	6.24	0.12	0.29	0.53	0	0	0	0.33	0.45	0	$ 7.96

165

the other months. Total annual cash costs are $131.73 per acre. Estimation of these monthly cash costs for enterprises provides part of the information required to develop projected cash flows for the farm business—one of the topics discussed in Chapter 6.

Although seasonal cash costs are estimated primarily for their use in projecting cash flows, they can also be used to estimate interest on the cash operating expenses. With the higher annual interest rates and prices of purchased inputs, interest on operating expense is a major cost item in today's farm business. The amount of interest charged for the cash expenses in each month can be obtained by multiplying the seasonal cash cost by the interest factor. The interest factor for any month depends on the month the crop is sold (or is transferred to the storage enterprise) and the annual interest rate. For instance, assume that the corn producer of Table 4.10 plans to sell his corn in November and that the annual interest rate to be charged for the use of operating capital is 12 percent. To simplify the calculations, assume all receipts and expenses occur on a given day of the month, such as the 15th. In the example, money invested in the previous November is invested for one full year. The interest factor is 1×0.12 and the amount of interest is $0.12 \times \$3.78$ or $0.45. However, the money spent in March is invested eight months. The interest factor for March is $2/3 \times 0.12$ or 0.08. The interest charge for inputs purchased in March is $78.01 $\times 0.08$ or $6.24. The interest factor for April is 7/12 of 0.12. It is 1/2 of 0.12, 5/12 of 0.12, and 1/12 of 0.12 for May, June, and October, respectively. The annual interest charge of $7.96 is obtained by summing the monthly interest charges in Table 4.10.

Table 4.11 Summary of Monthly Cash Costs per Group for Finishing Two Groups of 500 Feeder Pigs per Year

Months	Purchase of Pigs	Other Cash Inputs	Monthly Cash Operating Costs	Interest Factor	Interest Charge
January		$1,623.39	$ 1,623.39	0.01	$ 16.23
February		1,586.85	1,586.85	—	—
March	$25,675.20	1,972.04	27,647.24	0.04	1,105.89
April		1,762.39	1,762.39	0.03	52.87
May		1,762.39	1,762.39	0.02	35.25
June		1,623.39	1,623.39	0.01	16.23
July		1,599.33	1,599.33	—	—
August		0	0	—	0
September		0	0	—	0
October	21,408.00	1,972.04	23,380.04	0.04	935.20
November		1,972.39	1,972.39	0.03	59.17
December		1,762.39	1,762.39	0.02	35.25
Total			$64,719.80		$2,256.09

In some enterprises the product is sold two or more times per year. Estimates of the monthly cash costs, the interest factors, and the amount of interest annually is illustrated for a feeder pig enterprise in Table 4.11. In this example, the product, market hogs, is sold twice per year. Interest on the operating expenses for the first group of hogs is based on sale of the hogs on about February 15. One month's interest charge is included for capital invested in operating inputs in January, but receipts from the sale of some of the hogs are assumed to be available for payment of February operating costs. Interest on the operating inputs for the second group of hogs terminates when they are sold in July. Assuming a 12 percent annual rate and a four-month feeding period, the interest factor is 4/12 of 0.12 or 4 percent during the first month of the feeding period; similar calculations indicate rates of 3, 2, and 1 percent for subsequent months. The corn is assumed to be an intermediate product produced on the farm. It is not included in the cash flow since a cash outlay is not required. This is convenient since our purpose is to estimate costs and cash outflows for whole-farm planning. Total annual interest charges are $2,256.09.

ORGANIZING THE ESTIMATES INTO ENTERPRISE BUDGETS

An enterprise budget was defined in the previous chapter as a projection of average annual costs and returns for an enterprise. The enterprise budget normally includes an estimate of the physical resources required and products produced, their price, and the total value of each resource and product per unit of the enterprise for some future period of time. The format commonly used for enterprise budgets may contain eight parts: the title, livestock investment, receipts, operating costs, ownership costs, returns above costs shown, footnotes, and seasonal distribution of inputs. Each of these parts is discussed in turn. Three enterprise budgets, one each for corn, finishing feeder pigs, and a beef cow herd, are presented in Tables 4.12, 4.13, and 4.14 to illustrate the eight sections.

The Title

The purpose of the title is to name the product being produced, to indicate the unit for which the estimates have been prepared, and to describe the system of production. In some cases, it is useful to specify in the title how the product is to be marketed. For instance, corn produced for grain will probably not have the marketing system specified. However, a budget for the production of sweet corn should specify whether the sweet corn is for processing or for the fresh market, because this may affect some of the costs that are to be included in the enterprise budget. Crop budgets in the United States are typically prepared on an acre basis, and when they are, the unit is often not listed. The major soil types to which crop budget estimates apply should also be indicated in the title. If the enterprise budget is being prepared for a general area rather than an individual farm, it may be useful to indicate the geographic area by listing the section of the state or counties to which the estimates apply. If there are distinguishing features about the method of production, such as minimum tillage, irrigation, a unique method of harvesting, or a unique method of preparing the

Table 4.12 Corn for Grain Produced with Conventional Tillage on Webster Soils, Example Farm

Item	Unit	Quantity	Price or Cost/Unit	Value or Cost
1. Gross receipts				
Corn	Bu	120.00	$ 3.00	$360.00
2. Operating costs				
Preharvest				
Corn seed	Bags	0.30	50.40	15.12
Nitrogen	Lb	10.00	0.23	2.30
Anhydrous ammonia	Lb	134.00	0.16	21.44
Phosphate	Lb	60.00	0.25	15.00
Potash	Lb	45.00	0.11	4.95
Herbicides and insecticides	Acre	1.00	19.20	19.20
Crop insurance	Dol	360.00	0.025	9.00
Machinery	Acre	1.00	12.19	12.19
Labor	Hr	1.16	5.00	5.80
Harvest costs				
Machinery	Acre	1.00	32.53	32.53
Labor	Hr	1.14	5.00	5.70
Interest on operating capital	Dol		0.12	7.96
Total operating cost				$151.19
3. Income above operating costs				$208.81
4. Ownership costs				
Machinery depreciation, interest, insurance, and housing	Acre	1.00	54.37	54.37
Land taxes	Acre	1.00	12.00	12.00
Interest on land investment	Dol	2,000.00	0.04	80.00
Total ownership costs				$146.37
5. Total costs shown				$297.56
6. Net returns above costs shown				$ 62.44

Seasonal Distribution of Inputs

	Jan.	Feb.	Mar.	Apr.	May	Jun.	Jul.	Aug.	Sept.	Oct.	Nov.	Dec.
Labor (hr)				0.13	0.52	0.16				1.14	0.35	
Cash operating costs ($)			78.01	1.77	5.08	10.56				32.53	3.78	

Table 4.13 Finishing Feeder Pigs: Two Groups of 480 per Year, Fed in Solid Floor Confinement Facilities with Sales in February and July

Item	Unit	Quantity	Price or Cost/Unit	Value or Cost
1. Gross receipts				
230 lb market hogs in February	Hd	461	$51.45	$ 52,180.59
230 lb market hogs in July	Hd	461	55.10	55,882.42
Total receipts				$108,063.01
2. Operating costs				
40 lb feeder pigs purchased October	Hd	480	44.60	21,408.00
40 lb feeder pigs purchased July	Hd	480	53.49	25,675.20
Corn (10.1 bu/hd)[a]	Bu	9,696	3.00	29,088.00
Soybean meal[a]	Cwt	848.8	12.00	10,185.60
Mineral-vitamin supplement[a]	Cwt	169.0	7.00	1,183.00
Yardage and commission fees	Cwt	2,028.4	1.25	2,535.50
Veterinary and medicine	Hd	960	1.10	1,056.00
Utilities	Hd	960	0.10	96.00
Bedding	Tons	5.25	40.00	210.00
Insurance on feed and livestock	Dol	50,000	0.004	200.00
Miscellaneous expense	Dol			200.00
Truck operating expense	Hr	50	9.75	487.50
Machinery	Hr	175	7.56	1,323.00
Building repairs	Dol			160.00
Labor	Hr	720	5.00	3,600.00
Interest on operating capital	Dol			2,256.09
Total operating costs				$ 99,663.89
3. Income above operating costs				$ 8,399.12
4. Ownership costs				
Depreciation, interest, taxes, and				
insurance on buildings				5,801.00
Depreciation, interest, and insurance				
on machinery and truck				1,948.25
Total ownership costs				$ 7,749.25
5. Total costs shown				$107,413.14
6. Net returns above costs shown				$649.87

[a]The feed requirements assume a feed conversion of approximately 3.88 pounds of feed per pound of gain.

Seasonal Distribution of Inputs

	Jan.	Feb.	Mar.	Apr.	May	Jun.	Jul.	Aug.	Sept.	Oct.	Nov.	Dec.
Labor (hr)	72	96	48	72	72	72	96	0	0	48	72	72
Cash operating costs ($)	1,623	1,587	27,647	1,762	1,762	1,623	1,599	0	0	23,380	1,972	1,762

Table 4.14   Cow-Calf-Yearling Beef Cow Herd: 150 Cows Fed a Combination of Native Range, Small-Grain Pasture, and Sorghum Pasture, Rolling Plains Subregion of Texas and Oklahoma[a]

Livestock Investment		Head	Weight/Hd Cwt	Current Market Value Per Head	Total
Brood cows		131	10.0	$ 704.00	$ 92,224.00
Replacement heifers		19	8.5	579.00	11,001.00
Herd bulls		7	14.0	1,200.00	8,400.00
Replacement heifer calves		19	5.5	384.00	7,296.00
Total					$118,921.00

1.	Gross Receipts	Units	Number of Head	Weight	Price	Value	Value/Cow
	Steer calves	Cwt	22	4.42	$92.08	$ 8,953.86	
	Heifer calves	Cwt	20	4.35	77.59	6,750.33	
	Feeder steers	Cwt	41	5.80	92.74	22,053.57	
	Feeder heifers	Cwt	25	5.60	79.17	11,083.80	
	Cull cows	Cwt	16	8.75	49.47	6,925.80	
	Total receipts					$55,767.36	$371.78

2.	Operating Costs	Units	Number of Units	Price	Value	Cost/Cow
	Private range	Acre	1,051.0	$ 0.0	$ 0.0	$ 0.0
	Pasture rent/lease	Acre	473.0	2.93	1,385.89	9.24
	Small-grain pasture	Acre	598.0	4.41	2,637.18	17.58
	Sorghum pasture	Acre	105.0	30.52	3,204.60	21.36
	Hay aftermath	Acre	70.0	0.0	0.0	0.0
	Protein supplement	Cwt	296.2	9.35	2,769.47	18.46
	Grain (purchased)	Cwt	19.1	3.99	76.21	0.51
	Legume hay	Ton	15.8	69.88	1,104.10	7.36
	Hay (produced)	Ton	86.4	45.15	3,900.96	26.01
	Salt and minerals	Cwt	50.6	4.45	225.17	1.50
	Veterinary and medicine	Dol	451.0	1.00	451.00	3.01
	Marketing	Dol	541.0	1.00	541.00	3.61
	Hauling	Dol	118.0	1.00	118.00	0.79
	Range improvement	Dol	590.0	1.00	590.00	3.93
	Machinery fuel and lubrication				1,024.51	6.83
	Machinery repairs				726.17	4.84
	Equipment repairs				511.14	3.41
	Livestock labor	Hr	1,872.00	3.46	6,477.12	43.18
	Interest on operating capital	Dol	15,810.10	0.106	1,675.87	11.17
	Total operating costs				$27,418.39	182.79

3.	Income above Operating Costs				$28,348.97	$188.99

4.	Ownership Costs			
	Machinery		$4,042.22	$ 26.95
	Equipment		11,876.00	79.17

(*continued*)

Table 4.14   (*Continued*)

Livestock		
Depreciation on bulls		
(purchased)	630.00	4.20
Interest	5,946.05	39.64
Insurance and taxes	728.52	4.86
Land (interest and taxes)	1,000.60	6.67
Total ownership costs	$24,223.39	$161.49
5. Total Costs Shown	**$51,641.78**	**$344.28**
6. Net Returns above Costs Shown	**$4,125.58**	**$27.50**

[a]The budget assumes a calving rate of 88 percent, a replacement rate of 12.7 percent, cattle loss of 1.7 percent, and calf loss of 3.5 percent. The budget assumes spring calving, with fall selling of calves and yearlings sold the following spring. Year-round grazing using a combination of range, small-grain grazing, and forage sorghum pasture is assumed.

Source: Adapted to the above format from the Federal Enterprise Data System (enterprise code: 114411692). The budget was developed for Cow-Calf-Yearling Beef Cow Herds of 100 to 199 cows in size, for the Rolling Plains Subregion of Texas and Oklahoma for 1979.

Seasonal Distribution of Inputs

	Jan.	Feb.	Mar.	Apr.	May	Jun.	Jul.	Aug.	Sept.	Oct.	Nov.	Dec.
Labor (hr)	156	156	156	156	156	156	156	156	156	156	156	156

product for market, this information should be indicated in the title so that the user of the enterprise budget is aware of this special system of production that has been analyzed. Notice that the title of the enterprise budget in Table 4.12 indicates the crop, method of production, soil type, and farm to which it applies. The estimates have been prepared on a per acre basis. Thus, the unit is not mentioned.

Unlike crop production, it is very important to specify the unit of analysis in the title for livestock production. The unit for livestock production may be one cow, one sow, one litter, one ewe, or one hen. But a larger unit is frequently specified, given the economies of size that prevail in the use of facilities and labor for livestock production. It is common to specify in the enterprise budget the input requirements for a given size of enterprise, such as 100 beef cows, 480 feeder pigs, 48 sows, or 10,000 laying hens. The title of livestock budgets should also describe the housing, feeding, breeding, and marketing system assumed. Notice that the title of Table 4.13 lists the unit (two groups of 480), the type of facilities, and the expected marketing months. The budget in Table 4.14 was prepared for a general geographic area. The title indicates the products to be sold (calves and yearlings), the size of the enterprise (150 cows), the feeding program, and the geographic area to which the estimates apply.

Livestock Investment

Budgets for livestock enterprises often include a section to list the average number of animals in the breeding herd by type. Separate lines are normally maintained for mature females, replacement females, and male animals. The purpose of this section is to help describe the composition of the enterprise and to estimate the investment in breeding livestock so that appropriate ownership costs can be included. In Table 4.14, a beef cow enterprise budget, the breeding herd composed of 131 cows and 19 replacement heifers is listed first. Other entries are included for the bulls and replacement heifer calves. The value of offspring being fed for sale is not listed in the investment section since they will be sold within the year and are considered operating capital. For example, notice that the feeder pig finishing enterprise budget (Table 4.13) which has no breeding herd does not have a section for livestock investment.

Receipts

The receipts section normally lists each of the products that are expected to be marketed during a one-year period of time. A price is assigned to each of these products, and the total receipts are calculated per unit of the product. It is important to list all products that are marketed, minor as well as major. Thus, the amount of straw that is produced as well as grain should be listed for small-grain crops. The availability of aftermath grazing for corn production should be listed, providing time is available to make use of the aftermath grazing before the land is plowed for the following crop. The corn budget in Table 4.11 assumes fall plowing. Hence, time between harvest and plowing does not permit use of aftermath grazing, and it is not listed. In the case of livestock products, it may be of interest to list the value of the manure if costs have been included for its removal, transportation, and spreading in the field. The price assigned to each product should reflect the market price expected (including the appropriate seasonal adjustment) at the time the product is sold. The prices for hogs given in Table 4.12 have been seasonally adjusted. They are based on an average annual price of $50 per hundredweight. Seasonal adjustment of the major farm commodity prices is discussed in Chapter 7.

Operating Costs

Costs are divided into two sections, operating costs and ownership costs, as discussed in the previous chapter. Each of the variable or operating inputs should be listed in the operating costs section. The individual preparing the budget should list each operating input, give its unit of measure (pound, hundredweight, bushel, ton, gallon, etc.), as well as list the quantity, price, and value. It may be useful to list the operating inputs for crop enterprises under the two subheadings shown in Table 4.12—preharvest cost, and harvest cost. Notice that the fuel, lube, and repair cost for each of the machinery operations listed in Table 4.8 is summed and included under the operating costs section in Table 4.12. The quantity of seed and chemicals included has been selected using the economic principles discussed in the previous chapter. The levels of nitrogen and phosphate fertilizer are based on the analysis of fertilizer trials, such as those summarized in Table 3.8. A charge for insurance on the gross value of the crop has been included. The amounts of labor and interest on the operating capital determined in Tables 4.8 and 4.10 are also included. Labor can be either fixed (such as labor provided by the family or a full-time hired laborer), variable (such as labor hired by the day or on a piece-rate basis), or a combination of the two. We list it under the operating cost section. Whether it is variable, fixed, or a combination of variable and fixed must be determined on a case-by-case basis. Interest on the operating capital is a variable cost. It may be either a cash or a noncash cost (opportunity cost), and it should be listed under the operating cost section.

The procedure for listing operating costs for livestock budgets is essentially the same as for crops. Any purchased livestock are normally included in the operating cost section. Likewise, both purchased and home-grown feed required by the livestock enterprise are listed and a value assigned. The operating costs also include veterinarian and medical costs, marketing costs, utility costs, insurance premiums on livestock, and other expenses the operator would avoid if he or she did not have the livestock assumed in the budget. There is usually some machinery expense associated with livestock production. Machinery is required for feed hauling and manure handling, as well as for other operations in livestock enterprises. Thus, the machinery operating expense should also be listed. Insofar as building repairs are dependent on livestock occupying the buildings, they should be listed under the operating cost section. In some cases, a large proportion of the building repair and maintenance cost is essentially a function of time and is fixed. Labor and interest costs, as in the case of crop production, are normally listed under the operating cost section.

Ownership Costs

Ownership or fixed costs are those costs that will continue whether or not the enterprise is produced. In the case of crop production, this category normally includes the machinery and land ownership costs as shown in Table 4.12. The land charge is calculated in a manner analogous to the calculations for other capital assets discussed earlier. The appropriate land charge to include depends on whether the land is owned or rented. If the land is owned, then an interest charge on the current market value of the land plus a charge for the taxes is a reasonable estimate of the land charge (depreciation and insurance are zero). If charges for maintenance of drainage ditches, grass waterways, or other aspects of the land are required, these costs should also be included for owned land. If the land is rented on a cash rent basis, then the cash rent per acre can be listed as the land charge. If the land is rented on a crop share arrangement, no land charge should be included. The land charge is omitted in this case because the amount of production shown should be reduced by the amount the landlord is expected to receive.

Ownership or fixed costs for livestock production include the ownership cost for the buildings, land, and equipment used. The charge for the use of any owned land can be calculated in the manner suggested above for crop production. Expenditures for pasture rent, however, are normally listed under the operating cost section as shown in Table 4.14. Table 4.13 lists the appropriate building and machinery ownership costs for the feeder pig finishing enterprise.

The ownership costs associated with a breeding herd for livestock enterprises are normally included under the ownership or fixed category. The ownership costs include depreciation, interest on the money invested in the livestock, personal property taxes on the livestock, and insurance on the breeding herd.

Purchased breeding stock can be depreciated in a manner analogous to machinery. Dairy, beef, sheep, and swine breeding herds which raise replacement females and maintain a relatively constant age composition of the breeding herd normally include the cost of raising replacements (feed, veterinary expense, etc.) in the operating cost section as illustrated in Table 4.14. Thus, no depreciation needs to be calculated for the raised breeding stock. Interest, insurance, and personal property taxes are calculated as they are for machinery.

Returns above Costs Shown

Returns above costs shown are obtained by totaling operating and ownership costs, and subtracting the total cost shown from total receipts. This is a residual return to all of the factors for which a charge has not been included. A charge has been included for all of the variable cash costs, as well as the fixed resources—the operator's labor, land, machinery, and buildings—but a charge has not been included for management and other overhead expenses of the business. These other overhead expenses include office expenses, transportation, utilities, and perhaps other items depending on the type of business. Hence, this residual is a return to the operator's management and overhead expenses of the business. Returns above costs shown are $62.44 per acre for corn in Table 4.12, $649.87 for the swine enterprise budgeted in Table 4.13, and $4,125.58 for the cow-calf enterprise of Table 4.14. Enterprise budget calculations may result in a negative expected net return above all costs shown. A negative net return simply indicates that the enterprise is unable to make as great a contribution to the fixed resources as has been allocated. The operator should not necessarily stop production, however, as long as returns exceed variable costs.

Footnotes

Footnotes are an effective way of recording information that may help interpret the enterprise budget, but that cannot be entered directly into the budget itself. For instance, the particular chemicals used for herbicide and insecticide treatment might be listed in the footnote. The details of the crop share leasing arrangement might be given in the footnote to explain why certain categories of costs and returns are lower than they would be for an owner-operator. An explanation of certain harvesting and hauling expenses or the price received determined through some special marketing arrangement might also be mentioned in the footnote. In general, anything that is needed to understand the budget but that is not included in previous information can be included in the footnotes.

Seasonal Distribution of Inputs

The seasonal distribution of inputs is developed in the process of defining and specifying the production system as discussed in the introduction to this chapter. Budgets should normally summarize the seasonal use of the major inputs of concern in whole-farm planning; this includes labor and capital requirements as a minimum. Budgets might also show the seasonal distribution of building and facility requirements, machinery requirements, pasture requirements, irrigation requirements, or other items. Monthly labor and operating capital requirements are shown at the bottom of Tables 4.12 and 4.13.

These data on seasonal distribution are useful in whole-farm planning, and in developing an enterprise control system as we will discuss in Chapter 17. Estimating input requirements by period of the year will also be discussed more fully at that time.

Summary

To develop enterprise budgets, the concepts developed in the previous chapter must be combined with computational procedures. One of these procedures is the estimation of machinery costs. Machinery costs can be divided into two categories: fixed or ownership costs, including capital recovery or depreciation plus interest, housing, insurance, and personal property taxes; and operating costs which for owned machinery include fuel, lubrication, repairs and maintenance, and labor. Operating costs for leased equipment may be similar but will depend upon the lease agreement. To accurately estimate machinery costs, ownership and operating costs should be calculated based on current market or replacement prices. This approach is consistent with the concept that all costs and returns in an enterprise budget are at current market value.

Depreciation and interest are two key ownership costs. The purpose of including depreciation and interest allowances is to reflect the amount of money that should be set aside to cover the loss in value of the asset (depreciation) and interest on its remaining value. The traditional formulas based on average value, however, underestimate these costs; a more accurate but more complex computational procedure is the capital recovery method. The capital recovery method of estimating depreciation and interest costs consistently results in higher annual charges, but also more accurately reflects the true capital costs of a machinery purchase. Taxes, insurance, and housing represent the additional ownership costs and typically are a much smaller, but not insignificant, amount of total ownership costs.

Fuel costs are one of the key operating costs and are estimated based on hours of machine use, hourly fuel consumption, and fuel prices. Lubricant costs are traditionally calculated as 15 percent of fuel costs. Repair and maintenance costs are typically estimated, using engineering formulas, as a function of the list price of the machine and hours of utilization; the formulas used reflect increasing repair costs as utilization increases. Labor costs are estimated as the hourly wage rate times total hours of utilization increased by 20 percent to reflect lubrication and the time involved in servicing and transporting the machine.

To determine machine costs per acre, machinery accomplishment rates must be estimated using data on the size, speed, and efficiency of various machines. The cost per acre is determined by dividing hourly total ownership and operating costs by the accomplishment rate in acres per hour.

The annual costs of providing building services include depreciation, interest, repairs, taxes, and insurance. These costs are calculated using similar formulas to those used in determining machinery costs. For buildings that have multiple uses, it is necessary to allocate the total cost to various production enterprises.

Although enterprise budgets reflect annual costs and returns, the seasonality of input utilization and product flows can be extremely important in developing marketing as well as financial plans. Seasonal labor requirements can be estimated by determining the task that must be performed at various times of the year and the amount of time required to perform each task. Seasonal operating capital requirements depend upon when the operator pays cash for the various inputs. The seasonal flow of expenses is also important in estimating the interest cost on operating funds.

Once the basic cost and return data have been obtained, they can be organized into an enterprise budget. A typical enterprise budget should include a title describing the enterprise and any unique characteristics, receipts from the sale of the products of that enterprise, the operating costs categorized by specific cost item, the ownership costs also categorized by item, and an indication of the returns per unit of production above all costs. These data on enterprise costs and returns provide the basic component for doing whole-farm planning, which is the focus of the next chapter.

Questions and Problems

1. When should a manager estimate machinery costs?
2. Identify and discuss the specific fixed or ownership costs for farm machinery. Identify and discuss the specific operating costs for farm machinery.
3. Why are current replacement or market prices used as the base for estimating farm machinery costs?
4. What is the traditional formula for estimating annual depreciation on machinery?
5. What is the traditional formula for estimating annual interest on machinery?
6. Contrast the capital recovery method to the traditional depreciation and interest formulas for determining capital cost. Which method is more accurate in estimating cost?
7. Obtain the appropriate information for a specific farm machine and estimate the total ownership cost per hour of utilization.
8. Calculate the specific operating cost (fuel, lubricants, repairs and maintenance, and labor) for the machine you chose for question 7 and calculate total operating costs per hour of utilization.
9. What is the formula used to determine accomplishment rates for farm machinery? Using this formula, determine the cost per acre for the machine noted in questions 7 and 8.
10. What are the specific costs incurred in providing building services?
11. Collect data on a specific building and, using the appropriate formulas and procedures, estimate the specific costs of providing building services.
12. Why is it important to determine the seasonal flows of input use for a particular enterprise?
13. Select a livestock production enterprise and indicate how you would estimate the seasonal labor requirements.
14. How would you estimate the seasonal cash operating capital requirements and the operating capital interest cost for a crop production enterprise?
15. What are the major components of an enterprise budget? Discuss the items to include in each component.
16. Collect the data for a major crop or livestock product produced in your area and develop an enterprise budget for that product. What parts of the enterprise budget will be used if you are only concerned with short-run decisions? Long-run decisions?

Further Reading

Bowers, Wendell. *Fundamentals of Machine Operation: Machinery Management.* Moline, Ill.: Deere and Company, 1975.

Kletke, Darrel D. *Operation of the Enterprise Budget Generator.* Stillwater, Okla.: Oklahoma State University Agricultural Experiment Station Research Report P–790, August 1979.

Official Guide. St. Louis, Mo.: National Farm Power and Equipment Dealers Association. Published Fall and Spring annually.

Prentice, Paul T., and Lyle P. Schertz. *Inflation: A Food and Agricultural Perspective.* Washington, D.C.: U.S. Department of Agriculture, Agricultural Economic Report No. 463, 1981.

5

ECONOMIC CONCEPTS FOR WHOLE-FARM PLANNING

CHAPTER CONTENTS

The economic concepts we discussed in Chapter 3 are used to select the amount of variable input to use in producing one enterprise. Now we turn our attention to an economic concept that can be used to choose which combination of enterprises to produce. That is, we will consider the product-product model that can be used to help answer the question, "What combination of enterprises should be produced with a given group of resources?"

The product-product model provides a conceptual framework to allocate land, labor, and capital resources to alternative uses in a way that maximizes net returns to the manager's owned resources. Although maximizing net returns may not be the farm operator's only goal, achieving a high level of net income is inevitably one of the important goals, making it a useful starting point in selecting a farm plan. The conceptual model is illustrated with production, marketing, and financial alternatives, and the application of the model is illustrated with enterprise budgets (or gross margins). As in previous chapters, we will continue to assume that the relevant production functions are known with certainty and that the firm faces constant input and product prices.

THE PRODUCT-PRODUCT MODEL

The product-product frontier is a convenient way of summarizing the maximum possible production levels of two outputs for a given set of resources. More specifically, the product-product frontier is the technical relationship depicting the maximum amount of one product, Y_1, that can be produced for alternative levels of a second product, Y_2, with a specified set of resources. Thus, the product-product frontier shows the maximum quantities of output that a manager can produce with the resources available.

The firm's resource set may include both limited and unlimited resources. By unlimited we mean the manager can purchase as much of these resources as can be profitably used on the farm. The amount of unlimited resources to use in producing each enterprise can be determined by applying the rule to add additional amounts of the resource as long as the value of the marginal product exceeds the marginal factor cost. By applying this rule, a manager can use the optimum amount in each enterprise. Some resources are limited to the firm. By limited, we mean the manager cannot obtain enough of the resource to apply the profit-maximizing rule for an unlimited resource. Development of the product-product frontier is useful in allocating the limited resources and, in the process, selecting the level of each enterprise to produce.

Deriving the Product-Product Frontier

The product-product frontier can be derived from the production functions for the outputs. To illustrate, suppose that the production functions for corn in Table 5.1 apply to the two soil types on a farm. To simplify the illustration, suppose the farm's limited resources are composed of one acre of each type of soil and 100 pounds of nitrogen. In this case, the salable product is the same, but the product-product concept is applicable because two different processes are used. Although the limit on a variable input such as nitrogen may seem artificial, this type of situation frequently arises either because of limits on the availability of credit or because of a limited availability of the input from suppliers at the time it is needed. As we will see, the rule for allocating the limited resource can be applied broadly to allocating limited resources in farm planning.

The total physical product or yield of corn per acre is shown for alternative levels of nitrogen fertilizer on each soil type in Table 5.1. The difference in the water-holding capacity of the two soils causes the yield response to nitrogen fertilizer to differ in three general ways. First, the yield per acre is greater on soil type 2 for any nitrogen application rate per acre. Second, a positive response is achieved through 280 pounds of N on soil type 2, while the maximum yield is achieved at approximately 200 pounds of N per acre on soil type 1. While these two points may seem important, we will see that the third difference—the difference in the marginal physical product—is more relevant in the allocation of the 100 pounds of nitrogen.

Alternative ways of allocating the 100 pounds of nitrogen between corn on soil 1 and soil 2 are shown in Table 5.2. Combination 1 lists the yield of corn for an application of 100 pounds of nitrogen on soil 2 and no

Table 5.1 Yield Response of Corn to Nitrogen Fertilizer[a]

	Soil Type 1		Soil Type 2	
Nitrogen Fertilizer (lb/ac) X_1	Total Physical Product (bu/ac) Y_1	Marginal Physical Product (bu/ac) $\Delta Y_1/\Delta X_1$	Total Physical Product (bu/ac) Y_2	Marginal Physical Product (bu/ac) $\Delta Y_2/\Delta X_1$
0	30		90	
20	44	14	101	11
40	55	11	111	10
60	66	11	120	9
80	75	9	128	8
100	83	8	136	8
120	89	6	142	6
140	94	5	148	6
160	97	3	153	5
180	100	3	158	5
200	100	0	161	3
220	100	0	164	3
240	97	−3	166	2
260			167	1
280			168	1
300			167	−1

[a]The yield per acre is based on the following two functions:

$$Y_1 = 30.2 + 0.7N - 0.00175N^2$$
$$Y_2 = 90.4 + 0.55N - 0.00098N^2$$

where:

Y_1 = yield of corn on soil 1 in bushels per acre.
Y_2 = yield of corn on soil 2 in bushels per acre.
N = pounds of nitrogen applied per acre.

nitrogen on soil 1. Combination 2 lists the yields for an application of 80 pounds of nitrogen on soil 2 and 20 pounds on soil 1, and so on. Finally, combination 6 shows the yields on the two soils when all 100 pounds of the nitrogen are applied to soil 1.

The Choice Rule

The general rule to use in finding the profit-maximizing amount of each product or enterprise to produce is to continue substituting one product (enterprise) for another along the product-product frontier as long as the revenue added by the product being increased is greater than the revenue lost from the product being decreased.[1] The amount of product being

---

[1]We will derive this result graphically in Figure 5.1. For a mathematical derivation, see James M. Henderson and Richard E. Quandt, *Microeconomic Theory: A Mathematical Approach*, 2d ed. (New York: McGraw-Hill, 1971), pp. 89–98.

Table 5.2 Product-Product Frontier for Corn
with 100 Pounds of Nitrogen

	Pounds of Nitrogen Applied to Each Acre		Corn Yield on		Marginal Physical Product		Marginal Rate of Product Substitution
			Soil 1	Soil 2	Soil 1	Soil 2	$\dfrac{\Delta Y_2}{\Delta Y_1}$
Combination	Soil 1	Soil 2	Y_1	Y_2	$\Delta Y_1/\Delta X_1$	$\Delta Y_2/\Delta X_1$	
1	0	100	30	136	14	8	0.57
2	20	80	44	128	11	8	0.73
3	40	60	55	120	11	9	0.82
4	60	40	66	111	9	10	1.11
5	80	20	75	101	8	11	1.38
6	100	0	83	90			

added as one moves down Table 5.2 is the marginal physical product of 20
pounds of nitrogen on soil 1, $\Delta Y_1/\Delta X_1$, and the amount given up is the
marginal physical product of 20 pounds of nitrogen on soil 2, $\Delta Y_2/\Delta X_1$.
The most profitable combination is attained where the revenue foregone
equals the revenue added. This can be written as

$$\frac{\Delta Y_2}{\Delta X_1} \cdot P_{Y_2} = \frac{\Delta Y_1}{\Delta X_1} \cdot P_{Y_1} \tag{5.1}$$

where P_{Y_1} and P_{Y_2} are the net revenue per unit of Y_1 and Y_2, respectively.
This rule also can be written

$$\frac{\dfrac{\Delta Y_2}{\Delta X_1}}{\dfrac{\Delta Y_1}{\Delta X_1}} = \frac{P_{Y_1}}{P_{Y_2}} . \tag{5.1'}$$

This is usually written more simply by multiplying both numerator and
denominator of the ratio of marginal physical products by ΔX_1 and
obtaining

$$\frac{\Delta Y_2}{\Delta Y_1} = \frac{P_{Y_1}}{P_{Y_2}} . \tag{5.1''}$$

The ratio on the left is the marginal rate of product substitution (MRPS),
and the ratio on the right is the inverse ratio of net revenues.

The marginal rate of product substitution is shown in the right-hand
column of Table 5.2. If other inputs are varying with yield making the net
revenues $0.90 and $1.20 per bushel for Y_1 and Y_2, respectively, the ratio

P_{Y_1}/P_{Y_2} is 0.75. In applying the principle to discrete data, such as those shown in Table 5.2, an equality may not be found. In these cases, select the two adjacent substitution ratios, of which one is greater and the other is smaller than the price ratio. The combination of products between these two substitution ratios is the profit-maximizing solution. The profit-maximizing combination is 3, with 40 pounds of nitrogen on soil 1 and 60 pounds on soil 2.[2]

A Graphical Analysis

The same problem can be solved graphically. The product-product frontier for the data in Table 5.2 is plotted in Figure 5.1. This relationship is also referred to as either the production possibilities frontier or the isoresource frontier (because each point on the curve represents combinations of outputs that can be produced with an equal or isoamount of inputs). Regardless of which term is used, the relationship shows the maximum amount of Y_1 that can be produced for each level of Y_2 with the resources and technology available. The slope of the product-product frontier is given by the decrease in Y_2 required to achieve a unit of increase in Y_1. This is the MRPS or $\Delta Y_2/\Delta Y_1$. For example, the change from combination 1 to 2 is a decrease of 8 in Y_2 and an increase of 14 in Y_1, giving a slope of 8/14.

To find the profit-maximizing level of net return to the limited resources, we must develop a net revenue function with P_{Y_1} and P_{Y_2} representing the net revenue per unit of Y_1 and Y_2, respectively. This can be written as:

$$NR = P_{Y_1} \cdot Y_1 + P_{Y_2} \cdot Y_2 \tag{5.2}$$

To plot this net revenue equation on Figure 5.1, the relationship can be rewritten as:

$$Y_2 = \frac{NR}{P_{Y_2}} - \frac{P_{Y_1}}{P_{Y_2}} \cdot Y_1 \tag{5.3}$$

Notice that P_{Y_1} and P_{Y_2} are given. By specifying the level of NR, the equation, referred to as an isorevenue line, can be used to solve for all combinations of Y_1 and Y_2 that will provide an equal level of net revenue to the limited resources. For example, if P_{Y_1} is \$0.90, P_{Y_2} is \$1.20 and NR is \$90, the isorevenue line can be written as:

---

[2]A question frequently asked is, why should the net revenue per unit of product be used instead of the product price in selecting the optimum allocation of nitrogen to the two soils? The answer is that the objective is to maximize net returns to the limited resources—2 acres of land and 100 pounds of nitrogen in Table 5.2. The value of the variable inputs must be subtracted from gross receipts to obtain net revenue per unit of output. For example, the reader can verify that if the price per bushel of corn is the same on soils S1 and S2, *gross* returns are greater for fertilizer combination 4 than 3, but net revenue is greater for combination 3, which was selected.

$$Y_2 = \frac{\$90}{\$1.20} - \frac{\$.90}{\$1.20} \cdot Y_1 \tag{5.4}$$

$$= 75 - 0.75 \, Y_1 \tag{5.5}$$

The resulting isorevenue line is plotted in Figure 5.1. It has an intercept of 75 and a slope of minus 0.75. It is the locus of all combinations of Y_1 and Y_2 that result in exactly \$90 of net revenue. By increasing the level of net revenue, the isorevenue line which is tangent to the product-product frontier can be found. Inspection of Figure 5.1 indicates that the highest isorevenue line that is tangent or just touching the production possibilities frontier represents the maximum level of net revenue that can be attained with the resources, technology, and net revenue per unit of the enterprise. In Figure 5.1, this occurs at a net revenue of \$193.71 and a production

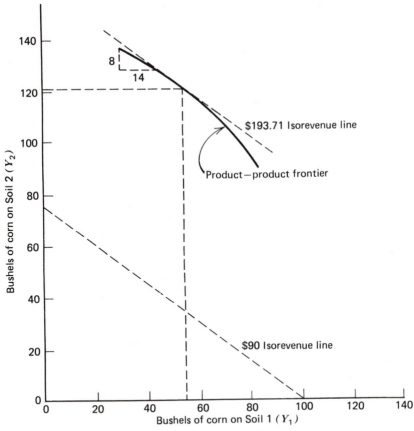

Figure 5.1.   Product-product frontier with one variable input.

level of 53.9 and 121 bushels of Y_1 and Y_2, respectively. This more precise solution would allocate 37.3 pounds of nitrogen to Y_1 and 62.7 to Y_2.

Figure 5.1 demonstrates that the rule to equate the MRPS and the ratio of net revenues maximizes net returns to the fixed resources. The slope of the product-product frontier and the slope of the profit-maximizing isorevenue line are equal at the point of tangency. The slope of the product-product frontier is indicated by $\Delta Y_2/\Delta Y_1$, and the slope of the isorevenue line is indicated by the ratio of net revenues P_{Y_1}/P_{Y_2}. Notice that both of these are negative. Thus, the sign is usually ignored for analyses such as those conducted in Table 5.2. One must either include the minus sign on both ratios or treat both ratios as positive values to apply the profit-maximizing rule outlined.

The Expansion Path

The product-product frontier is usually only of concern when the total amount of variable input is held constant at a particular value, making only one product-product curve relevant. When alternative amounts of the variable input (nitrogen in our example) are available, a product-product curve can be developed for each level. An analysis of varying amounts of the variable input provides the conceptual rule to use in allocating additional resources as a firm expands.

Alternative ways of allocating 200 pounds and 320 pounds of nitrogen to produce one acre of corn on soil 1 and one acre on soil 2 are shown in Table 5.3. The corresponding product-product frontiers for 100 pounds, 200 pounds, and 320 pounds of nitrogen are plotted in Figure 5.2. An isorevenue line with slope minus 0.75 is drawn tangent to each product-

Table 5.3 Product-Product Frontiers for Corn with 200 and 320 Pounds of Nitrogen

	200 Pounds of Nitrogen					320 Pounds of Nitrogen				
Combination	Y_1	Y_2	$\dfrac{\Delta Y_1}{\Delta X_1}$	$\dfrac{\Delta Y_2}{\Delta X_1}$	$\dfrac{\Delta Y_2}{\Delta Y_1}$	Y_1	Y_2	$\dfrac{\Delta Y_1}{\Delta X_1}$	$\dfrac{\Delta Y_2}{\Delta X_1}$	$\dfrac{\Delta Y_2}{\Delta Y_1}$
1	30	161				55	168			
			14	3	0.21			11	1	0.09
2	44	158				66	167			
			11	5	0.45			9	1	0.11
3	55	153				75	166			
			11	5	0.45			8	2	0.25
4	66	148				83	164			
			9	6	0.67			6	3	0.50
5	75	142				89	161			
			8	6	0.75			5	3	0.60
6	83	136				94	158			
			5	8	1.60			3	5	1.67
7	88	128				97	153			
			4	8	2.00			3	5	1.67
8	94	120				100	148			
			3	9	3.00			0	6	∞
9	97	111				100	142			
			3	10	3.33					
10	100	101								
			0	11	∞					
11	100	90								

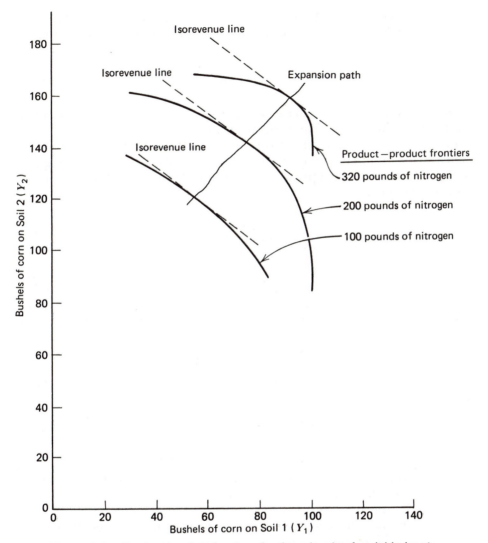

Figure 5.2.   Product-product frontiers for three levels of variable input.

product frontier. The most profitable allocation of 200 pounds of nitrogen is 80 pounds to soil 1 and 120 pounds to soil 2. The corresponding production levels are 75 and 142 bushels per acre. With 320 pounds of nitrogen, 131.3 should be allocated to soil 1 and 188.7 to soil 2. The corresponding yields are approximately 92 and 159.3 bushels per acre, respectively. The line connecting these maximum revenue points is referred to as an output expansion path. It shows the maximum revenue combination of outputs for alternative levels of the variable input and the P_{Y_1}/P_{Y_2} ratio specified.

Reconsider the condition that exists at the tangency between the product-product frontier and the isorevenue line. The condition was written as:

$$\frac{\Delta Y_2}{\Delta Y_1} = \frac{P_{Y_1}}{P_{Y_2}} \tag{5.6}$$

where the negative signs have been omitted on both ratios. We noted earlier that the limited input (nitrogen in our example) must be shifted from one product to the other to move along the product-product frontier. The discussion of equation 5.1 noted that the MRPS can be written as the ratio of the two marginal physical products, and consequently condition 5.6 can be written as:

$$\frac{\frac{\Delta Y_2}{\Delta X_1}}{\frac{\Delta Y_1}{\Delta X_1}} = \frac{P_{Y_1}}{P_{Y_2}} \tag{5.7}$$

This states that the ratio of the marginal physical products must equal the ratio of net revenues for the enterprise. By cross-multiplying, the previous equation can be written as:

$$P_{Y_1} \cdot \frac{\Delta Y_1}{\Delta X_1} = P_{Y_2} \cdot \frac{\Delta Y_2}{\Delta X_1} \tag{5.8}$$

or

$$VMP_{X_1 \cdot Y_1} = VMP_{X_1 \cdot Y_2} \tag{5.9}$$

This is read as the value of the marginal product of X_1 used in producing Y_1 must equal the value of the marginal product of X_1 used in producing Y_2. That is, the most profitable way to allocate a limited amount of a variable input is to allocate the input to its most profitable uses and equate the value of the marginal product in all of its uses.

The data in Table 5.1 indicate that the marginal physical product declines as more of the variable input is applied. As we move up the expansion path, the value of net revenue per unit of each output (P_{Y_1} and P_{Y_2}) remains constant, while the marginal physical product declines, causing the value of the marginal product to decline. However, at each point on the expansion path the value of the marginal product in the two uses is equal. The maximum profit point is achieved where the value of the marginal product in the two uses is equal *and* is equal to the price of the variable input. This can be written as

$$VMP_{X_1 \cdot Y_1} = VMP_{X_1 \cdot Y_2} = P_{X_1} \tag{5.10}$$

or, alternatively, dividing through by P_{X_1} we obtain the condition

$$\frac{VMP_{X_1 \cdot Y_1}}{P_{X_1}} = \frac{VMP_{X_1 \cdot Y_2}}{P_{X_1}} = 1 \tag{5.11}$$

EQUAL MARGINAL RETURN PRINCIPLE

The rule to allocate a limited amount of variable input to multiple outputs developed in the previous section is frequently referred to as the equal marginal return principle. This principle can be stated simply as follows: The returns from a scarce or limited resource are maximized when the input is allocated to its most profitable uses and the value added by the last unit of the resource is the same in each of its alternative uses.

The equal marginal return principle provides a basis to allocate limited resources in farm planning. It applies to the allocation of land, labor, capital, and irrigation water, as well as purchased inputs that can only be obtained in limited quantities. Thus, it is quite useful in making short-run decisions, such as how to allocate the limited amount of irrigation water among alternative crops during a particular period of the growing season. It can also be applied to allocate resources over a longer period, such as selecting the cropping program for future years. An application to the allocation of operating capital illustrates the concept. Other applications to the allocation of land and labor resources are presented in Chapter 6.

Suppose a manager is considering the allocation of a limited amount of capital among livestock production, nitrogen fertilizer on two soil types, and herbicides on the two crops. The manager has 500 acres of cropland equally divided between soil 1 and soil 2. To apply the equal marginal return principle, the operator develops the data shown in Table 5.4 which indicate the marginal value product from investing $1,000 increments of operating capital in each of the five alternative uses being considered on the farm. Investing the money in a certificate of deposit is the sixth alternative. For example, nitrogen fertilizer costs $0.20 per pound of nitrogen, and the cost of a 20 pound increment on each of 250 acres is $1,000. The data in Table 5.1 indicate that the first 20 pound increment increases yield 14 bushels per acre on soil 1. The net revenue per bushel on soil 1 is $0.90, making the marginal return for the first $1,000 of nitrogen equal to 250 × 14 × $0.90 = $3,150. The second 20 pounds of nitrogen increases yield 11 bushels per acre, making the marginal return for the second $1,000 spent on fertilizer for soil 1 equal to $2,475. Other entries for returns per $1,000 increment of fertilizer on soils 1 and 2 are calculated in an analogous manner from data in Table 5.1. Similar data (not shown) on the response to alternative herbicide applications were used to estimate the return per $1,000 spent on herbicides. Finally, a cattle feeding budget was calculated to estimate returns per $1,000 invested in cattle feeding.

The equal marginal return principle indicates that the first three

Table 5.4 Returns per $1,000 of Operating Capital in Each of Six Alternative Uses

$1,000 Increment of Operating Capital	Marginal Return per $1,000 Invested in					
	Nitrogen Fertilizer on Soil 1	Nitrogen Fertilizer on Soil 2	Herbicides on Soil 1	Herbicides on Soil 2	Cattle Feeding	Certificate of Deposit
1	$3,150	$3,300	$2,000	$2,250	$1,550	$1,150
2	2,475	3,000	2,000	2,250	1,550	1,150
3	2,475	2,700	2,000	2,150	1,550	1,150
4	2,025	2,400	1,600	1,750	1,500	1,150
5	1,800	2,400	1,100	1,150	1,450	1,150
6	1,350	1,800	800	850	1,400	1,150
7	1,125	1,800	500	450	1,400	1,150
8	675	1,500	150	90	1,400	1,150
9	675	1,500	—	—	1,400	1,150
10	0	900	—	—	1,400	1,150

$1,000 units of capital should be invested in nitrogen fertilizer on soil 2, soil 1, and soil 2, respectively. If the operator had $10,000 of operating capital to allocate, the most profitable allocation would be $3,000 for nitrogen fertilizer on soil 1, $5,000 for fertilizer on soil 2, and $2,000 for herbicides on soil 2. While the return to the marginal unit of capital is not equal in all of its uses, the returns in each one are greater than or equal to the marginal return that could be obtained in any other use under consideration. The returns from this allocation are $26,400, an average return of $2.64 for each dollar invested. This is higher than the return that can be achieved with any other allocation of the $10,000 using the data in Table 5.4. It should be noted that the operator would not want to allocate money to any alternative having a lower marginal return than the $1,150 that could be obtained by investing in a certificate of deposit.

The equal marginal return principle is sometimes referred to as the opportunity cost principle. The opportunity cost is defined as the return that can be achieved for the use of a resource in its most profitable alternative use. In the event a resource is allocated in a suboptimal manner, the opportunity cost will exceed the return achieved. For example, if the above operator allocates $3,000 to nitrogen fertilizer on soil 1, $5,000 to fertlizer on soil 2, and $2,000 to cattle feeding, the opportunity cost of the last $2,000 is the $4,500 that could be obtained from using the money for herbicides on soil 2. The concept emphasizes what is being given up or the opportunity cost ($4,500 in this example) to obtain the return achieved ($3,100 for the $2,000 invested in cattle feeding). Thus, an allocation based on the equal marginal return principle has an opportunity cost that is less than or equal to the return expected for the marginal unit of resource.

ALTERNATIVE PRODUCT-PRODUCT FRONTIERS

The product-product frontier can have one of several general shapes. The shape indicates the relationship between the two enterprises. This section describes the alternative relationships, referred to as competitive, supplementary, and complementary, and the conditions that typically foster them in production agriculture.

Enterprise 1 is competitive with enterprise 2 when increasing the output of enterprise 1 results in a decrease in the amount of output produced by 2 with a given set of resources. This describes the general product-product relationship developed and analyzed in Figures 5.1 and 5.2. The definitions assume a given and fixed set of limited or scarce resources. In farm planning, these typically include the land, livestock facilities, machinery, and labor supply.

The competitive relationship can be drawn with an increasing marginal rate of product substitution as illustrated in the nitrogen fertilizer example of Figure 5.1 and in Figure 5.3a. The increasing marginal rate of product substitution results from the diminishing marginal productivity as more of the limited resource is used in producing one of the products. This is illustrated numerically in the data in Tables 5.1 and 5.2, and shown graphically in Figure 5.1.

Figure 5.3b depicts a competitive relationship with a constant marginal rate of product substitution. In this situation, each unit of the limited resource that is shifted from production of Y_2 to Y_1 results in the same reduction in output of Y_2 and increase in output of Y_1, that is, constant marginal physical product for the input in the production of both enterprises. This relationship may exist when two crops can be produced on the same soil type and there is no yield increase from growing the crops in rotation. For example, suppose wheat has a yield of 35 bushels per acre, barley produces 50 bushels per acre, and there are no yield effects of growing any combination of the two small grains. Each acre shifted from wheat to barley would reduce wheat output 35 bushels and increase barley output 50 bushels.

The competitive relationship is the one most commonly dealt with. Crop enterprises and livestock enterprises using pasture on cropable land compete for the use of that land during a specified year or production period. Crop and livestock enterprises typically compete for the use of labor during at least some periods of the year. Crop enterprises may compete for the use of machinery services during certain periods, while livestock enterprises may require the same limited facilities during part or all of the production period. Frequently, capital resources available to the farm are limited, and all enterprises must compete for use of the available supply.

The profit-maximizing combination of competitive enterprises is

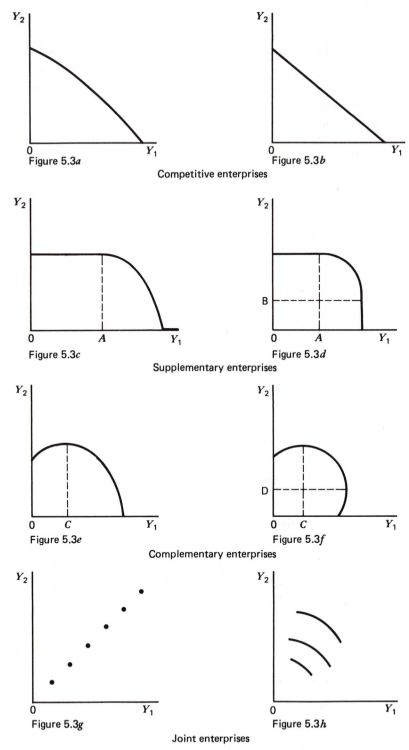

Figure 5.3. Alternative product-product relationships.

found by equating the MRPS with the inverse ratio of the net revenues as discussed in the previous section. A slight modification of that rule is required for enterprises having a constant marginal rate of product substitution, since no tangency can be defined between a linear product-product frontier and a linear isorevenue line. In this case, the most profitable alternative will occur at the maximum output of either Y_1 or Y_2. To determine which is the more profitable, remember that the profit-maximizing condition can be written as $\Delta Y_2/\Delta Y_1 = P_{Y_1}/P_{Y_2}$, or cross-multiplying, it can be written as $P_{Y_2}\Delta Y_2 = P_{Y_1}\Delta Y_1$. The slope of the product-product frontier is $\Delta Y_2/\Delta Y_1$. Given the net revenue, one can calculate $P_{Y_2}\Delta Y_2$ and $P_{Y_1}\Delta Y_1$. If $P_{Y_2}\Delta Y_2 > P_{Y_1}\Delta Y_1$, *producing all of Y_2* is more profitable, while $P_{Y_2}\Delta Y_2 < P_{Y_1}\Delta Y_1$ indicates producing all of Y_1 is profit maximizing. If $P_{Y_2}\Delta Y_2 = P_{Y_1}\Delta Y_1$, the slope of the product-product frontier and the isorevenue line are equal, making any combination of Y_1 and Y_2 equally profitable.

The second relationship illustrated in Figure 5.3 is supplementary. Enterprise 1 is supplementary to enterprise 2 when increasing the output of enterprise 1 has no effect on the amount of output produced by enterprise 2 with a given set of resources. This relationship is illustrated graphically in Figure 5.3c where an increase in Y_1 from O to OA does not change the amount of Y_2 that can be produced. Y_1 is supplementary to Y_2 within the range of O to OA. The two enterprises are competitive for larger outputs of Y_1. Figure 5.3d illustrates a supplementary relationship for both enterprises. Enterprise Y_1 is supplementary to Y_2 within the range of O to OA, and Y_2 is supplementary to Y_1 within the range of O to OB of Y_2.

The basis for this relationship is that the supplementary enterprise makes use of otherwise unused resources. Small poultry and livestock enterprises that utilize family labor and managerial skills that would otherwise be unutilized may be supplementary enterprises on small farms. Feeding lambs, feeder pigs, and cattle when labor and capital are not required for other uses may be supplementary enterprises. Two crops that can utilize the same equipment at different times of the year, such as the planting and harvesting of corn and soybeans in the Corn Belt, may be considered supplementary in terms of their use of machinery and labor.

All supplementary enterprises become competitive at some point. Increasing the level of a supplementary enterprise requires increasing amounts of the previously unused resource. When the supply of the unused resource is exhausted, the enterprises must compete for additional units of the resource. In agriculture, supplementary enterprises frequently become competitive at relatively low levels of output because of the management expertise required for planning, implementation, and control of the supplementary enterprises. While it may appear that underutilized facilities and labor exist on a farm, the management skills may not be available to add the enterprise. That is, the enterprises currently produced and those that might be supplementary in terms of using physical resources are competitive in their use of management ability, even at low levels of the supplementary enterprise. This implication seems to be frequently overlooked by advocates of more diversification in production agriculture.

The profit-maximizing combination of enterprises having supplementary ranges of output occurs where the MRPS equals the ratio of net revenues. That is, it occurs in the competitive range using the rule developed for competitive enterprises.

Complementary enterprise relationships are illustrated in Figures 5.3*e* and 5.3*f*. Enterprise 1 is complementary to enterprise 2 when increasing the output of enterprise 1 increases the amount of enterprise 2 that can be produced with a given set of resources. Y_1 is complementary to Y_2 within the range of O to OC of Y_1 in Figures 5.3*e* and 5.3*f*. The two enterprises are competitive for output levels to the right of C in Figure 5.3*e*. Figure 5.3*f* depicts two ranges of complementary relationships. Y_1 is complementary to Y_2 within the range of O to OC of Y_1, and Y_2 is complementary to Y_1 within the range of O to OD of Y_2. The two enterprises are competitive over the remainder of the range.

The basis for the complementary relationship is that one enterprise produces an input that enhances the output level of the other enterprise. The production of legumes to fix nitrogen in rotation with high nitrogen-using crops is a common example of a complementary relationship. During the first half of the twentieth century, a rotation of one year of legumes followed by three or four years of corn typically resulted in higher annual corn production on a given land area than planting the entire acreage to corn, even on very productive Corn Belt soils. In some cases, a higher proportion of legumes was required to reach output level OC of corn in Figure 5.3*e*. However, the development of methods to produce relatively inexpensive ammonia and nitrogen fertilizer from natural gas has enabled farmers to supply nitrogen fertilizer requirements without a legume crop and to avoid the complementary range. The development of resistant varieties, herbicides, insecticides, and fungicides has been effective in eliminating other examples where the use of rotations has been effective in controlling insect, disease, and weed problems that formerly resulted in complementary relationships between enterprises.

Notice that an increase in output per unit of limiting resource does not necessarily imply that these enterprises are complementary. This point is illustrated with hypothetical data in Table 5.5. The fixed set of resources includes 5 acres of land and the other resources required to produce any combination of 5 acres of crops A and B. The yields per acre are shown for alternative rotations of crops A and B. Notice that shifting from rotation 1 to rotation 2 reduces the acres in A by 20 percent and increases the yield of A 30 percent per acre. Inspection of the table indicates that when the proportionate increase in the yield exceeds the proportionate decrease in acreage, a complementary relationship exists. When yields of an enterprise

Table 5.5 Complementary Enterprises Illustrated

Rotation	Acres Produced of		Per Acre		Total for 5 Acres	
	Crop A	Crop B	A	B	A	B
1	5	0	100	—	500	0
2	4	1	130	14	520	14
3	3	2	180	13	540	26
4	2	3	190	12	380	36
5	1	4	200	11	200	44
6	0	5	—	10	0	50

increase by a smaller proportion than the reduction in acreage, the total production of the enterprise declines and a competitive relationship exists. Thus, enterprise A is complementary to B for rotation 1 through rotation 3. Enterprise A is competitive with enterprise B over the remainder of the range. The reader can verify this by graphing the product-product relationship using the total production figures in Table 5.5.

Complementary relationships become competitive at some level of output.[3] The profit-maximizing combination of complementary enterprises is found in the competitive range using the rule discussed earlier. It should be noted that the MRPS would be positive in the complementary range, that is, both ΔY_2 and ΔY_1 would be either positive or negative. Thus, it is important to be sure that the MRPS is in the competitive range when the signs are being ignored in finding the profit-maximizing level of outputs.

Joint products are illustrated in Figures 5.3g and 5.3h. Joint products result from the same production process. The production of meat and wool from sheep, meat and hide from cattle, and grain and stover from corn are examples of joint production. Figure 5.3g illustrates joint production in fixed proportions. In this case, no substitution between the two outputs is possible, such as the combination of meat and wool produced with one breed of sheep; the joint products can be considered as one output. But the joint products can often be produced in variable proportions. This relationship is illustrated in Figure 5.3h. For example, the combination of meat and wool produced can be varied by considering different

---

[3]This result follows from the assumption of diminishing marginal productivity for the use of the input produced. For example, the marginal productivity for the use of nitrogen produced by legumes in a legume-corn rotation will result in a competitive relationship between the two crops at some level of output.

breeds of sheep. In addition, the proportions of grain and straw produced with wheat can be varied by considering different varieties and fertilizer combinations. The profit-maximizing combination of joint products in variable proportions can be selected by applying the decision rule for competitive enterprises.

PRODUCT-PRODUCT FRONTIERS COMPOSED OF LINEAR SEGMENTS

An individual planning a farm business typically has neither the data available to estimate continuous production functions for the alternative enterprises and conditions nor the time required to obtain the data. Instead of estimating such functions and the corresponding product-product frontiers, data on the quantity of limited resources are obtained and used with enterprise budget data to evaluate alternative combinations of enterprises. This section develops the type of product-product frontier implicit in this approach—one composed of several straight-line segments—and describes the rule to use in finding the profit-maximizing combination of enterprises. This form of the product-product frontier is implicit in the planning procedures using budgeting described in Chapter 6.

Deriving the Product-Product Frontier

Enterprise budgets of the type developed in Chapter 4 can be treated as production functions that combine factors in fixed proportions. For example, the enterprise budget for corn in Table 4.12 combined 144 pounds of nitrogen and anhydrous ammonia, 60 pounds of phosphate, 45 pounds of potash, and the amounts of seed, pesticides, labor, and machinery services specified in the budget on one acre of land to produce 120 bushels of corn. Using double the amount of *all* inputs (288 pounds of nitrogen, 120 pounds of phosphate, 90 pounds of potash, and so on for the other inputs on two acres of land) would produce 240 bushels of corn. When the combination of inputs devoted to an enterprise is adjusted in fixed proportions, the marginal physical product per unit of the scarce or limited resource is constant. Consequently, the product-product frontier formed by one scarce or limited resource, such as land, is of the form shown in Figure 5.3b. Each unit of resource shifted from Y_2 (such as the corn above producing 120 bushels per acre) to Y_1 (such as soybeans on the same land that produce 35 bushels per acre) would decrease Y_2 by a fixed amount (120 bushels of corn) and increase the output of Y_1 by a given amount (35 bushels of soybeans). The limits imposed on the maximum production of two enterprises by several scarce or limited resources result in a product-product frontier composed of one or more straight-line segments. The following example, resulting in a product-product frontier composed of three line segments, illustrates the concept.

Table 5.6 Data to Prepare a Product-Product Frontier
Composed of Linear Segments

Item	Unit	Values per Acre		Values per Unit Output		Amount of Limited Resources Available
		Y_1	Y_2	Y_1	Y_2	
Yield	Units	100	50	1	1	
Price	Dol/unit	$ 2.00	$ 3.50	$2.00	$3.50	
Gross receipts	Dol	$200.00	$175.00	$2.00	$3.50	
Variable costs	Dol	$100.00	$ 50.00	$1.00	$1.00	
Net revenue	Dol	$100.00	$125.00	$1.00	$2.50	
Limited resources						
Land	Ac	1	1	0.01	0.02	300
Operating capital	Dol	16	8	0.16	0.16	$4,000
Labor 1	Hr	0	1	0	0.02	200
Labor 2	Hr	0.7	0.4	0.007	0.008	200

Suppose a manager wants to determine the combination of enterprises Y_1 and Y_2 that will maximize expected net returns to the land, family labor, and management available on the farm. Enterprise budgets for Y_1 and Y_2 are summarized in Table 5.6. Data in the first column indicate that enterprise Y_1 produces 100 units of output per acre and that the output is expected to sell for $2.00 per unit. Variable costs are $100, resulting in expected net revenue to the land, family labor, and management of $100 per acre. One acre of enterprise Y_1 requires one acre of land, $16 of operating capital, no labor during period 1, and 0.7 hour of labor during period 2. Comparable data are summarized for Y_2 in the second column of the table. The amount of each of the limited resources available on the farm is shown in the last column of Table 5.6. The limited resources that can be quantified include 300 acres of land, $4,000 of operating capital, and 200 hours of family labor in each period.

The product-product frontier is typically developed with the units of output rather than the number of acres on the axes. The cost, net revenue, and quantity of limited resources per unit of output are obtained by dividing the appropriate values per acre by the yield per acre. These values are shown for Y_1 and Y_2 in columns 3 and 4 of Table 5.6. For example, Y_1, which has a yield of 100 units per acre, requires 0.01 acre of land per unit of output. Other values are calculated in the same manner.

The requirements per unit of output and the corresponding amount of resources available can be used to develop an equation indicating the maximum amount of Y_1 and Y_2 that can be produced. For example, the following inequality represents the constraint imposed by land on the production of Y_1 and Y_2. This inequality states that 0.01 acres of land multiplied by the number of units of output Y_1 plus 0.02 acres of land multiplied by the number of units of Y_2 must be less than or equal to the 300 acres of land available.

$$0.01 \cdot Y_1 + 0.02 \cdot Y_2 \leq 300 \tag{5.12}$$

The inequality recognizes that production can be carried out without using all of the land (allowing some land to remain idle) if it is advantageous to do so, but that it is infeasible to consider a plan requiring more than 300 acres of land.

One inequality can be developed to represent the maximum limit placed on production of Y_1 and Y_2 by each of the limited resources. An equation to calculate the net revenue for alternative combinations of Y_1 and Y_2 can be developed using the net revenue per unit of Y_1 and Y_2. Thus, the problem being analyzed can be written as follows.

$$\text{Maximize } NR = 1Y_1 + 2.5Y_2 \tag{5.13}$$

$$\text{Subject to: } 0.01Y_1 + 0.02Y_2 \leq 300 \text{ acres land}$$
$$0.16Y_1 + 0.16Y_2 \leq \$4{,}000 \text{ operating capital}$$
$$0Y_1 + 0.02Y_2 \leq 200 \text{ hours labor in period 1}$$
$$0.007Y_1 + 0.008Y_2 \leq 200 \text{ hours labor in period 2} \tag{5.14}$$

This problem can be solved graphically. The four inequalities can be used to develop the product-product frontier for this problem. The net revenue equation can be used to form an isorevenue function in a manner analogous to the development in Figure 5.1.

The product-product frontier is to be developed with output of Y_2 on the vertical axis and Y_1 on the horizontal axis. Each of the constraints is treated as an equality representing the maximum level of Y_1 and Y_2 that can be produced with the amount of limited resource available. The land constraint can be written:

$$0.01Y_1 + 0.02Y_2 = 300 \tag{5.15}$$

To plot the equation on Figure 5.4, it is convenient to rewrite it in the following form.

$$Y_2 = 15{,}000 - 0.5Y_1 \tag{5.16}$$

This equation is plotted on Figure 5.4 with an intercept of 15,000 units of Y_2. The slope of minus 0.5 indicates that 0.5 unit of Y_2 must be given up for each unit increase in Y_1. The line indicates that either 15,000 units of Y_2, or 30,000 units of Y_1, or any combination on a straight line connecting the two points can be produced with the 300 acres of land available.

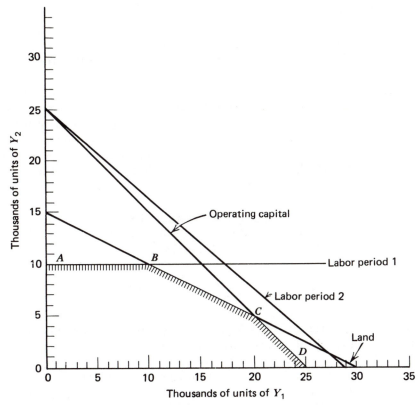

Figure 5.4. A product-product frontier composed of linear segments.

The land constraint divides Figure 5.4 into two parts. Any combination of Y_1 and Y_2 on or below the line can be produced with the 300 acres. However, the farm does not have sufficient land to produce any combination of Y_1 and Y_2 above the land constraint. It should also be noted that the marginal rate of product substitution is a constant 2 (ignoring the negative sign) on this constraint.

Each of the four resource constraints must be considered to obtain the product-product frontier for the problem. The constraints for operating capital, labor in period 1, and labor in period 2 can be written as:

$$Y_2 = 25,000 - 1Y_1$$
$$Y_2 = 10,000 - 0Y_1$$
$$Y_2 = 25,000 - 0.875Y_1 \tag{5.17}$$

The three equations are plotted on Figure 5.4. The combination of Y_1 and Y_2 selected may not use more of any of the limited resources than the amount available. Thus, the three line segments form the product-product frontier *ABCD* in this example. Line segment *AB* is part of the constraint imposed by period 1 labor, segment *BC* is part of the land constraint, and *CD* is imposed by the capital constraint. Notice that the fourth constraint, labor in period 2, does not form part of the product-product frontier and can be ignored.

The slope of the product-product frontier in Figure 5.4 indicates that more Y_2 must be given up to produce another unit of Y_1 as one increases the production of Y_1. The line segment *AB* has a slope of zero, indicating that Y_1 is supplementary to Y_2 within the range of 10,000 units of Y_2 and zero units of Y_1 (10,000, 0) to (10,000, 10,000). The slope changes to minus 0.5 from (10,000, 10,000) to (5,000, 20,000), and the slope of *CD* is minus 1 from (5,000, 20,000) to (0, 25,000).

Selecting the Product Mix

The net revenue function is used to obtain the profit-maximizing combination of Y_1 and Y_2. The net revenue function can be written as:

$$Y_2 = \frac{NR}{2.5} - \frac{1}{2.5}Y_1 = \frac{NR}{2.5} - 0.4Y_1 \tag{5.18}$$

The intercept is given by $NR/2.5$ and the slope is minus 0.4. The level of *NR* is adjusted until the isorevenue line just touches the product-product frontier *ABCD*. This occurs at point *B* in Figure 5.5. The solution has a net revenue of $35,000, producing 10,000 units of Y_2 and 10,000 units of Y_1. Using the yield per acre we find this solution involves producing 100 acres of Y_1 at a yield of 100 units per acre and 200 acres of Y_2 at a yield of 50 units per acre.

A tangency does not exist between the product-product frontier and the isorevenue line because the product-product frontier is not smooth at point *B*. However, the slope of the product-product frontier to the left of point *B* is less than the slope of the isorevenue line, which is less than the slope of the product-product frontier to the right of point *B*. Thus, the profit maximizing condition can be written as

$$\left[\frac{\Delta Y_2}{\Delta Y_1}\right]_L \geq -\frac{P_{Y_1}}{P_{Y_2}} \geq \left[\frac{\Delta Y_2}{\Delta Y_1}\right]_R \tag{5.19}$$

where the subscripts *L* and *R* refer to the left and right of the profit-maximizing combination.

The profit-maximizing condition can be analyzed by considering what it implies on each side of the profit-maximizing combination. Consider the first two terms to analyze the left side. The condition states that

$$\left[\frac{\Delta Y_2}{\Delta Y_1}\right]_L \geq -\frac{P_{Y_1}}{P_{Y_2}} \tag{5.20}$$

Cross-multiplying, we obtain

$$P_{Y_2} \cdot \Delta Y_{2_L} \geq -P_{Y_1} \cdot \Delta Y_{1_L} \qquad (5.21)$$

Since ΔY_2 is typically negative, we can multiply through by minus 1 to get rid of the minus sign and obtain

$$P_{Y_2} \cdot \Delta Y_{2_L} \leq P_{Y_1} \cdot \Delta Y_{1_L} \qquad (5.22)$$

This condition indicates that the change in the value of product 2 must be less than the change in the value of product 1 to the left of the profit-maximizing combination. Since ΔY_{2_L} is 0 for each one-unit change in ΔY_{1_L} in our example, the specific values for the above equation are:

$$2.5 \cdot 0 \leq 1.0 \cdot 1 \qquad (5.23)$$

or

$$0 < 1$$

Consider the condition to the right of the profit-maximizing point in the same manner.

$$-\frac{P_{Y_1}}{P_{Y_2}} \geq \left[\frac{\Delta Y_2}{\Delta Y_1} \right]_R \qquad (5.24)$$

Cross multiplying, we obtain

$$-P_{Y_1} \cdot \Delta Y_{1_R} \geq P_{Y_2} \cdot \Delta Y_{2_R} \qquad (5.25)$$

Remembering that ΔY_{2_R} is typically negative, we can multiply through by minus 1 and treat all quantities as positive values.

$$P_{Y_1} \cdot \Delta Y_{1_R} \leq P_{Y_2} \cdot \Delta Y_{2_R} \qquad (5.26)$$

This condition states that the value of producing an additional unit of Y_1 must be less than the value of the amount of Y_2 given up to the right of the profit-maximizing combination. The values to the right of the profit-maximizing solution in Figure 5.5 are

$$1 \cdot 1 \leq 2.5 \cdot 0.5 \qquad (5.27)$$

Thus, moving to the right of the profit-maximizing combination in Figure 5.5 would decrease the amount of Y_2 one-half unit for each unit of Y_1 added and reduce net revenue \$0.25 for each unit of Y_1 added.

The equilibrium condition for profit maximization can also be analyzed by remembering that the MRPS can be interpreted as a ratio of marginal physical products. The profit-maximizing condition to the left of the profit maximizing combination can be written as

$$\left[\frac{\frac{\Delta Y_2}{\Delta X_1}}{\frac{\Delta Y_1}{\Delta X_1}} \right]_L \geq -\frac{P_{Y_1}}{P_{Y_2}} \tag{5.28}$$

where ΔX_1 refers to a one-unit change in labor in the first period. Cross-multiplying results in

$$P_{Y_2} \cdot \frac{\Delta Y_{2_L}}{\Delta X_1} \geq -P_{Y_1} \cdot \frac{\Delta Y_{1_L}}{\Delta X_1} \tag{5.29}$$

This can be rewritten as

$$VMP_{X_1 \cdot Y_{2_L}} \geq - VMP_{X_1 \cdot Y_{1_L}} \tag{5.30}$$

The marginal physical product of X_1 can be obtained from Table 5.6. Output Y_2 requires one unit of labor in period 1, the yield is 50 units per acre, and $\frac{\Delta Y_2}{\Delta X_1}$ is 50. Output Y_1 requires 0 units of labor in period 1, and $\frac{\Delta Y_1}{\Delta X_1}$ is 0. Thus, the above relationship can be written as

$$2.5 \cdot 50 \geq -1.0 \cdot 0 \tag{5.31}$$

This states that the value of labor during the first period used in the production of Y_2 must be greater than the value of the same resource used in the production of Y_1 to the left of the equilibrium point.

The condition for profit maximization to the right of the profit-maximizing combination can be written as

$$-\frac{P_{Y_1}}{P_{Y_2}} \geq \left[\frac{\frac{\Delta Y_2}{\Delta X_2}}{\frac{\Delta Y_1}{\Delta X_2}} \right]_R \tag{5.32}$$

and cross-multiplying, we obtain

$$-P_{Y_1} \cdot \frac{\Delta Y_{1_R}}{\Delta X_2} \geq P_{Y_2} \cdot \frac{\Delta Y_{2_R}}{\Delta X_2} \tag{5.33}$$

Remembering that ΔY_2 is negative, we can multiply by minus 1 and treat all values as positive.

$$VMP_{X_2 \cdot Y_{1_R}} \leq VMP_{X_2 \cdot Y_{2_R}} \tag{5.34}$$

The value of MPP_{X_2} for one unit of X_2 (land) is 100 for Y_1 and 50 for Y_2. Thus, the numerical relationship for this example is

$$1.0 \cdot 100 \leq 2.5 \cdot 50 \tag{5.35}$$

and $\$100 < \$125$.

This condition states that the value of the marginal product of the limited resource forming the product-product frontier to the right of the profit-maximizing combination must be less for Y_1 than Y_2. If the opposite was true, it would be profitable to increase production of Y_1 and decrease Y_2.

The results of this example can be summarized verbally as follows. Two resources were found to limit the level of net returns. The value of the marginal product of the resource forming the product-product frontier to the left of the profit-maximizing combination must be less for Y_2 than Y_1, while the value of the marginal product for the resource forming the frontier to the right of the profit-maximizing combination must be less for Y_1 than Y_2. Another point that has not been emphasized but follows from inspecting Figure 5.5 is that minor changes in the amount of the other fixed resources (labor in period 2 and operating capital) would not affect the profit-maximizing combination.

The mechanics of the procedure to find the most profitable combination of enterprises can be summarized as: (1) identify the most limiting resources (in our example, labor in period 1 and land); (2) allocate as much of each limiting resource to the enterprise having the highest value of the marginal product for the use of that resource (Y_2 for labor in period 1 and Y_1 for land); and (3) ignore other limited resources in making this allocation.

These results suggest a method of applying the equal marginal return principle in choosing more profitable combinations of enterprises for a farm business. Data obtained on the available resources for a farm can be evaluated, and a judgment can be made on what is the most limiting resource or resources in developing a combination of enterprises. The enterprise budget data can be used to calculate net returns to the fixed resources, and an enterprise combination can be selected that maximizes net returns to the most limiting resources. Because many factors typically limit the most profitable plan for a farm business, mathematical programming (discussed in Chapter 10) is usually required to apply these procedures in a precise manner. However, this process can be used in a less precise manner

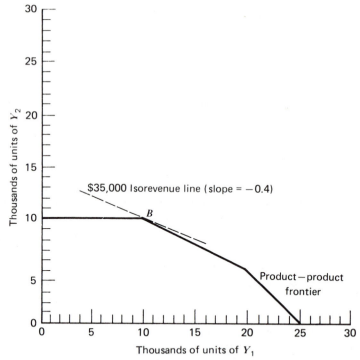

Figure 5.5. Profit maximizing solution for a product-product frontier composed of linear segments.

in whole-farm planning with budgeting. The application to an actual problem is illustrated in Chapter 6, in the section on selecting a long-run plan.

INCLUDING ADDITIONAL CONSIDERATIONS

The conceptual product-product model is sometimes dismissed as being of little value in guiding decision-making because some important considerations have not been included in the analysis. The purpose of this section is to illustrate how each of several additional considerations common in farm planning applications can be analyzed with the product-product model. In each case, the conceptual model indicates the direction of the adjustment in resource use. The additional considerations discussed here are differences in financing for the two outputs, marketing alternatives, and inclusion of income tax in the analysis.

Lenders may charge a higher rate of interest for credit used to finance enterprises that are perceived to be more risky than others. The financing cost is normally included as part of the variable cost in the enterprise budget, thus reducing the net revenue for those enterprises having high financing cost. For example, assume that enterprises Y_1 and Y_2 require C_1 and C_2 dollars of capital per unit of output, respectively. Suppose the financing charge (r) is increased to r' for enterprise Y_2. The net revenue ratio for the two outputs would be changed from

$$\frac{P_{Y_1} - rC_1}{P_{Y_2} - rC_2} \quad \text{to} \quad \frac{P_{Y_1} - rC_1}{P_{Y_2} - r'C_2}$$

This change reduces the denominator. Since the slope of the isorevenue line is negative, it rotates the isorevenue line to the right. This increases the relative profitability of Y_1 and tends to increase the amount of Y_1 included in the profit-maximizing solution.

Alternative methods of marketing and pricing products are frequently considered. Assume for the solution analyzed in Figure 5.5 that the operator has the opportunity either to build storage or to process the product which enables selling at a higher price. The additional costs of storage and the higher price can be estimated and added to the enterprise budget data to obtain a net revenue figure for the combined production and marketing strategy. For example, suppose that storage for product 1 in Figure 5.5 increases the variable cost of Y_1 by \$0.25 per unit and increases the price received to \$3.00. This would increase net revenue for Y_1 to \$1.75. The resulting net revenue equation would be

$$NR = 1.75 \cdot Y_1 + 2.5 \cdot Y_2 \tag{5.36}$$

The isorevenue equation could be written as

$$Y_2 = \frac{NR}{2.5} - \frac{1.75}{2.5}Y_1 = \frac{NR}{2.5} - 0.7Y_1 \tag{5.37}$$

The higher net revenue changes the slope of the isorevenue line and results in a new solution with net revenue of \$47,500. The most profitable combination of outputs shifts to 5,000 (or 100 acres) of Y_2 and 20,000 (or 200 acres) of Y_1. Thus, a change that affects the costs and/or price will adjust the net revenue per unit of output. The output that becomes relatively more profitable (Y_1 in this example) will tend to be increased.

Income tax effects can also be considered. Suppose the net revenue from both enterprises are taxed in the same manner. For example, assume that the revenues from both Y_1 and Y_2 are considered ordinary income and that the farmer is in the 50 percent tax bracket. The after-tax net revenues with the marketing alternative added are $\bar{P}_{Y_1} = 1.75 (1 - 0.5) = \$0.875$ and $\bar{P}_{Y_2} = 2.50 (1 - 0.5) = \$1.25$. The ratio of net revenue is

$$-\frac{\bar{P}_{Y_1}}{\bar{P}_{Y_2}} = -\frac{\$0.875}{1.25} \quad \text{or} \quad -0.7 \tag{5.38}$$

The slope of the isorevenue line is not changed, and the profit-maximizing enterprise combination remains at 5,000 units of Y_2 and 20,000 units of Y_1. However, the net revenue on an after-tax basis has been reduced by 50 percent.

While consideration of income taxes that affect the net revenue for both outputs in the same manner does not affect the most profitable combination of outputs, many tax provisions affect the net revenue received from alternative methods of producing one enterprise, as well as alternative enterprises, in different proportions. Differences in depreciation and investment credit may affect the after-tax costs of producing some products more than others. Likewise, part of the revenues from some enterprises may be considered capital gains that are taxed at a lower rate. For example, suppose the $1.75 net revenue per unit of Y_1 qualifies as capital gain income, but the $2.50 net revenue per unit of Y_2 is taxed as ordinary income. If we assume that 40 percent of capital gains income is subject to tax and the marginal tax rate is 50 percent, the after-tax net revenues are (0.6) (1.75) + (1.75) (0.4) (0.5) = $1.40 for Y_1 and $2.50 (0.5) = $1.25 for Y_2. The resulting isorevenue equation can be written as

$$Y_2 = \frac{35,000}{1.25} - 1.12Y_1 \tag{5.39}$$

With this price ratio, the after-tax net revenue is maximized by producing 25,000 units of Y_1 and no Y_2. The production of 25,000 units of Y_1 produces $35,000 of net revenue.

These examples illustrate that a wide variety of additional considerations can be analyzed with the product-product model. These considerations typically either affect the slope of the product-product frontier or the slope of the isorevenue line. In these cases, the simple product-product model can be used to analyze the effect of these additional considerations on the most profitable allocation of resources.

Summary

The product-product model of production theory provides the conceptual framework for whole-farm planning. The product-product model indicates the maximum possible production levels of two products for a given set of resources; this relationship is derived from the production functions for the two products. The optimal combination of products to produce, given a specific set and amount of limited inputs, is determined by equating the marginal rate of product substitution to the net revenue ratio for the two products. Note that the net revenue ratio recognizes potential differences in financing costs or income taxes for the products. An equivalent way of stating this optimizing condition is to equate the value of the marginal product in the two production processes, where the value of the marginal product is the marginal physical product of the inputs used in the production process times the net revenue for the specific product. This principle states that the returns from scarce or limited resources are maximized when the value added by the last unit of the resource in each of its alternative uses is the same. The equal marginal return principle can be used in making decisions about the allocation of labor, capital, or other inputs to alternative enterprises and the production of specific products.

The technical product-product relationship can exhibit various characteristics and shapes. The product-product relationship may be competitive in that to obtain a larger quantity of one product requires giving up part of an alternative product. This competitive relationship may exhibit increasing or constant marginal rates of product substitution. A complementary product-product relationship indicates that the production of one product results in increased production of the alternative; this relationship is rather rare in agriculture. Supplementarity occurs where the production of one product results in no increase or decrease in the quantity of the other product that can be produced; this relationship exists when the alternative product can make use of otherwise unused resources. Joint products are produced together in either fixed or variable proportions. In general, the optimal product mix will always occur in the competitive range of product-product substitution relationships.

Although much of the theory of production is developed assuming that the data are available to estimate continuous production functions, most real world applications utilize limited data that result in linear production functions and product-product relationships. The same basic concepts apply to linear or continuous product-product relationships; the only change in the logic is in the mathematics involved. In essence, with a linearly segmented product-product function, the optimal product mix cannot occur at a point of tangency between the curve reflecting the marginal rate of product substitution and the net revenue line, but must instead occur at a point where these two lines touch. This will always be a corner point on the product-product transformation curve.

Questions and Problems

1. Describe the product-product relationship. How is this relationship derived?
2. What is the marginal rate of product substitution? How is this concept used in determining the optimal or most profitable combination of products or enterprises to produce?

3. How is the expansion path defined? What is the economic usefulness and meaning of the expansion path?
4. What is the value of the marginal product? How can this concept be used in determining the optimal combination of products to produce?
5. State the equal marginal return principle. What is the economic usefulness and meaning of this principle?
6. What are the four different types of product-product relationships that can exist?
7. Collect data from an agronomist or animal scientist and illustrate the use of the equal marginal return principle in choosing the optimal combination of products to produce.
8. Identify two products that you expect might exhibit a complementary product-product relationship. Discuss the concept of complementarity.
9. Identify an example of joint products. Discuss the joint product concept.
10. Why is the optimal product combination always in the competitive range of the product-product relationship?
11. Identify and discuss the differences in the product-product optimizing conditions for a continuous product transformation curve compared to a linearly segmented curve.
12. Collect the appropriate budget data and develop a linearly segmented product-product frontier. Using relevant prices and net revenues for the products, determine the optimal combination of products to produce.
13. How can you incorporate finance charges and taxes in the product-product optimizing conditions? What impact will these factors have on the optimal product mix?

Further Reading

Doll, John P., and Frank Orazem. *Production Economics: Theory with Applications.* New York: John Wiley, 1984, Chapter 5.

Doll, John P., V. James Rhodes, and Jerry G. West. *Economics of Agricultural Production, Markets and Policy.* Homewood, Ill.: Richard D. Irwin, 1968, Chapter 6.

Grossack, Irvin Millman, and David Dale Martin. *Managerial Economics.* Boston: Little, Brown and Company, 1973, Chapters 5 and 6.

Heady, E. O. *Economics of Agricultural Production and Resource Use.* Englewood Cliffs, N.J.: Prentice-Hall, 1952, Chapters 7 and 8.

Longworth, John W., and Kenneth M. Menz. "Activity Analysis: Bridging the Gap Between Production Economics Theory and Practical Farm Management Procedures." *Review of Marketing and Agricultural Economics* 48 (April 1980): 7–20.

6

COMPUTATIONAL
PROCEDURES FOR
WHOLE-FARM PLANNING

CHAPTER CONTENTS

The successful manager of any activity must be able to complete the planning function. This is true of an artist painting a landscape, a military leader attempting to gain control of a certain geographic area, and the manager of a business firm. The individual managing the activity must establish goals for the activity and develop a plan to achieve the goals.

This chapter discusses a whole-farm planning procedure that managers can use to identify and evaluate alternative production, market, and financial plans. Budgeting techniques that integrate the tools discussed in Chapters 2 through 5 are developed to complete the analysis. This chapter emphasizes identification and evaluation of production plans. Some attention will be given to the method of marketing the products and financing the plan, but more thorough integration of market and financial planning considerations are reserved for Chapters 7 and 8.

AN OVERVIEW OF WHOLE-FARM PLANNING

The purpose of using a whole-farm planning procedure is to provide a means of systematically evaluating alternative farm organizations. It includes a means of projecting the more quantifiable aspects of alternative production, market, and financial plans, including the amounts of land, labor, and capital required, and the income expected. Completing the planning process also provides an opportunity for the manager to more thoroughly evaluate the less quantifiable aspects, including the management requirements, working conditions, seasonal work loads, and the stress related to high-risk and large debt levels. Completion of the whole-farm planning process provides comparable analyses of alternative plans, thereby enabling the manager to select the plan that appears to be the most consistent with the goals specified.

The major activities included in the planning function were identified in Table 1.1 of Chapter 1. Establishing long-run goals is perhaps the first of these activities to consider. Then attention turns to identifying the important aspects of the economic, social, and political environment within which the farmer operates, because these considerations set many conditions and constraints on the planning process. An overall plan to achieve the long-run goals within these constraints must be established. After a long-run plan has been developed, the planning process turns to the development of both goals and plans for the intermediate and short-run periods. Ideally, the planning procedures used should consider the effect of the interaction of production, marketing, and financing on the profitability of the business. Furthermore, the procedures should consider the impact of a change in one part of the plan on the profitability of the total business.

The activities discussed in Chapter 1 have been combined into a four-step procedure for whole-farm planning. The four steps are:

1. Specify the goals and inventory the available resources and planning restrictions.

2. Select a long-run plan.

3. Develop a short-run plan.

4. Develop intermediate-run plans.

The first step in the process requires the individual(s) operating the farm to specify the goals and objectives to be achieved over the short and longer run. The remainder of Step 1 is concerned with estimating the land, labor, and capital resources which the operator has to use, the restrictions placed on planning by the operator, and the social, political, and economic environment within which the farm must operate.

The second step is concerned with selecting a general long-run plan which the operator can use to achieve specific goals over a period of several years. The long-run plan specifies the general crop and livestock enterprises that are selected to achieve the operator's goals. The planning procedures used are rather general, partial budgeting procedures.

Detailed plans for the coming year are developed in the third step. Completion of this step requires development of production, market, and financial plans. The method of analysis includes the preparation of a projected monthly cash flow, a projected income statement, and a projected balance sheet.

The fourth step is concerned with developing specific, but somewhat less detailed, plans for the second, third, and perhaps fourth years into the future. The procedures used here typically include projected annual cash flows, projected income statements, and projected balance sheets for each of the years planned.

Following the four-step procedure provides a rather precise method for systematically evaluating alternative farm plans. Projecting the financial statements indicated in Steps 3 and 4 provides a specific picture of the expected effect of the plan on profitability, liquidity, and solvency of the business over the next several years. It may become apparent in moving through the steps that the goals cannot be achieved with the resources available and the restrictions placed on the planning process. In these cases, it is necessary to return to Step 1 and to reconsider the goals or resources and restrictions. It may then be necessary to revise the long-run plan selected and to proceed to Steps 3 and 4 again. Thus, completion of the four steps may be an iterative procedure rather than a simple progression.

Completion of all four steps is required only when significant changes are being contemplated in the business. Such changes typically occur when the farm is started, when a major expansion is made, when a partner is brought into the business, when the size of the business is reduced in anticipation of retirement, and when a change is needed to increase the profitability of the farm. However, Step 3 should be completed annually, and, in doing so, the operator may decide to go through one or more of the other steps as well. Thus, knowledge of the four-step procedure is useful each year in completing the planning process.

The four-step, whole-farm planning procedure described can be completed using several alternative analytical procedures. Budgeting procedures are emphasized here because they are the simplest and the most universally used farm planning procedures. Budgeting procedures also provide the basis to understand the application of linear programming and other more mathematically sophisticated techniques that are developed later.

GOALS, AVAILABLE RESOURCES, AND PLANNING RESTRICTIONS

The purpose of this step is to gather data on the goals, resources, and restrictions to use in the planning process. While this is the most difficult of the four steps to discuss in a specific manner, careful attention to this data-gathering activity is important in that it provides the basis for whole-farm planning. In addition to obtaining data on the goals and objectives, the inventory should include data on the land, labor, tangible working assets, capital position of the business, the institutional restrictions limiting the planning of the business, and managerial capacity of the operator.

Specifying the Goals and Objectives

An appropriate starting point for the planning process is to specify the goals and objectives of the unit managing the operation.[1] Clearly delineating the objectives to be achieved during the current year and over the longer run provides a basis to judge the acceptability of alternative plans. Without a set of longer run goals, the process may fall prey to the old adage, "If you don't know where you are going, any road will take you there."

The range of goals and objectives to be considered depends on the composition of the management unit. Many farms are operated as family businesses, which means that personal/family, as well as business goals and objectives, must be considered. The discussion here proceeds in this broad context of goal specification. The student should recognize that some of the personal/family considerations may be relatively unimportant for large commercial farms, while some of the business goals may be given less attention by those units emphasizing the way of life aspects of farming. For family operations, both the husband and wife should make separate lists of the things they want to achieve over the next year, the next five years, and the longer-run. In the case of a partnership, each partner and spouse should participate in this activity.

The goals should relate to both the personal/family and the business. These goals should be developed given the current status of the operation

---

[1]A distinction is frequently made between goals and objectives. Goals are usually considered to be more general guidelines of what is to be accomplished over a long period of time. The term "objective" is often reserved for short-run and more narrowly focused "targets" which, when accomplished, are useful in achieving the longer run goals. The terms are used largely interchangeably in this discussion.

and the business environment. Some of the categories that are usually important are:

	Next Year	*Within 5 Years*	*Beyond 5 Years*
Personal/Family Goals and Objectives			
Money withdrawn for family living			
Minimum required			
Other items			
Social and community			
Other			
Business Goals and Objectives			
Production			
Marketing			
Financial			
Annual net income			
Net worth level			
Minimum net capital ratio			
Minimum current capital ratio			

An appropriate way to initiate this process is to have each individual develop a list of personal/family goals and to compare the completed lists. An effort should be made to combine the lists and to rank the items in terms of the importance that the goal be achieved within the time frame indicated. This can be an effective means of determining the range in monetary withdrawals that may be placed on the business and the importance of these withdrawals to the individuals involved. It also provides information on the amount of time and effort to be devoted to nonbusiness activities and suggests how this may affect the business.

Then consider the business goals that are to be achieved within the next year, five years, and over a longer period. Plans for expansion and the type of expansion should be noted under the production and marketing areas. Other important goals to be considered at this point fall within the financial category. They include setting goals on net income and net worth. It may also be useful to set limits on borrowing stated as a minimum net capital ratio and a minimum current capital ratio.[2] The ratios specified imply something about the risk which those involved in the decision-making are willing to accept. It may also be important to control other aspects of risk by placing limits on the production and marketing plan. If so, these

---

[2]The current capital ratio is defined as total current assets divided by total current liabilities, as defined in Chapter 2. The net capital ratio is the total assets divided by total liabilities, the reciprocal of the debt/asset ratio defined in Chapter 2.

considerations should be included here. As in the personal/financial goals, the business goals should be ranked in terms of the importance that the goal be achieved within each of the time periods.

Completing the process highlights the areas of agreement and the divergence in personal, family, and business goals. While it is seldom possible to satisfy all goals, recognizing the differences and the conflicts at the start of the planning process may make it possible to satisfy a larger proportion of the important goals over time and to develop a useful plan for the farm business.

The Land Inventory

The land inventory is usually taken with two farm maps. The first map is a drawing approximately to scale showing the location and size of the physical features of the farm. The physical features include the size and shape of cropland areas, pasture land, woodland, lakes, ponds, and streams. The map should also show the location of man-made features, including the farmstead, fence lines, the location of contours, grass waterways, and irrigation wells. It may also be useful to indicate the location of crops presently being produced on the farm. Showing the current crop is particularly important if cropland is being used for perennial crops, such as alfalfa, orchards, and vineyards.

The second map required in an inventory of the land resources shows the soil types and the land capability classes. Information on soil types and land capability classes is important in developing the appropriate systems of crop production and in deciding on the appropriate arrangement of fields for the reorganized farming business. Capability grouping of soils prepared by the Soil Conservation Service of the U.S. Department of Agriculture is normally available for a farm from the local Soil Conservation Service office. The soils are grouped according to the limitations for field crop production and the way they respond to treatment. The capability system groups soils at three levels: the capability class, subclass, and unit. Capability classes are designated by Roman numerals I through VIII. Classes I through IV are considered suitable for cropping, with Class I soils having few limitations that restrict their use, while Class IV soils have severe restrictions that reduce their use for crops. Classes V through VII are soils limited to use as pasture, woodland, or wildlife, and Class VIII soils have limitations restricting their use to recreation, wildlife, water supply, or aesthetic purposes. Capability subclasses, designated by adding a small letter to the class numeral, indicate the major factor limiting the soil to the class. Subclasses commonly listed are "e" for erosion; "w" for water in or on the soil; "s" for shallow, droughty, or stony; and "c" for a climate that is too cold or too dry. Capability units, soil groups within the subclasses, are designated by adding an Arabic numeral to the subclass. In one symbol, such as III e-3, the Roman numeral designates the capability class or degree of limitation; the small letter indicates the subclass or kind of limitation; and the Arabic number identifies the severity of the limitation.

Specialists indicate that soils having the same class, subclass, and capability unit are similar enough to be suited to the same crops and pasture plants. Furthermore, they can be expected to require the same management and to have similar productivity and other responses to management practices.

Identifying the class, subclass, and unit for the major soil types on the farm is important for two major reasons. One, such data are useful in deciding on the field arrangement, and, two, the data are quite important in developing crop enterprise budgets. The size and shape of fields must be determined considering both the physical characteristics of the soils and the size required to efficiently perform tillage, planting, and harvesting operations. Data on the capability classes, subclasses, and units of soils help in developing reasonably homogeneous areas and the arrangement of fields for the farm.

One of the first decisions in developing crop enterprise budgets for farming operations is to combine the soils into a reasonable number of groups for farm planning purposes. The general rule is to combine soils that produce about the same yields with the same machinery operations and the same amount of variable inputs. Soils meeting this criterion will have the same cost and return estimates for a given crop. Data on the class, subclass, and unit provide a useful starting point to combine soils into groups meeting the above definition.

The Farm Labor Supply

The farm labor supply is composed of two major parts; the operator and family labor; and hired labor available to the farm. An inventory of the operator and family labor should consider the age and skill of both the operator and other adult workers in the family. The year should be divided into seasons based on the peak labor requirements for the major crop and livestock enterprises. Then the operator can estimate the hours of labor which each member of the family can provide by season of the year. Data on the approximate amount of family labor available for farm production activities can be estimated in this way. It is also important to distinguish between time that is available for direct labor and that used for management and overhead activities of the business. Farm planning procedures frequently estimate only the direct labor. Thus, it is important to be somewhat conservative in estimating the number of hours the operator can devote to direct labor tasks associated with crop and livestock production.

The willingness to use hired labor as well as labor availability and cost should also be determined. If the operator is willing to use hired labor, data on the availability by the day, week, and month should be obtained. In some areas, it may be necessary to hire labor on an annual basis to have labor when it is needed. Estimates of the wage rate, perquisites, bonuses, and incentives required to hire each type of labor and the tasks the labor could perform should also be obtained.

Tangible Working Assets

The major tangible working assets to be inventoried are the machinery, buildings, and breeding stock that are available on the farm. The purpose is to inventory the physical aspects—number, size, condition—of these assets that are of importance in farm planning. Their monetary value is included in the inventory of the capital position. The emphasis should be placed on the major items of machinery including the power units, harvesting equipment, planting equipment, and the major items of tillage and livestock equipment. It is particularly important to note if the machinery is inadequate for the current farm size, adequate, or more than adequate permitting an expanded operation without purchasing additional machinery. The availability of custom-hired machinery operations in the area should also be investigated to determine if the operator can hire a custom operator to perform operations for which machinery is not available on the farm. The space for livestock production, machinery storage, feed storage, and crop storage should be noted in inventorying buildings. Examples of the data to be obtained are a stanchion barn for 64 dairy cows, storage for 180 tons of dry forage, a concrete-stave silo 20′ × 50′, three grain bins with capacity of 8,000 bushels each, and a 40′ × 80′ pole building for machinery and supply storage. An inventory of the breeding stock including the number and age of dairy cows, beef cows, sows, ewes, and the related replacement animals should be made. It is also important to note the productive level of dairy cows for use in farm planning.

Capital Position

An inventory of the capital position includes two phases. The first involves obtaining a current balance sheet for the farm operator and the supporting information on the inventory of depreciable and nondepreciable assets. The balance sheet discussed in Chapter 2 lists the value of the current, intermediate, and long-term assets of the business. It also indicates the liabilities for which the operator is obligated. The balance sheet can be used to determine the liquidity, solvency, and use of debt and equity capital to finance the business. Thus, the balance sheet should indicate sources of equity capital which the operator has available for reinvestment in the business.

It is also useful to estimate the operator's ability and willingness to acquire debt capital. The ability to acquire debt capital depends on the operator's financial position as reflected on the balance sheet, the profitability of the operation, and the operator's credit rating. The proportion of the cost of current, intermediate, and long-term assets that can be borrowed varies from time to time depending on credit market conditions. The proportion that will be loaned also varies by type of farm. Many agricultural lenders prefer a current asset/current liability ratio of 1.5 or greater and a total asset/total liability ratio of 2.0 or more. The limits on availability of credit imposed by financial institutions can normally be obtained by discussing this matter with a major agricultural lender in the area.

The operator may wish to impose more severe restrictions on borrowing than those imposed by lending institutions. The minimum acceptable net capital ratio provided in the statement of business goals indicates the extent to which the operator(s) want to ration the use of debt capital. In the event a definite response was not obtained to the net capital ratio in the discussion of goals, it should be obtained here. Obviously, imposing a more severe restriction on credit use may limit the opportunities available for reorganization. But ignoring an internal capital rationing limit may result in evaluating a number of plans that are not acceptable because they require more debt capital than the operator is willing to borrow.

Institutional Factors

Many laws and operating procedures of regulatory agencies affect the best organization for the farm business. These can be discussed under the headings of government programs, marketing orders, zoning regulations, environmental restrictions, and long-term leasing arrangements. At times and for certain commodities, government programs designed to support farm income represent the most important institutional factor to consider in organizing the farm business. Limits on the number of acres of crops that can be planted and the corresponding support prices for products that are produced are extremely important considerations in the organization of the farm business. The provisions that apply vary from commodity to commodity and from year to year, making a thorough discussion of this topic impossible in the available space. However, data on the current provisions can be obtained from the local Agricultural Stabilization and Conservation Service Offices. In addition to affecting the production alternatives and the average price, these programs frequently provide direct payments to farmers in the event of low prices. Such programs may, therefore, affect the market risk. Methods to consider this aspect of government programs will be discussed in Chapter 11.

Marketing orders are authorized by Congress under the Agricultural Marketing Agreement Act of 1937 as amended, primarily as a tool for establishing and maintaining orderly marketing of designated agricultural commodities produced within specified geographic areas. Marketing orders for individual commodities and regions become effective only if the specified proportion of producers voting in a referendum approves the order. Once in effect, however, the order is binding on all producers of the commodity and may both impose quality regulations and place quantity limits on the marketing of the product. Many orders are currently in effect for fruits, vegetables, and specialty crops. Data on the quality standards and the base, set-aside, or other provisions to control quantity marketed should be obtained before planning production levels of these products.

Zoning regulations and environmental restrictions may limit the combination of enterprises and the systems of production that can be used. Although these regulations may be annoying to farm operators at times, they are typically imposed to preserve the environment by limiting the pollution of air and water, and to preserve the productivity of farmland. For example, restrictions may be placed on the amount of runoff from open feeding facilities to prevent pollution of public waters, and the size of open feeding facilities may be restricted to limit air pollution. Limits may be placed on the possibility of draining certain areas which are to be preserved for wildlife habitat. Restrictions on the spreading of manure, the utilization of certain chemicals, and other farming practices may be imposed in an effort to avoid the movement of effluent, chemicals, and eroded soil into public waters. Many of these restrictions are imposed by the state and local governmental units, making a general discussion of the current restrictions impractical for a large area. Information on the regulations and restrictions that apply to farms in an area can be obtained from the local extension service office.

Information on long-term leasing arrangements for land, pasture, livestock facilities, and breeding stock should be obtained. The provisions of the lease including length, payments, and opportunity for renewal should be obtained. Real estate is frequently rented on a year-to-year basis without a written agreement. In these cases, it is particularly important to evaluate the likelihood that the lease can be renewed and the probable terms for the near future.

Managerial Capacity

Perhaps the most difficult factor to judge is the managerial capacity of the individual(s) operating the business. However, some types of data can be obtained that are useful in planning the farm business. These data include a list of the enterprises the operator has produced and an indication of the level of the production efficiency achieved for each. Data on the yields per acre, pounds of feed per unit of meat, milk or eggs produced, and other measures of technical efficiency provide a basis to determine how the manager compares with other farmers operating under similar conditions. Data on the pricing procedures used and the average prices received for products provide an indication of marketing skills. The above data can be obtained from the operator's farm records. Furthermore, the operator should be asked what experience he/she has had supervising hired labor. While managerial capacity is difficult to judge, data on technical efficiency, marketing skills, experience in supervising hired labor, and a general discussion of changes that might be considered for the business provide an indication of the operator's managerial capacity that is useful in developing a farm plan.

LONG-RUN PLANNING

The data obtained in the first step of the whole-farm planning procedure provide the basis to begin developing and comparing alternative production, market, and financial plans for the future. Data on the goals indicate whether the business is to be expanded significantly, the enterprise mix maintained with some efforts made to increase the efficiency of labor and/or capital use, or whether the business is to be scaled back allowing the operator more time for other activities. Data on the physical resources (including the land, labor, tangible working assets, and capital), the institutional considerations, and the manageral capacity suggest the range in enterprises and systems of production to consider in a reorganization plan.

The next task is to develop the general plan the operator will follow over a period of several years. The concept of a long-run plan is the combination of production, market, and financing activities to be followed in three to five years that will accomplish the goals specified for the five-year period. This plan may involve a significant adjustment from the current organization to the plan selected. The process of adjustment will be considered during the third and fourth steps, after a long-run plan has been selected.

Completion of the second step in the whole-farm planning procedure involves identifying alternative long-run plans, budgeting the alternative plans, and selecting one of the plans for use. The comparison made considers the long-run profitability of the enterprises. Thus, the calculations are based on average prices that are expected to prevail over a period of five years or more and the net returns to resources that are fixed over the longer run. Average five-year planning prices for inputs and products are seldom available from the usual sources of outlook information (which specialize in predictions for periods up to one year into the future). However, such estimates typically are available from farm management extension specialists in the area. In addition to using long-run prices, it is important to calculate returns to the resources that are considered fixed over a period of five years or more. The list of resources considered fixed over that period is subject to the conditions on the farm being planned, but would normally include return to those resources already on the farm that could not be easily sold and moved off the farm if the firm went out of business. They normally include the owned land, the owned buildings and facilities, and the operator's labor and managerial ability. The machinery, equipment, and livestock normally could be adjusted over this period, and they are considered variable over the longer run period.

Identify Alternative Long-Run Plans

An analysis of alternative long-run plans should enable the operator to compare different ways of using available resources to accomplish the goals specified. Thus, the challenge is to identify more effective ways to use the land, labor, and capital available to achieve the goals. This can usually be accomplished by careful analysis of the plan currently followed and by applying the equal marginal return principle discussed in the previous chapter to allocate limited resources.

The Present Long-Run Plan   Budgeting the present long-run plan (the plan being followed at the current time) is usually appropriate when an existing farm business is being analyzed. The budget prepared for the present plan estimates the labor and capital required, as well as the expected net returns on a comparable basis to the estimates prepared for the other plans. The estimates for the present plan provide the basis to judge how much better (or worse) the alternative plans are. It may also suggest ways to improve productive efficiency and employ unused resources as noted below.

Improvements in Productive Efficiency   Working through the input combinations used may suggest changes in the production systems, such as changes in breeding programs to improve reproductive efficiency, changes in livestock rations to increase the amount of meat, milk, or eggs per unit of feed used, or changes in cropping practices that will increase the profitability of the business. It may also suggest that investments in drainage or machinery to complete planting or harvesting in a more timely manner would increase yields and net returns.

Employing Unused Resources   Unused resources available on the farm can be identified by budgeting the present plan. Livestock facilities, seasonal labor supplies, and unused borrowing capacity are the most common examples. The challenge is to find a productive use for these resources that increases the net returns of the business.

Selecting an Improved Combination of Enterprises   It may be possible to select a more profitable combination of enterprises using the data available on the farm's resources and the equal marginal return principle. Enterprise budgets are developed for the relevant crop and livestock alternatives as the basis to calculate returns to limited resources. One of the resources considered to be an important limiting resource in planning the business is chosen as the basis for enterprise selection, and the enterprise budgets are used to calculate residual returns per unit of the selected resource. The combination of enterprises is chosen by allocating the selected resource to its highest return use until one of the other resources is exhausted. Then consideration is given to allocating some of the selected resource to its second most profitable alternative, the third most profitable, and so on until the value of the marginal product is equal to the marginal factor cost or all of the selected resource has been allocated.

Selection of an appropriate limiting resource is essential to the successful application of this approach. For crop farms the amount of cropland or irrigation water is usually the appropriate limiting resource. The amount of forage available, or, more specifically, the amount available during those seasons of shortest supply, is often the appropriate limiting resource for ranching situations. The enterprise selection on combination crop and livestock operations should typically select the crop combination based on returns per acre of cropland and livestock enterprises based on returns per unit of livestock facilities.

Consider the example in Table 6.1. The operator has four enterprises (A through D) that require cropland and one enterprise, E, that can be thought of as either a confinement livestock enterprise, a marketing enterprise, or part-time employment off the farm. The gross returns, variable costs, and net returns, as well as the amount of each limited resource per unit of the enterprise, are shown.

Suppose the operator's limited resources include 100 acres of land and 200 hours of labor in each period with irrigation water being available in unlimited amounts. This is resource situation 1 in Table 6.2. Suppose the operator selects cropland as the most limiting resource. In this case, enterprises are selected based on net returns per acre of cropland (net returns to limited resources divided by acres of cropland). Enterprise D has the highest return per acre of cropland. Land is allocated to enterprise D

Table 6.1   Hypothetical Budget Data to Illustrate the Application of the Equal Marginal Return Principle in Selecting Enterprise Combinations

Item	Ent. A	Ent. B	Ent. C	Ent. D	Ent. E
Gross returns per unit	$200	$225	$275	$300	$20
Variable costs per unit	100	115	145	160	10
Net returns to limited resources	100	110	130	140	10
Limited resources					
Acres cropland	1	1	1	1	0
Units of irrigation water	0	2	8	10	0
Hours labor period 1	1	2	2	2	0.5
Hours labor period 2	0	0	3	4	0.5

Table 6.2 Resource Availability and Utilization Table for Plans Selected for Resource Situations 1 and 2

Resource Situation 1

Resources	Amount Available	50 Units Enterprise D		50 Units Enterprise B	
		Resources Required	Resources Remaining	Resources Required	Resources Remaining
Cropland	100 acres	50	50	50	0
Labor 1	200 hours	100	100	100	0
Labor 2	200 hours	200	0	0	0
Net returns	—	$7,000	—	$5,500	—

Resource Situation 2

Resources	Amount Available	75 Units Enterprise B		25 Units Enterprise A		50 Units Enterprise E	
		Resources Required	Resources Remaining	Resources Required	Resources Remaining	Resources Required	Resources Remaining
Cropland	100 acres	75	25	25	0	0	0
Irrigation water	150 units	150	0	0	0	0	0
Labor 1	200 hours	150	50	25	25	25	0
Labor 2	200 hours	0	200	0	200	25	175
Net returns	—	$8,250	—	$2,500	—	$500	—

until one of the other resources is exhausted. As shown in Table 6.2, 50 acres of D uses the 200 hours of labor in period 2. Enterprise C has the second highest net return per unit of cropland but cannot be included because it requires labor in period 2. The remaining 50 acres can be allocated to enterprise B, the enterprise with the third highest net return per unit of land. Since this exhausts the available land, labor in period 1, and labor in period 2, neither enterprise A nor E can be included. Thus, the plan for resource situation 1 is composed of 50 acres of enterprise B and 50 acres of enterprise D. It requires 100 acres of land, 200 hours of labor in period 1, and 200 hours in period 2, and has returns to the limited resources of $12,500.

Now consider the same five enterprises for resource situation 2 which has 100 acres of land, 200 hours of labor in each period, and 150 units of irrigation water. The greater returns per acre of enterprises requiring irrigation water and the very limited supply of water available suggest that irrigation water available may be the most limiting resource in selecting a farm plan. Thus, returns per unit of irrigation water is used to select the enterprises for the plan. Of the three irrigated enterprises, B has the highest net return per unit of water. Including 75 acres of B exhausts the water supply, leaving 25 acres of land, 50 hours of labor 1, and 200 hours of labor 2. Because all of the water has been allocated, another resource must be selected and returns per unit of that resource used to finish constructing the plan. Both enterprises A and E can be added to use some of the remaining resources. Because both require labor in period 1 (while they do not require either land or labor in period 2), the decision should be made on the basis of which provides greater returns per hour of labor 1. Net returns are greater for enterprise A ($100 \div 1$ for A versus $10 \div 0.5$ for E), and 25 acres of A are included. The remaining amount of labor 1 is allocated to enterprise E. Thus, the plan selected includes 25 acres of A, 75 acres of B, and 50 units of E. The plan requires 100 acres of land, 150 units of irrigation water, 200 hours of labor 1, and 25 hours of labor 2, and has net returns to the limited resources of $11,250. The reader can verify that other combinations of enterprises A, B, C, D, and E that do not exceed the amount of resources available would result in lower returns to the limited resources.

Selection of the appropriate limiting resource becomes more difficult as the number of limited resources (such as labor in each of many periods during the year, several qualities of land, livestock facilities of several types, and so on) increases. In many actual planning situations, several resources may limit the possible enterprise combination, making selection of the most profitable combination difficult with this simple approach. In these situations, a manager can select the two (or more) resources considered to be the more limiting and apply the procedure described above for each resource.

This results in one plan for each of the resources selected. The manager can compare the profitability of these plans and, by exercising some judgment, either modify the resulting plans or form combination plans that make more profitable use of the resources. It is unlikely that the most profitable combination of enterprises will be found in these complex situations without resorting to the mathematical programming procedures discussed in Chapter 10. However, the approach described focuses attention on the use of the appropriate rule—the equal marginal return principle—to allocate resources efficiently.

Improved Market and Financial Management   The list of ways to improve the long-run farm plan would be incomplete without mentioning the possibility of searching for ways to reduce input costs, increase product prices, and increase returns earned on debt and equity capital used in the business. However, discussion of these topics is reserved for Chapters 7 and 8.

Budgeting Procedures

The average annual profitability of the business under each of the alternative plans can be compared to determine which, if any, of the long-run plans warrant more detailed study in Steps 3 and 4. The procedures described here are more general and less time consuming to complete than those followed in subsequent steps. This analysis can be thought of as a sorting procedure to help identify the long-run plan which will be analyzed in more detail in Steps 3 and 4.

The budgeting procedures described here are used to estimate the differences in resource use and the average annual net farm income for one plan versus another. An annual measure of income is used that is analogous to the profit or loss reported on line 12 of the income statement (Figure 2.1). This represents the return to the unpaid family labor, the operator's labor, equity capital, and management. Thus, the budgeting procedures should consider the crop, livestock, and miscellaneous receipts, the cash farm operating expenses, the value of products produced on the farm that are consumed by the household, inventory changes, and changes in capital items. Two budgeting procedures, complete budgeting and partial budgeting, are commonly used to make these comparisons.

Complete Budgeting   A complete budget includes a listing of all production and income, and all inputs and expenses for the farm business. The complete budget, sometimes referred to as the total budget, includes an estimate of average annual cash operating receipts, cash operating expenses, inventory changes, and changes in capital items (depreciation on buildings, facilities, and machinery) to develop an estimate of average annual farm profit or loss. Inventories are usually considered constant. The relative profitability of alternative plans is compared by preparing a complete budget for each long-run plan being analyzed.

Consider an example. Andy has identified two alternative farm plans. One of these plans is composed of two crops, soybeans and corn, and one livestock enterprise, finishing 40 pound feeder pigs. Enterprise budgets have been prepared for each of these enterprises for the soils and conditions on Andy's operation. Long-run planning prices have been estimated for inputs and products and have been used in preparing the enterprise budgets. The dollar values and some of the physical data are summarized from the three enterprise budgets in Table 6.3.

The first task in developing the complete budget is to develop a resource availability and utilization table (Table 6.4). The availability of resources is determined in Step 1 of the whole-farm planning procedure. The inventory for Andy's farm indicates that the physical resources to be considered are 300 acres of owned land (which is of uniform quality), the amounts of operator and family labor by month shown, and livestock facilities to feed 480 feeder pigs at one time. The space for swine feeding can be utilized twice per calendar year. In addition to the physical resources, it is important to list any intermediate products so that the production plus purchases equals the use plus sales. In this example, corn is an intermediate product used by the livestock. The resources that may be limiting are listed in column 1, and the amounts available are recorded in column 2.

The amount of each resource required is calculated and entered in the column for the enterprise and the appropriate row for the resource. Plan 1 being considered includes 200 acres of corn, 100 acres of soybeans, and the finishing of two groups of 480 feeder pigs. The land and labor resources required per unit given in the enterprise budgets are multiplied by the number of acres to obtain the entries for corn and soybeans. The entries for swine are taken directly from the budget since "one unit" of that enterprise is being used. The uses or requirements of resources and intermediate products are recorded as positive numbers in columns 3 through 5, while the production of a resource or intermediate product is recorded as a negative value. Notice that the production of 200 acres of corn at 120 bushels per acre is recorded as negative 24,000, while the use of corn by swine is shown as a positive 9,696. The net balance in column 6 is the value in the resources available column minus the entry in each of the enterprise columns. The net balance is positive if the amount available plus production exceeds the amounts required by the plan. Excess resources may suggest ways to modify the plan that increase net farm income. Positive entries in the intermediate product rows indicate the amount of the product not required by other enterprises that is available for sale. In this example, the 200 acres of corn is expected to produce 14,304 bushels more corn than is required by the swine enterprise. This excess can be sold as cash grain. The negative entries in the net balance column indicate that the plan requires more resources than are available. A resource with a negative net balance indicates either that the plan must be changed to reduce the amount of resource required or that additional units of the resource must be obtained through purchasing, leasing, or trading work. Andy can hire a retired farmer in the neighborhood at $5.00 per hour to provide the additional labor required in May and October. He has ample supplies of the other resources.

The resources and intermediate products to include in the resource availability and utilization table depend on what appears to be limiting the farm organization. Ranches using native pasture or range may find it useful to include rows for the amount of grazing produced and consumed by season of the year. Farms producing pasture on cropland may include rows for pasture production and use by period of the year under intermediate products. An analysis of an irrigated operation should consider restrictions on the amount of water that can be applied per week or month throughout the irrigation season. The availability of storage space or machinery to perform a certain task in a timely manner may be an important consideration in planning some businesses, indicating this item should be included as a limited resource. For example, a line could be included showing the amount of storage space for silage, hay, or grain and the storage requirements for intermediate products to be used. Questions of field time in a particular period can be considered by including a row listing the hours or days available for the operation. Data are needed from the enterprise budget indicating the amount of field time required to perform the operation per acre. Then the balancing of time available and the amount required with the machinery complement budgeted can be considered within the framework of the resource availability and utilization table. The number of resources that should be included depends on the importance of each in planning the particular farm under consideration. Those resources that can be ignored, such as July labor, August labor, and storage space for corn in this example, reduce the time and effort required.[3]

The complete budget for plan 1 is shown in Table 6.5. The income and expense categories parallel those in the income statement shown in Figure 2.1, but have been combined to reduce the number of entries. Although income taxes are not considered in the complete budget, it is useful to show the division between ordinary income and capital gains to point out that potential tax advantage. A column is included for each

---

[3]It would be possible to include rows for cash receipts and expenses by month in the resource availability and utilization table. However, the projected cash flow requires more careful analysis than that devoted to the physical resources and intermediate products. The usual recommendation is, therefore, to evaluate plans at this level, assuming the financial resources can be obtained, and to reserve consideration of financial planning until the third and fourth steps of the whole-farm planning procedure.

Table 6.3 Summaries of Three Enterprise Budgets for Andy's Farm

Corn for Grain (figures are per acre)

	Value or Cost
1. Gross receipts	
Corn 120 bu @ $3.00	$360.00
2. Operating costs	
Seed	15.12
Fertilizer, herbicides, and insecticides	62.89
Crop insurance	9.00
Machinery operating expense	24.68
Labor 2.98 hr @ $5.00	14.93
Interest on operating capital	5.87
Total operating costs	$132.49
3. Income above operating costs	227.51
4. Ownership costs	
Machinery	53.79
Land taxes	12.00
Interest on land investment	88.00
Total ownership costs	$153.79
5. Total costs shown	$286.28
6. Net returns above costs shown	$ 33.72

Soybeans (figures are per acre)

	Value or Cost
1. Gross receipts	
Soybeans 40 bu @ $7.50	$300.00
2. Operating costs	
Seed	10.50
Fertilizer, herbicides, and insecticides	28.70
Crop insurance	7.50
Machinery operating expense	14.91
Labor 1.93 hr @ $5.00	9.63
Interest on operating capital	3.71
Total operating costs	$ 74.95
3. Income above operating costs	$225.05
4. Ownership costs	
Machinery	32.88
Land taxes	12.00
Interest on land investment	88.00
Total ownership costs	$132.88
5. Total costs shown	$207.83
6. Net returns above costs shown	$ 92.17

Finishing Feeder Pigs: Two Groups of 480 per Year

	Value or Cost
1. Gross receipts	
Market hogs in February	$ 52,180.59
Market hogs in July	55,882.42
Total receipts	$108,063.01
2. Operating costs	
480 feeder pigs in October	21,408.00
480 feeder pigs in March	25,675.20
Corn 9696 bu @ $3.00	29,088.00
Purchased supplement	11,368.60
Yardage and commission	2,535.50
Veterinary and medicine	1,056.00
Utilities	96.00
Bedding	210.00
Insurance on livestock	200.00
Miscellaneous	200.00
Truck operating expense	487.50
Machinery and tractor operating expense	1,323.00
Building repairs	160.00
Labor 720 hr @ $5.00	3,600.00

Monthly Cash Costs and Labor Requirements[a]

| | Enterprises | | | | | |
| | Corn | | Soybeans | | Swine finishing | |
Months	($/ac)	(hr/ac)	($/ac)	(hr/ac)	($/960 hogs)	(hr/960 hogs)
January					$ 1,623.39	72
February	$56.57				1,586.85	96
March			$39.20		27,647.24	48
April					1,762.39	72
May	5.07	0.66	5.52	0.75	1,762.39	72
June	32.45	0.41	8.34	0.16	1,623.39	72
July					1,599.33	96
August						
September						
October	14.38	1.56	5.33	0.67	23,380.04	48
November	3.22	0.35	3.22	0.35	1,972.39	72
December					1,762.39	72
Total	$111.69	2.98	$61.61	1.93	$64,719.80	720

	Interest on operating capital	2,256.09
2.	Total operating costs	$ 99,663.89
3.	Income above operating costs	8,399.12
4.	Ownership costs	
	Buildings	2,205.02
	Machinery	1,948.25
	Total ownership costs	$ 4,153.27
5.	Total costs shown	$103,817.16
6.	Net returns above costs shown	4,245.85

[a]The cash costs listed include all operating costs shown in the enterprise budget except the corn for feed normally produced on the farm, the labor and interest on the operating capital.

Table 6.4 Resource Availability and Utilization Table for Long-Run Plan 1

(1) Resources	(2) Amounts Available	(3) 200 Acres Corn	(4) 100 Acres Soybeans	(5) 2 Groups Feeder Pigs	(6) Net Balance (2) − (3) − (4) − (5) = (6)
Land					
Cropland	300 ac	200	100	0	0
Labor					
January-March	540 hr			216	324
April	200 hr			72	128
May	200 hr	132	75	72	−79
June	200 hr	82	16	72	30
July	200 hr			96	104
October	200 hr	312	67	48	−227
November	180 hr	70	35	72	3
December	180 hr			72	108
Livestock facilities					
Swine finishing	2 groups of 480 head			2 groups of 480 head	0
Intermediate products					
Corn grain	0 bu	−24,000		9,696	14,304

Table 6.5 Complete Budget for Long-Run Plan 1
for Andy's Farm

Item	200 Acres Corn	100 Acres Soybeans	2 Groups Feeder Pigs	Nonallocated Items	Total
Income					
Ordinary income	$42,912	$30,000	$108,063		$180,975
Capital gains					
Miscellaneous				$ 800	800
Total income	$42,912	$30,000	$108,063	$ 800	$181,775
Expenses					
Seed, crop supplies	3,024	1,050			4,074
Fertilizer and chemicals	12,578	2,870			15,448
Purchased feed			11,369		11,369
Purchased livestock			47,083		47,083
Breeding, veterinary, and livestock supplies			1,266		1,266
Marketing cost			3,023		3,023
Fuel, oil and utilities Machinery repairs	4,936	1,491	1,419		7,846
Other repairs			160	2,000	2,160
Insurance	1,800	750	200	800	3,550
Real estate taxes				2,100	2,100
Real estate rent					
Hired labor				1,530	1,530
Interest on operating loan	1,174	371	2,256		3,801
Interest on intermediate and long-term loan				17,000	17,000
Depreciation				18,699	18,699
Miscellaneous			200	1,800	2,000
Total expenses	$23,512	$ 6,532	$ 66,976	$43,929	$140,949
Farm profit or loss					$ 40,826

enterprise plan, one for nonallocated items and one for totals. The non-allocated column provides a place to record those items of income that are not a direct result of any enterprise listed. It is also used to record the overhead expenses of the business that are difficult to allocate to individual enterprises.

The sale of corn shown in Table 6.5 is the 14,304 bushels that is not required for swine feeding. It is sold at $3.00 per bushel. Soybean sales and swine sales are calculated directly from the enterprise budgets. Miscellaneous income includes gas tax refunds, dividends from cooperatives, minor amounts of custom work done for others, and the value of any products produced on the farm that are consumed by the family. Because we are calculating income for the average year, the average amount of production is included in sales and the change in inventory is zero. Thus, a calculation for an inventory change does not appear in the complete budget.

The expenses for the three enterprises are taken from the enterprise budgets in Table 6.3. The costs listed for each enterprise include the inputs listed in the operating costs section that are provided neither by resources available on the farm nor intermediate products produced on the farm. The expenses for corn and soybeans include all operating costs listed in the enterprise budgets except labor. The entries for swine include all operating costs except labor and the corn which is produced on the farm.

The ownership costs listed in the three enterprise budgets are *not* included under the enterprise columns of Table 6.5 for two reasons. The crop enterprises use a common set of machines, and the annual ownership costs (for depreciation, taxes, insurance, and interest) on this complement of equipment will probably not vary over a relatively wide range of the three enterprises unless the machines included in the complement change. Thus, it is more accurate to calculate the annual value of the appropriate overhead costs and to include them in the nonallocated column, since, as we pointed out in Chapter 4, any allocation of these costs to individual enterprises is largely arbitrary. The ownership costs listed in enterprise budgets include depreciation, taxes, insurance, and interest. The annual totals for the first three of these costs must be deducted in calculating farm profit or loss. However, only interest on debt (borrowed) capital is included as an expense in the complete budget because farm profit or loss is the return to unpaid labor, operator's labor, equity capital, and management.

Andy's nonallocated expenses include repairs for buildings, insurance on the buildings, machinery, real estate taxes on the farm, the cost of 306 hours of hired labor at $5.00 per hour, and miscellaneous overhead expenses for accounting, telephone expense, and farm share of the automobile and pickup truck expenses. The interest on the intermediate and long-term loans required by the plan and depreciation on the current buildings and complement of machinery are listed to complete the expenses.

Totaling the expenses and subtracting from total income results in a farm profit of $40,826. This is the average amount of money the operator is expected to have available from the farm business for payment of income taxes, family living, reinvestment in the business, and investment off of the farm.

A comparison of alternative plans with complete budgets requires development of a resource availability and utilization table (such as Table 6.4) as well as the complete budget (such as Table 6.5) for each plan. Comparison of the two tables (resource availability and utilization table and the complete budget) for each of two or more plans provides a comparison of resource use and the farm profit or loss. Thus, it provides an initial test of the desirability of the plan so that the operator can determine if more detailed planning is appropriate.

Partial Budgeting   Partial budgets are used to estimate the change that will occur in farm profit or loss from some change in the farm plan by considering only those items of income and expense that change. Partial budgets do not calculate the total income and total expense for each of two plans, but only list those items of income and expense that change to estimate the difference in profit or loss expected from the plans.

Partial budgeting is particularly useful in analyzing relatively small changes in the business such as the purchase of a piece of equipment to replace hiring a custom operator, considering a shift in the cropping program for the current year, or participation in a government program. Applications of partial budgeting to some of these problems will be illustrated in later chapters. It is possible to use the format to compare alternative long-run plans for the total farm business if one is careful to include all items of income and expense that change.

The general format for partial budgeting is presented in Figure 6.1. To interpret the headings in a consistent manner, the user of the partial budgeting procedure must first select one plan as the basis for comparison and the other as the alternative. Items listed under the first heading, additional income, are those items of income that would occur with the alternative plan that would not result from using the base plan. Entries under reduced expenses represent expenses for the base plan that are not incurred with the alternate plan. Reduced income and additional expenses are used to list items of income for the base plan that are not part of the alternate and additional expenses for the alternate plan, respectively. The subtotal of additional income and reduced expenses, minus the subtotal of reduced income and additional expenses, is an estimate of the difference in

---

Figure 6.1   Partial Budgeting Format

---

1. *Additional Income*
   This section lists the items of income from the alternate plan that will not be received from the base plan.
2. *Reduced Expenses*
   This section lists the items of expense for the base plan that will be avoided with the alternate plan.
3. *Subtotal* (1 + 2)
4. *Reduced Income*
   This section lists the items of income from the base plan that will not be received from the alternate plan.
5. *Additional Expenses*
   This section lists the items of expense from the alternate plan that are not required with the base plan.
6. *Subtotal* (4 + 5)
7. *Difference* (3 − 6)
   A positive (negative) difference indicates that the net income of the alternate plan exceeds (is less than) the net income of the base plan by the amount shown.

---

net farm income. A positive difference indicates that the alternate plan has higher expected net income than the base plan. A negative difference indicates that the base plan has the higher net income.

Consider how the partial budgeting format can be used to compare alternative long-run farm plans. Andy wants to compare the profitability of a second long-run plan that makes more complete use of the available labor supply throughout the year and that he feels may avoid some of the price risk associated with the finishing of feeder pigs. The second long-run plan includes 150 acres of corn, 150 acres of soybeans, and a 48 sow farrow-to-finish swine enterprise. The enterprise budgets for corn and soybeans in Table 6.3 apply to the second plan. Andy has estimated the investment costs for new farrowing and nursery facilities, the investment to remodel a pole building on the farm for gestation facilities, and the investment in new equipment to handle liquid manure. The additional investment for new facilities and equipment totals $49,948. The enterprise budget for the farrow-to-finish swine enterprise is shown in Table 6.6. The enterprise divides the 48 sows into three groups with approximately 16 litters farrowed every other month. Andy has selected farrowing periods so that sales of market hogs are expected in February, April, June, August, October, and December, as shown in the receipts section of the enterprise budget. Ownership costs on the new facilities and equipment were calculated using the procedures outlined in Chapter 4.

The resource availability and utilization table for long-run plan 2 is shown in Table 6.7. The net balance column shows that hired labor will be required in May, June, October, and November. Again, Andy feels he can hire the amounts available at $5.00 per hour. The net balance also indicates that 7,791 bushels of corn will be available for sale.

The partial budget in Table 6.8 considers long-run plan 1 as the base plan. Entries are presented considering long-run plan 2 as the alternative. The additional income from plan 2 is composed of the soybeans produced on an additional 50 acres and the sales from the proposed farrow-to-finish swine enterprise. The reduced expenses are the savings in operating expenses from reducing corn production 50 acres and from eliminating the finishing of feeder pigs. The operating expenses listed include all items in the operating costs section of the enterprise budget except the labor provided by the operator and the corn used as feed for the hogs which is produced as an intermediate product. Entries under part 4 of the partial budget include the reduced corn sales for plan 2 of 6,513 bushels (14,304 bu − 7,791 bu) and the loss of sales from eliminating feeder pig finishing. The increased expenses include the operating costs to produce the additional 50 acres of soybeans and the farrow-to-finish enterprise. More hired labor is required by plan 2. The additional 173 hours is included as an increased expense. The depreciation, taxes, and insurance in the additional buildings and machinery required, as well as interest on the additional money invested

Table 6.6 48 Sow Farrow-to-Finish High Investment, Six Litters Produced per Year

Item	Weight	Unit	Quantity	Price	Value
1. Receipts[a]					
Slaughter hogs					
February	2.2	Cwt	118	$ 51.45	$13,356.42
April	2.2	Cwt	118	46.65	12,110.34
June	2.2	Cwt	115	50.75	12,839.75
August	2.2	Cwt	118	54.70	14,200.12
October	2.2	Cwt	118	49.45	12,837.22
December	2.2	Cwt	118	47.60	12,356.96
Gilts—Nonbred	2.9	Cwt	7	48.00	974.40
Sows—Nonbred	3.6	Cwt	8	42.00	1,209.60
Sows cull	3.7	Cwt	18	40.50	2,697.30
Boars	4.5	Cwt	3	34.50	465.75
Total					$83,047.86
2. Operating costs					
Corn[b]		Bu	10,209.0	$ 3.00	$30,627.00
Soybean meal[b]		Cwt	1,000.5	12.00	12,006.00
Other purchased feed[b]		Dol			1,647.57
Veterinary and medicine		Dol	1,000.0	1.00	1,000.00
Electricity and fuel		Dol	1,300.0	1.00	1,300.00
Yardage and commission		Cwt	1,680.2	1.25	2,100.25
Young boars		Hd	3	400.00	1,200.00
Bedding		Tons	5.55	40.00	222.00
Miscellaneous		Dol	400	1.00	400.00
Livestock insurance			535		535.00
Truck and machinery operating expense		Dol			2,306.42
Building repairs		Dol			1,073.92
Labor		Hr	1,634	5.00	7,620.00
Interest on operating capital				0.12	1,367.47
Total					$63,405.63
3. Returns above operating costs					$19,642.23
4. Ownership costs					
Interest on livestock			10,185	0.05	$ 509.25
Depreciation, interest, taxes and insurance on:					
New buildings[c]					5,500.83
New machinery					1,665.00
Currently owned buildings, and equipment					3,468.70
Total					$11,143.78
5. Total costs shown					$74,549.41
6. Net returns above cost shown					$ 8,498.45

[a]Budget assumes a weaning rate of 8.0 pigs per litter and 7.4 pigs sold per litter. The calculations assume 6 pigs are saved as replacements and 4 die per 16 litters.

[b]Feed requirements are based on a feed conversion rate of 4.15 pounds of feed per pound of pork sold.

[c]Slatted floor farrowing house, partially slatted nursery and open front gestation. The ownership costs assume an investment of $21,850 in farrowing facilities, $12,000 in nursery facilities, $4,598 in remodeling an existing building for gestation facilities, $9,000 for new liquid manure equipment, $2,500 for a lagoon to control runoff, and use of existing finishing facilities.

239

Table 6.7 Resource Availability and Utilization Table for Long-Run Plan 2

(1) Item	(2) Resources Available	(3) 150 Acres Corn	(4) 150 Acres Soybeans	(5) 48 Sow Farrow- to-Finish Swine	(6) Net Balance (2) − (3) − (4) − (5) = (6)
Land					
Cropland	300 ac	150	150	0	0
Labor					
January	180 hr			115	65
February	180 hr			154	26
March	180 hr			128	52
April	200 hr			150	50
May	200 hr	99	113	113	−125
June	200 hr	62	24	147	−33
July	200 hr			134	66
August	200 hr			152	48
September	200 hr			115	85
October	200 hr	234	101	146	−281
November	180 hr	53	53	134	−60
December	180 hr			146	34
Livestock facilities					
Farrow-to-finish	48 sows			48 sows	0
Intermediate products					
Corn grain	0 bu	−18,000		10,209	7,791

Table 6.8 Partial Budget Estimating the Change
in Net Farm Income for Long-Run Plan 2

1.	Additional income	
	50 acres soybeans × 40 bu × $7.50	$ 15,000.00
	Sales from 48 sow swine enterprise	83,047.86
2.	Reduced expenses	
	50 acres corn at $117.56	5,878.00
	Finishing feeder pigs	66,975.89
3.	Subtotal (1 and 2)	$170,901.75
4.	Reduced income	
	Reduced corn sales 6,513 bu at $3.00	19,539.00
	Sales from feeder pig finishing	108,063.01
5.	Increased expenses	
	50 acres soybeans at $65.32	3,266.00
	Cash expenses from farrow-to-finish enterprise	25,158.63
	Additional hired labor (479 hr − 306 hr) = 173 hr @ $5.00	865.00
	Depreciation, interest, taxes, and insurance on new buildings	
	and machinery	7,165.83
	Interest on new investment in breeding stock	509.25
6.	Subtotal (4 and 5)	$164,566.72
7.	Difference (3 − 6)	$ 6,335.03

in buildings, machinery, and livestock, must be included as increased expenses for plan 2. Completing the calculation indicates a difference of $6,335.03. This positive difference indicates that long-run plan 2 has an annual average profit that is $6,335.03 greater than plan 1. The interested reader can develop a complete budget for long-run plan 2 and verify that the profit or loss is $47,161.14 ($40,826.11 + $6,335.03) for plan 2.

Thus, either complete or partial budgeting can be used to estimate the difference in net farm income between alternative farm plans. However, only the complete budget can be used to estimate the level of net farm income. It is possible to compare two plans and to indicate the difference in farm income with partial budgeting, but there is no assurance that either produces an acceptable net farm income. The advantage of the partial budgeting procedure is that it typically requires fewer computations and avoids the need to list many overhead expenses which are identical for the alternative plans compared. As illustrated in the comparison of plans 1 and 2, it is often appropriate to develop a complete budget for one plan to establish the level of net income and use partial budgeting to quickly compare alternative plans.

Evaluating Alternative Long-Run Plans

The budgeting procedures described provide a relatively quick method of balancing resource requirements with the amount available. This helps insure the feasibility of carrying out the plan with the resources available and provides useful information on seasonal requirements for some resources, such as labor, pasture, and irrigation water. Some plans may be considered infeasible because of relatively high seasonal requirements for labor or other inputs which would be difficult to manage, if not impossible to obtain. The operator can contrast the resource use and the difference in net farm income to narrow the list of plans for more detailed planning in Steps 3 and 4. Frequently, the process of comparing the results suggests other plans which may be more effective in satisfying the goals specified. If this is the case, the new alternatives can be compared using the procedures described in this step of the whole-farm planning process. When the operator feels the relevant long-run plans have been considered, one is selected that appears to be the most consistent with the goals. More detailed planning is carried out on this plan in Steps 3 and 4.

SHORT-RUN PLANNING

The development of a workable production, marketing, and financial plan for the coming year is referred to as short-run planning. The process involves specifying the timing of input purchases, production activities, product sales, and investments for the coming year. It also includes the development of a financial plan to carry on the year's activities.

The procedures described in this step should be used annually when only minor adjustments are to be made in last year's plan, as well as when a major change is planned for the business. Short-run planning determines whether the physical resources are available to carry out the plan in the allotted time and evaluates the profitability and financial progress anticipated for the year. When the short-run plan is satisfactory in these respects, it serves as the basis to arrange for purchasing inputs and financing. As the operator moves into the year, the plan also provides a basis for control of the operation. While a year's business activity will probably never turn out exactly as planned, the effect of changes can be managed more easily based on a carefully developed and detailed plan.

The process described has four parts. (1) Decide on the enterprise combination, the systems of production to be used, the timing of production and marketing activities, the timing of any investments, and the method of financing the operation. (2) Project a monthly cash flow for the coming year based on these decisions. (3) Develop a projected income statement for the first year to estimate the profitability of the plan. (4) Project the balance sheet for the end of the year's operation and compare it to the balance sheet at the start of the year to estimate the effect of the plan on the structure of assets and liabilities. Consider each of these four parts for Andy's next year of operation.

The Detailed Plan

Andy has decided to develop a short-run plan for long-run alternative 2. This plan includes 150 acres of corn, 150 acres of soybeans, and a 48 sow six-litter farrow-to-finish swine enterprise. He has selected this plan because it provides more constant employment for his labor, does not require him to hire much more labor than his current organization, and appears to provide greater net farm income. Furthermore, he feels there is less risk associated with farrow-to-finish swine enterprises than the buy-sell livestock enterprises he has had in the past.

After some study, Andy outlines the production, marketing, and financial aspects required to implement this organization during the current year. The crop organization can be shifted to the proposed plan without difficulty. The machinery, equipment, and crop storage space required are available on the farm. He plans to follow the production system used to develop the enterprise budgets in Table 6.3. The input requirements, yields, and production costs listed in those budgets are appropriate for planning for next year.

Implementation of the farrow-to-finish swine enterprise requires building new farrowing and nursery facilities, as well as remodeling an existing building to house the breeding stock. Andy develops a schedule that coordinates facility construction, the purchase of breeding stock in February, April, and June, and the first farrowings in August, October, and December. Because the pigs from the new enterprise will not use the finishing facilities until later, he decides to purchase one group of feeder pigs on about March 15 and to finish them. Thus, his production and investment plans include production of 150 acres of corn and 150 acres of soybeans, finishing one group of feeder pigs beginning about March 15, developing the new swine enterprise, and having a contractor remodel and build the necessary facilities.

The production and investment decisions are made considering the prices expected to prevail during the coming year, the cash flow needs of the business, and the income tax implications of the decision. Andy plans to sell the remaining soybeans he has in storage in early January and the corn (not required for feed) in March and June. He also plans to sell most of the year's soybean production in December because sales and taxable income are expected to be low during the current year. Andy also estimates sales of slaughter hogs in July and cull breeding stock later in the year.

Andy recently consolidated some personal real estate loans, obtaining a new real estate loan of $170,000. The interest rate is 10 percent, and the loan is to be repaid with annual payments of $22,355 per year for 15 years. The first payment is due April 1. He also has a current operating loan for $30,000 with an interest rate of 12 percent. The interest on the operating loan is paid through the end of the previous calendar year. Andy plans to use the line of credit he has to finance the operating expenses of the business. He plans to obtain a new intermediate-term loan to provide the cash for the new facilities, equipment, and breeding stock.

Projected Cash Flow

The cash flow is a recording of all sources and uses of cash throughout the production period. The production period for short-run planning is usually one year in length, and the sources and uses are grouped into monthly, bimonthly, or quarterly periods as desired. Monthly periods are the most appropriate when the individual sources and uses of cash fluctuate by relatively large amounts from one month to the next, which is the typical situation in agricultural production. However, farms with a relatively uniform monthly pattern of receipts and expenses may find a bimonthly or quarterly cash flow satisfactory for planning.

Several types of data are needed to complete the projected monthly cash flow. Much of this data results from the detailed planning just described. Completion of the cash receipts section requires data on the quantity and timing of crop sales, livestock sales, and the sale of any capital items planned for the year. These estimates should be based on prices expected during the current year. The amount and timing of any miscellaneous income from custom work, government programs, gas tax refunds, and other sources must be estimated.

Many of the data on the quantity and timing of cash operating expenditures for crop and livestock enterprises are included in the enterprise budgets. Again, it is important to use price data for the current year in estimating the cash outflows. Data on overhead cash costs, including utilities, repairs, property taxes, insurance on buildings and equipment, rents, and hired labor, must be estimated. Data on the timing and dollar value of capital expenditures are required. Monthly proprietor withdrawals for family living, savings, and payment of income and social security taxes must be specified. Finally, interest and principal payments required on long-term and intermediate-term loans are necessary to complete the cash outflow section. Data on the cash available at the start of the year, the outstanding loan balances at the start of the year, and the interest rate for the operating loan are needed to complete the flow of funds summary and the loan summary.

The projected monthly cash flow for the first year of Andy's operation illustrates the data required to prepare such a projection (Table 6.9). The cash receipts section is completed using decisions on the amount and timing of crop and livestock sales and price projections for the current year. The sources of miscellaneous income from government programs, gas tax refunds, custom work for other farmers, etc., are estimated. Andy does not plan any capital sales during the year, but such items would be included on line 6 before totaling cash receipts by month.

Enterprise budgets include much of the data on quantity and timing of cash operating expenses required to complete the cash outflow section. Again, it is important to use estimated input prices for the current year. Andy feels that the input quantities and prices for corn and soybeans used in the enterprise budgets are appropriate for the first year. The operating expense for these two enterprises is obtained by multiplying the cash outflow per acre in Table 6.3 by 150 acres in each case. Andy estimates that the purchase price of feeder pigs will be $38 per head in March rather than the long-run price budgeted in Table 6.3. He bases the estimate of other cash expenses for the finishing of 480 feeder pigs purchased in March on the cash outflow shown in the enterprise budget. Andy uses the enterprise budget in Table 6.6 to prepare estimates of monthly cash expenses for the farrow-to-finish enterprise.

The cost of utilities, other repairs, real estate taxes, and insurance on buildings and equipment was based on the cost of these items last year, anticipated changes in the amount of use, and expected changes in the price of these items. Hired labor costs are based on the quantities estimated in Table 6.7 and a cost of $5.00 per hour.

The capital expenditures by month for the new facilities, equipment, and purchase of the breeding stock required for the farrow-to-finish swine enterprise are shown on lines 18 and 19. Andy estimates that they will need to withdraw $12,000 for family living over the year, plus an additional $5,500 in January for payment of taxes and social security on last year's income. Scheduled payments on previous debt include $180 of accounts payable and a payment of $22,355 on the land loan in March.

The flow of funds summary is completed on a month-by-month basis. Andy's beginning cash balance on January 1 is $2,083. The January column is completed by adding the January cash receipts from line 7 and subtracting the total January cash outflow (line 24) to obtain the cash difference of −$3,862. The negative cash difference indicates that other sources of funds will be required to pay the outflows listed. Andy plans to draw on an intermediate loan to pay for the capital purchases; the amount of borrowing on the intermediate loan each month equals the planned capital purchases for the month. He plans to borrow additional money on the operating loan as needed in $100 increments to maintain an ending cash balance of $2,000 or more. So in January Andy borrows $5,000 on the intermediate-term loan and $900 on the operating loan. No payments are made on the operating loan in January, leaving an ending cash balance of $2,038. This is also the beginning cash balance for February.

Andy updates the debt outstanding for the end of January by adding the $5,000 and the $900 of new borrowings to the ending balance of the intermediate and operating loans. Finally, the accrued interest on the operating loan is calculated.

Projected cash receipts are less than projected cash outflow during

Table 6.9 A Projected Monthly Cash Flow for the First Year of Andy's Operation with Alternative Plan 2

Item	Jan.	Feb.	Mar.	Apr.	May	Jun.	Jul.	Aug.	Sept.	Oct.	Nov.	Dec.	Total
Cash Receipts													
1. Corn			30,000			6,000							36,000
2. Soybeans	5,985											42,500	48,485
3. Feeder pig enterprise							53,753						53,753
4. Farrow-to-finish swine					568		568		1,171		946		3,253
5. Miscellaneous income			200			200			200			200	800
6. Capital sales													0
7. Total cash receipts	5,985	0	30,200	0	568	6,200	54,321	0	1,371	0	946	42,700	142,291
Operating Expense													
8. Corn production			8,486		760	4,868				2,157	483		16,754
9. Soybean production			5,880		828	1,251				800	483		9,242
10. Feeder pig enterprise • Purchased pigs			18,240										18,240
11. • Other cash expense			1,972	1,762	1,762	1,623	1,599		562	708	794	1,321	8,718
12. Farrow-to-finish swine		181	89	153	146	146	350	549	562	708	794	1,321	4,999
13. Utilities	250	250	250	100	100	100	100	100	150	200	200	250	2,000
14. Other repairs	180	200	200	200		200				200	200		2,000
15. Taxes, insurance, and rent							400	400	400				2,180
16. Hired labor				1,050						1,850			2,900
17. Other					625	165				1,405	300		2,495
Capital Expenditures													
18. New facilities	5,000	5,000			17,000			12,000	11,000				50,000
19. Purchase breeding stock		4,400		3,200		3,520		960		960	960	960	14,000
Proprietor Withdrawals													
20. Living expenses and taxes	6,500	900	800	800	800	800	800	1,200	1,200	1,000	1,200	1,500	17,500

Payments on Previous Debt

	1	2	3	4	5	6	7	8	9	10	11	12	Total
21. Intermediate													
• Principal													
• Interest													
22. Long-term													
• Principal													5,355
• Interest													17,000
23. Other													
24. Total cash outflow	11,930	10,931	58,022	7,265	22,021	12,673	3,249	15,209	13,312	9,080	3,660	4,031	171,383

Flow-of-Funds Summary

	1	2	3	4	5	6	7	8	9	10	11	12	Total
25. Beginning cash balance	2,083	2,038	2,007	2,085	2,020	2,067	2,014	2,081	2,032	2,091	2,071	2,057	2,083
26. Cash receipts	5,985	0	30,200	0	568	6,200	54,321	0	1,371	0	946	42,700	142,291
27. Cash outflow	11,930	10,931	58,022	7,265	22,021	12,673	3,249	15,209	13,312	9,080	3,660	4,031	171,383
28. Cash difference	(3,862)	(8,893)	(25,815)	(5,180)	(19,433)	(4,406)	53,086	(13,128)	(9,909)	(6,989)	(643)	40,762	(27,009)
29. Borrowing this month													
• Intermediate and long	5,000	9,400	0	3,200	17,000	3,520	0	12,960	11,000	960	0	960	64,000
• Operating	900	1,500	27,900	4,000	4,500	2,900	0	2,200	1,000	8,100	2,700	0	55,700
30. Payment on operating loan													
31. • Principal							49,600					36,100	85,700
• Interest							1,405					1,412	2,817
32. Ending cash balance	2,038	2,007	2,085	2,020	2,067	2,014	2,081	2,032	2,091	2,071	2,057	4,174	4,174

Debt Outstanding	Beginning Balance	1	2	3	4	5	6	7	8	9	10	11	12	Total
33. Long term	170,000	170,000	164,645	164,645	164,645	164,645	164,645	164,645	164,645	164,645	164,645	164,645	164,645	XX
34. Intermediate loan	0	5,000	14,400	14,400	17,600	34,600	38,120	38,120	51,080	62,080	63,040	63,040	64,000	XX
35. Operating loan	30,000	30,900	32,400	60,300	64,300	68,800	71,700	22,100	24,300	25,300	33,400	36,100	0	XX
36. Accrued interest on operating loan	0	309	633	1,236	1,879	2,567	1,405	221	464	717	1,051	1,412	0	XX

each of the first six months. The operating loan balance increases from $30,000 on January 1 to $71,700 at the end of June. The planned sale of the finished feeder pigs in July should provide money to pay the accumulated interest on the operating loan and to repay some of the principal. Andy's projection includes payment of $1,405 of interest and $49,600 of principal during July, reducing the operating loan balance to $22,100 at the end of July. The operating loan is projected to increase monthly from August through November. Andy's projected sale of soybeans in December should provide the cash receipts required to repay the remaining interest and principal on the operating loan during December, leaving an ending cash balance of $4,174 on December 31.

Two additional points should be noted about the debt outstanding section of the projected cash flow. The amount of long-term and intermediate-term debt is decreased by the amount of principal payments on those loans, as illustrated by the annual payment on the long-term loan in March. Finally, only the interest accrued on the operating loan is listed on line 36. Interest also accrues on the long-term and intermediate-term loans, but the scheduled principal and interest payments on this debt are included on lines 21 and 22. An individual using this cash flow should recognize that interest has accrued on $164,645 of long-term debt from April 1 through the end of the year, and interest has accrued on the intermediate loan throughout the year. These interest costs were not cash items during the year; like other noncash items, such as depreciation and inventory change, they are not recorded on the projected cash flow.

The total column serves as a check on the computation. It can also be interpreted as a projected annual cash flow. The rows in the cash receipts and cash outflows sections simply record the annual totals for the item. The total or annual flow-of-funds summary is completed by using the starting cash balance for the year ($2,083) and calculating the cash difference in the usual manner. The annual borrowing is obtained by summing the monthly totals. Similarly, the annual payments of interest and principal on the operating loan are the sum of the monthly totals. Completing the annual flow-of-funds summary in the usual manner results in the ending cash balance of $4,174 for the end of the year, the same value obtained for the monthly projection. A comparison of the monthly and the total column illustrates the relationship of the monthly cash flow to the annual cash flow.

Development of a projected monthly cash flow provides a framework to record the expected financial consequences of the production, marketing, and financial plans. This affords an opportunity to evaluate whether the plans appear to be technically feasible and consistent with the managerial capacity available. It should be obvious that a reliable cash flow projection will enable an operator to determine the timing and magnitude of borrowing, as well as the timing and magnitude of loan repayment. Basing the financing plan on a projected cash flow provides information on liquidity and loan repayability that is very important for both the lender and the farm operator.

Projected Income Statement

The projected income statement to measure profitability (Table 6.10) can be prepared by combining the appropriate cash items from the projected cash flow with the noncash sources of income and expense for the year. The farm operating receipts and expenses are transferred from the projected cash flow to the income statement. Notice that purchases and sales of breeding stock, machinery and equipment, and buildings are recorded in the section on adjustments for changes in capital items. Andy completes the ending inventory line by projecting the December 31 value of corn and soybeans, the three groups of pigs that will be on feed, supplies, and prepaid expenses and accounts payable. It is important to include accrued interest at the beginning and end of the year to more accurately estimate the net farm income for the year. In Andy's first year, accounts payable at the beginning of the year include $180 on account at the machinery dealer, plus $12,750 accrued interest on the real estate loan ($170,000 at 10 percent for 9 months). Andy projects no accounts payable on December 31, but estimates that accrued interest on the intermediate loan and real estate loans will be $5,285 and $12,348 ($164,645 at 10 percent for 9 months), respectively. Projected farm profit or loss for the first year is $48,785. Andy does not project any off-farm income for the year. Subtracting the $17,500 of proprietor withdrawals results in an addition to retained earnings of $31,285.

Projected Balance Sheet

Andy also develops the projected balance sheet for the end of the first year's operation. The balance sheet for both the start of the year and the end of year 1 is shown in Table 6.11. Notice that the sections and line numbers are the same as the business portion of the balance sheet in Figure 2.1. The personal part can be added if it is important to do so. The asset section is completed by using the ending cash balance from the projected cash flow and the ending values for inventory and capital assets on the projected income statement. One of the issues is whether to include appreciation of machinery, equipment, and real estate in the ending value of assets. Our recommendation is to use current market value of these assets for the current balance sheet and current market value less reasonable depreciation to estimate the ending value. This permits a comparison of how the structure of assets and liabilities is expected to change over the year, based on the changes in external financing and the use of earnings generated by the business. Then the expected appreciation is considered in

Table 6.10 Projected Income Statement
for Andy's Operation with Alternative Plan 2

Period Covered:	*January 1 Year 1 to December 31 Year 1*

Farm Operating Receipts				Farm Operating Expenses		
Livestock and livestock products				Corn enterprise	$ 16,754	
	Units					
Slaughter hogs	( )	$ 53,753		Soybean enterprise	9,242	
	( )					
	( )					
	( )			Finishing feeder pigs	26,958	
	Subtotal	$ 53,753	(1)			
Crop sales				Farrow-to-finish	4,999	
Corn	( )	$ 36,000		Utilities	2,000	
	( )			Other repairs	2,180	
Soybeans	( )	48,485		Taxes, real estate, sales	2,100	
	( )			Insurance	800	
	Subtotal	$ 84,485	(2).	Rents		
Other operating receipts				Trucking and market		
Miscellaneous		$ 800		Hired labor	2,495	
				Farm interest paid		
				($2,817 + $17,000)	19,817	
	Subtotal	$ 800	(3)	Other		
Gross farm operating				Gross farm operating		
receipts (1) + (2) + (3)		$139,038	(4)	expense	$ 87,345	(5)
				Net cash operating income		
				(4) − (5)	$ 51,693	(6)

Value of Farm Products Produced on the Farm and Consumed by the Household	$ 0	(7)

Adjustment for Changes in Inventory

	Feed and Grain	Market Livestock	Accounts Receivable	Supplies and Prepaid Expenses	Accounts Payable		
Ending inventory	$65,200	$22,440	$ 0	$ 2,000	Begin $ 12,930		
Beginning inventory	63,000	0	0	913	End 17,633		
Net adjustment	2,200	22,440	0	1,087	(4,703)	$ 21,024	(8)

(continued)

Table 6.10   *(Continued)*

Adjustments for Changes in Capital Items

	Breeding Livestock	Machinery and Equipment	Buildings and Improvements		
Ending inventory	$ 8,464	$ 104,742	$ 103,300		
Plus sales	3,253	0	0		
Subtotal	11,717	104,742	103,300	(9)	
Beginning inventory	0	108,191	71,500		
Plus purchases	14,000	9,000	41,000		
Subtotal	14,000	117,191	112,500	(10)	
Net capital adjustment (9) − (10)	(2,283)	(12,449)	(9,200)		$(23,932) (11)
Farm profit or loss (6) + (7) + (8) + (11) (returns to unpaid labor, operator's labor, equity capital, and management)					48,785 (12)
Off-farm income					0 (13)
Total net income (12) + (13)					48,785 (14)
Proprietor withdrawal					17,500 (15)
Addition to retained earnings (14) − (15)					31,285 (16)

the section on summary of liquidity, solvency, and profitability that follows, to estimate the effect of appreciation on these financial characteristics of the firm.

The liabilities are obtained from the ending loan balances listed on the cash flow and from the ending accounts payable on the income statement. The amount of principal to be repaid during the second year is estimated for both the real estate and the intermediate loans, and is recorded as a current liability. The accrued interest on the two loans, shown in the ending accounts payable section of the income statement, also is recorded as a current liability. The deferred portion of the two loans is recorded on the appropriate lines, and the balance sheet can be completed.

Contrast of the Three Financial Statements

After listening to a discussion of how to project the three financial statements, a student once commented, "It looks to me like all you are doing as you go from one statement to the next is rearranging the same numbers!" This statement emphasizes the need to keep the purpose of each of the statements in mind as they are being completed and to understand some basic differences in the data used to prepare them.

The projected cash flow provides an indication of liquidity (whether sufficient cash inflows will be available to meet the cash outflows month by month throughout the production period) and loan repayability. It identi-

Table 6.11 The Projected Balance Sheet
for Andy's Operation with Alternative Plan 2

Assets	Actual 1/1/Year 1	Projected 1/1/Year 2	Projected 1/1/Year 3
Current Business			
1. Cash and checking account	$ 2,083	$ 4,174	$ 29,051
2. Farm notes and accounts receivable	0	0	0
3. Livestock held for sale	0	22,440	22,440
4. Crops held for sale	63,000	65,200	65,200
5. Value in growing crops	0	0	0
6. Farm supplies	913	2,000	2,000
7. Prepaid expenses	0	0	0
8. Other	0	0	0
9. Total current assets	65,996	93,814	118,691
Intermediate Business			
10. Machinery, equipment, and vehicles[a]	108,191	104,742	103,093
11. Breeding livestock	0	8,464	8,464
12. Movable farm buildings	0	0	0
13. Securities not readily marketed	0	0	0
14. Other	0	0	0
15. Total intermediate assets	108,191	113,206	111,557
16. Total current and intermediate assets (9 + 15)	174,187	207,020	230,248
Long-Term Business			
17. Farmland[a]	480,000	480,000	480,000
18. Permanent buildings and improvements	71,500	103,300	92,050
19. Other	0	0	0
20. Total long-term assets	551,500	583,300	572,050
21. Total business assets (16) + (20)	725,687	790,320	802,298
Liabilities			
Current Business			
27. Accounts payable	180	0	0
28. Notes payable within 12 months	30,000	0	0
29. Principal payments on longer term debt due within 12 months	5,355	12,233	13,583
30. Estimate accrued interest	12,750	17,633	18,826
31. Estimate accrued tax	0	0	0
32. Accrued rent	0	0	0
33. Total current liabilities	48,285	29,866	32,409
Intermediate Business			
34. Deferred principal owed	0	57,658	50,555
35. Deferred accounts payable	0	0	0

(*continued*)

Table 6.11   (*Continued*)

Liabilities (continued)	Actual 1/1/Year 1	Projected 1/1/Year 2	Projected 1/1/Year 3
36. Deferred notes payable	0	0	0
37. Contingent income tax liabilities[a]	0	0	0
38. Total intermediate	0	57,658	50,555
39. Total current and intermediate (33) + (38)	48,285	87,524	82,964
Long-Term Business			
40. Deferred principal on-farm real estate	164,645	158,754	152,274
41. Other	0	0	0
42. Contingent capital gains tax liability on real estate[a]	52,700	52,700	52,700
43. Total long-term liabilities	217,345	211,454	204,974
44. Total business liabilities (39) + (43)	265,630	298,978	287,938
45. Net worth (21) − (44)	460,057	491,342	514,360
46. Total business liabilities and net worth	725,687	790,320	802,298

[a]Projected appreciation is not included.

fies the amount and timing of all cash items, including new borrowing, principal payments, and proprietor withdrawals, as must be done to answer questions on operating loan requirements and the availability of funds for debt repayment. However, it is not adequate to estimate profitability because it does not include the noncash items.

The purpose of the projected income statement is to measure the expected profitability of the business. The measure used is farm profit or loss (also referred to as net farm income), the before-tax residual return to unpaid family labor, the operator's labor, equity capital, and management. To project net farm income, the statement includes all expenses to be incurred (whether or not they are paid during the production period) and all receipts to be earned (whether or not they are converted to cash during the production period). Thus, the income statement includes the change in the value of inventories, the change in accounts payable and receivable, the value of items produced on the farm that are consumed by the household, and the change in the value of the capital assets. The income statement considers neither the amount of funds borrowed nor principal payments, because these cash flows as such do not represent income earned and expense incurred. The interest paid for the use of borrowed funds is an expense for the use of debt capital, and it is deducted in calculating net farm income. Given the definition of net farm income, it follows that proprietor withdrawals for income tax, social security payments, family living, and savings are not subtracted in calculating the residual return. Therefore, the income statement projects profitability but includes several noncash sources of income and expense, and omits several cash flow items required to estimate liquidity and loan repayability.

The purpose of the projected balance sheet is to show the effect of the production, marketing, and financing plan on the structure of assets and liabilities and on the solvency of the business. It estimates the value of the assets, liabilities, and the net worth (or equity capital) for the business at a point in time. In planning, the projected balance sheet is prepared for the end of each production period. The comparison of current assets and liabilities at the beginning and end of the first year indicates changes in the liquidity position of the firm over the first year. Changes in total assets, total liabilities, and net worth also provide information on changes in solvency to be expected over time.

Summary of Liquidity, Solvency, and Profitability

In addition to the information on liquidity during year 1 provided by the projected cash flow, the estimate of current capital (current assets − current liabilities) provides data on the change in the firm's liquidity position. Andy's current capital increased from $17,711 to $63,948 over year 1. The current capital ratio shown in Table 6.12 is projected to increase from 1.37 to 3.14. This change results largely from Andy's plan to use equity capital generated during the year to repay the operating loan and to borrow the money required for the new swine enterprise on an intermediate-term loan. This will have the effect of reducing current liabilities while the inventory of hogs on feed is increasing, making a very sizable change in both the amount of current capital and the current capital ratio.

Information on solvency is provided by the change in net worth and the net capital ratio. It is useful to calculate this both with and without appreciation of capital assets in order to determine the change in solvency that can be attributed to operation of the business and the change that can be attributed to ownership of the assets. Prior to any adjustment for appreciation, Andy's net worth increases from $460,057 to $491,342. However, the relative increase in liabilities is greater than the relative increase in assets, and the net capital ratio declines slightly from 2.73 to 2.64 (line 4 of Table 6.12). Andy expects the value of land to increase about 10 percent, with the value of other capital assets remaining constant. An increase of $48,000 in the value of land will increase Andy's contingent tax liabilities approximately $9,600, making the increase in net worth with land appreciation $69,685 ($529,742 − $460,057) and the end-of-year net capital ratio 2.72.

The profit or loss estimated on the net income statement provides one measure of profitability. This can be divided between a return to labor, equity capital, and management using the procedures presented in Chapter 2. Andy calculates equity capital for year 1 as the average of beginning and ending net worth without land appreciation. This averages $475,700

Table 6.12 Summary of Liquidity, Solvency, and Profitability Analysis for Andy's Plan 2

Item		Jan. 1 Year 1	Jan. 1 Year 2	Jan. 1 Year 3
1.	Current capital	$ 17,711	$ 63,948	$ 86,282
2.	Current ratio	1.37	3.14	3.66
3.	Net worth without projected appreciation	$460,057	$491,342	$514,360
4.	Net capital ratio	2.73	2.64	2.79
5.	Appreciation of capital assets	XX	$48,000	$100,800
6.	Total assets with appreciation	XX	$838,320	$903,098
7.	Increase in contingent tax liability	XX	$9,600	$20,160
8.	Total liabilities with appreciation	XX	$308,578	$308,098
9.	Net worth with appreciation	XX	$529,742	$595,000
10.	Net capital ratio with appreciation	XX	2.72	2.93

Item		Year 1	Year 2
11.	Profit or loss	$48,785	$40,518
12.	Interest on projected net worth in business at 6 percent	$28,542	$30,171
13.	Opportunity cost of unpaid family labor	0	0
14.	Opportunity cost of operator's labor	$15,000	$15,000
15.	Residual return to operator's labor and management	$20,243	$10,347
16.	Operator's labor and management earnings per month	$1,687	$862
17.	Residual return to equity capital	$33,785	$25,518
18.	Rate of return on equity capital (percent)	7.10	5.07

for year 1. Andy recognizes that the cash rate of return on capital invested in real estate is quite low. He calculates a weighted average opportunity cost rate of return on his equity capital as 6 percent, making the interest on equity capital $28,542. The residual return to the operator's labor and management is $20,243 or $1,687 per month. Alternatively, Andy finds that when he assigns an opportunity cost of $15,000 per year to his labor, the residual return to $475,700 of equity capital is $33,785, an average annual rate of return of 7.10 percent.

Andy will want to compare these indications of profitability with the values from his farm records for previous years. He may also want to consider the sensitivity of his plan to changes in prices or yields by analyzing the magnitude of a change in one of these items on the cash flow, the net farm income, and the net worth of the business. If the levels and variability projected are not consistent with Andy's goals, he may want to go through this process with another plan.

INTERMEDIATE-RUN PLANNING

When the plan for the first year is considered acceptable, the operator must consider whether projections should be made for one or more subsequent years. Enough years should be planned to move through any transitions in production and financing that may cause significant fluctuations in year-to-year cash flows. If the cash flows for the last year planned are expected to prevail in subsequent years, the data for the last year indicate the general level of profitability and the trend in liquidity and solvency expected in subsequent years.

Again, the first task is to develop a production, marketing, and financing plan for the year and to complete the projected cash flow. This can be done by projecting a quarterly or annual cash flow. Given the improvement in the current capital ratio, an annual cash flow should be adequate for Andy. Andy makes his projection for year 2 in Table 6.13 based on the budgets and price conditions he expects to prevail during the second year. He decides to project receipts based on expected production levels and average long-run planning prices. However, he expects the costs of certain inputs (fertilizer, chemicals, and hired labor) as well as some of the overhead costs to increase in year 2. He also estimates $12,000 will be needed to replace machinery. He calculates the interest and principal payments for the intermediate loan (financed over 7 years at 12 percent) and real estate loans. He estimates that a line of operating credit for $30,000 will be required and that interest on the operating capital will total $2,000 for year 2. The ending cash balance for year 2 is $29,051, somewhat more than for year 1, indicating that Andy should be in a position to repay the intermediate- and long-term loans more rapidly than scheduled.

Andy also completes the projected income statement (in Table 6.14) for year 2. The operating receipts and operating expenses are taken from the projected cash flow for year 2. Andy projects no change in inventories of grain or livestock for year 2, but the accrued interest on the two loans is $1,193 greater at the end of year 2 because of the new intermediate loan. Thus, inventory adjustments are negative. Capital adjustments include depreciation and capital purchases. The net adjustment is a minus $19,551. Projected farm profit is $40,518 during the second year, and the addition to retained earnings is $23,018. Andy also completes the balance sheet shown in Table 6.12 for the end of year 2 (January 1, Year 3).

The implications of the year 2 projections for Andy's liquidity, solvency, and profitability are summarized in the right-hand column of Table 6.12. The liquidity position at the end of year 2 as indicated by the current capital ($86,282) and the current ratio (3.67) is even better than that pro-

Table 6.13 Projected Annual Cash Flow for Andy in Year 2 with Plan 2

Item	Year 2
Cash Receipts	
Livestock sales	$ 83,048
Crop sales	68,973
Miscellaneous income	800
Capital sales	0
Total cash receipts	152,821
Cash Outflows	
Corn production	17,602
Soybean production	9,830
Farrow-to-finish swine	22,791
Utilities	2,100
Other repairs	1,900
Taxes, insurance, and rent	3,100
Hired labor	2,744
Capital expenditures	12,000
Proprietor withdrawals and income taxes	17,500
Interest and long-term loan payments	
• Principal	12,233
• Interest	24,144
Total cash outflow	125,944
Flow-of-funds summary	
Cash balance beginning	4,174
Cash receipts	152,821
Cash outflow	125,944
Cash difference	31,051
New borrowing	30,000
Payment on operating loan	
• Principal	30,000
• Interest	2,000
Cash balance end	29,051
Loan Balances End of Year	
Long term	158,754
Intermediate	57,658
Operating	0

jected for the end of year 1. Net worth also shows an increase during the second year, and the net capital ratio at the end of year 2 is somewhat higher than at the current time. Including the appreciation in capital assets increases net worth more rapidly and improves the net capital ratio.

The analysis of profitability indicates that projected net farm income

Table 6.14   Projected Income Statement for Andy in Year 2

Period Covered: *January 1 Year 2 to December 31 Year 2*

Farm Operating Receipts				*Farm Operating Expenses*		
Livestock and livestock products				Corn enterprise	$ 17,602	
	Units			Chemicals and other		
Swine sales	(____)	$ 77,700		crop supplies		
_____	(____)	_____		Soybean production	9,830	
_____	(____)	_____		Storage		
_____	(____)	_____		Farrow-to-finish swine	22,791	
	Subtotal	$ 77,700	(1)	Feeder livestock bought	_____	
Crop sales				Breeding	_____	
Corn	7,991 bu	$ 23,973		Veterinary	_____	
_____	(____)	_____		Livestock supplies	_____	
Soybeans	6,000 bu	45,000		Fuel and oil	_____	
_____	(____)	_____		Utilities	2,100	
	Subtotal	$ 68,973	(2)	Machinery repairs	_____	
Other operating receipts				Other repairs	1,900	
Miscellaneous sales		$ 800		Taxes, real estate, sales	3,100	
_____		_____		Insurance	_____	
_____		_____		Rents	_____	
	Subtotal	$ 800	(3)	Trucking and market	_____	
Gross farm operating				Hired labor	2,744	
receipts (1) + (2) + (3)		$147,473	(4)	Farm interest paid	26,144	
				Other	_____	
				Gross farm operating		
				expense	$ 86,211	(5)
				Net cash operating income		
				(4) − (5)	$ 61,262	(6)

Value of Farm Products Produced on the Farm
and Consumed by the Household $____0   (7)

Adjustment for Changes in Inventory

	Feed and Grain	Market Livestock	Accounts Receivable	Supplies and Prepaid Expenses		Accounts Payable	
Ending							
inventory	$65,200	$22,440	$ 0	$ 2,000	Begin	$ 17,633	
Beginning							
inventory	65,200	22,440	0	2,000	End	18,826	
Net adjust-							
ment	0	0	0	0		(1,193)	$ (1,193) (8)

(continued)

Table 6.14 (*Continued*)

	Period Covered:	*January 1 Year 2 to December 31 Year 2*

Adjustments for Changes in Capital Items

	Breeding Livestock	Machinery and Equipment	Buildings and Improvements			
Ending inventory	$ 8,464	$ 103,093	$ 92,050			
Plus sales	5,347	0	0			
Subtotal	13,811	103,093	92,050	(9)		
Beginning inventory	8,464	104,742	103,300			
Plus purchases	0	12,000	0			
Subtotal	8,464	116,741	103,300	(10)		
Net capital adjustment						
(9) − (10)	5,347	(13,648)	(11,250)		$(19,551)	(11)

Farm profit or loss (6) + (7) + (8) + (11) (returns to unpaid labor, operator's labor, equity capital, and management)	40,518	(12)
Off-farm income	0	(13)
Total net income (12) + (13)	40,518	(14)
Proprietor withdrawal	17,500	(15)
Addition to retained earnings (14) − (15)	23,018	(16)

is lower in year 2 than 1. With a lower net farm income and increasing net worth, it is not surprising that both the residual return to operator's labor and the rate of return on equity capital are lower in year 2 than year 1. As Andy analyzes why this is the case, it is obviously because he is projecting the price of hogs and soybeans to be somewhat lower in year 2 than year 1, while projecting higher farm expenditures for some inputs, reducing net farm income. While this is somewhat bothersome, it simply implies that somewhat lower gross margins are being projected for year 2 rather than suggesting a basic problem with his production, marketing, and financing plan. Furthermore, the analysis of liquidity and solvency does not suggest that the lower gross margins will cause Andy any particular difficulty during the second year.

USING COMPUTER PROGRAMS FOR WHOLE-FARM PLANNING WITH BUDGETING

The four-step procedure described for whole-farm planning with budgeting requires many calculations. The individual using this procedure to plan a farm business must understand both why and how the calculations are performed so that the input data are prepared properly and the results can be evaluated. Completing the calculations for several plans by hand can be time consuming and provides the opportunity for arithmetic errors. These problems may cause the individual to focus on the computations rather than on preparation of the input data and evaluation of the results.

Computer programs that largely parallel the computational procedures described in the four steps have been made available for use in many areas of the world through the agricultural extension service, other public agencies, and some private concerns. Use of these programs requires development of data on the appropriate goals, the resources available, and the planning restrictions discussed in Step 1. They also require the development of enterprise budgets and the selection of alternative whole-farm plans for comparison. The appropriate data on the resources, enterprise budgets, and farm plans to be compared are inputed and used to prepare complete budgets, projected cash flows, projected income statements, and projected balance sheets.

The speed and arithmetic accuracy with which the computations can be completed are beneficial in several respects. They permit the individual to concentrate on the preparation of the appropriate input, including the development of plans for comparison, and evaluation of the results. The speed with which the computations can be completed permits studying the desirability of a larger number of plans and testing the sensitivity of the results to changes in production levels, price levels, credit availability, and other factors of interest in selecting a plan.

Summary

Whole-farm planning involves the systematic evaluation of alternative farm organizations. Through the whole-farm planning process, the manager can evaluate various production, marketing, and financial plans and select the plan that is the most consistent with the goals specified. Whole-farm planning involves four steps: (1) specify goals and inventory the available resources and planning restrictions; (2) select a long-run plan; (3) develop a short-run plan; and (4) develop intermediate-run plans. Various computational procedures can be used in the whole-farm planning process, but the simplest and most universally accepted procedure is that of whole-farm budgeting.

The starting point for the planning process is specifying the goals and identifying which goals are most important over the next year, the next five years, and a longer period of time. The resource inventory includes the available land, the supply of operator, family, and hired labor, and the available tangible working assets, including machinery, buildings, and breeding stock. A summary of the availability of equity and debt funds that can be used to finance the business is also necessary. Finally, the institutional restrictions and regulations (laws, regulatory agencies, environmental restrictions, etc.) as well as the managerial capacity must be realistically appraised.

The second step in whole-farm planning is to identify and evaluate alternative long-run plans and select one of those plans for implementation. Since this step considers long-run profitability, the evaluation should be based on prices and efficiency factors that will prevail over a longer period of time such as five years rather than current conditions. The process of selecting plans to evaluate should focus on ways to improve productive efficiency and employ unused resources. More profitable enterprise combinations often can be identified by selecting limiting resource(s) and allocating the limiting resource(s) using the equal marginal return principle. For crop farms, the amount of cropland or irrigation water is typically the appropriate limiting resource; for livestock operations it may be livestock facilities. The budgeting procedures used in the second step of whole-farm planning are complete budgeting and partial budgeting.

Once the long-run plan has been developed, short-run planning that focuses on the timing of input purchases, production activities, product sales, and investments during the coming year must be completed. The steps involved in short-run planning include (1) determining the enterprise combination and production systems to be used along with the timing of production, marketing, and financial activities, (2) projecting monthly cash flows for the upcoming year, (3) developing a projected income statement for the first year to estimate the profitability of the plan, and (4) projecting a balance sheet to estimate the effect of the plan on the structure of assets and liabilities. The techniques used to develop these various accounting statements were discussed in Chapter 2.

Intermediate-run plans that project financial consequences and implications for two or three years into the future also can be developed using the same types of budgeting techniques. Because the computations can be tedious and time consuming, computerized budgeting techniques may be used to accomplish whole-farm planning in a more efficient fashion.

The four-step procedure provides a flexible method of systematically evaluating alternative farm organizations. Step 3 must be completed each year. In some cases, the operator may want to expand the business and follow the same general plan. For these situations, completing Steps 1, 3, and 4 may be sufficient. In other cases, the operator is interested in the broader question of what organization appears best for the farm. Following the entire four-step procedure is appropriate for these cases.

The farm planning procedure provides a method to simultaneously consider the effect of production, market, and financial planning. The discussion in this chapter has emphasized production planning, but the effect of market and financial management concepts presented in the next two chapters can be analyzed in combination with production strategies using the procedures discussed in this chapter.

Questions and Problems

1. What is the purpose of whole-farm planning?
2. Identify and discuss the steps of whole-farm planning.
3. How might you go about the process of developing an inventory of the land for a farm operation?
4. What other resources should be inventoried in developing whole-farm plans? How might you develop this inventory?
5. What concepts and procedures can be used to identify alternative long-run plans?
6. Discuss the complete budgeting procedure. When should you use this procedure in whole-farm planning?
7. Collect the appropriate data and complete a whole-farm budget for a farm operation.
8. Discuss the partial budgeting procedure. When might you use this procedure in whole-farm planning?
9. Collect the appropriate data and apply the partial budgeting procedure to a whole-farm planning decision.
10. What are the four parts of the short-run whole-farm planning process? Why develop short-run plans?
11. What is the purpose of intermediate-run planning?

Further Reading

Barnard, C. S., and J. S. Nix. *Farm Planning and Control.* Cambridge: Cambridge University Press, 1973, Chapters 13 and 14.

Carson, E. E., and others. "Farm Planning and Financial Management." ID-68 Cooperative Extension Service, Purdue University, West Lafayette, Ind., rev. 1975.

Harsh, Stephen B., Larry J. Connor, and Gerald D. Schwab. *Managing the Farm Business.* Englewood Cliffs, N.J.: Prentice-Hall, 1981, Chapters 9 and 10.

Herbst, J. H. *Farm Management: Principles, Budgets, Plans.* Champaign, Ill.: Stipes Publishing, 1976, Chapter 4.

Milligan, Robert A., Stuart F. Smith, and Eddy L. LaDue. *Monthly Cash Flow Planning.* Ithaca, N.Y.: Cornell University Department of Agricultural Economics A. E. Ext. 76–2, 1976.

Osburn, D. D., and K. C. Schneeberger. *Modern Agricultural Management.* Reston, Va.: Reston Publishing Company, 1978.

Thomas, Kenneth H., Richard O. Hawkins, Robert A. Luening, and Richard N. Weigle. *Managing Your Farm Financial Future.* St. Paul, Minn.: North Central Regional Extension Publication 34, 1973.

MARKET PLANNING

CHAPTER CONTENTS

Although marketing and market planning are essential for successful management of the farm business, many farmers do not take a systematic approach to making marketing decisions. In the past, prices of inputs and products were relatively stable (at least in comparison to current times), and pricing and delivery options were limited. A large proportion of agricultural commodities were priced at the time of delivery, and delivery occurred when the farmer needed cash to purchase other inputs, the storage facilities were needed for other crops, or the livestock had reached market weight. With relatively stable prices and fewer marketing options, the profit opportunities from spending management time on marketing decisions were small compared to production decisions. Consequently, market planning received little emphasis in farm management decision-making, and probably rightly so.

But modern-day agriculture faces a new market environment. First, fluctuations in both commodity and input prices are much larger than in the past. For example, during 1980, hog prices (seven major Midwest markets) fluctuated from $28.97 per hundredweight in April to $48.40 per hundredweight in August; ten years earlier in 1970, hog prices ranged from $15.70 in December to $28.73 in February. Likewise, corn prices

(Chicago) ranged from $2.54 per bushel in January to $3.54 in December in 1980, while in 1970 they varied from $1.24 in March to $1.54 in December.[1] Input prices have also become more variable. Interest rates have fluctuated dramatically in recent years, and prices of commercial fertilizer and supplement for hogs and cattle have also changed significantly even from month to month. Furthermore, the number of pricing and delivery options available to the farm manager has increased. Opportunities to forward contract the sale price and conditions of grain and livestock products or to use the futures market to price such products are more readily available now than in the past. Choices among quality differentials and weight discounts and premiums in livestock production must be evaluated to determine the most profitable weight and grade of livestock to market. Alternative market and delivery outlets must be analyzed to determine whether the product should be delivered to local elevators, country buyers, or shipped to terminal or regional markets. The alternative dealers for inputs such as fertilizer, seed, and chemicals must also be evaluated as to reliability, quality, price, and service. Market planning is, therefore, an integral part of farm management in modern agriculture, because marketing decisions can have a significant impact on profits.

Market planning must be integrated with production and financial planning if the management process is to be successful. In particular, the long-run production planning process, which involves the choice of enterprises that will be included in the business, requires long-run projections as to the market prices of products and inputs and evaluation of long-run contracting and pricing alternatives. The year-to-year adjustments in production plans that are part of intermediate-run planning require market information on price cycles and changing market outlets and services. Annual production and financial planning cannot be accomplished without information on seasonal price patterns as well as marketing decisions as to when and how to price inputs and outputs.

In this chapter, we will briefly outline the key components of a market plan and the decisions that must be made in the market planning process. Our discussion will include market decisions with respect to both inputs and outputs used in the production of crops and livestock. Short-run planning will be emphasized, but marketing considerations in intermediate- and long-run planning will also be reviewed. Clearly, a discussion of such a broad topic in such a few pages will not allow us to evaluate all of the marketing options and alternatives in depth. Our purpose here is to systematize the market planning process and to identify and briefly describe some of the various marketing options; detailed evaluation of those options is left for other discussions.

---

[1]USDA-ERS, *Agricultural Prices,* various years.

MARKETING DECISIONS

Development of a marketing plan for the farm business requires answers to the following six questions: (1) When to price? (2) Where to price? (3) What form, grade, or quality to deliver or purchase? (4) What services to provide or acquire? (5) What method to use in pricing? and (6) When and how to deliver? Each of these six decisions will be discussed in brief here; the last section of this chapter will discuss in detail alternative strategies with respect to each of these questions.

1. *When to price?* This decision requires determining the time when the price for a particular product or input will be established. In the past, most farmers priced their commodities or inputs at delivery. However, with the opportunities for forward contracting and the use of futures markets, the time of pricing can be significantly different from the time for delivery. For example, a producer can sign a contract in January that will establish the price for grain he expects to plant and produce during the growing season and deliver to the buyer in December. A cattle feeder can contract at a set price for future delivery both his feeder animals and the market livestock. Or a grain producer can price his fertilizer through a contract agreement many months prior to delivery. With the wide fluctuations of input and commodity prices noted earlier, the decision as to when to price inputs and outputs is an important component of the market plan.

2. *Where to price?* Although the number of market outlets for some products and inputs has declined in some geographic regions of the United States, in many cases, particularly when regional as well as local outlets are included, the options as to where a product or input can be bought or sold have increased. Should grain farmers sell their crops at the local elevator or at a regional outlet? Should they deliver the crops directly to a processor or exporting firm? Should livestock producers deliver directly to the packer or use local or terminal livestock marketing facilities? What about selling through cooperatives versus noncooperative firms? Should inputs be bought from local dealers or from wholesalers? These questions as to location or where to price products and inputs are also part of the market plan.

3. *What form, grade, or quality?* For most grain and livestock products, a price differential is established for different qualities or grades. Given this price differential, a producer must determine the profit opportunities for producing products with various qualities. For example, should cattle producers attempt to market primarily "choice" cattle, or should they market cattle that will grade "good"? And at what weight should they market— 1,000 pounds or 1,200 pounds? Should hog producers market their animals at 200 pounds or 240 pounds given the discounts and premiums for different weight categories? Should commercial supplement be purchased in bulk or bag form? As to fertilizer, should dry or liquid fertilizer be used? Particularly in fresh and frozen fruits and vegetables and livestock products, the quality differentials may be substantial, and a choice must be made as to which quality will be produced and marketed.

4. *What services to use?* This marketing decision is particularly important with respect to purchasing inputs. Various dealers or suppliers may provide different types and qualities of services, including free pickup and delivery, extended credit terms, quality repair services, maintenance contracts, and warranties. With respect to products, some buyers will offer cash, whereas others may only offer delayed payment terms. Some may be able to provide long- or short-term storage. Services may not be a major consideration in marketing products compared to price. However, for inputs, particularly capital items such as machinery and equipment, the repair or other services offered may be very important in choosing among various marketing alternatives.

5. *How to price?* This marketing decision essentially involves choosing among such alternatives as cash sales, forward contracting, hedging with futures markets, delayed sales contracts, basis contracts, and formula contracts; these and other alternatives will be discussed in detail later. The methods available to price a product or input are numerous, and the economic advantages in terms of increased profits as well as reduced risks will be different for each method and producer. In many cases, pricing a product on the date of delivery, which has been the traditional marketing method for many farmers, may result not only in lower profits but also in significantly higher risks for the producer. Analysis of the alternative methods for pricing is important in today's volatile agricultural environment.

6. *When and how to deliver?* In some cases, particularly if a product is priced when it is delivered to the market, the choice of when and how to deliver may already have been made. If futures markets are used to price a commodity, however, the choice of a delivery place and time is separated from the pricing decision. The product could be delivered to satisfy the futures contract, although rarely does delivery on such contracts occur. More typically, the futures contract is offset, and the product is delivered at a local or regional market. When and where the product will be delivered depends to a significant degree upon the relationship between the cash price and the nearby futures contract price. Even when a product is priced in the cash market at a local outlet, it may be desirable to deliver the product to the regional elevator or the processing plant or to take delivery of inputs at the manufacturer's or the processor's plant, depending upon the cost that will be incurred for transportation, distribution, or setup. Therefore, in some cases the delivery decision is separate from the pricing decision.

STEPS IN MARKET PLANNING

As with production and financial planning, market planning requires a systematic procedure for specifying goals and analyzing alternative methods to accomplish those goals. The steps in market planning will be reviewed here; then alternative methods and strategies to implement this plan will be discussed.

1. *Specify goals.* Explicit specification of marketing goals is often over-looked in planning. In Chapter 6, we indicated that the overall goals of the operator(s) must be specified to adequately evaluate a farm plan; here we concentrate on the specific goals of the market plan. Implicitly, many farm managers appear to have the goal of selling their commodities at the highest possible price or buying inputs at the lowest price. For most producers, however, this goal of hitting the high or low is unrealistic, particularly given the volatility and uncertainty of prices in today's agriculture. Rarely is one able to consistently predict market price movements to accomplish this goal, and an individual with this objective will almost certainly be frustrated. This frustration may lead to unpredictable and emotional behavior rather than an analytical and systematic approach to marketing products and buying inputs. Alternative goals would appear to be more realistic.

One alternative marketing goal would be to price a product or input to meet a particular profit or rate of return target. For example, a producer might strive to obtain a specified rate of return in his or her grain operation (for example, 12 percent) or a particular gross margin or profit per head on his or her hog or cattle enterprise as discussed in Chapters 4 and 5. Given this target, the marketing goal will be to price the product when this minimum rate of return or profit per head is reached. Such an approach requires estimation of the cost of production and the breakeven prices for the various commodities being priced so that the profit or rate of return can be determined. A profit or rate of return goal could be implemented with various strategies, including forward contracting, hedging, or even possibly cash sales.

This approach of pricing a commodity or purchasing an input when a particular profit or rate of return goal has been achieved is consistent with the concept of integrating the marketing decisions and plans with production and financial decisions and plans. However, it is typically necessary to combine this goal with other goals, so that if prices do not rise sufficiently to meet the target profit or rate of return, a marketing strategy can still be implemented. Thus, contingency plans that recognize alternative goals must be included in the marketing plan.

A third possible goal of the marketing plan would be to sell above the average or median price, or to sell with an average price in the top one-third of the price range during the marketing year. Such a goal recognizes the unpredictability of prices and the inability to pick highs and lows; instead, the objective is to sell at an average price for the entire volume of production that is "acceptable." Note that if this goal or the profit goal identified earlier is to be achieved, adjustments must be made in selling prices to account for any storage, handling, or carrying charges that may be associated with various marketing alternatives. Thus, the goal should be to obtain a specified profit above all marketing costs or a *net* selling price above the average, median, or in the top one-third of the price range in the marketing year—the selling price must be adjusted for storage, handling, and other carrying charges.

For some farmers, the marketing objective is to sell commodities when cash is needed for other purposes and to buy inputs when cash is available. Meeting the cash flow of the business is an important consideration in market as well as financial and production planning. However, choosing a marketing strategy based only on cash flow ignores the difference between pricing a commodity and delivery of that commodity for cash payment. As noted earlier, it may be possible to set the price for a commodity or for an input prior to delivery through use of futures contracts or forward contracting techniques; delivery could then be specified when cash is needed or is available. Furthermore, the opportunity to use borrowed funds to cover cash flow needs and to hold on to commodities or schedule the purchase of inputs until more favorable prices can be obtained must be evaluated. In many cases, it may be desirable to use credit to meet cash flow needs rather than have cash flow be the major and only determinant of the marketing plan.

Finally, a key concern in marketing is that of risk. Price volatility is a major source of risk for many farm businesses, and the successful farm manager must manage risk through use of various production, financial, and marketing strategies. Using various marketing strategies to reduce risk may enable a producer to operate a larger business and to use more leverage without increasing the total financial and business risk of the farm operation. Risk due to price volatility can be managed, and astute producers will incorporate risk reduction as part of their marketing plan.

Clearly, a producer will want to consider a combination of the goals noted above. Although one goal may be predominant, other goals should be considered in specifying the market plan. Regardless of the goals chosen, a reasonable and effective market plan cannot be developed and implemented unless realistic goals have been specified.

2. *Determine production schedule and volume.* The marketing plan must be integrated with the production and financing plan. This integration occurs through estimation of the volume of various products to be produced and the timing or schedule of production as specified by the production plan. The detailed production plan developed for short- and intermediate-run planning in Chapter 6 provides estimates of the quantities of various inputs that must be purchased and the volume of various products that will be available for sale. This plan also indicates the time when inputs will be applied or utilized as well as when products will be available for sale.

The volume and schedule of inputs required and products produced must be estimated well ahead of the actual implementation of production plans if a reasonable marketing plan is to be developed and implemented.

As we will note later, there are seasonal fluctuations in the prices and availabilities of various inputs, and unless the volume, quality, and quantity of these inputs are known well ahead of their utilization, the opportunity to take advantage of lower prices cannot be exploited. For example, if producers must purchase all of their feed for livestock production, they may want to price corn, commercial supplement, or other feed ingredients well ahead of when they will actually be used; in particular, they may want to price grain in the fall of the year when such commodities are frequently at their seasonal low prices. Or grain producers may want to price fertilizer well ahead of when they actually plan to apply it. Without a complete set of production plans, it is difficult, if not impossible, to determine the quantity of inputs needed in the upcoming production period and the volume that could be priced prior to delivery.

Similarly, an estimate of the volume of production is essential to develop a marketing plan that includes the use of hedging, forward contracting for any marketing procedure where the pricing decision is made prior to delivery of the product. For example, grains can be priced up to almost a full year before they will be available for delivery through the use of hedging or forward contracting arrangements. Most analysts suggest that such forward pricing procedures be utilized only for a portion of the expected production because of the uncertainty due to weather, disease, pests, or other risks of obtaining the expected yield. Without an estimate of future expected production, it is extremely difficult to implement any forward pricing alternatives, whether they call for selling all or only a portion of the crop or livestock product to be produced. In essence, projection of future production schedules, the quantity of inputs needed to implement production plans, and the expected volume of output is necessary to implement any marketing strategy that involves pricing the product or input prior to delivery. And if one limits one's marketing alternatives to cash sales or purchases at delivery, the opportunities to increase income through using profitable marketing strategies are severely limited.

3. *Estimate production costs.* To implement a marketing plan, particularly a plan that includes profit or returns as a goal, the cost of producing the product must be estimated. Production costs are not only essential in determining what price is needed to meet a particular profit goal on selling crop and livestock products, but also useful in evaluating the breakeven price that can be paid for various inputs, such as feeder cattle or feeder pigs. Estimating the cost of production should be relatively straightforward if the producer has developed reasonable production plans. In fact, the production plan should indicate the expected quantity of inputs required and products produced, and by specifying prices for the various inputs, production costs per unit of output can be easily determined.

Two concepts of production costs may be useful in developing the marketing plan. The first concept includes total costs of all resources, whether purchased for cash or contributed by the farmer and his or her family. This concept includes both fixed and variable costs as discussed in Chapter 3 and is essential to implement a marketing plan where the goal is to generate a particular profit per unit of production or a specified rate of return. To illustrate, the total costs of production for corn in central Iowa are represented in Table 7.1. Farmers with these costs would need to sell their corn at an average price of $3.36 to generate a 20 cent profit per bushel. Alternatively, if they wanted a 5 percent profit margin above all costs, they would have to sell their corn at an average price of $3.32.

A second concept of cost that may be useful in implementing the marketing plan is that of cash flow costs. This concept of cost recognizes the cash flow requirements of producing a particular commodity and the fact that these cash costs may be higher or lower than the total economic cost depending on the producer's financial and tenure situation. For example, if corn producers rent all of their land on a cash rent basis and have substantial debt obligations against their machinery and operating inputs,

Table 7.1 Economic and Cash Costs of Producing Corn (per acre)

Item	(1) Total Economic Costs	(2) Cash Costs Owner-Operator (no land or machinery debt)	(3) Cash Costs Owner-Operator (50% debt on land and machinery)
Seed	$18.00	$18.00	$18.00
Fertilizer	50.00	50.00	50.00
Herbicide and insecticide	24.25	24.25	24.25
Miscellaneous	10.50	10.50	10.50
Interest on operating capital[a]	22.00	22.00	22.00
Labor[b]	25.94	25.94	25.94
Variable machinery and equipment costs (fuel, repairs, etc.)	43.12	43.12	43.12
Fixed machinery and equipment costs	69.96	3.00[c]	74.35[e]
Land charge	115.00	12.00[d]	196.11
Total	$378.77	$208.81	$464.27
Per bushel (120 bu yield)	$3.16	$1.74	$3.87

[a]Operating capital funds must be borrowed in all cases, resulting in a cash interest cost on operating capital.

[b]Labor is included as a cash cost, even if the operator does all of the work, to reflect the per acre cash requirements for family living.

[c]Insurance and taxes only.

[d]Property taxes only.

[e]Includes per acre annual principal and interest payments on machinery debt plus insurance and taxes.

their cash flow requirements will be substantially higher than those of farmers who own all of their land and machinery and have no debt obligations to worry about.

The cash flow production costs for two farmers with vastly different financial arrangements are illustrated in columns 2 and 3 of Table 7.1. Note that for farmers who rent their land and have sizable debt obligations, the cash flow cost of production is substantially higher than the economic cost of production of column 1. This occurs because the annual debt servicing requirements for borrowed funds used to acquire capital items is usually larger than the depreciation allowances used in determining economic costs of production. The cash costs for the full owner are much lower than the economic costs of production. Consequently, producers with high leverage or rented land may need higher market prices for their products to cover cash costs and debt and rent obligations compared to producers with fewer borrowed funds and no rented land. Even though the economic costs in the long run may be the same for both producers, the market prices needed to cover financial obligations may be quite different for the highly leveraged renter compared to the low-leveraged owner-operator. Thus, these two producers would be expected to choose different marketing strategies for their products. Estimating both economic costs of production and cash flow production costs is essential to develop a reasonable marketing plan.

4. *Project market prices.* As noted in detail later, projecting the prices of products and inputs requires a basic understanding of the concepts of supply and demand as well as systematic price variations including seasonal, cyclical, and trend changes in prices. Price projections involve the use of outlook information on production, placements, yield, disappearance, inventories, carryover, etc., which can be obtained from various public and private sources. No matter how accurate the data or system used, some method for estimating future prices must be utilized to implement marketing as well as production and financing plans.

5. *Make the marketing decisions.* Decisions concerning when to price, where to price, what form, grade, or quality, what services to utilize, how to price, and when and how to deliver encompass the marketing plan.

Most marketing plans do not include only a single alternative or method for each of these decisions. For example, the marketing plan may include pricing grain at four different times of the year with various methods, including futures contracts, forward contracting, and cash sales. It may involve delivery of part of the crop to local elevators and the remainder to a regional outlet. The marketing plan should also include "backup" or contingency plans so that, for example, if prices do not reach the levels necessary to implement one part of the plan, an alternative is available. Thus, market planning is a complex process, but systematic analysis using the steps outlined here should significantly improve the marketing process for most farm managers.

PROJECTING PRICES

What will prices for cattle be this fall? What about corn prices next year? What will I have to pay for feeder cattle this winter? These are common questions asked by farmers. Many producers find it difficult to obtain answers to these questions, and predicting prices two or three years into the future is even more difficult. Since they cannot predict future prices with certainty, some producers do not develop marketing plans. No one can predict the prices of products and inputs in the future with complete certainty. Even when prices are supposedly set by the manufacturer, such as in the automobile industry, actual prices may vary dramatically from the specified prices due to dealer discounts, rebates, and other adjustments that result from changing supply and demand conditions.

Farm managers should not aspire to predict future prices with complete accuracy. However, they should expect to project a range of future possible prices and be able to attach some probability or chances to specific prices within this range. We will discuss procedures to do this in Chapter 11. Furthermore, if managers understand the forces and factors that influence prices, they will have the basis to make adjustments in their expectations or chances of a particular price occurring in the future when changes in the overall economy, supply conditions, foreign exchange rates, weather in the United States or other countries, or export limitations occur.

We will not present specific price forecasting techniques and procedures here but instead will review some of the basic factors and forces that influence prices as well as systematic price variations for various commodities. In essence, our discussion will focus on what factors influence prices, what information can be obtained about these factors, and how much impact a specific change in these factors will have on prices. The discussion will emphasize product prices, although similar arguments apply to prices of inputs as well.

Factors That Influence Prices

The price of any commodity or product in the long run is determined by supply and demand conditions. Thus, our discussion will first review those factors and forces.

Demand   The demand curve for agricultural products reflects the quantity of that product that will be purchased at different prices given specified levels of population, consumer incomes, tastes and preferences, and the availability of substitutes. The demand curve is downward sloping; with lower prices a larger quantity of the product will be taken from the market. The demand for agricultural products at the farm gate is derived from the consumer demand for food products. Because a minimum quantity of food is needed to have a nutritionally balanced diet, and concern about excess weight as well as biological capacity limits the maximum amount that can be consumed, the quantity of food demanded is not as responsive to price as are some other products such as automobiles, jewelry, or luxury items. In general, a dramatic decline in food prices will not result in substantial increases in the quantity demanded, and dramatic increases in prices will only result in limited reductions in the quantity demanded. The demand curve for food in general is inelastic, and thus the demand for agricultural products at the farm gate is also inelastic.[2] Note, however, that the demand for certain food items is more responsive to price than for other items; for example, the demand for meat products is more responsive to prices than the demand for bread and potatoes. The demand curve then reflects the relationship between price and quantity demanded of a particular product; the steepness of the slope of the demand curve reflects the responsiveness of quantity demanded to price changes.

Various factors may result in shifts in the demand curve: some forces will shift the demand out or increase the quantity demanded at a given price, and other forces will result in a leftward shift of the demand curve with a lower quantity of the product demanded at a given price. An important shifter of the demand curve for farm products is population growth. With increased population (assuming that this increased population has the money to make the purchases), the quantity of food and agricultural products demanded at any particular price will increase. Worldwide population growth has been a major explanation for the increased demand for U.S. agricultural products in export markets. This increased demand has shown up primarily in food grains such as wheat, but growing exports of feed grains including corn and soybeans have also been a result in part of growing population worldwide.

The demand for agricultural products is also significantly influenced by consumer incomes. This is particularly the case for fresh fruits and vegetables, meat products and feed grains, and oilseed crops. Rising real incomes of U.S. consumers during the decades of the 1960s and 1970s resulted in substantial increases in the demand for meat products. To illustrate, the price response of pork and beef to increased consumer incomes is shown in Table 7.3 presented later in this chapter; as personal income increases by 1 percent, beef and pork prices are expected to increase from 0.80 to 1.10 percent. A large part of the increased export demand for feed grains and soybeans can also be explained by rising real

[2]William G. Tomek, and Kenneth L. Robinson, *Agricultural Product Prices* (Ithaca, N.Y.: Cornell University Press, 1972), Chapter 3.

incomes of consumers in Europe, the U.S.S.R., Japan, and other foreign countries; these higher real incomes resulted in increased demand for meat products, which, in turn, increased the demand for feed grains and soybeans to use in livestock production.

As one might expect, the demand for certain agricultural products is much more responsive to changes in income compared to other products; clearly, the demand for fresh fruits and vegetables and meat products is affected more by changes in incomes than the demand for wheat. This income elasticity of demand is an important consideration in evaluating potential shifts in the demand curves for such products as well as the feed grains and oilseed products used in livestock production.

Tastes and preferences also influence the demand for specific agricultural products. Because of religious or other moral beliefs, some groups of consumers do not eat certain meat products. Some individuals simply prefer some food items to others; for example, some people prefer beef to pork, whereas others prefer fish to poultry. And recent concerns about the nutritional and health dimensions of various agricultural products have resulted in some shift in demand, particularly for products that have a high fat or cholesterol content. One might expect that, at least in the United States, concerns about the nutritional and health characteristics of food products will become increasingly important as a determinant of the demand for agricultural commodities.

Finally, and most importantly for many agricultural commodities, the price and availability of substitutes will significantly affect the quantity of a particular product demanded at a specified price, that is, they will shift the demand curve. The impact of the price and availability of substitutes can clearly be seen in the demand for cotton; with the increased availability and declining price of synthetic fibers, the demand for cotton and cotton products decreased. A similar phenomenon may have occurred in the livestock and meat products sector during the late 1970s. With substantially lower prices of poultry and pork products compared to beef, it appears that consumers have shifted consumption away from beef to pork and poultry. Thus, the price and availability of substitutes can significantly influence the demand for specific agricultural commodities; this substitution concept may be particularly important in understanding adjustments to the meat animal and fiber sectors of the agricultural industry.

Supply   Whereas the demand curve reflects the quantity of a product that will be demanded at different prices, the supply curve reflects the quantity of that product that will be available at different prices given specified levels of technology and input prices. The supply curve slopes upward to the right, reflecting the fact that to supply additional quantities of a product to the market, higher costs must be incurred requiring higher prices. The shape of the supply function is basically determined by the shape of the marginal cost curves of the producers of a product.[3] If additional production can occur without substantial increases in cost, then the supply curve will be relatively flat and the quantity supplied will be very responsive to price changes. However, if production adjustments require dramatic changes in costs (for example, bringing marginal lands into production or adding significantly to the productive capacity of agriculture through massive outlays of capital), then the supply curve will have a relatively steep slope and the quantity supplied to the markets will not be as responsive to price changes.

Major shifts in the supply curve will occur because of variations in uncontrollable events such as weather and disease and because of changes in the technology and prices of inputs used in the production process. If new technology improves the efficiency of agricultural production, which has been the case in past years, the supply curve will shift to the right; i.e., an increased quantity of products will be available at the same cost and thus price because of the improved efficiency and productivity of the inputs used in the production process. A decline in the prices of various inputs would also result in a shift of the supply curve to the right, since lower input prices would encourage producers to use more inputs, and thus they would be able to produce a larger volume of output at the same cost. Rising input prices would be expected to shift the supply curve to the left, resulting in reduced supplies at the same cost or price.

Supply and Demand Balance   We will briefly consider several specific supply and demand considerations and procedures used in price forecasting for grains and oilseeds and livestock products. For grains and oilseeds, the key forecasting tool is the balance sheet of supply and utilization. The "balance sheet" summarizes the supply from carryover or inventory, production, and imports for a specified marketing year, as well as the utilization for livestock feed, food, industrial and other uses, and exports. The difference between supply and utilization results in an ending carryover which is extremely important in determining the balance between supply and demand and thus price for a particular year. The balance sheet of supply and utilization can be generated on a worldwide basis or for a particular country. In projecting future prices for U.S. grain and oilseed products, a balance sheet for the United States is typically used. Example balance sheets provided by the U.S. Department of Agriculture for corn for the years 1978–82 are presented in Table 7.2. Note the projected U.S. seasonal average price of corn based on this estimate of supply and utilization. The local price is obtained by adjusting the U.S. price for transportation and other marketing charges to move the product from the local market.

[3]Tomek and Robinson, *op. cit.,* Chapter 4.

Table 7.2 Corn Balance Sheet (millions of bushels)

	1978–79	*1979–80*	*1980–81*	*Projected 1981–82*
Supplies				
Carryover, October 1	1,111	1,304	1,617	1,034
Production and imports	7,269	7,940	6,646	8,202
Total	8,380	9,244	8,263	9,236
Utilization				
Feed	4,368	4,519	4,139	4,250
Food, industrial, and seed	575	675	735	785
Exports	2,133	2,433	2,355	2,125
Total	7,076	7,627	7,229	7,160
Carryover, October 1	1,304	1,617	1,034	2,076
U.S. seasonal average price	$2.25	$2.52	$3.11	$2.40–2.55

Source: USDA-ERS, *Outlook and Situation Reports,* 1982.

The data used to develop a balance sheet of supply and utilization of U.S. grains and oilseeds are collected by the U.S. Department of Agriculture through its worldwide information system. Estimates of acreage, yields per acre, and total production are obtained from sampling procedures and farmer questionnaires in the United States as well as from official reports and personal visits to foreign countries. Domestic utilization by livestock is estimated from livestock numbers and feeding rates adjusted for weather conditions, finished weights of livestock, etc. Exports are determined by monitoring growing conditions in countries that purchase agricultural commodities from the United States as well as conditions in those countries that compete with the United States in the export markets. This information on supply and demand conditions is reported in various forms, including monthly crop reports showing the current year's potential production of major crops, grain stock reports indicating the supplies available at a particular point in time, weather and crop condition reports summarizing the current status of growing crops as well as moisture conditions, and world production and trade reports indicating international movements of grains.

A key entry in the balance sheet of supply and utilization is the size of the carryover projected for the end of the year, which is calculated as total supplies minus total utilization. If the carryover is expected to decrease at the end of the market year compared to previous years, prices typically will rise; if carryover stocks increase substantially, prices will likely decline if these excess supplies remain on the open market and are not acquired through government purchases or other actions such as the reserve program. If carryover stocks are near the minimum levels needed by the trade for business transactions, prices will be expected to be much more responsive to changes in the projected size of the carryover than if burdensome carryover stocks currently existed.

The balance sheet of supply and utilization is used primarily to assess the general outlook for the price of a particular commodity for the marketing season. Variations around this average price would be expected to occur depending upon seasonal price patterns, as will be noted later. One of the important ways that this balance sheet is used to analyze future prices is to update periodically changes expected in supply as growing conditions change throughout the production period, and then determine what impact these changing production and crop conditions will have on expected future prices. In general, assuming that other market factors and conditions remain unchanged, a 1 percent change in total corn supplies is expected to change the seasonal average price by about 2 to 3 percent in the opposite direction. Thus, for example, if crop conditions suggest in July that the corn crop will be 2 percent lower than expected at the beginning of the production period, seasonal average corn prices will be expected on average to increase by about 4 to 6 percent. For soybeans, a 1 percent change in supplies will result in approximately a 2 to 2½ percent change in prices in the opposite direction. For wheat, a 1 percent change in supplies will result in about a 1 to 1½ percent change in prices.

Changes in export sales or domestic consumption of grains can be analyzed in a similar fashion. For example, if new information suggests that increased exports will account for approximately 2 percent more of the total supply of corn, market prices will be expected to increase by approximately 4 to 6 percent assuming other market factors remain constant. Hence, the balance sheet of supply and utilization provides an indication of supply and demand conditions and can be used to estimate seasonal average prices for grain products.

Our discussion of projecting livestock prices will focus on supply considerations, although changes in demand cannot be ignored in determining farm gate prices of livestock. In addition, because of the biological processes involved and the delay in supply adjustment, livestock prices typically exhibit cyclical as well as seasonal patterns as will be discussed in the following section.

To obtain an estimate of livestock supplies, the first step is to project the size of the breeding herd as well as the expected production of that herd as reflected by calf crop, lamb crop, or pigs per litter. Then replacement rates must be estimated to determine what portion of the breeding herd will be liquidated for slaughter and how much of that herd will be replaced by younger animals. Weather conditions that might influence the carrying capacity of pastures, and thus the need to liquidate breeding herds, as well as expected prices of grain and thus the placement of cattle in feedlots at lighter or heavier weights, must also be included as part of the supply and estimation process for cattle. Such factors as interest rates may also be important in estimating supplies, since higher interest costs result in higher costs of production and the slowing of placements on feed as well as expansion of breeding herds. Interest cost may also influence the demand on the part of packers and retailers, because it increases their costs of storage and holding livestock products in inventory, thus decreasing demand in the short run. Data on breeding inventories; placements; calf lamb, and pig crops; pasture and range conditions; weekly marketings; livestock on feed, etc., are published by the U.S. Department of Agriculture.

Once estimates of future supplies are obtained, a price forecast can be generated using the price flexibility coefficients in Table 7.3 adjusted for any seasonal and cyclical effects. These coefficients indicate the expected change in seasonal average price if supplies increase or decrease by a given percentage. For example, the coefficients of Table 7.3 show that a 1 percent change in the supply of pork would be expected to have a 1.7 to 2.3 percent change in the opposite direction on live hog prices. Likewise, a 1 percent change in the supply of beef would affect beef prices in the opposite direction by 1.6 to 2.0 percent. The data in Table 7.3 also indicate the substitution relationship between various livestock products. Thus, for example, a 1 percent increase in beef supplies is estimated to result in a 0.40 to 0.60 percent decline in hog prices; with larger supplies of beef, beef

Table 7.3 Price Response Coefficients[a]

Change in Hog Price from a 1 Percent Change in:	*Percent*
Pork production	−1.70 to −2.30
Beef production	−0.40 to −0.60
Chicken production	−0.30 to −0.60
Total personal income	+0.90 to +1.10

Change in Fed Cattle Price from a 1 Percent Change in:	
Beef production	−1.60 to −2.00
Pork production	−0.30 to −0.50
Chicken production	−0.10 to −0.30
Total personal income	+0.80 to +1.00

[a]A negative sign indicates that the price effect is in the opposite direction of the change in supply; a positive sign indicates that the price effect is in the same direction as the change in income.

Source: Gene Futrell (ed.), *Marketing for Farmers* (St. Louis, Mo.: Doane-Western, 1982, pp. 222–223).

prices will decline and consumers will substitute beef for pork, which reduces the demand and consequently the price of pork.

Although the discussion presented here does not provide a foolproof way to predict future prices, it does identify the major factors and considerations that must be taken into account in developing price expectations. Farm managers must develop their skills to evaluate market forces and information so that they can continually make better price projections to use in market planning.

Systematic Price Variations

Although the terminology "systematic price variations" implies that prices of farm products vary in a very predictable fashion, any student of agriculture will quickly recognize that is not the case. We use the term "systematic" here in an imprecise fashion. Our interest is in general tendencies for prices to move in a particular direction over time such as in cycles or a general upward or downward trend. The systematic price changes discussed here reflect general movements of prices, and one should not expect prices to follow these systematic patterns precisely.

The previous discussion has indicated the factors and forces that must be taken into account in determining the seasonal average price or price range for particular commodities. This seasonal average price must be adjusted for systematic price movements that may occur during the year or over years. One systematic movement is a trend—a general upward or downward movement of prices. In general, prices of agricultural products, like those of most other commodities and products in the U.S. economy, have trended upward during the last two or three decades, in part reflecting general price increases or inflation. Thus, for example, during the 1965–69 period, the annual average price of corn at Chicago ranged from \$1.14 to \$1.34, whereas during the 1975–79 period, corn prices ranged from \$2.26 to \$2.81.[4] The trend in agricultural prices, however, is much more erratic than that of many other products such as steel, automobiles, and refrigerators. Because of the biological nature and competitive characteristics of farming, prices in agriculture typically are more variable than those in most other industries. Therefore, even though the trend of agricultural prices has been upward, the price variations from this trend have been quite substantial, particularly in recent years.

In addition to the trend over time, prices of livestock products exhibit a cyclical movement that includes sequential years of higher prices followed by years of lower prices and then a repeat of the cycle. In most cases, these cyclical price patterns are a function of the biological processes associated with breeding and reproduction of the livestock species. Essentially, price cycles and production cycles are the reverse of each other; when produc-

[4]USDA, Feed Outlook and Situation, various years.

tion volume or supplies are at their peak, prices typically are at the trough of the cycle. Similarly, when production or supplies are at the trough of the cycle, prices are at or near the peak.

The most predominant price cycle in the past has been in the cattle industry where cycles range from 9 to 14 years from peak to peak or trough to trough as shown in Figure 7.1. In the typical cattle price cycle, large beef supplies and low prices result in reductions of breeding stock and, in the short run, even lower prices because of the increased supplies added to the market. But with slaughter of part of the breeding herd, the size of calf crop is reduced, beef supplies begin to decline, and prices begin to rise. After a period of profitable prices, producers begin to expand herd size by holding back replacement heifers rather than shipping them for slaughter. The short-run impact of this expansion is to reduce the market supplies further and to increase prices even more, but as the size of the breeding herd expands and the calf crop increases, beef supplies also increase and prices begin to decline. After a period of declining or low prices, farmers again cut back on breeding herd size, and the cycle starts again. Although the beef cycle has become less predictable and more uncertain than in the past, it still is important to be aware of the current phase of the cycle in making marketing and long-run production plans.

Hog prices have also followed a cyclical pattern of about three to four years in length as noted in Figure 7.1. Similar explanations exist for the hog cycle as have been discussed for the cattle cycle; the hog cycle is shorter because of the shorter gestation and production period for hogs. Some analysts have argued that the hog cycle has been less predictable in recent times because of the increased dominance in the hog industry of large production units that will remain in production in both profitable and unprofitable years. However, there still appears to be some cyclical pattern to hog production and prices, and again it is important to determine the current phase of the price cycle when making market and production plans.

Numerous forces can influence whether the regular production adjustments suggested by price and production cycles occur or whether a cycle might be truncated or extended. For example, a period of growing real consumer income may result in the prices of hogs or cattle remaining relatively high even with expanding supplies, thus extending the peak of the price cycle beyond what is expected. Alternatively, rising input costs such as interest rates or feed costs may result in declining production and supplies of hogs and cattle, even though prices are relatively high, because producers are losing money in spite of the higher prices and are thus cutting back production. One must, therefore, interpret cycles carefully in light of the other forces that may be influencing demand, supply, and prices.

A third systematic price variation that is particularly useful in adjust-

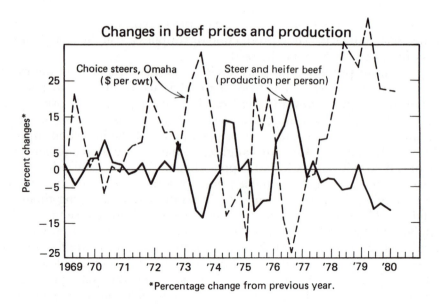

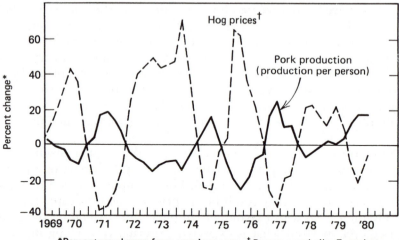

Figure 7.1. Hog and beef production and price cycles. *(Source: USDA Livestock Market News.)*

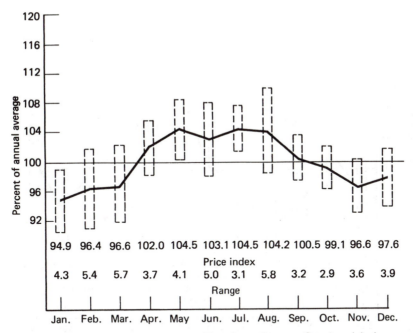

Figure 7.2a. Seasonal price index: Slaughter Steers Omaha (choice grade), 1970–79. (*Source:* Gary Stasko and Gene Futrell, Cooperative Extension Service, Iowa State University.)

ing the seasonal average price is the seasonal price pattern. Seasonal price patterns occur because of the patterns of production and consumption during the year. For many livestock products, particularly cattle and hogs, the seasonal pattern relates explicitly to annual breeding and production decisions; for example, seasonal patterns in cattle production indicate that supplies are typically lower in late fall and early winter. Consequently, cattle prices typically follow a seasonal pattern as noted in Figure 7.2a, with higher prices in the spring and summer months and lower prices in the fall and winter months.[5] For hogs, midsummer and midwinter are typically periods of smaller supplies, while production increases seasonally during the late fall and late winter and early spring months. Consequently, the seasonal pattern of Figure 7.2b exists for hog prices with higher prices in the midsummer and winter months than during the other months of the year. Seasonal price patterns for feeder steers, lambs, eggs, and milk are also shown in Figures 7.2c through 7.2f.

Although these seasonal patterns are still part of livestock production, seasonality is becoming less dominant in agriculture as products are pro-

---

[5]This figure and the others that illustrate seasonal price patterns also indicate the price range in the last ten years for each month.

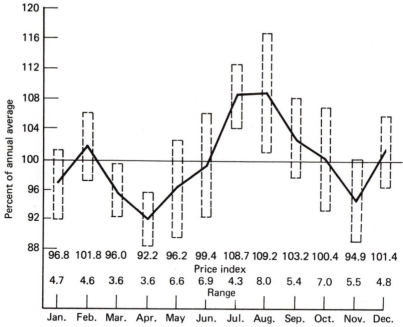

Figure 7.2b. Seasonal price index: Slaughter hogs, seven markets, 1970–79. (Indianapolis, Kansas City, Omaha, St. Louis, Sioux City, St. Joseph, and South St. Paul) (*Source:* Gary Stasko and Gene Futrell, Cooperative Extension Service, Iowa State University.)

duced over a wider geographic area and are thus less subject to weather influences. In addition, seasonality is reduced with production of more livestock in larger units that may operate at a constant or level output compared to smaller, less well-capitalized farming operations.

Seasonality also exists in grain prices. The seasonal patterns of corn, wheat, and soybean prices are identified in Figures 7.3a through 7.3c. Note that both corn and soybeans exhibit seasonal lows during the fall harvesting period and then rise through the winter months until summer where a seasonal decline frequently occurs. Wheat exhibits a similar pattern, with seasonal lows occurring in the summer immediately after harvest and then increases in prices through the fall and early winter months until the expected size of the upcoming crop and its impending availability result in declining prices during the late winter and spring months. With grains, the major seasonal price patterns reflect availability at harvest and then the cost of storage following harvest, as well as the anticipated availability of the upcoming crop during the growing season.

Trends, cycles, and seasonal patterns in prices can be useful in projecting future prices of products and inputs to use in developing the production and marketing plan. In particular, price trends can be useful in devel-

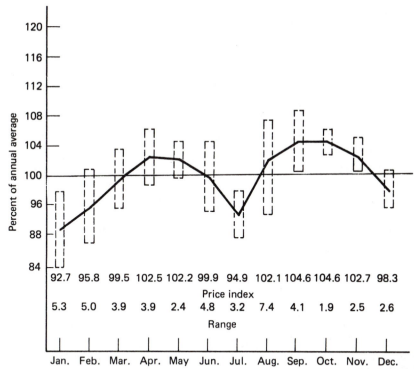

Figure 7.2c. Seasonal price index: Feeder steers Omaha, 500–700 lb (all grades), 1970–79. (*Source:* Gary Stasko and Gene Futrell, Cooperative Extension Service, Iowa State University.)

oping long-run plans as to what products to produce; cyclical price patterns can be used in making the year-to-year adjustments that are part of intermediate-run planning; and seasonal patterns must be considered when putting together the current year's production and marketing schedule. These systematic price movements will not always occur, however, and other forces in the markets, for example, changes in government policy, worldwide weather and export potential, or domestic production patterns, may counter these systematic movements and result in prices different from those expected using the best price projection methods. That is why a good marketing plan includes a combination of strategies to take advantage of favorable prices as well as to protect the firm from unfavorable prices.

Market Information and Projections
As noted earlier, the U.S. Department of Agriculture provides a substantial amount of information which can be used in developing projections of commodity prices. Although some farmers criticize the accuracy of these

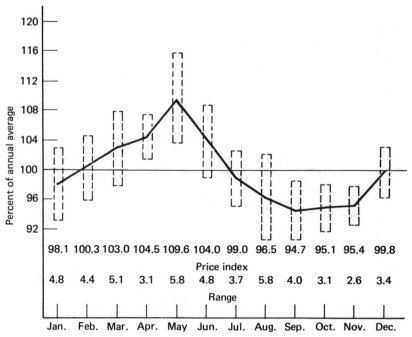

Figure 7.2d. Seasonal price index: Slaughter lambs Iowa-So. Minn. (choice grade), 1970–79. (*Source:* Gary Stasko and Gene Futrell, Cooperative Extension Service, Iowa State University.)

data, with the decentralized agricultural industry of the United States, even the best statistical procedures will not guarantee exact results. All that can be expected is a reasonable estimate, and an improved estimate as new information is obtained about actual supply and demand prospects such as the growing conditions for crops or the placements of livestock. In addition, some government reports such as planting intentions and livestock placements may result in changes in the future behavior of farmers as a group, and thus have a significant impact on future prices. Hence, even if government reports were 100 percent accurate at the time they were published, one would still not expect the projections of prices based on them to come true because of changes in farmer behavior in response to the reports.

Statistics from government reports are utilized by government and university analysts, companies and firms in the marketing channel, private forecasters, and farmers to analyze expected future conditions and the impact these conditions may have on prices. Rather than do their own price forecasting, many farmers rely on outlook information provided by university extension economists or commercial market advisory services. Most state extension services provide weekly, biweekly, or monthly outlook newsletters that attempt to interpret current forces and factors in the mar-

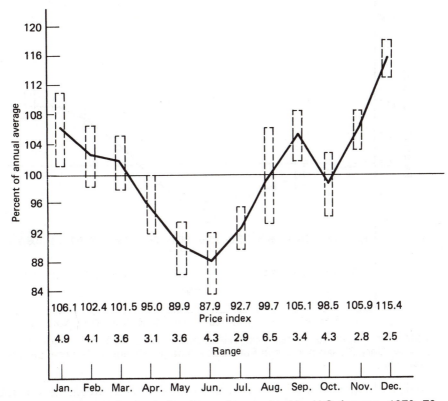

Figure 7.2e. Seasonal price index: Egg price received by U.S. farmers, 1970–79. (*Source:* Gary Stasko and Gene Futrell, Cooperative Extension Service, Iowa State University.)

kets and to indicate what these forces may suggest as to changes in commodity prices. Numerous commercial market advisory services have developed throughout the country to provide not only price forecasts, but also recommendations concerning when to price, how to price, and where to price. Some of these advisory services use sophisticated mathematical equations to project future prices of various products, while others use more traditional judgmental approaches to interpret the potential impact of changing supply and demand conditions. At the current time, there is no concrete evidence that the sophisticated mathematical approach is superior to the traditional judgmental and analytical approach to market price forecasting.

Some producers view the commodity futures markets as a good forecaster of future prices. Volatility of commodity futures prices leads one to be cautious in using futures prices (adjusted for the basis) as forecasts of what cash prices will be at a future date. Recent research suggests that futures prices may not be any better or worse than some of the sophisti-

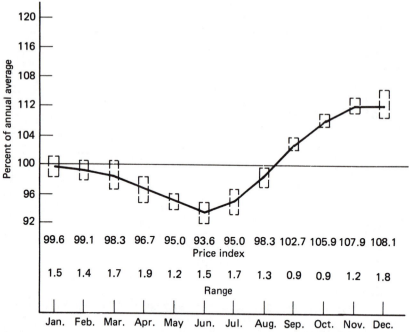

Figure 7.2f. Seasonal price index: All milk price received by U.S. farmers, 1970–79. (*Source:* Gary Stasko and Gene Futrell, Cooperative Extension Service, Iowa State University.)

cated mathematical methods used to forecast prices.[6] Regardless of their use as a forecasting tool, futures prices can play a key role in developing and complementing the market plan, because they provide the base for forward contract bids for most elevators and processors as well as the opportunity to establish a price through the hedging process.

WHEN TO PRICE

As noted earlier, the pricing decision for a commodity or input can be separated from the delivery decision. A manager may want to separate pricing from delivery to take advantage of the seasonal pattern of price variations noted earlier. For grain products and other storable commodities, prices can be set prior to and during the production period, at harvest, or after harvest during the storage period. For most production inputs, livestock products, and other perishables that cannot be stored, prices can be set prior to or at delivery. Regardless of the commodity or input, numerous opportunities for determining the time of pricing exist.

[6]Richard E. Just, and Gordon C. Rausser, "Commodity Price Forecasting with Large-Scale Econometric Models and the Futures Markets," *Am. J. Ag. Econ.* 63, (May 1981): 197–208.

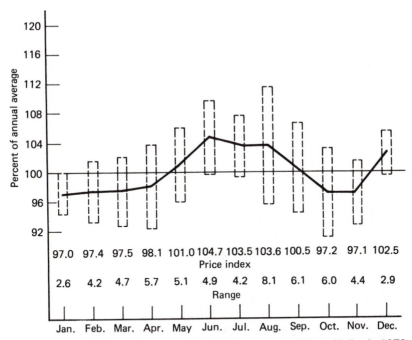

Figure 7.3a. Seasonal price index: Corn Chicago cash (No. 2 Yellow), 1970– 79. (*Source:* Gary Stasko and Gene Futrell, Cooperative Extension Service, Iowa State University.)

Pricing grain and livestock products as well as inputs prior to production or delivery is becoming a more common practice in the agricultural sector. These pricing decisions are typically implemented through the use of hedging or forward contracting methods which will be discussed later. Key considerations in pricing a commodity or input prior to production or delivery include expected changes in prices between the date of pricing and the date the product will be available or delivered, the financial and legal arrangements necessary to guarantee the price, the risk associated with expected production quantities and scheduling, and the financial risk that the firm can bear if prices should be higher or lower than current levels at the delivery date. If prices of inputs are expected to rise or products decline as the delivery date approaches and a more favorable price can be obtained by pricing prior to delivery, it may be desirable to forward price part of the crop or inputs. Even if prices are not *expected* to be significantly different at the delivery date, the potential risk of an *unexpected* rise in input prices or a decline in commodity prices may suggest that forward pricing is desirable.

If forward pricing requires a substantial financial commitment in the form of a deposit or prepayment for inputs, then these costs must be considered in determining the economic benefits of the pricing alterna-

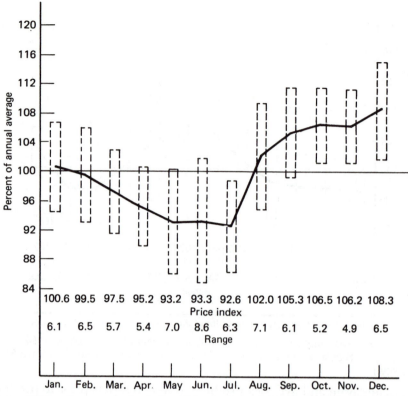

Figure 7.3b. Seasonal price index: Wheat price received by U.S. farmers, 1970–79. (*Source:* Gary Stasko and Gene Futrell, Cooperative Extension Service, Iowa State University.)

tives. And certainly the risk associated with future production activities must be considered in evaluating forward pricing alternatives. Forward pricing of a particular input may involve a commitment to a particular form or quality that may not be as profitable when the production plan is actually implemented. For example, forward pricing of lightweight choice feeder cattle will result in a commitment to feed those cattle even though heavier weight Holstein steers may appear to be more profitable when the cattle are actually placed in the lot. Likewise, pricing a commodity that has not yet been produced involves the risk that weather, disease, or other factors may result in lower production and less actual output than has been priced or contracted. Thus, many producers forward price only a portion of their expected future output so that they do not have to purchase the commodity on the open market to fulfill forward pricing or contracted obligations.

For most perishable commodities as well as a large portion of commodities that are storable, the price is established at harvest or when the production cycle or process ceases. The choice of whether to price at har-

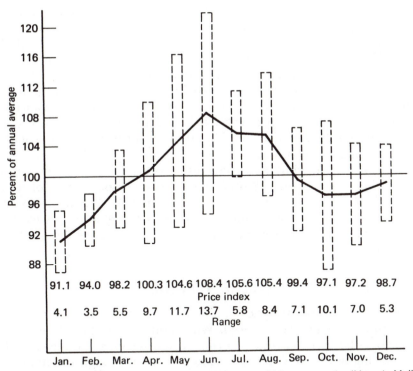

Figure 7.3c. Seasonal price index: Soybeans Chicago cash (No. 1 Yellow), 1970–79. (*Source:* Gary Stasko and Gene Futrell, Cooperative Extension Service, Iowa State University.)

vest or later again depends to a large degree on expectations of future prices, as well as the cost incurred in carrying the product until the future delivery date. However, an astute farm manager should choose harvest date pricing based on a thorough analysis of alternative pricing dates rather than by default because alternatives have not been investigated or the opportunities for storage of the commodity are not available.

For storable commodities, the opportunities to price the product after harvest must also be investigated. Considerations here again include expected future trends and seasonal price patterns, potential variability of future prices and the risk that can be taken by the firm, and the carrying and storing charges that will be incurred. Carrying and storage charges include not only the handling cost that will be incurred to move the commodity into and out of storage and the cost that must be paid for the storage process itself (either the cash cost of storage paid to the elevator or the operating cost associated with on-farm storage), but also the shrinkage and possible deterioration that occur when storing commodities and the interest charges of carrying the commodity in inventory. With current high interest rates and the costs associated with commercial storage, significant price rises after harvest are typically required to cover storage costs. These costs have frequently been recouped by selling the commodity at higher prices during the postharvest period.

With the flexibility to price a commodity prior to, during, and after the production process, the opportunities for timing the pricing decision are enormous. For example, the marketing period for grain can extend over 24 months or more from approximately 12 months prior to harvesting a crop when futures and forward contracts are typically available to 12 or more months after the crop has been harvested when it can still be priced from storage. A manager will usually want to consider a combination of pricing periods; for example, part of expected output may be priced prior to or during the production process, another portion at harvest or when the production process has been complete, part of it after three months of storage (if it is a storable commodity), and the rest of it after 6 months of storage. Numerous other pricing strategies exist and must be evaluated to answer the question of when to price the product or input.

WHERE TO PRICE

The choice of an input supplier or a market outlet depends on a combination of price and services offered. With modern transportation systems, particularly the vast network of primary and secondary roads as well as interstate highways, most farmers find that, even though the number of local suppliers or market outlets has declined, the transportation system has made available a number of nonlocal outlets. Farmers can purchase inputs from a local dealer, a wholesaler, or directly from the manufacturer or processor. Livestock products can be shipped to terminal markets, directly to the packing plant, and to country buying stations, or priced at the farm gate with the purchaser covering the cost of transportation. Grain can also be sold to local, private, or cooperative elevators, to regional shippers, or directly to the processing plant.

A farmer should become a comparison buyer and/or seller just like consumers are comparison shoppers when it comes to purchasing various brands of food in the supermarket or cars from the automobile dealer. Make a list of the alternative market outlets or dealers for a particular product within a 5, 10, 25, or 30 mile radius of your production unit. Most producers are surprised at the number of outlets or dealers on their list. Then consider the cost that will be incurred in transporting the product or input to or from each of these outlets. Compare the price bids within each of the different mileage bands; substantial differences frequently exist in prices offered by different outlets that are only a few miles apart. These differences reflect local supply and demand conditions as well as the ability of various market outlets to handle, transport, or process the commodity. Grain prices may vary within 3 to 5 percent within the same market area on the same day, and livestock prices may be as much as a dollar or two per hundredweight higher at one outlet compared to another.

With rising energy prices, the costs of transportation have become increasingly important in choosing a market outlet. A manager should calculate the cost that will be incurred to deliver a product to various outlets compared to the prices that will be received. In some cases, the prices received on central markets for livestock or at the processing plant for grains and oilseeds may not be sufficient to justify the additional transportation cost that will be incurred to deliver the product there compared to a local market. In other cases, a regional shipper, particularly in the grain business, may provide higher on-farm bids than could be obtained from local outlets, because he can transport the grain directly from on-farm storage by truck to barge or rail-loading facilities.

In addition to transportation costs, the other factors that should be considered in choosing a market include the fees or marketing charges that may be assessed, the marketing services that are available including grading, quality, and quantity discounts, the availability of alternative methods to price the product including cash and contract opportunities, and the price competitiveness of the market. If commercial storage is used, the choice of a particular market outlet must be combined with the choice of a storage facility, since it usually is quite costly to move products from one storage facility to a different market outlet. Check the competitiveness of the price bids as well as the cost of storage at various outlets before you make a decision as to where to price.

The financial condition of the market outlet has become increasingly important in choosing among alternative outlets in recent years. Grain elevators and packing plants can go bankrupt. And frequently, those farmers who have delivered grain or livestock to such firms find that they are treated by the bankruptcy courts as unsecured creditors and thus do not have a very high priority in the bankruptcy settlement process. The direct risk that a farmer faces is large enough without having to take additional risks caused by an unscrupulous dealer or poorly managed market outlet.

With livestock as well as with some crop products, a major consideration in choosing the location of a market is the shrink or weight loss that will occur during the transportation and shipment process. Livestock incur two types of shrink: fill shrink and tissue shrink. Fill shrink occurs from the normal waste and water excretion processes of the animal; such shrink is hard to control directly, and the amount of weight loss attributable to fill shrink can be reduced only by the timing and conditions of weighing the animals. If the payweight for the animals is based on weights at the farm, fill shrink will be substantially less than if the payweight is determined once the animals have arrived at the processing or packing plant.

Tissue shrink is moisture loss from the animal tissue. Tissue shrink is primarily influenced by the time between the last opportunity to have access to feed and water and the time of slaughter; if the animal is slaughtered within approximately 12 hours of when it last had access to feed and water, tissue shrink should be minimal. However, substantial stress in the sorting, loading, or shipment process as well as no access to feed or water for an extended period of time can result in tissue shrink. Tissue and fill shrink can amount to 1 to 2 percent of the body weight of the animal, depending on the way the animals are transported and handled as well as the timing and conditions of determining animal payweight.

WHAT SERVICES TO ACQUIRE OR UTILIZE

Marketing firms, particularly input supply firms, offer different services along with the product. Although most decisions concerning the choice among different outlets will be primarily influenced by price, services can be very important in some instances. With respect to machinery and equipment, the quality of the repair and maintenance services that are offered can be extremely important in the choice among various brands or even dealers for the same brand. A dealer who stocks the major parts or can obtain hard-to-get parts with overnight service; has well-trained mechanics and service personnel; will make personnel available for on-farm adjustments and servicing of major items; or can provide a temporary replacement if a major power or harvesting unit must be laid up for repairs during the critical seasons would probably be preferred by most managers to a dealer who cannot provide reasonable parts and repair services. Likewise, a dealer who may be willing to rent or lease equipment on a short-term basis so that potential customers can obtain some experience with the machinery prior to purchase may be able to attract additional business.

The credit terms that are available from the dealer in purchasing operating inputs such as feed, seed, fertilizer, and chemicals as well as machinery and equipment may be important; delayed payment terms, interest-free credit periods, and extended open accounts may be services that will influence a farmer's choice among products or dealers. Services such as soil testing by fertilizer companies, ingredient analysis by feed companies, and irrigation schedule and pest management recommendations by equipment and chemical suppliers may also be of interest to various producers. Some manufacturers may have price adjustments if the input is purchased in a different form (for example, bulk versus bag for feed), and certain dealers may be able to handle various forms of a product such as liquid, dry, and anhydrous fertilizers. The delivery services may be significantly different, with some dealers offering immediate delivery to the farm at no charge and others charging for delivery services or requiring a number of days' notification prior to delivery. And convenience can be particularly important to a producer, particularly if convenience can result in reduced time spent in purchasing and taking delivery of input items. A related characteristic of reliability, which means that the dealer will typically have the desired input in stock when it is needed, can also be extremely important for some producers.

The services offered by product purchasing firms may also be important in deciding where to sell the product. Some firms will offer a variety of marketing methods including cash sales and forward contracts. This flexibility may be important for some producers in choosing one marketing outlet over another. Payment terms may also be important; some firms will pay immediately in cash or a cashier's check, whereas others may delay payment for one or two weeks. Some firms will provide on-farm pickup service at a nominal fee or no charge, whereas others will require delivery charges to be paid by the producer. In livestock marketing, some firms will buy only on a liveweight basis, whereas others will offer liveweight or grade and yield options. The grade and yield alternative may be particularly important for producers who are interested in improving the quality and performance of their livestock.

Even though it may not be considered explicitly as a service, many farmers have the opportunity to sell their products to private companies or a cooperative. In these circumstances, the patronage distribution and equity redemption program of the cooperative may be important in the choice among cooperatives or between a cooperative and noncooperative market outlet.

Even such considerations as the hours when the input supply or market outlet are open to transact business may be important in choosing a dealer or market outlet. The ability to handle delivery without long time delays has become extremely important in recent years, particularly in the Midwest where some grain elevators do not have adequate handling facilities and may require delays of two to three hours to unload during the critical harvesting period.

Numerous other illustrations of services that are available or can be acquired could be mentioned. The relative importance of services compared to price in choosing between various dealers or market outlets will depend upon the individual manager, but clearly services are one of the considerations in developing a marketing plan and choosing a market outlet.

WHAT FORM, GRADE, OR QUALITY

The form, grade, or quality of input or product may have a significant impact on the price paid or received. With respect to inputs, certain forms may be more readily available and lower priced than others; for example, anhydrous fertilizer will frequently be priced lower than dry fertilizer, and bagged commercial supplement may be higher priced than the same product in bulk form. Feed additives may be differently priced if purchased separately than as part of a pre-mix. In livestock production, feeder heifers are typically priced lower per hundredweight than feeder steers, and heavy feeders or yearlings have a lower price per hundredweight than light calves. In addition, the quality of the feeder animal will significantly influence its price.

The differential price of inputs by grade, form, or quality will typically change over time; therefore, producers cannot choose a particular quality of input and not make adjustments with changing prices if they want to implement the most profitable production and marketing strategy. This is particularly true with feeder livestock where the expected price of the finished animals as well as the relative availability of various qualities, weights, and grades will influence the relative prices of steers versus heifers, light feeder pigs versus heavy pigs, etc. Flexibility to handle different qualities, grades, or forms of inputs can be important in taking advantage of favorable prices in the input markets.

In some sense, a similar decision, particularly with respect to quality, must be made in purchasing capital items. With the rapid rate of technological advance in agriculture, the choice must be made between less expensive facilities and equipment that may wear out more quickly compared to more expensive equipment that has a longer useful life but may become technologically obsolete before it wears out. This is particularly the case in decisions concerning the construction of new livestock facilities for the confinement feeding of hogs and cattle.

The issue of quality, grade, and form is also important in the product market. The manager in a fruit and vegetable operation must choose to produce for the fresh, the frozen, or the canned market. Likewise, in citrus operations the choice between the fresh fruit or canned juice market must be made. In dairy operations, a price differential is paid for Grade A compared to Grade B milk, and some processors now provide premiums and/or discounts depending upon the butterfat content. These market differentials affect production planning as well; for example, the type of production facility and the rations fed can have a significant impact on milk quality as well as fat content.

The grade and quality of livestock and grain products to sell must also be determined. In recent years, substantial discounts have existed for over-fat cattle, and the price premium paid for prime compared to good to choice cattle has narrowed significantly. In hog production, some packers will pay premiums for hogs that are marketed at various weights, whereas others will discount selected weights, particularly heavyweight hogs, substantially. Opportunities to sell corn from the field prior to drying to livestock producers who can handle wet corn must be compared to the alternative of selling number two corn at 14.5 percent moisture content. And for very large producers, the opportunities to blend different qualities of grain to take advantage of the minimum acceptable levels of foreign matter, test weight, etc., should be evaluated. Thus, a producer must evaluate the cost compared to the benefits of alternative weights, grades, and forms of livestock and grain products to sell; the partial budgeting procedure presented in Chapter 6 can be very useful in this analysis.

HOW TO PRICE

There are three basic ways to price any commodity: cash sales at delivery, hedging with futures markets, and forward contracting. Each of these options will be discussed in turn. Although the discussion will emphasize various methods of selling products, most inputs can also be priced with different methods, and the following comments are equally applicable to the pricing of inputs as well as products.

Cash Sales

As noted earlier, a large majority of agricultural products are priced at delivery for cash. For livestock and fruits and vegetables that are perishable and difficult to store, cash sales typically occur when market weights are achieved or at harvest. Even if the marketing strategy includes only cash sales when the production process is complete, the timing of sales can still be important. For some products, livestock products in particular, prices may be more favorable during certain days of the week. For example, in the Midwestern cattle markets, most of the sales occur early in the week in order to minimize the chances that the packer will have to carry an inventory over the weekend. Since local market conditions vary, local prices should be checked to determine if a weekly pattern exists.

For storable commodities such as grains, opportunities for cash sales after harvest should be evaluated by comparing the expected seasonal patterns in prices to the carrying cost of storage. Although storage is not always the best option, in a large majority of the cases the higher prices received after harvest will more than compensate for storage costs, at least the variable costs of storage.

As noted earlier, it is virtually impossible to hit the top of the market when selling farm commodities because of the uncertainty and volatility of commodity prices. In fact, some studies have shown that almost two-thirds of the grain sold by farmers is sold in the lower one-third of the seasonal price range. To improve the chances of receiving an average or higher price and to reduce the risk of receiving well below-average prices, some managers are using a periodic selling plan to market commodities. One such plan would be to sell approximately one-fourth of the crop during four predetermined months—say, December, March, July, and September. Alternatively, some producers attempt to market a portion of their crop each month. Similarly, larger livestock producers may be able to market animals on a monthly basis. Spreading sales rather than one-time selling may not only improve the average price received, but also reduce the variability of prices received over time.

Another method that can be used to improve the chances of marketing the total crop at a higher than average price is a "scaling up" procedure. A "scaling up" plan calls for the pricing of a certain amount or proportion of the total crop as prices hit successively higher targets. For example, one "scaling up" procedure would be to sell one-fifth of the corn crop at a price of $2.90 and one-fifth of the crop at each $0.10 rise in price. If we used this procedure, the total crop would be sold at prices ranging from $2.90 to $3.30 with an average price of $3.10. If such a "scaling up" procedure is used, the producer must decide when to begin scale-up selling and the size of the increments to activate additional sales. Estimates of the cost of production can be useful in determining when to begin selling, and historical ranges of price variability may provide some indication of the total range in prices expected, and thus the increments to use in activating additional sales. The producer must also develop plans to complete sales in case markets do not move high enough to activate the total scale-up plan. Thus, "scaling up" plans are only part of the total plan for pricing a commodity; contingency plans must also be developed in case market prices do not rise sufficiently to activate the entire plan.

A variation of the scale-up plan is to sell increasingly larger quantities

Table 7.4 Alternative Scale-up Selling Procedures

Price	Bushels Sold	Dollar Sales
$2.90	10,000	$ 29,000
3.00	10,000	30,000
3.10	10,000	31,000
3.20	10,000	32,000
3.30	10,000	33,000
	50,000	$155,000

Average price per bushel = $3.10

Price	Bushels Sold	Dollar Sales
$2.90	2,500	$ 7,250
3.00	5,000	15,000
3.10	10,000	31,000
3.20	15,000	48,000
3.30	17,500	57,750
	50,000	$159,000

Average price per bushel = $3.18

of the commodity as prices rise. For example, if producers are planning to market 50,000 bushels of corn with a scale-up procedure, their plan may call for selling 2,500 bushels at $2.90, 5,000 bushels at $3.00, 10,000 bushels at $3.10, 15,000 bushels at $3.20, and 17,500 at $3.30. If prices rise such that the full scale-up plan can be implemented, the average price in this situation will be higher than would occur if an equal amount of the commodity were marketed at each price level as shown in Table 7.4.

Hedging

A second method that can be used to price a commodity is that of hedging with the commodity futures market. Through the use of hedging, the pricing decision is separated from the decision of how and when to deliver. Although the technicalities of hedging and the various strategies that can be used to determine when to place and lift hedges are relatively complicated, a simple example can be used to illustrate the basic concepts.

Assume a farmer produces 20,000 bushels of corn, and during April the December corn futures price is observed to be $3.80 per bushel. After adjusting the futures price for the basis of $0.50 (the expected difference between cash prices and futures prices in the delivery month) and commissions and interest of $0.10, the $3.80 futures price would result in a $3.20 per bushel local price. Since $3.20 is acceptable to the producer and he or she does not want to take a chance that corn prices will be substantially lower after harvest, he or she decides to "lock-in" this price for half the crop by hedging on the futures market. The transactions for this hedge are shown in Table 7.5. Two 5,000 bushel December corn contracts are sold in April, and then when the crop is harvested and sold at a local market in November for $3.02, the December futures contracts are bought back, thus closing out the futures market position at a net profit of $0.18 per bushel. Even though the 10,000 bushels of corn sold brings only $3.02 on the cash market, the gain in the futures contract of $0.18 per bushel results in a net price of $3.20. If the cash market for corn had been higher, say $3.31, then

Table 7.5 Using the Futures Markets to Price Corn

Date	*Cash Market*	*Futures Market*
April		Sell two 5,000 bushel futures contracts, December option, at $3.80 ($3.20 equivalent local price)
November	Sell 10,000 bushels at $3.02 per bushel	Buy back two 5,000 bushel futures contracts, December option, at $3.52
Net price	$3.02 per bushel	$0.18 per bushel ($3.80 − [$3.52 + $0.10])
Total price per bushel (cash plus futures markets) = $3.20		

the futures contract would most likely have been closed out at a loss of approximately $0.11 per bushel, again resulting in an average net price of approximately $3.20 per bushel.

Hedging works because cash prices and futures prices tend to fluctuate together. The futures price quotation reflects the price at an acceptable delivery point and time such as December delivery in Chicago. Transportation, storage, carrying charges, and interest are required to deliver a commodity from the local market to the accepted delivery point. These costs are reflected in the basis which is the difference between local cash prices and futures prices. Because of changing supply and demand conditions as well as transportation, interest, and storage costs, the basis may fluctuate throughout the crop season so that the perfect hedge illustrated in Table 7.5 will not always occur. But basis fluctuations are typically much smaller than price fluctuations during the year. Therefore, even though hedging does not eliminate all the risk associated with pricing a commodity, it does substitute the fairly small basis risk for the fairly large price risk.[7] Thus, hedging can be used not only to lock in a profitable price, but also to reduce price risk. However, by reducing downside price risk with a hedging strategy, the possibility of gains from an increase in market prices is also eliminated.

To use hedging, a capital or cash reserve is required to meet margin calls. Margin calls are deposits to restore the margin account or performance bond to its original level after a loss in the futures position has occurred. Consequently, while hedging may reduce price risk, it may increase financial risk when margin calls occur if funds are not available to meet these calls.

Various rules can be used to establish hedges. Producers who are hedging their crops should evaluate basis patterns and estimate their cost of production so that they know what price is needed to hedge a profit rather than a loss. Some producers use the periodic or scale-up pricing strategies discussed earlier in determining when to place hedges in the futures market as well as when to make cash sales.

Hedging can be used to price products prior to or after harvest for storable commodities. However, if a hedging procedure is used to price a commodity prior to or during the production process, care must be exercised not to price more product than will be produced, since this would put the individual in a speculative position. When hedging is used, decisions must also be made as to when the hedge will be lifted as well as when it will

[7]The concept of the basis and fluctuations in the basis are very important in developing hedging strategies and in determining when to place and lift hedges. A thorough discussion of this concept is beyond the focus of this brief review of hedging; for a more detailed discussion, see Merrill Oster, "Commodity Futures for Profit: A Farmer's Guide to Hedging," Investor Publications, Cedar Falls, Iowa, 1979, and Thomas A. Hieronymus, "Economics of Futures Trading," Commodity Research Bureau, One Liberty Plaza, New York, New York, 1971.

be placed. For a true hedge position, the initial position should be offset when the cash commodity is priced. Thus, for example, if a November soybean futures contract is sold in May to establish a hedge position and the soybeans are sold for cash in October following harvest, a November contract to liquidate the futures market position should be bought on the same day in October that the beans were sold for cash.

Hedging can also be used to lock in prices of purchased inputs as well as prices of commodities that will be sold. For example, some livestock producers use futures markets to lock in the purchase price of feeder cattle or of corn and soybean meal. The opportunities to use financial futures markets to lock in interest rates have also been discussed recently, although the sizes of the contracts in interest rate futures are typically so large that many farmers find it impossible to use this method as a way of protecting themselves from interest rate volatility. If producers use futures markets to lock in the purchase price of inputs, they in essence have committed themselves to follow through with the production plan that will require those inputs; at least changes in production plans may become more difficult and costly. For example, if producers lock in a price on feeder cattle, and if conditions in the future change such that they do not want to place cattle in the lot, they may have to incur significant costs to offset their hedges on purchased inputs.

Forward Contracting

The third alternative for pricing a product or input is to forward contract it. Forward contracting has become increasingly important as a method of pricing commodities in recent years, particularly in the grain markets. Forward contracts are used to take advantage of favorable prices that may occur prior to production or utilization of the product or input, to lock in a price and thus reduce price risk, and to guarantee a market outlet. An additional benefit of forward contracting may be access to additional borrowed funds, particularly for a producer who is highly leveraged. By forward contracting part of their production, highly leveraged farmers may reduce their risk and increase their credit-worthiness and ability to borrow additional funds. In general, forward contracting has important advantages over hedging; problems such as potential margin calls, an unstable basis, or the minimum size of the contract do not exist with forward contracting. A contract sale is less flexible than hedging, however, because delivery is required on a contract, while a hedge can be lifted at any time. Thus, any penalties associated with nondelivery or delivering products of different quality specifications than those indicated in the contract should be thoroughly evaluated before a forward contracting arrangement is used. This decrease in flexibility may be advantageous, however, to producers who are tempted to speculate by lifting their futures market hedge when margin calls are received.

A number of forward contracting options are available; we will discuss some of the options used in the grain market to illustrate the kinds of possible arrangements. A *cash forward contract* is an agreement between the producer and marketing outlet concerning the terms of future sale and delivery of a product or purchase of an input. Such contracts usually include specification of the quality or type of commodity and discounts for deviations from these specifications, the quantity of product or input, the specific price, the location and date of delivery, and procedures for settling any disagreements. Although some cash forward contract arrangements are based on oral agreements, recent problems in enforcing such agreements suggest that a written contractual arrangement should be utilized. Depending upon the product, additional terms may be included in the contract such as payment arrangements and responsibility for delivery and grading cost. Forward contracting arrangements are commonly used in pricing fruit and vegetable products and to some extent in marketing livestock products as well as grain.

A second forward contracting alternative is the *delayed or deferred pricing contract* arrangement. With this option, all the terms of the contract agreement other than the exact price are specified, and producers are given some flexibility as to when and how they will set the price for the commodity. In grain marketing, two common types of pricing formula are used in delayed pricing contracts: a fixed basis formula and a service charge formula. With a fixed basis contract, the specific price received by the seller is determined as the futures price on the date of the seller's choice less the basis between cash and futures price on the date the commodity was delivered to the buyer. With this arrangement, sellers fix the basis or discount they will get from the futures price on the date of delivery, but they have the flexibility to wait until a later date to set the gross price from which this basis will be subtracted to determine the net sale price. Producers who use this arrangement will want to deliver the crop when the basis is narrow to reduce the size of the discount from the futures price; if they deliver when the basis is wide, the discount from the futures price will be larger.

With a fixed basis contract, title for the grain usually passes from the buyer to the elevator when it is delivered, and the elevator frequently sells the grain and purchases a futures contract to protect its obligation to the seller. Consequently, sellers may receive part of their proceeds in cash shortly after the grain is delivered, with the remaining proceeds received when the grain is actually priced at a later date. Note, however, that since the seller no longer has title to the grain, he or she is really an unsecured creditor of the purchaser. If for some reason the elevator encounters financial difficulties, the seller may not be in a particularly strong financial position to receive the remainder of his or her payments. Consequently, one should thoroughly investigate the financial strength of the purchaser before entering into fixed basis forward contracts.

The *service charge contract* agreement is quite similar to the fixed basis contract in that the seller has the opportunity to delay the pricing of grain until some point after delivery, but is charged a service charge or fee for each month following delivery he or she chooses to delay pricing. With a service charge contract, the net price received by the seller is the local cash price on the date he or she chooses to price the commodity minus accumulated service charges. Again, since the purchaser takes title to the grain and frequently will sell it and protect his or her obligation with a futures contract, the concerns noted earlier with respect to the financial strength of the buyer are also a consideration in deciding whether to use a service charge contract. A farmer who is considering selling with a service charge contract should evaluate the monthly service charges compared to the cost of commercial storage; if the service charges are the same or larger than commercial storage costs, the producer may be better off to place the crop in storage and price it for cash at a later delivery date.

A provision that is frequently included in various types of forward contracting arrangements (and can be used with various cash sales arrangements as well) is that of delayed or deferred payments. For producers who use the cash accounting system, the delayed payment provision may allow them to better manage their income tax liability. In essence, such a provision allows the producer to defer the receipt of income until a future tax year. Although the Internal Revenue Service (IRS) has treated such provisions with skepticism, recent rulings indicate that deferred payment contracts will not be subject to as much IRS scrutiny in the future. However, a producer must evaluate the tax savings of such provisions compared to the interest or earnings foregone by not having use of the funds during the deferral period, as well as the risk of receiving no payment at all if the buyer encounters financial difficulties between the time when the product is delivered and payment is to be received.

WHEN AND HOW TO DELIVER

If a product is priced through the use of a forward contract or cash at delivery, the choice of where to price and how and when to deliver is made simultaneously. If, however, the futures market is used as the mechanism to price a commodity, then the pricing decision has been separated from the delivery decision and the considerations noted earlier must be evaluated to determine when, where, and how to deliver the product. Remember that if hedging methods are being used to price the product, the timing of delivery must be coordinated with the timing of lifting the hedge if hedging is to reduce risk.

As noted earlier, determining how to deliver a product will depend primarily on the availability of transportation facilities, the cost of transporting the product compared to on-farm pickup, and the convenience of various delivery methods. In recent years, the delays encountered in delivering grain to local elevators during the harvest period have encouraged many producers to consider alternative delivery procedures as well as on-farm storage or delivery to alternative market outlets.

SUMMARY OF THE MARKETING PLAN

As with any plan, it is desirable to put together a written summary of the key components of the marketing plan. This summary will not only facilitate implementing the plan and reduce the possibility of emotional and hasty day-to-day decisions based on incomplete analysis, but it will also enable the manager to monitor performance in the marketing area by comparing actual marketing decisions and results to the plan.

Table 7.6 illustrates a commodity sales schedule that might be used to summarize the product sales component of the marketing plan. For each product produced by the firm, a summary of the supply and usage should be developed to determine the marketable surplus. This is particularly important for storable commodities and those such as feed grains that might be used on-farm to produce other commodities. The schedule of Table 7.6 includes unhedged cash sales, forward contract sales, and hedged sales using the futures market. Planned sales are summarized by month with entries for the quantity sold, the price per unit, and the gross sales. In addition, the delivery location for cash sales, the date contracted and delivery location for contract sales, and the futures contract month, month sold, selling price, and expected basis for hedged sales are recorded. Note in Table 7.6 that the marketing plan calls for cash sales of corn in February and June (2,000 and 6,000 bushels, respectively), forward contract sales of 6,000 bushels in April, and hedged sales of 2,000 bushels in March and 5,000 bushels in July. Proceeds from corn sales are expected to total $22,520 from cash sales, $17,640 from contract sales, and $19,060 from hedged sales. The data for hog sales can be interpreted in a similar fashion. This monthly summary of expected sales by marketing alternative not only provides documentation of this component of the marketing plan, but can also be used in completing the cash flow budget of the financial plan.

In addition to the commodity sales schedule, a schedule of input purchases could be developed to document expected prices and quantities of inputs to buy as well as to use in completing the cash flow budget. Furthermore, it is recommended that the key objectives and assumptions of the marketing plan be summarized so that a permanent record of the underlying analysis for the plan is available. This summary will be particularly useful in assessing later during the year whether fundamental market con-

Table 7.6   Commodity Sales Schedule

				Unhedged Cash Sales			
Crop	Inventory		Month[a]	Quantity	Price/ Unit ($)	Gross Sales ($)	Delivery Location
Corn	Beginning carryover	28,500	January				
	Production	42,700	February	2,000	2.65	5,300	Mallard Cooperative
	Total	71,200	March				
	Usage on farm	25,200	April				
	Ending carryover	25,000	May				
	Total	50,200	June	6,000	2.87	17,200	Mallard Cooperative
	Marketable surplus	21,000	July				
			August				
			September				
			October				
			November				
			December				
			Total	8,000	—	22,520	—
Soybeans	Beginning carryover		January				
	Production		February				
	Total		March				
	Usage on farm		April				
	Ending carryover		May				
	Total		June				
	Marketable surplus		July				
			August				
			September				
			October				
			November				
			December				
			Total				
Hogs	Beginning carryover	380	January				
	Production	1,000	February	280	110	30,800	Diamond Pack
	Total	1,380	March	100	110	11,000	Diamond Pack
	Usage on farm	0	April				
	Ending carryover	350	May				
	Total	350	June				
	Marketable surplus	1,030	July				
			August				
			September				
			October				
			November				
			December				
			Total	380	—	41,800	—
Cattle	Beginning carryover		January				
	Production		February				
	Total		March				
	Usage on farm		April				
	Ending carryover		May				
	Total		June				
	Marketable surplus		July				
			August				
			September				
			October				
			November				
			December				
			Total				

(*continued*)

[a]The month for forward contract and hedged sales refers to the month when the commodity will be delivered and cash received.

Table 7.6 (*Continued*)

	Forward Contract Sales					Hedged Sales						
	Quantity	Price/ Unit ($)	Gross Sales ($)	Date Contracted	Delivery Location	Quantity	Contract Month	Month Sold	Price/ Unit ($)	Expected Basis ($)	Net Price/ Unit ($)	Gross Sales ($)
J												
F												
M						2,000	March	Dec.	3.08	0.45	2.63	5,260
A	6,000	2.94	17,640	Sept.	Osage							
M					Cooperative							
J												
J						5,000	July	Jan.	3.16	0.40	2.76	13,800
A												
S												
O												
N												
D												
T	6,000	—	17,640	—	—	7,000	—	—	—	—	—	19,060
J												
F												
M												
A												
M												
J												
J												
A												
S												
O												
N												
D												
T												
J												
F												
M												
A												
M												
J	300	115	34,500	Jan.	Butler Food							
J	100	120	12,000	Jan.	Butler Food							
A												
S												
O												
N												
D						250	Dec.	July	126	7	119	29,750
T	400	—	46,500	—	—	250	—	—	—	—	—	29,750
J												
F												
M												
A												
M												
J												
J												
A												
S												
O												
N												
D												
T												

ditions have changed from those assumed earlier and whether major changes should be made in the marketing plan.

Summary

In addition to production and financial planning, farmers must develop market plans. With the increased variability in commodity and input prices and the higher risk faced by many farmers, market planning is much more important than it has been in the past. Developing a marketing plan requires answers to six basic questions: (1) when to set the price, (2) where to price, (3) what form, grade, or quality to deliver or purchase, (4) what services to provide or acquire, (5) what method to use in pricing, and (6) when and how to deliver.

The steps in market planning include (1) specification of goals, (2) determination of the production schedule and volume, (3) estimation of production costs, (4) projection of market prices, and (5) implementation of marketing decisions in the six areas noted earlier.

Projecting prices of agricultural commodities and inputs is not an easy task. Prices in the long run are determined by basic supply and demand conditions. These conditions are typically summarized for grain and oilseed products in the "balance sheet" which indicates the supply and utilization of agricultural commodities. The size of the carryover expected at the end of the year from this balance sheet is extremely important in projecting future prices; an increase in the carryover suggests declining future prices, whereas a decline in the carryover (unless it has been burdensome) suggests rising commodity prices. For livestock products, supplies are estimated using projections of the size of the breeding herd and expected productivity of that herd. For both crop and livestock products, price flexibility coefficients can be used to estimate the price response from changes in supply and demand conditions. Agricultural commodity prices also exhibit seasonal and cyclical price patterns which should be considered in developing the marketing plan.

Farmers have many options as to when and how to set the price of their commodities. In many cases, the price is set at the date of delivery with that date being at maturity or harvest, or at a later date with commodities that are storable. For some commodities, opportunities for forward pricing through the use of contracts and hedging with futures market transactions are also available. Hedging requires knowledge of the basis (the difference between cash prices and futures prices in the delivery month) and the ability to meet margin calls if required. Various types of forward contracting options are available including a cash forward contract, the delayed or deferred pricing procedure, and a service charge contract. Determining where to price includes an analysis of the competitiveness of various market outlets along with the transportation and marketing charges associated with that outlet.

When purchasing inputs, farmers must choose the services as well as the product. Support and maintenance services, credit arrangements, reliability, and delivery services are all important considerations in choosing among various suppliers. The marketing, credit, and payment practices of various product purchasing firms may also influence the choice of a particular market outlet. The choice of the grade or quality of input to acquire and product to market is another important marketing decision.

Questions and Problems

1. List and discuss the steps involved in market planning.
2. Identify the various goals that might be specified for the market plan and discuss when each of these goals might be appropriate.
3. What are the major factors that result in a shift in the demand curve?
4. What determines the supply curve for a commodity? What factors result in a shift in this supply curve?
5. Locate the U.S. Department of Agriculture balance sheet for a major agricultural commodity (corn, wheat, soybeans, etc.) and determine the projected carryover for the upcoming marketing year. What would you expect to happen to market prices if this carryover increased by 2 percent?
6. Choose a major agricultural commodity and discuss the cyclical and seasonal trends in the price of that commodity. Why do these systematic price variations occur, and under what circumstances would you expect deviations from these cyclical or seasonal price patterns?
7. Using the latest U.S. Department of Agriculture and other sources of information (including those from any private sector forecasters), develop a price forecast for the next one, three, and six months for an agricultural commodity.
8. What are the major factors that should be considered in determining where to price a product or input?
9. Identify two specific competing input supply firms (two feed dealers, machinery dealers, etc.) and compare the services they offer.
10. What factors should you consider in determining whether to price a product at delivery, use a forward contract, or hedge on the futures market?
11. Describe how you might use a "scale-up" procedure as part of your marketing plan.
12. What are the advantages and disadvantages of hedging as a way to set the price for a product?
13. Compare and contrast the cash forward contract, the delayed or deferred pricing contract, and the service charge contract methods of forward pricing.

Further Reading

Branson, Robert E., and Douglass G. Norvell. *Introduction to Agricultural Marketing.* New York: McGraw-Hill, 1983.

Futrell, Gene A., editor. *Marketing for Farmers*. St. Louis, Mo.: Doane Western, 1982.

Kohls, Richard L., and Joseph N. Uhl. *Marketing of Agricultural Products*. 5th ed., New York: Macmillan, 1980.

Purcell, Wayne. *Agricultural Marketing: Systems, Coordination, Cash and Futures Prices*. Reston, Va., Reston Publishing Company, 1979.

Rhodes, V. James. *The Agricultural Marketing System*. 2d ed., New York: John Wiley and Sons, 1983.

Shepherd, Geoffrey S., and Gene A. Futrell. *Marketing Farm Products*. 7th ed., Ames, Iowa: Iowa State University Press, 1982.

Tomek, William G., and Kenneth L. Robinson. *Agricultural Product Prices*. 2d ed., Ithaca, N.Y.: Cornell University Press, 1981.

INVESTMENT, FINANCIAL, AND TAX MANAGEMENT CONSIDERATIONS IN PLANNING

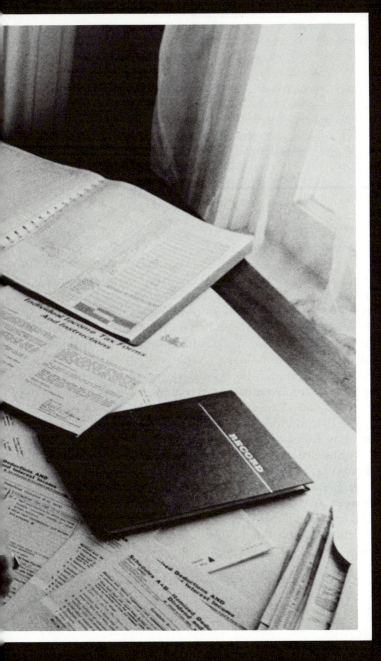

CHAPTER CONTENTS

Since the purchase of capital inputs is usually necessary to carry out production plans over time, the evaluation of alternative investments and their impact on profitability and cash flow is a crucial task in the farm planning process. For many businesses the long-run production plan calls for expansion to obtain more income and a better standard of living. Only through the use of the capital budgeting and financial analyses procedures discussed here can alternative investments be evaluated and profitable long-run plans be developed and implemented. If credit is used to finance capital purchases, the effects of different proportions of debt and equity on the risk and return of the firm should be analyzed. Tax obligations are an important expense which influence the after-tax income of various enterprises as well as the profitability of various capital investments. Furthermore, tax liabilities can be reduced or minimized with appropriate planning. This chapter will first develop the concepts and tools of capital budgeting and investment analysis, and then discuss the principle and implications of financial leverage. Finally, the concepts of tax management will be reviewed, with emphasis on integrating tax considerations in the farm planning process.

INVESTMENT ANALYSIS

Capital investment decisions that involve the purchase of durable inputs such as land, machinery, buildings, or equipment are among the most important decisions undertaken by the farm manager. These decisions typically involve the commitment of large sums of money, and they will affect the farm operation over a number of years. Furthermore, the funds to purchase a capital item must be paid out immediately, whereas the income or benefits accrue over time. Because the benefits are based on future events and the ability to foresee the future is imperfect, considerable effort should be made to evaluate investment alternatives as thoroughly as possible. This evaluation may include analysis of the decision under alternative futures with respect to prices, productivity, and cost, for once the decision is made and an alternative is chosen, the direction and operation of the firm will be affected for a number of years.

There are four major steps involved in the evaluation of capital expenditure proposals:

1. *Identify all possible profitable investment opportunities.* This step should be taken to insure that *the most profitable*—not just *a profitable*—investment is chosen.

2. *Evaluate the economic profitability and financial feasibility of the various investment opportunities.* Evaluating *economic profitability* involves determining the capital outlay required for each alternative and the earnings or benefits that will likely result from each alternative, and comparing the outlay to the benefit stream. Different methods of comparison or criteria for evaluating economic profitability are available and will be discussed shortly. *Financial feasibility* involves a comparison of the cash inflows generated by the investment project with the principal and interest payments that are due on any borrowed funds used to purchase the capital item.

3. *Reevaluate the decision under different price and yield assumptions.* Since the investment decision involves projections into the future and a major commitment over time, it is desirable to evaluate the economic profitability and financial feasibility of an investment alternative under different sets of future prices and productivity. This should provide information to the manager concerning the desirability of an investment when prices or conditions are unfavorable as well as in "normal" times.

4. *Choose an alternative based on the economic and financial evaluation as well as other factors that would influence the investment decision.* As with any managerial decision, judgment must be combined with the economic analysis to select an alternative. Note that these steps parallel those of the decision-making process presented in Chapter 1.

The most important task of investment analysis is that of gathering the appropriate data. Even though the proper procedures are used to evaluate the decision, inaccurate or incomplete data will most certainly negate an otherwise thorough and complete analysis.

Economic Profitability

The purpose of the economic profitability analysis is to determine whether the investment project will contribute to the long-run profits of the firm. Various criteria can be used in this analysis as will be discussed shortly. Even if an alternative is economically profitable, however, it may not be financially feasible, i.e., the cash flows may be insufficient to make the required principal and interest payments. Consequently, financial feasibility analysis must be completed before a final decision is made to accept or reject a particular project. Financial feasibility analysis will be briefly discussed following the review of profitability analysis. A more detailed discussion of financing alternatives (including leasing), and the procedures for analyzing these alternatives will be discussed in Chapter 14.

Four common criteria used in evaluating the profitability of alternative capital investments are the payback period, the simple rate of return, net present value and the internal rate of return.[1] We will briefly discuss these alternative criteria or methods of capital budgeting, and then develop the net present value procedure in detail using an example. The Appendix to this chapter illustrates the computations for all four methods using this example.

Payback   The payback period is frequently used in investment analysis as evidenced by the often-heard statement, "That combine looks like a good investment, it will pay for itself in about three years." The payback procedure essentially looks at the time or number of years it takes to recover the initial outlay from the earnings generated by the investment. A simple formula summarizing the calculations of the payback period is presented in equation 8.1.

$$P = \frac{O}{I} \tag{8.1}$$

where P is the payback period in years, O is the capital outlay required to purchase the item, and I denotes the average annual after-tax earnings (before depreciation) expected from the investment. Capital items that have a shorter payback period are more desirable than those with a longer payback period.

Table 8.1 illustrates the computations of the payback period for two alternative investments. Project A requires a $10,000 capital outlay and generates an income stream of $2,500 for six years, whereas Project B

---

[1]The discussion of the payback and rate of return criteria follows closely the arguments presented by Richard D. Aplin, George L. Casler, and Cheryl P. Francis, *Capital Investment Analysis Using Discounted Cash Flows,* 2d ed. (Columbus, Ohio: Grid, 1977.)

Table 8.1 Comparison of Payback and Simple Rate of Return Investment Criteria

Item	Investment Alternatives	
	Project A	Project B
Capital outlay	$10,000	$10,000
Average after-tax earnings		
Year 1	2,500	4,000
Year 2	2,500	4,000
Year 3	2,500	4,000
Year 4	2,500	—
Year 5	2,500	—
Year 6	2,500	—
Total	$15,000	$12,000
Annual depreciation	$1,667	$3,333
Payback period: $P = \dfrac{O}{I}$	$\dfrac{10,000}{2,500} = 4.0$ yr	$\dfrac{10,000}{4,000} = 2.5$ yr
Rate of return: $R = \dfrac{I - D}{O}$	$\dfrac{2,500 - 1,667}{10,000} = 8.33\%$	$\dfrac{4,000 - 3,333}{10,000} = 6.67\%$

requires the same capital outlay but generates $4,000 of annual income for only three years. Note that the payback period for Project A is four years, whereas Project B has a payback period of 2.5 years. Consequently, B would be the chosen alternative using the payback criterion.

Although the payback period does provide some evidence of the desirability of alternative investments, it does have serious shortcomings as a measure of economic profitability. First, the payback period ignores the economic life of the investment or the income earned after the initial capital outlay has been recovered. In reality, the payback period is not a measure of profitability, but one of liquidity since it measures how quickly investment dollars will be recouped. A second problem with the payback period is that there is no logical basis for the criterion or cutoff between acceptance and rejection of an investment. For example, assume that four alternative investments are being evaluated which have payback periods ranging from three to ten years, and sufficient funds are available to choose more than one alternative. Which alternatives should be chosen? It may be true that a shorter payback period is more desirable than a longer period, but should all investments with payback periods shorter than six years be chosen, or only those with payback periods shorter than four years? And what is the logical basis for selecting four years, six years, or any other number of years as the cutoff criterion?

Finally, the payback period fails to consider the timing of the cash inflows and outflows generated by the investment. Thus, it ignores the concept of the time value of money, or the fact that a dollar in hand today is more valuable than a dollar that will be received sometime in the future.

Simple Rate of Return A second criterion that might be used in investment analysis is the simple rate of return. Although there are variations of this analysis procedure, a commonly computed simple rate of return can be determined by the formula of equation 8.2:

$$R = \frac{I - D}{O}$$

(8.2)

where R is the annual rate of return, I is the average annual after-tax earnings before depreciation expected from the investment, D is the annual average depreciation for the investment, and O is the capital outlay that is required. The rate of return for various projects can be computed with this formula and then compared to a goal established by management such as a minimum acceptable return of 10 percent. Moreover, alternative investments can be ranked from most to least desirable using the rate of return criterion with those investments exhibiting the higher rate being more desirable.

The simple rate of return for the two investment alternatives identified earlier is also summarized in Table 8.1. Note that the rate of return for investment A is 8.33 percent, and for investment B the rate is 6.67 percent, indicating that investment A is preferred to B. This is the opposite ranking suggested by the payback criterion discussed earlier.

The simple rate of return criterion is superior to the payback procedure because it considers the earnings of the investment over its entire expected life. Thus, income that is generated after the initial outlay is returned is important in determining whether an investment is desirable. Furthermore, the simple rate of return can be more easily compared to external criteria to determine the desirability of an investment. For example, a rate of return of 8 percent for a particular capital item can be compared to the return generated from investing the funds in savings instruments or certificates of deposit to determine whether or not a particular investment is desirable.

The simple rate of return is not without criticism, however. First, numerous methods can be used to compute the rate of return, and some of these methods will result in rates that are not directly comparable to the figures used in the financial world, such as interest rates or yields on stocks and bonds. In addition, the rate of return also does not consider the timing of the capital outlays and the benefit stream from an investment—the time value of money.

It should be noted that the technique discussed earlier of partial budgeting can also be used in investment analysis, but this procedure calculates the average annual cost and returns over the life of the investment and thus is only a variation of the computation of the simple rate of return. Thus, cost and return or partial budgeting procedures have similar deficiencies as the simple rate of return in choosing among alternative investments in capital budgeting.

Net Present Value   A third method that can be used to evaluate investment alternatives is that of net present value or discounted cash flow. This analysis procedure is one of the more desirable methods to use in capital budgeting because it reflects the opportunity cost of having funds tied up in capital items and the time value of money. The basic concept of the discounted cash flow (net present value) procedure is that a dollar in hand today is worth more than a dollar to be received sometime in the future. A dollar is worth more today than tomorrow because today's dollar can be invested and can generate earnings. In addition, the uncertainty of receiving a dollar in the future and inflation make a future dollar less valuable.

To reflect the fact that "money begets money," or that funds invested in capital items have an opportunity cost because they could be earning a return in some other investment, a discounting procedure is applied to money flows. This discounting procedure converts the money flows that occur over a period of future years into a single current value so that alternative investments can be compared on the basis of this single value. This conversion of flows over time into a single figure via the discounting procedure takes into account the opportunity cost of having money tied up in the investment.

The concept of the time value of money can be illustrated by the following example. Assume that a farmer can earn a 10 percent annual return on funds invested in his or her business. Based on this return, only $621 needs to be invested today to obtain $1,000 in five years as shown below. Thus, $1,000 received in five years is worth only $621 today if the annual return on invested funds is 10 percent.

Year	Value Beginning of Year	Annual Interest	Amount at End of Year
1	$621	$621 × .10 = $62	$ 683
2	683	683 × .10 = 68	751
3	751	751 × .10 = 75	826
4	826	826 × .10 = 83	909
5	909	909 × .10 = 91	1,000

Adjusting money for its time value can be accomplished with either a compounding or a present value formula. The compounding formula was used in the above example to obtain the future value of a current principal sum. The compounding formula can be stated as:

$$V_n^f = V_n^p (1 + i)^n \tag{8.3}$$

where V_n^f is the future value of the investment, V_n^p is the current amount or principal sum invested, i is the rate of earnings on the investment, and n is the number of annual periods over which the principal sum is to be compounded. Although the term $(1 + i)^n$ can be computed, it is easier to use a compound factor table, such as Appendix Table I. The compound factor can be obtained by entering the table for the appropriate rate and year.

The discounting process is the reciprocal (inverse) of compounding; its purpose is to find the current or present value of income received in the future. Since discounting is the inverse process of compounding, the formula for discounting a future amount can be derived from the compounding formula by dividing both sides of equation 8.3 by $(1 + i)^n$. Thus

$$\frac{V_n^f}{(1 + i)^n} = \frac{V_n^p(1 + i)^n}{(1 + i)^n} \quad \text{or} \quad V_n^p = \frac{V_n^f}{(1 + i)^n} \tag{8.4}$$

where V_n^p is the present value of the future sum, V_n^f is the future amount of income, and i and n are as previously defined. Again, the term $1/(1 + i)^n$ can be calculated, but the present value or discount table (Appendix Table II) is much more convenient to use.

To illustrate the present value computations, assume that a farmer earning 10 percent on his or her capital is to receive $1,000 at the end of each year for the next five years. The discount factor for money received at the end of the first year assuming a 10 percent rate is 0.909 (Appendix Table II). Hence, the $1,000 received at the end of the first year has a present value of only $909 ($1,000 × 0.909). In similar fashion, the present value of the $1,000 received at the end of years 2, 3, 4, and 5 can be calculated using the discount factors from Appendix Table II for 10 percent and the appropriate years. The present value of this flow of money is then determined as the sum of the annual present values, or

Year	Cash Flow	Discount Factor	Present Value
1	$1,000	0.909	= $ 909
2	1,000	0.826	= 826
3	1,000	0.751	= 751
4	1,000	0.683	= 683
5	1,000	0.621	= 621
		Total	$3,790

So the present value of an annual flow of $1,000 for each of five years (assuming a 10 percent discount rate) is only $3,790. The farmer would be equally well off if he or she were to receive a current payment of $3,790 or the annual payment of $1,000 per year for five years assuming a 10 percent discount rate.

In equation form, the net present value for a particular investment can be calculated as:

$$N = \sum_{n=1}^{K} \frac{I_n}{(1 + d)^n} - O \qquad (8.5)$$

where N denotes net present value, n denotes the time period with K indicating the last period an inflow is expected, Σ denotes a summation of all n periods, I_n denotes the net cash inflow in period n, d the rate of discount, and O the cash outlay required to purchase the capital asset. The specific computational steps required to use this procedure are discussed next.

Computation Steps The steps of the net present value analysis procedure are relatively straightforward:

Step 1. Choose an appropriate discount rate to reflect the time value of money. The discount rate is used to adjust future flows of income back to their present value. The discount rate chosen essentially indicates the minimum acceptable rate of return for an investment; it represents the "cutoff criterion" in judging whether or not an investment is desirable. Consequently, the selection of the correct discount rate is of crucial concern.

The discount rate can be based on the cost that the firm will incur to finance the investment. This approach, commonly referred to as the cost of capital approach, argues that any investment should return at least the cost of the debt and equity funds that must be committed or acquired by the firm to purchase the asset.[2]

How should the combination of debt and equity funds used to finance an investment be determined? In the long run, the funds a firm uses to acquire any capital item will come from both debt (borrowed funds) and equity (the owner's financial contribution to the firm) sources. Therefore, the cost of capital should be based on the combination of debt and equity capital used in the "long run" to finance the operation, not the specific combination of debt and equity that may be used to finance a particular purchase. Even though a high proportion of debt may be used to finance current investments, using this debt now will reduce the firm's ability to use credit in the financing of future investments.

[2]For a thorough discussion of the concept of the cost of capital, see James C. Van Horne, *Financial Management and Policy,* 5th ed. (Englewood Cliffs, N.J., Prentice-Hall, 1980).

The objective is to evaluate investment alternatives based on the long-run optimal capital structure of the firm—the capital structure or combination of debt and equity which the entrepreneur expects to maintain over a number of years. To determine the long-run cost of capital (based on this optimal capital structure) for the firm, the cost of debt funds and the cost of equity funds must be weighted by the long-run proportions of debt and equity that will be used to finance the farm operation. This results in a weighted cost of capital as summarized by equation 8.6.

$$d = k_e W_e + k_d(1 - t)W_d \qquad (8.6)$$

where d is the discount rate, k_e is the after-tax cost of equity funds (rate of return on equity capital), W_e is the proportion of equity funds used in the firm, k_d is the cost of debt funds (interest), t is the marginal tax rate, and W_d is the proportion of debt funds in the firm.

The purpose of the weighted cost of capital formula is to obtain a discount rate which accurately reflects the long-run direct cost of debt funds and the opportunity cost of equity funds, along with the long-run proportions of debt and equity that will be used in the firm. Note that the cost of equity funds is best estimated as the opportunity cost (income foregone) of committing equity to this particular investment compared to other investments. The best way to specifically measure this cost is to look at the after-tax rate of return being generated by the equity capital currently being used in the firm. This rate of return can be calculated as the sum of the cash return after-tax plus the after-tax gain in asset values divided by net worth or equity (market value basis) as measured on the balance sheet.[3] The annual cash return is calculated as annual net income from the income statement (as discussed in Chapter 2) minus expected income taxes. Annual capital gain is determined by comparing the market value of assets of the firm (particularly farmland) this year to the value last year; these data can be obtained from the balance sheet as discussed in Chapter 2. Anticipated capital gains tax must again be subtracted to obtain the after-tax gain. Also note that because interest (the cost of debt funds) is tax deductible, thus reducing the tax liability of the firm, the true cost of debt is the rate of interest on debt funds minus the tax savings. Equivalently, the true after-tax cost of debt can be calculated as the interest rate (k_d) times one minus the marginal tax rate $(1 - t)$.

Equation 8.6 indicates that the costs of equity and debt funds are multiplied by the respective proportions of equity and debt in the firm to obtain the long-run cost of capital or the discount rate. These proportions

---

[3]A farm firm with no real estate will typically have a high rate of cash return and no appreciation, whereas a firm that includes a substantial amount of real estate will have a lower rate of cash return and a higher appreciation component.

of debt and equity can be obtained from the balance sheet where W_d is calculated as total liabilities divided by total assets, the debt/asset ratio defined in equation 2.4 (market value basis), and W_e is calculated as equity divided by total assets (market value basis). Alternatively, since total liabilities plus net worth equal total assets, W_e can be calculated as $1 - W_d$. If the current balance sheet does not reflect the desired or expected long-run mix of equity and liabilities, adjustments in W_d and W_e should be made. For example, the current amount of debt in the balance sheet may be higher than the operator desires or plans to have in the long run because a farm was purchased recently on contract with a low downpayment and the operator plans to pay the contract off as soon as possible. In this case, the value of W_d calculated from the current balance sheet should be reduced to reflect the proportion of debt that the operator expects to have in the long run. Thus, if the long-run optimal or desired capital structure will include more or less debt than is currently reflected in the balance sheet, appropriate adjustments may be required.

It should be recognized that this approach to the computation of the discount rate essentially reflects the method of financing for the investment and tax deductibility of interest. Thus, in the later computation of the cash outlay required for a particular investment (Step 2), the method of financing or the issue of the source of funds need not be considered.[4]

In addition, note that *market* rates of interest and return on equity capital are used in the cost of capital calculation. These market rates include the expectation of the market participants as to the rate of inflation in the overall economy. Consequently, to be consistent in the computations, similar expectations concerning price and cost increases or inflation must be incorporated in the calculation of the annual cash income as discussed in Step 3.

By using the cost of capital as the estimate of the discount rate in the net present value computation, the farm manager is evaluating the returns for a particular investment compared to the cost of the debt and equity funds committed to that investment. Consequently, a particular investment is desirable only if it will return more income than the expenses that will be incurred to finance its purchase.

Step 2. Calculate the present value of the cash outlay required to purchase the asset. In most cases, the present value of the cash outlay will be equal to the purchase price of the asset since all the capital must be commit-

---

[4]An alternative (and typically less accurate) procedure that may be used to reflect the source of financing (debt versus equity) and the tax deductibility of interest in the analysis is to specify the discount rate as the cost of equity funds only. The cash inflows and outflows of Steps 2 and 3 would then include explicitly the downpayment, annual principal and interest payments, and the tax savings attributable to interest payments. We do not advocate this procedure, and we suggest that the weighted cost of capital approach to obtaining the discount rate enables us to evaluate the benefits of alternative financing arrangements as discussed in Chapter 14.

ted at the time the purchase is made. In some cases, however, an additional capital outlay will occur in future years in order, for example, to replace equipment that wears out before the end of the useful life of a building or facility. In this situation, these future capital outlays must be discounted to the present and added to the initial outlay. Note that whether or not money will be borrowed to purchase the investment need not be considered in this step, since the source of funds (debt versus equity) and the tax deductibility of interest have already been taken into account in the cost of capital calculation of Step 1. If the fact that payments will be made over time to purchase a particular asset because of debt financing is reflected in this step and also in the calculation of the discount rate, double counting of the tax deductibility of interest and the benefits of the financing arrangement would result.[5]

In the computation of the capital outlay for a particular investment, it is important to include all additional outlays that may be required. For example, if a farmer were evaluating the construction of a new farrow-finish facility for his or her hog enterprise, the capital outlay would include not only the purchase price of the building and equipment, but also the cost of any additional breeding stock or additional feed and veterinary supply inventories that might be required to support the larger hog operation. In essence, these additional working capital commitments will be necessary to operate the larger hog facility and must be considered as part of the capital outlay for the new investment.

Step 3. Calculate the benefits or annual net cash flow for each year from the investment over its useful life. As suggested by the term "discounted cash flow," the benefits to be included are the increased net *cash flows* that result from a particular investment. These cash flows should be calculated on an after-tax basis. Since depreciation is not a cash flow but only an accounting entry to allocate the cost of a capital item over its useful life, it does not enter directly in the computation of annual net cash flows. Instead, depreciation enters the calculations only as it influences the tax liability or the tax savings of a particular investment. In addition, since the discount rate reflects current expectations of inflation because the data used in the calculation come from current market rates, the estimation of future cash income and cash expenses should also reflect expected price increases for inputs and outputs.

The annual net cash flows for an investment can be computed with the following formula:

$$ANCF_n = CI_n - CE_n - T_n + TSIC_1 + S_K \qquad (8.7)$$

---

[5]If the discount rate reflected the cost of equity funds only, the outlays of this step would include the present value of the downpayment plus principal and interest payments.

where $ANCF_n$ is the net cash flow in year n, CI_n is the cash income in year n, CE_n is the cash expenses in year n, T_n is the income tax in year n, $TSIC_1$ is the tax savings from investment credit (available in the first year only with values for subsequent years equal to zero), and S_K is the salvage value of the asset in the last (K^{th}) year of its useful life. Cash income for a particular investment would be calculated as product sales from that particular investment times the expected prices, whereas cash expenses would include the cash costs of the inputs used in production. Note that interest on debt used to finance the investment is not included as a cash expense since it has already been included in the computation of the cost of capital (Step 1).[6]

Income taxes are computed as:

$$T_n = [CI_n - (CE_n + D_n)] \, TR_n \qquad (8.8)$$

where D_n is depreciation and TR_n is the marginal tax rate, both in year n. The tax variable (T_n) reflects the additional taxes that will be paid on income generated by the particular investment. Note that depreciation enters in this computation of the tax liability. As shown in equation 8.7, the additional tax liability will reduce the net cash flow from a particular investment. Also note that the tax savings from investment credit ($TSIC_1$) is included for the first year to reflect the reduced tax liability that will be incurred if qualified investment credit property is purchased. The 1983 investment credit rules are discussed later in this chapter. Any other tax credits such as the energy credits or state investment credits should also be included in the analysis.

The salvage value of a particular machine or a piece of equipment should be included as a positive cash flow in the last year if it is to be sold or traded on a new item. The salvage value is part of the cash benefit stream that will be received if the machine is sold. Or if it is traded for a new machine, the salvage value reflects the reduced cash outlay that will be incurred to purchase the new machine.

The impact of inflation on the discount rate and the cost and returns from an investment cannot be ignored in the practical use of the net present value procedure. Some analysts argue that all computations should be done in "real" terms. Thus, the discount rate obtained from the market rates of interest and return on equity capital would be reduced by the amount of expected inflation that is reflected in these market rates, and anticipated price increases for products and inputs would be ignored in the computation of annual net cash flows. The procedures discussed here sug-

---

[6]However, if the alternative computation procedure of footnote 4 were used, interest would be included as a cash expense in this calculation and in the computation of taxes in equation 8.7.

gest that inflation or expected price increases should be built directly into both the net cash flow stream and the discount rate. Since the market rates of interest and return on equity capital already reflect current expectations of inflation, no adjustment need be made to build inflation into the discount rate. By considering future increases in product prices and input costs in the calculation of the cash flow stream, inflation and its impact on net revenues are easily reflected.

A major advantage of measuring both the cash flow stream and the discount rate in "nominal" dollars is that the financial feasibility analyses to be discussed shortly compare the annual net cash flow in each year to the required principal and interest payments. If the cash flows have been calculated on a "nominal" basis for the economic profitability phase of the capital budgeting process, the same numbers can be used in evaluating the financial feasibility of the project. If "real" cash flows were used in the calculations, they would have to be adjusted to "nominal" values to complete the financial feasibility analysis. Regardless of whether "real" or "nominal" values are used in the net present value calculation, one thing is essential and that is consistency—*both* the discount rate and the net cash flow must be measured either in "real" terms or in "nominal" terms.

Equation 8.7 is used to calculate the annual net cash flow *for each year* of the useful life of the asset. Thus, a series of annual net cash flows is computed in this step.

Step 4. Calculate the present value of the annual net cash flows. In Step 3, the annual net cash flow stream for the entire useful life of the asset was calculated. Now we want to convert this stream into a single figure which represents the current or present value of such a stream of income over time. As has been suggested earlier, the *present value* of income that will be received sometime in the future can be determined by multiplying the annual income times the discount factor for the appropriate discount rate and year. By multiplying the annual net cash flow for each year times the discount factor and then summing the discounted annual net cash flows, a single present value figure can be obtained.

The discount factors to be used in this computation are obtained from the discount table (Appendix Table II). The factor for each year is determined by entering the table for the appropriate discount rate as computed in Step 1 and the appropriate year. For example, if a discount rate of 12 percent was calculated in Step 1, the discount factor for year 1 to be used in the computation of the present value of the annual net cash flows would be 0.893. Likewise, the discount factor for year 2 (12 percent) would be 0.797.

Step 5. Compute the net present value. Net present value is simply computed as the present value of the net cash flows obtained in Step 4 minus the present value of the cash outlay to purchase the investment of Step 2.

Step 6. Accept or reject the investment. The criterion for acceptance or rejection of an investment is simple—for mutually exclusive alternatives accept an investment if it has a positive net present value and reject that investment if it has a negative net present value. This simple criterion is possible because when the benefit stream or the annual net cash flows for a particular investment are discounted with the cost of capital, the resulting figure represents the maximum amount that the manager could afford to pay for the investment and expect to just "break even." Therefore, a net present value of zero indicates that the particular investment is generating a return exactly equivalent to the cost of capital or the cost of debt and equity funds that have been used to finance the investment. A positive net present value indicates that the particular investment is generating a benefit stream larger than the cost of the funds used to finance the investment; hence, the investment is a profitable one. In essence, the additional return adjusted by the time value of money is larger than the additional cost of the investment. In contrast, a negative net present value indicates that the increased income received from the investment will be less than the cost of funds required to support that investment. Thus, the investment is undesirable, and the funds should be committed to some alternative investment that will generate a return at least equivalent to their cost.

In some cases, the decision may not be one of accepting or rejecting a particular investment but of choosing among a number of alternative investments. In this situation, the investment alternatives can be ranked in order of preference based on their net present values, with the alternative having the highest net present value being ranked first and the one with the lowest net present value ranked last. The decision-maker would then implement all those alternatives with a positive net present value if the funds were available to do so. If the funds to acquire the alternative investments were limited, one would choose that combination of projects that generates the largest total net present value with the limited funds.

An Example   To illustrate the use of the net present value procedure in capital budgeting, it will be applied to the decision of purchasing a new combine by a custom operator. The operator has priced a 4–30″ row combine with corn head only at $76,420 (no trade). He expects to custom harvest 900 acres of corn per year with the machine at a rate of $25.00 per acre. The combine has an eight-year life and an after-tax salvage value at the end of eight years of $19,586. Cash operating costs are projected to amount to $3.08 per acre for repairs, $1.63 per acre for fuel and lubricants, $2.83 per acre for labor, and $1.49 per acre for insurance, taxes, and housing, for total cash costs of $9.03 per acre. To simplify the analysis, both cash costs and custom rates are assumed to increase by 5 percent per year.

The operator is in the 28 percent tax bracket and expects to finance his or her custom combining business with 30 percent debt and 70 percent equity in the long run. The cost of equity capital (after-tax rate of return on equity funds currently being used in the business) is 13.5 percent, and debt funds can be borrowed at 12 percent.

Step 1. Compute the discount rate. The discount rate is computed as:

$$d = k_e W_e + k_d (1 - t) W_e$$
$$= (0.135 \times 0.70) + (0.12 \times 0.72 \times 0.30)$$

$$= .094 + .026$$
$$= .12 \tag{8.9}$$

Step 2. Calculate the present value of the cash outlay. The purchase price of the combine is $76,420. No additional working capital will be required, and the total outlay must be committed immediately. So the present value of the cash outlay is $76,420.

Step 3. Calculate annual net cash flows. The annual net cash flows are computed as:

$$ANCF_n = CI_n - CE_n - T_n + TSIC_1 + S_K \tag{8.10}$$

where:

$$T_n = [CI_n - (CE_n + D_n)] TR_n \tag{8.11}$$

Annual computations would be as follows:

Year	Cash Income[a]	Cash Expenses[b]	Taxes[c]	Investment Credit[d]	Salvage Value	Net Cash Flow
1	$22,500	$ 8,127	$ 975	$7,642	$—	$21,040
2	23,625	8,533	−246 [e]	—	—	15,338
3	24,806	8,960	168	—	—	15,678
4	26,046	9,408	390	—	—	16,248
5	27,348	9,878	623	—	—	16,847
6	28,715	10,372	5,136	—	—	13,207
7	30,151	10,891	5,393	—	—	13,867
8	31,659	11,436	5,662	—	19,586[f]	34,147

[a]900 acres times $25.00 per acre for the first year; custom rates increase 5 percent per year for each subsequent year.

[b]900 acres times $9.03 per acre for the first year; costs increase 5 percent per year for each following year.

[c]Cash income minus the sum of cash expenses plus depreciation times the tax rate. Depreciation was calculated according to the Accelerated Cost Recovery System (ACRS) schedule for 5-year life items as 15 percent of the basis ($76,420 minus 50 percent of the investment credit of $7,642) in the first year, 22 percent in the second year, and 21 percent in years 3, 4, and 5. For the first year, taxes total [$22,500 − ($8,127 + $10,890)] · 0.28 = $975.

d$76,420 times 0.10.

*e*The deductions exceed the income in year 2 by $880 [$23,625 − ($8,533 + $15,972)] resulting in a loss. This loss can be used to offset other income which would be taxed in the 28 percent bracket, thus resulting in tax savings (or negative taxes) of $246.

*f*This after-tax salvage value is based on the remaining value of 24.1 percent of the purchase price reported in Table 4.3 adjusted for increases of 5 percent per year for 8 years minus income taxes computed on the total sales value (the depreciated value or tax basis is $0) at the 28 percent bracket. The value is thus [($76,420 × 0.241) · (1.05)^8] · 0.72 = $19.586.

Step 4. Calculate the present value of the net cash flows. The present value of the net cash flows is computed as the sum of the discounted annual net cash flows (net cash flow times the discount factor).

Year	Annual Net Cash Flow	Discount Factor	Present Value of Annual Net Cash Flow
1	$21,040	0.893	$18,789
2	15,338	0.797	12,224
3	15,678	0.712	11,163
4	16,248	0.636	10,334
5	16,847	0.567	9,552
6	13,207	0.507	6,696
7	13,867	0.452	6,268
8	34,147	0.404	13,795
	Present value of the net cash flows		$88,821

Step 5. Compute the net present value. Net present value is computed as the present value of the net cash flows minus the present value of the cash outlay.

$$\$88,821 - \$76,420 = \$12,401$$

Step 6. Accept or reject. Based on the positive net present value of Step 5, the custom operator would purchase the combine (assuming that other investment alternatives have a lower net present value). The combine will generate $12,401 more income than the cost of the funds used to finance the purchase during its eight-year life. In fact, the operator could afford to pay an additional $12,401 for the machine if necessary, and it would generate enough income to just pay the cost of capital. If the operator was required to pay more than $88,821 for the combine, it would not generate sufficient income to pay for the cost of the funds used to finance the purchase, and so it would be rejected as an undesirable investment.

Two common problems are frequently encountered in practical application of the net present value procedure, particularly when it is used to compare investment alternatives that are mutually exclusive. The first problem is that of differences between the two alternative projects in the size of the capital outlay. For example, one alternative investment may require a capital outlay of $20,000, whereas the other alternative requires a $26,000 outlay. The issue is, "Does the net present value procedure provide an accurate comparison between these two investments with different capital outlays?"

Fortunately, the size of the initial outlay is not important when comparing the net present values for two projects. Thus, the investment with the larger net present value is preferred, regardless of the size of the capital outlay. This occurs because the net present value of excess cash (for example, the $6,000 difference between the two investment projects noted earlier) is zero. Note that a net present value of zero does not imply that the extra cash is not being invested to generate earnings. What is implied is that the rate of earnings on the excess cash is exactly equal to the rate of discount. When the earnings from the extra cash plus the receipt of the principal sum at the end of the investment period are discounted at a rate equal to the earnings rate, and then subtracted from the initial cash outlay, the net present value is zero. Since the net present value of the difference in the outlays between two projects is zero, this difference can be ignored and the project with the highest net present value can be chosen regardless of the size of the capital outlay.

A second problem frequently encountered in the practical use of net present value in comparing alternative investments involves projects with different useful lives. For example, machine A may have a useful life of five years and machine B a useful life of eight years. In this case, a direct comparison of net present values is not appropriate, and an adjustment must be made so that both projects have the same useful lives.

One adjustment procedure would be to assume a common terminal date as specified by the useful life of the longest life investment, which in the above case would be eight years. Then, it is assumed that the shorter life investment is replaced with an identical machine at the end of its useful life (five years in this case), and sold at the common terminal date, with the salvage value of the replacement machine included as a cash inflow in that year (year 8 in our example).

A second procedure to obtain equal lives is to assume that each of the machines is replaced with identical models until they both reach the end of their useful lives in the same year. In our example, the common denominator would be 40 years, which would require machine A to be replaced seven times and machine B to be replaced four times. This solution procedure is not usually very practical.

A third procedure, and often the preferred method of adjusting for differences in useful lives of the investments, is to use the annuity concept and convert the net present values into annual equivalent values. For example, assume that machine A has a net present value of $420 and machine B a net present value of $480. The annual equivalent of these net present values can be calculated by using the amortization factors from Appendix Table IV, assuming the appropriate discount rate and useful life of the asset. For example, a net present value of $420 over five years (machine A), assuming a 10 percent discount rate, is equal to an annual equivalent value of $110.80 per year ($420 × 0.2638). Likewise, the annual equivalent net present value of machine B would be $89.95. Thus, machine A would be preferred to machine B because it has the higher annual equivalent net present value.

Internal Rate of Return A fourth criterion that may be used in investment analysis is the internal rate of return. Essentially, the internal rate of return is determined as the rate of discount that equates the present value of the cash inflows with the cash outlays. Thus, the internal rate of return is r in the following equation.

$$\sum_{n=1}^{K} \frac{I_n}{(1 + r)^n} = O \qquad (8.12)$$

where n again denotes the time period, K the last period in the analysis, Σ the summation over all K periods, I_n the net cash inflow in each period, and O the cash outlay. This internal rate of return, r, can be compared to alternative rates on other investments to determine if the particular investment is desirable. If r exceeds the rates on other investments or the minimum acceptable rate (usually specified as the cost of capital), the project is accepted; those with lower rates are rejected.

The net present value and internal rate of return criteria use similar data and computation procedures and will generate the same results in most practical cases.[7] In fact, the internal rate of return is that rate of discount which results in a net present value of zero. If the internal rate of return for a project exceeds the minimum acceptable rate of return, and this minimum acceptable rate is the cost of capital used to discount the annual cash flows using the net present value procedure, then the net present value of the project will be positive. Thus, in both cases the project

---

[7]Differences between the two criteria in terms of the "reinvestment rate" assumption—the rate at which net cash inflows are assumed to be reinvested—may result in different and conflicting answers from the two procedures in unique situations. In essence, the net present value procedure presumes that net cash inflows are reinvested or generate earnings at the discount rate, whereas the internal rate of return procedure presumes that they are reinvested at the internal rate of return offered by the project. Thus, a high internal rate of return may be partially a result of the high rate of return that is presumed to be earned on reinvested net cash flows, and may, therefore, be unrealistic because, in reality, this cash could not generate this high rate in other parts of the business.

would be accepted. Similarly, if the internal rate of return is less than the minimum acceptable rate, the net present value will be negative and the project will be rejected with both criteria.

The internal rate of return can only be obtained through an iterative process of solving equation 8.12 using different values of r until the equality holds. Without the aid of a computer, this can be a tedious and time-consuming task, and since the net present value and internal rate of return criteria result in identical results in most practical situations, the net present value procedure is typically used.[8] To illustrate the calculations, the computation of the internal rate of return for the combine example discussed earlier is presented in the Appendix to this chapter.

Financial Feasibility

Once the profitability of various investments has been analyzed and an alternative chosen, its financial feasibility should be evaluated. The purpose of financial feasibility analysis is to determine whether or not the investment project will generate sufficient cash income to make the principal and interest payments on borrowed funds used to purchase the asset. If the purchase is to be made with equity funds and a loan is not required, then financial feasibility analysis is unnecessary. Furthermore, the discussion of financial feasibility presented in this chapter presumes that a particular financing arrangement will be used; it does not consider the issue of the optimal financing arrangement. Procedures to evaluate alternative financing arrangements (including leasing) are discussed in Chapter 14.

The first step in financial feasibility analysis is to determine the annual net cash flows for the project. Fortunately, these annual flows have already been calculated as part of the economic profitability analysis if the net present value or internal rate of return procedures are used (see equation 8.7). Next, the annual principal and interest payments must be determined based on the loan repayment schedule. Since the annual net cash flows are after-tax and the payment schedule is before-tax, this payment schedule must be adjusted to an after-tax basis by calculating the tax savings from the deductibility of interest and subtracting this savings from the payment schedule. Then, the annual net cash flow is compared to the after-tax annual principal and interest payments to determine if a cash surplus or deficit will occur. If a cash surplus results, the investment project will generate sufficient cash flow to make the loan payments, and the project is

---

[8]A fifth criterion, the profitability index or benefit-cost ratio, is calculated as the present value of the cash inflows divided by the cash outlay. If this ratio exceeds the value one (1), the project is accepted; if the index is less than one (1), the project is rejected because a value less than one (1) implies that the additional benefits of the project are less than the initial outlay. This index or ratio gives answers consistent with net present value procedures when only one project is being evaluated. But it may be inconsistent with net present value and incorrect when two or more mutually exclusive projects are compared. See Van Horne, *op. cit.,* Chapter 5.

financially feasible as well as economically profitable. If a cash deficit results, the project is not financially feasible—it will not generate sufficient cash income to make the loan payments. Cash deficits do not mean that the investment is unprofitable or should not be made; they simply mean that loan servicing problems will likely be encountered.

Cash deficits can be reduced or eliminated in a number of ways. Extending the loan terms (i.e., more years to repay the principal) will result in lower annual debt servicing requirements, thus reducing the deficit. Increasing the amount of the downpayment will reduce the size of the loan and the annual principal and interest payments. Possibly, the net cash flow from the project could be increased by controlling expenses more carefully or increasing utilization as in the case of a combine that could be used for custom work. If the deficit cannot be reduced or eliminated, then the project must be subsidized with cash from some other source such as other livestock or cropping enterprises, or maybe even off farm employment, to make it financially feasible. By completing a financial feasibility analysis, the size of the subsidy needed can be estimated.

To illustrate the use of financial feasibility analysis, it will be applied to the earlier example of the decision to purchase the combine. Recall that the economic profitability analysis had resulted in a decision to make the purchase. Now the question is, "Will it generate enough cash to make the loan payments?" We will assume that the lender has agreed to a four-year loan for the full purchase price with four equal annual principal payments and 13 percent interest on the outstanding balance.

The data used in the financial feasibility analysis are summarized in Table 8.2. The annual net cash flow (not the *discounted* net cash flow) was calculated earlier in the economic profitability analysis (see Step 3 of the example on page 328). The loan repayment schedule calls for $19,105 of principal each year, plus interest on the outstanding balance ($9,935 of interest in year 1, $7,451 in year 2, $4,967 in year 3, and $2,484 in year 4).

Table 8.2 The Financial Feasibility of a Combine Purchase

| Year | Annual Net Cash Flow | Payment Schedule | | | Tax Savings from Interest Deductibility | After-Tax Payment Schedule | Surplus (+) or Deficit (−) |
		Principal	Interest	Total			
1	$21,040	$19,105	$9,935	$29,040	$2,782	$26,258	$ −5,218
2	15,338	19,105	7,451	26,556	2,086	24,470	−9,132
3	15,678	19,105	4,967	24,072	1,391	22,681	−7,003
4	16,248	19,105	2,484	21,589	696	20,893	−4,645
5	16,847	—	—	—	—	—	+16,847
6	13,207	—	—	—	—	—	+13,207
7	13,867	—	—	—	—	—	+13,687
8	34,147	—	—	—	—	—	+34,147

The tax savings from interest deductibility are then calculated as the interest payment times the marginal tax bracket (28 percent in this example), resulting in an after-tax payment schedule as noted in Table 8.2. As indicated in the last column of Table 8.2, a cash deficit occurs in the first four years, and then cash surpluses occur for the remaining life of the combine. Consequently, the custom operator knows that, even though the combine is economically profitable, it will not generate sufficient cash in the first four years to make the loan payments. These cash deficits might be reduced by extending the term of the loan or increasing the acreage harvested each year. Alternatively, the operator could plan to subsidize the combine purchase with cash from other sources; the subsidy required is clearly documented by the financial feasibility analysis as $5,218 in the first year, $9,132 in the second year, $7,003 in the third year, and $4,645 in the fourth year. After year 4, the combine will generate excess cash which could be used to subsidize the later purchase of other capital assets.

The analysis of capital investment alternatives is an essential part of long-run whole-farm planning. By using the procedures of investment analysis, the various expansion alternatives available to the business can be evaluated. For example, the farm operator may have the opportunity to purchase an additional piece of farmland or a larger combine or tractor, or to expand his or her hog finishing facilities. The expected contribution of each investment alternative to annual net income can be obtained by using whole-farm or partial budgeting. Then to obtain an estimate of the lifetime profitability of these capital items, investment analysis procedures that reflect the time value of money can be applied. In this fashion, the annual data from whole-farm or partial budgeting are used as input into the investment analysis procedure. Thus, capital budgeting procedures are a key component of long-run planning and build on the whole-farm and partial budgeting procedures discussed earlier.

LEVERAGE

A discussion of financial analysis and planning would not be complete without a review of the concept of financial leverage. Financial leverage essentially involves the use of credit or debt funds with a fixed cost in the form of interest in the anticipation of increasing the return to the equity capital of the business.[9] The amount of financial leverage a firm is using can be measured by the ratio of debt to equity. Thus, Firm A of Table 8.3 has a leverage ratio of zero since it is using no debt and all equity. Firm B has a leverage ratio of 1.0 ($100,000 of debt and $100,000 of equity), and

---

[9]The interest cost is fixed in the short run regardless of whether the interest rate is fixed or variable.

Table 8.3 The Impact of Leverage

Item	Firm A	Firm B	Firm C
Debt (12 percent)	$ 0	$100,000	$150,000
Equity	200,000	100,000	50,000
Total capital	200,000	200,000	200,000
Leverage ratio	0	1.0	3.0
All Capital Returns 15 Percent			
Returns to total capital	$ 30,000	$ 30,000	$ 30,000
Interest on debt (12 percent)	0	12,000	18,000
Return to equity capital	$ 30,000	$ 18,000	$ 12,000
Percentage return to equity	15	18	24
All Capital Returns 8 Percent			
Returns to total capital	$ 16,000	$ 16,000	$ 16,000
Interest on debt (12 percent)	0	12,000	18,000
Return to equity capital	$ 16,000	$ 4,000	$−2,000
Percentage return to equity	8	4	−4

Firm C has a leverage ratio of 3.0 ($150,000 of debt and $50,000 of equity). An increase in the leverage ratio will result in an increase not only in the potential return to equity capital, but also in the potential losses that can occur if prices or productivity are below expectations. Increased leverage, therefore, results in an increase in the potential return as well as in financial risk. These risk and return implications of financial leverage can be combined with the earlier concepts of cash flow and debt repayment capacity to determine the acceptable level of credit to use in the farm business.

The specific impact of leverage on the potential income or losses of the firm can best be illustrated with the examples of Table 8.3. Note that all three firms of Table 8.3 include the same amount of total capital or assets ($200,000), but Firm A includes all equity and no debt, Firm B includes equal amounts of equity and debt, and Firm C has $150,000 of debt and $50,000 of equity. Firm A might be representative of an older, well-established farmer who has paid off all debt, whereas Firm C may be indicative of a young farmer who is getting established and has a high debt obligation in relation to equity, and thus a high leverage ratio.

The data in Table 8.3 indicate that if all capital earns a 15 percent rate of return, the return to total capital will be $30,000 for each of the three firms. However, Firm A includes no debt so the fixed interest obligation will be zero, leaving a return to equity capital of $30,000. In contrast, Firms B and C must pay interest on the debt capital at 12 percent, resulting in payments of $12,000 and $18,000, respectively. Thus, the return to equity capital is only $18,000 for Firm B and $12,000 for Firm C. The percentage return to equity capital (return to equity capital divided by the amount of equity capital invested in the firm) is 15 percent for Firm A ($30,000 ÷ $200,000), 18 percent for Firm B ($18,000 ÷ $100,000), and 24 percent for Firm C ($12,000 ÷ $50,000). When the rate of return on those assets purchased with borrowed funds is greater than the rate of interest on borrowed funds, increased leverage will result in an increase in the rate of return on equity capital.

But what if low prices or productivity result in a rate of return on total capital of only 8 percent? If we use the same computational procedure as before, the returns to total capital now amount to $16,000 for each of the three firms. When we subtract the fixed interest expense from the return to total capital, the return to equity is $16,000 for Firm A, $4,000 for Firm B, and −$2,000 for Firm C. The rate of return on equity capital is then 8 percent for Firm A, 4 percent for Firm B, and −4 percent for Firm C. Thus, the return to equity capital declines sharply with higher leverage if the rate of return on total capital is less than the interest that must be paid on borrowed funds.

Increased leverage or the use of increasing amounts of debt with a given level of equity is, therefore, a two-edged sword. If the interest rate paid on debt is less than the rate of return on capital, the margin is earned by the holder of equity capital. Thus, a higher leveraged firm can accumulate equity more rapidly and expand its operations at a faster rate if it is generating a higher return on borrowed funds than the interest rate. Because of the fixed nature of the interest payment, a higher leveraged firm will also *lose* more money if the rate of return on capital is less than the interest rate on borrowed funds. While increased leverage increases the possibility of a higher return to equity capital, it also increases the size of the possible loss or the risk associated with generating that return.

Furthermore, the principle of increasing risk indicates that as leverage increases, the potential losses that can occur will be much larger than the potential benefits that can be generated. This asymmetric impact of leverage is a result of the fixed interest obligation on credit which must be paid regardless of the rate of return. To illustrate, note in the examples of Table 8.3 that when all capital generates a 15 percent rate of return, increasing leverage from 0 to 3.0 increases the rate of return by 9 percentage points (15 percent from Firm A to 24 percent for Firm C). When all capital generates only an 8 percent rate of return, however, the increased leverage results in a 12-point reduction in the rate of return, from 8 percent for Firm A to −4 percent for Firm C. Thus, the potential benefit of increasing the leverage ratio from 0 to 3.0 is a 9 percent increase in the rate of return to equity capital, but the potential reduction in return is 12 percent as the leverage ratio increases.

TAX MANAGEMENT

The income tax liability can be a major cash expenditure for successful farm businesses. As noted in the earlier discussion of capital budgeting procedures, income taxes can have a significant impact on the cash flows and thus the economic profitability and financial feasibility of a capital investment.

Since most farmers (approximately 97 percent) use the cash accounting method for determining income tax obligations, the tax liability can be effected by judicious tax management. The goal of tax management is to maximize the income after taxes over a period of years. Although a number of tax management alternatives could be discussed, we will concentrate on methods of income leveling, deferring current tax liabilities, capital gains tax provisions, and investment credit. Because of the progressive structure of the income tax rates, wide variations in income from year to year will result in a larger tax bill over the years compared to level annual incomes. Therefore, income leveling and deferring taxes are particularly important issues when incomes are expected to be substantially higher or lower than in previous years.

The tax management procedures presented here are based on U.S. income tax regulations as of 1983. Since revisions in the income tax regulations have occurred rather frequently in recent years, the astute farm manager should keep up to date concerning new tax laws. The discussion assumes that the cash accounting system is being used since it allows the most flexibility in tax management. The major differences between the cash and accrual systems of accounting are summarized in Chapter 2. Moreover, the tax implications of different methods of business organization (sole proprietorship, partnership, corporation) in terms of tax rates and special rules are reviewed in Chapter 9.

Estimating Tax Due

The first step in tax management is to make an estimate of the tax liability for the current year. This can be done with the aid of a tax estimate worksheet (Figure 8.1). At the top of this worksheet, all receipts to date and an estimate of receipts for the rest of the year are entered in the first two columns and totaled in the last column. Likewise, farm expenses to date and an estimate for the rest of the year are entered on the worksheet and totaled in the last column. The next item, depreciation, can be estimated by adding the amount of depreciation on capital purchases made during the year to the actual depreciation deduction taken in the previous year. Gains from the sale of capital items, itemized deductions, and nonbusiness deductions and exemptions must also be projected. Finally, the income and self-employment tax can be estimated, and credits can be deducted to obtain

Income Tax Estimate Worksheet

	Amount to Date	Estimated Rest of Year	Estimated Year's Total
RECEIPTS:			

Sales of products raised* and miscellaneous receipts:

	Amount to Date	Estimated Rest of Year	Estimated Year's Total
Cattle, hogs, sheep and wool, etc.	$_____	_____	_____
Poultry, eggs and dairy products	$_____	_____	_____
All crop sales	$_____	_____	_____
Custom work, prorations and refunds agriculture program payments	$_____	_____	_____
Total sales and other farm income . . . . (1)	$_____	_____	_____
Sales of purchased market livestock . . . . †$_____		_____	_____
Purchase cost (subtract) . . . . ‡$_____		_____	_____
Gross profits on sales of purchased livestock . . . . . . †(2)	$_____	_____	_____
Gross farm profits (Item 1 + 2) . . . . (3)	$_____	_____	_____

FARM EXPENSES:

Labor hired	$_____	Veterinary, medicine .	$_____
Repairs, maintenance	_____	Gasoline, fuel, oil . . .	_____
Interest	_____	Storage, warehousing	_____
Rent of farm, pasture	_____	Taxes	_____
Feed purchased . .	_____	Insurance	_____
Seed, plants purchased	_____	Utilities	_____
Fertilizers, lime . .	_____	Freight, trucking . . .	_____
Machine hire . . .	_____	Conservation expenses	_____
Supplies purchased	_____	Other	_____
Breeding fees . . .	_____	Other	_____

	Amount to Date	Estimated Rest of Year	Estimated Year's Total
Total cash farm expenses . . . . (4)	$_____	_____	_____
Depreciation on machinery improvements, dairy and breeding stock . . . . (5)	$_____	_____	_____
Total deductions (Item 4 + 5) . . . . (6)	$_____	_____	_____
Self employment farm income (Item 3 less item 6) . . . . . ‡(7)	$_____		
Net taxable gain from Schedule D (Sales of dairy and breeding stock, machinery and other capital exchanges) . . . . (8)	$_____	_____	_____
Taxable non-farm income . . . . (9)	$_____	_____	_____
Adjusted gross income (Item 7 + 8 + 9) . . . . (10)	$_____	_____	_____
Less: zero bracket amount or itemized deductions§ . . . . $_____		_____	_____
$1000 x _____ personal exemptions** . . . . $_____			
Total non-business deductions and exemptions . . . . . . (11)	$_____	_____	_____
Taxable income (Item 10 less item 11) . . . . (12)	$_____	_____	_____
Estimated income tax (calculated from applicable tax computation table or rates) . . . . (13)	$_____	_____	_____
Estimated self-employment tax (Item 7 x current rate) . . . (14)	$_____	_____	_____
TOTAL TAX (Item 13 + 14) . . . . (15)	$_____	_____	_____
Less Credits: allowable investment credit and carryover, gas tax, income tax withheld and estimated tax paid . . . (16)	$_____	_____	_____
Estimated tax due (Item 15 less item 16) . . . . (17)	$_____	_____	_____

Last year's marginal tax bracket _____ %
This year's estimated marginal tax bracket _____ %
Next year's expected marginal tax bracket _____ %

† Omit for accrual method.
‡ For accrual method adjust for change in inventory and new livestock purchases.
§ Use itemized deductions if larger.
* For accrual method include sales of all livestock.
** Exemption for 1980, see current tax regulation for subsequent years.

Figure 8.1. Income tax estimate worksheet. (*Source:* R. N. Weigle, R. Edward Brown, Jr. and Robert S. Smith, "Income Tax Management for Farmers," North Central Regional Publication No. 2, Oct. 1980 (revised).)

the tax due this year. Once this estimate is obtained, the question arises as to how to influence the tax liability if it is too high or too low.

Income Averaging

One method that can be used to reduce the income tax liability in a year of excessively high income is that of income averaging. Income averaging allows part of an unusually large amount of taxable income to be taxed in lower brackets. The result is a reduction in the overall amount of tax due. To qualify for income averaging, averageable income for the tax year must be in excess of $3,000. Averageable income is calculated as the amount by which adjusted taxable income for the computation year exceeds 120 percent of the average taxable income for the four preceding tax years. Income averaging computations and reporting are accomplished on Schedule G of Form 1040 (Figure 8.2).

The benefit of income averaging can be illustrated by a simple example. Assume a farmer (married, filing a joint return) had taxable income in 1978 through 1981 of $10,000, $9,500, $10,000, and $10,500, respectively. In 1982, the current tax year, taxable income is $30,000. If we use the format of Schedule G (Figure 8.2), the averageable income for 1982 amounts to $18,000, which qualifies the farmer for income averaging. The tax due under income averaging amounts to $4,673. If the farmer had not used income averaging for his 1982 tax return, the tax liability would have been $5,607. Thus, income averaging would save $934 of tax payments, a savings of 16.7 percent.

Tax Deferral

It is frequently desirable to defer the payment of income taxes to future years, particularly if income levels are anticipated to be lower in the future. Income deferral enables the taxpayer to move income out of the higher tax brackets in a year of high income into lower tax brackets in the future when income may be lower. A number of methods can be used to defer income taxes to future years. We will only briefly review three such methods— accelerated depreciation; soil and water conservation and land clearing expenditures; and prepayment of ordinary business expenses or delayed sales.

With passage of the Economic Recovery Tax Act of 1981, the accelerated cost recovery system (ACRS) established a new set of rules with respect to the tax depreciation of capital items. In general, the ACRS system allows property to be classified into one of four categories for purposes of determining the annual depreciation allowance:

1. 3-year property—includes such items as breeding hogs, light trucks, and automobiles.

Schedule G
(Form 1040)
Department of the Treasury (X)
Internal Revenue Service

Income Averaging

▶ See instructions on back. ▶ Attach to Form 1040.

OMB No. 1545-0074

1982
20

Name(s) as shown on Form 1040 | Your social security number

Step 1 Figure your income for 1978–1981

1978	1	Fill in the amount from your 1978 Form 1040 (line 34) or Form 1040A (line 10)	**1** 13,000		
	2	Multiply your total exemptions in 1978 by $750	**2** 3,000		
	3	Subtract line 2 from line 1. If less than zero, enter zero		**3**	10,000
1979	4	Fill in the amount from your 1979 Form 1040 (line 34) or Form 1040A (line 11)	**4** 13,500		
	5	Multiply your total exemptions in 1979 by $1,000 . . .	**5** 4,000		
	6	Subtract line 5 from line 4. If less than zero, enter zero		**6**	9,500
1980	7	Fill in the amount from your 1980 Form 1040 (line 34) or Form 1040A (line 11)	**7** 14,000		
	8	Multiply your total exemptions in 1980 by $1,000 . . .	**8** 4,000		
	9	Subtract line 8 from line 7. If less than zero, enter zero		**9**	10,000
1981	10	Taxable income. Fill in the amount from your 1981 Form 1040 (line 34) or Form 1040A (line 12). If less than zero, enter zero		**10**	10,500
Total	11	Fill in all income earned outside of the United States or within U.S. possessions and excluded for 1978 through 1981		**11**	--
	12	Add lines 3, 6, 9, 10 and 11		**12**	40,000

Step 2 Figure your averageable income

	Multiply the amount on line 12 by 30% (.30)			
13	Write in the answer	**13**	12,000	
14	Fill in your taxable income for 1982 from Form 1040, line 37	**14**	30,000	
15	If you received a premature or excessive distribution subject to a penalty under section 72, see instructions	**15**	--	
16	Subtract line 15 from line 14	**16**	30,000	
17	If you live in a community property state and are filing a separate return, see instructions	**17**	--	
18	Subtract line 17 from line 16. If less than zero, enter zero	**18**	30,000	
19	Write in the amount from line 13 above	**19**	12,000	
20	Subtract line 19 from line 18. This is your averageable income	**20**	18,000	

If line 20 is $3,000 or less, do not complete the rest of this form. You do not qualify for income averaging.

Step 3 Figure your tax

	Multiply the amount on line 20 by 20% (.20)			
21	Write in the answer	**21**	3,600	
22	Write in the amount from line 13 above	**22**	12,000	
23	Add lines 21 and 22	**23**	15,600	
24	Write in the amount from line 17 above	**24**	--	
25	Add lines 23 and 24	**25**	15,600	
26	Tax on amount on line 25 (from Tax Rate Schedule X, Y, or Z)	**26**	1,937	
27	Tax on amount on line 23 (from Tax Rate Schedule X, Y, or Z) . .	**27** 1,937		
28	Tax on amount on line 22 (from Tax Rate Schedule X, Y, or Z) . .	**28** 1,253		
29	Subtract line 28 from line 27	**29** 684		
	Multiply the amount on line 29 by 4			
30	Write in the answer		**30**	2,736
	If you have no entry on line 15, skip lines 31 through 33 and go to line 34.			
31	Tax on amount on line 14 (from Tax Rate Schedule X, Y, or Z) . .	**31** --		
32	Tax on amount on line 16 (from Tax Rate Schedule X, Y, or Z) . .	**32** --		
33	Subtract line 32 from line 31		**33**	--
34	Add lines 26, 30, and 33. Write the result here and on Form 1040, line 38. Be sure to check the Schedule G box on that line		**34**	4,673

G

For Paperwork Reduction Act Notice, see Form 1040 instructions.

Figure 8.2. Schedule G (Form 1040) income averaging.

2. 5-year property—includes most machinery and equipment used in farming plus storage facilities, fences, tile lines, water systems, paved drives and feeding floors, depreciable parts of dams and ponds, and single-purpose livestock and horticultural structures.

3. 10-year property—real property, including buildings with an asset depreciation range midpoint life of 12.5 years or less.

4. 15-year property—real property, including buildings with an asset depreciation range midpoint life of more than 12.5 years.

Most farm personal property falls in the 5-year ACRS category. A taxpayer can choose to use depreciation methods other than ACRS, but the optional periods in general are longer and result in lower depreciation deductions in the early years of the life of an asset.

Once the property has been placed in the correct ACRS category, the rate of write-off for each year is obtained from Table 8.4. The recovery rate for 15-year ACRS property depends upon the month placed in service; the numbers in Table 8.4 presume that the property is placed in service in January. Note that, with the new ACRS provisions, the salvage value of the property is ignored. Thus, the write-off rate of Table 8.4 is applied to the full basis of the property which for new property (assuming

Table 8.4 ACRS Schedule for Property Placed in Service during 1981 and Thereafter

Recovery Year	Applicable Percentage for the Class of Property			
	3 year	5 year	10 year	15 year[a]
1	25	15	8	12
2	38	22	14	10
3	37	21	12	9
4	—	21	10	8
5	—	21	10	7
6	—	—	10	6
7	—	—	9	6
8	—	—	9	6
9	—	—	9	6
10	—	—	9	5
11	—	—	—	5
12	—	—	—	5
13	—	—	—	5
14	—	—	—	5
15	—	—	—	5
	100	100	100	100

[a]Assumes property is placed in service during the first month of the fiscal year.

no trade-in) is the purchase price. As is noted later, however, provisions implemented in 1982 require a reduction in the basis of property if the full investment tax credit is taken.

In addition to the ACRS method of calculating depreciation, the new 1981 law initiated "expense method depreciation." This method allows up to 100 percent of the purchase price of tangible depreciable property to be deducted currently—up to a 100 percent write-off for the year the property is purchased. The amount of property eligible for expense method depreciation is $5,000 in 1982 and 1983, $7,500 in 1984 and 1985, and $10,000 in 1986 and thereafter.

It should be recognized that ACRS and the faster methods of depreciation available prior to 1981 such as double declining balance and sum-of-the-years-digits only defer the payment of income taxes into future years. The amount of money saved through tax deferral in the current year can be invested, however, and will yield earnings that will accrue in future years. The full economic benefits of ACRS and faster depreciation methods can be evaluated using the present value procedures discussed earlier. The appropriate discount rate to use in this analysis is again the cost of capital.

To determine the benefits of fast depreciation, the additional depreciation and tax savings or losses that occur must first be calculated. If we assume a $10,000 machine with a 12-year useful life and a farmer in the 32 percent tax bracket, the tax savings of ACRS assuming a 5-year life compared to straight-line depreciation over 12 years are listed in Table 8.5. If

Table 8.5 Net Present Value of ACRS (5 years) Compared to 12-Year Straight-Line Depreciation

Year	Straight-Line Depreciation[a]	ACRS 5 Year	Additional Depreciation with ACRS	Tax Savings from ACRS	Discount Factor (10%)	Present Value of Tax Savings
1	$833	$1425	$ 592	$ 189.44	0.909	$ 172.20
2	833	2090	1257	402.24	0.826	332.25
3	833	1995	1162	371.84	0.751	279.25
4	833	1995	1162	371.84	0.683	253.97
5	833	1995	1162	371.84	0.621	230.91
6	833	0	−833	−266.56	0.564	−150.34
7	833	0	−833	−266.56	0.513	−136.74
8	833	0	−833	−266.56	0.467	−124.48
9	833	0	−833	−266.56	0.424	−113.02
10	833	0	−833	−266.56	0.386	−102.89
11	833	0	−833	−266.56	0.350	−93.30
12	833	0	−833	−266.56	0.319	−85.03
Net present value						+462.78

[a]Assumes a zero salvage value.

the farmer has a discount rate of 10 percent, the tax savings for each year can be multiplied times the appropriate discount factors and summed to obtain the net present value of the additional depreciation deductions. This net present value totals $462.78 for the $10,000 machine, indicating that the economic benefits of ACRS are significant even with only a $10,000 outlay.

In addition to using depreciation deductions as a means of reducing taxable income and tax liabilities, farmers on the cash accounting system may be able to schedule the sales of products and purchases of supplies to reduce taxable income. For example, sales of crops or livestock might be deferred until after the end of the year, so that these receipts would show up in the next year's taxable income. A contract for delivery of certain crops or livestock after the end of the year—but signed in December—would allow a farmer to guarantee himself a price, and yet the income would not be reported until the following year. Another method that can be used to reduce taxable income is to purchase supplies such as feed, fertilizer, and chemicals for use next year before the end of the current year. Recent court rulings uphold the right of farmers to prepay expenses as long as there is a binding commitment to make the purchase (as contrasted to a deposit), the purchase is for a business purpose (lower price, guarantee supply, etc.), and it does not materially distort income. There has been substantial litigation in this area of prepayment of expenses by farmers. A producer should, therefore, pay for and take delivery of the supplies or have a binding contract for a specified quantity and type of input before December 31, and not just sign a contract for the delivery of supplies when they are needed.

Farmers who are in their early sixties and are approaching retirement may want to give particular attention to the ideas of income deferral. After the age of 65, the personal exemption is doubled and an additional credit (the credit for the elderly) is available. Additional tax savings will occur because income deferred until after the age of 65 will be partially offset by additional exemptions and credits. The total tax liability for retiring farmers can also be reduced by selling property, particularly real estate, on an installment basis so that any capital gains can be spread over a number of years rather than being reported in one year. Farmers close to retirement age may want to make sure they maintain income sufficiently high (if possible) to receive full social security benefits at retirement. And any taxpayer should plan to have at least enough income in each year to utilize all personal exemptions and deductions, plus the zero bracket amount, since unused personal deductions and exemptions cannot be carried forward to future years.

If farmers desire to increase income in a particular year, they can accelerate the sale of crops and livestock or delay the purchase of supplies. However, they should consider expected changes in product and input prices before they make these types of deferred or accelerated sale and input purchasing decisions. Product sales should not be delayed if prices are expected to decline after the end of the tax year, and supplies should normally not be purchased just for tax purposes if they are expected to cost less next year.

A high-income year may also be the time to take advantage of the lucrative tax deductions that can be taken for soil and water conservation and land clearing expenses. A farmer may deduct, as current expenditures, outlays for soil and water conservation improvements up to a maximum of 25 percent of the *gross income from farming*. Outlays in excess of the 25 percent maximum can be carried forward as deductions in succeeding years. However, if the land is sold in subsequent years, these deductions may be recaptured as ordinary income. If the sale occurs within five years of the date of acquisition of the land (not the date when the improvement was made), all soil and water conservation deductions are recaptured, whereas no recapture occurs if the land is sold ten years after it is acquired. The recapture proportions for land sold in six, seven, eight, or nine years after the date of acquisition are 80, 60, 40, and 20 percent, respectively. If soil and water conservation expenses are not deducted in the year they are incurred or carried forward, they are added to the basis of the land and recovered at the time of sale.

Expenses for clearing land to make it suitable for farming can also be deducted in the year they are incurred. The land clearing deduction is limited to 25 percent of the *taxable income from farming* up to a maximum of $5,000. Excess land clearing deductions cannot be carried forward, but are added to the basis of the land and recovered at the time of sale. If the land is subsequently sold, the recapture rules discussed earlier for soil and water conservation expenses also apply to land clearing expenses. These soil and water conservation and land clearing expenses can be valuable deductions in years of abnormally high income.

Income Sharing

Another method that can be used to reduce income tax liabilities is to share income with family members. Wages can be paid to children for farm work as long as they are reasonable and the children are true employees in the farm operation. Although these wages must be included as income for the child, and he or she may be required to pay income tax, the children will typically be in a lower tax bracket. Furthermore, an unmarried child can earn $2,300 without a tax liability, and as long as the child is under 19 or regularly enrolled in school or an on-farm training program, and the parents pay for over one-half of the child's support, they can also receive a $1,000 personal exemption because the child is a dependent.

Sharing income among family members can also be accomplished by transferring income-producing property such as land, livestock, or machinery to the children through a gift or sale. The children then report the income received from this property in their tax returns, and it is again typically taxed at a lower marginal tax rate.

Capital Gains

The increase or decrease in value of capital items may receive preferential treatment under the income tax regulations. If capital assets such as land or breeding stock are held for an appropriate period of time, the increase in value is treated as long-term capital gain, and only 40 percent of this gain is taxed as income. In general, capital assets must be held for more than 12 months to be considered for long-term gain treatment. To qualify for long-term capital gains, however, breeding cattle and horses must be held for 24 months and other breeding livestock for 12 months.

The capital gains provisions should be considered in scheduling the sale of machinery and equipment, land, and breeding livestock. By holding an item for the appropriate holding period, the tax liability may be reduced by 60 percent. For example, if raised breeding stock such as sows were held for 12 months (one or possibly two litters), income received when they were culled from the herd would be capital gain. If cull breeding sows are selling for $150 per head, qualifying the cull stock for capital gains treatment would save $28.80 per sow in income taxes (assuming the breeding stock has a zero basis, which is typically the case for raised breeding stock under the cash accounting system) for a farmer in the 32 percent tax bracket [($150 × 0.32) − ($60 × 0.32)]. And if the sows are kept for fewer litters and replaced by raised gilts, a higher proportion of the income from hog sales would be capital gain rather than ordinary income. To determine how to manage the breeding herd, however, this tax benefit must be compared to the lower reproductive efficiency and litter size that typically occur with younger breeding stock.

The special tax treatment of capital gains compared to ordinary income should also be considered when completing an investment analysis using the net present value or other procedures discussed earlier. If part of the salvage value or annual income qualifies for capital gains tax treatment, the tax liability will be reduced significantly, thus increasing the after-tax cash flow from the project and its profit potential.

Investment Credit

Another method that is available to reduce income tax liabilities is investment credit. Currently, a direct credit is allowed against the income tax due in the amount of 10 percent of the value of qualifying property that has a useful life in the hands of the taxpayer of at least five years under the ACRS rules discussed earlier. For property with a three-year life under the ACRS rules, the credit is 6 percent. Investment credit can be taken on both new and used property. In general, the credit is limited to the income tax liability, or $25,000 plus 85 percent of the tax liability in excess of $25,000, whichever is less. If the taxpayer sells or disposes of the property before the end of the specified ACRS life, the investment credit must be recomputed and a recapture of this credit occurs in the later computation year. The recapture depends upon the number of years the property has been owned. In general, the credit accrues at the rate of 2 percent per year; thus, three-year property disposed of before the end of three years will require a recapture at the rate of 2 percent per year for early disposition. For property with a five-year or greater life, recapture at the rate of 2 percent per year for early disposition during the first five years is required; property held for five years or more is not subject to investment credit recapture.

Legislation passed in 1982 requires a taxpayer to choose between the full investment credit and the full ACRS deduction. In essence, the taxpayer can choose either to reduce the investment credit by two percentage points and use the full basis of the property (for example, the purchase price for new property) for depreciation or cost recovery purposes, or take the full investment credit and reduce the tax basis used in the depreciation or cost recovery calculations by 50 percent of the amount of investment tax credit taken.

All tangible personal property used in the farm business qualifies for investment credit. This includes machinery, equipment and vehicles used in the business, livestock held for breeding purposes other than horses, fences used in connection with raising livestock and keeping livestock and other animals out of cultivated areas, paved barnyards, and handling facilities, waterwells, drain tiles on agricultural land, groves, orchards, and vineyards in the productive stage, and storage facilities such as grain storage bins, corn cribs, and silos. Furthermore, single-purpose livestock or horticultural structures used, for example, in confinement production of hogs or cattle qualify for investment credit.

Investment credit must be taken in the year the property is placed in service, or it is lost. Property is placed in service when it is placed on the depreciation schedule or when it is in a condition of readiness and availability for a specified function, whichever is earliest. Thus, a farmer can purchase an item that will qualify for investment credit in one year and not take delivery until the following year, as long as there is a binding contract and the item is set up and ready for operation on the dealer's lot by the end of the current year.

The use of investment credit can result in substantial tax savings. The investment credit on a $10,000 machine amounts to a $1,000 tax credit or tax reduction. For the taxpayer in the 22 percent tax bracket, a total of $4,545 of additional depreciation or other deductions would be required to generate this same tax savings. Investment credit is also an important consideration in investment analysis. To illustrate, in the combine example used earlier to demonstrate the use of net present value analysis, the investment credit amounted to $7,642 which has a present value of $6,824 ($7,642 × 0.893). If investment credit was ignored or the combine did not qualify for this credit, the net present value would be reduced by 7.7 percent from $88,821 to $81,997. Not considering investment credit in the analysis clearly reduces the profitability of the combine, and in some cases may make the difference between a profitable and unprofitable investment. Thus, investment credit is an important means of reducing taxes and should certainly not be ignored in making investment decisions and computing tax liabilities.

Summary

An integral part of the planning process is that of developing investment and financial plans for the future. Financial planning includes not only the evaluation of alternative investments and the financing of the purchase of capital assets, but also the management of the tax liability of the farm business.

Analyzing investment alternatives (frequently referred to as capital budgeting) involves four major steps: (1) identify possible investment opportunities, (2) evaluate the economic profitability and financial feasibility of each opportunity, (3) reevaluate the alternatives under different price and yield assumptions, and (4) choose an alternative. Different methods can be used in determining the economic profitability of an investment; such criteria include the payback period, the simple rate of return, the internal rate of return, and net present value. Of these four criteria, the net present value procedure is preferred because it is relatively simple to calculate and it recognizes explicitly the time value of money.

The computational steps of the net present value analysis procedure are: (1) choose an appropriate discount rate to reflect the time value of money, (2) calculate the present value of the cash outlay required to purchase the asset, (3) determine the benefits or annual net cash flow from the asset over its useful life, (4) calculate the present value of the annual net cash flows, (5) compute net present value by subtracting the capital outlay from the cash flows, and (6) accept investments with a positive net present value and reject those with a negative net present value. If the discount rate reflects the weighted cost of capital for the firm, a positive net present value indicates that the investment is generating an income stream larger than the cost of funds used to finance the investment; consequently, the investment is profitable. If the net present value is negative, the income generated by the investment is less than the cost of funds and the money would be better invested elsewhere.

Financial feasibility analysis answers the basic question: Will the asset pay its own way? To complete a financial feasibility analysis, the after-tax annual net cash flows are compared to the principal and interest payment schedule on loans incurred to buy the asset. Annual surpluses indicate that the investment should generate sufficient funds to service the debt, whereas deficits indicate that adjustments may be needed in the repayment schedule or a cash subsidy will be required from other parts of the farming operation.

When farmers borrow money and combine it with their own equity capital to finance the business, they are using leverage. The amount of leverage a firm is using can be measured by the ratio of debt to equity, with a higher ratio reflecting increased leverage. Leverage is a two-edged sword—if the interest rate paid on debt is less than the rate of return on capital, the firm can accumulate equity more rapidly with higher leverage. But a higher leveraged firm will also lose more money if the rate of return on capital is less than the interest rate on borrowed funds. Furthermore, the principal of increasing risk indicates that as leverage increases, the potential losses that can occur will be much larger than the potential benefits that can be generated.

Income taxes can be a major cash expenditure for the farm business; thus, tax management is an important component of farm and financial management. The goal of tax management is to maximize the income after taxes over a period of years. Because of the progressive nature of the income tax rate structure, this goal is best accomplished by using tax management techniques to obtain approximately equal taxable incomes over time. Such tax management procedures include use of income averaging; deferral of income through rapid depreciation methods such as the accelerated cost recovery system; delay of product sales to future years; prepayment of expenses; use of deductions for land clearing and soil and water conservation expenses; and various methods for sharing income among family members. Tax management strategies should also recognize the important differences in the tax treatment of capital gains (only 40 percent of any long-term capital gain is taxable). Furthermore, investment tax credit provisions should be explicitly considered when making any type of investment or capital acquisition decision.

APPENDIX

Application of Various Capital Budgeting Procedures

To illustrate the practical application of the various capital budgeting procedures (payback period, simple rate of return, net present value, and internal rate of return), they will be applied to the combine example presented earlier in the chapter. Table 8.6 summarizes the annual after-tax net cash flow or earnings from the combine. As indicated, *average* annual earnings are $15,848 per year.

PAYBACK PERIOD

If we assume a capital outlay of $76,420 to purchase the combine, the payback period is calculated as:

$$P = \frac{O}{I} = \frac{\$76,420}{\$15,848} = 4.8 \text{ years} \qquad (8.13)$$

Thus, if we assume average annual earnings, the capital outlay will be recovered in 4.8 years. This payback period can be compared to that of other machines, and the one with the shortest payback period will be the preferred choice.

Table 8.6 Annual After-Tax Cash
Earnings from a 4-30″ Row Combine

Year	Earnings
1	$ 21,040
2	15,338
3	15,678
4	16,248
5	16,847
6	13,207
7	13,867
8	14,561
Total	$126,783
Average earnings =	$\frac{\$126,783}{8} = \$15,848$

SIMPLE RATE OF RETURN

To use the simple rate of return method, the average annual depreciation must be determined. If we assume a 10 percent salvage value ($7,642) and an eight-year life, annual depreciation on the combine is calculated as:

$$D = \frac{\$76,420 - \$7,642}{8} = \frac{\$68,778}{8} = \$8,597 \qquad (8.14)$$

Thus, the simple rate of return is calculated as:

$$R = \frac{I - D}{O} = \frac{\$15,848 - \$8,597}{\$76,420} = 0.0949 \text{ or } 9.49 \text{ percent} \quad (8.15)$$

This rate can be compared to the rate of return calculated in a similar fashion for other investment alternatives, and the one with the highest simple rate of return would be the preferred choice.

NET PRESENT VALUE

Figure 8.3 summarizes the cash flows that are used in the net present value and internal rate of return calculations using a time line. The stream of income above the time line reflects the annual cash earnings plus the salvage value of the combine. The entry below the time line reflects the initial outlay to buy the combine. As indicated in the earlier discussion, the net present value of this stream of cash inflows and outflows (assuming a 12 percent discount rate) is $12,401. The magnitude of this net present value is in part a function of the discount rate used in the computations. With a higher discount rate, the net present value is reduced as reflected in Figure 8.4, and as the discount rate continues to rise the net present value will decline to zero and then become negative. For the combine, a discount rate of 16.36 percent results in a net present value of zero. As we will show shortly, this rate where the net present value is zero indicates the internal rate of return for the combine.

INTERNAL RATE OF RETURN

To calculate the internal rate of return for the combine requires use of an interative process to determine the rate of discount that equates the dis-

Table 8.7 Calculation of the Internal Rate of Return for the Combine

Year	Annual Net Cash Flow	Discount Factors (16%)	Present Value of Annual Net Cash Flow (16%)	Discount Factors (17%)	Present Value of Annual Net Cash Flows (17%)
1	$21,040	0.8621	$18,139	0.8547	$17,983
2	15,338	0.7432	11,399	0.7305	11,204
3	15,678	0.6407	10,045	0.6244	9,789
4	16,248	0.5523	8,974	0.5337	8,672
5	16,847	0.4761	8,021	0.4561	7,684
6	13,207	0.4104	5,420	0.3898	5,148
7	13,867	0.3538	4,906	0.3330	4,618
8	34,147	0.3050	10,415	0.2849	9,728
Present value of net cash flows			$77,319		$74,826
Capital outlay			$76,420		$76,420
Net present value			$+899		$-1,594

Interpolation between 16% and 17%:
    Internal rate of return $= 16\% + [899/(899 + 1594)](17 - 16) = 16.36$

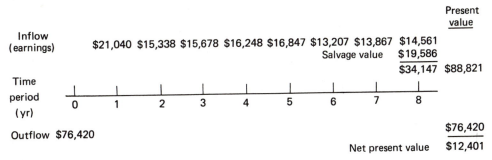

Figure 8.3. Time line of net cash flows for the combine.

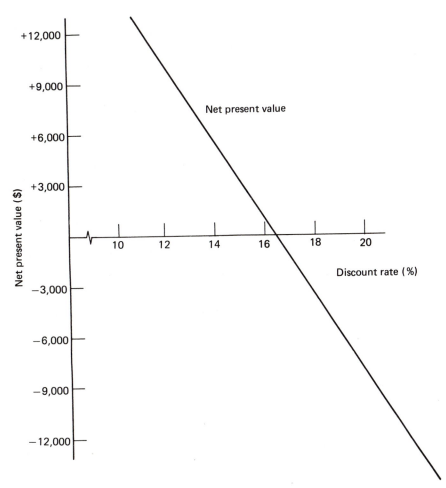

Figure 8.4. Net present value of the combine at various discount rates.

counted net cash flows with the outlay. Table 8.7 illustrates the manual calculation of the internal rate of return using an interpolation procedure. Note that the internal rate of return using this technique is estimated as 16.36 percent. If the maximum acceptable rate of return as specified by the cost of capital is 12 percent, this combine investment would be an acceptable purchase using the internal rate of return procedure.

Questions and Problems

1. List and discuss the four steps of capital budgeting.
2. Distinguish between the concepts of economic profitability and financial feasibility.
3. Summarize and discuss the payback period method of capital budgeting. What are the advantages and disadvantages of using this method?
4. How would you calculate the internal rate of return for a particular capital investment? What are the advantages and disadvantages of using this method of capital budgeting?
5. Why does money have a time value?
6. What are the steps of the net present value procedure of capital budgeting? Identify a specific capital investment project, collect the data, and apply this procedure to determine whether it is an acceptable investment.
7. What is the formula for determining the weighted cost of capital? Discuss how you would calculate the cost of capital for a particular farm business. What are the advantages and disadvantages of using the weighted cost of capital as an estimate of the discount rate?
8. Assume you have evaluated a capital investment and it has a positive net present value. Should you make the investment or not? What can you say about the rate of return of this particular investment? How much could you afford to pay for this investment?
9. How might you recognize or incorporate the concept of inflation in the net present value analysis procedure?
10. Discuss the computations required to complete a financial feasibility analysis of an investment project.
11. Collect the data and complete a financial feasibility analysis for a specific investment.
12. What is leverage? What are the advantages and disadvantages of high leverage?
13. What is the principle of increasing risk? What does it suggest about the risks and returns of higher leverage?
14. What is the goal of tax management? What management strategies can be used to accomplish this goal?
15. What is income averaging? How might it be used to reduce a farmer's tax liability?
16. Describe the accelerated cost recovery system method of depreciation. Calculate the annual write-offs for a $10,000 machine which is categorized as five-year property.
17. What methods might farmers use to defer income taxes?

18. Describe the current investment tax credit provisions. Choose a specific farm machine and calculate the amount of the credit. Why is it important to consider investment tax credit in the calculation of net present value?

Further Reading

Aplin, Richard D., George L. Casler, and Cheryl P. Francis. *Capital Investment Analysis Using Discounted Cash Flows*. 2d ed., Columbus, Ohio: Grid, 1977.

Barry, Peter J., John A. Hopkin, and C. B. Baker. *Financial Management in Agriculture*. 2d ed., Danville, Ill.: Interstate Printers and Publishers, 1979.

Bierman, Harold, Jr., and Seymour Smidt. *The Capital Budgeting Decision*. 4th ed., New York: Macmillan, 1975.

Harris, Phillip, R. Edward Brown, Jr., and W. A. Tinsley. "Income Tax Management for Farmers." North Central Regional Publication No. 2, October 1982.

Lee, Warren F., Michael D. Boehlje, Aaron G. Nelson, and William G. Murray. *Agricultural Finance*. 7th ed., Ames: Iowa State University Press, 1980.

Markowitz, H. M. *Portfolio Selection: Efficient Diversification of Investments*. New York: John Wiley, 1959.

Penson, John B., Jr., and David A. Lins. *Agricultural Finance*. Englewood Cliffs, N.J.: Prentice-Hall, 1980.

Schall, Lawrence D., and Charles W. Haley. *Introduction to Financial Management*. New York: McGraw-Hill, 1977.

Sharp, William F. "Capital Asset Prices: A Theory of Market Equilibrium Under Conditions of Risk." *Journal of Finance* 19(September): 425–42, 1964.

Sharp, William F. *Portfolio Theory and Capital Markets*. New York: McGraw-Hill, 1970.

Solomon, Ezra. *The Theory of Financial Management*. New York: Columbia University Press, 1963.

U.S. Department of the Treasury, Internal Revenue Service. *Farmer's Tax Guide*. Publication 225 (Rev. October 82), Washington, D.C.

Van Horne, James C. *Financial Management and Policy*. 5th ed., Englewood Cliffs, N.J.: Prentice-Hall, 1980.

9

PLANNING THE FARM BUSINESS ARRANGEMENT

CHAPTER CONTENTS*

The family farm is a basic institution in U.S. agriculture. Many parents want their businesses to continue; thus, their objective is to pass on their farm to the next generation. A growing number of farm families are con-

*This chapter is adapted from Kenneth H. Thomas, and Michael D. Boehlje, "Farm Business Arrangements: Which One for You?," North Central Regional Extension Publication 50, Revised 1982.

fronted with the problem of how best to get a son or daughter started and established in farming and to successfully transfer both property and management control to their heirs.

The older farmer without relatives to take over the business faces similar problems in bringing an unrelated young person into the business as a working manager or partner. Because of the high capital and management requirements of modern agriculture, most beginning farmers recognize that they initially have to "piggyback" on an existing farm operation.

Selecting a business arrangement under these circumstances is an important decision that is not to be taken lightly or done hastily. In most cases, the decision involves the economic security of the older generation, the future career potential of the new entrant into farming, and overall family good-will. It is, therefore, important that the arrangement best meet the needs of all individuals involved.

Since family farming dominates U.S. agriculture, choosing a farm business arrangement is most frequently encountered when planning for the transfer of ownership between generations. The decision must be faced any time a new business is started, whether one person or many are involved, or whether or not the individuals are members of the same family. Because of the importance of this decision in the family context, our discussion will focus on choosing a business arrangement to facilitate the transfer of the farm between generations. However, many of the same factors must be considered in developing arrangements between those who are not members of the same family.

Most farm businesses are managed by an individual operator rather than a management team. The 1978 Census of Agriculture indicated that 85.7 percent of U.S. farms with sales over $25,000 per year were operated as sole proprietorships, 11.2 percent were partnerships, 2.7 percent were incorporated, and 0.4 percent were in estates and trusts. The growing capital and management requirements of modern farming and the resultant entry and transfer problems mean that the number of individually operated businesses will likely decline, while the number of joint family operations involving the use of partnerships and corporations will tend to increase in the future.

FARM AND FAMILY LIFE CYCLES

The individually operated farm business has a traditional life cycle paralleling that of the farm operator. The typical farm operator and firm pass through at least three stages during a farming career: entry, growth, and exit stages (Figure 9.1).

The entry stage involves the two major processes of testing and establishment. The operator first evaluates the opportunities in farming and determines whether or not to choose farming as a career. He or she must

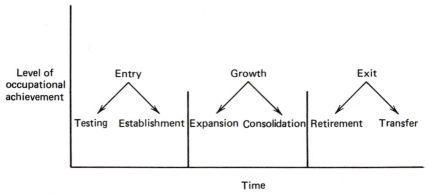

Figure 9.1. Typical family/farm life cycle of individually operated business.

then acquire sufficient capital resources and managerial ability to establish a viable economic unit that will generate a competitive income and be capable of growth.

The growth stage involves the processes of expansion and consolidation. During this stage, the operator first attempts to extend the resource base through purchase or lease. Capital requirements tend to escalate and the operator usually must utilize debt capital (borrowed funds). The operator then tends to shift emphasis from expansion to consolidation of previous gains and stabilization of income.

The exit stage involves consideration of retirement and intergenerational transfer. The farm operator attempts to reduce management responsibilities while maintaining sufficient control of farm assets to generate adequate retirement income. He or she should also develop estate plans that will implement lifetime or at-death transfers of property and the associated managerial responsibility to the next generation.

The growing capital and management requirements of modern farming and the resultant increasingly complex and costly establishment and transfer problems have caused farm families to look toward the merger of farm businesses or farming generations. The emphasis here will be on the merging of generations. The merger of businesses from the same generation should be approached in a similar fashion.

Merging the life cycle of a second-generation farm family with the typical life cycle of the parents must be approached with caution and candor and explored carefully. Each situation is different for at least three sets of reasons. One reason is the level of achievement and objectives of the parents. Some parents have put together a very substantial business, one that would easily accommodate one or more "next generation" farmers (Situation 1, Figure 9.2). Others have provided adequately for their family

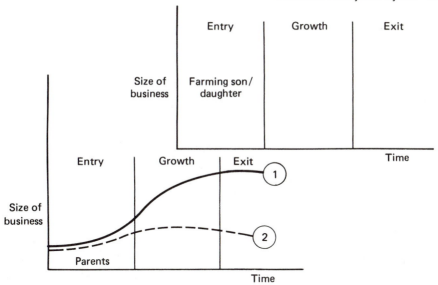

Figure 9.2. Merging family/farm life cycles of parents and farming son or daughter.

but find themselves with a one or a one-and-a-half person business (Situation 2, Figure 9.2).

Parents' goals, objectives, values, and priorities also vary. Some may want their businesses to grow and prosper; others may want to consolidate their gains and slow down. Some may want to begin transferring large amounts of property, while others are concerned with enjoying the fruits of their labor. Some may favor the farming son or daughter, while others want to treat all of their heirs "equally."

A second set of reasons relates to the child's managerial ability and the timing of his or her entrance into farming relative to the parent's retirement. Prospective farmers vary widely in their present managerial skills as well as their commitment to a farming career. Moreover, the timing of their entrance into a farming situation where the parent plans to farm for 20 more years creates a much more complex situation than one where the parent is within five years of retirement.

A third set of reasons relates to the number of other children involved and the degree of interest they have in farming. The situation is much easier to deal with when there is only one child than when there are eight children and three want to farm.

Thus, the merging of the life cycles of two farming generations can be quite simple under the right set of circumstances. Such a case might be the father who is five years from retirement, has a very good business, and has only one child. At the other extreme is the 45-year-old father with less than a two-person business, two children who want to farm, and five other children.

DECISION FRAMEWORK AND BUSINESS ARRANGEMENT ALTERNATIVES

Many other situations could be described. However, it now should be apparent that from a decision-making standpoint the farm family—both generations—should, first, carefully analyze their current situation and identify their objectives. Second, they should make a preliminary appraisal of the possibilities of developing a successful joint operation (Figure 9.3).

If a joint operation is decided upon, then both parties normally should enter a so-called testing stage. The purpose of this testing stage is twofold: to help the child decide whether he or she really wants to farm, and to help the parties involved determine whether they can make a joint business venture work from both personal and business points of view. Some families may find that they can jump over this testing stage and go to more complex joint operating arrangements. But these cases are generally the exception rather than the rule.

At the end of the testing period, the family must decide what future course they wish to follow. In some cases, they may decide to go their separate ways. This may involve the child getting out of farming altogether or moving to another farm with no direct ties to the home farm. In other situations, the child may establish his or her own separate farming unit but share labor or machinery with the parent in a so-called spin-off arrangement. In some cases, the parents may desire to retire from active farming and may sell or rent the farm to the on-farm heir. Or in cases of a larger business, establishment of a multiownership, multimanagement firm may be in order.

The major business arrangement alternatives for organizing a farm business should now be recognized. The sole proprietorship is a form of business organization in which the farm is operated by one individual and

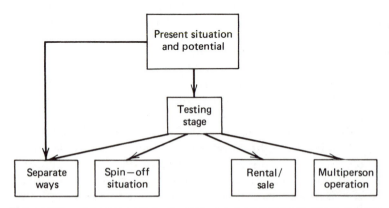

Figure 9.3. A framework for business arrangement decision-making.

exists until the owner retires or dies. Under current conditions, individual sole proprietorships often become "modified" by various types of contractual arrangements, such as a modified sole proprietorship established when a son, daughter, or unrelated party is first taken into the business. Wage-incentive, wage-share, enterprise and rental agreements, and various joint-venture arrangements are examples of modified proprietorships.

A general partnership is an aggregation of owners. Two or more persons contribute their assets to the business and share with each other the management responsibility, profits, and losses. Each partner pledges faith in the other partners and stands liable for the actions of all partners within the scope of partnership activities.

A limited partnership is a special form of partnership permitted by state law in which the liability of one or more partners for partnership debts and obligations is limited to their investment in the business. A limited partner is just an investor; a limited partner who participates in management becomes liable for all partnership obligations as a general partner. A limited partnership must have at least one general partner who handles the management of the business and is fully liable for all partnership debt and obligations.

A corporation is an artificial entity created under state law. It is a separate business entity distinct from its owners, who are called shareholders because they own shares of interests in the corporation. The major characteristic of the corporate form of business organization is this sharp line of distinction between the business and the owners. The corporation is a separate legal entity as well as a separate taxpayer.

A tax-option corporation (Subchapter S) is a creation of federal tax law. It is a corporation in all respects except that the corporate entity pays no income tax; instead, each shareholder-owner reports a share of corporate income for income tax purposes.

Other arrangements such as the trust can be considered. All of these business arrangements can be combined in various ways, such as incorporating part of the business and holding the remainder as a sole proprietorship or limited partnership.

How each of the above business arrangements tends to fit into the decision-making framework described in Figure 9.3 will now be illustrated (Figure 9.4). Most farm families will be starting with an individually operated sole proprietorship. As they move to the testing stage, one of the various types of "modified" sole proprietorships, ranging from wages to some type of joint working arrangement, will probably be selected. These are simple arrangements; yet they provide the opportunity to test the new entrant's true desires regarding farming and the compatibility of the participants.

After this testing stage experience, some families will decide to go their separate ways. Dad likely will continue in his sole proprietorship, and

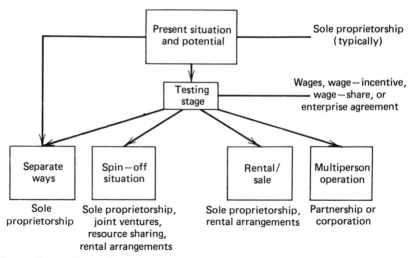

Figure 9.4. Alternative business arrangements and the decision framework.

the son or daughter will establish his or her own operation (sole proprietorship) or farm with someone else. Upon dad's retirement, the home farm business will likely be discontinued, with the land rented out or sold. In other cases, the son or daughter may desire to establish his or her own farming unit, but in a spin-off situation relative to the home farm. They may end up exchanging labor for machinery and in some cases becoming involved in a joint-venture situation. This may involve joint ownership of machinery or may mean becoming involved in one or more enterprises together. Dad may rent some of his land to the son or daughter and even co-sign some notes. If dad is at or near retirement, then the plan may call for the establishment of a rental arrangement, with dad serving as the landlord and the farming heirs serving as tenants. Eventually, dad may sell part or all of the farm to them. When the business is deemed adequate and the parties can get along well together, then farming together in a multiperson operation using either a partnership or corporation becomes the viable business arrangement.

APPRAISING YOUR PRESENT SITUATION AND POTENTIAL FOR JOINT OPERATION

Before beginning any type of joint operation, the parties involved should examine their present situation carefully and make a preliminary judgment of their potential for developing a successful joint operation.

Do You Really Want to Farm? Together?

Here, you are trying to size up the human factor—usually the most important item in successful business arrangements. Some of the key questions to be answered include: Is the heir and his or her spouse really committed to farming or is it just a passing fancy? If they really want to give it a try, will the family members be able to live, work, and manage together?

It is essential that the participants in a joint operation relate well to each other in a personal as well as a business sense. All parties (participants and their spouses and families) must be tolerant and understanding as well as have the ability to overlook each other's faults. More business arrangements are dissolved because of disagreements over trivial things than over major issues. The ability to compromise is essential. Many parents are conservative, their children venturesome. Spouses must be kept informed, be interested in the business, and be compatible.

All parties should work toward similar business objectives to make the business succeed. When goals and values differ, care must be taken to arrive at a reasonable compromise. Joint participation in managerial decisions is a must. Most parents understand the importance of transferring property ownership to the operating farm heir. Gradual transfer of management control can be just as vital. A check that you might make as to your ability to live, work, and manage in a joint business arrangement setting is the test contained in "Test Yourself—Would You Be a Good Partner?" (Figure 9.5).

Is the Business Profitable Enough?

Can the business provide security for the parents and opportunity for the child? No matter how equitable an agreement may be and how well the parties can get along, a business arrangement will not be successful if business earnings are inadequate. The income must be sufficient to support the owners by providing an adequate standard of living and compensating each individual for his or her capital resources. This may require (1) expansion in the business; (2) improvements in production, marketing, and financial skills; and/or (3) shifts in the amounts of resources supplied by the various parties. If there appears to be any doubt as to the adequacy of the business, it is recommended that a more detailed financial analysis be made.

Some of the budgeting techniques discussed earlier can be useful in assessing the financial feasibility of a specific expansion or joint-venture plan. For example, the minimum number of acres or units of livestock required to generate a specified income level can be determined using enterprise cost and return budgets. The corn enterprise budget of Table 9.1 indicates that the net return above costs including land charges would be $11.23. This is a return to management, overhead, and risk. Adding in the labor charge would result in an estimate of the income per acre received by the farmer for his labor, management, overhead, and risk. In this case, the corn enterprise is charged $25.94 ($9.10 + $16.84) per acre for labor. Thus, the income per acre received by the operators would amount to $37.17 ($11.23 + $25.94). If a farmer and his son wanted to start a joint-

TEST YOURSELF...

Would you be a good partner?

By CLAUDE W. GIFFORD

Partnerships more often stumble over human relationships than over business arrangements.

Here is a farm partnership test that will tell you your chances of success and happiness in a father-son partnership.

After scoring yourself, turn page to see how you rate.

Rating Scale

If your answer to the question is:	Give yourself these points:
A firm "yes"	10
Yes	9
Yes, but barely	8
Maybe, with "conditions"	7
Could live with it	6
Possibly	5
Very difficult	4
Isn't likely	3
Highly improbable	2
A firm "no"	1

Partnership test for fathers:

Points (See rating scale top right):

_____ Are you willing to work up a partnership agreement with your son, or son-in-law—now—and put it in writing?

_____ Can you usually "suggest" advice to your son, or son-in-law, as a business partner—rather than giving him commands or correcting him as a son?

_____ Can you willingly "give in" to your son's, or son-in-law's, wishes when you see that it means a great deal to him, his wife and family? Even though it might cost you money?

_____ Are you willing to turn over definite areas of important responsibility to your son, or son-in-law, for him to make the management decisions? And then live by his decisions without grumbling?

_____ Can you discuss family and business affairs with your son, or son-in-law, without getting emotional, angry or upset?

_____ Are you willing to cut down on your personal spending a bit, or even go deeper into debt to expand your farm, or livestock operation, to provide money for two families?

_____ If your son, or son-in-law, makes a mistake that you might honestly have made at the same age, can you mark it up as a "useful experience" in helping him grow—rather than scold him about it, or brood over it yourself?

_____ Are you willing to have a heart-to-heart talk with your son, or son-in-law, so that he knows fully now what his future is on the farm?

_____ Are you willing to accept the fact that your son's wife and his family—and his future—are rightfully his No. 1 concern?

_____ Are you willing to make a contract now that protects your son's, or son-in-law's, investment of time and money in the farm in case of an untimely death for you?

Your total score

For mothers:

Points:

_____ Can you accept the fact that your son or daughter has married a person of his or her choice and that this partner and their future together is his or her No. 1 concern?

_____ Can you willingly accept the fact that it may be necessary for you and your husband to go deeper into debt in order to provide more income for two families?

_____ Are you willing to go along with your husband now in giving your son, or son-in-law, and his family, a clear picture of their future on the farm?

_____ Can you take pride in your daughter's, or daughter-in-law's, clothes and her home furnishings for the pleasure it brings her and her family—rather than comparing them with what you have or being upset when yours are not as new?

_____ Can you refrain from giving advice to your daughter, or daughter-in-law, raising her children—yet enjoy them as your grandchildren?

_____ Can you accept the fact that young couples are likely to be more carefree, spend money more lavishly and be more irresponsible than couples of your age?

_____ Can you willingly compliment your son or daughter about his or her married partner's good qualities once in a while—and refrain from dwelling on their shortcomings?

_____ Can you let your young "in-laws" have their own life—with couples their own age—and be free to come and go without comment while you stay a reserved distance away, even though they live on the same farm (or even in the same house)?

_____ Do you believe that a farm partnership is a business arrangement with advantages to both families—rather than a favor for the young family?

_____ As you strive to get along in the partnership would other people say that you are kind and considerate; that you are discreet about what you say; and that you are a person who controls her anger?

Your total score

REPRINTED FROM THE FEBRUARY 1968 ISSUE OF **Farm Journal** PHILADELPHIA, PA.

U. S. Copyright 1968, Farm Journal, Inc.

Figure 9.5. Test yourself . . . Would you be a good partner?

Rating Scale

If your answer to the question is:	Give yourself these points:
A firm "yes"	10
Yes	9
Yes, but barely	8
Maybe, with "conditions"	7
Could live with it	6
Possibly	5
Very difficult	4
Isn't likely	3
Highly improbable	2
A firm "no"	1

If you find that you don't score as high as you thought you might, don't despair. The couples who helped set up the test say that a good partnership doesn't just happen—it is developed by people who "put themselves out" to make it work.

If you decide after the test that a farm partnership isn't for you, there's nothing wrong with that. After all, not everyone is cut out to be a business partner. Many successful farmers would be unhappy in a partnership.

Your chances for success as a good farming partner

Total points	
91-100	You're a fine partner
81-90	You should make a go
71-80	It may be rough
61-70	Barely "passable"
51-60	I think of another setup
41-50	You'll be very unhappy
31-40	You couldn't stand it
21-30	Please don't do it
10-20	Not a chance in the world

If the combined total points reach 320 or more for the four people (two couples) in the partnership, it should be a success. If the combined total is 240 or less, the partnership will have major troubles.

For sons and sons-in-law:

Points: (See rating scale top left):

_____ Can you accept advice from your father, or father-in-law, with an open mind—believing that it might have merit? Are you willing to "give it a try" when it is important to him, even when you think the advice might not work?

_____ Are you patient enough to take time to "grow into" the farm business that your father, or father-in-law, has spent a lifetime building up?

_____ Since you have the advantage of youth, strength and stamina, are you willing to do more than your share of the physical work on the farm without complaining or feeling resentment?

_____ Can you appreciate with understanding that your father, or father-in-law, may have spent many years running the farm, making the decisions, perhaps even dealing with you as a boy—and that it will take some time for him to get used to your coming in as a decision-making business partner?

_____ Can you feel that you may be getting a much faster start in life, with much more certainty, as a result of the partnership—and that this "debt" is something you owe to the partnership?

_____ In return for the advantages that the partnership offers you, are you willing to take on the prospect that maybe you will be primarily responsible for caring for your parents or "in-laws" in later years?

_____ Can you refrain from pressing your advice on your parents or "in-laws" regarding their personal or family affairs—keeping in mind that they may enjoy things that you wouldn't?

_____ Can you willingly reserve for your father, or father-in-law, an important area of responsibility in the farm business, even in his advancing years? Can you avoid imposing your will on him, in this area, even though he may "hold back" the farm business some?

_____ Are you willing to keep a good set of record books of partnership expenses and income so that you and your father, or father-in-law, can see what is making money and what isn't?

_____ In situations that call for judgment affecting the partnership and your father, or father-in-law, can you willingly lean over backward so that others would judge you as being calm, fair and considerate?

Your total score

For daughters and daughters-in-law:

Points:

_____ Can you accept without resentment the fact that your mother, or mother-in-law, has spent a lifetime raising a family and making the farm go, and that she is entitled to rest, travel, good furniture—things that you may not be able to afford at this stage of life?

_____ Can you appreciate the faster start in life that you may be getting as a result of the partnership; and be truly thankful without resenting your dependence on the older couple?

_____ Can you "make do" with the house you have, the furniture, the car, the conveniences—without complaining that the older couple is responsible?

_____ Can you accept the fact that married "spats" are normal and then keep them to yourself without burdening the older couple or expecting them to take your side; and without blaming them or "taking it out" on them because of the partnership?

_____ Can you use good judgment in not imposing too much on your mother, or mother-in-law, to take care of the children, prepare meals, baby-sit, and the like?

_____ Can you teach your children to enjoy the grandparents' attention and their home when the grandparents want the children; but otherwise keep the children from "having the run" of the grandparents' house?

_____ Can you be discreet around your children so they aren't "carrying stories" between the two homes?

_____ Are you willing to make and keep a budget of household expenses that will help you plan and get along on what may be "short funds" due to two families living on the same farm?

_____ Can you refrain from "egging" on your husband to get him to make more of the farm decisions, ask for a bigger share of money, or do less of the work or chores—when this makes the partnership more difficult and your husband more uncomfortable?

_____ If you're a daughter-in-law of the older farm couple, are you willing to take time to "grow into" your husband's family? If you are a daughter of the older farm couple, can you make your husband the "confidant" of your thoughts; work out your "troubles" with him; and not lean on your daughter-family relationships so that your husband feels like an outsider?

Your total score

venture farming operation that would generate at least a $15,000 labor and management return for each of them ($30,000 total labor and management for the farm), they would need to have approximately 807 acres of corn ($30,000 ÷ $37.17).

A similar procedure could be used to determine the number of feeder

Table 9.1   Corn Enterprise Budget

Item	Unit	Quantity	Price or Cost/Unit	Value or Cost
1. Gross receipts				
Corn	Bu	120.00	$ 3.25	$390.00
2. Operating costs				
Preharvest				
Seed	Acre	1.00	18.00	18.00
Nitrogen	Lb	160.00	0.14	22.40
Phosphate	Lb	60.00	0.265	15.90
Potash	Lb	70.00	0.11	7.70
Lime	Ton			4.00
Herbicides and insecticides	Acre	1.00	24.25	24.25
Hail insurance	Dol	390.00	0.025	9.75
Machinery	Acre	1.00	19.41	19.41
Labor	Hour	1.40	6.50	9.10
Harvest costs				
Machinery	Acre	1.00	24.46	24.46
Labor	Hour	2.59	6.50	16.84
Interest on operating capital	Dol		0.175	22.00
Total operating cost				$193.81
3. Income above operating costs				$196.19
4. Ownership costs				
Machinery depreciation, interest, insurance, and housing	Acre			69.96
Land charge	Acre			115.00
Total ownership costs				$184.96
5. Total costs shown				$378.77
6. Net returns above costs shown				$ 11.23

pigs required to generate a specified level of income. As shown in Table 9.2, the net return to management and risk plus labor charge in feeder pig production is budgeted at $11.07 plus $3.00 for a total per unit of $14.07. Consequently, in order to have a sizable enough feeder pig operation so that it will generate sufficient income to support one of the partners ($15,000), a total of 1,066 feeder pigs must be purchased and raised ($15,000 ÷ $14.07). Thus, enterprise budgets can be useful in determining the minimum size required for a viable joint-venture farming operation.

The enterprise budgets can also be used to determine the resource requirements necessary to generate additional income from a planned expansion. For example, the hog budget of Table 9.2 indicates that the return above variable costs minus labor totals $20.84 [$116.60 − ($98.76 − $3.00)] per pig and that each pig requires 0.6 hours of labor. Consequently,

Table 9.2 Feeder Pig Enterprise Budget

Item	Unit	Quantity	Price or Cost/Unit	Value
1. Gross receipts				
220 lb market hog	Hd	1	$53.00	$116.60
2. Operating costs				
40 lb feeder pig	Hd	1.03[a]	43.04	44.33
Corn	Bu	10.5	2.34	24.57
Supplement	Lb	99.0	0.1465	14.50
Veterinary and medicine	Hd	1.0	1.25	1.25
Bedding, utilities, and miscellaneous	Hd	1.0	4.75	4.75
Hauling and commission	Hd	1.0	2.00	2.00
Labor	Hr	0.6	5.00	3.00
Interest	Dol		0.175	4.36
Total operating cost				$ 98.76
3. Income above operating costs				$17.84
4. Ownership costs				
Depreciation, interest, taxes, and insurance on buildings and equipment				$ 6.77
5. Total costs shown				$105.53
6. Net returns above costs shown				$ 11.07

[a]Assumes 3 percent death loss.

if adequate facilities are available, an additional $1,000 of income could be generated with 28.8 hours [($1,000 ÷ $20.84) × 0.6] of additional labor. In a similar fashion, the additional capital and other resources needed to generate additional income from expansion can be estimated using the enterprise budgets.

Can a Workable Transfer Plan Be Developed?

The younger generation must have the opportunity to invest, thus increasing equity and control of the business as well as taking over management responsibility as time goes by. The basic question is, can you develop a workable transfer plan? Specific issues include:

Is dad willing to shift managerial responsibility over time? Development of the child's management abilities is critical to the future of the business. If the parents are not willing to shift responsibilities on a reasonable timetable, all of the other transfer plans may be for naught. Moreover, the child will get frustrated and probably leave.

Are the parents able and willing to transfer ownership of property over time? Here, a careful analysis must first be made to determine whether the parents are in a financial position to make gifts of property to the younger generation. As a guideline, if the value of the parents' estate (in 1983 dollars) is below $300,000, then the prospect of their estate being able to transfer much property during their lifetime is not great. Parents in this situation should be concerned primarily with their own welfare, with little or no transfer of property contemplated. Parents with estates between $300,000 and $600,000 probably have an adequate estate for their life expectancy needs. However, they should be cautious in terms of gifting property—particularly at this point. Thus, families with either of these two situations should be quite open about the fact that the parents are not in a good position to transfer property and should not be expected to. Protecting the parents' security is the number one priority in business arrangement decision-making, not getting the child established.

Families with larger estates—$600,000 or more—are facing a potentially serious estate tax problem. With even moderate inflation rates, such an estate could grow to $1 to 2 million in 10 to 15 years. Parents in this situation are usually able to begin the transfer process. The question is, are they willing to do so? If not, they should tell the child right away so he or she can make other career plans. The situation regarding other heirs must also be explored. How do they fit into the transfer plans? How can they be best treated fairly? After answering these questions, one should be able to reach some tentative conclusions about farming together.

THE TESTING STAGE: DO YOU REALLY WANT TO FARM? TOGETHER?

Most existing family farm units are sole proprietorships. Most families have had little or no experience in operating a business together. The son or daughter may be uncertain about farming. Therefore, it is recommended that the parent and child first enter a testing stage. The family's objectives at this point should be to determine whether the son or daughter really wants to farm and whether the parties can work together.

Alternative Business Arrangements for the Testing Stage

Business arrangements adapted to the testing stage can be grouped into two broad categories. The first category includes various types of wage, wage-incentive, and wage-share arrangements in which the child contributes primarily labor and possibly some management. The second category includes various joint working agreements that may at times border on being a partnership. With these latter arrangements, the child will be supplying at least some personal property as well as labor and management.

Employer-Employee Type Arrangements First, alternative arrangements in which the child contributes primarily labor and some management are discussed.

Many families find the easiest way to start a child in the business is to pay wages. This plan may also include a year-end bonus and benefits. This is probably a good place to start the testing process. Two notes of caution are in order, however. First, the wage should be a reasonable one. Second, this type of arrangement should be considered very temporary. Since the child does not have a direct interest in the success of the business, he or she will lose interest in farming if forced to work on a wage basis for an extended period of time. One approach is to pay well in excess of a hired man's wage. This would keep the child motivated, test the financial adequacy of the business, and insure that the parent would not prolong the arrangement unduly.

Wage-incentive plans often are used to encourage a farming child to take a more active interest, and additional responsibility, in the farm business. For the employee, the compensation should be in addition to his or her basic wage, not a substitute for a reasonable wage, agreeable working conditions, and adequate housing. It is assumed that the result of an incentive plan will be increased returns to parent as well as child.

The details and characteristics of various income and physical efficiency incentive programs are discussed thoroughly in Chapter 12 (Labor Acquisition and Management). These same programs can be used for family employees or in the testing stage prior to the development of a permanent business arrangement for farming together. In a sound incentive program, the worker should be able to influence the size of payment received by the work performed. Payments should be sufficiently large and attainable to encourage extra effort. A written agreement should be developed describing the purpose of the arrangement, employee responsibilities, method of calculating and making the payment, and provisions for arbitration.

A wage- and income-sharing plan is particularly well adapted as a beginning agreement when the child is not sure of continuing in business on the home farm or does not want to become too involved financially. From a legal standpoint, a wage- and income-sharing but not loss-sharing plan establishes an employer-employee relationship rather than a partnership. The employee and employer would thereby avoid some of the liability aspects associated with the general partnership.

Under this plan the parent typically furnishes the farm, housing for the child, all farm personal property, and his or her own labor and management. The parent also pays all farm expenses. The child works on the farm full-time and receives a guaranteed monthly wage and a share of net farm income. The wage rates could be comparable to current wages for similar services by hired help in the area.

Some of the advantages of wage-based types of business organizations include: (1) determining if the child really wants to farm and if the parent and child can get along well when farming together; (2) giving the child experience and "know-how" about farming; and (3) starting and stopping easily, as there is no jointly held property, etc. Among the disadvantages are: (1) the child's major interest in the farm may be the monthly paycheck; (2) payment in wages does not encourage savings, nor does the arrangement permit the child to gain an equity in the business; and (3) such plans are often kept in force long after the child is ready to become a full-fledged member of the business.

Joint Working Agreement   The testing period may extend long enough or relationships may develop fast enough that the child begins to contribute personal property to the farm business along with his labor and management. Similarly, a so-called holding pattern may be in order in those cases where the parent is very close to retirement. Rather than expanding the business, establishing a partnership, etc., the parties may decide to extend the testing stage and become involved in some type of joint working agreement. Two types of joint working agreements will be explored at this point, although there obviously could be many others.

With an enterprise-type joint working agreement, the farming child may furnish some personal property (for example, livestock and machinery) and some management in addition to labor. He or she normally would not make whole-farm decisions, but might make most of the major decisions for one enterprise.

The child may buy into a given enterprise such as the dairy herd, beef cow herd, or hog enterprise on a partial or complete basis, either paying the parent for the use of feed, buildings, and pasture or working out a livestock lease arrangement whereby part of the production is given to the parent. Any resultant agreement should be put in writing and cover such topics as job responsibilities, contributions, distribution of income, method of settling disputes, and dissolution.

Enterprise working agreements normally should be regarded as only temporary. Most farms are too small to be subdivided into separate enterprises. With this type of arrangement, there is a tendency for the child to concentrate on his or her enterprise at the expense of the overall farming operation. In addition, the child is exposed to considerable risk when he or she depends on one source of income. Recordkeeping may be difficult.

There may be situations in which the parent and child operate in a joint-venture agreement and share labor and machinery, but own or rent their own land, facilities and livestock. Normally, the family should have made some long-range decisions before entering such agreements and thus would have moved out of the so-called testing stage into the spin-off stage.

Length of Testing Stage

The length of the testing period depends on the family's situation, objectives, and progress toward deciding which future route to follow. Two to three years normally should suffice for the testing period. Delaying a decision beyond this time should be viewed critically, particularly if the parties still are involved in a wage-type arrangement. Remember, the purpose of the testing period is to determine whether the child wants to farm and whether the parties can work together. Once these issues have been resolved, be prepared to move out of the testing stage.

FARMING SEPARATELY OR TOGETHER: SOME POSSIBLE ROUTES AND ARRANGEMENTS

This section discusses some possible routes and types of business arrangements to consider after the testing stage.

Going Your Separate Ways

You may decide to dismantle any joint arrangements that may have developed during the testing stage and go your separate ways. Such a separation may be caused by an inability to work together or the parent's unwillingness to transfer property. Or it may have been part of the plan all along because the original business was too small.

Full and frank communications are important if such a separation is likely to occur. It is unfortunate when parents have assumed that the child would eventually farm on his or her own, but the child was under the impression that he or she would operate the home farm. The sooner both parties can agree on their future course, the better.

If it appears that a split is to occur, care should be taken to insure that good records are kept of how much each party has contributed and who owns what property. An answer eventually must be given to the question: do you wish to keep the home farm in the family in the future? If the answer is no, then the decision becomes a parental one of whether the farm is to be disposed of before or after death and how. If the farm is to be kept in the family, more complex decisions must be made about who will eventually own it and how it will be operated.

Spin-Off Situation

When the home farm is too small or differences surface between the parent and child about how the business should be developed, a spin-off situation might prove desirable. The child might rent or buy a separate farming unit. Machinery might be owned jointly, or owned separately and used jointly. They normally would exchange work and possibly operate a livestock operation together. Such joint working arrangements can take on many forms and can get a child established in farming without jeopardizing the security the parent has built up in his or her own business. It also permits the child to better exhibit management know-how and to establish independence because he or she has a separate business entity. The key to this arrangement is the child's ability to obtain a farming unit sizable enough to generate profits for an adequate standard of living and net worth improvement, and a unit with future business growth potential.

Rental or Sale of Home Farm from Parent to Child

The parent may be at that stage in his or her career when he or she is ready to rent the farm to the child. Usually, the parent wholly or partially retires from the day-to-day work and management of the farm. If there is adequate housing, parents often continue to live on the home farm and receive part of their living, especially food and housing, from it.

Whether the rental arrangement should be a cash or share lease depends on the financial circumstances of the parent and child. A cash lease shifts the risk of low prices and yields to the child, who has to pay the parent a fixed rent regardless of price and crop conditions. A flexible cash lease could be developed to reduce some of the risk assumed by the child. With a share lease, the parent shares at least part of the financial ups and downs of the farm business.

It usually is recommended that at the beginning of the rental period the child acquire full ownership of livestock, feed, crops, and supplies. The parent will be relieved of the worry of managing these assets, and the child will be in a much better financial and managerial position to take the final step towards attaining ownership of the home farm. Ownership of the machinery should be worked out with an accountant to avoid investment credit recapture.

The rental arrangement is a fairly satisfactory agreement when the parent wants to retire but needs a source of retirement income. The child may find it to his or her advantage to rent rather than buy at this time. However, it is necessary to decide on some type of arrangement to eventually transfer at least partial ownership, as the child's future in farming may be closely tied to acquiring control of the real estate and other business assets.

Multiperson Operation

When the parties involved can get along well together and the existing farm business is adequate in size or can be expanded without jeopardizing the parents' financial position, then farming together in a formal multiperson operation seems in order. This usually involves the use of a partnership or corporate type of business arrangement.

MULTIPERSON OPERATIONS: PARTNERSHIP OR CORPORATION?

When it appears that the parties involved are ready for a multiperson operation, two questions remain. First, should we be in a partnership or corporation? Second, how do we go about forming such arrangements? We shall discuss each of these questions in this section.

Should We Be in a Partnership or Corporation?

Of course, there is no perfect form of doing business, nor is there likely to be one that will meet all one's needs. As can be seen in Table 9.3, both the partnership and corporation have their advantages and disadvantages, their costs and benefits. Therefore, farm families will need to work closely with their accountant, attorney, or management consultant in deciding which arrangement can best meet their specific needs. The following discussion is designed to provide a general feel of where each of these tools tends to fit best.

Corporation   When one looks to the business world, it appears that the corporation is the ultimate form of business organization. Just as most major businesses are organized as corporations, a similar pattern appears to be emerging in farming. The corporate structure tends to be utilized in those situations where (1) there is a large, growing business with continuing income tax problems, and (2) continuity, control, and transfer of the business are of key concern to the owners.

The corporate form has several characteristics that make it attractive from an income tax management standpoint. As can be seen in Table 9.3, the corporation is a separate legal entity, which means that another taxing unit has been created. This permits a splitting of income between the corporation (retained earnings) and the individuals (salaries, rents, interest). Also note that the corporate form has very favorable tax rates relative to those available to individuals at similar income levels.

To take advantage of these features, however, the after-tax earnings of the corporation must be retained in the corporation and reinvested there. Otherwise, they will be subject to an accumulated earnings tax or double taxation if distributed as dividends. Thus, it is generally recommended that the corporation be considered in those situations where there is a relatively severe tax problem ($30,000 to $35,000 or more of taxable income) and a growing business where retained earnings can be reinvested for an acceptable business purpose.

It should also be noted that since the corporation is a separate legal entity, the owner/operators of the corporation become employees of the corporation. As a result, they become eligible for a wider range of fringe benefits than would be true under the self-employed status of the partnership. The corporation does not have as favorable tax treatment from the standpoint of capital gains, however. Part of this disadvantage can be avoided by keeping certain capital gains property, such as breeding stock, out of the corporation.

Two key estate planning features of the corporation are the ease with which property can be transferred (shares of stock versus property items) and the fact that control of the business can be exercised as long as the parents (or, later, the farming heir) retain at least 51 percent of the voting stock. Therefore, a key concern in its formation is what property should be

Table 9.3 Comparison of Farm Business Organization Alternatives

Nature of Entity	Sole Proprietor: Single Individual	Partnership: Aggregate of Two or More Individuals	Corporation: Legal Entity Separate from Owners
Business Establishment and Growth Considerations			
Source of capital	Personal investment; loans	Partners' contributions; loans	Contributions of shareholders for stock, sale of stock, bonds, and other loans
Liability	Personally liable	Each partner liable for all partnership obligations	Shareholders may not be liable for corporate obligations.
Limits on business activity	Proprietor's discretion	Partnership agreement	Articles of incorporation and state corporation laws.
Management decision	Proprietor	Agreement of partners	Shareholders elect directors who manage business through officers elected by directors.
Income tax	Income taxed to individual: 60 percent deduction for long-term capital gains	Partnership files an information return but pays no tax. Each partner reports share of income or loss, capital gains, and losses as an individual	Regular corporation—corporation files a tax return and pays tax on income; salaries to shareholder-employees deductible. Capital gains offset by capital losses; no 60 percent deduction for capital gains. 1983 tax rates: 15 percent on first $25,000, 18 percent on next $25,000, 30 percent on third $25,000, 40 percent on fourth $25,000, 46% on excess above $100,000. Shareholders taxed on dividends paid. Tax-option corporation—corporation files an information return but pays no tax. Each shareholder reports share of income, operating loss, and long-term capital gain.

Transfer Considerations

Business life continuity	Terminates on death; business liquidates	Agreed term; terminates at death of partner. Liquidation or sale to surviving partners	Perpetual or fixed term of years. Death may have little effect on corporation. Stock passes by will or inheritance.
Transfer of interest	Terminates proprietor-ship	Dissolves partnership; new partnership may be formed if all agree	Transfer of stock does not affect continuity of business—may be transferred to outsiders if no restrictions.

Source: Adapted from North Central Regional Extension Publication No. 11, *The Farm Corporation*, revised 1979.

put into the corporation—particularly, should land be included? If considerable land is kept outside of the corporation (for tax and other reasons), the ease of transfer benefit is reduced markedly. A second concern is how the corporation is structured financially. For example, if the corporation were capitalized with preferred stock or debentures (interest-bearing notes due in 10 or more years) going to the parents or to the off-farm heirs, this would reduce the amount of common stock which the farming heir would have to own to control the business. It would also "freeze" that portion of the parents' estate, saving future estate taxes or the cost of acquiring the property by the farming child. However, a carefully structured limited partnership, combined with a general partnership, can to a marked degree accomplish the same estate planning features.

Other legal features that may or may not be an advantage of the corporation include business continuity and limited liability. The corporation can be set up for a perpetual life, whereas the partnership must be reformed at the departure of any of the partners. Remaining corporate members must have the managerial skills and financial wherewithal to control and operate the business for it to continue, regardless of the form of business organization.

Limited liability becomes an important issue when the owner-operators have personal investments outside the business. The sole proprietor and general partners risk personal investment when obligations against the business are greater than the value of capital within the business. As a general rule, corporate shareholders and limited partners are not subject to unlimited personal liability. The corporate business organization has an advantage in this regard. However, there is no absolute guarantee of limited liability. Even corporate shareholders may be subject to liability greater than their investment or their commitment to invest, especially when shareholders assume personal liability for debt obligations. This frequently occurs in farming. Adequate liability and property insurance should be carried to protect the business capital. The liability issue should be checked carefully with an attorney as well.

The corporation also has some negative characteristics in comparison to the partnership. Generally, it costs more to set up and maintain a corporation. There are also some possible adverse tax costs in the formation, operation, and dissolution of the corporation. At formation there may be a tax liability if a tax-free incorporation is not possible. There also may be a possible recapture of investment credit if part of the real estate or other property is withheld from the corporation. During operation, social security taxes will be higher with the corporation. In addition, a corporation gets no deduction for capital gains. Retained earnings can also be a problem if the business is not growing fast enough to provide a place to reinvest earnings. Since incorporation usually is made as a tax-free exchange, termination holds the potential for a large capital gains tax.

In a closely held corporation, majority stockholders, who also may be employees of the organization, can vote against minority stockholders' interests. Most of the profits could be taken as salaries or rent rather than being paid out as dividends, leaving little or no cash return on the capital investment of minority stockholders. There may be little or no market value for the stock of a closely held corporation. Therefore, minority stockholders may not receive annual income on their stock, nor will they receive anything if the stock is offered for sale. Minority stockholders in a corporation may be locked in unless the articles of incorporation and bylaws are written to specify minority stockholder rights and buy-sell agreements protecting their investment. In a partnership, a minor partner, under the terms of the partnership agreement, can typically sell or dispose of his or her share in order to get out of the partnership agreement.

Partnership   A general partnership is the simplest form of business association of two or more persons. In structure, it can be likened to an "overgrown" sole proprietorship. It is easy to create, as no written agreement is required. With few exceptions, the arrangement can become whatever the partners agree to. Therefore, it can range from very simple to complex. Other key characteristics of the partnership are noted in Table 9.3.

As a result, the general partnership tends to fill the gap between the sole proprietorship and the corporation. It can represent a stepping stone to incorporation, remain a continuing means of doing business, or revert to a sole proprietorship upon the retirement, death, or departure for some other reason of one of the partners. It serves a useful role for many midsized businesses where taxes are not a problem. When combined with a limited partnership it can overcome much of the advantage of the corporation as far as estate transfer is concerned.

A partnership is based on the following general conditions: (1) each member contributes at least part of his or her time to the operation, (2) each individual contributes or rents resources or property to the partnership, (3) management decisions are made jointly by the individuals concerned, and (4) profits and losses are shared according to each individual's contribution to the partnership.

Numerous profit and loss income-sharing methods can be used in a partnership, but there are three basic methods: (1) income is shared in proportion to the total capital and labor contributions of each partner, (2) return is shared in proportion to the labor contribution after capital is compensated at the agreed rate, (3) profit (or loss) is shared equally after labor and capital are compensated at the agreed rate. The use of each of these methods to share partnership income will be illustrated.

The following preliminary steps are necessary for all partnership income sharing procedures.

1. Establish a mutually agreeable wage rate for contributed labor and an interest rate for contributed capital.

    Example: Wage rate—$5.00/hour
    Interest rate—9 percent

2. Determine the amount and value of each partner's capital and labor contribution.

    Example

	Amount of Contribution		
	Partner 1	*Partner 2*	*Total*
Capital	$250,000	$50,000	$300,000
Labor	700 hr	2,800 hr	3,500 hr

	Value of Contribution		
	Partner 1	*Partner 2*	*Total*
Capital (@ 9%)	$ 22,500	$ 4,500	$ 27,000
Labor (@ $5.00/hr)	3,500	14,000	17,500
Total	$ 26,000	$18,500	$ 44,500
Percentage of total contribution	58.4%	41.6%	100%

3. Calculate the farm income to be shared as gross receipts minus operating expenses.

    Example

Gross receipts	$142,000	
Operating expenses	86,000	
	$ 56,000	(returns to the partners' land, labor, fixed capital, overhead, and management)

Method 1 Income is shared in proportion to the total capital and labor contributions of each partner.

	Total	*Partner 1*	*Partner 2*
Income	$56,000	—	—
Percentage of total contribution	100%	58.4%	41.6%
Total compensation	$56,000	$32,704	$23,296

Method 2 Return is shared in proportion to the labor contribution after capital is compensated at the agreed rate.

	Partner 1		Partner 2	
	Contribution	*Compensation*	*Contribution*	*Compensation*
Capital (@ 9%)	$250,000	$22,500	$ 50,000	$ 4,500
	(Labor return = $56,000 − ($22,500 + $4,500) = $29,000)			
Labor (percentage of total labor contribution)	20%	$ 5,800	80%	$23,200
Total compensation		$28,300		$27,700

Method 3 Profit (or loss) is shared equally after labor and capital are compensated at the agreed rate.

	Partner 1		Partner 2	
	Contribution	*Compensation*	*Contribution*	*Compensation*
Capital (@ 9%)	$250,000	$22,500	$ 50,000	$ 4,500
Labor (@ $5.00/hr)	700 hr	$ 3,500	2,800 hr	$14,000
Subtotal		$26,000		$18,500
	(Profit = $56,000 − ($26,000 + $18,500) = $11,500)			
Profit shared equally		$ 5,750		$ 5,750
Total compensation		$31,750		$24,250

It is important to evaluate the income each partner receives under various arrangements to determine whether it is adequate to meet debt commitments and obligations as well as provide the desired, level of living.

Forming a Partnership or Corporation: Some Checklists

The partnership agreement should be put in writing once the desired share arrangement has been agreed upon. It should specify who is contributing what to the partnership; how it is to be operated; how profits and losses are to be shared; and the duties, powers, and limitations of the partners. Since transfer and control of a farming operation are so critical today, provision also must be made should the partnership be dissolved because of death, disability, or for other reasons. It should include options-to-buy, buy-sell agreements, and a statement showing that the spouses involved agree to the proposed settlement procedures. Property titles, wills, and other documents should be altered where necessary to reflect the desired partnership relationship. The "Checklist for General Farm Partnerships" (Appendix 9A) should prove helpful in developing a partnership agreement. Legal help should be secured in making up the final agreement.

Forming a corporation involves considerations similar to those discussed for a partnership, but the structure by law is required to be more formal. If a farm corporation is to be formed, then the "Checklist for Farm Incorporation" should prove helpful (Appendix 9B).

Two key issues that can have a major impact on future tax burdens and the ease with which property and control are shifted to the heirs must be faced at the time of corporate formation. These are: (1) what property should be put in the corporation, and (2) what type of capital structure should be used? Therefore, a farm family should not form a corporation unless they can secure the services of an accountant and attorney skilled in the formation process.

TRANSFERRING THE FARM TO THE NEXT GENERATION

The issue of transfer of the farm to a younger generation is a difficult one for everyone concerned. The parents have owned the farm for years. They have worked hard to build and maintain a viable business. It has become a part of their lives and, as such, is difficult to part with it. A question of security and independence also is involved. As long as the parents have the farm, they have a means of supporting themselves and of being independent of others for their support.

The children, on the other hand, are trying to build a place for themselves and their families. They are seeking opportunities to build a profitable business with a future and are anxious to try out their wings in farm management decisions. One of the main objectives of an estate transfer plan is to integrate these different goals into a plan that will meet the needs of all involved.

A sound transfer arrangement should be tested on several different points. First, and perhaps most important, does it provide the parents with a reasonable degree of security during retirement and old age? Second, does it provide a reasonable degree of security for the managing child? A third aspect of a sound transfer arrangement is equitable treatment of nonfarm heirs. And finally, the arrangement must be based on suitable legal advice. The legal consequences of different ways of owning property (joint tenancy, tenancy in common, life estate, etc.) and the use of various transfer arrangements such as wills, gifts, and trust arrangements are important considerations in choosing a transfer plan. And estate taxes and settlement costs can still result in a substantial estate shrinkage during the transfer process in spite of recent changes in tax laws that have reduced the estate tax burden. Transfer problems are far too complex for farm families to draw up their own "legal" documents.

When to make the farm ownership transfer depends on the characteristics of both the parents and the farming heir. The ages of both parents and heir are usually key factors. The farm transfer should be made when the parent or parents, completely or partially, retire from the day-to-day work and other responsibilities of the farm operation, and/or when the parents have adequate sources of income. The farm transfer should be seriously considered when the farming heir (1) attains maturity with respect to farm experience, managerial competence, and business judgment; (2) is certain he or she will farm; and (3) has sufficient capital and income to support additional debts and other ownership responsibilities. A cash flow budget should be prepared to determine whether the child has the potential to take over the farm business.

Continuation of a parent-child partnership or corporation with the child purchasing a neighboring farm may be the answer when the parent is not old enough to retire and his or her income needs are still high, but the child has built a financial reserve for a farm purchase.

Families who are considering an intrafamily farm transfer are most likely to achieve their goals and effect a successful transfer if the following conditions exist: (1) the farm is large enough and productive enough to be an efficient unit; (2) income from the farm transfer or from other sources can be expected to provide a reasonable degree of financial security for the parents following retirement from the operation; and (3) the transfer, or at least part of the transfer, of the property occurs at a reasonably early period in the life of the child who is to be the future operator and owner of the farm.

With careful planning, transfer of the farm business to the next generation can become a reality under normal circumstances. However, life and circumstances are not always normal. Families also should plan for the unexpected, such as premature death of one of the parents or a breakup of a business arrangement. The parties involved should decide how they want the transfer to occur should something unexpected happen and specify this in their business agreements and wills. Options to buy, buy-sell agreements, and the funding of such agreements are important issues, and the partners, spouses, and other heirs should be fully informed of the nature and implications of these agreements.

Summary

Selecting the appropriate farm business arrangement is an important decision in the planning process. Although most farm businesses have been organized as sole proprietorships in the past, other forms of organization including the partnership, corporation, and various types of joint ventures will be more important in the future with the growth in the number of multioperator farming units.

Appraising the potential for two parties to develop a successful joint operation requires answers to the following questions: (1) Do you really want to farm, and farm together? (2) Is the business profitable enough to provide a reasonable standard of living for both parties? (3) Can a workable plan be developed to eventually transfer ownership from one party to another (particularly in the case of father-son or father-daughter joint operations)? Whole-farm and enterprise budgeting techniques discussed in earlier chapters can be very useful in assessing the financial feasibility of a specific joint-venture plan and the implications of various expansion and/or financing alternatives.

Various types of legal arrangements can be used to formalize a joint venture. Employer-employee types of arrangements including the wage agreement, wage-incentive plan, and wage- and income-share plans are particularly useful in the early stages. A logical next step may be a joint working arrangement such as an enterprise agreement whereby both parties furnish some of the capital and management resources for a particular enterprise and share the income and expenses. Similar arrangements may be developed for the total farm as well as a specific enterprise. At some point, particularly in the case of father-daughter or father-son farming operations, the farm may be rented or sold from the parent to the child. In other cases, a multiperson operation involving a partnership or corporation may be formed.

Choosing between a partnership or corporation for a multiperson operation involves a number of considerations. The corporate form is frequently advantageous from the perspective of minimizing income taxes and in gradually transferring the business between the parties. Furthermore, the corporation facilitates the transfer of ownership without the loss of control as long as 51 percent of the stock is retained. Limited liability may be an additional advantage of the corporate structure in certain circumstances. However, the partnership arrangement is typically less costly to set up and maintain than a corporation and can usually be dissolved without severe tax consequences. In many cases, the partnership may represent a stepping stone between the sole proprietorship and the corporate structure.

One of the key decisions that must be made in any business arrangement, and that can be a particularly difficult problem in a partnership, is the basis upon which income is shared. Three income-sharing methods can be used in the partnership arrangement: (1) income shared in proportion to the total capital and labor contributions of each partner, (2) return shared in proportion to the labor contribution after capital is compensated at the agreed rate, or (3) profit shared equally after labor and capital are compensated at the agreed rate. These different sharing arrangements will result in the parties having different risks and returns from the joint farming operation.

In the case of family members, the issue of transferring the farm to the next generation should be an important consideration in developing the business arrangement. The business organization and transfer plan developed should give the parents a reasonable degree of security during retirement and old age, provide an opportunity for the younger generation to acquire control of the farming operation, and treat any nonfarm heirs in an equitable fashion. The problems are sufficiently complex that the advice of an attorney, accountant, insurance representative, and other experts should be obtained in developing a business organization and transfer plan. One final note—it is usually best to document the business organization and estate transfer plan with some form of written agreement.

APPENDIX 9A
CHECKLIST FOR GENERAL FARM PARTNERSHIPS*

PRELIMINARY STATEMENTS

1. What are the names and addresses of the partners?

2. What name (if any) has been selected for the partnership?

3. Where will the partnership have its principal place of business?

4. In what farm (nonfarm) activities will the partnership be permitted to engage?

5. When will the partnership agreement take effect? How long will it run? Will it be automatically renewable?

6. When and how must termination be given?

7. An annual review of the agreement will be held.

CAPITAL CONTRIBUTIONS—CASH, PERSONAL PROPERTY, REAL ESTATE

1. A schedule indicating the initial capital contribution of each of the partners should be prepared, indicating its value and how contributed—whether outright, use only, or by lease.

Source: Kenneth H. Thomas and Michael D. Boehlje, "Farm Business Arrangements: Which One for You?," *North Central Regional Extension Publication* No. 50, revised 1982, pp. 29–30.

2. Under what conditions can a partner make additional contributions to the partnership? Leave profits or annual use charges in the partnership?

3. If major improvements are made on a given partner's property, who will own them? How will they be paid for?

4. Under what conditions can a partner withdraw part of his or her contribution?

CONTRIBUTIONS OF LABOR AND MANAGEMENT: DUTIES AND LIMITS

1. Indicate the amount and value of each partner's labor and management contribution to the business.

2. What nonpartnership work, if any, may a partner engage in?

3. Under what conditions will the partners not have an equal voice in management, and what will the arrangement be?

4. How will disagreements be settled?

5. How will management responsibilities be divided?

6. What will happen if a partner is unable to perform his or her assigned duties?

7. What limitations will be placed on the partners relative to the use and disposition of existing property? The acquisition of property and debt? The management of personnel and dealings with other parties?

RECORDS AND ACCOUNTS

1. Where will the partnership establish its checking account? Who will be empowered to sign the checks?

2. What accounting system will be used? Who will keep the accounts? Will they be paid?

3. Who will do the tax work? Develop the capital account?

4. What periodic and annual financial statements will be provided to the partners?

PARTNERSHIP INCOME, EXPENSES, ANNUAL SETTLEMENT

1. Except for the following, all other income shall be considered partnership income: _____

   _____

2. Except for the following, all other business-related expense shall be paid by the partnership _____

   _____

3. Cash withdrawals—what amounts will each partner be able to withdraw monthly for living expenses, debt servicing, etc.? How can this amount be changed?

4. Are inventory changes to be considered on an annual basis or for the period of the agreement?

5. How will profits and losses be shared?

6. When are wages, salaries, rents, interest, or other types of compensation to partners to be paid?

PARTNERSHIP DISSOLUTION: BUSINESS LIQUIDATION OR CONTINUATION

1. What will happen to the partnership business if the partners agree to dissolve the partnership or the term of the partnership expires?

2. What will happen to the partnership business if a partner withdraws or retires? What limitations or penalties shall be placed on a withdrawing partner?

3. What will happen to the partnership business if one of the partners becomes incapacitated or dies?

4. If the business is liquidated, how will the person be selected to handle the winding up of the partnership assets? What will be the priorities for distributing partnership assets? How will assets in the capital account be shared?

5. If the business is to continue, what agreement has been reached as to the buy-out of the departed partner's interest by the remaining partners? Will the buy-out be mandatory or voluntary?

6. How will the value be established and payments made to execute the buy-out provision? Will life insurance be used to fund part or all of the buy-out? Who will own and pay for the insurance?

ARBITRATION AND MISCELLANEOUS PROVISIONS

1. How will disagreements be arbitrated?

2. How will partners' housing be provided?

3. How will partners' vacations and time-off be handled?

4. Is provision for admitting a new partner desired?

5. Have the spouses been involved in developing the agreement? Do they understand it and consent to it?

APPENDIX 9B

CHECKLIST FOR FARM INCORPORATION*

1. *Name.* What is to be the corporation name? Consider application to reserve corporate name.

2. *Duration.* Will the corporation be organized to exist perpetually? Or for a term of years?

3. *Purpose.* What are the purposes of the corporation? Narrowly defined or broadly stated?

4. *Stock and debt capital structure.*
    a. How many classes of stock will be authorized? How many shares of stock will be issued? What are the characteristics of each class as to:
        (1) Voting rights—voting stock, nonvoting stock, proxy voting, cumulative rights.
        (2) Dividend rights.
        (3) Preference on liquidation.
        (4) Conversion rights, if any.
        (5) Par value. (Consider low par value to minimize annual fee on stated capital.)
        (6) Fair market value on issuance.
        (7) Preemptive rights.
    b. Is debt capital to be used? (Watch tax-free incorporation limitation.)
        (1) Type of debt security (note, bond, debenture) and amount.
        (2) Time of maturity.
        (3) Conversion to stock.
        (4) Interest rate.
        (5) Priority on liquidation.

5. *Stock transfer restriction.* What type of restriction will be used (consent, first option, buy-sell agreement)? Method of stock valuation (book value, appraised value, periodically renegotiated fixed value)? Arrangements for payment by purchasers?

6. *Shareholders.* Names and addresses? Date of annual meeting? Place of annual meeting? Voting requirements? Quorum requirements? Pooling agreements? Voting trusts? Shareholders' agreements? For minor shareholders, consider using Uniform Gifts to Minor Act custodianship. Custodian should be someone other than donor.

Source: Kenneth H. Thomas and Michael D. Boehlje, "Farm Business Arrangements: Which One for You?" *North Central Regional Extension Publication* No. 50, revised 1982, pp. 30–31. Prepared by Neil E. Harl, Professor of Economics, Iowa State University and member of Iowa Bar.

7. *Board of directors.*   Number of directors on board? Names of first directors? Voting requirements? Quorum requirements? Arrangements for meetings. Director fees? Is preincorporation agreement desirable?

8. *Officers.*   What offices will be authorized? Who is expected to be elected to each office? What salary will be authorized for each officer? Is corporation to pay entire social security tax or only one-half? Will a bonus policy be authorized? What authority are officers to have in terms of borrowing money, signing negotiable instruments, executing contracts, or signing other documents? Explain the proper format for signatures on corporate documents.

9. *Other employees.*   What individuals are to be employed by the corporation in addition to the officers? What are terms of employment? Is an employment contract to be drafted? Arrangements for compensation? Is corporation to pay entire social security tax or only one-half?

10. *Assets to be owned by corporation.*   What property is to be transferred to the corporation?
    a. Prepare inventory for each transferor and list each by name of owner, description of asset, income tax basis, fair market value, indebtedness, and holding period. Preserve copies to be submitted with income tax returns. Watch gifts between and among transferors of property. Note insurance carried on assets and assets under special registration.
    b. Is transfer to be tax-free or taxable? Check eligibility requirements for one desired.
    c. Who will value assets?
    d. Have the property taxes been paid by transferors to date of incorporation?
    e. Documentary stamp taxes on land transferred?
    f. Abstracts of title?
    g. Prepare deeds and bills of sale.

11. *Assets to be leased by corporation.*   What property will be leased to the corporation? List each item by name of lessor, description of property, and rental to be charged. Prepare leases.

12. *Bank.*   What bank will be the depository bank? Resolution of officer authority to borrow money and sign negotiable instruments should be prepared and sent to bank.

13. *Income taxation.*   What method of income taxation will be followed?
    a. Regular. File Form 1120 annually.
    b. Subchapter S—review eligibility requirements for election; prepare Form 2553; if corporation has operated previously as regular corporation, check operating loss carryover, investment credit carryover, and recapture of investment credit. File Form 1120-S annually.

14. *Identification number.*   Prepare and submit Form SS4, "Employer's Application for Identification Number."

15. *Registered office.*   What is the address of the registered office of the corporation?

16. *Registered agent.* Who is to be the registered agent of the corporation?

17. *Notice of incorporation.* If required by state law, as in Iowa, prepare notice of incorporation, forward to publisher of eligible newspaper, and, where required, send affidavit of publication to secretary of state.

18. *Incorporation kit.* Order corporate kit, specifying type of seal, if any; number and type of stock certificates (have stock transfer restriction printed thereon or type restriction on certificate when received); minute book.

19. *Loans, mortgages.* What loans or mortgages are to be assumed or taken subject to by corporation? Give special attention to Federal Land Bank, Production Credit, Farmers Home Administration loans.

20. *Basis.* Determine corporation's income tax basis of assets for purposes of depreciation and sale. Calculate and make a record of shareholders' basis for stock and securities received. Because of "galloping basis," repeat every year for Subchapter S corporations.

21. *Fiscal year.* What is to be the corporation's fiscal year? Consider fiscal year other than calendar year for Subchapter S corporations.

22. *Method of accounting.* Is the corporation to be on the cash or accrual basis? How are inventories to be valued?

23. *Special elections.* Check on elections for treatment of commodity credit loans, soil and water conservation expenses, and land clearing expenses.

24. *Residences.* All houses to be transferred to corporation? Reasonable rental to be paid by occupants? Or occupants to report value of occupancy as additional income? Or rely on I.R.C. §119?

25. *Motor vehicles.* What vehicles will be transferred to corporations? Insurance arrangements? Title transfer? What vehicles will be individually owned? Rate of compensation for business use? Insurance coverage for accidents involving employee-owned vehicles within scope of employment?

26. *Recapture.* If corporation is not a mere change in form of doing business, will depreciation and investment credit be recaptured? If Subchapter S taxation is elected after operation as a regular corporation, file shareholder consent to be responsible for recapture investment credit with last Form 1120.

27. *Fringe benefits.* What fringe benefits are to be provided? Check health and accident plan, group term life insurance (10 or more employees or "baby group" plan), sick pay, and deferred compensation for retirement.

28. *Doing business in other states.* Will the corporation be doing business in another state? How much? Necessary to qualify to do business as a foreign corporation?

29. *Minorities.* Will stock be permitted to pass to off-farm shareholders? Consider assuring management rights, current income, and market for stock in planning for protection of minority shareholders.

30. *Wills.*   Do wills and estate plans of shareholders need to be updated by codicil or completely rewritten? Consider provisions to direct executor to consent to Subchapter S election and to comply with restrictions on stock transfer. For holders or potential holders of Subchapter S of corporation stock, consider substitute provisions in lieu of trusts, for example, legal life estate rather than marital deduction trust.

31. *Memberships.*   What about memberships in cooperatives? Farm organizations? Breed associations?

32. *Insurance.*   Check on casualty insurance, liability insurance, workmen's compensation election, and motor vehicle liability.

Questions and Problems

1. Discuss the three stages of the typical life cycle of the family farm.
2. Why is the process of merging the life cycle of children and parents within a single business operation frequently so difficult?
3. List the major business arrangement alternatives that might be used to organize a farm business. Briefly describe each arrangement.
4. How might you determine if a farm business is large enough to generate an adequate living for two families? If a business is too small to provide an adequate standard of living, what adjustments might be made?
5. How might the wage incentive or wage and income-share plan be used as part of the testing stage in choosing a farm business arrangement?
6. Describe the enterprise joint working arrangement and discuss how it might be used to facilitate joint farming operations.
7. Under what conditions might rental or sale arrangements be the best alternative rather than multiperson arrangements such as a partnership or corporation?
8. What are the major advantages and disadvantages of using the corporate structure for joint farming operations?
9. What are the major advantages and disadvantages of using the partnership for joint farming operations?
10. Describe the three basic income-sharing arrangements that can be used in a partnership. What are the advantages and disadvantages of each arrangement?
11. Why should estate planning and transfer of the farm business be considered when chosing the legal form of business arrangement?

Further Reading

Harl, Neil E. *Farm Estate & Business Planning.* 7th ed., Skokie, Ill.: Century Communications, 1982.

Harl, Neil E., and John C. O'Byrne. "The Farm Corporation." North Central Regional Extension Publication No. 11, Revised January 1983.

Henderson, P. A. "Fixed and Flexible Cash Rental Arrangements for Your Farm," North Central Regional Extension Bulletin No. 75. Lease Form, North Central Regional Extension Bulletin No. 76.

Hepp, R. E., and M. P. Kelsey. "General Partnership for Agricultural Producers." Extension Bulletin E–731, Michigan State University, Rev., September 1975.

Pretzer, Don D. "Business Organization of the Family Farm." MF 279, Kansas State University, October 1980.

Pretzer, Don D. "Partnerships as a Farm Business Arrangement, Part II." Kansas State University, C-590, January 1978.

Smith, R. S., and R. N. Weigle. "Taxmanship in Buying or Selling a Farm." North Central Regional Extension Bulletin No. 43, September 1975.

Thomas, Kenneth H., and Michael D. Boehlje. "Farm Business Arrangement: Which One for You?" North Central Regional Extension Publication No. 50, University of Minnesota, St. Paul, Minnesota, Revised 1982.

Thomas, Kenneth, and Michael Boehlje. "Farm Estate and Transfer Planning: A Management Perspective." North Central Regional Extension Publication No. 139, July 1983.

Thomas, Kenneth, P. Kunkel, and D. C. Dahl. "Minnesota Farm Business Partnership: Legal and Economic Considerations." Economic Report ER81–7, Department of Agricultural Economics, University of Minnesota, St. Paul, Minnesota, 1981.

10

FARM PLANNING WITH
LINEAR PROGRAMMING

CHAPTER CONTENTS

Various tools and techniques that can be used to analyze the efficient utilization and allocation of scarce resources within the farm business have been discussed, including the techniques of partial and whole-farm budgeting. Another procedure that has been used extensively in planning the farm business organization will now be discussed—the technique of linear programming. The discussion will emphasize the relationships between linear programming and budgeting as well as the concepts of building a programming model of the farm firm, solving that model, and interpreting the results.

LINEAR PROGRAMMING—WHAT IS IT?

Linear programming is essentially a mathematical technique for solving a problem that has certain characteristics. The essential characteristics of a linear programming problem are that there is a function or objective to be maximized or minimized, there are limited resources to be used in the satisfaction of this objective, and numerous means of using the resources are available. Most resource allocation decisions faced by farm managers have these characteristics and can be evaluated through the use of linear programming. For example, farmers may have land, machinery and equipment, and labor resources that they can use to produce corn, soybeans, wheat, hay, oats, and other crops, and they would like to combine these resources in such a fashion as to maximize net income. Alternatively, cattle feeders may have numerous ingredients such as corn, protein supplement, wheat, corn silage, grain sorghum or milo, cottonseed meal, and hay that they can feed to cattle, and they want to combine these ingredients so as to satisfy the minimum and maximum energy, protein, vitamin, and mineral requirements of the cattle and to do so at the least cost per pound of gain. Such problems as the optimal whole-farm organization, the appropriate combination of hogs and cattle to produce, or the least cost feed ration for various types of livestock can be evaluated with this technique. Furthermore, linear programming can be used to determine whether minimum tillage technology is more profitable than conventional tillage for a specific farm situation, or whether confinement facilities should be substituted for an open lot in a cattle feeding operation. These are but a few examples of the allocation problems faced by the farm manager that can be evaluated with linear programming. Although most farm management applications of linear programming involve short-run planning where resource supplies are fixed, long-run planning where additional resources can be acquired over time can also be done with linear programming.

The technique of linear programming was originally developed for use during World War II to evaluate alternative routes for shipping supplies to Allied troups and to determine the optimal combination of scarce labor and capital resources to produce war materials. Applications in the industrial sector have been numerous, with examples in the transportation, petroleum, and meatpacking industries, and other sectors involving production, transportation, and distribution of goods and services. The linear programming technique has also been an important tool in agricultural economics research during the past two decades. The technique is also being used extensively in the planning of individual farm businesses through the use of prestructured linear programming packages that are available through many extension services. Consequently, the tools and techniques of linear programming have been well tested and proven applicable to a wide variety of problems in the farm and business management area.

As has been indicated, linear programming is a set of mathematical rules for solving specific problems. Thus, it is not an economic theory. Instead, it is the procedure used to solve an economic problem. We will not discuss the mathematics underlying the linear programming procedure or the mechanical rules for solving a programming problem. Standard computer programs are available to accomplish these tasks. Instead, our emphasis will be on the economic implications and interpretation of the results of a linear programming analysis of farm management problems.

WHY STUDY LINEAR PROGRAMMING?

Since we have already discussed numerous planning tools, the student might ask, "Why study another?" The reasons for studying linear programming are numerous. First, the procedure is applicable to almost any resource allocation problem faced by the farm manager. This ability to handle different issues with one analysis procedure means that time allocated to understanding the procedure can be "spread over many potential uses." Second, the procedure can handle more complex problems than budgeting or marginal analysis. Although the data and input requirements for linear programming are similar to those of other techniques, the computations required to obtain a solution are much more complex and tedious using budgeting and other techniques. Consequently, these techniques are usually applied to relatively simple problems because of the cost and complexity of obtaining a solution. With linear programming, we can specify more complex, realistic problems without concern for the cost or feasibility of obtaining an answer.

Third, linear programming provides not only information on the best or optimal way of allocating resources and the best production-marketing-financial plan, but also additional information concerning the value of various resources used in that plan. Thus, a computational byproduct of the programming procedure is information concerning what resources are limiting the income potential of the farm operation, what resources are in excess, and how much it is worth to acquire additional units of the limiting resources—the marginal value product of these resources. For example, linear programming would indicate how much family labor is used and how much is unused in a whole-farm planning problem, and if labor is limiting the potential for growing additional crops and increasing income, how much the operator could pay for additional hired labor. Although this information could be obtained from other analysis procedures as well, it would be computed only with great difficulty and significant effort.

A fourth attribute of using programming in farm management analysis is that it can easily be used to evaluate how the results would change if changes occurred in product prices or technical efficiency—the sensitivity or stability of the farm plan. For example, the linear programming procedure indicates how the whole-farm organization will change as the relative prices of products increase or decrease. In addition, adjustments that should be made as changes occur in grain yields or rates of gain in livestock production can be readily evaluated. Other questions of the "what would happen if . . . ?" variety, such as what would happen to income if additional labor was available and how would it be best used; what additional enterprises would be produced on an additional 80 acres of land; and how would additional credit or capital be allocated and how much additional income would it generate, can also be easily evaluated.

A final reason for studying and using linear programming in farm management analysis is the ease with which it handles the issues of "opportunity cost." As has been indicated earlier, opportunity costs reflect the income forgone in using a resource in an alternative enterprise—for example, using labor in the hog enterprise compared to the corn enterprise. By using budgeting procedures, the opportunity cost as measured by the marginal value product of a resource in another enterprise within the business is not easily determined. Therefore, it becomes extremely difficult to determine how the reallocation of resources among enterprises will influence the profits of the firm. The process of pricing resources in the production of various products, based on the income-generation capacity or opportunity cost of that resource in alternative uses, is the heart of the programming procedure. Thus, linear programming is one of the few techniques available that can solve a realistically defined farm management problem using mathematical procedures consistent with the economic concepts of marginal analysis discussed earlier. Because of this adaptability to many problems and consistency with economic theory, linear programming has become a key tool in quantitative economic and business analysis.

THE CONCEPT OF A PROCESS (ACTIVITY)

The concept of a process (activity) is basic to the understanding of linear programming. A process is a method of transforming resources or inputs into a specific output. Although this definition sounds quite similar to that of a production function, a process is much more restrictive in that the relative proportion of all inputs and outputs is fixed in a process. Thus, a process can be represented as a single ray (*OA*) in the factor-product space as illustrated in Figure 10.1. A different process with a different input-output relationship would be represented by a separate ray (*OB*) in the factor-product space. Two productive events are instances of the same process if they consume the same resources in the same proportions and

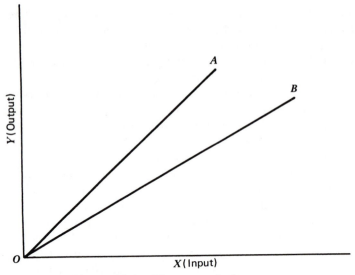

Figure 10.1. The concept of a process.

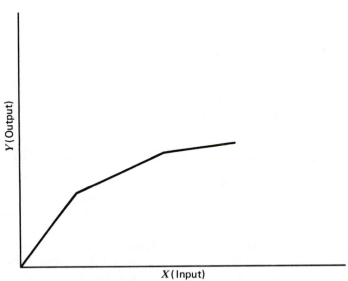

Figure 10.2. A combination of processes resulting in a non-linear production function.

result in the same output. Hence, a process can only vary in size or scale; any changes in the relative amounts of inputs or outputs result in a new process.

A production function, then, is really a family of processes. Any two points on a production function will define different processes unless the proportionality condition specified above holds. If this condition does hold, we have a linear production function. A nonlinear production function would be represented by many different processes as indicated in Figure 10.2. Thus, the concept of a process implies that there are many alternative ways to produce a single product in a linear programming model. In addition to determining the optimal combination of products or enterprises to include and the allocation of scarce resources to those enterprises, the linear programming procedure can also indicate the optimal technology or method of production that should be used.

ASSUMPTIONS OF LINEAR PROGRAMMING

Four basic assumptions are necessary to use linear programming. These assumptions can be used to determine whether linear programming is applicable to a particular problem and whether it will provide a meaningful and precise answer. These assumptions also provide the basis for comparing linear programming to other analysis procedures which we will do in the next section.

Additivity and Linearity

The additivity assumption specifies that the total amount of resources used by two or more processes must be the sum of the amount of resources used by each process. The same assumption applies to products produced—the total output of two processes is equal to the sum of the output from each process. The implication of this assumption is that interactions between processes are not allowed. Thus, the complementary relationships discussed earlier between corn and alfalfa in a crop rotation, for example, cannot be reflected *in* the corn and alfalfa production processes of a linear programming model. Instead, a new process must be defined that includes a corn-alfalfa crop rotation. This new process can, in fact, reflect the complementarity between the two products. Potential interactions between processes are, therefore, handled by defining a new process that includes these interactions.

The assumption of linearity follows directly from that of additivity. Linearity implies that multiplying all inputs used in a process by a constant results in a constant change in the output of that process. Thus, the production function for a process is linear. To reflect nonlinear production relationships, these relationships are approximated by linear segments with each linear segment representing a process. This approximating procedure is illustrated in Figure 10.2. Therefore, the assumptions of linearity and additivity refer to relationships between processes. They do not restrict the definition of a process in such a fashion as to prohibit complementary and antagonistic relationships in production or nonlinear production relationships.

Divisibility

This assumption specifies that all resources and products can be produced in fractional amounts. Thus, we cannot expect to feed exactly 100 head of cattle or rent exactly 80 acres of land. Because the mathematical procedure requires complete divisibility of inputs and outputs, a practical interpretation of the results requires the judgment of the analyst. "Rounding off" the answer to the nearest integer or whole number when implementing the plan generated by a linear programming model usually does not affect the income significantly. However, in some cases where resources come only in a critical mass, like a 40- or 80-acre tract of land or one tractor, it may be necessary to be more ingenious in model specification or interpretation. Problems associated with and methods to relax this assumption of divisibility will be discussed later.

Finiteness

This assumption specifies a limit to the number of alternative processes and resource restrictions that can be included in the analysis. Such a practical assumption is not unrealistic in evaluating real world problems, even if the number of alternatives seems infinite. Farm managers can only allocate so much time to the evaluation of alternatives and must restrict their analysis to a subset of the possible production and marketing alternatives available to them. Current programming routines and computer packages make this assumption somewhat academic since they will handle literally thousands of processes and restrictions.

Single-valued Expectations

The single-valued expectations assumption essentially eliminates the important dimension of risk from linear programming analysis. This assumption specifies that resource supplies, input-output coefficients, and commodity and input prices must be known with certainty. Although many would argue that this assumption is unrealistic and makes the results suspect, all is not lost. First, the same assumption is required of almost all other analysis procedures used by farm managers, including budgeting and marginal analysis. Second, prices and production coefficients can be easily varied in the linear programming framework, and this "sensitivity analysis" can illustrate the resource allocation and income impacts of alternative sets of prices and production efficiencies. This information can be quite useful in evaluating the implications of price or production variability.

It should be noted that these four assumptions apply to linear programming only, and not to the broader range of analysis tools and techniques commonly referred to as mathematical programming procedures. Three of the four assumptions just specified can be relaxed with these procedures. Nonlinear, separable, and quadratic programming techniques can be used to handle nonlinear functions. Integer programming can be used in situations where fractional amounts of inputs or outputs are not feasible from a technical as well as a practical point of view; and stochastic and quadratic programming procedures can be used to incorporate risk dimensions in the analysis, replacing the single-valued expectations requirement. Thus, techniques are available to relax all of the important assumptions of linear programming if they seem too restrictive. However, the cost of using these techniques is that of increased complexity of model construction, reduced model size, and higher cost for model solution. Furthermore, as we shall discuss in the next section, the assumptions of linear programming are no more restrictive than, and are quite similar to, those required in traditional farm management and economic analysis using budgeting or marginal analysis.

LINEAR PROGRAMMING AND BUDGETING

The discussion thus far provides the basis to compare linear programming with the budgeting procedure discussed earlier as a tool in farm planning. Both techniques use the same set of assumptions—linearity of production relationships, completely divisible outputs and inputs, a finite set of production alternatives, and constant, single-valued expectations of prices and production efficiencies. Furthermore, both techniques require the same type of information—input-output coefficients, price and cost relationships and relative profitability of various alternatives, and resource availability as well as personal preferences to reflect physical and economic restrictions on production. For a given planning problem, the budgeting and linear programming procedures will, therefore, give almost identical answers in terms of resource use and income generated. In fact, it can be argued that linear programming is only a formalization of the familiar budgeting procedure.

What, then, is the criterion to be used in choosing linear programming or budgeting to solve a particular farm management problem? Essentially, the criterion is one of computational ease. With the budgeting procedure, each specific plan to be evaluated must be identified and the detailed calculations completed. For a complex farm business, the computations become tedious and burdensome. Furthermore, only by luck will one of the plans chosen for budgeting analysis prove to be the optimal or most profitable plan. In contrast, the linear programming procedure utilizes a set of "optimizing rules" to evaluate successively more profitable plans until the optimal plan is determined. Given the specification of input-output coefficients, prices and costs of production alternatives, and resource availabilities, the final solution of a linear programming model indicates the most profitable way to organize the farm business and allocate the scarce resources.

Hence, the choice between budgeting and linear programming is not based on differences in the underlying assumptions, but on the complexity of computations. Programming can be used to handle problems of a more complex nature (usually reflecting a more realistic statement of the problem), generates an optimal plan for the problem specified, and provides additional information concerning the sensitivity of the results to price and cost changes as well as the income-generation capability of additional limiting resources. Furthermore, as indicated earlier, linear programming easily handles the opportunity cost problems frequently encountered in organizing the farm business—the problems of allocating limited resources to different enterprises based on their contribution to profit in that enterprise compared to the profit contribution or "income forgone" from other enterprises.

LINEAR PROGRAMMING AND MARGINAL ANALYSIS

The principles and concepts of marginality or marginal analysis have been the basis of conventional production theory and farm management decision-making. Thus, the question must be asked, "Is linear programming consistent with marginal analysis?"

One way to answer this question is to again compare the assumptions required for linear programming and marginal analysis. Both techniques require single-valued expectations of input-output coefficients and prices, as well as a finite number of alternatives to analyze. Marginal analysis requires complete divisibility of inputs and output as does linear programming, but it also requires that the functional relationships involved in the analysis be continuous. Programming requires the functional relationships to be linear. Thus, the major difference between the two analysis procedures in terms of theoretical content is the assumption concerning linear compared to nonlinear relationships.

At first glance, this linearity requirement of the programming approach seems quite restrictive and unrealistic. This may not be the case, however. Nonlinear functional relationships can be estimated through the linear approximation method identified in Figure 10.2 whereby, for example, a different process is used to identify various segments on a nonlinear production function. These processes have different input-output relationships and are included in the programming model as separate activities. By restricting the range of these processes or activities within acceptable bounds, the programming model can choose a specific process which reflects a particular level of efficiency on a nonlinear production function.

Furthermore, the linearity assumption is also used in marginal analysis with respect to returns to scale in most applications of conventional marginal theory. The nonlinear production function discussed earlier and in most texts on production theory results from the law of variable proportions—the fact that increased quantities of one input used with a fixed quantity of other inputs will result in increasingly smaller increments in output. However, if *all* inputs are varied at the same time by a constant amount, the output may increase by that same constant amount. This phenomenon is known as constant returns to scale and is the implicit assumption of most applications of marginal analysis as well as linear programming. In fact, empirical estimation of agricultural production functions has verified the constant returns to scale concept for many farm firms. Thus, linear programming and marginal analysis are based on essentially the same set of assumptions, focus on the same types of problems, and utilize the same types of analytical concepts. The decision of which analytical procedure to use again comes down to which technique will handle a specific problem with the least computational burden. As with budgeting, the techniques of marginal analysis are not easily adaptable to complex, real world problems. The mathematical tool of calculus is used to solve marginal analysis problems, and solution procedures become very unwieldy and difficult to use with a problem of realistic size. In contrast, the solution algorithm of the linear programming model can handle sizable, real world problems. Consequently, linear programming has an advantage over marginal analysis in obtaining numerical answers to real world problems.

It should be pointed out again that linear programming is not an economic theory. Linear programming is simply a set of mathematical rules to solve a particular problem—it is completely devoid of economic content. In this context, linear programming is much like the mathematical procedure of calculus. Calculus is used to solve the continuous model utilized in marginal analysis. Likewise, linear programming is the mathematical tool used to solve the linear or discontinuous model of economic analysis. In both cases, the economic concepts, principles, and content are included in the model, not in the solution procedure. Just as calculus is not an economic concept and has no economic content, linear programming also is not an economic theory or concept.

A LINEAR PROGRAMMING EXAMPLE

The basic concept of the linear programming procedure is to maximize or minimize a specific outcome variable that is influenced by and dependent on decisions made by the decision-maker, subject to a set of restrictions or constraints limiting the decisions that can be made. For example, cattle feeders may desire to obtain the highest profit from feeding a combination of yearlings or calves, but are limited in the number of cattle that can be fed by feedlot space, feed supply, labor supply, or the amount of money they can borrow from a lender. Furthermore, their profit will be influenced by the rate of gain and feed efficiency of different cattle, quality of cattle and feed supply, health problems and death and preventive medicine costs, costs of supplement, and response to feed additives and environment (weather conditions and the existence of shelter or other facilities to protect the cattle).

This simple problem can be specified in a linear programming framework by using the following general algebraic notation:

$$\text{Max } \pi = c_1 X_1 + c_2 X_2 \ldots + c_n X_n \tag{10.1}$$

subject to:

$$a_{11} X_1 + a_{12} X_2 + \ldots + a_{1n} X_n \leq b_1$$

$$a_{21} X_1 + a_{22} X_2 + \ldots + a_{2n} X_n \leq b_2$$

$$a_{m1} X_1 + a_{m2} X_2 + \ldots + a_{mn} X_n \leq b_m \tag{10.2}$$

$$X_1, X_2 \ldots X_n \geq 0 \tag{10.3}$$

If we use the standard summation notation $\left(\sum_{j=1}^{n} \right)$, the model can also be written as:

$$\text{Max } \pi = \sum_{j=1}^{n} c_j X_j \tag{10.1'}$$

subject to:

$$\sum_{j=1}^{n} a_{ij} X_j \leq b_i \text{ for } i = 1 \ldots m \tag{10.2'}$$

$$X_j \geq 0 \text{ for } j = 1 \ldots n \tag{10.3'}$$

where:

X_j = the level of the jth production process or activity,

c_j = the per unit return to the unpaid resources (b_i's) for the jth
   activity,

a_{ij} = the amount of the ith resource required per unit of the jth
   activity,

b_i = the amount of the ith resource available.

Equations 10.1 and 10.1' specify the objective function as the maximization of the summation of the net returns (c_j) from a set of processes or activities (type of cattle fed, for example) times the number of units of each process in solution (X_j). Equations 10.2 and 10.2' indicate the limits on how many units of each process or activity can be in the final solution. These limits are specified as the summation of the resources required to produce one unit of each process (a_{ij}) times the number of units of each process produced (X_j) cannot exceed the resources available (b_i), where b_i refers to feed or labor in the earlier example. In an actual model of a farm operation, a separate equation is included for each restriction or limit imposed on the processes. Equations 10.3 and 10.3' reflect the practical reality and mathematical necessity that processes or activities cannot take on negative values or that negative production is not permissible.

To illustrate graphically the solution procedure of linear programming, assume that a cattle feeder has provided the data in Table 10.1 on capital availability and feed supply and wants to know if it is more profitable to feed calves, yearlings, or some combination of the two. The data in Table 10.1 should be familiar as they are obtained from enterprise budgets as discussed in Chapter 4. This simple example could be solved by whole farm budgeting as discussed in Chapter 5, but is used here to illustrate the concepts of the linear programming procedure.

The total supply of feed and capital limit the number of calves and yearlings that can be finished. For simplicity we have abstracted from the possibility of hiring more labor or buying more feed. The linear programming procedure can readily handle these additional extensions. If we evaluate each constraint separately, with the 20,000 bushels (509.1 metric tons) of corn, the feeder could finish up to 289 head of yearlings, 198 head of calves, or a linear combination of yearlings and calves. This is graphed in

Table 10.1 Resource Requirements and Availability
in Cattle Feeding

Item	Calves	Yearlings	Total Supply Available
Net return	$ 20	$ 18	—
Feed requirement	101 bu	69 bu	20,000 bu
(corn equivalent)	(2,565.5 kg)	(1,752.70 kg)	(509.1 metric tons)
Capital requirement	$150	$250	$50,000

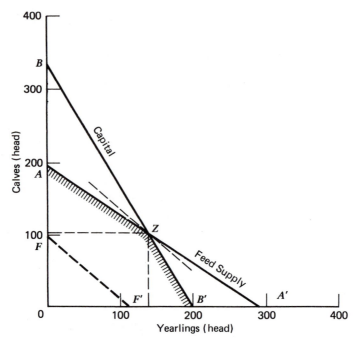

Figure 10.3. Graphical representation of the linear programming example.

Figure 10.3 as line *AA'*. Alternatively, the feeder could finish up to 333 head of calves or 200 head of yearlings with the $50,000 of capital, or a linear combination as indicated in Figure 10.3 by line *BB'*. If we consider both constraints simultaneously, any points inside the area or on the boundary of *OAZB'* are technically feasible combinations of calves and yearlings to finish. This area (*OAZB'*) is known as the "feasible set."

The most profitable combination of calves and yearlings depends on their relative profitability. The data in Table 10.1 indicate that calves will generate $20 per head net return, whereas yearlings will generate $18 per head. Based on these figures, 1.11 yearlings will generate the same net return as one calf. If the relationship denoting equal net returns generated by calves and yearlings is graphed, the line *FF'* (Figure 10.3) would result. As line *FF'* is moved parallel to the right, higher levels of net returns are reflected. Since the cattle feeder wants to generate the highest level of net return possible, the equi-net return line (*FF'*) is moved to the right *until* it just touches the edge of the feasible set (*OAZB'*). Production beyond this point is not technically feasible because adequate quantities of one or both the capital and feed resources are not available. Below this point of tangency, it is possible to move upward and obtain more total net returns. So the final point of optimal production is denoted by the tangency between the feasible set and the equi-net return line at point *Z*. If we move parallel from point Z to each axis of the graph, the analysis indicates that 137 head of yearlings and 104 head of calves will maximize net return. This combination of yearlings and calves will generate $4,546 of profit [(137 × $18) + (104 × $20)]. As more constraints and production activities are added, the example becomes a multidimensional problem that would be impossible to represent graphically. However, this simple example indicates the basis behind the optimization procedure of linear programming.

THE SOLUTION PROCEDURE

Conveniently, a precise set of mechanical rules have been developed to solve a linear programming model, and these rules are the same no matter what the problem. The mechanical rules specify each step that is to be followed in the solution process. In one sense, these rules really are a trial-and-error procedure for problem-solving. However, they have been constructed in such a way that each trial results in an improved answer. The rules also guarantee that, if an optimal value exists, it will be found in a finite number of steps. Because of their mechanical nature, these rules have been computerized so that the farm manager need only develop the model of the problem and submit that model to the computer for "number-crunching."

The mechanical rules for solving linear programming problems are collectively known as the "simplex method." We will not devote any time to discussing the simplex method in detail. The interested student can obtain numerous references which provide detailed discussions of this procedure along with examples that can be solved by hand.[1] However, the previously discussed characteristics of a linear programming problem and the example of Figure 10.3 can be used to provide a basic understanding of the simplex procedure.

Returning to Figure 10.3, it was shown earlier that the optimal combination of yearlings and calves to produce occurred on the boundary of the feasible set at the point of intersection between the feed and capital resource constraints. This intersection is referred to as a "corner point." In fact, because of the linearity assumptions used in the programming procedure, it can be proven mathematically that the optimal solution will always be at a corner point. Thus, to determine an optimum, the only points that need to be investigated in the trial-and-error process of the simplex method are the corner points on the boundary of the feasible set. This is exactly how the mechanical rules of the simplex procedure operate: they

[1]For a detailed discussion of these rules, see Earl O. Heady and Wilfred Candler, *Linear Programming Methods* (Ames: Iowa State University Press, 1958); or R. C. Agrawal and Earl O. Heady, *Operations Research Methods for Agricultural Decisions,* (Ames: Iowa State University Press, 1972).

search the corner points of the boundary of the feasible set in a sequential fashion. For example, the procedure would start at the origin corner point of Figure 10.3 and move along the yearling axis to the corner point denoted by B'. Profit would be evaluated at that point and then the next corner point of the feasible space, corner point Z, would be investigated. Once corner point Z is evaluated, the simplex method would investigate the possibility of moving to corner point A. Since corner point A has a lower objective function value than corner point Z, the procedure would stop and declare corner point Z the optimum. The mechanical rules of the simplex method are structured so that each corner point will result in a higher value for the objective function. Once a corner point is reached that has a lower value, further investigation is unnecessary because no other corner point in the feasible set has a higher value for the objective function and an optimal solution has been found.

MODEL-BUILDING

Building a linear programming model of a farm business is partly an art and partly a science—additional expertise always comes with practice. The purpose of this discussion is to review some of the basic concepts of building a linear programming model. The discussions will emphasize building whole-farm planning models. Students should appreciate that even if they do not intend to build such a model, an understanding of these concepts will facilitate their ability to interpret the results of linear programming analyses and determine whether a particular model is appropriate and will give realistic answers to a particular problem.

A Matrix—The Basic Structure

The basic structure of the linear programming model is a matrix. Figure 10.4 illustrates such a matrix for a simple farm management problem. Down the left-hand side of the matrix are identified the resource restrictions or constraints that summarize the resources available in the farm business as well as personal and institutional constraints on how the business can operate. Resource restrictions include such items as land availability (possibly by month or season as well as annual), machinery capability, capital and credit availability, the capacity of fixed capital such as a farrowing house, and personal restrictions on certain processes such as the maximum number of dairy cows that will be milked. Across the top of the matrix are identified the various processes or activities in which the farm business can be involved. Processes or activities include such items as the production of corn, soybeans, and other crops, the production of beef cows or hogs, the purchase of feed or other inputs, and the sale of livestock and crop products. We will discuss the different types of constraints or re-

Figure 10.4   A Linear Programming Matrix of a Farm Firm

Constraints or Restrictions	Produce Corn	Produce Hogs	Produce Cattle	Sell Corn	Rent Land	Constraint Type	Resource Availability
Objective function ($)	−81	71	20	3	−100		
Land (acres)	1				−1	≤	300
Labor (hours)	5	16	7			≤	2,000
Capital ($)	60	200	300			≤	100,000
Corn transfer (bu)	−120	105	61	1		=	0
Hog facilities (head space)		1				≤	40
Cattle minimum (head)			1			≥	100

strictions and processes or activities incorporated in linear programming models of farm management problems in more detail later.

The first row of numbers in the body of the matrix is the objective function that has been specified for the problem—the variable that is to be maximized or minimized. For example, if the objective is to maximize income above variable cost, the income figures for each activity or process will be included in the objective function. Thus, for the corn-selling activity of Figure 10.4, the income figure of $3 per bushel sold is entered in the objective function row. For activities that reduce income, a negative figure is inserted in the objective function as is illustrated with the land rental activity of Figure 10.4. The income that will be generated by producing corn on this land is included in the objective function through the entry for the corn production and selling processes of Figure 10.4. In general, the numbers in this row of the matrix represent the c_j coefficients of equations 10.1 and 10.1′; they provide the basis to graph a line similar to FF' of Figure 10.3.

The right-hand column of Figure 10.4, the "Resource Availability" column, indicates the amount of resources that can be allocated to the various production processes. For example, the figure indicates that this farm has 300 acres of land, 2,000 hours of labor, $100,000 of capital, and 40 units of hog facilities. The numbers in this column represent the b_i coefficients of equations 10.2 and 10.2′. Again, the details concerning specification of these constraints will be discussed later. Note that the column headed "Constraint Type" includes a mathematical symbol for each constraint. For example, the symbol "≤" denotes a "less than or equal to" constraint. This means that the sum of the values on the left-hand side of this sign cannot exceed the value on the right-hand side. Therefore, all of the uses of labor in crop and livestock production must be less than or equal to or cannot exceed the amount available—2,000 hours.

The remaining cells of the matrix (frequently referred to as the body of the matrix) reflect input-output relationships between the processes and the constraints. These are physical relationships that indicate how much each process uses or contributes to each of the constraints of the model. For example, producing corn requires 1 acre of land, 5 hours of labor, and $60 of capital. The other coefficients in the body of the matrix reflect similar relationships between processes and constraints. In general, the numbers in these cells represent the a_{ij} coefficients of equations 10.3 and 10.3'; these coefficients along with the resource availability coefficients ($b_i's$) can be used to graph relationships similar to those of lines AA' and BB' of Figure 10.3.

With this basic understanding of the structure of the linear programming matrix, we will now discuss in more detail the characteristics of constraints and processes. First, our discussion will emphasize the mathematical and model-building characteristics of constraints; then it moves to the specification of processes or activities.

Specification of Constraints

Mathematical Type   From a strictly mathematical viewpoint, a linear programming model can contain four different types of constraints or restrictions. The mathematical type refers to whether the constraint is a maximum constraint, a minimum constraint, an equality constraint, or nonconstraining. The maximum constraint is specified as a less than or equal to constraint ($\leq$). This constraint type indicates that all uses of a particular resource must be less than or equal to the amount of that resource available. As is shown in Figure 10.4, the mathematical type is specified by the sign between the input-output coefficients in the body of the matrix and the value in the resource availability column. The land, labor, capital, and hog facility constraints of Figure 10.4 all reflect the fact that resource use cannot exceed the amount available. Mathematically, this relationship can be summarized (using the data of the labor row) as:

$$\text{acres of corn} \cdot (5) + \text{number of hogs} \cdot (16) + \text{number of cattle} \cdot (7)$$
$$\leq 2,000 \text{ hours of labor.}$$

A second mathematical type of constraint is a greater than or equal to constraint ($\geq$). This constraint type reflects the fact that all uses of a particular resource must exceed a minimum amount, and so this constraint type is frequently called a minimum restriction or constraint. Examples may include the specification that a minimum amount of alfalfa must be produced or that a dairy herd must be of a minimum size. Such a constraint is illustrated in Figure 10.4 with the cattle minimum constraint.

A third constraint type is the equality ($=$). This mathematical type indicates that all uses of resources must be equal to the amount available. Consequently, an equality constraint does not allow any resources to remain idle, even if it would be more profitable to do so. Equality constraints are frequently used with transfer rows to transfer resources from one process such as corn production to another process such as the production of hogs. Transfer rows will be discussed in more detail later.

The fourth constraint type is a nonconstraining possibility. This constraint type is most frequently used for the objective function. This row is specified as nonconstraining because we want it to take on the highest or lowest possible value, depending on the problem being analyzed.

Economic Type   Four decisions must be made with respect to each restriction or constraint included in a linear programming model: (1) Its name and economic meaning. This decision reflects the purpose or rationale for including the restraint in the model. (2) The units of measurement. Is the resource constraint measured in hours, acres, animal units, dollars, or some other unit? (3) Mathematical type. This decision involves specification of a minimum constraint ($\geq$), a maximum constraint ($\leq$), an equality ($=$), or nonconstraining. (4) The resource availability. This decision also is frequently referred to as specification of the "right-hand side." It involves specification of the amount of the resource available or the value that is to be placed on the right-hand side of the equality or inequality row.

From an economic point of view, three general types of constraints or restrictions can be defined—real constraints, institutional constraints, and accounting constraints. Each of these constraint types will be discussed and illustrated in detail.

Real Constraints   This category refers to physical resource restrictions. The major types of physical resource restrictions include land and pasture, labor, capital, and water.

Land   A separate constraint is needed for each quality of land that is available. Similarly, each different type or quality of pasture requires a separate constraint. Figure 10.5 illustrates the specification of land constraints that might be included in a linear programming model and the four decisions that must be made for each constraint.[2] For example, the first row indicates that the model will include land with silty-clay soil structure. The units for this land constraint will be acres, and the mathematical type will be a less than or equal to constraint indicating that there is a maximum quantity of this land available. This maximum quantity is specified under the "Availability (RHS)" heading as 125 acres. Each of the land

---

[2]The use of and interactions between this and other types of constraints and activities as part of a total model are illustrated later in this chapter.

Figure 10.5. Land and pasture constraints.

Name	Units	Type	Availability (RHS)
Silty-clay	Acres	≤	125
Loam	Acres	≤	230
Clay-loam	Acres	≤	110
Improved pasture	Acres	≤	20
Native pasture	Acres	≤	115

classes in Figure 10.5 reflects a different level of productivity or fertility level, resulting in different crop or forage yields.

Labor A programming model must include a separate labor constraint for each period when timeliness is important. Thus, the labor periods should reflect the ability to accomplish certain tasks in sequence within a limited amount of time. In the Midwest, timely operations in the critical planting and harvesting period are important, and labor restrictions should reflect the limited amount of time available to do the planting and harvesting jobs. Labor periods need not all be the same length, and they also need not cover the entire year if there obviously is going to be excess labor available in one season. Consequently, weekly or biweekly periods may be necessary for the planting and harvesting seasons for a crop farmer, but restrictions would not be included during the winter period if excess labor was going to be available.

Figure 10.6 illustrates labor constraints for a typical whole-farm programming model. Note again the specification of the four items—name, units, mathematical type, and resource availability—for each constraint.

Capital The capital that can be used in the farm business can be measured in two ways—the amount of monetary funds that are available, or the physical capacity of various buildings and equipment. Both methods of measurement might be included in one model, the first to reflect operating funds available for various enterprises and the second to reflect the

Figure 10.6. Labor constraints.

Name	Units	Type	Availability (RHS)
January–April	Hours	≤	860
May 1–15	Hours	≤	100
May 15–30	Hours	≤	110
June–August	Hours	≤	640

Figure 10.7.   Capital constraints.

Name	Units	Type	Availability (RHS)
Monetary Funds			
Operating capital	Dollars	≤	20,000
Investment capital	Dollars	≤	40,000
Physical Capacity			
Combine	Acres	≤	300
Cattle barn	Head	≤	100
Hog facilities	Sow and two litters	≤	40

physical capacity of the current fixed livestock facilities and machinery. Figure 10.7 illustrates the specification of both monetary funds and physical capacity constraints for a programming model. Note that the monetary funds are separated into those that can be used for short-term operating purposes and those that are available for new investment in additional capital items such as machinery and/or buildings. The values in the "Availability (RHS)" column could include the amount of funds the farmer has available to commit for operating or investment purposes as well as the amount that could be borrowed from various lenders. Alternatively, the amount borrowed can be transferred into these constraints from other borrowing restrictions as will be discussed later. Constraints on the physical capacity of various machinery and equipment items are also illustrated in Figure 10.7.

Water   Water availability is an important restriction on farm production in many parts of the United States where irrigation is practiced. If water availability is unrestricted and the only concern is the cost of pumping, the irrigation costs will be included in the objective function and constraints on water availability need not be included in the model. However, if only a specified amount of water can be obtained in any critical period of plant production because of a limit in the capacity of the distribution system, natural conditions of the water source, or legal restrictions on its use, water constraints should be included in the model. Figure 10.8 illustrates water restrictions that might be included in a programming model for irrigated areas.

Figure 10.8.   Water constraints.

Name	Units	Type	Availability (RHS)
June 1–30	Acre-inches	≤	300
July 1–31	Acre-inches	≤	400

Figure 10.9 Institutional Constraints

Name	Units	Type	Availability (RHS)
Peanut allotment	Acres	≤	50
Borrowing	Dollars	≤	50,000
Rented land	Acres	≤	200
Soybean maximum	Acres	≤	150
Beef cow minimum	Head	≥	40
Hog maximum	Sow and 2 litters	≤	80

Institutional Constraints Institutional constraints include restrictions imposed on the firm by two sources: outside institutions that limit the activities of the business, and personal preferences of the farm operator. With respect to the limits imposed on the farm business by outside institutions, governmental regulations such as allotments and Environmental Protection Agency requirements may restrict the size of particular enterprises. In addition, lenders may impose limits on the amount of funds that can be borrowed, and resource ownership and market conditions may be such as to limit the amount of additional land that can be rented. These types of constraints are illustrated in Figure 10.9. Also illustrated are restrictions that are a result of the farm manager's personal preferences. For numerous reasons, some of which may not be based on economic rationale, farmers may have minimum or maximum quantities of certain enterprises which they will impose on the farm business. For example, they may specify maximum soybean acreage because of erosion problems, or their landlords may specify maximum size for a particular livestock enterprise. These types of restrictions are easily incorporated in a programming model as illustrated in Figure 10.9.

Accounting Constraints Accounting constraints are used primarily to keep track of resources transferred between processes, to collect information on resources or processes, and to provide structure for the model. For example, an accounting constraint would be used to transfer the corn from a corn production process to the hog feeding process. The accounting

Figure 10.10 Accounting Constraints

Name	Units	Type	Availability (RHS)
Corn transfer	Bushels	=	0
Hay transfer	Tons	=	0

constraint is most frequently used in the ease of intermediate products as is illustrated in Figure 10.10. The purpose of the equality (=) mathematical type and the zero (0) right-hand side is to force everything that enters this transfer row or constraint through production out through consumption or utilization. This guarantees that none of the resource or product is lost in the transfer process.

Specification of Processes or Activities

As with constraints, each process or activity included in a linear programming model requires four decisions: (1) The type of activity to include and its economic content. This decision essentially requires an economic rationale for why the particular process is included in the model. (2) Units of measurement. The units decision must be made with consideration for consistency of model structure as well as ease of interpretation of the results. For example, livestock production processes can be measured in number of head or number of hundredweights produced. There are no rules on unit specification for a process except those of consistency and ease of interpretation. (3) Objective function value (C_j). The contribution of each activity or process to the objective function must be specified. In most cases, this is the net return per unit, the price per unit, or the cost per unit of a particular process. (4) Input-output coefficients (a_{ij}). This involves specification of the amount of each resource or constraint that is used by each process. These input-output coefficients reflect the physical production relationships between products and resources, and are typically obtained from farm records (if available) or agronomists, animal scientists, and other production specialists.

From an economic or biologic viewpoint, four basic types of activities or processes are included in most linear programming models: production processes, resource supply processes, product marketing processes, and transfer processes. Each of these types of processes will be discussed in turn.

Production Processes   This category refers to those processes that utilize various resources to generate physical outputs. Examples include the production of wheat on silty-clay land, the production of corn on loam land, the feeding of cattle, and the farrowing and feeding of market hogs. In most whole-farm planning models, the objective function value for production processes will include returns to the unpaid resources of land, family labor, fixed capital, risk, and management. The input-output coefficients for each production process reflect the amount of various resources or constraints utilized for each unit that process produced. The standard linear programming format is to reflect resource use within the body of the matrix with a positive coefficient, and resource augmentation or addition to a particular constraint with a negative coefficient. An easy rule to re-

Figure 10.11. Production processes.

Item	Corn Growing and Harvesting (1 Acre)	Soybean Growing and Selling (1 Acre)	Feeder Pig Finishing (1 Head)	Cattle Finishing (1 Head)
Objective function ($)	−81	170	26	56
Land (acres)	1	1		
May labor (hours)	1.2	1.0	0.1	0.5
October labor (hours)	0.4	0.6		0.5
Capital ($)	68	42	40	510
Livestock facilities (sq. ft.)			8	20
Corn transfer (bu)	−120		10	65

member in structuring a linear programming matrix of the maximization type is that negative input-output coefficients *add* to the resource constraint and positive coefficients *subtract* from the resource constraint. With the values for the objective function, the standard format that negative values reflect costs and positive values reflect returns is utilized.

Examples of the specification of production processes or activities for a whole-farm linear programming model are illustrated in Figure 10.11. Note that producing soybeans generates a $170 per acre return above variable cost as reflected in the objective function. Soybean production requires 1 acre of land, 1.0 hour of May labor, 0.6 hours of October labor, and $42 of operating capital. In this example, the soybeans are produced and sold by the same process. In the case of corn, the grain is produced by one process and either fed to cattle or hogs or marketed through a separate process. Consequently, an accounting constraint in the form of the corn transfer row is included in the model, and the yield per acre (120 bushels) is reflected in this constraint under the corn growing and harvesting process with a negative sign. Since the corn growing and harvesting activity does not generate any direct revenue, the objective function coefficient reflects the variable cost of production ($81).

Resource Supply Processes   These processes are used to acquire additional inputs and make them available for use in the production processes. Examples include the hiring of labor, the borrowing of capital, or the renting of additional land. Figure 10.12 illustrates the specification of resource supply processes. For example, the labor hiring process reduces the objective function by $5.50 for each hour of labor hired, adds to the amount of total labor available for production, and reduces the amount of labor available to hire. Similarly, borrowing long-term capital at 12 percent reduces the objective function by $0.12 for each dollar borrowed, increases the amount of long-term capital available for production by $1, and reduces the amount of long-term capital available to borrow by $1.

Figure 10.12. Resource supply processes.

Item	Hire Labor (1 Hour)	Borrow Long-Term Capital (1$)	Rent Land (1 Acre)
Objective function ($)	−5.50	−0.12	−100
Labor (hours)	−1		
Capital ($)		−1	
Land (acres)			−1
Hire labor (hours)	1		
Capital borrowing ($)		1	
Land rental (acres)			1

Marketing Processes   These processes are included in a model to sell the products or commodities produced by the production processes. In some cases, the production process may encompass both the production and the sale of a particular commodity. However, if alternative ways of marketing a product such as direct sale on a cash basis, delayed sale on a contract basis, or feeding to livestock are to be examined, it is necessary to separate the production and marketing processes or activities. In this situation, specific processes would be included for the sale of corn, the sale of soybeans, and the sale of cattle or other specific commodities. The relationship between production processes and marketing processes is shown in Figure 10.13. The marketing processes included in this figure are the corn selling and cattle selling processes. Corn production and cattle feeding are both production processes, whereas corn buying is a resource supply process. Note that Figure 10.13 also includes an accounting constraint or row for corn—the commodity that has alternative uses. Corn produced by the corn production process is added to the corn accounting constraint (−120). In addition, purchased corn is added to this constraint (−1). These processes reflect only the variable cost of production ($81) or purchase price ($3) in the objective function. Thus, corn can either be fed to cattle as reflected in the cattle feeding process or sold as reflected in the corn selling process. The two marketing processes, sell corn and sell cattle, subtract corn and cattle from the appropriate accounting constraint and have positive values

Figure 10.13. Marketing processes.

Item	Produce Corn (1 Acre)	Buy Corn (1 bu)	Feed Cattle (1 Head)	Sell Corn (1 bu)	Sell Cattle (1 cwt)
Objective function ($)	−81	−3	−740	2.90	75
Corn (bu)	−120	−1	65	1	
Cattle (cwt)			−11		1

in the objective function reflecting the net return from corn or cattle selling. Hence, a marketing process essentially takes a commodity from an accounting constraint and increases the objective function by the sale price per unit of that commodity.

Transfer Processes   The purpose of transfer processes in a linear programming model is to transfer resources or commodities from one constraint to another. For example, if a model includes seasonal operating capital constraints, it is usually desirable to make operating capital not used in one period available in the following period. Figure 10.14 illustrates an example of such a transfer process. Note that the transfer process in this case has no value in the objective function and simply provides the mechanical linkage between two seasonal operating capital constraints. Other examples of the use of transfer processes may involve the transfer of commodities between certain restrictions, the transfer of fixed costs into the objective function, the payment of income taxes, or the transfer of resources such as physical capital capacity or funds between periods in a multi-period model.

Special Cases
Some unique problems encountered in the construction of linear programming models of the farm firm should be noted here.

Complementarity and Joint Products   It has been indicated that complementary relationships such as exist in crop rotations cannot be reflected between processes in a programming model but can be included *within* a process. The typical procedure is to specify a rotation process or activity to reflect the complementary relationship. For example, a corn-corn-oats-meadow rotation might be included in a programming model to reflect the fact that corn following meadow may not require as much commercial

Figure 10.14.   Transfer processes.

Item	Capital Transfer (1 $)	Corn Transfer (1 bu)
Objective function	0	0
Operating capital—summer ($)	1	
Operating capital—fall ($)	−1	
Corn inventory—July (bu)		1
Corn inventory—August (bu)		−1

fertilizer or other inputs to obtain the same yield level, and that erosion may be reduced if these crops are produced in rotation rather than as separate processes. The units for this rotation could be defined as four acres, and all resource requirements and the objective function value could reflect the cost, returns, and requirements for this four-acre unit. Alternatively, the units could be specified as one acre, and the input-output coefficients and objective function value would reflect one-fourth of the returns, costs, and resource requirements for each crop in the rotation.

A similar problem of joint production arises in livestock enterprises. When constructing livestock activities for a linear programming model, it is important to include all products that will be sold. For sheep, this will include the wool as well as the animal itself. If a breeding herd is included in the model such as a sow or beef-cow herd, the costs and returns for the maintenance and sale of cull breeding stock must be included in the model as well as the production and sale of the offspring.

Nonlinear Relationships   As has been indicated earlier, nonlinear relationships can be approximated through linearly segmented functions. For example, to include a production relationship that reflects decreasing marginal productivity, the relationship would be approximated with a set of linear segments and a separate process would be included in the programming model for each linear segment. Constraints would also be included in the model to restrict the feasible values for each process to the relevant range for each linear segment of the relationship. This procedure can be used to handle the traditional nonlinear production function that exhibits declining marginal products. However, it cannot be used with a production function that has increasing marginal products.

Share Rents   Many farm operations include share rental agreements as well as full ownership or cash rental of land. To analyze the farm business from the tenant's viewpoint, a new process must be identified for the farming of share rental land separate from the process for farming owned land or cash-rented land. This new process would include only the tenant's share of the income received and costs incurred in the objective function. Furthermore, only the tenant's share of resources used is included in the input-output coefficients of the matrix. This approach to modeling share rental arrangements may also require constraints on land use to reflect any limits imposed by the landlord. For example, the landlord may require a minimum acreage of a certain crop on share rented land, so a minimum restriction would be included in the model for this process. If the farm management problem is being viewed from the position of the landlord, then the resources that he or she must contribute and his or her income and costs are reflected in the coefficients for the share rental process. Therefore, the key issue with respect to share rental arrangements is to determine whether the model is being constructed for the landlord or the tenant, and then reflect only the appropriate costs, returns, and resource requirements in the matrix.

A SIMPLE FARM PLANNING MODEL

To illustrate the development of a linear programming model of a whole-farm planning problem and the interpretation of the results, a simple case example will be used. An example of using linear programming to develop least cost feed rations is discussed in the Appendix to this chapter.

The Farm Situation

A farm operator wants to evaluate alternative methods of organizing his or her farm operation. The farm consists of 300 acres of land capable of producing row crops, a labor supply of 2,000 hours, and hog facilities that can be used to produce 150 litters per year. He or she estimates fixed costs for buildings, equipment, land and other resources, and income taxes to total $45,200. Because of the soil types on the farm and the use of a pasture-oriented hog system, the operator wants a minimum of ten acres of oats to be included as part of the farm organization. Alternative enterprises that are being considered include corn and corn-soybean rotations, oats, and hogs. The farmer is willing to buy and sell all commodities produced on the farm and hire additional labor if needed. Her or his objective is to maximize net return above variable and fixed costs.

A linear programming matrix that can be used to evaluate the optimal organization for this farm operator is shown in Figure 10.15. This matrix includes physical resource constraints (labor, land, and hog facility restrictions), institutional and personal constraints (the minimum oat acreage restriction), and accounting constraints (corn balance, soybean balance, oat balance, straw balance, and fixed cost and income tax payment restrictions). With respect to processes or activities, the matrix includes production processes (corn growing and harvesting, corn-corn-beans rotation growing and harvesting, oats growing and harvesting, and swine farrow-finish and selling activities), marketing processes (corn selling, soybean selling, oats selling, and straw selling activities), input purchasing processes (corn buying, oats buying, straw buying, and labor hiring activities), and a transfer process (fixed cost and income taxpaying activity).

The crop production processes utilize labor and land resources and contribute products to the corn, soybean, oats, and straw accounting rows or constraints. The coefficients in the objective function for the crop production processes reflect the cash costs involved in production. Thus, for example, corn growing and harvesting has an operating cost of $82.41, utilizes 4.5 hours of labor, 1 acre of land, and contributes 120 bushels of corn to the corn accounting row.

The swine farrow-finish and selling process is a combination production and marketing process. This process requires 34 hours of labor, 2

Figure 10.15. Linear programming matrix for the case example.

Constraints	Corn Growing and Harvesting (1 Acre)	Corn-Corn-Beans Rotation Growing and Harvesting (3 Acres)	Oats Growing and Harvesting with Straw (1 Acre)	Swine Farrow-Finishing with Selling (1 Sow / 2 Litters)	Corn Selling (1 bu)	Soybean Selling (1 bu)	Oat Selling (1 bu)	Straw Selling (1 Ton)	Corn Buying (1 bu)	Oat Buying (1 bu)	Straw Buying (1 Ton)	Labor Hiring (1 hr)	Fixed Cost Paying (1 $)	Math Type	Availability (RHS)
Objective function ($)	-82.41	-201.25	-34.12	612.21	2.90	7.90	1.52	45.00	-3.00	-1.58	-48.00	-5.75	-1	—	—
Labor (hrs)	4.5	13.5	4	34	0.005	0.005	0.005	0.6	0.005	0.005	0.6	-1		$\leq$	2,000
Land (acres)	1	3	1											$\leq$	300
Hog facility (head space)				2										$\leq$	150
Corn balance (bu)	-120	-250		175	1				-1					$\leq$	0
Soybean balance (bu)		-35				1								$\leq$	0
Oat balance (bu)			-60	42			1			-1				$\leq$	0
Straw balance (ton)			-0.90	0.5				1			-1			$\leq$	0
Fixed cost ($)													1	$\leq$	45,200
Oats minimum (acres)			1											$\geq$	10

units of facility space (the process unit is one sow and two litters), 175 bushels of corn, 42 bushels of oats, and 0.5 tons of straw. Each unit of the process (a sow and two litters) generates $612.21 of return above variable cost which is reflected in the objective function.

The commodity selling processes (corn selling, soybean selling, oats selling, and straw selling) indicate the income and resource requirements for each of the marketing activities. For example, selling a bushel of corn results in a $2.90 contribution to the objective function and requires 0.005 hours of labor. Each bushel of corn sold reduces the amount of corn available in the corn balance row or constraint by one bushel. Similar explanations can be applied to the other commodity marketing processes.

In addition to selling commodities, the farmer is willing to purchase commodities as reflected in the corn buying, oat buying, and straw buying processes. These processes add to the respective accounting constraints as reflected by the negative values in each respective row, and they reduce the objective function by their respective purchase prices per unit. The purchase of commodities also requires labor as reflected by the positive coefficients in the labor constraint. Note that the selling price for each commodity is less than the purchase price for the same commodity. This reflects the realistic margin for handling and transportation.

The labor hiring process enables farmers to increase their labor supply. This process has a negative coefficient in the objective function, reflecting the wage rate per hour of labor ($5.75), and a negative coefficient (-1) in the labor constraint, reflecting the addition of one hour of labor for each unit of the hiring process utilized.

The final activity of the model transfers the required fixed cost and income tax payments into the objective function. The coefficients indicate that each dollar of fixed costs and taxes paid reduces the objective function value by one dollar. The equality mathematical type for this constraint forces the fixed cost and taxpaying activity into the solution at a value equal to $45,200. Consequently, the full $45,200 of fixed costs and taxes will be transferred into the objective function and subtracted from the income above variable costs generated by the other production, marketing, and input supply processes.[3]

---

[3]This formulation of the model requires an estimate of fixed costs and income taxes prior to obtaining a solution. A more accurate specification of the model would include taxable income accounting constraints and detailed taxpaying activities. To implement this specification, a taxable income accounting constraint would first be added to the model. Then a separate constraint would be added to the model for each income tax bracket. Income would be transferred from the taxable income accounting constraint into the appropriate constraint for each tax bracket, and then the tax liability would be transferred into the objective function as a cost. For more details on building income tax provisions into linear programming models, see J. M. Vandeputte, and C. B. Baker, "Specifying the Allocation of Income Among Taxes, Consumption, and Savings in Linear Programming Models," *American Journal of Agricultural Economics* 52, No. 4 (November 1970) 521–527.

Generating a Solution

Once the matrix for the programming model has been developed, the numbers are transferred onto data sheets and eventually onto cards for solution. The mechanics of this process will not be reviewed in detail, but the procedures are typically straightforward.[4] Computer packages that can be used to solve the linear programming model once it has been constructed and put on cards are available from various computer hardware and software manufacturers. Each computation center uses its own format for job setup and submittal, but most linear programming computation routines utilize similar input formats and procedures and provide similar types of output information.

Interpretation of Results

To illustrate the type of information that can be provided by the linear programming procedure, we will discuss the output that was generated for the whole-farm programming model of Figure 10.15. This output is summarized in Table 10.2. Note that the output contains three major sections: a section that indicates the value of the objective function, a section that presents data on the optimal processes or activities, and a section that summarizes the status of the rows or constraints in the optimal solution. Each of these sections will be discussed in turn.

The first item of the output indicates the value of the objective function or the amount of farm income generated by the optimal enterprise combination. For the resource situation, cost, and physical efficiencies of Figure 10.15, the optimal combination of enterprises would generate an income above variable and fixed costs of $33,416.32. The optimal combination of enterprises and the allocation of the various resources to produce this level of income are provided in the next sections of the linear programming output as summarized in Table 10.2.

The optimal organization for our simple case example includes 96.7 units (3 acres per unit) of the corn-corn-soybean rotation, 10 acres of oats, and 15.6 sow and 2 litter units of hogs (31.2 litters total). A total of 21,439 bushels of corn, 3,383 bushels of soybeans, and 1.2 tons of straw are sold in the optimal solution. To support the hog enterprise, 55 bushels of oats are purchased. A total of $45,200 of fixed costs are paid as part of the optimal solution.

Unless the various production activities generate income that will be treated differently by the tax rules (such as capital gain rather than ordinary income), the inclusion of the detailed taxpaying activities noted above will have no impact on short-run optimal enterprise choice and the optimal short-run farm plan. It may, however, more accurately reflect the income tax obligation and thus the amount of income available for family living or reinvestment in the firm.

[4]See Raymond R. Beneke and Ronald Winterboer, *Linear Programming Applications to Agriculture,* (Ames: Iowa State University Press, 1973) for a thorough discussion of entering the data on data sheets for further processing.

Table 10.2   The Output for the Case Example

I.   *Farm Income—$33,416.32*

II.   *Processes*

Process	Units	Cost or Return	Reduced Income
Corn growing and harvesting (1 acre)		−$82.41	$1.22
Corn-corn-soybeans growing and harvesting (3 acres)	96.7	−201.25	
Oats growing and harvesting with straw (1 acre)	10.0	−34.12	
Swine farrow-finishing with selling (1 sow, 2 litters)	15.6	612.21	
Corn selling (1 bu)	21,439.3	2.90	
Soybean selling (1 bu)	3,383.3	7.90	
Oat selling (1 bu)		1.52	0.06
Straw selling (1 ton)	1.2	45.00	
Corn buying (1 bu)		−3.00	0.10
Oat buying (1 bu)	54.6	−1.58	
Straw buying (1 ton)		−48.00	3.58
Labor hiring (1 hr)		−5.75	5.27
Fixed cost paying (1 $)	45,200.0	−1.00	

III.   *Constraints*

Constraint	Amount Used	Amount Unused (Slack)	Lower Limit	Upper Limit	Resource Value
Labor (hrs)	2,000.0		None	2,000.0	$0.48
Land (acres)	300.0		None	300.0	264.36
Hog facility (head space)	31.17	118.83	None	150.00	
Corn balance (bu)			None		2.90
Soybean balance (bu)			None		7.90
Oat balance (bu)			None		1.58
Straw balance (ton)			None		44.71
Fixed cost ($)	45,200.0		45,200.0	45,200.0	1.00
Oats minimum (acres)	10.0		10.0	None	−165.22

In addition to this information concerning the optimal combination of processes or enterprises, the linear programming output also summarizes other characteristics of the optimal solution. The column in Table 10.2 entitled "Cost or Return" indicates the value used in the objective function for each of the processes included in the model. These data can be used along with the size of each activity to determine the relative importance of each process in generating income or incurring costs.

The programming procedure provides additional information on the amount of income that would be lost if one of the processes not in the optimal solution were forced into solution. This information is summarized in the "Reduced Income" column of Table 10.2. For example, continuous corn was not included in the optimal farm organization. The figures in the "Reduced Income" column indicate that the objective function or farm income would have been reduced by $1.22 for each acre of continuous corn that was forced into the enterprise organization. An alternative interpretation of the "Reduced Income" data is that the cost of producing continuous corn would have to decrease by only $1.22 per acre before this activity would be competitive with the corn-corn-soybean rotation. This small income penalty suggests that continuous corn and corn-corn-soybean rotation are very similar in terms of profitability. A similar interpretation applies to the "Reduced Income" values for the other processes. For example, no corn, oats, or straw are purchased in the optimal solution; in fact, farm income would have been reduced by $0.10 for each bushel of corn, $0.06 for each bushel of oats, and $3.58 for each ton of straw if these commodities would have been purchased. For corn and oats these values represent the difference between the buying and selling prices inputted into the analysis; if corn or oats were purchased, it would not be consumed but sold at a loss of $0.10 and $0.06 per bushel, respectively. No additional labor is hired, and farm income would have been reduced by $5.27 for each hour if extra labor had been hired.

The third section of Table 10.2 summarizes the status of the constraints in the optimal solution. The "Amount Used" column reflects the total amount of each resource or constraint that is utilized in the various processes. For example, the entire 2,000 hours of labor and 300 acres of land are utilized in the production processes. Only 31.2 units of the hog facility space are used. The fixed costs are paid, and the minimum of 10 acres of oats is also included in the optimal enterprise organization. When some of the resources are not completely utilized as is the case with the hog facility space, this is reflected in the "Amount Unused (Slack)" column of the third section of Table 10.2. Thus, 118.8 units of hog facility space are not utilized. Such information indicates the availability of excess resources and which resources or constraints are limiting production.

The "Lower Limit" and "Upper Limit" columns of the constraints section of Table 10.2 reflect the amount of the resources available or the restrictions included in the model. For example, an upper limit of 2,000 hours was placed on the labor restriction and 300 acres on the land restriction. Since the fixed cost restriction was specified as an equality, both the upper limit and lower limit are equal to the amount of fixed cost that must be paid—$45,200. The lower limit of 10 acres on the oat minimum constraint reflects the minimum size of this enterprise in the optimal solution.

An additional item of information generated by the linear programming model is the value of the resources used in production. This information is summarized in the "Resource Value" column of the third section of Table 10.2. The information in this column indicates the amount of income that would be given up if one less unit of each of the resources was available. For example, the value for the land resource in this column is $264.36. This number indicates that if only 299 acres of land were available rather than 300, income above variable costs would be reduced by $264.36. Alternatively, the value of the last acre of land or the marginal value product of land at the margin is $264.36. Note the value in the "Resource Value" column for the oat minimum constraint; this value indicates that reducing the oat minimum by one acre would have *increased* income by $165.22. The minimum specified by the farm operator on oat acreage is actually reducing his or her income because that land is not available to use in corn and soybean production. Thus, one way for the farmer to increase his/her income from the specified set of resources would be to reduce the acres of oats produced, increase the number of acres devoted to corn and soybean production, and buy all the oats and straw needed for the hog enterprise.

In addition to this information, the linear programming output can provide additional detailed data concerning the sensitivity of the results to changes in costs, prices, and input-output coefficients as well as the "ranges" over which the "Resource Value" and "Reduced Income" values are valid or appropriate. This information can be quite useful in further interpretation of the optimal solution and its practical application to the farm business under varying conditions and changes in price and production relationships.

PRESTRUCTURED LINEAR PROGRAMMING PACKAGES

Constructing a linear programming model is not an easy task. It requires an understanding of basic mathematics as well as the concepts of linear programming. To facilitate the use of the programming procedure for those who do not have this knowledge, numerous "prestructured" linear programming models have been developed for use in analyzing farm management problems. Such models are available for whole-farm planning, machinery scheduling and selection, and least cost ration formulation, for example. These prestructured packages require farm managers to provide information on their resource availability, alternative processes they are willing to consider and the cost and returns associated with each process, and physical input-output coefficients that relate each process to resource availability. In some cases, the prestructured packages will include suggested values for the cost and input-output coefficients so that farm managers can evaluate these coefficients and make adjustments if the coefficients do not seem to reflect their cost or efficiency. Output information from a prestructured model usually is organized in a readable and concise set of reports.

Prestructured packages do not reduce the data or information requirements that are necessary to use linear programming. The input information is almost identical whether a prestructured package is used or managers construct their own unique model. The advantage of a prestructured package is that managers can concentrate their time on obtaining the data necessary to solve the problem and interpreting the results rather than structuring the programming model. A significant disadvantage of the prestructured model is that it may not have the flexibility desired by managers to analyze the unique and specific characteristics of their management problems. Examples of prestructured models include the Top Farmer programs available from Purdue University, Crop-Opt from Iowa State University, LP-FARM from Oklahoma State University, Tel-plan programs available from Michigan State University, and CMN programs available from Virginia Polytechnical Institute and State University. Many other land grant institutions have or are developing prestructured programming models, some of which are to be solved on micro- or personal computers. The farm manager and student should, therefore check with their state land grant institution concerning the availability and usefulness of the models for their particular management problem.

A FINAL COMMENT

We have discussed in detail the potential use and application of linear programming in farm management analysis along with the problems of model construction and interpretation of the results. It is important to recognize that linear programming cannot solve all problems associated with farm planning. Specifically, linear programming cannot help the manager determine what prices to expect in the future or what the physical production relationships will be on his or her farm. And it may be difficult to specify in detail the constraints that are needed or the activities or processes that should be included for an accurate model of the management problem. Moreover, the programming formulation does not include the important dimensions of risk in the farm planning process, nor can it handle relationships that involve decreasing costs. Thus, the procedure is not a panacea for farm planning. However, it is a significant improvement over the traditional farm planning methods of budgeting and marginal analysis for most farm firms where complex interrelationships between enterprises and resource availability must be evaluated. The primary constraint in using linear programming to solve farm management problems is the time available for model construction and interpretation. Many of the prestructured linear programming models can reduce the time required in model construction significantly and also facilitate interpretation of the output. As these prestructured models are improved and updated to reflect new technology and changing cost and price relationships, linear programming will become an increasingly important planning tool for many farm firms.

Summary

Linear programming can be a very useful tool in planning the farm business organization. Four basic assumptions are essential to use linear programming: (1) additivity and linearity—inputs and products are additive: (2) divisibility—resources and products are divisible: (3) finiteness—a finite number of alternative activities and constraints can be specified, and (4) single-valued expectations—prices, resource supplies, and input-output coefficients are known with certainty. Since similar assumptions underlie the budgeting and marginal analysis procedures, linear programming results should be similar to those obtained using other analytical procedures. The advantage of using linear programming is that it can be used to solve more complex problems with less computation burden, and it typically will provide more information concerning resource use and valuation.

The basic concept of linear programming is to choose a set of activities or processes that maximize or minimize the specific outcome variable, subject to a set of constraints that limit the decisions that can be made. A set of mathematical rules known as the simplex method are used to solve linear programming problems with the aid of a computer.

Building linear programming models is part art and part science. The basic structure of any linear programming model is a matrix, with the columns in that matrix being the processes or activities and the rows of the matrix being the resource restrictions or constraints. Three general types of constraints or restrictions are included in most farm management models: (1) real constraints which limit the physical resource availability, (2) institutional and subjective constraints which reflect limits imposed by outside institutions or personal preferences of the operator, and (3) accounting constraints which are used to keep track of resources or will provide structure to the model. Four basic types of activities are included in most linear programming models: (1) production activities which generate physical output, (2) resource supply activities which acquire inputs and make them available for use in the production process, (3) marketing activities which sell products or commodities, and (4) transfer acitivites which transfer resources or commodities from one constraint to another. A special row in the linear programming model, called the objective function, includes the costs and/or returns that are to be maximized or minimized; and a special column, called the right-hand side, reflects the endowment of resources that are available or the limits on the production process.

The output from a linear programming model indicates not only the optimal set, but also the most efficient allocation of resources and the value of those resources in the production process. Additional information on the sensitivity of the results to changes in resource availability, activity prices or costs, and input-output coefficients is also generated by the linear programming model.

To facilitate the use of the linear programming procedure and the construction of programming models, numerous "prestructured" models have been developed for use in analyzing farm management problems.

APPENDIX

Least Cost Rations

The least cost ration problem is frequently solved using a linear programming model. In this application, the objective is to minimize the cost of satisfying the nutritional requirements of a particular type of livestock using alternative feed ingredients. This procedure is used frequently by feed manufacturers as well as larger livestock producers who are blending their own rations.

To illustrate, assume that a beef feedlot operator wants to formulate the least cost ration to increase animal weight from 300 kg (660 pounds) to 400 kg (880 pounds). The ration will be formulated in terms of kilograms of various ingredients to feed per head per day to meet daily nutritional requirements.

Table 10.3 Nutritional Requirements per Head per Day

Nutrient		Amount
Digestible protein	Minimum	0.58 kg
Metabolizable energy	Minimum	19 Mcal
TDN	Minimum	5.3 kg
Calcium	Minimum	26 g
Phosphorus	Minimum	19 g
Carotene	Minimum	39.5 g
Dry matter intake	Maximum	8 kg

Table 10.4 Nutrient Characteristics
of Various Feed Ingredients

Item	Prices ($/kg)	Carotene (mg/kg)	Phosphorus (g/kg)	Calcium (g/kg)	TDN (%)	Metabolizable Energy (Mcal/kg)	Digestible Protein (kg/kg)	Dry Matter (%)
Alfalfa hay	0.077	29.70	1.96	12.00	0.517	1.87	0.108	0.892
Ground ear corn	0.07		2.70	0.43	0.78	2.82	0.04	0.87
Cracked shelled corn	0.11	1.78	3.10	0.18	0.81	2.92	0.067	0.89
Bone meal	0.05		123.50	258.00				0.945
Oats	0.12		3.47	0.979	0.676	2.44	0.088	0.89
Soybean meal	0.288		6.67	3.20	0.72	2.60	0.389	0.89
Cottonseed meal	0.258		11.98	1.55	0.636	2.47	0.333	0.915
Barley	0.107		4.18	0.80	0.73	2.67	0.087	0.89
Sorghum-milo	0.087			0.35	0.71	2.57	0.063	0.89
Linseed meal	0.24		8.30	4.00	0.69	2.50	0.31	0.91
Corn silage	0.024		0.80	1.08	0.28	1.01	0.0188	0.40

Figure 10.16 A Linear Programming Matrix of a Least Cost Feed Ration Problem

Constraints	Alfalfa Hay[a]	Ground Ear Corn	Cracked Shelled Corn	Bone Meal	Oats	Soybean Meal	Cottonseed Meal	Barley	Sorghum-Milo	Linseed Meal	Corn Silage	Math Type	Nutrient Requirements
Objective function ($)	0.077	0.07	0.11	0.05	0.12	0.288	0.258	0.107	0.087	0.24	0.024		
Dry matter (%)	0.892	0.87	0.89	0.945	0.89	0.89	0.915	0.89	0.89	0.91	0.4	$\leq$	8
Digestible protein (kg)	0.108	0.04	0.067		0.088	0.389	0.333	0.087	0.063	0.31	0.0188	$\geq$	0.58
Metabolizable energy (Mcal)	1.87	2.82	2.92		2.44	2.6	2.47	2.67	2.57	2.5	1.01	$\geq$	19
Total digestible nutrients (%)	0.517	0.78	0.81		0.676	0.72	0.636	0.73	0.71	0.69	0.28	$\geq$	5.3
Calcium (g)	12	0.43	0.18	258	0.979	3.2	1.55	0.8	0.35	4	1.08	$\geq$	26
Phosphorus (g)	1.96	2.7	3.1	123.5	3.47	6.67	11.98	4.18		8.3	0.8	$\geq$	19
Carotene (mg)	29.7		1.78									$\geq$	39.5

[a]Units for all activities are kilograms (kg).

431

Table 10.5 Least Cost Daily Ration for
Feeding Cattle from 300 kg (660 Pounds)
to 400 kg (880 Pounds)

Ingredient	Amount (kg)
Alfalfa hay	3.10
Ground ear corn	.26
Cracked shelled corn	—
Bone meal	.018
Oats	—
Soybean meal	—
Cotton seed meal	—
Barley	—
Sorghum-milo	—
Linseed meal	—
Corn silage	12.46

The daily nutritional requirements per head per day are summarized in Table 10.3. The potential feed ingredients include commonly available feedstuffs in the Southwestern states; the nutrient characteristics of these ingredients are summarized in Table 10.4 along with the market price per kilogram. It is assumed that all ingredients are available in unlimited quantities at the specified price.

The linear programming matrix for this simple least cost feed formulation problem is summarized in Figure 10.16. Note that only alfalfa hay and ground yellow corn provide carotene; thus, one or both of these ingredients will be included in the least cost ration. In addition, the constraints with one exception are all minimum constraints (i.e., the least cost ration must include a minimum of the various nutrients). The only maximum constraint is dry matter intake which restricts the amount of total feed intake of the animal to 8 kg per day.

The least cost daily ration for this set of prices and input-output coefficients is summarized in Table 10.5. The daily ration includes 3.10 kg of alfalfa hay, 0.26 kg of ground ear corn, 0.018 kg of bone meal, and 12.46 kg of corn silage. The cost of feeding this ration would be $0.56 per head per day. Including shelled corn in the ration would increase the cost by 2 cents per head per day for each kilogram included, and using milo would increase the cost by 1.5 cents per head per day for each kilogram used. The ration formulated includes excess calcium, carotene, and metabolizable energy, but no other ration can meet the daily nutrient requirements of the animal at a lower cost.

Questions and Problems

1. What are the advantages of linear programming compared to other analysis techniques discussed earlier?
2. Describe the basic structure of a linear programming model.
3. Identify and discuss the four basic assumptions of linear programming.
4. What is the difference between a linear programming process and a production function?
5. Discuss the relationship between linear programming and budgeting? Between linear programming and marginal analysis?
6. Specify a simple linear programming model with 2 activities and 3 constraints. Then solve the model graphically.
7. What are the four mathematical types of constraints that might be included in a linear programming model? Provide an illustration of each of these mathematical types.
8. Identify the four decisions that must be made with respect to each restriction or constraint in a linear programming model.
9. Differentiate between real constraints, institutional and subjective constraints, and accounting constraints. Provide an illustration of each of these types of constraints.
10. What are the four decisions that must be made with respect to each process or activity in a linear programming model?
11. Differentiate between production, resource supply, product marketing, and transfer processes or activities. Provide an illustration of each of these types of processes or activities.
12. How can complementarity and joint products be handled in a linear programming model?
13. How can nonlinear relationships be handled in a linear programming model?
14. Obtain the output from a linear programming model and analyze the solution results.
15. Discuss the type of information that can be obtained from the solution of a linear programming problem.

Further Reading

Agrawal, R. C., and Earl O. Heady. *Operations Research Methods for Agricultural Decisions*. Ames: Iowa State University Press, 1972.

Baumol, William J. *Economic Theory and Operations Analysis,* Englewood Cliffs, N.J.: Prentice-Hall, 1965.

Beneke, Raymond R., and Ronald Winterboer. *Linear Programming Applications to Agriculture*. Ames: Iowa State University Press, 1973.

Heady, Earl O. "Simplified Presentation and Logical Aspects of Linear Programming Techniques." *Journal of Farm Economies* 36, No. 5 (December 1954): 1,035–51.

Heady, Earl O., and Wilfred Candler. *Linear Programming Methods*. Ames: Iowa State University Press, 1958.

McCorkle, Chester P., Jr. "Linear Programming as a Tool in Farm Management Analysis." *Journal of Farm Economics* 37, No. 5 (December 1955): 1,222–36.

Naylor, Thomas H. "The Theory of the Firm: A Comparison of Marginal Analysis and Linear Programming." *Southern Economic Journal* 32, No. 3 (1966): 263–74.

Wagner, Harvey. *Principles of Operations Research.* Englewood Cliffs, N.J.: Prentice-Hall, 1969.

11

CONSIDERATION OF RISK AND UNCERTAINTY

CHAPTER CONTENTS

The discussion of farm planning procedures in previous chapters has assumed either perfect knowledge of input-output and price relationships or the use of expected values. The factor-product and factor-factor models discussed in Chapter 3, as well as the product-product model presented in Chapter 5, assume that the output level(s) for any combination of inputs can be determined. The analysis of cost-minimizing input levels and the profit-maximizing output level(s) was based on known input and product prices. The budgeting and linear programming procedures were described using expected input requirements, output levels, and prices.

This chapter describes procedures that can be used in analyzing decisions considering uncertainty. The manager of a farm business seldom, if ever, has complete knowledge of the input-output relationships and prices involved in making decisions. Typically, farmers have some information on the possible outcomes and some feeling for those that are more likely to occur. In this situation, appropriate decisions must consider both the possible outcomes *and* the individual's attitude towards bearing risk. Both of these aspects of decision-making under uncertainty are considered in this discussion.

The procedures described are offered as methods to make good decisions given the information available at the time the decision is made. Thus, the procedures lead to "good decisions" from an *ex ante* point of view. The reader should understand that these procedures do not necessarily provide good decisions after the fact, or from an *ex post* point of view. For example, farmers who insure their crops against hail may conclude at the end of the growing season that the money spent on the premium could have been saved because no hail storm occurred. After the fact, the decision to buy hail insurance might be considered unwise because it reduced the operator's income. However, if loss of the crop to hail would have created significant liquidity or solvency problems for the business, this decision might be considered a good decision from an *ex ante* perspective.

The only way to guarantee good decisions after the fact is to have knowledge of the outcome for each alternative the decision-maker has available. We advocate obtaining more information as a basis for decisions when the information is available and the expected value of the information exceeds the cost of obtaining it. In most decisions it is impractical, if not impossible, to obtain perfect knowledge. Managers must be prepared to make decisions with less than perfect knowledge. For some decisions farmers have a great deal of information on the possible outcome. In others they have very little. The procedures described are based on modern decision theory and can be used with as much or as little information as the decision-maker has available.

RISK, UNCERTAINTY, AND PROBABILITY

Traditional analyses of decision-making with less than perfect knowledge were divided into two classes based on the pioneering work of Frank H. Knight.[1] Knight divided decision-making situations into risk and uncertainty. He defined the risk situation as one in which the decision-maker knows both the alternative outcomes and the probability associated with each outcome. These known probability distributions are referred to as objective probabilities. Objective probabilities can be defined as the limits of relative frequencies from a very large number of outcomes. A farmer in

[1]Frank H. Knight, *Risk, Uncertainty and Profit* (Boston: Houghton Mifflin, 1921).

Knight's risk situation with respect to crop yields for alternative fertilizer levels would know the probability distribution of the random variable, crop yield, for each fertilizer level that could be applied. For example, the farmer in the risk situation would know that the probability of corn yields between 111 and 120 bushels is 0.2, that yield in the 121 to 130 bushels range is 0.15, and so on for other yield levels. According to Knight's definitions, uncertainty exists when the decision-maker has less information about the alternative outcomes and their probability of occurrence. Under uncertainty the farmer does not know the probability of alternative outcomes, such as the probability of alternative corn yields above. Furthermore, the decision-maker may or may not know the alternative outcomes that can occur.

Modern decision theory recognizes that the objective probabilities required by traditional analysis for risk seldom, if ever, exist. Games of chance involving the use of coins, dice, cards, roulette wheels, and other devices played under carefully controlled conditions may result in relative frequencies closely approximating the objective probabilities. However, business decisions are made a limited number of times, and the environment is different each time the decision is made. It is sometimes argued that historical yield data, or at least weather data, can be used to provide objective probabilities of outcomes that are relevant for agricultural production decisions. However, the historical yield data have been generated over years with differing soil moisture and temperature conditions at planting time, different varieties (or hybrids), different cultural practices, and changes in other factors that make the historical yields irrelevant for use as relative frequencies in the current year. Even the use of historical data on rainfall amounts and other weather variables requires the presumption that the system generating the rainfall has not changed over the time and that the selection of the number of years of data to use in calculating the frequencies is long enough to result in constant frequencies for longer periods. The frequencies for virtually any weather variable, such as rainfall, will be somewhat different if one uses 10, 20, 30, or some other number of years of data to estimate the frequencies. While it may be very appropriate to consider such data in forming personal beliefs about the likelihood of alternative outcomes, the decision to use such data involves subjective judgment.

Modern decision theory is based on the use of personal probabilities. Personal or subjective probabilities represent the degree of belief or the strength of conviction that an individual has about the occurrence of an outcome. To be useful, subjective probabilities must be consistent with the axioms and rules of probability. They also must be consistent with the decision-maker's degree of belief. As the term "personal probability" implies, the individual making the decision should develop his/her own subjective probabilities. The decision-maker may use historical data, expert advice, and other data in forming personal probabilities. A decision-maker's degree of belief about a certain occurrence frequently changes as the individual's experience and knowledge expand. Two individuals, such as two farmers operating adjacent farms, may hold different degrees of belief about some event, such as the price of a crop at a future date. However, the degree of belief of two or more individuals usually corresponds more closely as their accumulation of common experience increases.

Modern discussions of agricultural decision-making, like modern discussions of decision theory, do not distinguish between the terms "risk" and "uncertainty." The two words are used interchangeably throughout this book. By either risk or uncertainty, we mean that the action a decision-maker selects has alterative outcomes. The decision-maker may not know the full range of alternative outcomes, but the approach presented presumes the available data are used to identify possible outcomes and estimate personal probabilities of their occurrence. The approach described can be used quite widely in decision-making for nonfarm and farm decisions alike. Within the farm business it can be used for decision-making for planning, implementation, and control. The applications presented here are to planning problems, but other applications will be noted in later chapt' rs.

TYPES OF RISK OR UNCERTAINTY

The types of uncertainty that must be considered depend in part on the outcome of interest. The possible outcomes we might analyze include physical production (such as yield per acre or milk production), gross returns, the cash difference between cash receipts and cash outflow, net farm income (or farm profit or loss), and addition to retained earnings. Since our emphasis is on the profitability and financial success of the business, it is appropriate to evaluate the uncertainty of the contribution which enterprises make to the overhead expenses of the business. The contribution for an enterprise can be calculated as the gross receipts less the variable cash costs, or the gross margin. An analysis of uncertainty for the total farm business should also consider the payment of cash overhead expenses and financing costs. The farm profit or loss (net farm income) on the projected income statement (Figure 6.10, line 12) is an appropriate measure of this outcome for the total farm business.[2]

With the emphasis focused on financial outcomes, the risks farmers face can be divided into two broad types: business and financial.

---

[2]If the amount of income tax paid is expected to vary widely for alternative outcomes, farmers may prefer a measure such as farm profit or loss (net farm income) less income taxes. The worksheet to estimate income taxes illustrated in Chapter 8 can be used to convert the net farm income figures developed here to an after-tax basis.

Business Risk

Business risk or uncertainty is commonly defined as the inherent uncertainty in the firm independent of the way it is financed. Thus, business risk includes those sources that would be present with 100 percent equity financing. The major sources in any production period are price and production uncertainty, although a number of other sources may affect the price and production variation, particularly over a period of time.

Price Risk   All of the factors leading to unpredictable shifts in the supply and demand of inputs and products are sources of price uncertainty. Movements of a seasonal, cyclical, and trend nature are predictable to some extent, but the inability of the operator to predict these prices accurately in making decisions represents a source of business risk. Many governmental actions including trade agreements, embargoes, and fiscal and monetary policy affect the uncertainty of prices.

Production Risk   The variation in the production level resulting from factors beyond the manager's control—including weather, pests, genetic variation, changes in regulations on use of pesticides and feed additives, and timing of production practices—represents a major source of business risk. Production risk is reflected in the variability of yields per acre, weaning weights, rate of gain, death loss, and other variables used to measure the amount of physical production.

Financial Risk

Financial risk or uncertainty is defined as the added variability of net returns to owner's equity that results from the financial obligation associated with debt financing. This risk results from the concept of leverage discussed in Chapter 8; leverage multiplies the potential financial return or loss that will be generated with different production and price levels. Furthermore, there are other risks inherent in using credit. Uncertainty associated with the cost and availability of credit is reflected partly in the interest rates for loans and partly through nonprice sources. Uncertain interest rates reflect uncertain financial market conditions and represent an uncertain input price—the price of debt capital. The nonprice sources, a type of institutional uncertainty, include differing loan limits, security requirements, and maturities, depending on the availability of loan funds over time. Thus, financial risk also includes uncertain interest rates and uncertain loan availability.

These three sources of uncertainty parallel the three areas of farm management described in Chapter 1—production, marketing, and finance. We emphasize production and market (or price) risk in the examples in this chapter. The uncertainty of interest rates, which is only part of financial risk, is considered in the application to whole-farm planning. This illustrates that the same methods of analysis and decision rules can be used to consider the combined effect of price, production, and financial risk in whole-farm analyses.

RISK ATTITUDES

Much of the discussion in this chapter refers to alternative attitudes toward bearing risk. Here we define the terms used to describe risk attitudes and provide a method that an individual can use to evaluate his/her risk attitude for alternative sizes and probabilities of gains and losses.

The type of risk attitude is not a reflection on the individual's management ability. There are no right or wrong risk attitudes, and this discussion does not suggest that one attitude is more appropriate than another. The emphasis is placed on knowing one's risk attitudes and using them in decision-making. Experience suggests that an improved understanding of risk attitudes can help in analyzing investments as well as in making day-to-day management decisions.

Three Categories

Risk attitudes can be divided into three types: risk averse, risk preferring, and risk neutral. It is common to place an individual in one of these categories, although as we will see below, any one individual may not be in the same category for all decisions. Risk averters or avoiders are characterized as more cautious individuals with preferences for less risky sources of income and investment. In general, they will sacrifice some amount of expected income to reduce the probability of low income and losses. Risk preferrers or takers are more adventuresome with a preference for more risky business alternatives. In comparing decisions with similar expected values, risk preferrers select the alternative with some probability of a higher outcome, even though they must also accept some probability of a lower outcome. Risk neutral is the limiting case between risk averse and risk preferring. The risk-neutral individual chooses the decision with the highest expected return, regardless of the probabilities associated with alternative levels of gain and loss.

Attitudes toward risk vary depending on the individual's objectives and financial resources. Some farmers are willing to accept more risk than others. They apparently experience little stress in making and living with business decisions having the probability of large financial losses, while other farmers with the same financial resources making the same decisions experience sleepless nights and other forms of anxiety. A simple way to characterize these differences is to recognize that farmers have different objectives. We will present procedures that can be used by farmers with alternative risk attitudes, rather than discuss the causes of this difference.

Risk attitudes also are logically related to the financial ability of the individual to accept a small gain or loss. A farmer with a net worth of $200,000 may certainly be adversely affected by a $30,000 loss in one year, but is unlikely to be forced out of business by this one unfavorable outcome. However, his or her neighbor with a net worth of less than $50,000 may be forced out of business by a similar loss. In general, farmers and other decision-makers tend to exhibit more risk-averse behavior when the financial consequences of an unfavorable outcome are very severe.

Given both the individual's objectives and financial situation, the farmer's attitude about risk also may vary depending on the probabilities and size of gains and losses. An individual may react as a risk preferrer or as being risk neutral when considering a decision with relatively minor financial consequences. However, the same individual may be risk averse when considering a decision with more severe consequences of either a personal or a financial nature.

The Certainty Equivalent and the Risk Premium

The three risk attitudes can be defined more precisely by introducing some additional concepts. Assume a decision-maker has been offered the opportunity to participate in a venture that will pay either $14 or minus $10. The outcome is to be determined by the flip of a coin, making the probability 0.5 for either event. Suppose the decision-maker is asked to specify the amount of money for certain that makes him/her indifferent between the amount for certain and participation in the venture. This amount for certain is referred to as the certainty equivalent.

Suppose we ask three individuals, A, B, and C, to specify the certainty equivalent for this venture and they respond with values of $1, $2, and $3, respectively. We can compare the certainty equivalent (CE) with the expected payoff or expected monetary value (EMV) of the venture. The expected monetary value is the probability weighted sum of the possible monetary outcomes. In this venture the EMV is 0.5 ($14) + 0.5 (−$10) = $7 + (−$5) = $2. The EMV minus the CE is called the risk premium. The risk premium for the first decision-maker, individual A, is the EMV ($2) minus the CE ($1) or $1. This indicates that individual A must receive a positive premium to accept risk. Individual A and all individuals requiring a positive risk premium are categorized as risk averse. On the other hand, individual C has a negative risk premium (the EMV − CE = $2 − $3 = −$1), indicating C is willing to accept risk for the chance of receiving a higher payoff. This individual and all other individuals willing to accept a negative risk premium are risk preferring. Individual B has a zero risk premium (the EMV − CE = $2 − $2 = 0) and is risk neutral. Obtaining data on the certainty equivalent and calculating the risk premium is a relatively simple method of determining a decision-maker's risk attitude considering the probabilities and sizes of gains and losses.

The relative size of the risk premium and even the classification as risk averse, risk neutral, or risk preferring may change depending on the size of the gains and losses. For example, suppose individual C, who has a certainty equivalent of $3 when the venture has a probability of 0.5 of $14 and 0.5 of minus $10, indicates a certainty equivalent of $200 for a venture with a probability of 0.5 of $1,400 and 0.5 of minus $1,000. The certainty equivalent is equal to the expected monetary value for the second venture, indicating risk neutrality. The individual's financial situation is expected to be important in determining the response as the gains and losses are increased. Decision-makers with larger financial reserves may continue to respond with approximately the same relative risk premium (such as $1 for the first venture and $100 for the second venture above), but many decision-makers will require a larger risk premium as the potential loss increases. Suppose individual C has a certainty equivalent of −$1,000 for a venture with a probability of 0.5 of $14,000 and 0.5 of −$10,000. In this case the risk premium is $3,000 [EMV − CE = $2,000 − (−$1,000)] and clearly indicates risk aversion. Of course, there is no reason the risk premium must decline as the size of the gains and losses increases. One can imagine that a wealthy individual might become more interested as the possible gains and losses from the venture become large enough to make some difference to the individual. Such an individual might shift from risk neutrality at low levels of gains and losses to risk preferring as the gains and losses increase.

Risk attitudes may also change over time. As we discussed in Chapters 1 and 6, it is natural for managers to change their goals over the life cycle of the farm firm. They may adjust their risk attitude in ways that the manager feels will help achieve goals, such as firm survival, providing a desired standard of living, growth of the firm, and providing funds for retirement. The risk attitude of individuals may change as their financial situation changes over time. An individual who becomes less risk averse for gains and losses of a specified size as net worth increases may become more risk averse following a year of large losses and resulting declines in net worth.

The changes in risk attitudes for alternative probabilities and sizes of gains and losses, the change in risk attitudes over time, and the change due to the changing financial situation make it difficult to predict how an individual will respond to risky decisions over time. However, it is clear that knowing one's attitudes about accepting uncertain gains and losses can aid in making management decisions.

Consideration of the difference in farmers' response to risk can often explain the difference in the decisions made. Suppose two farmers approximately the same age, farming similar farms, and having approximately the same financial structure are bidding on a cash lease to rent an additional 80 acres of land that is adjacent to both of their farms. Suppose one farmer bids $6,400 and the other $8,000 per year as the cash rent payment. While many explanations of the difference in bids could be hypothesized, one is that the two individuals appraised the risk in different ways. This could have occurred because the individual making the lower bid was more risk averse and required a larger risk premium. However, it is possible that the two farmers' appraisal of the probabilities and sizes of gains and losses from renting the land were quite different. The farmers could have quite similar attitudes for the same probabilities and sizes of gains and losses, but the farmer making the higher bid could have based his or her decision on more favorable probabilities and sizes of gains and losses. This emphasizes the importance of not only knowing one's risk attitudes, but also developing estimates of the consequences carefully. Developing such estimates is the topic emphasized in the remainder of this chapter.

COMPONENTS OF A RISKY DECISION PROBLEM

The enterprise budget for corn shown in Table 6.3 is based on an application of 144 pounds of nitrogen. This includes 10 pounds of N as starter fertilizer and 134 pounds applied prior to planting as anhydrous ammonia. Andy has data on the response of corn on his soils for four alternative levels of N (90 pounds, 110 pounds, 144 pounds, and 160 pounds) and alternative weather conditions. He decides to reconsider the amount of nitrogen to use given uncertain weather and corn prices.

The decision problem has several components. The alternative actions available are to apply: (1) 90 pounds of N, (2) 110 pounds of N, (3) 144 pounds of N, and (4) 160 pounds of N. The *events* or *states* that can occur are alternative weather conditions and corn price levels. A review of the yield data indicates that the production uncertainty can be represented by considering three events: (1) favorable, (2) normal or typical, and (3) unfavorable. The yields in bushels per acre for each of Andy's four actions and the three weather events are given in Table 11.1. Andy also decides to consider corn prices ranging from $2.25 to $3.25 per bushel by $0.25 intervals. The *payoffs, outcomes,* or *consequences* are the gross margins per acre (gross return minus variable costs). The element of chance or uncertainty is represented by the probability Andy estimates for each of the events or states. Andy is willing to develop subjective *prior probabilities* based on his knowledge and experience expressing his degree of belief that each event or state will occur. He is also interested in combining price *predictions* made available by his advisor with his estimates to form revised or *posterior probabilities* about the degree of belief of each event or state. Finally, Andy wants to consider alternative *choice criteria* that might be used in selecting a course of action. Thus, this decision problem under uncertainty, like almost any problem of decision-making under uncertainty, may have as many as seven components: (1) actions representing the choices available to the decision-maker, (2) events or states representing alternative levels of variables, such as weather and prices in this example, that the decision-

Table 11.1 Corn Yields in Bushels per Acre for Four Levels
of Fertilizer and Three Weather Events on Andy's Farm

Weather Events θ_i	Actions			
	a_1 90 lb N	a_2 110 lb N	a_3 144 lb N	a_4 160 lb N
Favorable	90	115	140	145
Normal	85	110	120	120
Unfavorable	75	90	90	90

maker cannot control, (3) payoffs or consequences, (4) prior probabilities, (5) predictions obtained from consultants, experts, or through the use of some additional predictive device, (6) posterior probabilities which combine the prior probabilities and the data on their accuracy of the prediction, and (7) choice criteria used to select a course of action. The shorthand notation used to refer to the components is a_j for the jth action, θ_i for the ith event, O_{ij} as the outcome for the ith event and the jth action, and $P(\theta_i)$ for the prior probability of θ_i.

The first task is to describe Andy's problem showing the possible courses of action, events, and the consequences for each action-event combination. Two methods are commonly used to diagram the decision problem—the decision tree and the payoff matrix. There are three steps in building either a decision tree or a payoff matrix: (1) list the alternative actions that are relevant, (2) list the possible events that could occur, and (3) budget the payoff for each action-event combination and enter it in the appropriate space.

We will simplify the presentation initially by considering only uncertain yield. This treats corn price as known and reduces the number of events that must be considered. The decision tree for Andy's fertilizer problem is shown in Figure 11.1. The diagram has one branch for each of Andy's four actions. The three alternative weather events are shown for each of Andy's actions. Andy uses historical weather and yield data to estimate the prior probabilities of the alternative events. The prior probabilities, 0.3 for favorable, 0.5 for normal, and 0.2 for unfavorable, are listed on each event branch.

Andy uses the data in Table 6.3 to budget the variable costs for each action-event combination for the coming year. To do this he must decide which items of income and variable cost will change with alternative actions and events. Andy decides to budget the gross receipts for each action-event combination using the yields in Table 11.1 and a price of $3.00 per bushel. The major cost items that will change with the action selected are the cost of fertilizer and the amount of insurance purchased on the crop. The amount of drying and handling cost depends on the yield and will differ for each

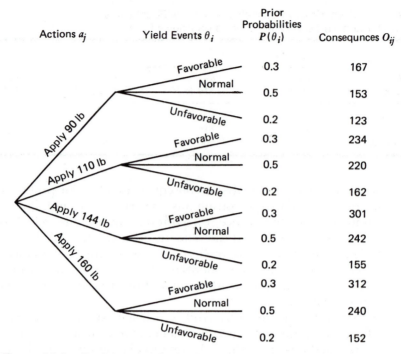

Figure 11.1. Decision tree for fertilizer decision considering uncertain yield.

action-event combination. The interest on the operating capital depends on the money required for the other variable inputs. He feels other costs will be essentially constant for each of the action-event combinations. If we use the data in Table 6.3, the gross margin for 144 pounds of N and normal weather is \$242.44 (\$3.00·120 bu − \$117.56 variable costs) and is rounded off to the nearest dollar before it is listed in Figure 11.1. The consequence for a 90 pound application of N and unfavorable weather conditions was estimated as \$123.46. Gross receipts are \$3.00 per bushel times 75 bushels. Fertilizer costs in Table 6.3 are reduced by \$8.64 (54 lb N at \$0.16). Drying and handling costs are reduced by \$4.05 (45 bu at \$0.09), and insurance based on the gross receipts per acre with normal weather is reduced \$2.62. The reduction in operating capital required lowers the interest on operating capital \$0.71 per acre. Thus, Andy's estimate of the gross margin for a 90 pound application of N and unfavorable weather conditions is \$123.46. This value is rounded to the nearest dollar and is listed as the consequence in Figure 11.1. Other consequences were estimated in the same manner.

An alternative method of displaying the data is with the use of a payoff matrix. The payoff matrix for Andy's fertilizer problem is shown in

Table 11.2 Payoff Matrix for Fertilizer Decision Considering
Uncertain Weather

| | | *Per Acre Returns Above Variable Cash Costs* | | | |
| | | *Actions (Amount of N Applied per Acre)* | | | |
Weather Events θ_i	*Prior Probabilities* $P(\theta_i)$	*a_1* *90 lb*	*a_2* *110 lb*	*a_3* *144 lb*	*a_4* *160 lb*
Favorable	0.3	$167	$234	$301	$312
Normal	0.5	153	220	242	240
Unfavorable	0.2	123	162	155	152
Expected value		151.20	212.60	242.30	244.00

Table 11.2. The decision-makers' actions are usually listed across the top of
the table, and events or states are typically listed along the left side of the
table. The prior probabilities are shown to the right of each event, and the
consequence for each action-event combination is recorded in the appropriate cell.

The decision tree and payoff matrix are two alternative ways of depicting actions, events, probabilities of the alternative events, and the consequences of alternative action-event combinations. Some individuals prefer one method of presenting the data and some another. The important
point to note here is that either method can be used to summarize the same
data.

In this example, the action with the highest payoff differs for the
three events. The largest gross margin under unfavorable weather condition ($162) results from applying 110 pounds of N per acre, whereas the
largest consequence for normal and favorable weather conditions occurs
from applying 144 and 160 pounds of N, respectively. Thus, the choice of
the appropriate level of N to apply is not obvious. The expected value using
Andy's prior probabilities $P(\theta_i)$ for each event-action combination O_{ij} is
calculated as

$$E(O_j) = \sum_{i=1}^{m} P(\theta_i)O_{ij} \qquad (11.1)$$

For example, the expected value for action a_3 is $E(O_3) = (0.3)(301) +$
$(0.5)(242) + (0.2)(155) = \$90.30 + \$121.00 + \$31.00 = \$242.30$. The expected value in this example is the largest for action a_4. However, Andy
notes that the gross margin with a_3 exceeds the gross margin for a_4 for both
unfavorable and normal years, and that the gross margin in favorable years
is only $11 greater for a_4 than a_3. Thus, he decides to consider uncertain

corn price levels and some alternative decision criterion before making a decision on the level of fertilizer to select.

This simple example has illustrated several components of the decision problem under uncertainty: actions, events, consequences, and prior probabilities. We will develop a more complete example of these components for Andy's fertilizer decision before introducing the use of alternative choice criteria, a fifth component. The final two components, predictions and the development of revised or posterior probabilities, are incorporated into the fertilizer decision problem in the Appendix to this chapter.

APPLICATION TO ENTERPRISE DECISIONS

We begin the development of a more realistic analysis of Andy's problem by discussing some important considerations in developing actions, events, probabilities, and payoffs. Then a more complete payoff matrix will be developed for Andy's problem.

Specifying Alternative Actions

A large number of alternative actions can be specified for consideration in most decision problems. The time required for analysis can be reduced by including only the more relevant alternatives, but it is important to remember that only the actions included in the payoff table can be selected. The suggested approach is to identify a limited number of actions for consideration that appear to include the most relevant alternatives available. To be relevant, the actions must contribute to the decision-maker's goals and be consistent with the resources available. For example, if one of Andy's goals is to make more money, there is no reason to include additional actions like a_1 in Table 11.2 which has a lower return for all events. Furthermore, it is important to determine that the actions included are feasible with the amount of land, labor, capital, and mangerial capacity available. Finally, it is useful to consider a wide range of alternative actions which may involve shifting several factors under the manager's control. For instance, Andy should consider shifting the level of other inputs, such as row spacing, plant population, and other inputs, when adjustments in nitrogen are made. Andy considers these points and concludes that the actions listed in Table 11.2 are appropriate.

Identifying the Events

It is desirable to limit the large number of events that can be specified for most problems to reduce the complexity of the analysis. The first consideration is to include only the most important sources of uncertainty in the analysis. The most important sources are those with a large potential impact on the payoff and a significant chance of very favorable and/or unfavorable outcomes occurring. Given these considerations, Andy decides to consider production uncertainty (resulting from uncertain weather and pests) and product price uncertainty. These are the two he considers to be the most important, although he recognizes that the uncertainty of input prices and many other factors has been ignored. It should be noted that Andy has insured his crop against loss from hail and storm losses. He includes a cost for the insurance and ignores the risk of loss from those sources in the analysis.

After the important sources of uncertainty have been selected, it is necessary to identify the events to be considered in the analysis. The events must be mutually exclusive and collectively exhaustive. By mutually exclusive we mean that no more than one event can occur during this period. Collectively exhaustive means that one of the events must occur, i.e., no possible events have been omitted.

Andy selects three weather events which he characterizes as favorable, normal, and unfavorable. He feels the price of corn will be in the range of $2.10 to $3.40 per bushel. He selects five levels of corn price—$2.25, $2.50, $2.75, $3.00, and $3.25—to represent this range. Because the yield of corn is based on local weather conditions and the price of corn is dependent largely on international supply and demand considerations, Andy feels it is possible for each corn price to occur with each of the three yield events. This results in the fifteen events shown in the payoff matrix in Table 11.3. Specifying a larger number of yield and price events might contribute to

Table 11.3 Payoff Matrix for Fertilizer Decision Considering Both Uncertain Yield and Uncertain Price

Events			Actions			
		Prior	a_1	a_2	a_3	a_4
Yield	Price	Probability	90 lb	110 lb	144 lb	160 lb
Favorable	$3.25	0.03	$190	$263	$336	$349
	3.00	0.06	167	234	301	312
	2.75	0.09	145	206	266	276
	2.50	0.06	122	177	231	240
	2.25	0.06	100	148	196	204
Normal	3.25	0.05	174	247	272	270
	3.00	0.10	153	220	242	240
	2.75	0.15	131	192	212	210
	2.50	0.10	110	165	182	180
	2.25	0.10	89	137	152	150
Unfavorable	3.25	0.02	142	184	178	175
	3.00	0.04	123	162	155	152
	2.75	0.06	105	139	133	130
	2.50	0.04	86	117	110	107
	2.25	0.04	67	94	88	85

more precise analysis, but it would also add to the computational burden. Likewise, additional sources of uncertainty, such as input costs, can be included by specifying a mutually exclusive and exhaustive set of events for the additional source of uncertainty and adding to the list of events. For example, considering three levels of input prices for each of the yield and corn price levels in Table 11.1 would increase the total number of events to 15 times 3 or 45.

Estimating Subjective Probabilities

As we noted earlier, subjective probabilities represent the degree of belief or the strength of conviction that an individual has about the occurrence of an outcome. To be useful in decision-making, these personal or subjective probabilities must both follow the rules of probability and be consistent with the decision-maker's degree of belief. This section briefly reviews the rules of probability and one simple method of eliciting subjective probabilities.

Several basic axioms or rules of probabilities should be remembered. First, probabilities must be within the range 0 through 1. An event with a probability of 0 is certain not to occur, while an event with a probability of 1 is certain to occur, making these logical limits for the numerical value of probability. Second, the probability that two or more mutually exclusive events will occur is the sum of their respective probabilities. For example, the probability of either unfavorable or normal yield in Table 11.2 is 0.2 + 0.5 or 0.7. Third, the probability of the mutually exclusive and collectively exhaustive events must equal 1. This simply states that if we have listed all possible events, then the outcome must fall within one of the event categories. If θ_i, $i = 1, \ldots, m$, denotes the ith mutually exclusive event of a collectively exhaustive set of m events, the above three rules can be stated as:

$$0 \leq P(\theta_i) \leq 1 \qquad (11.2)$$

$$P(\theta_i \text{ or } \theta_j) = P(\theta_i) + P(\theta_j) \qquad (11.3)$$

$$\sum_{i=1}^{m} P(\theta_i) = 1 \qquad (11.4)$$

These three rules must be followed in developing probabilities for each source of uncertainty. Many methods of obtaining subjective probabilities have been proposed. Consider one method that has been used frequently with agricultural producers.[3]

[3]See R. L. Winkler, "The Assessment of Prior Distribution in Bayesian Analysis," *Journal of the American Statistical Association* 62, No. 320 (1967): 776–800 and "Probabilistic Prediction: Some Experimental Results," *Journal of the American Statistical Association* 66, No. 336

(continued)

Table 11.4 Assessing Subjective Probabilities
with Conviction Weights

Corn Price in $ per Bu		Andy's Conviction That the Price Will Be In the Given Range (0 through 100)	Subjective Probabilities
Range	Midpoint		
Less than 1.88		0	0.00
1.88–2.12	2.00	0	0.00
2.13–2.37	2.25	67	0.20
2.38–2.62	2.50	67	0.20
2.63–2.87	2.75	100	0.30
2.88–3.12	3.00	67	0.20
3.13–3.37	3.25	33	0.10
3.38–3.62	3.50	0	0.00
More than 3.62		0	0.00
		334	1.00

The method of assigning conviction weights can be used either to estimate one's own subjective probabilities or to elicit the subjective probabilities of another individual about some uncertainty. The procedure is illustrated with data on corn prices for the fertilizer example, but it can be used to elicit subjective probabilities of yields, changes in government policies, and other uncertainties as well.

The first step in estimating subjective probabilities is to specify a range of outcomes for the uncertainty (the price of corn in our example) and divide the range into intervals. Experience suggests that the total range should be sufficiently wide so that the decision-maker is not limited in assigning strength of conviction weights to each interval. Then the total range can be divided into discrete intervals in a manner that is meaningful for the decision-maker. For example, if a wheat producer thinks about wheat yields in 5 bushel intervals, it will be meaningful to develop discrete ranges with 5, 10, 15, etc., bushels as the midpoint. If the same producer considers prices in $0.25 intervals, it will be meaningful to specify $0.25 intervals centered on $3.50, $3.75, $4.00, etc. Then the decision-maker should understand that a price of $3.50 really means a range from $3.38 through $3.62. For the example, Andy specifies the $0.25 price intervals for the price of corn shown in the left-hand column of Table 11.4. He also lists the midpoint of each range in the next column.

(1971):675–85 for a discussion and appraisal of alternative methods of estimating subjective probabilities. A more complete discussion of the techniques used to elicit subjective probabilities for agricultural producers is provided by A. Gene Nelson, George L. Casler, and Odell L. Walker, *Making Farm Decisions in a Risky World: A Guidebook* (Oregon State University Extension Service, 1978), Chapter 4, pp. 5–13; and Jock R. Anderson, John L. Dillon, and Brian Hardaker, *Agricultural Decision Analysis* (Ames: Iowa State University Press, 1977), pp. 19–26.

Then the decision-maker must assign a number 0 through 100 to each range indicating his/her belief that the uncertain event will fall in the range. An appropriate starting point is to select the range the individual feels has the greatest probability of containing the actual event and assigning a numerical value of 100 to it. The decision-maker than assigns a value of 0 through 100 to each of the other ranges to indicate the strength of conviction (relative to the most likely range) the actual event will fall in that range. In Table 11.4, Andy feels that a corn price in the range of $2.63 to $2.87 has the highest probability. He assigns a conviction weight of 100 to this range and smaller weights to the others. Notice that zeros were assigned to several intervals, indicating Andy does not feel the price of corn will fall in these ranges. The conviction weight in each row is divided by the sum of the conviction weight to obtain the subjective probability for each $0.25 range.

Notice that the subjective probabilities estimated satisfy the three rules of probability. If Andy has developed them carefully, they should also reflect his degree of belief that corn prices will fall in each of the ranges. Development of subjective probabilities that reflect the decision-makers' degree of belief requires people to be honest with themselves. If they discover they have made mistakes, they must correct them.

The probabilities estimated can be graphed in alternative ways. One method is to graph the discrete distribution as a histogram as shown in Figure 11.2. This is particularly useful for the decision-maker to appraise if the resulting shape of the distribution is consistent with his/her beliefs. The probabilities can also be graphed as a cumulative distribution function showing the probability of receiving a corn price less than or equal to a specified level. The cumulative distribution is particularly useful if one wants to determine the probability of prices for ranges other than those estimated. The probability estimates for each interval in Figure 11.2 are summed across intervals from left to right to calculate the cumulative prob-

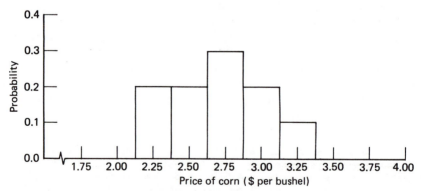

Figure 11.2. Andy's subjective probability distribution for price of corn.

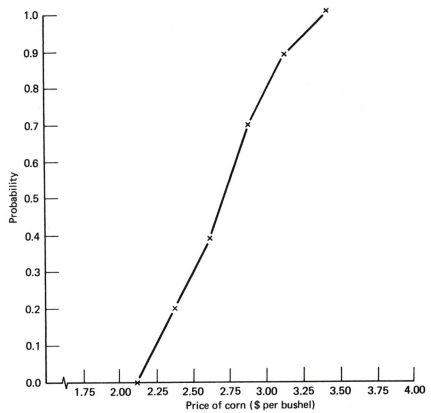

Figure 11.3. Andy's subjective cumulative distribution function for price of corn.

ability at the upper extreme of each interval and plotted in Figure 11.3. The resulting points can either be connected with straight-line segments as illustrated, or a free-hand curve can be drawn through the points to approximate a continuous cumulative distribution function.

To this point we have been discussing what might loosely be called the unconditional probabilities of one event. In actuality, the elicited price distribution above is conditional on the data and information Andy has available. This means his degree of conviction weights for the alternative intervals may change as more information becomes available, requiring reassessment. The subjective probabilities elicited are unconditional in the sense that we have not asked Andy to assign degree of conviction weights given some additional information, such as the size of the national corn crop, the price of soybeans, or other information that might have a specific effect on his assessment. The same procedure described above can be used to elicit subjective probabilities conditional on other information, but doing so for each of several levels of the conditional event is often a more difficult task than providing the responses for unconditional elicitation.

Decision-making often requires specification of joint probabilities for two or more events, such as yield level and price in the fertilizer example. Thus, it is important to review the rules of probability for joint events and to discuss how they can be applied to decision-making at the applied level. Given two possible events A and B, the rule of calculating the probability of the joint event A and B, written as $P(A,B)$, is

$$P(A,B) = P(B)\ P(A|B) \tag{11.5}$$

Here we are discussing the probability of A *and* B, that is the probability of both events occurring. (Earlier we discussed the probability of one *or* the other and added the probabilities.) This statement indicates that the probability of A and B occurring is equal to the probability of B multiplied by the conditional probability of A given that B has occurred. Since A and B are not specified, we can also write

$$P(A,B) = P(A)\ P(B|A) \tag{11.6}$$

To make this more operational, suppose there are three levels for A (say soybean yields) and three levels for B (say corn yield). Then attaching subscripts 1, 2, and 3 to each letter to reflect the yield level, we can write

$$P(A_1,B_1) = P(A_1)\ P(B_1|A_1)$$
$$P(A_1,B_2) = P(A_1)\ P(B_2|A_1)$$
$$\cdot$$
$$\cdot$$
$$\cdot$$
$$P(A_3,B_3) = P(A_3)\ P(B_3|A_3) \tag{11.7}$$

to indicate how the nine joint probabilities can be calculated.

Two special or limiting cases are worth noting. In some applications, the events may be independent, indicating the outcome of A does not affect the outcome of B, or alternatively, the factors affecting the outcome of A are largely independent of the factors affecting the outcome of B. Farmers may be willing to assume that the yield of a crop on a specific farm is independent of its price, when the crop, such as corn or wheat, is grown over a wide geographic area. In this case, the conditional probabilities are the same as the unconditional probabilities. We have

$$P(A|B) = P(A) \text{ and } P(B|A) = P(B) \tag{11.8}$$

and we can write

$$P(A,B) = P(A) \cdot P(B) \qquad\qquad \textbf{(11.9)}$$

A second limiting case occurs when the outcome of A is perfectly correlated with (or is directly related to) the outcome of B. In this case, the conditional probability is 1, making the joint probability equal to the unconditional probability for either A or B. Whether this situation exists in actual decision-making is questionable. However, given the difficulty of assessing conditional probabilities for many applications, producers may be willing to assume this is the case when they feel the conditional probabilities are near 1.[4]

Let us return to the fertilizer problem. Andy is now in a position to complete the more detailed payoff matrix. He feels it is reasonable to assume that the price he receives for corn in January is independent of the conditions affecting the yield on his farm. Thus, he calculates the joint probability of yield and price by multiplying the appropriate unconditional probabilities as indicated in equation 11.9. For example, the joint probability of favorable yield and a corn price of $2.50 is $(0.3) \cdot (0.2) = 0.06$. The joint probabilities are given in Table 11.5. They are also listed as the prior probabilities for the joint yield-price states in the payoff matrix presented in Table 11.3. Notice that the joint probabilities derived comply with the three basic rules of probability. After the probabilities were filled in, the budgeting procedures described earlier were used to estimate the gross margin for each action and joint yield-price event to complete Table 11.3.

---

[4]The assessment of joint probabilities for three or more events is frequently of interest in agricultural decision-making, particularly in assessing the uncertainty of net farm income or the addition to retained earnings for a whole-farm plan. The rules to estimate the joint probabilities for three or more events follow from the sequential application of the rule to calculate joint probabilities for two events. For three events A, B, and C we have:

$$\begin{aligned} P(A,B,C) &= P(A,B)\ P(C|A,B) \\ &= P(A)\ P(B|A)\ P(C|A,B) \end{aligned} \qquad\qquad \textbf{(11.10)}$$

For four events A, B, C, and D we have:

$$\begin{aligned} P(A,B,C,D) &= P(A,B,C)\ P(D|A,B,C) \\ &= P(A)\ P(B|A)\ P(C|A,B)\ P(D|A,B,C) \end{aligned} \qquad\qquad \textbf{(11.11)}$$

Empirical assessment of conditional probabiliites for one event given two others, $P(C|A,B)$, and one event given three others $P(D|A,B,C)$ is usually too time consuming for routine application by agricultural producers. Except in those rare cases where the extension service or some other advisory service has prepared appropriate estimates of the conditional probabilities, producers considering the combined uncertainty of several events, such as yields, prices, and financing costs for a whole-farm plan, will need to prepare joint probabilities directly. We will return to this topic in the section on whole-farm applications.

Table 11.5 Joint Probabilities of Independent Yield
and Price Events

		Price Events				
		$3.25	$3.00	$2.75	$2.50	$2.25
Yield Events	*P(θ)*	*0.1*	*0.2*	*0.3*	*0.2*	*0.2*
Favorable	0.3	0.03	0.06	0.09	0.06	0.06
Normal	0.5	0.05	0.10	0.15	0.10	0.10
Unfavorable	0.2	0.02	0.04	0.06	0.04	0.04

Decision Rules

Previous sections of this chapter have discussed considerations in estimating the sizes and probabilities of possible gains and losses for decision problems under uncertainty. This section considers some alternative criteria that a farmer can use in selecting a course of action based on the data summarized in the payoff matrix.

Decision theory suggests that maximization of expected utility or satisfaction is the appropriate criterion. This criterion is indeed attractive when a utility function describing the satisfaction an individual derives from alternative sizes and probabilities of possible gains and losses is available. Unfortunately, our ability to estimate such utility functions is limited. An equivalent approach is to have the decision-maker assign a certainty equivalent to each action. The action with the highest certainty equivalent is the preferred choice. For example, if Andy can assign a certainty equivalent to each action in Table 11.3, then selection of the action with the highest certainty equivalent is the decision that maximizes expected utility. Experience indicates that it is very difficult to assign certainty equivalents to actions with more than a few possible outcomes, particularly when the events have a wide range in probability values. Thus, many authors have suggested alternative rules that decision-makers might use.

A number of rules have been advanced which rank alternative actions. A decision-maker must decide to what extent each decision criterion is appropriate given his/her goals, the sizes and probabilities of possible gains and losses, and his/her financial ability to bear risk. We consider three groups of rules. The first group does not require probability estimates. These criteria may be of interest when decision-makers feel particularly uncomfortable about estimating the probabilities of alternative outcomes. The second group requires probability estimates and results in selection of a specific action. While the first and second group can be referred to as decision criteria because they result in the selection of a specific action, the third group can be referred to as efficiency criteria. Efficiency criteria simply sort those actions that should be considered from those that should not be for specific groups of decision-makers. The efficiency criteria eliminate those actions that are dominated by other actions being considered. The decision-maker is then left with a smaller set of actions to consider in making the decision. Although efficiency criteria may not result in a definite choice, they may be preferred by decision-makers who feel the assumptions of the decision criteria are too restrictive to represent their ordering of risky choices.

Decision Criteria That Do Not Require Probability Estimates   Three decision criteria that do not require information on the probability distribution of alternative events are mentioned.[5] Situations where the manager has no information on the probability of alternative events may be rather rare, but these criteria have been advanced by various scholars of decision-making and provide a basis of comparison for criteria that utilize probability estimates.

Maximin   Abraham Wald suggests that a decision-maker examine the worst outcome for each action and then select the action that maximizes the minimum gain. The minimum monetary gain from each fertilizer application in Table 11.3 is recorded in the first row of Table 11.6. The minimum value for each action occurs with the lowest price and unfavorable yield. Following the Wald criterion, one would select action a_2 (and apply 110 pounds of N per acre) because it has the largest minimum outcome.

  This is a very conservative or pessimistic approach in that it only considers the worst outcome. It does not consider the higher income that might be achieved with actions a_3 and a_4 under normal and favorable weather conditions. However, proponents note that it may be appropriate for situations where the consequences of the worst outcome are particularly undesirable, such as situations that can have very dire financial consequences or health consequences for the decision-maker.

Maximax   This is a logical opposite to the maximin criterion. The rule is to identify the most desirable outcome for each action and select the action with the most desirable return. The maximum monetary gain for each fertilizer application in Table 11.3 is recorded in the second row of Table 11.6. The maximax rule would select action a_4, an application of 160 pounds of N per acre.

  This is a very optimistic rule that only considers the most desirable

[5]See R. Duncan Luce and Howard Raiffa, *Games and Decisions, Introduction and Critical Survey* (New York: John Wiley and Sons, 1957), Chapter 13, for a detailed discussion of these three criteria as well as several other criteria that do not require probability estimates.

Table 11.6 Values Used to Rank Alternative Actions by Some of the Decision Criteria

| Item | Actions | | | |
	a_1 90 lb	a_2 110 lb	a_3 144 lb	a_4 160 lb
Minimum value for action	$ 67.00	$ 94.00	$ 88.00	$ 85.00
Maximum value for action	190.00	263.00	336.00	349.00
Simple average value for action	126.93	179.00	203.60	205.33
Expected value with prior probabilities	125.82	180.26	206.36	207.73
Return exceeded with probability = 0.92	86.00	117.00	110.00	107.00
Return exceeded with probability = 0.70	105.00	148.00	178.00	175.00
Variance with prior probabilities	887.25	1,686.57	3,482.93	3,959.34
Standard deviation with prior probabilities	29.79	41.07	59.02	62.92

outcome. It can be criticized on the same basis as the maximin criterion—it only considers a small part of the information contained in the payoff table.

Principle of Insufficient Reason   The principle of insufficient reason argues that if the decision-maker does not know the probabilities of the alternative events, a reasonable approach is to treat all events as if they were equally likely and select the action with the most desirable average outcome. To apply this criterion to the actions in Table 11.3, the simple average of the 15 net returns was calculated for each action and recorded in the third row of Table 11.6. This is equivalent to weighting the events with a uniform probability distribution. Action a_4 is selected with this criterion.

Unlike the maximin and maximax, the principle of insufficient reason uses the payoffs for all events in calculating the desirability of each action. Like the two previous criteria, it ignores any belief the decision-maker has about some events being more likely than other. Furthermore, the criterion does not consider the dispersion of outcomes except insofar as it affects the average value for the action. Thus, two actions with equal average values would be considered equally desirable, even though one had a wider variation of outcomes than the other.

Decision Criteria That Require Probability Estimates   Two decision criteria that consider subjective probabilities are examined here. They are maximizing expected monetary payoffs and a safety-first criterion.

Maximizing Expected Monetary Value   It is understandable that a decision-maker would like to have one statistic that summarizes each action and that can be used to compare alternative actions. Each of the decision criteria discussed above is based on a single statistic, but they have been criticized because they do not utilize all of the information available—all outcomes and the probabilities of the outcomes. The expected monetary value is a single statistic that utilizes all of the outcomes and the probability information available.

The expected value of a variable is equal to the probability-weighted sum of the possible values of the variable. If we have m possible events for the jth action with the ith event denoted as θ_i, having outcome O_{ij} and probability P_i, then the expected monetary value of the outcome is given by

$$E(O_j) = P_1 O_{1_j} + P_2 O_{2_j} + \ldots + P_m O_{mj}$$

$$= \sum_{i=1}^{m} P_i O_{ij} \qquad (11.12)$$

The expected monetary values were calculated for the four fertilizer actions in Table 11.3. The expected value recorded in Table 11.6 is $207.73 for a_4, higher than for any of the other actions, although it is not much greater than the expected outcome for a_3.

Selecting the action with the highest expected monetary value is the utility-maximizing decision for managers who are risk neutral, that is, neither risk averse nor risk preferring. However, the expected monetary value provides no information about the variability of the outcome. Individuals who are either risk averse or risk preferring also will want to consider information on the variability of monetary outcomes in making the choice. The remaining rules consider dispersion in making the choice.

Safety-First Criterion   Safety-first is a multiple-objective criterion that places a constraint on maximizing expected monetary value. The rule is to maximize expected monetary value subject to a specified probability of exceeding a minimum level of net income. The first step in applying this rule is to specify both the minimum income level and the probability by which an action must exceed this income level. The minimum income level is usually based on the cash required to meet business overhead expenses, family living, and debt repayment. The probability level depends on the decision-maker's financial situation to meet cash requirements from other sources if they cannot be provided from the business. The second step is to select those actions that provide the minimum outcome with the specified probability. The third step is to consider the actions identified in Step 2 and select the action with the highest expected monetary value.

To illustrate this with the data in Table 11.3, suppose Andy wants to select a fertilizer strategy that maximizes gross margins subject to a probability of 0.92 (approximately 11 years out of 12) that the gross margin will exceed \$109 per acre. Inspection of the payoffs for action a_1 in Table 11.3 indicates that the lowest payoffs are \$67 and \$86 with a prior probability of 0.04 for each level. Thus, the data indicate the prior probability that returns will exceed \$86 is 0.92. If we consider the other actions, the probability is 0.92 that returns will exceed \$117 for a_2, \$110 for a_3 and \$107 for a_4. These values are recorded in Table 11.6. If Andy's rule is to maximize expected returns subject to a probability of 0.92 that returns exceed \$109 per acre, actions a_1 and a_4 will be eliminated because they do not meet the condition. The choice will be between a_2 and a_3. Action a_3 has the higher expected value (\$206.36) and is selected.

Now consider an example using the data in Table 11.3 which maximizes expected monetary value subject to the prior probability of 0.70 that returns will exceed \$160 per acre. Inspection of action a_1 indicates a probability of 0.04 of \$67, 0.04 of \$86, 0.10 of \$89, 0.06 of \$100, and 0.06 of \$105, or a proability of 0.30 of a return of \$105 or less. Thus, there is a proability of 0.70 of achieving a return greater than \$105. The comparable returns are \$148 for a_2, \$178 for a_3 and \$175 for a_4. Notice that both a_3 and a_4 exceed the \$160 minimum with a probability of 0.7. Action a_4 would be selected because it has a higher expected value.

The safety-first criterion is intuitively appealing because it only requires the decision-maker to specify the disaster level and the probability of exceeding it. Some have argued it is a satisfactory approximation to the use of conventional utility functions. This argument is strengthened by considering what the decision-maker would do if none of the alternatives met the minimum income provisions. In this case, the decision-maker would inspect the alternatives and select the best action without specifying her/his objectives. However, the requirement to specify a minimum income level and probability level may lead to excluding actions that almost meet the constraint and have much higher expected outcome. Thus, criteria that consider the trade-off between expected outcome and dispersion in a more general manner would be useful.

Efficiency Criteria   The efficiency criteria consider the trade-off between the expected income and the dispersion of the outcomes. Consideration of the dispersion reduces our ability to select a preferred action. However, we may be able to eliminate some actions that are dominated by (always less preferred than) other actions. This reduces the number of actions a decision-maker must compare in making a choice.

Expected Value-Variance   The most commonly used summary statistic of dispersion is the variance. If a decision-maker wants to consider both the expected outcome and the dispersion of the possible outcomes, consideration of the expected value and the variance is a relatively simple way of considering both in a mathematically precise manner.

The variance is the sum of the probability-weighted squares of the difference between the possible outcomes and the expected value. If we let $V(O_j)$ represent the variance of outcome for the jth action, the method to calculate the variance can be written as:

$$V(O_j) = E[O_{ij} - E(O_j)]^2$$
$$= P_1[O_{1j} - E(O_j)]^2 + P_2[O_{2j} - E(O_j)]^2 + \ldots + P_m[O_{mj} - E(O_j)]^2$$
$$= \sum_{i=1}^{m} P_i[O_{ij} - E(O_j)]^2 \qquad\qquad \textbf{(11.13)}$$

For example, the variance for action a_1 in Table 11.3 is calculated as:

$$V(O_1) = 0.03[190 - 125.82]^2 + 0.06[167 - 125.82]^2$$
$$+ \ldots + 0.04[67 - 125.82]^2$$
$$= 0.03[4119.0072] + 0.06[1695.7924]$$
$$+ \ldots + 0.04[3459.7924]$$
$$= 123.5722 + 101.7475 + \ldots + 138.3917$$
$$= 887.25$$

Since both the probability and the squared deviation must be nonnegative, the sum of the probability-weighted squared deviations must be nonnegative. If all of the possible outcomes are the same value, each of the deviations will be zero and the variance will be zero. This is the case of perfect knowledge or certainty. As the dispersion of possible outcomes increases, the squared deviations increase in magnitude, leading to a positive variance. In general, a larger variance indicates more dispersion of outcomes and more uncertainty.

The positive square root of variance is known as the standard deviation. This measure is sometimes used in place of the variance because it is smaller and more convenient. However, both measures have the same implications about variation and uncertainty.

The variance and the standard deviation are listed for each of the four actions in Table 11.6. Notice that both the expected value and the variance increase as we move to increasingly higher levels of nitrogen fertilizer.

The relative change in expected income and variance can be visualized more easily by plotting the values for the four fertilizer actions on a graph such as Figure 11.4. The relationship connecting the four points is an estimate of the expected value and variance for intermediate levels of nitrogen that might be applied. If we have defined the production systems with the least variation for each level of expected income, this relationship

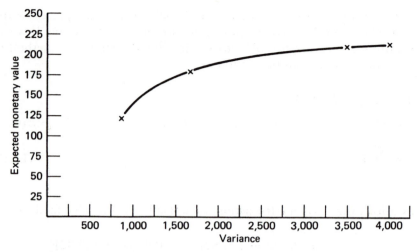

Figure 11.4. The efficiency frontier for the fertilizer decision.

can be referred to as an efficiency frontier. An efficiency frontier indicates the maximum expected income for any given level of variance, or alternatively, it shows the minimum level of variance for any level of expected income. Thus, it would be possible to find other production systems having an expected-value variance combination below the frontier, but not above it.

The strength of the expected value-variance (E-V) analysis and the development of the efficiency frontier is in defining efficient alternatives for consideration in decision-making. If we are comparing a number of alternative production systems, involving alternative seeding rates, tillage systems, pesticide programs, and other variables, the expected gross margin and the variance of gross margin could be calculated for each production system. The E-V analysis would be an effective means of identifying the systems having the greatest expected value for any level of variance. In general, decision-makers who are either risk neutral or risk averse will prefer a production system on the frontier to any below the frontier.[6] The risk neutral decision-maker considers only the expected outcome and will select the action with the highest expected value regardless of the variance. This is the extreme right end of the efficiency frontier in Figure 11.4.

The efficiency frontier is an effective way to summarize the data for risk averse decision-makers. Notice that the efficiency frontier in Figure

---

[6]To be technically correct for risk averse decision-makers, this statement also requires either that the outcomes are normally distributed or that the decision-maker's utility is a function of only the mean and variance. When skewness and higher moments are considered, the risk averse decision-maker need not find the most desirable outcome on the efficiency frontier. We deal with this more general case in second degree stochastic dominance below.

11.4 increases at a decreasing rate as we move from left to right. This indicates that the marginal change in variance (the change in variance per dollar of change in expected income) increases as one moves from left (low levels of N) to right (higher levels of N). Risk averse decision-makers can evaluate the alternatives along the efficiency frontier and make their choice on an inspection basis.

The E-V criterion considers both the expected value and the variance. It is possible to have many alternative forms of the probability distribution of outcomes that have the same expected value and variance. For example, distributions that have the same probability for all outcomes (referred to as uniform distributions), normal distributions, skewed distributions, and bimodal distributions can be constructed for a given expected value and variance. For this reason, decision-makers frequently prefer to use criteria that consider the total distribution rather than just one or two summary statistics. Two of the simpler criteria that consider the total distribution are first and second degree stochastic dominance.[7]

First Degree Stochastic Dominance   The stochastic dominance criteria introduced here can be discussed most conveniently using cumulative distribution functions of the type we introduced in Figure 11.3. First degree stoachastic dominance rests on the very reasonable assumption that decision-makers prefer more to less. That is, if gross margin or net income is the outcome specified, the application of first degree stochastic dominance is relevant, providing the decision-maker prefers more gross margin or net income to less.

Consider the continuous cumulative distribution function for actions F and G in Figure 11.5. F dominates G by first degree stochastic dominance if the cumulative probability of F at all outcomes (on the horizontal axis) is less than or equal to the cumulative probability of G, with the inequality holding for at least one outcome level. Notice that the cumulative probability of action F is less than G at all outcome levels in the figure, and we can say that F dominates G by first degree stochastic dominance. Graphically, this means that the cumulative distribution of the dominant function (F in the figure) cannot have any points to the left of the dominated function (G in Figure 11.5).

This property is transitive, meaning that if F dominates G and G dominates another cumulative function K (not shown), then F will also

---

[7]See Anderson, Dillon, and Hardaker, *op. cit.,* pp. 282–88 and R. P. Zentner, D. D. Green, T. L. Hickenbotham, and V. R. Eidman, *Ordinary and Generalized Stochastic Dominance: A Primer,* Department of Agricultural and Applied Economics Staff Paper P81–27, University of Minnesota, St. Paul, September 1981, pp. 15–27, for a more rigorous discussion of first and second degree stochastic dominance. The basic work elaborating and formalizing first and second degree stochastic dominance is also referenced in these publications.

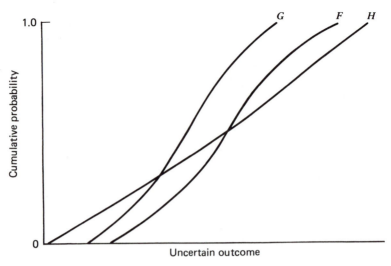

Figure 11.5. Illustration of first degree stochastic dominance (*F* dominates *G*, but not *H*).

dominate K. Distributions that are dominant in the first degree will be preferred by all decision-makers meeting the requirement that they prefer more to less. However, this criterion provides little help in ranking distributions that neither dominate another distribution nor are dominated by another distribution. For example, the distributions F and H in Figure 11.5 cannot be ranked by first degree stochastic dominance. A decision-maker could reduce the set of actions from F, G, and H to F and H based on first degree stochastic dominance. The choice between F and H would have to be made based on inspection of the distributions or another decision criterion that would rank the distributions.

Discrete, as well as continuous distributions, can be analyzed with this rule. Cumulative probability distributions for the gross margin data in Table 11.3 can be plotted as either continuous or discrete distributions. We present the cumulative distributions in discrete form here and leave plotting them in continuous form (analogous to Figure 11.3) as an exercise for the reader. The cumulative probabilities for actions a_1, a_2, and a_3 are plotted in Figure 11.6 as discrete distributions, and both actions a_2 and a_3 dominate a_1 in terms of first degree stochastic dominance. However, neither a_2 nor a_3 dominates the other. As we noted earlier, the gross margin (shown in Table 11.3) is less for a_1 than the gross margin for the same event with either a_2, a_3, or a_4. First degree stochastic dominance is a generalization of the dominance concept we noted earlier.

The concept of first degree stochastic dominance is easily understood, and its underlying assumptions make it very widely applicable. However, it cannot rank actions, such as a_2 and a_3, whose cumulative distributions

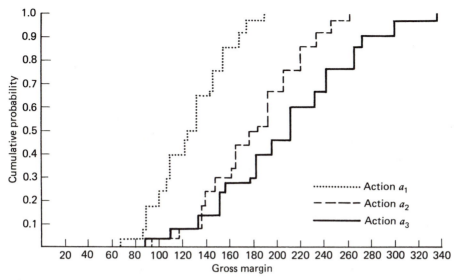

Figure 11.6.   Discrete cumulative distribution functions for actions a_1, a_2, and a_3.

cross. Experience with decision-making under uncertainty suggests that distributions which cross are the rule rather than the exception. We turn our attention to second degree stochastic dominance which permits ranking certain actions whose distributions cross.

Second Degree Stochastic Dominance   The selection rules for second degree stochastic dominance provide the basis for eliminating risky prospects from the stochastically efficient set developed under first degree stochastic dominance. In addition to preferring more to less, strict second degree stochastic dominance assumes that the decision-maker is risk averse. Thus, the application of second degree stochastic dominance would be appropriate to select a set of actions for a risk averse decision-maker to consider, and then choose one act from the set by inspection. However, the application of second degree stochastic dominance might exclude some actions preferred by risk neutral and risk preferring decision-makers, making it an inappropriate method to select an efficient set of actions for these individuals.

The decision rule for second degree stochastic dominance can be stated in terms of the area under the cumulative distribution curves. Given two uncertain actions F and G, F is preferred to G by all decision-makers who are risk averse if the area under the cumulative distribution function of F never exceeds and somewhere is less than the area under the cumulative distribution function of G. Consider the illustration in Figure 11.7. Distribution I dominates distribution J if the area under distribution I is

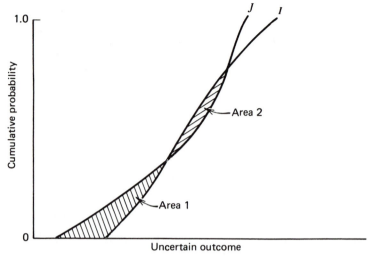

Figure 11.7. Illustration of second degree stochastic dominance (*I* dominates *J*).

less than or equal to the area under *J* for all outcome levels, with the inequality holding for at least one monetary level. The difference in the area under the cumulative distribution functions for *I* and *J* is equivalent to considering the difference in the two shaded areas 1 and 2 in Figure 11.7. If we start at low levels of the outcome, *I* lies to the right and clearly dominates *J*. As we move to higher levels of outcome, *I* crosses *J* and lies to the left of *J*. Area 2 is clearly smaller in size than area 1, indicating that the sum of area 1 minus that part of area 2 up to each outcome is positive for all monetary outcomes. The cumulative distribution function for *I* lies to the right of *J* for outcomes to the right of area 2. Thus, *I* dominates *J* by second degree stochastic dominance.

Consider the previous actions a_1, a_2, a_3, and a_4 in Table 11.3. We already determined by inspection that a_2 and a_3 dominate a_1 by first degree stochastic dominance. Although a_4 was not graphed in Figure 11.6, inspection of the outcomes in Table 11.5 indicates that the cumulative distribution function for a_4 will also lie to the right of the distribution function for a_1. Furthermore, we can tell by inspection of Figure 11.6 that any action which dominates another by first degree stochastic dominance will also dominate the other by second degree stochastic dominance. Thus, we need not consider a_1 in selecting the efficient set for second degree stochastic dominance. As we compare the cumulative distribution functions for a_2 and a_3, notice that a_3 has some outcomes at the lower monetary values to the left of a_2. Likewise, the outcomes in Table 11.3 indicate that a_4 has some monetary outcomes to the left of either a_2 or a_3. Thus, it is clear that the area under a_3 will be greater than the area under a_2 at very low outcomes, indicating that a_3 will not dominate a_2 by second degree stochastic dominance. What about the reverse? Will a_2 dominate a_3? Inspection of the two cumulative functions indicates that the total area under a_2 will exceed the area under a_3 at higher outcomes. Thus, a choice cannot be made between a_2 and a_3 with second degree stochastic dominance. A similar pairwise comparison of a_2 and a_4 (not plotted in Figure 11.6), as well as a_3 and a_4, indicates that none of the three strategies can be eliminated with second degree stochastic dominance. Thus, the efficient set of strategies for both first degree and second degree stochastic dominance is composed of a_2, a_3, and a_4.

Does this result seem intuitively reasonable? Remember that second degree stochastic dominance should eliminate any actions which a risk averse decision-maker would not consider. Action a_3 has some lower outcomes than a_2, and a_4 has some lower outcomes than either a_2 or a_3. It is clear that some risk averse decision-makers may prefer a_2, while others may prefer a_3 or a_4, making the result intuitively reasonable.

Now assume Andy has an opportunity to introduce a new marketing strategy that will permit him to contract the sale of the corn he produces at $2.75 per bushel. Signing this contract will eliminate the price uncertainty. If Andy applies 144 pounds of N per acre, the events, probabilities, and payoffs for this new action, a_5, are as shown in Table 11.7. This action could be added to the payoff matrix in Table 11.3 by including another column for a_5 and listing an outcome of $266 for each of the five favorable yield events, $212 as the outcome for each of the five normal yield events, and $133 for each of the five unfavorable yield events.

The cumulative distribution functions for a_3 and a_5 are plotted in Figure 11.8. For a_5 to dominate a_3 by second degree stochastic dominance, the cumulative area under a_5 must be less than for a_3 for each monetary outcome. With a discrete cumulative function, the total area under the two curves only has to be calculated at each change in the cumulative probabilities.

The calculations to compare the area under the cumulative distributions for a_3 and a_5 are given in Table 11.8. Notice that the cumulative area

Table 11.7 Events, Probabilities, and Payoffs for Action a_5, Contracting for $2.75 per Bushel

Events		Prior	Action
Yield	Price	Probability	a_5
Favorable	$2.75	0.3	$266
Normal	2.75	0.5	212
Unfavorable	2.75	0.2	133

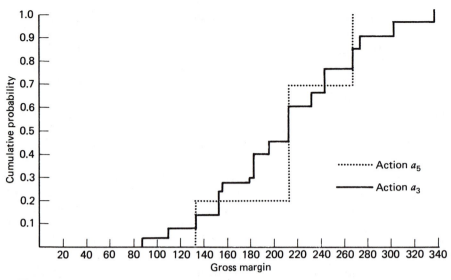

Figure 11.8. Discrete cumulative distribution functions for actions a_3 and a_5.

on each line of the table is the total cumulative area under the distribution to the left of the monetary level. Thus, the area under the distribution for a_3 at \$110 is 0.88, the area under the "first step." The area under the cumulative function for a_3 to the left of \$133 is 2.72, the area under the "first and second steps," while the area under the cumulative distribution for a_5 to the left of \$133 is 0. Comparing the area under the two cumulative distributions indicates that the area under the cumulative function for a_5 is less than for a_3 at each of the 15 "steps." This indicates that action a_5 dominates a_3 by second degree stochastic dominance.

Does a_5 dominate either a_2 or a_4? We can be sure that neither a_2 nor a_4 dominates a_5, because a_5 has a larger minimum outcome. However, we must complete the computational procedure comparing a_5 with a_2 and a_5 with a_4 to determine if a_5 dominates one or both of these actions.

Summary

The procedures described can be used to analyze a wide variety of decisions at the enterprise level when uncertainty of outcome is an important consideration. Decision problems, including an evaluation of alternative crop and livestock production systems, alternative tillage methods, alternative fallow systems for dryland farming, alternative irrigation strategies, alternative marketing or pricing strategies, and an evaluation of several sizes of machines to purchase (where a larger machine has higher costs but performs the task in a more timely manner) can be analyzed with the approach illustrated.

Table 11.8 Calculation of Area Under the Cumulative
Distribution Functions for Second Degree Stochastic
Dominance Analysis

Outcomes O_i	Cumulative Probabilities		Change in Outcomes ΔO_i	Area Under the Cumulative Curve[a]	
	a_3	a_5		a_3	a_5
88	0.04	0.00	—	0.00	0.0
110	0.08	0.00	22	0.88	0.0
133	0.14	0.20	23	2.72	0.0
152	0.24	0.20	19	5.38	3.8
155	0.28	0.20	3	6.22	4.4
178	0.30	0.20	23	12.66	9.0
182	0.40	0.20	4	13.86	9.8
196	0.46	0.20	14	19.46	12.6
212	0.61	0.70	16	26.82	15.8
231	0.67	0.70	19	38.41	29.1
242	0.77	0.70	11	45.78	36.8
266	0.86	1.00	24	66.42	53.6
272	0.91	1.00	6	71.58	59.6
301	0.97	1.00	29	97.97	88.6
336	1.00	1.00	35	131.92	123.6

[a]The area under the cumulative distribution function at each outcome is calculated as

$$\sum_{i=2}^{m} A_j(O_{i-1})\Delta O_i.$$

For example, the area under a_3 at $110 is 0.04(22) = 0.88 and the area under a_3 at $133 is 0.88 + 0.08(23) = 0.88 + 1.84 = 2.72, and so on. The values calculated represent the cumulative area under a_3 at $110 and $133, respectively.

The procedures discussed have been applied to a wide range of problems and reported in the literature.[8] Some of the applications reported include additional considerations that were not illustrated in this chapter, such as calculating the outcomes on an after-tax basis. While space limitations prevent the inclusion of a wider range of applications, the guidelines discussed to define the relevant actions and events can be applied to planning and control problems of operating farms. The number of events in actual applications will typically be greater than the examples presented. However, the methods discussed can be used to estimate the prior probabilities and the payoff matrix for each action-event combination. The procedures illustrated in Chapter 8 can be used to prepare the estimates on an after-tax basis. Calculating the payoffs for a large number of event-action

[8]For a partial list of applications, see Anderson, Dillon, and Hardaker, *op. cit.* pp. 155–59, 225–80, and 319–21.

combinations by hand would be tedious, but such calculations can be finished easily and quickly with microcomputers, making such applications routine and practical. The appropriate decision rule can be selected from those presented based on the decision-maker's objectives and financial situation as well as the size and probabilities of gains and losses. As noted earlier, the appropriate rule may differ depending on the size of probabilities and outcomes involved.

The analysis in this section considered five of the seven components of a decision problem under uncertainty—actions, events, outcomes, prior probabilities, and decision criteria. The remaining components—predictors of uncertain events and development of revised probabilities of the outcome given the prediction (posterior probabilities)—are presented in the Appendix to this chapter. An application to an enterprise example is also included.

APPLICATION TO WHOLE-FARM PLANNING

The procedures outlined for decision-making under uncertainty are as applicable to whole-farm planning as they are to individual parts of the business. The increased number of variables—the prices and production levels for each of several products, the risk of financing costs, and the cost of other inputs—makes it difficult to complete the computations in a reasonable amount of time without the aid of computing equipment. We illustrate two analyses that can be made to aid in the evaluation of the risk associated with alternative whole-farm plans. The procedure to develop a payoff matrix and more complete analysis is described at the end of the section.

The first task is to select the outcome of interest. For enterprises we evaluated the uncertainty of the contribution which enterprises make to the overhead expenses of the business. At the firm level, it is appropriate to concentrate on the uncertainty of the return to the firm's unpaid resources. The appropriate outcome for long-run planning is the farm profit or loss (or net farm income) which is the before-tax return to unpaid family labor, operator labor, equity capital, and management. This measure considers the overhead expenses that must be paid in cash on a year-by-year basis, as well as the depreciation that must also be paid over a period of time. Some managers may prefer to estimate the uncertainty of the addition to retained earnings for long-run analyses. This measure recognizes the withdrawal for family living and income tax payments and may provide very relevant information for comparison of alternative farm plans, particularly when the taxes paid are expected to vary greatly between plans. However, we will only consider before-tax estimates here to avoid introducing the complication of making the estimates on an after-tax basis.

The appropriate outcome for short-run planning depends on whether the manager wishes to focus on the uncertainty of liquidity or profitability. Some managers concerned with short-run liquidity may prefer to estimate the uncertainty of the cash difference on the projected annual cash flow (line 28, Tables 6.9 and 6.13).[9] Those less concerned with meeting liquidity requirements may prefer to evaluate the uncertainty of profitability and estimate the distribution of farm profit or loss.

The procedures to estimate the variability of either measure resulting from uncertain prices, production, and interest rates are essentially the same. Our example considers variability of farm profit or loss for the long-run.

Alternative procedures can be used to provide information on the uncertainty associated with alternative farm plans. We consider three alternatives—sensitivity analyses, estimation of the variance of farm profit or loss, and the construction of the payoff matrix. Each successive method is more complex to prepare, but also provides more information for decision-makers.

Sensitivity Analyses

Sensitivity analysis is used to estimate the effect of a specified change in a price, a production level, an interest rate, or combinations of changes in prices, production levels, and interest rates on the outcome of concern. The analysis can be completed easily either on an enterprise or a whole-farm basis. The sensitivity analysis is more meaningful for decision-making when the level of events relates to the decision-maker's subjective probabilities for such events.

Consider the application of sensitivity analysis to the estimation of farm profit or loss for Andy's long-run Plan 1 and Plan 2 in Table 6.5. Andy could specify the levels of corn, soybean, and hog prices as well as the levels of production for each and calculate the effect of various price and production combinations on farm profit or loss. However, he observes that total cash receipts from the sale of crops and livestock typically have been within 10 percent of his planned values, and he decides to analyze the effect of a ± 10 percent change in total cash receipts for corn, soybeans, and slaughter hogs on farm profit or loss. Cash receipts for these items (Table 6.5) total $180,975 for Plan 1 and $151,421 for Plan 2. The effect of a 10 percent price increase and decrease on farm profit or loss is shown in the second row of Tables 11.9 and 11.10 for Plans 1 and 2, respectively. Andy also feels that interest rates may fluctuate within a ± 2 percent range. He estimates the effect of a 2 percent change in the interest rate on all debt

---

[9]More technically, we would want to use the cash difference (line 28) less the interest on operating capital (line 31) when variability of interest costs is one of the sources of risk being considered in the analysis. We will deal with this in the section entitled Additional Considerations.

Table 11.9 Sensitivity of Farm Profit or Loss for Long-Run Plan 1 to Changes in Cash Receipts and Interest Rates

Change in Interest Rates	Cash Receipts		
	−10%	As Budgeted	+10%
+2%	$18,694	$36,792	$54,890
0	22,782	40,826	58,924
−2%	26,762	44,860	62,958

capital for both plans. An increase of 2 percent would increase expenses $4,034 per year for Plan 1 and $5,070 per year for Plan 2. The resulting farm profit or loss for ± 2 percent change is shown in the first and third rows of Tables 11.9 and 11.10.

Sensitivity analysis is a quick and convenient method to determine the approximate impact of the change in one or more variables on the outcome. However, the entries may provide little more than a general indication of the more important variables for consideration in risk analysis, unless the manager can relate approximate subjective probability levels to the level of events analyzed. For example, the entries in Tables 11.9 and 11.10 will be of limited use unless the manager estimates the likelihood that gross cash receipts will be within 10 percent of the budgeted value.

Estimating Variance of Whole-Farm Returns

The variance of income for whole-farm plans can be estimated by treating the enterprise gross margins and input costs that vary (such as interest costs) as random variables and using the statistical properties of jointly distributed random variables. This discussion considers the preparation of estimates of the expected farm profit or loss (net farm income) and its variance. However, the procedures to estimate the variability of other measures of whole-farm income are very similar.

The procedure of estimating farm profit or loss with an income state-

Table 11.10 Sensitivity of Farm Profit or Loss for Long-Run Plan 2 to Changes in Cash Receipts and Interest Rates

Change in Interest Rates	Cash Receipts		
	−10%	As Budgeted	+10%
+2%	26,949	42,091	57,233
0	32,019	47,161	62,303
−2%	37,089	52,231	67,373

ment can be stated in equation form by considering Y_j as the random gross margin of the jth enterprise and K_j as the number of units of the jth enterprise, such as the number of acres or head. If interest rates are uncertain, the uncertain cost of operating capital related directly to the enterprise is one of the causes (along with price and production uncertainty) of uncertain gross margins. The uncertainty of interest costs for fixed debt commitments can be recognized as a variable interest cost per unit of debt capital, I, multiplied by the number of units of capital, d. Finally, K_o represents the constant and fixed overhead costs of the business. In making the calculation for farm profit or loss, K_o includes general building maintenance not related to the amount of use, casualty insurance, property taxes, depreciation, accounting costs, and other overhead costs considered constant for the analysis. Then the expected profit or loss, $E(P \text{ or } L)$, for a farm with n enterprises can be stated as:

$$\begin{aligned} E(P \text{ or } L) &= E(K_1Y_1 + K_2Y_2 + \ldots + K_nY_n - dI - K_o) \\ &= K_1E(Y_1) + K_2E(Y_2) + \ldots + K_nE(Y_n) - dE(I) - K_o \end{aligned} \quad \textbf{(11.14)}$$

since the expected values of the constants (the Ks and d) are equal to themselves. Equation 11.14 only formalizes the calculation we have made using the income statement and identifies the random variables to be considered in the analysis. It is not intended to change the calculation.

Once we have the expected gross margin per unit $E(Y_1)$ and the variance per unit σ_1^2, the variance of the gross margin for K_1 units is:

$$V(K_1Y_1) = K_1^2\sigma_1^2 \quad \textbf{(11.15)}$$

For example, the gross margin is \$206.36 per acre for corn produced with 144 pounds of nitrogen on Andy's farm, and the variance of the gross margin per acre is \$3,482.93 (Table 11.6). The expected gross margin for 200 acres of corn for long-run Plan 1 is:

$$200 \cdot \$206.36 = \$41,272$$

The variance of the gross margin for 200 acres of corn is:

$$(200)^2(\$3,482.93) = (40,000)(\$3,482.93) = \$139,317,200$$

The variance of the sum of two random variables Y_1 and Y_2 is given by:

$$V(K_1Y_1 + K_2Y_2) = K_1^2\sigma_1^2 + K_2^2\sigma_2^2 + 2K_1K_2\sigma_{12} \quad \textbf{(11.16)}$$

where σ_{12} is the covariance, a measure of the way Y_1 and Y_2 move together. The covariance is estimated as:

$$\hat{\sigma}_{12} = \underset{Y_1 Y_2}{\Sigma \, \Sigma} \, [Y_1 - E(Y_1)][Y_2 - E(Y_2)][P(Y_1, Y_2)]$$

(11.17)

where $P(Y_1, Y_2)$ is the joint probability of Y_1 and Y_2 as defined in equation 11.5. If the variables tend to move together, such as the yields of two crops grown during the same season of the year in a given area, the covariance is positive. If the variables tend to move in opposite directions, so that one is high when the other is low, the covariance is negative. When there is no relationship, the two variables are independent. In this case the joint probability $P(Y_1, Y_2)$ equals the product of the unconditional probabilities $P(Y_1) \cdot P(Y_2)$ as we noted in equation 11.9, and the covariance is zero.

The magnitude of the covariance depends on the unit used to measure the outcome. For example, the magnitude of the covariance for two crop yields depends on whether the yields are measured in pounds or kilograms. It is often useful to standardize the relationship between two random variables by calculating the correlation coefficient given by:

$$r_{12} = \frac{\sigma_{12}}{\sigma_1 \cdot \sigma_2}$$

(11.18)

The value of r is always bounded by:

$$-1 \le r \le 1$$

Whenever Y_1 and Y_2 have a perfect positive relationship, r takes a value of $+1$. This will occur when all points in Figure 11.9 are on a straight line. If there is a perfect negative relationship such that Y_1 is high when Y_2 is low and vice versa (Figure 11.10), then r will be -1. The value of r is zero when Y_1 and Y_2 are independent as illustrated by the scatter of dots in Figure 11.11.

If we use the correlation coefficient, equation 10.16 can be written as:

$$V(K_1 Y_1 + K_2 Y_2) = K_1^2 \sigma_1^2 + K_2^2 \sigma_2^2 + 2 K_1 K_2 r_{12} \, \sigma_1 \sigma_2$$

(11.19)

Stating the covariance in terms of the correlation and the two standard deviations is advantageous when the correlations can be estimated more easily than the covariance. We will return to the problem of estimating correlations in the example that follows.

Before writing the variance expression for the profit and loss, we should note that the variance of the difference between two random variables such as Y_1 and I in equation 11.14 can be written as:

$$V(K_1 Y_1 - dI) = K_1^2 \sigma_1^2 + d^2 \sigma_I^2 - 2 K_1 d r_{1I} \sigma_1 \sigma_I$$

(11.20)

The variances are added, but the covariance is subtracted. Thus, if higher

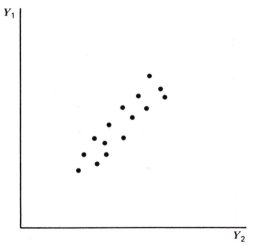

Figure 11.9. Positive correlation.

interest rates are positively correlated with favorable gross margins making $r_{iI} > 0$, then the variance will be smaller than the sum of the two variance terms. Of course, the opposite is true if $r_{iI} < 0$.

Using relationships 11.19 and 11.20, we can write the variance of farm profit or loss as:

$$V(P \text{ or } L) = K_1^2\sigma_1^2 + K_2^2\sigma_2^2 + \ldots + K_n^2\sigma_n^2 + d^2\sigma_I^2$$
$$+ \sum_i\sum_j K_iK_jr_{ij}\sigma_i\sigma_j - \sum_i K_idr_{iI}\sigma_i\sigma_I \qquad \textbf{(11.21)}$$

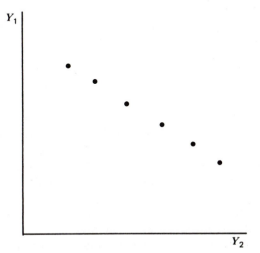

Figure 11.10. Correlation of -1.

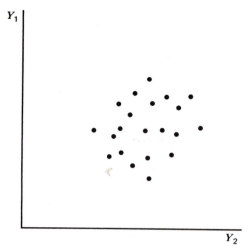

Figure 11.11. Correlation of zero.

Equation 11.21 can be used to estimate the variance of profit or loss for alternative farm plans, such as Andy's farm Plan 1 and 2. The K_i's are the number of units of enterprise i and d is the number of units of capital. These values are available from the data used to develop the complete budget for long-run planning and from the projected income statement for short- and intermediate-run planning. The variance of the enterprises and interest rates can be estimated using the procedures illustrated with the data in Table 11.3 and equation 11.13. The probability of the outcome levels should relate to the time period being analyzed. The distribution of prices and yields for the long run should be used to estimate variances for long-run plans. Probabilities for next year should be used to estimate variances for that period.

Estimates of the correlation coefficients are the remaining data required to calculate the variance of farm profit or loss with equation 11.21. One approach to obtaining correlation coefficients for such calculations is to assume that the physical and economic environment that has prevailed in the recent past will continue to prevail in the future. If the manager is willing to make this assumption, correlation coefficients can be estimated from historic data. Estimates of the correlation of gross margins based on historic data are frequently prepared by the Agricultural Experiment Station and are made available to managers in the area. When such data are not available, the manager can estimate the correlation coefficients using either gross margins from records for the farm or estimated gross margins calculated from average yield and price data reported by the state Crop and Livestock Reporting Service or other agency reporting agricultural statistics for the geographic area.[10] A second source of such estimates is to elicit the appropriate conditional probabilities and estimate the correlation coefficients from the elicited probabilities and gross margin estimates of the type in Table 11.3 for the two enterprises. A description of this procedure is beyond the scope of this discussion.[11]

Each of the two long-run plans analyzed in Chapter 6 has three enterprises. Plan 1 is composed of 200 acres of corn, 100 acres of soybeans, and finishing two groups of 480 feeder pigs. Plan 2 includes 150 acres of corn, 150 acres of soybeans, and a 48-sow farrow-to-finish swine enterprise. The appropriate equation to estimate the variance of farm profit or loss is

$$
\begin{aligned}
V(P \text{ or } L) = {} & K_1^2\sigma_1^2 + K_2^2\sigma_2^2 + K_3^2\sigma_3^2 + 2K_1K_2r_{12}\,\sigma_1\sigma_2 + \\
& 2K_1K_3r_{13}\,\sigma_1\sigma_3 + 2K_2K_3r_{23}\,\sigma_2\sigma_3 + \\
& d^2\sigma_I^2 - 2K_1dr_{1I}\sigma_1\sigma_I - 2K_2dr_{2I}\sigma_2\sigma_I \\
& -2K_3dr_{3I}\sigma_3\sigma_I
\end{aligned}
\tag{11.22}
$$

where subscripts 1, 2, and 3 refer to the corn, soybean, and swine enterprise, respectively. Andy estimates the variances of gross margins using the procedures illustrated earlier in the chapter. He develops estimates of correlations between enterprise gross margins based on his farm record data. He estimates the correlation between corn and soybeans is 0.56, but the estimated correlations between the hog enterprises and the crops is approximately zero. Thus, he decides to assume they are zero for the analysis.

Andy feels that higher interest rates are correlated with lower product prices, suggesting r_{1I}, r_{2I}, and r_{3I} may be negative. However, he finds neither estimates of these correlations nor data to make his own estimates. He decides to assume these correlations are zero, but he recognizes that doing so will bias his estimate of the $V(P \text{ or } L)$ downward if the correlations are in fact negative. Andy's estimates of the appropriate coefficients are shown in Table 11.11. Notice that the estimates of the variance and the standard deviation are on a per acre basis for the crops, for the total enterprise for swine, and per $1,000 for borrowed capital. If we use these data, the variance for long-run Plans 1 and 2 is estimated as:

---

[10]The formulas to use in estimating the variance and covariance from sample data can be written as:

$$
\hat{\sigma}_1^2 = \frac{1}{n-1}\sum_{i=1}^{n}[Y_{1i} - E(Y_1)]^2
$$

$$
\hat{\sigma}_{1,2} = \frac{1}{n-1}\sum_{i=1}^{n}[Y_{1i} - E(Y_1)][Y_{2i} - E(Y_2)]
$$

[11]Procedures to elicit subjective probability data and to estimate correlation coefficients for enterprise gross margins based on such estimates are presented by Anderson, Dillon, and Hardaker, *op. cit.*, pp. 21–37.

Table 11.11 Andy's Subjective Estimates of Variance and
the Standard Deviation of Enterprise Gross Margin

Enterprise	Unit	Variance	Standard Deviation
Corn	Acre	$ 3,482.93	$ 59.02
Soybeans	Acre	3,178.23	56.38
Finish feeder pigs	960 hd	50,346,735.00	7,096.00
Farrow-to-finish	48-sow	37,350,457.00	6,111.00
Intermediate- and long-term financing cost	$1,000	160.00	12.65

$$
\begin{aligned}
V(P \text{ or } L)_1 &= (200)^2(3,482.93) + (100)^2(3,178.23) + \\
&\quad (1)^2(50,346,735) + (2)(200)(100)(0.56)(59.02)(56.38) + \\
&\quad (170)^2(160) \\
&= 139,317,200 + 31,782,300 + 50,346,735 + \\
&\quad 74,537,066 + 4,624,000 \\
&= 300,607,301 \\
\sqrt{V(P \text{ or } L)_1} &= 17,338
\end{aligned}
$$

$$
\begin{aligned}
V(P \text{ or } L)_2 &= (150)^2(3,482.93) + (150)^2(3,178.23) + \\
&\quad (1)^2(37,350,457) + 2\ (150)(150)(0.56)(59.02) \\
&\quad (56.38) + (23.4)^2(240.25) + (230)^2(160) \\
&= 78,365,925 + 71,510,175 + 37,350,457 + \\
&\quad 83,854,200 + 8,464,000 \\
&= 279,544,760 \\
\sqrt{V(P \text{ or } L)_2} &= 16,720
\end{aligned}
$$

Table 11.12 Estimated Farm Profit or Loss, the Variance,
and the Level of Returns at Several Points on the Farm Profit
or Loss Distribution

Method of Calculating	Plan 1	Plan 2	Farm Profit Will Exceed This Level with Probability Shown If the Distribution Is Normal
Mean + 1.645 standard deviation	$69,347	$74,665	0.05
Mean + 1.284 standard deviation	63,088	68,629	0.10
Mean + 0.846 standard deviation	55,494	61,306	0.20
Mean	40,826	47,161	0.50
Mean − 0.846 standard deviation	26,158	33,016	0.70
Mean − 1.284 standard deviation	18,564	25,693	0.90
Mean − 1.645 standard deviation	12,305	19,657	0.95
Standard deviation	17,338	16,720	

If we take the square root of the variance, the standard deviation is $17,338 for Plan 1 and $16,720 for Plan 2.

The data we have developed on the expected farm profit or loss and the variability for farm Plans 1 and 2 are summarized in Table 11.12. The farm profit or loss estimated for each plan is considered the mean or expected value. The standard deviation calculated above is listed in the last row. Plan 2 has both a higher expected farm profit or loss and a smaller standard deviation (and variance). Thus, Plan 2 is preferred to Plan 1 based on the E-V criterion. To apply the other criteria that make use of probability data, we must know (or assume) the shape of the distribution of farm profit or loss. If the enterprise returns are jointly normally distributed, then any linear combination of the enterprise returns, such as farm profit or loss in equation 11.14, is also a normally distributed random variable.[12] Managers and those advising managers are frequently willing to assume that this condition holds. By doing so, we can estimate the returns at alternative points on the cumulative probability distribution. Five percent of the area under a normal curve is to the right of the ordinate at the mean plus 1.645 standard deviations. Thus, by assuming that farm profit or loss is normally distributed, we can say that the probability it will exceed $69,347 [$40,826 + 1.645($17,338)] for Plan 1 is 0.05. Given the symmetric property of the normal distribution, the probability that farm profit or loss will be less than the mean minus 1.645 standard deviations is also 0.05. Based on the assumption of normality, the levels of farm profit or loss that will be exceeded with a probability of 0.05, 0.10, 0.20, 0.50, 0.70, 0.90, and 0.95 are listed in Table 11.12.

The assumption of normality along with the estimates of the mean and standard deviation can be used to apply the safety-first criterion as well as first and second degree stochastic dominance. For example, suppose Andy wants to maximize expected farm profit or loss subject to a probability of 0.95 that farm profit or loss will exceed the $12,000 required for family living. In this case, both plans exceed the minimum income level, and Plan 2 is selected because it has the higher expected value. The cumulative distribution could be plotted using the data in Table 11.12. Notice that the income at each probability level is higher for Plan 2 than Plan 1. Thus, Plan 2 dominates Plan 1 by both first and second degree stochastic dominance.

Constructing a Payoff Matrix

A third method of comparing alternative plans is to construct a payoff table for whole-farm plans in a manner analogous to that used for the corn

---

[12]See a statistics text such as Thomas H. Wonnacott and Ronald J. Wonnacott, *Introductory Statistics for Business and Economics*, 2d ed., (New York: John Wiley and Sons, 1977), p. 151, for a discussion of the theorem.

enterprise in Table 11.3. Developing the payoff matrix showing the probability of achieving alternative levels of farm profit or loss for each plan would provide more data on the total distribution, avoiding the need to assume that the distribution is either normal or of some other specific form. Completion of the payoff matrix would permit the application of the decision rules discussed for enterprise decisions to the selection of the whole-farm plan.

Consider how such a payoff matrix could be developed. The alternative farm plans, such as long-run Plans 1 and 2 analyzed in Chapter 6, would be the acts in the payoff matrix. The plans could be composed of production, marketing, and financing alternatives selected using the procedures outlined in previous chapters.

The events would be defined to reflect the combinations of alternative levels of uncertain prices, production, interest rates, and other sources of uncertainty considered important for the analysis. The obvious problem is that a large number of events is required if the price and production level are considered uncertain for each of several enterprises. For example, whole-farm Plans 1 and 2 have three enterprises. If three levels (high, expected, and low) are considered for both the price and production level of one enterprise, nine events are required. Considering the same number of events for each of three enterprises results in $9 \times 9 \times 9$ or 729 events. Adding three levels of interest rates would increase the number of events to be listed to $729 \times 3 = 2,187$. In general, the number of levels to be considered for each event, h, and the number of sources of uncertainty, n, can be combined to calculate the number of events as h^n. In the above example three levels for each of three prices, three production levels, and one interest rate results in $3^7 = 2,187$. The large number of events make this approach rather tedious without the help of a computer to calculate the outcomes for each event and make the other computations desired.

Perhaps the estimation of the joint probabilities for the events is an even more difficult problem than delineating the events and calculating the outcome for each event. Unless the events are independent, as we assumed for price and yield of corn in Table 11.2, conditional probabilities must be estimated. Assuming the seven events in the above example are independent is analogous to assuming all of the covariance terms in equation 11.22 are zero. The estimate of variance of profit or loss summarized in Table 11.12 assumed that only the covariance between corn and soybeans was nonzero. An analogous treatment in constructing the payoff table is to estimate the conditional probability of the gross margin for soybeans given the gross margin for corn, but assuming all other sources of uncertainty are independent.

After the payoff matrix is constructed, the data can be used to form cumulative distributions in the same manner as illustrated with the enterprise example in Table 11.3. Then each of the decision rules can be applied. The information gained through developing the payoff matrix instead of just estimating the variance of each plan would be more detailed specification of the farm profit or loss distribution. However, development of the payoff matrix is time consuming, and the previous methods—sensitivity analyses and estimating the variance of whole-farm plans—although less precise are more commonly used.

Additional Considerations

The discussion has focused on preparing estimates of variability for long-run farm profit or loss. Two additional points on estimating the risk of whole-farm plans are noted here. One is the difference in preparing estimates of variability for annual versus long-run plans. The other is the change in procedures required to estimate risk of the cash difference (less the interest on operating capital) of the annual cash flow instead of the risk of farm profit or loss.

The first consideration in estimating long-run versus short-run risk is to decide if there are more sources of uncertainty to consider in the longer run. The sources considered in the example—price, production, and financial—are likely to be important in both the short and long run. There may be other sources, such as a change in the cost and availability of rented land, that should also be considered in the long run. A variable can be added to equations 11.14 and 11.21 for the cash leasing cost of rented land or other variables considered important. A second consideration in estimating long-run versus annual risk is that the appropriate range and the probabilities for each source of uncertainty may be different. Farmers typically have more information on supply and demand conditions for the coming year than years further in the future. Likewise, they may have more information on soil moisture conditions and the likelihood of pest problems for the coming year. Thus, it is common for the subjective probabilities for price and production to be more concentrated for the next year than for the longer run, although the procedures to estimate the variability of gross margins are the same for both periods.

We have suggested throughout the discussion that procedures similar to those described can be used to estimate the variability of the net cash difference (less interest on the operating capital). Making this estimate is of the most use when the decision-maker is concerned with the effect of risk on liquidity. First, notice that the major sources of uncertainty—price, production, and interest costs—are the same for either outcome and that a given dollar change in these sources of uncertainty will have the same dollar effect on both outcomes, providing all of the other items included in calculating the two income items are constant. Referring to Table 6.10, item K_o in equation 11.14 is the algebraic sum of miscellaneous receipts, utilities, other repairs, taxes, insurance, hired labor, value of farm products produced on the farm and consumed by the family, and adjustments for changes in inventory and in capital items. If equation 11.14 were used to calculate the expected net cash difference for the annual cash flow (Table 6.9), the constant term would be the algebraic sum of miscellaneous income, utilities, other repairs, taxes, insurance, hired labor, capital expenditures, living expenses and taxes, principal payments on intermediate and long-term debt, and the beginning cash balance. Providing the items listed as part of the constant term are considered constant as enterprise gross margins and interest rates vary, they affect the expected outcome (equation 11.14) but not the variability of the outcome (equation 11.21). In this case, the expected values will differ, but the variance will be the same in both cases. Thus, in preparing estimates for the coming year, it is possible to use the same variability estimates for both the cash difference and profit or loss, providing the analysis is limited to the same sources of uncertainty.

Finally, it should be emphasized that the cash difference on line 28 (Table 6.9 and other cash flows) does not include the interest cost on operating capital. Thus, it is convenient to use the cash difference (line 28) less interest on the operating loan (line 31) as the value having approximately the same variability as farm profit or loss under the conditions noted above.

STRATEGIES TO USE IN DEALING WITH UNCERTAINTY

Farmers rely on a variety of strategies to deal with risk and uncertainty. These strategies can be divided into three broad categories: those strategies designed to reduce uncertainty, those selected to shift some of the risk to another firm, and those that rely on reserves to carry one through low-income or loss periods. We will comment on several strategies that can be included under each of these headings.

A number of strategies can be listed which may reduce uncertainty. The selection of certain systems of production for use in producing a product may have less variability than others. Irrigated production is often considered to have less variability of yield and gross margins than nonirrigated production on soils with low available water-holding capacities. Likewise, some enterprises, such as dairy and farrow-to-finish swine enterprises, are considered to have less risk than certain types of specialty crop production and cattle feeding. It is important to note that, while statements such as these may be true in general, they need not be true in all time periods and all locations. The procedures described can be used to estimate and compare the risk for alternative enterprises.

Diversification is commonly advanced as a method of reducing risk. Two ways of diversifying are noted—by adding resources and by diversifying with existing resources. The two cases result in quite different answers. It is important to notice that the variance of returns for two enterprises is the sum of the variances of each enterprise plus the covariance of the two. Thus, the opportunities to reduce variance by increasing resources and adding enterprises are not very great unless enterprises can be added that have either less variance per dollar of return or a negative correlation with the included enterprises. For example, consider what would happen to the variance of enterprise gross margins (ignoring the financial risk) if Andy had only one enterprise, corn, and planted the 300 acres to corn. The variance of the enterprise gross margins is:

$$K_1^2\sigma_1^2 = (300)^2(3,482.93) = 313,463,700$$

If Andy diversified by adding another 100 acres of land and planting the 100 acres to soybeans, the estimated variance of the gross margins on the 400 acres would be:

$$K_1^2\sigma_1^2 + K_2^2\sigma_2^2 + 2K_1K_2r_{12}\sigma_1\sigma_2$$
$$= (300)^2(3,482.93) + (100)^2(3,178.23) + (2)(300)$$
$$(100)(0.56)(59.02)(56.38)$$
$$= 313,463,700 + 31,782,300 + 199,686,460$$
$$= 544,932,460$$

Thus, adding the 100 acres and a second enterprise increases the variance of enterprise gross margins. Reduction of the variance by adding the 100 acres of land and planting the second crop would require a negative covariance that more than offsets the variance of the 100 acres of soybeans. Enterprises with low variance often have low returns, making them less attractive in terms of expected income. We also find that crops grown during the same season tend to have positively correlated yields. There is also a tendency for the major crop and livestock prices to be positively correlated. Thus, the opportunity to combine negatively correlated enterprises is not great on many farms.

Diversifying on a given resource base, however, may be an effective means of reducing the variance of gross margins, even when the gross margins from the enterprises are positively correlated. In this case, the variance is normally reduced by diversification because the variance is weighted by $(K_i)^2$ and the covariance is weighted by (K_iK_j). For example, we noted above that the enterprise variance of planting Andy's 300 acres to corn is 313,463,700. However, the variance of the gross margins from planting 200 acres to corn and 100 to soybeans is

$$(200)^2(3,482.93) + (100)^2(3,178.23) + 2(200)(100)(0.56)$$
$$(59.02)(56.38)$$
$$= 139,317,200 + 31,782,300 + 74,537,066$$
$$= 245,636,566$$

Thus, diversifying on the existing land base reduced total variance of the cropping plan. In general, diversifying with a second enterprise that has a lower variance than the first will reduce total variance even if the correlation between the two is $+1.0$. If the added enterprise has greater variance, then diversification may or may not reduce the gross margin variance of the diversified plan, depending on the magnitude of the correlation coefficient[13] Of course, the variance resulting from financing would have to be considered to obtain variance of profit or loss.

Perhaps one of the more effective means of reducing production and price uncertainty is to monitor the operation and to adjust plans in response to changing production and price conditions. Although it is difficult to estimate the effect of exercising the control function on business risk, casual observation suggests it is rather great. We will discuss control procedures in the last three chapters of the book.

Several strategies can be listed which tend to shift some of the risk to another individual or firm. Purchasing insurance of various kinds—fire, wind, all-risk crop insurance, life, and health—shifts certain risks to other firms or agencies. Price risk can be shifted for many commodities by selling on contract or through hedging on the futures markets. In each of these cases there is a cost to shift the risk, either in the form of a premium for insurance or the forgone opportunity for a higher price.

An important strategy to use in living with uncertainty is to maintain reserves. Some farmers maintain reserves of physical resources in the form of feed to use during shortage periods. Most farmers maintain financial reserves in the form of liquidity and solvency to carry the business through low-income or loss periods. The procedures described in this chapter can be used to analyze the appropriate size of feed reserves as well as the required financial reserves to carry the business through low-income periods.

Summary

Most farm management decisions must be made in a risky environment. The concept of risk suggests that future events are not known with certainty and that decision-makers can assign probabilities or chances (at least subjectively) to possible future events. These probabilities become an integral part of the process of making decisions in a risky environment.

The risks that farmers face can be classified into two categories—business risk and financial risk. Business risks are a function of the unpredictability of future prices and production levels due to market forces, weather, government policy, and institutional changes. Financial risk is the added variability of net returns that results from the financial obligations associated with debt or lease financing. Financial risk results from the

---

[13]See Anderson, Dillon, and Hardaker, *op. cit.,* pp. 185–99, for an additional discussion of these conditions.

concept of leverage as well as the unpredictability of the future cost and availability of nonequity funds.

The farm manager can exhibit three different attitudes toward risk. The risk averse attitude is commonly exhibited by managers—risk averters prefer less risky alternatives to those that will result in more risk. A risk neutral manager ignores risk in the decision-making process; this individual makes decisions based only on expected returns or values. The risk preferrer is more adventuresome and will choose the more risky business alternative. A person's attitude toward risk can be determined by evaluating the risk premium associated with various risky decision alternatives. Risk attitudes depend on a manager's objectives and financial resources and may change over time. Risk attitudes are not a reflection of management capability.

To evaluate a risky problem, it should be broken down into its components. The components include the alternative actions that the decision-maker might take, the states or possible future events that might occur, but that are not known with certainty, the payoff or outcomes that indicate the consequences for each action and state or event combination, and the probabilities or chances (odds) associated with each event. Given these components of the risky decision problem, choice and efficiency criteria can be used to aid in selecting a particular action to implement. Alternative criteria include maximizing expected values, maximizing expected utility, maximizing the minimum gain, maximizing the maximum gain, safety-first (or maximizing expected income subject to a probability of exceeding a minimum level), and stochastic dominance. Each of these decision criteria may reflect different attitudes and perceptions of risk.

Decision tree and payoff matrices that summarize the consequences of alternative actions for different events can be very useful in understanding and analyzing risky problems. Similarly, sensitivity analyses in whole-farm budgeting using manual or computer techniques can be used to illustrate the potential outcomes for different prices, yields, or other stochastic events. By estimating both the expected value and the variance or standard deviation of returns from a specific farm organization, the potential risk as well as the return from a particular plan can be evaluated.

If farm managers determine that they are facing too much risk, various risk management strategies might be adopted including diversification, flexible facilities and adjustable enterprises, insurance, maintaining financial and resource reserves, forward pricing or hedging of products and inputs, contingency plans, and closer monitoring or control of the farm organization.

Development of the payoff matrices and application of the decision criteria require many calculations for application to the planning and control problems of operating farm businesses. Such applications require calculating payoffs for a large number of events and making the calculations on an after-tax basis using the procedures described in Chapter 8. Fortunately, the necessary arithmetic can be completed efficiently with a microcomputer, making routine application of the decision and efficiency criteria practical for operating businesses.

APPENDIX

Revising Probabilities as Additional Information Becomes Available

We have discussed procedures that can be used to estimate subjective probabilities at a point in time. Now consider the problem of revising probabilities as additional information becomes available. Many agricultural producers appear to revise their subjective probabilities in an informal manner as they receive weather reports, national production estimates, data on domestic use and exports, price predictions, and other data that may affect the net income of their operation. Such probability revisions can be accomplished in a logical and mathematically correct manner by applying Bayes' theorem. Bayes' theorem is an elementary theorem of probability derived during the eighteenth century by the English clergyman Thomas Bayes. This theorem is normally developed in introductory statistics courses, and its logical validity is demonstrated in many books on decision theory.[14] We will simply state the theorem and illustrate its use here.

Revising Probabilities

Suppose that the decision-maker can obtain a predictor of the events (θ_i). Consider the kth level of the predictor as Z_k. Since predictors of uncertain phenomena such as price and yield levels for agricultural production are less than perfect, it is important to consider the accuracy of the predictor in revising probability estimates. The likelihood of obtaining a particular forecast given the event that occurred $P(Z_k|\theta_i)$ can be obtained by utilizing data on previous forecasts (Z) and the actual outcomes (θ). Then Bayes' theorem can be used to combine the prior probabilities of the decision-maker, $P(\theta_i)$, and the data on the accuracy of the predictor, $P(Z_k|\theta_i)$, to

---

[14]For example, see Albert N. Halter and Gerald W. Dean, *Decisions Under Uncertainty* (Chicago: Southwestern Publishing Company, 1971), pp. 26–27, or Anderson, Dillon and Hardaker, *op. cit.*, pp. 50–64.

estimate the posterior probabilities, $P(\theta_i|Z_k)$. The posterior probabilities indicate the probability that an event will occur given the prediction that has been made. Bayes' theorem can be expressed as:

$$P(\theta_i|Z_k) = \frac{P(\theta_i Z_k)}{P(Z_k)} = \frac{P(\theta_i)P(Z_k|\theta_i)}{\sum_i P(\theta_i)P(Z_k|\theta_i)} \tag{11.23}$$

The expression to the right of the first equal sign follows from the definition of conditional probability and states that the posterior probability equals the joint probability of θ and Z divided by the unconditional probability of Z. The second formula makes this operational by substituting the product of the prior probability and the likelihood probability for the joint probability in the numerator. Notice that the unconditional probability of Z_k in the denominator can be calculated by summing the product of the prior probability and the likelihood over all i for the kth value of Z.

Consider the application of Bayes' theorem in the revision of Andy's subjective probabilities. Andy subscribes to a market outlook service. Based on past history, he determined the accuracy of the predictions [the $P(Z_k|\theta_i)$ values shown in Table 11.13] for the service. The data indicate, for example, that when Z_4 (weak demand and moderate supplies) was predicted in the past, prices of \$3.25, \$3.00, and \$2.75 (θ_1, θ_2, and θ_3 respectively) occurred 10 percent of the time, a price of \$2.50 (θ_4) occurred 50 percent of the time, and a price of \$2.25 (θ_5) occurred 20 percent of the time. The values in other columns of the conditional probability matrix are interpreted in a similar manner.

Andy wants to combine the predictions he receives with his prior probabilities using Bayes' theorem. The joint probabilities required for the numerator of Bayes' theorem are calculated and recorded in the upper right portion of Table 11.13. For example, $P(\theta_2) P(Z_4|\theta_2) = (0.2)(0.1) = (0.02)$. After completing the calculation of the joint probabilities, the denominator can be calculated by summing each column. For example, $P(Z_4) = P(\theta_i) P(Z_4|\theta_i) = 0.01 + 0.02 + 0.03 + 0.10 + 0.04 = 0.20$. Notice that summing the $P(Z_k)$ over all values of k equals 1.

Following Bayes' theorem, the posterior probabilities can be calculated by dividing the joint probabilities by the unconditional probability of Z_k. For example, $P(\theta_2|Z_4) = 0.02/0.20 = 0.10$.

The posterior probabilities for prices replace the prior probabilities estimated in Table 11.2. However, we have five sets of posterior probabilities, one for each level of predicted price (Z_k). Each of the five sets must be combined with the prior probabilities for yield to form five sets of posterior joint probabilities for combined yield-price events. Again, we assume independence of the posterior price and the prior yield probabilities in calculating the revised probabilities listed in Table 11.14. The

Table 11.13 Derivation of Posterior Price Probabilities

Corn Price Events (θ_i)	Conditional Probabilities $P(Z_k\|\theta_i)$					Prior Probabilities $P(\theta_i)$	Joint Probabilities $P(\theta_i)P(Z_k\|\theta_i)$				
	Observations						Observations				
	Z_1	Z_2	Z_3	Z_4	Z_5	$P(\theta_i)$	Z_1	Z_2	Z_3	Z_4	Z_5
3.25	0.6	0.2	0.1	0.1	0.0	0.1	0.06	0.02	0.01	0.01	0.00
3.00	0.2	0.5	0.2	0.1	0.0	0.2	0.04	0.10	0.04	0.02	0.00
2.75	0.1	0.1	0.6	0.1	0.1	0.3	0.03	0.03	0.18	0.03	0.03
2.50	0.1	0.1	0.1	0.5	0.2	0.2	0.02	0.02	0.02	0.10	0.04
2.25	0.0	0.1	0.1	0.2	0.6	0.2	0.00	0.02	0.02	0.04	0.12
$P(Z_k)$							0.15	0.19	0.27	0.20	0.19
$P(\theta_1\|Z_k)$							0.400	0.105	0.037	0.050	0.000
$P(\theta_2\|Z_k)$							0.267	0.527	0.148	0.100	0.000
$P(\theta_3\|Z_k)$							0.200	0.158	0.667	0.150	0.158
$P(\theta_4\|Z_k)$							0.133	0.105	0.074	0.500	0.210
$P(\theta_5\|Z_k)$							0.000	0.105	0.074	0.200	0.632

payoffs for actions a_3, a_4, and a_5 also are listed in Table 11.14. Actions a_3 and a_4 represent the application of 144 pounds and 160 pounds of N per acre, respectively. Both a_3 and a_4 assume the corn is priced at delivery. Action a_5 represents the same fertilizer application as a_3 (and the same yield variability), but a_5 assumes the crop is sold under contract for $2.75 per bushel.

The next step is to calculate the expected returns for each action using each set of posterior joint probabilities. The results are shown in Table 11.15. When Z_1 ($3.25 per bushel) is predicted, the expected gross margin is $240.38 per acre with a_3 and $242.18 per acre for a_4. Notice that the predicted price does not affect the expected gross margin for corn sold under contract, a_5, and the expected gross margin is the same for all values of Z. Further inspection of Table 11.15 indicates that a_4 has the highest expected gross margin for price prediction of Z_1 ($3.25), Z_2 ($3.00), and Z_3 ($2.75), while a_5 has the highest expected return when Z_4 ($2.50) and Z_5 ($2.25) are predicted. Thus, we can say the optimal strategy for a risk neutral decision-maker is (a_4, a_4, a_4, a_5, a_5), meaning the decision-maker will maximize expected utility by selecting action a_4 when Z_1, Z_2, and Z_3 are predicted, but selecting a_5 when either Z_4 or Z_5 are predicted.

It should be noted that the posterior joint probabilities and the payoffs in Table 11.14 can be used with each of the decision rules we discussed in this chapter. In this case, we need to apply the appropriate rule for each set of posterior probabilities (each predicted price level). Then an optimal

Table 11.14 Payoff Matrix for Fertilizer-Pricing Decision Considering Both Uncertain Yield and Price

Events		Prior Probabilities	Posterior Joint Probabilities with Prediction of					Actions		
Yield	Price		Z_1	Z_2	Z_3	Z_4	Z_5	a_3 144 lb Open Mkt	a_4 160 lb Open Mkt	a_5 144 lb Contract
Favorable	$3.25	0.03	0.120	0.032	0.011	0.015	0.000	336	349	266
	3.00	0.06	0.080	0.158	0.045	0.030	0.000	301	312	266
	2.75	0.09	0.060	0.047	0.200	0.045	0.047	266	276	266
	2.50	0.06	0.040	0.032	0.022	0.150	0.063	231	240	266
	2.25	0.06	0.000	0.032	0.022	0.060	0.190	196	204	266
Normal	3.25	0.05	0.200	0.052	0.018	0.025	0.000	272	270	212
	3.00	0.10	0.134	0.264	0.074	0.050	0.000	242	240	212
	2.75	0.15	0.100	0.079	0.334	0.075	0.079	212	210	212
	2.50	0.10	0.066	0.052	0.037	0.250	0.105	182	180	212
	2.25	0.10	0.000	0.052	0.037	0.100	0.316	152	150	212
Unfavorable	3.25	0.02	0.080	0.021	0.007	0.010	0.000	178	175	133
	3.00	0.04	0.053	0.105	0.030	0.020	0.000	155	152	133
	2.75	0.06	0.040	0.032	0.133	0.030	0.032	133	130	133
	2.50	0.04	0.027	0.021	0.015	0.100	0.042	110	107	133
	2.25	0.04	0.000	0.021	0.015	0.040	0.126	88	85	133

Table 11.15 Expected Value of Action Using Posterior Probabilities

Value Predicted	Actions		
	a_3	a_4	a_5
Z_1	$240.38	$242.18	$212.40
Z_2	225.05	226.62	212.40
Z_3	212.36	213.77	212.40
Z_4	191.34	192.54	212.40
Z_5	168.15	169.11	212.40

strategy, indicating which action to select for each predicted price, could be defined for the rule selected by the decision-maker.

To this point we have only considered incorporating one forecast in revising the probabilities. Bayes' theorem provides a logical basis to aggregate any number of independent pieces of probabilistic information. In each case, the posterior distribution estimated from one revision becomes the prior distribution for the next revision. While it may seem heroic to assume a farm operator will devote the time required for repeated application of Bayes' theorem, its most important feature is that it provides a logical mechanism for the consistent processing of additional information.

Value of the Predictor

It is reasonable to ask if the use of the predictor will increase the expected gross margin. Furthermore, there may be a charge for the market service. A manager would like to know if the increase in expected gross margin will exceed the cost of the service. We can answer these questions by comparing the expected return using the optimal strategy with the predictor to the expected return for the optimal action without the predictor.

Consider the computations for the risk neutral decision-maker maximizing expected gross margin per acre. This individual's optimal action without the predictor is a_5 with an expected gross margin of $212.40. Action a_5 has a higher expected gross margin than any of the actions listed in Table 11.6. It is the action with the highest expected gross margin based on the prior probabilities, making it the optimal action without the predictor. The expected value of the optimal strategy with the predictor is given by multiplying the expected payoff from the actions by $P(Z)$, which represents the probability the action will be used. For the data in Table 11.15 this is $(0.15)(242.18) + (0.19)(226.62) + (0.27)(213.77) + (0.20)(212.40) + (0.19)(212.40) = $36.33 + 43.06 + 57.72 + 42.48 + 40.36 = 219.95. The expected value of the predictor is $219.95 - $212.40 = 7.55 per acre, or $1132.50 on the 150 acres of corn Andy plans to produce. Purchasing the service would increase Andy's expected gross margin from the corn enterprise if the service cost less than $1,132.50 per year.

Decision-makers selecting the optimal strategy based on other decision rules may have a different optimal action based on the prior probabilities as well as a different optimal strategy with the predictor. However, the same type of analysis could be repeated using the expected gross margin of their prior optimal act and of each action in the optimal strategy to calculate the value of the predictor.

The optimal strategies based on predictors and posterior probabilities are frequently referred to as Bayes strategies. Bayes strategies imply utility-maximizing strategies. We have not converted the dollar values to expected utility indexes and thus have refrained from referring to these as Bayes strategies.

The procedures to develop optimal strategies have many applications in calculating the value of information provided by public and private sources. These procedures can be used to estimate the value of frost warnings, outlook information, and other sources of public information. A discussion of these applications is presented elsewhere.[15]

Questions and Problems

1. What is a subjective or personal probability? How can such probabilities be derived?
2. Identify the types of risk or uncertainty faced in farming. What is the difference between business and financial risk? What are the various sources of business and financial risk?
3. Identify and discuss the three basic risk attitudes. How does one determine a particular farmer's risk attitude?
4. What are the components of a risky decision problem? Identify a specific farm management problem that encompasses risk and identify the specific components of that risky problem.
5. Choose a farm management problem such as making a specific marketing or production decision where risk is a consideration and develop a decision tree of that problem.
6. Develop a payoff matrix for the specific decision problem identified in question 5.
7. How should one go about the process of estimating subjective probabilities? Estimate the subjective probabilities for the various events of the decision problem noted in question 5.
8. What are the three mathematical rules that must be followed in developing subjective probabilities? Why are these rules important?

[15]See Halter and Dean, *op. cit.*, Chapter 11, and Anderson, Dillon and Hardaker, *op. cit.*, Chapter 5.

9. What is the maximin decision rule? Discuss its usefulness in making mangement decisions.
10. What is the maximax decision rule? How might it be used in making farm management decisions?
11. Discuss the principle of insufficient reason and its application to farm management decisions.
12. Under what conditions would maximizing expected monetary value be a reasonable decision criterion for a risky problem? What is this decision criterion?
13. What is the safety-first criterion for making risky decisions? When might it be applicable in farm management decisions?
14. Discuss the concept of an E-V efficiency frontier. How useful is this concept in evaluating risky decisions?
15. What is meant by stochastic dominance? Contrast and compare first degree and second degree stochastic dominance.
16. Collect the data for a particular whole-farm plan and compute the expected value of income as well as the variance of income for that plan. How might you use the variance of income in analyzing the risk of the farm plan?
17. What kind of risk management strategies might a farm operator use to control risk?

Further Reading

Anderson, Jock R., John L. Dillon, and Brian Hardaker. *Agricultural Decision Analysis.* Ames: Iowa State University Press, 1977.

Eidman, Vernon R., Gerald W. Dean, and Harold O. Carter. "An Application of Statistical Decision Theory to Commercial Turkey Production." *Journal of Farm Economics* 49(November 1967):852–68.

Halter, Albert N., and Gerald W. Dean. *Decisions Under Uncertainty.* Chicago: South-Western Publishing Company, 1971.

Halter, Albert N., John L. Dillon, and J. P. Makeham. *Best-Bet Farm Decisions.* Professional Farm Management Guidebook Number 6. Sydney: Southwood Press, 1969.

Luce, R. Duncan, and Howard Raiffa. *Games and Decisions, Introduction and Critical Survey.* New York: John Wiley and Sons, 1957.

Nelson, A. Gene, George L. Casler, and Odell L. Walker. *Making Farm Decisions in a Risky World: A Guidebook.* Oregon State University Extension Service, July 1978.

Wald, Abraham. *Statistical Decision Functions.* New York: John Wiley & Sons, 1950.

Winkler, R. L. "The Assessment of Prior Distributions in Bayesian Analysis." *Journal of the American Statistical Association* 63, No. 320 (1967):776–800.

Winkler, R. L. "Probabilistic Prediction: Some Experimental Results." *Journal of the American Statistical Association* 66, No. 336 (1971):675–85.

IMPLEMENTATION

12

LABOR ACQUISITION
AND MANAGEMENT

CHAPTER CONTENTS

As the trend to larger farm units continues, many farms are becoming multiple operator units. In some cases the business includes only family members, but in many situations additional seasonal or full-time labor is hired. Labor management is becoming as important as capital and other management skills in the success of many farming units.

The tasks of the farm manager with respect to labor or personnel acquisition and management are fourfold (Figure 12.1): (1) Obtain the proper kind and number of personnel. This task requires analysis of the quality and type of manpower required (full-time or seasonal, managerial or labor skills) and the recruitment and selection of employees. (2) Develop the skills necessary for proper job performance. This task involves appropriate training sessions to insure adequate performance and safety of the employees, as well as encouragement and support for the employees to

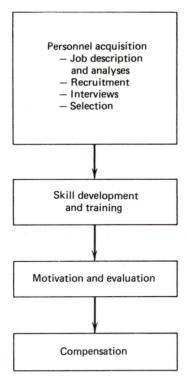

Figure 12.1.   The personnel management process.

improve their skills and capabilities. (3) Motivate employees to perform to the best of their capabilities. This task requires the farm manager to evaluate the working environment in terms of physical work conditions as well as personnel relationships and to discuss with the employees their performance on the job. Praise for a job well done as well as suggestions for improvement in performance are important in motivating people. (4) Develop appropriate and equitable compensation plans for the employees. Although working conditions and employer-employee relationships are important in maintaining employee morale, development of compensation plans is one of the most important functions of labor management. The compensation program will typically include a cash salary plus bonuses and may include various types of incentives to improve performance.

The following discussion delineates principles and ideas which the farm manager can use to perform these four tasks of labor or personnel management. The concepts are equally applicable to employees with management skills as well as labor skills ("middle managers"), and the discussion will emphasize both workers and management personnel.

HUMAN NEEDS

Successful personnel management requires an understanding and appreciation of basic human needs. Maslow (1954) has suggested that individuals have the following hierarchical need structure: (1) physiological needs, (2) safety needs, (3) needs for belonging and love, (4) esteem needs, (5) needs for self-actualization, and (6) needs for cognitive understanding. Physiological needs include the basic requirements of food, water, and air; they are the physiological basis of life. Safety needs include the avoidance of pain and danger and the comfort that comes from not actually incurring or being threatened with physical damage. The need for belonging and love refers to the human desire to be part of a group and the sense of security and comfort that comes from an intimate relationship with other people. Self-esteem involves the desire to be accepted by others, and the drive for status and recognition combined with the perception by the individual that he or she does have status and is accepted. Self-actualization refers to the desire to understand one's potential and capabilities and to grow and develop the skills that will utilize those capabilities. Furthermore, this need includes having reasonable and attainable expectations as to one's capabilities so as to minimize the frustration and stress of underachievement. Finally, the need for cognitive understanding suggests that humans are aware of and desire to assess the meaning of their own personality, their life, and the environment in which they live. These needs provide the basis to understand and explain the motivations and behavior of individuals with respect to their personal life as well as their employment.

RESPONSIBILITY, AUTHORITY, ACCOUNTABILITY

Given a basic comprehension of people's needs, the farm manager must understand the concepts of responsibility, authority, and accountability and the relationship between these concepts in order to be successful at labor and personnel management.

Responsibility

Responsibility is the obligation of individuals to perform assigned duties or functions to the best of their ability. To obtain responsible performance, it is desirable to assign certain tasks or jobs to various employees. If possible, these tasks should be assigned based on their similarity so that the employee can become somewhat specialized. Thus, if a farmer has two employees and both hog and cattle enterprises in the farming operation, major responsibility for the cattle feeding activity should be assigned to one employee and the hog feeding activity to a different employee. This process facilitates specialization, which should result in improved job performance as well as giving each employee specific responsibilities. Responsibility should be clearly defined, and there should be no gaps in assignments or the work may not get done. Assigning responsibility to an employee requires delegation on the part of the farm manager—delegation of authority as well as responsibility.

Authority

Authority is the right to decide what should be done and how the job should be accomplished. Authority enables employees to "get the job done" and is derived from the responsibilities that have been assigned to them. If employees have responsibility for certain tasks, commensurate authority should also be delegated to them so that their tasks can be carried out. This relationship between responsibility and authority is one of the most frequently overlooked by the farm manager. In all too many cases, the manager expects an employee to do a certain job but is unwilling to give the authority to do it in the way the employee feels is best. Certainly, some employees must be guided more than others—told not only what to do, but how to do it. However, a good training program and the hiring of competent, skilled employees will enable the manager to delegate authority along with responsibility.

Accountability

Accountability is answering for one's performance. If employees have been given the responsibility and authority to do certain tasks, they should be held accountable. Conversely, employees should not be accountable for tasks in an area in which they have no responsibility and authority. Accountability is the basis for employee evaluation because it requires the specification of performance standards—whether they be formal or informal, quantitative or qualitative—and the assessment of how well the employee has performed compared to these standards. Thus, the process of evaluating an employee's performance and determining not only compensation, but also whether that employee should be given additional responsibilities and decision-making authority, is based on his/her accountability for current areas of responsibility and authority.

THE PROCESS OF PERSONNEL MANAGEMENT

The following discussion will focus on the personnel management tasks identified earlier and the concepts useful in accomplishing them.

Personnel Acquisition

Job Analysis and Description   The first question to be answered in the area of labor or personnel acquisition and management is what kind of employee is needed. Is the farm operation large enough to support a full-time employee, or is seasonal labor required? Will the employee only perform physical tasks, or will he or she be relied on for day-to-day management decisions?

The labor skills needed in farming have changed, along with the complexity and capital intensity of the industry, from physical and brute force to the ability to make day-to-day operating decisions. If the employee is going to be working alone and assume responsibility for part of the operation, she or he must have the ability to identify and evaluate problems and make decisions to correct those problems. A farmer with a large livestock operation may want an employee who knows animal nutrition and can adjust the ration depending on changes in weight of the livestock or the nutrient content of the feedstuffs. A large cash grain farmer will need an employee who understands fertilizer and chemical recommendations and can operate and service equipment such as the combine or grain dryer without supervision. Thus, the first task is to evaluate the job to be performed by the employee and the qualifications that are needed.

A job description can often prove useful in personnel acquisition. In general, the job description identifies the types of responsibilities and authorities that will be delegated to the employee and the duties to be performed. It may also include the specification of working conditions in terms of hours and fringe benefits and relationships to other employees and the supervisor or farm manager. The overall purpose of the job description is to make it clear to the employee and the farm manager what is expected of the worker. Thus, the job description is useful in recruiting, selecting, and hiring people with appropriate skills.

Once an employee has been hired, the job description can be used to orient him or her to the responsibilities and duties of the position. It may also provide the basic document in job evaluation or assessing accountability. By identifying areas of responsibility and authority, the employee and employer have a better chance for agreement on the standards of accountability. The job description may even specify in quantitative or qualitative terms the performance to be achieved by the employee during a certain period of time which often provides the basis for job evaluation. Whether written or oral, a job description provides a means of communication between the employer and employee and a basis for assessment of the employee's performance. Figure 12.2 illustrates a job description for a Midwest farming operation with a sizable livestock enterprise.

Recruiting, Interviewing, and Selection   The next task is that of selecting and hiring the appropriate people for the position. This task involves recruiting individuals and then using various screening devices such as the application form, reference checks, and the interview to determine their qualifications as outlined and specified in the job description. If the screening devices are used correctly, selection of the best applicant for the job

Figure 12.2. An example job description.

Job Description
Cattle Herdsman   Name _____

Goodacre Farms

This job description can be changed at any time by management to reflect changing responsibilities and authorities.

Responsibilities

1. Feed mixing and feeding of cattle.
2. Monitor health and comfort of cattle.
3. Notify manager of serious health problems.
4. Maintain work area with emphasis on safety and cleanliness.
5. Provide a quarterly inventory of livestock to assistant manager (number, type, and approximate weight).
6. Notify assistant manager of needed repairs and supplies for the cattle operation.
7. Assist in crop production activities when necessary.
8. Assist in other livestock production activities when necessary.
9. Assist in maintenance and repair of machinery.

Authorities

1. Administer approved medication.
2. Call the veterinarian in consultation with assistant manager.
3. Change rations in consultation with manager.
4. Purchase health and medical supplies within budget as needed.

should result. Unfortunately, many farm managers do not take a systematic approach to the selection and acquisition of labor, and consequently, problems of poor job performance and unsatisfactory labor relations frequently result.

Recruitment   Recruitment is the process of searching for prospective employees. Some farmers use an informal recruitment process and obtain names of possible candidates by asking around the community or checking with other farmers who have had experience in hiring labor. However, as the skills required of farm employees increase and more "middle-management" personnel are required in agriculture, the recruitment process may involve the traditional sources of employees used by many nonfarm industries.

Federal or state employment agencies are available in almost any county and are an excellent source of information on employees who have basic labor skills. In addition, vocational agriculture instructors, bankers, credit personnel, and county extension directors also are a source of information on prospective employees. Advertising in the local newspaper is another method of recruiting for personnel with limited management skills.

Many community colleges now offer specialized two- and four-year courses in agricultural technology and management and are an excellent source of personnel for positions that may require additional training or management expertise. For people who will be placed in full-time management positions, four-year land grant institutions are also an excellent source of personnel. Many land grant institutions and community colleges offer summer "internship" programs whereby students work for a farmer to gain on-the-job experience. These internship programs may be a way to obtain additional seasonal help in the summer periods as well as to evaluate the potential of a specific individual to become a full-time employee following graduation. Most universities and community colleges also provide placement services, and the farm manager may find the college placement officer quite effective in identifying prospective employees who have the appropriate combination of training, skills, and attitudes that will fit within the farming operation. In addition, there are some commercial placement services that, for a fee, will assist the farmer in recruiting employees. In most cases, commercial placement services are involved primarily in the recruitment and placement of more highly skilled employees.

Interviewing and Selection   As indicated earlier, the selection process involves the use of screening devices such as the application form, reference checks, and interviews to select that one applicant who best fits the qualifications of the position as outlined in the job description. Few farmers will use an application blank, although a simple form requesting basic information on the applicant and her/his prior work experience could certainly prove useful in the interview and reference checks.

The interview typically provides the opportunity to obtain detailed information concerning the prospective employee's qualifications and personal characteristics. In all too many cases, the interview turns into a casual personality evaluation at best. During the interview, the employer should strive to obtain precise information about previous work experience, technical knowledge, personal goals, family background, willingness to learn and adapt to new methods, and attitudes concerning employer-employee relationships. Consequently, it is important to plan before the interview occurs. This planning will include selection of a comfortable, quiet place for the interview where there will be few interruptions and a set of questions that cover all important points. As much information as possible should be gathered about the applicant prior to the interview so that valuable time is not wasted in reviewing personal factors. The interview should start with general questions to develop a relaxed atmosphere, but should move quickly into issues of previous experience, goals and plans, and skills that relate to the job. Questions should be asked in such a fashion that the applicant is encouraged to talk about his/her capabilities and background, but the interviewer should not lose control of the interview process. Applicants should be asked why they want the job, what they expect to get from the job, and what they consider are their strengths and weaknesses. If interviewing for a management position, it is also important to evaluate the applicant's leadership qualities and ability to communicate and manage others as well as to identify problems and make decisions. A checklist of the questions or topics that are to be covered with each applicant may prove useful; the same checklist may be used to guide the discussion with the applicant's references. An evaluation of the applicant's performance should be completed during or immediately after the interview with each applicant. Figure 12.3 illustrates a checklist and evaluation form that might be utilized in the interview process.

The purposes of the interview are twofold: (1) to evaluate potential candidates as to their qualifications for the job, and (2) to inform the applicants of the type of position and the opportunities available if they take the job. Consequently, it is important to explain not only the work that will be required but also the financial compensation, benefit program, and opportunities for advancement that will be available to the employee. If the

Figure 12.3.   Interview evaluation form.

Name _____

Characteristics	*Low*				*High*
1. Leadership qualities	1	2	3	4	5
2. Ability to work with others	1	2	3	4	5
3. Receptiveness to receiving directions	1	2	3	4	5
4. Motivation to learn	1	2	3	4	5
5. Willingness to work (physical labor)	1	2	3	4	5
6. Training and background in:					
Animal nutrition	1	2	3	4	5
Animal health	1	2	3	4	5
Crop production	1	2	3	4	5
Livestock production	1	2	3	4	5
Mechanical skills	1	2	3	4	5
Management concepts	1	2	3	4	5
Finance	1	2	3	4	5
Marketing procedures	1	2	3	4	5
7. Ability to manage others	1	2	3	4	5
8. Personal goals	1	2	3	4	5
9. Initiative and imagination	1	2	3	4	5
10. Motivation	1	2	3	4	5
11. Determination	1	2	3	4	5
12. Ability to compromise	1	2	3	4	5
13. Ability to identify problems	1	2	3	4	5
14. Ability to make a decision	1	2	3	4	5

applicant has numerous other opportunities and is definitely qualified for the position, it is important to make him or her aware of the opportunities and advancement potential in the farm operation. Thus, the applicants should be given an opportunity to ask questions about the farm business, how it is operated, and plans for the future so that they can evaluate whether or not they will fit in the organization.

The references provided by the applicant are also a useful screening device. References should be consulted and specific questions asked concerning the applicant's capabilities and personality. A telephone or personal conversation with the references may be very enlightening as to previous work and how well the applicant got along with previous employers. Naturally an applicant will probably provide the names of people who will provide favorable comments; these references should, therefore, be used to obtain the names of other people who know the applicant or for whom the applicant worked. These secondary references may be more willing to identify and discuss the applicant's potential weaknesses and any problems that may have been encountered. Don't be hesitant to ask pointed questions of the references. If there is a problem, you should know about it. If there is no problem, a reference should not be offended.

Training

Regardless of previous experience, each new farm employee should receive initial on-the-job training to become acquainted with the particular farm and the methods used to do various jobs. An on-the-job training program will pay high benefits in terms of increased productivity, better employee morale, reduced supervision, and fewer accidents.

Each new employee should receive instructions or training in the characteristics and capabilities of any machine that will be operated, including the machine maintenance procedure and schedule. A few minutes spent in instructing the employee on operating procedures may save valuable hours and dollars in "downtime" during the critical harvest or planting season due to improper maintenance or operation of the machine by the employee. In similar fashion, employees who will be responsible for the feeding and handling of livestock should be given a thorough orientation and training in feeding procedures, health and preventive medicine practices, and other day-to-day livestock management responsibilities.

A thorough review of operating procedures and an adequate training program will contribute to maximum labor efficiency in the farm business. Training may require an investment of time by the farm manager, but the long-term benefits offset the initial cost. On-the-job training programs for new employees also reduce the potential for personal injury and liability claims, and may enable the employee to handle additional responsibilities and "take over" for a short period if the manager should become ill, sustain injuries, or go on vacation.

Training should be an on-going process. It may be possible to send the employee to vocational agriculture or other night school classes where new skills can be learned. Extension or industry-sponsored conferences, workshops, or conventions such as those sponsored by state and national commodity groups (Annual Pork Producers Conference or Cattle Feeders Day) may also prove worthwhile. Management-type employees should be given an opportunity and encouraged to improve their skills and capabilities.

Motivation and Evaluation

Management is sometimes defined as the ability to get things done through people and other resources. To be successful at managing employees, farm managers must understand what their workers are looking for in a job and motivate them to perform to the best of their ability. Clearly, appreciation for people's basic human needs as discussed earlier is important in understanding what employees aspire to obtain from their jobs.

Employees are typically seeking a number of items from their employment environment. (1) Pay. Adequate compensation is essential and a major objective of the employee. (2) Job security. Constant concern about the possibility of being fired or laid off will not allow the employee to concentrate on doing the best job possible. (3) Pleasant co-workers. Employees who cannot get along with each other will not function as a team and will usually be dissatisfied with the job. (4) Credit for work done. Recognition for a job well done through verbal praise or monetary reward provides support for the self-esteem of the employee. (5) A meaningful job. Every employee wants to feel that he/she is accomplishing something and that his/her work is making a contribution to the objectives of the farm business. (6) Opportunity to advance. Few employees will be satisfied with a "dead-end" job with little opportunity to improve their financial situation or be assigned additional authority and responsibility. (7) Comfortable, safe, and attractive working conditions. These needs may require the use of more modern machinery and livestock facilities to replace some of the drudgery of farm work. In addition, safe working conditions are a must in any job. (8) Competent and fair leadership. Employees perform best when they have respect for the employer and feel a sense of pride in the employer's accomplishments. (9) Reasonable orders and directions. A manager who is reasonable and fair in his/her demands and clearly identifies the responsibilities and work activities for the employee will usually have a more satisfied as well as more productive work force. An understanding of these wants and needs will help the farm manager motivate the employees to improved performance.

Motivation occurs when the employee judges that a particular type of behavior or doing a particular job will help satisfy one of the wants identified above. Thus, farm workers spend extra time checking the feed bunks because they expect the manager to reward them (verbally or monetarily) for less feed waste and more efficient gains on the cattle. If employees judge that this additional effort will not be recognized by the manager, they will not be motivated to do the task.

Since motivation is an important part of employee relations and employee productivity, the farm manager should be aware of various tools that can be used to motivate employees. An awareness of basic human needs is again useful in understanding employee motivation techniques and procedures. Certainly, praise and credit for work well done is an important motivation tool that will satisfy one of the basic needs of the employee. To be an effective motivational tool praise must be handed out judiciously and only for superior performance (measured by mutually agreed upon standards as noted later). Good work should not be rewarded by silence at the same time that bad work always results in a reprimand.

The expression of sincere concern and interest in the employee as an individual and in the employee's family is also a useful motivational tool. Occasional comments concerning the academic accomplishments of the employee's son or a brief discussion of the employee's hobbies and interests indicate interest in the worker as a human being. Competition between employees can also be a useful motivational tool when carefully handled; when not properly handled, competition can reduce productivity as employees begin to undercut each other's work and cease functioning as a team in order to "look better" than their competitors.

Another motivational tool that can be used is that of pride—pride not only in the work of each employee but also in the success of the overall organization. Public recognition of good performance can instill a sense of pride in the employee. The simple task of calling farm workers employees rather than "hired hands" recognizes the employee's contribution to the farm operation and will instill a sense of pride in the position and employment role. Delegation of additional responsibility if the employee can handle it is another way to provide motivation. It indicates that the manager trusts the employee and feels that he or she has the capability to handle additional responsibility and authority. Finally, money can be used as a motivational tool. We will discuss the use of compensation and incentive plans to motivate employees in more detail in the next section.

Consistent, high-level performance by employees requires that the employer establish clear-cut policies, procedures, and performance standards. Performance standards indicate the performance levels that are expected of the employee and provide a yardstick against which actual performance can be measured. They provide a benchmark for appraising the employee and her/his job performance. If possible, these performance standards should be quantitative and should identify goals that the employee should attempt to achieve, such as the number of acres to be plowed in a good working day or the time required to do chores or feed the cattle.

In addition to performance standards, a work schedule is also useful in determining priorities and getting the farm work done efficiently. A maintenance checklist for all machinery and equipment is a good idea, not just for new employees, but also for experienced workers. Likewise, a schedule for the hog operation is useful in assuring the timely performance of such operations as clipping needle teeth, giving iron shots, weaning, and the start of creep feeding. Checklists and schedules not only help get various jobs done regularly and on time, but they also relieve the farmer of some of the everyday management responsibilities so that greater attention can be given to important long-term planning and decisions.

Each employee's performance should be reviewed periodically to evaluate any problems and provide guidance concerning points of weakness or where improvement is needed. This appraisal process should include commendations for good performance as well as constructive suggestions concerning weaknesses. Discussion of the employee's expectations and whether the goals set by the employer are realistically achievable may also occur during the appraisal process. In some cases, it may be preferable to have appraisal sessions with the employee on a quarterly or semiannual basis rather than once a year. This will enable the manager to make suggestions and discuss problems that are fresh in both parties' minds rather than trying to recall what happened almost a year ago. More experienced employees may not need as frequent appraisal sessions, but they must be evaluated at least once a year to determine appropriate salary or wage increases.

As suggested earlier, it is important that employers establish and display a considerate attitude toward their employees. Today's farmers must recognize that most farm workers are employees with an ability to think. Employees' opinions and ideas concerning specific tasks should be solicited. This gives them a feeling of usefulness and self-confidence while providing a second and often useful view of how the farm operation might be improved. Each worker's complaints and ideas should be listened to thoroughly and objectively, and criticism should be constructive rather than degrading. In addition, employees should be criticized privately, but praised in public to protect their self-esteem.

The employee should also be given both the responsibility and authority for completing assigned tasks. By giving the workers some authority, they learn to make decisions and become more productive in the operation. If the manager retains all authority, sooner or later the capable worker will move to a more responsible position, while the others will let the farmer not only make the decisions, but also do most of the work. A delegation of authority implies freedom for employees to carry out their own ideas and methods. In determining how a task should be done, it is frequently preferable to follow the method suggested by the persons who will ultimately carry it out. This encourages the employees to prove that their ideas are, in fact, workable. The amount of authority delegated must vary according to individual ability.

Two final points concerning motivation should be made. Frequently, farm employees leave for nonfarm jobs because their families are unhappy—particularly with the hours. The hours are typically long and tiring, especially during the planting and harvesting seasons. The spouse and family often feel neglected and mistreated because the employee spends little time at home during these seasons. It is important for the farm manager to compensate for the long hours during rush seasons by allowing the workers some additional time with their families during slack times. A spontaneous display of good-will, such as an afternoon off to go to the fair or attend to some personal matters, helps maintain good employee morale. A further means of recognizing and maximizing an employee's productivity is to encourage self-improvement activities, such as farmer's night school meetings, field days, and professional meetings. It may be desirable to give the employee some time off to attend these sessions and maybe even pay part or all the cost for participation in them. The payments are a tax deductible farm expense and may be a low-cost method of improving employees' skills and also displaying confidence in their capabilities. Providing subscriptions to farm publications is another inexpensive means of recognizing the employee's desire for self-improvement and of motivating them to higher levels of performance.

Compensation

A fourth major task in farm labor management is employee compensation. A salary and fringe benefit package that is competitive with alternative jobs is necessary to attract and keep good farm employees. If workers are not monetarily rewarded for their ability to skillfully operate machinery and work with livestock as well as make day-to-day decisions on the farm, they will take their talents elsewhere.

The determination of adequate and reasonable compensation is a difficult, judgmental task. A number of factors affect the compensation decision; the farm manager's awareness of these factors may make her/his decision easier. (1) Labor supply and demand. Farmers typically cannot set wages independently of the number of employees available and the demand for those employees. They must pay a price that will bring forth the supply or "the going wage rate." Supply and demand for farm labor, therefore, influence wage rates just as supply and demand influence the price the farmer receives for corn, soybeans, and cattle. (2) Labor unions. Although unionized labor does not comprise a major component of the farm labor force, there is a trend to increased unionization of farm labor. The main thrusts of labor unions are to increase the wage rate or the benefits received by employees, and to obtain better working conditions.

Farmers will not be able to ignore the unionization movement and will be required to become knowledgeable concerning collective bargaining and negotiation on wages and working conditions. (3) Ability to pay. A farm or business that is marginally profitable will find it much more difficult to pay competitive salaries or wages for their employees than one that has reasonable profits. (4) Productivity. This factor should be the primary basis for compensation decisions, particularly increments in compensation for various employees. Rewarding high productivity and giving fewer or no rewards for low productivity provide a monetary incentive for improved performance. (5) Cost of living. With the high rates of inflation experienced during the past few years, many are advocating cost of living adjustments in wages to keep real wages from declining. However, it should be recognized that cost of living adjustments do not and should not reflect differences in productivity. (6) Government regulations. Minimum wage regulations specify the minimum wage that employees in certain industries can be paid. The specific regulations that affect agriculture will be discussed in the next section on governmental regulations. Even if agriculture was not directly affected by minimum wage legislation, the specification of minimum wages for firms and industries that compete for the same labor skills would have a significant indirect impact on the supply and wage rates paid for farm employees.

Compensation packages for farm employees typically include an hourly, daily, or monthly wage rate plus various fringe benefits including insurance programs, housing, electricity, and meat. Many farmers are now finding it advantageous to pay overtime for hours worked after a certain time of day. Overtime pay helps compensate for the employees' additional effort and the burden placed on their families by the extra hours. It also motivates employees to work the longer hours that may be required—hours that have a high value to the employer during the critical harvesting and planting seasons.

An attractive fringe benefit program including such things as life, hospital, and accident insurance and a paid vacation may also be part of the compensation package offered to employees. However, problems may arise with the provision of these fringe benefits and other prerequisites such as housing, electricity, or meat because of different estimates of their dollar value by the employer and employee. The employee characteristically does not value fringe items as highly as the employer. Thus, the worker should be informed of the cost to the employer of insurance and other benefit programs and the value assigned to housing and other prerequisites. Furthermore, it may be desirable to offer the employee the option of taking a housing allowance or meat allowance in cash rather than "in-kind."

So that employees know how much their total compensation package costs and what value the employer is placing on fringe benefits, an em-

Figure 12.4 Employee Compensation Summary Form

	Name _____	
	Compensation Summary	
	Monthly	*Annual*
Cash wages	_____	_____
Social security		
Employee contribution	_____	_____
Employer contribution	_____	_____
Worker's compensation	_____	_____
Unemployment insurance	_____	_____
Health and medical insurance	_____	_____
Retirement contribution	_____	_____
Housing	_____	_____
Meat and other farm products	_____	_____
Other	_____	_____
Total	_____	_____

ployer may want to provide a compensation summary to each employee. An illustration of such a summary is provided in Figure 12.4. This summary can be useful to remind the employees of the other components of their compensation package in addition to the cash salary or wage and the value of fringe benefits as measured by the employer's cost of providing these benefits.

Some farmers supplement the base wage and the fringe benefit package with various types of bonus and incentive programs. If the purpose of such programs is to improve the employee's productivity, an incentive program will usually be more effective than a bonus. Although the bonus may be used to keep an employee through a critical season, it may also have negative effects. If a bonus is given every year, the employees may feel that it is a part of their base pay and that they will receive it no matter what. Then, if the bonus is not given in a specific year, they may feel that their wage has been reduced and become dissatisfied or disillusioned.

An incentive program compensates workers based directly on their productivity in a particular enterprise or in the entire farming operation. The purposes of an incentive program are (1) to promote higher levels of performance by the employee, (2) to attract better qualified employees and encourage them to develop their skills and capabilities, (3) to encourage the employee to continue in the present job and provide the tenure that will be helpful to both the employer and employee, and (4) to provide adjustments in salaries to reflect in part the profit situation of the business.

LABOR INCENTIVE PROGRAMS

A number of principles should be considered in the development of a labor incentive program.[1]

1. The program should be simple and easily understood by the employee. There is a danger that oversimplification may lead to uneconomical practices, but reasonable balance is needed.

2. The program should be based on factors largely within the employee's control. This may be hard to attain, but some degree of control is necessary.

3. The program should aim at rewarding work that is in the best interests of the employer. A good program is designed so that outstanding performance benefits both the employer and employee.

4. The program should provide a cash return large enough to provide motivation for improved performance. Individuals in industry have found that 15 to 20 percent of an employee's wage should be in the form of an incentive payment if an incentive program is to encourage better performance.

5. The incentive payment should be made promptly or as soon after the completion of the work as possible.

6. The incentive program should be written, contain provisions for arbitration of misunderstandings, and indicate the duration of the program. Written copies of the program which are provided to both parties will help minimize misunderstandings from the beginning.

7. The incentive program should set forth employee responsibilities and be administered equitably.

8. The incentive payment should not be considered as a substitute for competitive base wages and good labor relations.

Types of Incentive Programs

Incentive programs can be grouped into four basic types: production incentives, livestock incentives, crop incentives, and percentage of income incentives.

Production Incentives Production incentives provide a means of rewarding an employee for performance which increases production or sales

[1]Phil Carroll, *Better Wage Incentives* (New York: McGraw-Hill Book Company, 1957); Paul Wesley Harrison Weightman, "Financial Incentive Plans for Farm Labor in New York State," Unpublished Ph.D. dissertation, Cornell University, Ithaca, New York, 1966; and John Wolfe, Michael Boehlje, and Vernon Eidman, "An Economic Evaluation of Wage Rates and Incentive Agreements for Full-Time Hired Labor on Oklahoma Farms," Oklahoma Agricultural Experiment Station, Oklahoma State University, Bulletin B–713, April 1974.

Table 12.1 Bonus/Incentive/Share Guide

The following examples of incentive programs should be used only as guides and be adapted to your situation. They should be tied to work responsibilities carried out by the employee and over which he/she has some control.

Suggested Incentives	*Type of Employee Status*		
	Semi-Skilled	*Skilled*	*Supervisory/Management*
Normal incentive should be equal to:	2–5 percent of cash wages	4–10 percent of cash wages	5–15 percent of cash wages
General farm	End of year bonus = $100 to $400 per year plus $50 for each year of service	End of year bonus = $200 to $600 per year plus $75 for each year of service	End of year bonus = $300 to $1,000 per year plus $100 for each year of service
Small farm	Weekly bonus of 1½ times cash wage rate for each hour worked over 60 hours per week	Weekly bonus of 1½ times cash wage rate for each hour worked over 60 hours per week	2–5 percent of net cash income
Large farm	Weekly bonus of 1½ times cash wage rate for each hour worked over 48 hours per week	Weekly bonus of 1½ times cash wage rate for each hour worked over 48 hours per week	1–4 percent of net cash income
Crop farm	$1–2/hour tractor driven after 7:00 P.M. (paid weekly) $2–3/hour tractor driven after 11:00 P.M. (paid weekly)	$1–3/hour tractor and/or combine driven after 7:00 P.M. (paid weekly) $2–5/hour tractor and/or combine driven after 11:00 P.M. (paid weekly)	2–6 cents per bushel of corn produced over county average 5–15 cents per bushel of soybeans produced over county average
Dairy	$1–3 for each cow detected in heat *Calving interval* 14 months = $50 13.5 months = $150 13 months = $300 12.5 months = $500 $3–5 per calf weaned if death loss kept below 15 percent $5–10 per calf weaned if death loss kept below 10 percent		*Herd milk production avg.* 12,000# = $100/year 14,000# = $400/year 16,000# = $800/year 18,000# = $1,600/year

(continued)

514

Table 12.1 *(Continued)*

The following examples of incentive programs should be used only as guides and be adapted to your situation. They should be tied to work responsibilities carried out by the employee and over which he/she has some control.

	Type of *Employee Status*		
Suggested Incentives	*Semi-Skilled*	*Skilled*	*Supervisory/Management*
Hogs	$0.50–1.00 for each sow detected in heat	*Pigs saved per litter* 6.5 = $50/year 7.0 = $150/year 7.5 = $300/year 8.0 = $500/year 8.5 = $900/year	*Feed conversion farrow-to-finish* 450# = $100/year 400# = $200/year 350# = $400/year 300# = $700/year
Beef	$5–10 for each feeder detected sick, treated, and recovered	*Calf crop sold* 80% = $100/year 85% = $200/year 90% = $400/year 95% = $700/year 100% = $1,100/year	Same as other two categories

Source: Kenneth H. Thomas and Michael D. Boehlje. *Farm Business Arrangements: Which One for You?* North Central Regional Extension Publication 50. University of Minnesota, St. Paul, 1982.

of an enterprise. The incentive payment should be based on a measure of production that will insure an increase in net income of the entire farm operation rather than an increase in one enterprise at the expense of others. Production incentives are frequently used to make growth or expansion in an enterpise more acceptable to the employee.

Livestock Incentives   A livestock incentive program gives an employee the opportunity to raise a limited number of livestock and receive a share or all of the income from the sale of those livestock. This is sometimes referred to as an equity accumulation program designed to retain a good employee. The program may require the employee to purchase the livestock and pay a minimal fee for forage and feed. Alternatively, the employee may receive the animals as the incentive, with all operating costs paid by the employer.

Crop Incentives   A crop incentive program gives an employee the opportunity to grow a specified acreage of crops and receive a share or all of the income from the sale of those crops. Crop incentives usually require the employees to pay for some part of the operating expenses; they also receive some part of the income and possibly government payment. The employee may grow the same crop on the same acreage, or the program may allow the employee to select a crop and choose one of several alternative locations specified by the employer.

Percentage of Income Incentives   With this incentive program, the employee receives a percentage of the farm income. Gross income or net income may be used to calculate the payment. This program can be used with an enterprise or the whole farm if the payment is based on profits. The program usually considers all operating expenses as costs when determining profits. However, taxes, depreciation, and operator salaries may not always be treated as operating expenses.

Examples of various types of incentive programs for crops, livestock, and the total farm operation are provided in Table 12.1. Suggestions as to the percentage of normal cash wages that the incentive payment should comprise are shown at the top of the table for different skill levels, with specific payment procedures and schemes that would result in the suggested percentage payment listed below. These suggested procedures must necessarily be adjusted to different farm operations as well as changes in farm wage rates and commodity prices over time. The impact of a specific incentive payment procedure on both the income of the employee and the profits of the employer should be evaluated before a particular alternative is chosen.

Estimating the Profitability of an Incentive Program

The profitability to the employer of a labor incentive agreement depends on its combined effect on productivity and costs. The technique of partial budgeting as discussed in Chapter 6 can be used to estimate the effect of an incentive agreement on profitability. Only those items of income and expense that change as a result of using the incentive agreement are considered, making the computations relatively simple.

The seven parts of the partial budgeting format and an explanation of the entries to be made in evaluating an incentive program are given in Figure 12.5. The additional receipts and reduced expenses include the items that increase income and thus make adoption of the labor incentive agreement more profitable. Summing the entries in these two parts gives total credits attributed to the agreement. The reduced receipts and additional expenses include the items that reduce income and make adoption of the labor incentive agreement less profitable. The sum of the entries in these two parts equals total debits. Subtracting total debits from total credits gives the difference. If the difference is positive, adoption of the labor incentive agreement is expected to increase the net returns of the business. A negative difference indicates that the agreement will decrease net returns.

The application of the partial budgeting procedure to a situation involving one enterprise is summarized in Figure 12.6. Assume a dairy operator has one employee and is presently milking 100 cows. The average production per cow is 14,000 pounds of milk each year, which is sold for $13 a hundredweight. The employer feels that over the next few years the employee could be instrumental in raising the herd average to 15,000 pounds per year and increasing the herd size to 120 cows. The employer is considering offering the employee a production incentive of $1 per hundredweight to be paid monthly on total production over the present annual average of 1.4 million pounds (100 × 14,000 pounds). The employer is interested in determining the incentive payment the employee will receive if the goals are met as well as her/his own gain from the program.

A typical dairy budget is used to estimate costs and returns of the dairy enterprise. The additional receipts from the increased production and herd size would be 400,000 pounds per year [(100 cows × 1,000 pounds) + (20 cows × 15,000 pounds)]. Since there are no reduced costs assumed, total credits would be $52,000 per year. Additional costs are the incentive payment and expenses involved with more cows and higher production. The incentive payment will be $4,000 ([400,000/100] × $1). Additional expenses include the overhead costs of 20 additional cows estimated by the farmer to be $7,400 per year and a variable cost of $9 per hundredweight for additional milk. Since the change does not affect other enterprises, there are no reduced receipts.

Completing the calculations indicates that the benefits to the employer

Figure 12.5. Evaluating an incentive program with partial budgeting.

Additional Receipts

Those items of additional gross receipts expected when an incentive plan is used are listed here. The increased production per unit (e.g., per cow or acre), additional units to be included in the plan (e.g., more cows or acres), and higher selling prices due to a better quality product are possible sources of additional receipts resulting from adoption of a labor incentive agreement.

Reduced Expenses

Items of cost that will be avoided or reduced when an incentive plan is used are listed here. Reduced expenditures for additional hired labor during busy seasons, more efficient feed use, lower veterinary bills, and reduced repair costs are possible sources of reduced expenses resulting from the adoption of a labor incentive agreement.

Total Credits

This is the sum of all items of ADDITIONAL RECEIPTS and REDUCED EXPENSES.

Reduced Receipts

Those items of income that will be reduced or no longer received when a labor incentive agreement is adopted are listed here. Additional livestock production under a labor incentive agreement may require hay and grain formerly sold to be fed on the farm. Likewise, specialization may result in the elimination of some enterprises and hence of some sources of gross receipts.

Additional Expenses

Items of additional cost that will be required when an incentive plan is used are listed here. The incentive payments and the cost of additional nonlabor inputs required under the incentive agreement, such as feed, are possible sources of additional expense.

Total Debits

This is the sum of all items of REDUCED RECEIPTS and ADDITIONAL EXPENSE.

Difference

This is TOTAL CREDITS minus TOTAL DEBITS. A positive (negative) DIFFERENCE indicates that adoption of the labor incentive agreement is expected to increase (decrease) the net return of the farming operation.

---

total $4,600; the estimated total credits exceed the estimated total debits by this amount. Thus, the incentive program will result in average monthly incentive payments to the employee of $333.33 ($4,000 ÷ 12 months) and an increase in the employer's net returns of $4,600 per year. Normally, the employer will want to go through the calculations several times using different values for production levels, the price of milk, and feed costs before deciding on the desirability of the program and on the level of incentive to offer.

Figure 12.6. Evaluating a production incentive for a dairy enterprise.

Additional Receipts		
4,000 cwt of milk per year @ $13	$52,000	
Reduced Expenses		
None		
Total Credits		$52,000
Reduced Receipts		
None		
Additional Expenses		
Incentive payment: 4,000 cwt × $1	$4,000	
Overhead costs: $370 per cow × 20 cows	$7,400	
Variable costs: 4,000 cwt × $9	$36,000	
Total Debits		$47,400
Difference		$4,600

The partial budgeting procedure can also be used to evaluate other types of incentive programs. Evaluation of a percentage of income is usually more complex than the above illustration because all enterprises affect net farm income. Consequently, the effect of a labor incentive agreement on the costs and returns of each enterprise must be considered in the evaluation.

Crop and livestock incentives can also be analyzed with the same basic concepts. With these incentive programs the operator incurs two kinds of costs: the direct cost of inputs used on the employee's enterprise and the opportunity cost of not receiving any profits from those resources which the employee operates. The additional receipts of these programs are also the result of increased labor productivity.

REGULATIONS AFFECTING FARM EMPLOYEES

Numerous government regulations affect the employment and working conditions of farm employees. These regulations include social security, worker's compensation and employee liability, occupational safety and health regulations, child labor regulations, minimum wage requirements, and unemployment insurance. Since regulations on farm employees are constantly being revised by both federal and state governmental agencies, only the general characteristics of farm labor legislation will be reviewed.

Minimum Wage Regulations

State and federal minimum wage regulations specify the minimum wage that must be paid to employees. With passage of the Fair Labor Standards Act of 1974, minimum wage rules have been in effect for farmers. Minimum wage regulations apply to farm employers with at least 500 man-days of hired work during any quarter of the preceding year. The regulations define a man-day of work as one worker doing one hour or more of work during any day. As of July 1, 1982, the minimum wage rate is $3.50 per hour; if farmers are subject to minimum wage rules, they must pay minimum wages to all employees whether paid weekly, monthly, or hourly.

Hourly rates less than the minimum wage can be paid to the employer's immediate family, "hand-harvest" workers paid on a piece rate basis who commute daily and are employed less than 13 weeks per year, and employees caring for range livestock who must be available 24 hours a day. The minimum wage legislation does not require overtime pay for farm workers. To comply with minimum wage and other labor regulations imposed by the Fair Labor Standards Act, farmers should keep records concerning the pay, number of hours worked, any additional compensation paid, and other details concerning their employees.

Social Security Taxes

Farm employees are subject to social security taxes if they are paid more than $150 in cash wages during the year or have been employed for 20 or more days during the year. If a worker is subject to social security taxation, the employer is required to withhold 6.70 percent of his/her cash wages on the first $35,700 of wages paid (1983 regulations). This 6.70 percent is matched by an equal amount paid by the employer. The social security withheld from the employee's paycheck and that contributed by the employer must be deposited in an authorized commercial bank or Federal Reserve Bank on either a monthly or yearly basis, depending on the total amount of social security tax withheld. The employer must keep records of the tax withheld and also prepare a W-2 form, and a Wage and Tax Statement, for each employee whose wages have been subject to social security tax. Farm employees are exempt from income tax withholding for federal income purposes, but some states do require withholding for state tax purposes.

Child Labor Regulations

The Fair Labor Standards Act of 1974 specifies a minimum age for employees. In general, sixteen years is the minimum age for employment in agriculture during school hours. Outside of school hours, the minimum age for employment is fourteen with two exceptions; children twelve or thirteen can be employed with written parental consent, and children under twelve can work on their parents' farm. Children under the age of twelve cannot be hired for farm work except by their parents. Currently, any violation of the child labor provisions is subject to a maximum of $1,000 fine per violation.

Additional regulations apply to jobs that are classified as "hazardous." Such jobs include working with agricultural chemicals and anhydrous ammonia, driving and operating most farm machinery such as tractors, combines, corn pickers, and mowers, and working with breeding stock such as a bull or a boar. Children under the age of fourteen cannot be employed in hazardous jobs, and fourteen- or fifteen-year-old employees must be certified for these jobs. Certification can occur by taking extension or vocational agricultural safety training and machine operation courses. These courses will result in the employee receiving a "Certificate of Training" which will then qualify the employee to do work designated as hazardous.

Occupational and Safety Health Act (OSHA)

OSHA regulations are in a constant state of flux, but they do affect many farms and farm employees. The purpose of the OSHA regulations is to provide safe working conditions and to eliminate potential accidents on the job. Current OSHA regulations require the following from farm employers: (1) A notice must be posted that explains the "rights and responsibilities of employees" under the OSHA regulations and the responsibilities of the employer as well. (2) Slow-moving vehicle (SMV) signs must be displayed on the back of tractors and implements driven by employees who travel 25 miles per hour or less on public roads. Even though some farms may not explicitly be covered under the OSHA regulations, many states have also adopted the SMV sign and require it as a matter of state law. (3) Detailed records must be kept on any work-related accident if seven or more employees are hired, and fatal accidents must be reported to the nearest OSHA office within 48 hours. (4) Each employee is to be trained in the safe performance of any task she or he is assigned, and adequate safety equipment is to be provided for the employee's use. (5) Rollbar protection must be provided on most tractors (except low profile tractors), and the employee must be informed of the major safety regulations affecting tractor utilization. (6) All shields and guards must be kept in place on machinery. In addition, electrical outlets must be fitted with appropriate devices to eliminate electrocution. The employer has the general responsibility to keep the workplace safe and free from hazards.

Worker's Compensation and Employee Liability

In many states farm employees are covered by worker's compensation regulations. Under these regulations, employees can receive benefits for job-related injuries or illnesses that result in permanent or temporary, partial, or total disability or death. Benefits are received regardless of whether the employee or the employer was at fault. The benefits received by the employee include income maintenance during the treatment and recovery period as well as medical care and rehabilitation if required. Because the employer is financially responsible, regardless of fault, for payments under the worker's compensation regulations, it is usually advisable to participate in the state insurance program or to purchase an insurance policy from a private carrier to cover such financial risks. Premiums for workman's compensation policies typically amount to 5 to 8 percent of the employee's salary but cannot be deducted from wages.

Even in states where worker's compensation does not result in strict liability for employee accidents, an employer may still face liability for injuries to employees. In general, the employer does have the responsibility to provide safe working conditions and safe equipment and tools with which to work, to warn employees of potential dangers and instruct them in the use of dangerous equipment, and to encourage them to wear protective clothing and other devices when handling chemicals such as herbicides and pesticides. However, under civil law liability regulations, the employer must be negligent and that negligence must be the cause before an employee can recover for an on-the-job injury.

To maintain good employee relations and reduce the financial hardships that may be associated with on-the-job injuries, many employers carry hospital, medical, and various types of life insurance policies on their employees. In addition, liability insurance to protect an employer in case of an accident is also an important part of a total insurance program.

Unemployment Insurance

Unemployment insurance is a controversial topic in many farm communities, and state and federal regulations are in a constant state of flux, but current rules require farm employers who hire large amounts of labor to pay for part or all of the cost of unemployment insurance through a combination of state and federal taxes.

Under federal law, a farm employer must pay a tax for unemployment insurance of 3.4 percent of the first $6,000 of wages for each employee if total wages are in excess of $20,000 in any calendar quarter in the current or preceding year, or if they employ 10 or more workers in 20 different weeks in the current or preceding year. A state tax in addition to the federal tax may also be imposed. The unemployment insurance program provides payments for a certain period of time to an employee who is terminated and unable to obtain other employment.

In general, farm managers can expect additional governmental regulations to influence the working conditions, pay, and environment for farm employees. In some states, rules concerning the collective bargaining rights and union representation of farm workers and the regulation of migrant worker employment and working conditions are becoming increasingly important and will affect the employment conditions and cost of hiring labor. The successful farm manager who intends to hire farm employees must obtain the latest information concerning regulations on minimum wage, social security, occupational safety and health, workman's compensation, and collective bargaining and unionization.

Summary

Many farmers hire part- or full-time employees; consequently, labor management skills are important to the successful operation of the farm. Labor or personnel management involves four important functions: (1) obtain the proper kind and number of personnel, (2) develop the skills that are necessary for proper job performance, (3) motivate employees to perform to the best of their capabilities, and (4) develop appropriate and equitable compensation plans.

Job analysis and descriptions that identify the responsibilities, authorities, and duties to be performed by the employee can be very useful in assessing the type of personnel needed in the farm operation and in communicating with the employee his or her job activities and the basis for assessing performance. Recruitment, interviewing, and selection of employees are tasks that must be accomplished in a systematic fashion to obtain qualified employees with the appropriate skills. Once an employee has been hired, he or she should participate in an on-the-job training program to enhance employee productivity and morale, and reduce the chances of work-related accidents.

A major task of personnel management is employee motivation and evaluation. Employees usually are looking for a number of sources of satisfaction from their employment environment including (1) pay, (2) job security, (3) pleasant co-workers, (4) credit for work done, (5) a meaningful job, (6) opportunity to advance, (7) comfortable, safe, and attractive working conditions, (8) competent and fair leadership, and (9) reasonable orders and directions. Periodic review sessions should be used to evaluate performance and provide guidance concerning areas where improvement is needed.

Compensation of farm employees must be competitive with alternative jobs to attract and keep high-quality personnel. The compensation package in many cases will include a salary plus a set of fringe benefits such as insurance, housing, electricity, and other perquisites. Various types of incentive programs might also be used to enhance labor productivity and share the benefits of this increased productivity between an employer and employee.

Employers must continually be aware of the changing state and federal regulations concerning employment and working conditions. These regulations include social security, worker's compensation and employee liability, minimum wage requirements, Occupational Safety and

Health Act regulations, child labor regulations, and unemployment insurance. These regulations not only influence the working environment, but they also can have a significant impact on the cost of hiring labor.

Questions and Problems

1. Identify the four major functions of labor or personnel management.
2. Discuss the relationships between the concepts of responsibility, authority, and accountability.
3. Write a job description for a swine herdsman in a large farming operation. What authorities and responsibilities would you include in such a job description?
4. List some of the questions you might ask in an interview with a prospective farm employee.
5. Discuss the factors an employer should consider in motivating and evaluating employees.
6. How can policies, procedures, and performance standards be used in the motivation and evaluation process?
7. How would you determine the appropriate salary to pay a farm employee?
8. What kind of fringe benefits might be considered as part of the benefit package for farm employees?
9. List and discuss the principles that should be considered in developing a labor incentive program.
10. How would you evaluate the profitability of an incentive program for the employer and the employee?
11. What are current federal minimum wage regulations for farm employees?
12. Summarize current Occupational Safety and Health Act regulations that employers must comply with to provide safe working conditions for their employees.

Further Reading

Bishop, C. E. "Dimension of the Farm Labor Problem." *Farm Labor in the United States.* New York: Columbia University Press, 1967, pp. 1–17.

Brown, Lauren H. *Making Farm Employment Competitive.* Rural Manpower Center Special Paper No. 1. East Lansing, Michigan State University, 1967.

Carroll, Phil. *Better Wage Incentives.* New York: McGraw-Hill Book Company, 1957.

Downey, W. D. "People Problems on the Farm." *Purdue Farm Management Report.* Cooperative Extension Service, Purdue University, June 1976.

Fuller, Earl I. "Increasing Labor Productivity." Agricultural Extension Service, University of Minnesota, 1977.

Knorr, Lawrence A., and Joachim G. Elterich. *Analysis of Delaware's Full-time Hired Farm Labor Situation.* Agricultural Experiment Station Bulletin 385. Newark: University of Delaware, 1971.

Robbins, Paul R. *Keeping Good Hired Farm Labor.* Cooperative Extension Service EC–306. Lafayette, Ind.: Purdue University, 1966.

Schaffer, Harry W., George L. Casler, and Robert S. Smith. *Incentive Payment Plans for Hired Men.* Agricultural Economics Extension No. 49. Ithaca: New York State College of Agriculture, 1959.

Smith, Richard B., and Earl O. Heady. "Paradox of Farm Labor." *Iowa Farm Science* 24, No. 12 (June 1970): 3–5.

Thomas, Kenneth H., and Michael D. Boehlje. *Farm Business Arrangements: Which One For You?* North Central Regional Extension Publication No. 50. St. Paul: University of Minnesota, 1982.

Weightman, Paul Wesley Harrison. "Financial Incentive Plans for Farm Labor in New York State." Unpub. Ph.D. dissertation, Cornell University, Ithaca, New York, 1966.

Wolfe, John, Michael Boehlje, and Vernon Eidman. "An Economic Evaluation of Wage Rates and Incentive Agreements for Full Time Hired Labor on Oklahoma Farms." Bulletin B–713, Oklahoma Agricultural Experiment Station, Stillwater, Oklahoma, April 1974.

13

LAND ACQUISITION
AND MANAGEMENT

CHAPTER CONTENTS

Land is a basic resource for most agricultural production, and for many farmers a key concern is how to obtain and maintain access to an adequate land base. From a business viewpoint, land is a basic factor of production in the farming operation. However, there are numerous intangible dimensions associated with the ownership of land, and the ownership of farm real estate plays an important role in the social structure of the community. Ownership or control of land provides an opportunity for employment and a place to raise the family. Therefore, land has social, community, and family dimensions and implications as well as being a source of income. Furthermore, compared to many other resources used in farming, land is relatively fixed in supply and is a nonrenewable resource. Consequently, stewardship and maintenance of the productivity of this resource are important considerations for the farm manager and the population as a whole.

In this chapter we will discuss various methods of obtaining the services of land, including ownership as well as leasing and rental arrangements. The discussion will emphasize the economic considerations of land valuation and appraisal, the financial feasibility of purchasing land, and alternative methods of transferring property between individuals. The economic, legal, and control implications of various types of leasing arrangements such as the cash lease, the crop share or livestock share lease, and the flexible cash lease will also be discussed. Finally, the economic dimensions of conservation programs and land improvements such as tiling and waterways, terracing, and clearing will be reviewed along with the important tax considerations and "public" versus "private" costs of such investments. Although the emphasis of the discussion in this chapter will be on the economic and financial dimensions of land acquisition and management, the social and community dimensions of farm real estate should not be ignored—particularly when evaluating ownership compared to rental of farm real estate.

CONTROL OF LAND

Acquiring control of land need not involve ownership. The services of land can be acquired through rental or leasing arrangements as well; in fact, almost 50 percent of the land farmed in some Midwestern states is leased or rented and farmed by a tenant-operator. A key issue faced by many farmers is whether they should buy or rent land.

The choice between buying or renting land depends on the unique characteristics of each farm operation. Ownership of at least some real estate enables farmers to have a base for their farm operation. This base not only gives them a minimum acreage to farm, but it also provides the home and farmstead for raising their families and a location for intensifying the farming operation through expansion of the livestock facilities. Ownership also eliminates the risk of losing a lease and thus not having an adequate land base for the livestock operation and efficient employment of the machinery complement.

In recent years, an additional advantage of ownership has been the price appreciation of real estate. Unlike most resources used in farming, land does not depreciate or deteriorate if managed properly. Although the farmer has not received the financial benefits of price appreciation in a cash form that is available for direct consumption, appreciation has increased net worth. This increased net worth can be used as a financial base for borrowing funds to expand the farm operation as well as a cushion or reserve against short-term financial losses that may require "refinancing."

Thus, land ownership has important income, capital appreciation, and risk-reduction dimensions for the farm operator as well as the social and family dimensions of a permanent home and residence for the farm family.

The price that must be paid for these attributes of ownership is the substantial capital outlay needed to purchase land. In many parts of the Midwest, land is selling for $3,000 to $4,000 an acre, and ownership of a quarter- or half-section involves a capital outlay of $500,000 or more. Most farmers, particularly beginning farmers, do not have sufficient capital for the down payment required for land acquisition and have enough left for machinery and equipment purchases and working capital. The financial requirements of purchasing land can drain valuable funds away from other investment alternatives. The basic question, therefore, becomes one of which method of land acquisition has the highest financial payoff compared to alternative uses of the farmer's funds, and which alternative is "financially feasible" or within the financial capability of the farm operator. We will discuss the issues and attributes of land ownership first, followed by the important dimensions of land rental or lease.

OWNERSHIP OF REAL ESTATE

To evaluate the advantages and disadvantages of owning rather than renting farmland, the farm manager must understand the concepts of land valuation and appraisal, the financial feasibility of land ownership, and alternative methods of purchasing land or transferring real estate from buyer to seller. Each of these topics will be discussed in turn.

Land Valuation and Appraisal

As a factor of production, economic theory would suggest that farmland should be valued on the basis of the value of the products produced from the land. However, land is a unique resource and, as has been suggested earlier, has more than purely economic characteristics. Consequently, the value of land is not based soley on its economic productivity. For the owner-operator the land provides not only a source of income, but also a residence with certain characteristics including location and style of life. For the absentee landlord, land may be solely a source of investment income or capital gain. He or she may hold real estate primarily as a hedge against inflation. Hence, different individuals will place different values on a particular piece of land depending on their intended use of the real estate. Some will value it as a place to live. others as a means of making a living or as an investment, while others will value it based on the style of life and the community (schools, churches, social groups, etc.) of which it is a part.

Although each individual parcel of land has unique characteristics and unique value, some generalizations concerning land values can be made.

Analysts and professional appraisers agree that good land has a tendency to be underpriced and poor land overpriced. This occurs in many cases because of incomplete information and an inability to recognize the differences in productivity between various types of land. Furthermore, there is a natural price resistance on the part of buyers of land of higher quality and thus higher price.

A second generalization is that tracts that include improvements typically do not sell for a significant premium above unimproved land. Thus, buildings and improvements can typically be obtained more economically by purchasing them with land rather than separately. This has become particularly true in many areas of the Midwest that are in transition from livestock or general farms to cash grain farms. Improvements add value to real estate only when they are in demand, are adaptable to future use, and can be incorporated as part of the overall farm operation. If the unit is being purchased as a headquarters for a farming operation, then improvements can be expected to add to the value of the farm real estate. In the case of the "add-on" unit where farmers are expanding the size of their operation, improvements may not add significantly to the value of the property.

It also appears that large farms tend to sell for a lower price and small farms for a higher price per acre. This probably occurs because there are more potential bidders for the smaller acreage. For many larger farms, the capital outlay necessary to make a purchase may exclude many potential buyers from actively participating in the bidding process. Furthermore, for the large farm operation, the residence and improvements will usually amount to a lower cost per acre compared to a smaller farm, thus contributing to the lower value per acre of the larger unit.

Appraisal Methods

In spite of the earlier generalizations concerning land values, each parcel does have unique characteristics that merit specific evaluation and appraisal. In many cases, this evaluation will be done by a professional real estate appraiser. The appraiser will typically use a combination of three approaches in determining the value for a particular parcel of real estate: the market value method, the cost method, and the income method. Although each of these three methods will be briefly reviewed, the income approach to land valuation and appraisal will be emphasized because of its consistency with the net present value method of evaluating investments discussed earlier. In essence, this approach to valuation determines the long-run profitability of a land investment.

The Market Approach   The market approach to valuing real estate essentially attempts to determine what the property would bring if sold. According to Murray, et. al., this approach involves four basic steps: (1) defining the kind of sale method to be used (cash, contract, etc.); (2) selecting and analyzing nearby sales; (3) determining the comparability of the nearby sales and the tract being appraised; and (4) adjusting for value changes between the time of sale of comparable properties and the time of the present appraisal.[1]

Only bona fide sales between a willing buyer and willing seller are used in the market approach to valuing a particular tract of real estate. Sales between family members at concessionairy prices, or distress sales because of sickness, death, or financial difficulties are not included in the comparable sales data. Judgment is an important component of the valuation procedure using the comparable sales approach. Adjustments in comparable sales data to reflect differences in location, soil type and fertility, type and condition of improvements and residence, size of farm and ease of access, and even the terms of the sale must be made. For example, a contract sale with an interest rate below the "market rate" must be adjusted to eliminate the price premium that is paid for the lower rate of interest. And even after all of these adjustments have been made for differences in quality characteristics, an additional adjustment may be necessary to reflect a time lag since the comparable sales occurred. In order to use the market approach to land valuation, the farm manager or appraiser must, therefore, be knowledgeable about real estate market conditions and changes that have occurred in that market since the comparable sales occurred, as well as the physical characteristics of farmland including soils, yield potential, erosion problems, and the characteristics of improvements.

The Cost Approach   A second and somewhat antiquated approach to value is that of cost. The basic philosophy of the cost approach is to inventory the various resources of the farm, estimate their cost, and then sum these costs to obtain a total value. Because of the extremely difficult task of associating a cost with land, this approach is quite difficult to use for unimproved acreages. Tracts with a major set of improvements may utilize the cost approach to valuation more effectively, particularly in the valuation of the buildings.

The typical approach in applying the cost method is to determine the use that will be made of a building, its physical condition, and the outlay that would be required to construct a similar structure at today's prices. The cost method reflects the replacement cost for a facility or structure in the same condition, not the initial cost minus depreciation. Estimating replacement cost was discussed in detail in Chapter 2. Note that this replacement cost approach will not result in the same depreciated value that is reflected in the seller's tax records.

A further difficulty encountered with the cost approach to appraisal or valuation is that it does not recognize basic supply and demand factors

---

[1]See William Murray, Duane Harris, Gerald Miller, and Neill Thompson, *Farm Appraisal and Valuation.* Ames: Iowa State University Press, 1983 for further discussion of these steps.

that may exist in the market. With the current interest in "add-on" units in cash grain areas, bidders frequently prefer unimproved land that is completely tillable to improved tracts with buildings. Consequently, even though the buildings may be in good physical condition, and when valued at replacement cost would add to the value of the tract, buyers may not be willing to pay a premium for this improved land compared to unimproved real estate. Thus, the cost approach is most useful in valuing buildings and improvements by providing some indication of the outlay that would be required to construct comparable structures, but it may not be an accurate indication of their value in the marketplace.

The Income Approach   The third method of resource appraisal and the one that provides the best economic base for determining the value of a tract of real estate is the income approach. This approach involves determining what the present value of the income stream from the tract of land would be for a landlord or an owner-operator. Essentially, a whole-farm budget is developed for the farm, including the crop rotation and livestock program. Projected prices are assigned to the various livestock and crop outputs, and expenses are determined. The income and expense items to be included in the computation of net income depend on whether the buyer will be a landlord or owner-operator. Once a net income figure has been determined, this income is capitalized at a chosen rate of return to obtain the income value for the farm. This procedure is in reality a modified application of the net present value procedure discussed in Chapter 8. Because the income approach and capitalization process assume a perpetual life of the asset and a perpetual income stream, it is important to use long-run estimates of prices, yields, expenses, and the capitalization rate in the computations rather than being swayed by recent trends or short-run price and yield fluctuations.

In more explicit fashion, the steps involved in the income or earnings approach to appraisal, assuming an owner-operator is the potential purchaser, are as follows: (1) select a long-run average set of prices for each crop and livestock product that is realistic in terms of future planning and price expectations; (2) estimate total cash farm receipts based on the productivity capacity of the land and the size and capacity of buildings and structures, using typical yields from crops and livestock; (3) estimate cash farm operating expenses, excluding principal and interest payments on real estate; (4) calculate a net cash balance by subtracting cash operating expenses from cash farm receipts; (5) adjust net cash flow by deleting the value of the operator's unpaid family labor if not charged as a cash expense, and subtracting depreciation on buildings and improvements as well as machinery and equipment. If not included in cash expenses, interest on nonreal estate capital is also deducted. The final result should be a residual return to the real estate. (6) Divide this return by the capitalization rate to arrive at the earnings value of the farm.

The simple formula for determining the income or earnings value of real estate is

$$V = \frac{R - E - L - I}{r} \tag{13.1}$$

where V = earnings value, R = total cash farm receipts, E = total cash farm expenses, L = the value of the operator's and unpaid family labor, I = interest on nonreal estate capital, and r = the capitalization rate.[2] As has been suggested earlier, the basic data and information used in the earnings approach to valuation can be generated by developing whole-farm budgets for the farms as discussed in Chapter 6.[3]

One of the most difficult decisions required in using the income approach to valuation is choosing the appropriate capitalization rate. From a conceptual viewpoint, the capitalization rate should reflect the cost of capital or the cost of funds committed to the purchase of real estate as discussed in Chapter 8. However, adjustments are necessary to reflect differences in the risk associated with land compared to alternative investments, the expected rate of growth in income (and thus price appreciation) from farm real estate, and the historical rates of return that have been acceptable by participants in the real estate market. If the real estate investment appears to be a much more risky venture compared to alternative uses of funds, then a risk premium must be added to the base capitalization rate to reflect a higher minimum acceptable rate of return from the more risky land investment. Conversely, if the buyer anticipates an increase in the income stream of the asset over time which has not been explicitly recognized in the cash flow calculations, she or he should be willing to accept a lower current cash rate of return because of the growing income stream. In this case, the capitalization rate would be reduced to reflect the lower rate of current cash return that is acceptable because of the financial gain expected to accrue through a growing income stream and the resulting price appreciation of the asset.[4] For example, if the farmer expects the income

---

[2]For an asset with an infinite life and perpetual income stream, the net present value formula of Chapter 8 mathematically collapses into the simple formula noted in equation 13.1.

[3]This simple formula does not consider income taxes; both the income stream and the capitalization rate are calculated on a before-tax basis. If taxes are included as a cash expense, then the capitalization rate must also be reduced to an after-tax rate. For assets with an infinite life, where all the income is taxed as ordinary income, the before-tax and after-tax analysis procedures will generate the same results. If part of the return is taxed as capital gain, the resuls may be different and depend on the specific marginal tax bracket of the purchaser. A potential purchaser with a higher marginal tax bracket will in general be able to pay a higher price for the land than a purchaser with a lower tax bracket.

[4]Emanuel Melichar, "Capital Gains Versus Current Income in the Farming Sector," *American Journal of Agricultural Economics* (December 1979): 1085–92.

from real estate to increase by an average of 4 to 5 percent per year (which is the historical average) and has a minimum acceptable rate of return for investment purposes of 10 percent per year, his or her cash income capitalization rate would be 5 to 6 percent. This adjustment essentially reflects the fact that the farmer will receive a 5 to 6 percent rate of return on the investment in a cash form and a 4 to 5 percent return in the form of increased value of the asset or price appreciation.

A third adjustment may be required in the capitalization rate to reflect the minimum acceptable rate of return by participants in the farm real estate market. Because farming offers nonmonetary benefits as well as a place to live and a source of employment, it has been suggested that farmers are willing to accept a lower rate of return in farming than they might be able to obtain by investing their funds outside of the agriculatural sector. In fact, prior to the early and mid-1970s, the relationship between current net income earned by farmers and current market value of farm property suggested an average rate of return of 4 to 5 percent. Thus, even though a higher return could be obtained by investing outside of agriculture, the nonmonetary benefits of farming and expectations of a growing income stream were substantial enough to make a 5 percent current monetary rate of return acceptable to farmers. Thus, some would argue that the capitalization rate should not reflect the opportunity cost of a farmer investing his or her funds outside of agriculture, but the rate of return on investment in agriculture as reflected by current net income compared to current farm property values in the market.

The choice of a capitalization rate is not an easy task and requires sound judgment and an assessment of the motivations of the participants in the real estate market. The best advice is to base the capitalization rate on the opportunity cost of the funds that are being invested in the real estate, and to make appropriate adjustments to reflect differences in risk, differences in expected rates of growth in income (and thus price appreciation), and differences in the noneconomic benefits associated with the alternative investments. To illustrate, a potential buyer who has a cost of capital of 12 percent and judges that the risk adjustment should be 1 percent, that the expected growth and appreciation rate is 6 percent, and that a noneconomic benefit of 1 percent is justified will use a capitalization rate of 6 percent $(12 + 1 - 6 - 1)$.

Financial Feasibility

In addition to an analysis of the economic profitability or the price that can be paid for a particular parcel of land, the financial feasibility of the land purchase must also be evaluated. As noted in Chapter 8, financial feasibility essentially involves evaluating the cash flow generated from the land compared to the principal and interest payments required on the real estate loan. If a cash purchase is made, financial feasibility is not a problem.

Alternatively, the installment land contract can result in severe cash flow problems for the farm firm because of the low downpayment, the short term, and the resulting high annual interest and principal payments on land purchases.

The steps involved in evaluating the financial feasibility of a land purchase are quite simple and utilize many of the same data that have been developed for the profitability or valuation analysis. First, the cash expenses and cash income are estimated, and a net cash flow per acre or for the farm as a unit is calculated. Because principal and interest payments must be made with cash and do not come from noncash items such as depreciation, the net cash flow is not adjusted for noncash items as is the case with the valuation procedure discussed earlier.

Next, the annual principal and interest payments required to amortize the contract or mortgage are determined. By comparing these cash requirements for debt servicing with the cash availability, the ability of the land to "make its own payments" can be evaluated. If a cash deficit occurs, cash subsidies from some other enterprise or part of the farm business will be necessary to make the land purchase financially feasible. Alternatively, different loan terms that include a larger downpayment, a longer term, or possibly a balloon payment at the end of the contract may be required.

Thus, as with any investment decision, the analysis involves two phases: (1) the economic profitability of the investment, which indicates the maximum value of the real estate or the price that can be paid for the investment, and (2) the financial feasibility of the investment or the ability of the asset to generate sufficient cash to cover all cash expenses plus pay principal and interest on the loan.

An Example

A specific example will be used to illustrate the application of these concepts to a real estate purchase. The farm includes 527 acres: 291 acres of high-quality land with less than 4 percent slope, 147 acres of medium-quality land with slopes in excess of 6 percent, 47 acres of pasture land, and 42 acres of buildings, roads, and waste. The long-run cropping plan is to produce 146 acres of corn and 145 acres of soybeans on the higher productivity land, 132 acres of corn and 15 acres of alfalfa hay on the medium-quality land, and to use the pasture in a cow-calf operation. Table 13.1 summarizes the yield and price expectations for the crops and the estimation of gross income from the farm. Using 1981 data on cost of production per acre, Table 13.2 indicates the crop expenses that would be incurred by an owner-operator who owned the farm. These data include both the operating expenses and any fixed expenses (including machinery depreciation) that might be incurred in the operation of the farm. The computation of net farm income is summarized in Table 13.3. Note that net income above all expenses would total $59,667 using the specified long-term

Table 13.1 Gross Farm Sales

Land Description	Acres	Crop	Yield	Price	Gross Income per Acre
High-quality cropland (<4%)	146	Corn	117 bu	$3.00	$351.00
High-quality cropland (<4%)	145	Soybeans	45 bu	7.50	337.50
Medium-quality cropland (>6%)	132	Corn	90 bu	3.00	270.00
Medium-quality cropland (>6%)	15	Alfalfa hay	4 ton	75.00	300.00
Pasture	47	Native	100 animal days	0.30	30.00
Buildings, roads, and waste	42				
Total	527				

prices, yields, and costs. Finally, alternative values for this farm based on different capitalization rates are summarized in Table 13.4. The values per acre range from $1,132 to $2,831 depending on the capitalization rate. Thus, with a higher growth rate, decreased risk, or a lower minimum acceptable rate of return, the land has a higher value. For example, reducing the capitalization rate from 8 to 6 percent results in a 33 percent increase in the value of land. A similar relationship exists in other cases, indicating the sensitivity of land values to the capitalization rate.

The financial feasibility analysis for this land purchase is illustrated in Table 13.5. If we assume a sale price of $1,875 per acre, a 25 percent downpayment, and a 25-year loan at 12 percent, the annual principal and interest payments amount to $94,489. Given the cash income and expense projections utilized in the valuation or appraisal analysis, a $15,000 living

Table 13.2 Total Farm Expenses

Item	Corn	Soybeans	Hay	Pasture
Production Expenses (Per Acre)				
Preharvest machinery	$ 26.10	$ 35.10	$ 7.60	$10.00
Harvesting and storage	51.80	23.15	54.00	—
Labor	17.60	18.40	22.00	—
Seed, chemical, and fertilizer	79.40	71.00	60.80	—
Total per acre	$174.90	$147.65	$144.40	$10.00
Fixed Expenses (Total for the Farm)				
Property taxes	$5,000			
Depreciation on improvements	4,000			
Insurance on improvements	400			
Total	$9,400			

Table 13.3 Net Farm Income

Land Description	Acres	Crop	Gross Income per Acre	Total	Operating Expenses per Acre	Total
High-quality cropland (<4%)	146	Corn	$351.00	$ 51,246	$174.90	$25,535
High-quality cropland (<4%)	145	Soybeans	337.50	48,938	147.65	21,409
Medium-quality cropland (>6%)	132	Corn	270.00	35,640	174.90	23,087
Medium-quality cropland (>6%)	15	Alfalfa hay	300.00	4,500	144.40	2,166
Pasture	47	Native	30.00	1,410	10.00	470
Total				$141,734		$72,667
Income above operating expense						$69,067
Fixed expense						$ 9,400
Net farm income						$59,667

allowance, and the $94,489 principal and interest payment, the data in Table 13.5 indicate a negative net cash flow of $31,576 will result. The data in Table 13.6 illustrate the impact of different loan terms and arrangements on the financial feasibility of the purchase. Increasing the downpayment to 35 percent and extending the loan to 35 years would still not result in a positive annual net cash flow. Column 4 in Table 13.6 indicates the impact of increasing the price expectations on corn to $3.60; these higher price expectations increase the net cash flow and make the purchase financially feasible.

This Midwest example and the data in Tables 13.1–13.6 illustrate the type of analysis that must be completed for an appropriate assessment of the economic profitability or value of a tract of land and the financial feasibility of the purchase.

Methods of Transferring Real Estate

Numerous arrangements are available to transfer the ownership of real estate between buyer and seller. These methods include the cash, mortgage financed, or contract sale, the exchange, the gift, or the transfer of property at death. A number of factors must be considered in choosing among these transfer methods. These factors include inflation, life expectations, costs of transfer, tax regulations, cash flow, estate planning objectives, and security interest.

1. *Inflation.* Inflation has become a major consideration in any investment or disinvestment decision. If buyers expect land to appreciate at a rate similar to the rate of inflation as has occurred in the past, they can

Table 13.4 Earnings Value at Different Capitalization Rates

Cost of capital (required rate of return) (%)	12	12	12	12	12	12
Rate of income growth and appreciation (%)	2	3	4	5	6	7	8
Capitalization rate (%)	10	9	8	7	6	5	4
Net farm income ($)	59,667	59,667	59,667	59,667	59,667	59,667	59,667
Farm value ($)	596,670	662,967	745,838	852,386	994,450	1,193,340	1,491,675
Value per acre ($)	1,132	1,258	1,415	1,617	1,887	2,264	2,831

Table 13.5 Financial Feasibility of the Farm Purchase

Sales price	$988,125 ($1,875 per acre)	
Downpayment	25 percent	
Interest rate	12 percent	
Term	25 years	
Annual payment (even total payment plan)	$94,489	
Cash flow		
Cash income		$141,734
Cash operating expenses[a]	$58,421	
Property taxes and insurance	$5,400	
Family living and income taxes	$15,000	
Annual land payment	$94,489	
Total	$173,310	$173,310
Net cash flow		−$31,576

[a]$14,246 of depreciation and other fixed costs are not included as cash expenses.

expect to pay more for the land at some future date. Consequently, if they have adequate financing and want to expand their land base, it may be desirable to make the land purchase now rather than wait. For the seller, particularly the retiring farmer, inflation is also an important consideration. Retiring farmers must be careful not to lock themselves into fixed or constant income investments where the income stream and the investment principal do not adjust with inflation or increase with the general price level.

Historically, real estate has increased in value at an equivalent or faster rate than the rate of inflation. For example, the rate of price appreciation in Iowa farmland from 1965 to 1980 averaged approximately 13 percent

Table 13.6 Financial Feasibility with Alternative Financing Arrangements and Price Assumptions

Item	Extend Loan to 35 Years	Increase Downpayment to 35 Percent	Extend Loan to 35 Years and Increase Downpayment to 35 Percent	Extend Loan to 35 Years, Increase Downpayment to 35 Percent and Corn Price of $3.60
Cash income	$141,734	$141,734	$141,734	$159,111
Cash operating expenses	58,421	58,421	58,421	58,421
Property taxes and insurance	5,400	5,400	5,400	5,400
Family living and taxes	15,000	15,000	15,000	15,000
Annual payment	90,636	81,891	78,551	78,551
Net cash flow	−$27,723	−$18,978	−$15,638	+$1,739

per year. This rate of increase in the value of farm real estate compares to an average inflation rate of approximately 7 to 8 percent during this period. Thus, real estate was an effective inflation hedge during the 1960s and 1970s.

2. *Life expectations.* This is a particularly important consideration from the viewpoint of the seller. During the past 20 years in particular, medical discoveries have significantly increased the life expectancy of older people. Medical breakthroughs with respect to heart disease and other age-related illnesses may not be far away and will increase the life expectancy of retirement-age people even more. With the possibility of living more years, retiring farmers must be careful that they do not outlive their "income stream."

3. *Costs of transfer.* Substantial costs may be incurred in the process of transferring property, particularly through the sale method. These costs include the real estate brokerage fees plus the abstracting, legal, and appraisal costs that may be incurred during the transfer process. Depending on the type of real estate and its location, these fees may amount to as much as 4 to 10 percent of the fair market value of the property. This shrinkage in the amount of proceeds received at the time of sale of real estate cannot be ignored by the seller and may influence the price the seller will try to extract from the buyer.

4. *Tax regulations.* Both income (ordinary and capital gains) and estate tax regulations influence the type of transfer method chosen. The details of their impact for different transfer methods will be discussed later. However, it should be cautioned here that, although tax regulations are important, they should not be the overriding factor or consideration in choosing a particular transfer method. In fact, exploiting all tax advantages may result in the sacrifice of other important goals of the retiring farmer.

5. *Cash flow.* The current cash needs of the seller and the amount of cash the buyer can use in the purchase will influence the transfer method used. For example, a farmer who plans to invest the sale proceeds elsewhere will find that the installment contract method of sale which spreads the payments out over a number of years will probably not be appropriate. Alternatively, if a retiring farmer desires to have a constant and fixed annual inflow during the retirement years, a contract sale may be desirable. If the farmer has other assets and current income is not required, a gift or exchange may be appropriate. For the buyer who has limited cash, a contract with a low downpayment, low annual payment, and a balloon at the end of the contract may be appropriate.

6. *Estate planning objectives.* The objectives held by the farm family with respect to estate planning and intergenerational transfers are critical in choosing a particular real estate transfer method. For example, if continuation of the firm is planned with one of the children taking over the farm business, it may be desirable to transfer some of the ownership in the real

estate to the on-farm heir during the lifetime of the retiring farmer. With this procedure the heir can obtain some experience in managing the property while the father is available to provide advice and counsel. A lifetime transfer would also give the on-farm heir(s) some financial security since they would own an interest in the farm that could not be wrestled from them at the death of the parents. At the same time, the parents may not want to transfer so much of their real estate or farm assets to the heirs that they become dependent on the children during their retirement years. Thus, the parents can protect their financial security by maintaining ownership in a substantial portion of the farm real estate.

7. *Security interest.* A final concern, particularly if the sale method of transfer is used, is the security interest the buyer and seller have in the transferred property. Different types of security interests will influence the cost in terms of time and money that must be incurred by the seller to obtain redress if the buyer defaults. The type of security interest will also create the possibility of the buyer losing the land if he or she does not perform all contractual obligations in purchasing the property.

Real estate transfers can occur during the lifetime of the owner or at death. The three most common methods of lifetime transfer of real estate are the sale, the exchange, and the gift. In some cases, real property is also transferred in exchange for a private annuity. In a private annuity, the recipient of the property promises to make periodic payments to the transferor for a specified time—frequently until her or his death. Developing the terms of a private annuity requires serious consideration of the estate, gift, and income tax regulations as well as the equity of the arrangement to both parties.

The Sale   The sale is the most popular method of transferring real estate. With the sale, complete ownership in the property is transferred to the buyer. A sale can be financed through three different types of financial arrangements: (1) the buyer obtains credit from a third party; (2) the seller takes back a mortgage for part or all of the selling price; or (3) the property is sold on contract, with the seller again providing the financing.

The Cash Sale   With a cash sale, the buyer pays the entire purchase price in exchange for the deed to the real estate. The seller receives the full purchase price of the property in cash which can then be used for personal consumption or for reinvestment elsewhere.

If the buyer has insufficient cash to finance the land purchase, he or she may execute a separate mortgage or take over the seller's current mortgage. If the buyer "assumes" the seller's mortgage, responsibility for the mortgage payments is transferred to the buyer who also agrees to personal liability for the mortgage debt. If the property is purchased "subject to" the mortgage, the buyer agrees to make the payments on the outstanding debt but he or she is not agreeing to personal liability for the debt obligation. Consequently, if the buyer defaults on the loan, the seller can foreclose on the property, but the buyer's other assets are not available to settle the debt.

If a cash sale is used, the seller is required to pay capital gains tax on the amount of the gain (selling price minus adjusted basis) in the year of sale. This tax liability can result in a substantial reduction in the amount of cash realized by the seller. There is no concern with the seller's security interest when a cash sale is used since the seller obtains the full purchase price and thus complete performance has occurred.

Retained Mortgage Sale   If the seller retains a mortgage on the property, a substantial downpayment is usually required before the deed will be transferred to the buyer. The buyer has basically the same rights and obligations as when third-party financing is used. With this financing arrangement, the seller only obtains the downpayment in cash form, and the mortgage is equivalent to a fixed payment (income) contract that does not adjust or change in value with inflation or deflation. This type of transaction may qualify for installment reporting of gain for tax purposes if the regulations to be discussed shortly are met. If the buyer defaults in making payments on the mortgage, the seller must typically go through foreclosure procedures to obtain redress. These procedures can be costly as well as time consuming; they may take up to two years to complete and may entail some reduction in productivity because of disagreements as to ownership of the property.

Installment Contract Sale   The installment contract sale is an increasingly popular method for financing the transfer of real estate. With the contract sale, the buyer obtains possession of the property for the consideration of a downpayment and the execution of a contract for the remaining amount of the purchase price. The seller typically retains the deed or legal title to the property until all contract payments have been made.

The advantages and disadvantages of using the installment contract arrangement are numerous. The buyer who uses a contract can usually purchase real estate with a lower downpayment than with a conventional mortgage sale. A buyer also has better security of tenure and begins building equity in the real estate earlier compared to renting the land. However, the buyer may lose possession of the farm very quickly upon default. In addition, the annual payments on the contract may be high compared to the income from the land. If the balloon payment arrangement discussed shortly is used, later refinancing may be required at higher interest rates. Note that although most discussions of the installment sale contract emphasize the tax advantages of this method of transfer, numerous other advantages as well as disadvantages can be identified.

From the seller's viewpoint, the low downpayment that is typical with the contract may increase the number of potential buyers of the property. The contract payments provide a constant, steady flow of income during the retirement years that will not be reduced in amount or value during periods of deflation, and they can typically qualify for installment reporting of capital gain for tax purposes. Although the low downpayment may increase the risk of default on the part of the buyer, the seller can frequently regain complete ownership of the property within a relatively short period—often 30 days—depending on the forfeiture provision under state law. From the seller's viewpoint, a major disadvantage of the installment method of sale is the reduced liquidity and the fact that the contract does not adjust with inflation.

One arrangement frequently used with installment sale contracts is to combine partial amortization over a period of time and a balloon payment. For example, the contract might call for a 29-percent downpayment and the payment of an additional 31 percent of the principal plus interest over a 10-year period, with a balloon payment at the end of the tenth year of the remaining 40 percent of the purchase price. This procedure makes it possible to keep the annual payments on the contract relatively low, thus enabling the buyer to purchase the real estate without as large a cash subsidy from sources other than the land itself. At the end of the tenth year, the buyer has usually obtained sufficient equity to refinance the property and make the balloon payment using a third-party mortgage.

Installment Reporting of Gain Under Internal Revenue Code provisions, sellers of property can spread any capital gain over the life of a contract rather than report the gain for tax purposes in the year of sale. Sellers can delay tax payments and spread out the gain over many years rather than reporting it all in one year and be taxed in a higher marginal tax bracket. If the seller uses this installment treatment of gain, the annual contract proceeds are reported partly as interest taxed at ordinary income rates, partly as recovery of capital (and thus not taxed), and partly as capital gain. If this gain is a long-term capital gain (which is typically the case with real estate), then 40 percent of it is reported as income and taxed at the ordinary income tax rates.

Because of the incentives for the seller to use a contract with a low interest rate (and thus higher price per acre) to convert ordinary income into capital gain, the installment payment of tax provisions include an unstated interest rule. Essentially, this rule specifies that if the interest rate on an installment contract is below the "test rate," it will be recomputed at a higher rate and the interest received by the seller will be taxed as ordinary income at this higher rate. Regulations passed in 1981 set the test rate at 9 percent and the recomputation rate at 10 percent compounded semiannually, with the exception of the first $500,000 of value on sales between family members where the rate can be as low as 6 percent.

Prior to October 1980, for a contract to qualify for installment reporting, the contract payments in the year of sale could not exceed 30 percent of the selling price, and the payments had to be spread over at least two taxable years and involve two or more payments (Internal Revenue Code Section 453[b] before amendment by Public Law 96-471, 94 Statute 2247 [1980]). Recent legislation has dramatically altered these rules and the qualifications for reporting gain on an installment basis (Public Law 96-471, 94 Statute 2247 [1980]). The legislation eliminates the 30-percent limit on the downpayment and the two-year, two-payment requirement. It also restricts the potential for a seller to report the gain on an installment basis if the buyer is related to the seller and sells or transfers the property to a third party—thereby obtaining the same family liquidity as could be obtained in a cash sale. If such a transaction occurs within two years after the original sale, the original seller must report any remaining gain and be taxed thereon at the time of the subsequent sale. The legislation also changes the tax treatment of the property when a buyer on contract inherits the contract obligation at the death of the seller. This frequently occurs with contract sales between family members. In essence, the new provisions no longer allow the contract to be inherited tax-free; thus, the remaining untaxed gain is taxed to the estate, and an income tax must be paid in addition to the estate tax.

An example comparing an outright sale and an installment sale over a five-year period will clarify the tax differences. The following assumptions are made:

a. Sale price of the farm is $300,000; adjusted tax basis is $100,000.

b. Owner is married, files joint return, and both owner and spouse are under 65 years of age.

c. Regular taxable income without the sale of the farm after taking the standard deduction and exemptions is $7,000.

d. Under the installment sale, 20 percent of the sale price is received in the year of sale and 20 percent is received per year during each of the following four years, with 10 percent interest on the unpaid balance.

e. Interest income on the unpaid balance is ignored as it is assumed that if the sale was on a cash basis, the proceeds would have been invested and equivalent interest received.

All income received in one year from the sale of the farm.

Gain on sale of farm: $300,000 − $100,000 = $200,000
(40 percent of gain is taxable)

Taxable income:

Farm sale	$80,000	
Regular	7,000	
Total taxable income		$87,000
Income tax (1980 rates)		34,328

Farm sold on the installment basis.

Twenty percent of $300,000 or $60,000 received in year of sale and $60,000 per year received for subsequent four years.

In order to determine what portion of the installment payment is gain, the gross profit percentage must be figured. The gross profit percentage is calculated by dividing the profit by the contract price. This percentage is then multiplied by the annual payment to obtain the reportable gain.

Gross profit percentage: ($300,000 − $100,000)/$300,000 = 0.666
Annual reportable gain: $60,000 × 0.666 = $39,960
(40 percent of gain is taxable)

Taxable income:

Farm sale	$15,984	
Regular	7,000	
Total taxable income		$22,984
Income tax (1980 rates)		4,053

If the seller's taxable income exemptions and tax rate remained the same each year for five years, the total income tax would be as follows:

Years	Cash Sale	Installment Sale
1	$34,328	$4,053
2	534	4,053
3	534	4,053
4	534	4,053
5	534	4,053
Total 5-year	$36,464	$20,265

Thus, the installment sale results in a tax savings of $16,199 compared to an outright sale. If the differences in the timing of tax payments are taken into account, the savings of the contract will be even greater because of the delay in paying the taxes. If the regular taxable income or capital gain is greater, savings will be even larger using the installment procedure. The savings would also be greater if the installments were spread over a larger number of years.

One of the frequently overlooked characteristics of the installment sale contract or the seller-financed mortgage sale is that part of the cash proceeds from the property is tied up in a fixed payment contract. Even if the interest on alternative investments is identical to that on the contract, the contract or mortgage itself is not increasing in value or adjusting with inflation. In this respect, the contract or mortgage is identical to an annuity.

To illustrate the importance of this concept, let us return to our earlier example. Assume that if a cash sale occurred, the net proceeds after taxes would be invested in other real estate or personal property that increases in value at the rate of 7 percent per year. For the cash sale, the net after-tax proceeds from the sale of the farm are $266,206 ($300,000 − $33,794 of tax attributable to the farm sale). Based on a 7-percent rate of appreciation, the $266,206 of proceeds would increase in value to $373,381 by the end of the fifth year. This is an increase in value above the original $300,000 sale price of the farm due to price appreciation alone (ignoring interest or cash return annually) of $73,381 during the five-year period.

Alternatively, assume that the annual payments on the installment sale contract are also invested in assets that increase in value at the rate of 7 percent per year. The after-tax proceeds received from each payment amount to $56,481 ($60,000 − $3,519 of tax attributable to the installment sale of the farm). However, since the seller only receives one-fifth of the total sale proceeds during the first year, it is not possible to invest the entire proceeds in "inflation hedge" property immediately. If each payment is invested when it is received, the value of the property that can be purchased with the after-tax contract proceeds plus the appreciation of that property amounts to $347,545 at the end of five years. Thus, the increase in value due to price appreciation for the contract proceeds is only $47,545 ($347,545 − $300,000) compared to $73,381 for the cash sale. Alternatively, if the general rate of inflation is 7 percent, the contract sale sacrifices $25,836 in reduced purchasing power due to dollar shrinkage during the five-year period. It should be noted, however, that the purchasing power of the proceeds of a fixed income contract would actually increase during periods of recession and deflation.

Sale of the Farm Residence   Capital gain that occurs upon the sale of the residence receives special tax treatment under Internal Revenue Code regulations. If a taxpayer sells his or her residence, recognition of any gain can be postponed to the extent proceeds are reinvested in another residence. Reinvestment of the proceeds must occur within 24 months before or after the sale of the old residence. If the new residence costs more than the old residence, the entire gain is postponed. However, if the new residence costs less than the old residence, gain is recognized in the amount of the difference between the adjusted basis of the old residence and the cost of the new residence.

For farmers a major problem in postponing any gain on the sale of the residence is apportioning the basis and the selling price of the property to the residence and to the farm, or specifying that part of the farm that qualifies as a residence. Internal Revenue Service regulations are not very helpful in determining how many acres of a farm qualify as a residence. As few as 5 acres and as many as 65 acres have qualified in specific cases. However, the residence cannot include any part of the premises used for business purposes, such as a garage that houses a truck or pickup used in the farm business.

If a taxpayer is over the age of 55 at the time of sale of her or his residence, all or part of the gain may be *excluded* from the taxable income rather than recognition postponed. To qualify for this exclusion, the residence must have been the principal residence for at least three of the past five years. In addition, the taxpayer must be 55 or over at the time of the sale; the exclusion is available only once during the taxpayer's or spouse's lifetime. The amount of gain that can be excluded from taxable income is limited to $125,000.

The Exchange The tax-free exchange method of transferring real estate may offer substantial advantages over the sale or gift method. The exchange allows a reciprocal transfer of property or receipt of one item in return for another regarded as an equivalent. If the exchange is qualified as nontaxable, property with a fair market value that is greater than its basis can be exchanged for other property without recognition of any gain or the paying of any tax.

To qualify as nontaxable, the transaction must be an exchange and not a sale and separate purchase. However, this requirement does not eliminate the possibility of multiple exchanges, which would enable all of the parties to obtain the type of property they desire without the consequences of a sale. A second requirement for the exchange to be nontaxable is that it involve "like-kind" property. The "like-kind" requirement specifies that the property be similar or related in service or use, not in grade or quality. Thus, for example, rental urban real estate can be exchanged for a farm and qualify as a "like-kind" exchange. Improved real estate is considered to be "like-kind" with unimproved real estate for exchange qualification purposes. However, the exchange of real estate for personal property does not qualify as a nontaxable transaction. Real estate and personal property are not considered "like-kind" according to Internal Revenue Code regulations.

The third requirement for a nontaxable exchange is that the property be investment or business property and not property for resale. Thus, for example, a residence that is considered personal property cannot be exchanged tax-free for a farm that is considered business property. In addition, property must not be sold soon after an exchange, for this may be an indication that the property was being exchanged for resale rather than investment or business purposes.

Table 13.7 summarizes the consequences of using the exchange method of transferring real estate. Assume that Johnson has a farm with a fair market value of $400,000 and a tax basis of $100,000. Rather than sell this farm and be taxed on the $300,000 gain, Johnson decides to exchange this farm for one owned by Kramer which also has a fair market value of $400,000 and a basis of $125,000. The figures in Table 13.7 (Example 1) indicate that, by using a qualified exchange, Smith is able to postpone the recognition of any capital gain through appropriate adjustments in the basis of the new property he receives in the exchange transaction.

The exchange transaction need not involve property that has equal fair market values. In many cases, a mortgage will also be transferred with the property or cash (boot) received by one of the parties in the exchange transaction. If cash is involved in the exchange, gain is recognized and taxed to the extent of any cash *received*. Furthermore, the basis of the property received in the exchange is adjusted by decreasing the basis carried over from the property transferred by the amount of cash received

Table 13.7   Tax Consequences of an Exchange

		Example 1	*Example 2*
Johnson farm	Fair market value	$400,000	$400,000
	Basis	$100,000	$100,000
Kramer farm	Fair market value	$400,000	$350,000
	Basis	$125,000	$125,000
Computation of Johnson's Capital Gain			
Value property received (Kramer farm)		$400,000	$350,000
Cash received		—	$50,000
Total		$400,000	$400,000
Less cost basis of property transferred		$100,000	$100,000
Potential gain		$300,000	$300,000
Gain taxed to extent cash received		—	$50,000
Tax Basis of Johnson's New Property			
Basis of old property		$100,000	$100,000
Plus gain recognized		—	$50,000
Minus cash received		—	$50,000
Basis		$100,000	$100,000
Market Value of Johnson's New Property		$400,000	$350,000

and increasing that basis by the amount of gain recognized. Example 2 in Table 13.7 summarizes an exchange whereby $50,000 of cash (boot) is *received* by Johnson. In this example, Johnson must recognize $50,000 of gain during the year the transaction occurred. The basis of the property received in the exchange is adjusted for the cash payment and gain recognized. Note that if Johnson had sold his property which had a fair market value of $400,000 and a basis of $100,000, a gain of $300,000 would have been recognized. Through exchanging this property for other property with a fair market value of $350,000 and receiving $50,000 of cash, Smith has been required to recognize only $50,000 of gain. The recognition of the remaining $250,000 of gain has been postponed, but will be recognized and taxed at a subsequent sale of the property received in the exchange, since this new property is now valued at $350,000 and has a $100,000 adjusted basis. A mortgage that is assumed by the other party in the exchange transaction must be treated by the taxpayer as equivalent to cash received unless an equal-sized mortgage is taken back in the transaction.

Even though the qualified exchange transaction does not require the recognition of capital gain, if depreciable property is involved, recapture of excess depreciation may be required. This recapture will typically be taxed at ordinary income tax rates. Thus, the qualified exchange transaction may involve some tax consequences, but not the taxation of capital gains.

Gifts A third method that can be used to transfer real estate during one's lifetime is a gift. Essentially, a gift involves transferring complete ownership in property to a donee (recipient). The recipient of the property has the right to sell, transfer, use, or dispose of the property in any way he or she desires. Since a gift must be complete to qualify for preferred tax treatment, the donor (giver) not only disposes of property when a gift is given, but the stream of income associated with that property is also transferred to the recipient. Gifts are particularly useful in the development of estate plans when it is desired to transfer property to an heir who is involved in the farm business while the parents are still alive.

The tax treatment of gifts will only be briefly reviewed here. A donor may transfer $10,000 *each* to an unlimited number of recipients annually without paying a gift tax (the annual exclusion). If the spouse consents in the gift, this annual exclusion is increased to $20,000 per year. With respect to capital gains tax, property transferred by gift has a "carryover" basis or the same basis in the hands of the recipient as that of the donor. Thus, if the recipient sells the property, capital gains tax must be paid on the difference between the sale value and the carryover basis. However, capital gains will not be recognized if the recipient subsequently exchanges the property in a tax-free exchange.

Frequently, the gift and sale transfer methods can be combined to implement various property transfer plans. One alternative method of combining the gift and sale is to qualify part of the real estate being transferred for the gift tax exemption, with the remainder being treated as a sale. In this case, capital gains tax must be paid on only that part of the property that is sold. The basis for the gift portion of the property is "carried over" from the donor to the recipient, whereas the purchased portion of the property receives a new basis at the time of sale.

A second means of combining the gift and sale procedure is to use the installment sale contract with a cancellation of part or all of the installment payments as they come due or returning the payments to the buyer once they have been made. This enables the seller to transfer property to an heir, for example, without having to divide or portion the property to meet the gift tax exclusions. For example, a contract on a 40-acre parcel of land might be signed that requires annual principal and interest payments of $10,000. When the annual payments become due, the seller can collect the payment, pay the appropriate taxes, and return the payment to the buyer as a gift. Caution should be exercised to make sure that the return of the payment is well documented as a gift. Furthermore, a definite or fixed plan for returning the payments should not be developed at the time the sale is consummated, or the entire transaction will be considered a present gift at the time of transfer and gift taxes may be due. If the transaction is treated in a business-like manner as a sale, it will usually be treated by the Internal Revenue Service as a sale. One additional benefit of this procedure is that, if the transaction is treated as a sale, the buyer's basis is the fair market value or purchase price of the property. Furthermore, a reduction in this basis is not required as gifts are received when payments are due. Thus, this procedure provides a way to transfer property through the gift mechanism and obtain a new basis. Using this procedure is not costless since the tax on any capital gain must be paid by the seller-donor.

Transfers at Death   Real estate can be transferred at death as well as during life. In fact, in some cases it may be desirable to transfer property through a will or by the laws of descent at death. When the property is transferred upon death, the real estate can be held as an inflation hedge during the retirement years. In addition, the financial security of the owner is not reduced as might occur with a gift. The real estate receives a new or "stepped up" basis at the time of a death transfer; hence, capital gains tax would be minimized at the time of subsequent sale. Because real estate may be valued more conservatively in an estate than stocks or bonds that are traded on an established market, the estate taxes that must be paid may be lower if the real property is held until death rather than being sold and the proceeds invested in a portfolio of traded stocks and bonds that generate similar earnings. Furthermore, real property may qualify for reduced valuation under the special use value procedures contained in the estate tax law.

Maintaining complete ownership in the farm real estate during life and transferring it at death may be undesirable if an heir is active in the farm business. This procedure not only makes it impossible for the farm heir to obtain security in the real estate, but it also may limit the opportunity to assume the financial and managerial responsibilities of the farm. Maintaining ownership until death may also be unsatisfactory if the cash rent from the farm will not satisfy the current cash needs of the farm couple during their retirement years.

The Legal Dimensions of Real Estate Transfers

Although there are numerous legal dimensions to the real estate transfer process, only three of these dimensions will be discussed here: the evidence of transfer or conveyance and ownership, the legal description, and methods of holding title to real estate.

Evidence of Property Transfer and Ownership The instrument used to convey a right, title to, or interest in real estate from one person to another (for example, from a seller to a buyer) is called a deed. There are two commonly used deeds—a warranty deed and a quit claim deed. A warranty deed is used to convey clear and merchantable title from the seller to the buyer. Such a deed not only conveys title in the property to the seller, but it also warrants or guarantees that the seller or grantor (1) possessed good and merchantable title, (2) had the right to transfer such title, (3) will defend the title against all lawful claimants, and (4) has not placed liens and encumberances on the real estate except those noted in the deed or abstract.

In contrast, a quit claim deed conveys any right, title, or interest in the property held by the seller, but does not warrant or purport that the seller has any right, title, or interest to transfer. A quit claim deed, therefore, conveys whatever interest the seller might have, but does not guarantee that she or he has an interest in the property. Thus, the usual purpose of the quit claim deed is to clear a title or to "cure a defect" in the chain of title for property. For example, it might be used to terminate the spouse's interest in real estate. In this case, the husband or wife would give a warranty deed to the buyer, and a quit claim deed might be requested of the spouse which indicates that she or he is giving up any ownership interest in the property.

The deed must meet certain formal requirements, including the names of the grantor (seller) and grantee (buyer), a legal description of the property, a warranty (if a warranty deed), and any exceptions or title defects that might exist. Furthermore, in order to be effective in most states it must be dated, signed, and delivered to the buyer or a designated representative.

To protect the property interests of both the buyer and seller, many states require the deed to be recorded as a means of serving public notice of the transfer and the rights of the various parties. Recording is done at the County Recorder's Office or some such designated authority, typically at the county level. Recording the deed is an important step in finalizing the real estate transaction, because the laws in many states indicate that the person who records the deed first has priority. Thus, for example, if a deed is transferred but not recorded and a lien or judgment against the previous owner is attached to the real estate through the recording process, the purchaser of the property may in fact become liable for the judgment because it was not indicated publicly in the real estate records that he or she was the new owner of the property.

The Legal Description   The legal description is used to identify the location of a particular tract of real estate. Two systems of describing real estate are used in the United States—metes and bounds, and the rectangular survey. The metes and bounds system is used primarily in the eastern part of the United States and uses natural monuments or topographic features in the description system. The rectangular survey system is a unique American development and utilizes the well-known township and county geographic and political subdivisions in the property description. A property description is essential to determine the unique location of each parcel of land and is a necessary element of any valid warranty deed or other method of conveying real estate. Furthermore, property descriptions are also utilized in most leasing arrangements to identify the location of the real estate being leased.

Property Ownership   Five types of property ownership are common in most parts of the United States. *Sole or fee simple* ownership refers to the ownership of property by one person with all legal rights to dispose of the property in any way desired. Ownership of property as *tenants-in-common* exists when two or more individuals have undivided interests in the property. Each tenant-in-common has the right to sell, mortgage, assign, or convey by any legal means, including a will, his or her interest in the property. *Joint tenancy* is a type of property ownership between two or more individuals which includes the right of survivorship. Thus, if one joint tenant dies, his or her interest in the property is immediately and automatically conveyed equally to the other joint tenants. A joint tenant does have the right to sell, mortgage, or assign an interest in jointly held property during his or her lifetime, but the will is ineffective with respect to jointly held property. In some states, real estate owned by a husband and wife may be owned as *tenants-by-the-entireties,* and the survivorship rights apply upon the death of either spouse. Property held in tenancy-by-the-entirety cannot be sold, mortgaged, assigned, or conveyed by one co-owner alone, except upon divorce.

Ownership of property either in fee simple or as tenants-in-common rather than as joint tenants or tenants-by-the-entirety may have a substantial influence on the freedom to manage and transfer the property as one desires. Joint decisions are required to mortgage or transfer the property if it is owned by two or more people. Furthermore, when a person dies, joint-tenancy property is transferred immediately to the surviving joint tenants. Because of the automatic nature of the transfer of property held in joint tenancy, the opportunity to make transfers to other family members and the flexibility to adjust the estate transfer plan to changing family and estate conditions are not available.

The fifth type of property ownership is the *life estate*. The purpose of a life estate is to divide the ownership interest in property between two parties, one party possessing a current interest and the other party a remainder interest. The party possessing the remainder interest can exercise the rights of ownership and take possession of the property only after the death of the party that has the current interest. Thus, title in property can be held as a life estate in one or both parents with the children having a remainder interest. The effect of this type of title division is that the parents have the right to use the property and receive the income from it during their lifetime, but the children will receive the property at the death of the parents. The children also have an interest in the property that can be sold, mortgaged, or conveyed in any way to anyone.

The use of a life estate may accomplish lifetime and estate transfer goals as well. A remainder interest held by an heir(s) who is operating the farm guarantees eventual ownership of the property. Thus, the heir(s) may be willing to undertake necessary improvements and maintenance as well as make needed additions to the real estate and buildings without fear of these improvements going to another heir at the death of the parents. In addition, a life estate held by the parents or a surviving spouse provides lifetime security for the parents. However, if an heir who has a remainder interest in property dies, leaving this interest to his or her minor children, it may become difficult for the parents to mortgage or convey the property because of the restrictions on the transfer of property by minors.

A sixth method of property ownership, *community property*, is available only in eight states—Arizona, California, Idaho, Louisiana, Nevada, New Mexico, Texas, and Washington. In general, in these states all property acquired during marriage is community property; one-half is presumed to be owned by each spouse, and at the death of a husband or wife, his or her one-half interest in community property is subject to federal estate tax and is transferred by the terms of the will.

RENTING OR LEASING LAND

Because of the large capital outlay required to purchase all of the real estate included in the farm business, many farmers rent all or part of the land they operate. In fact, in various areas of the Midwest as much as two-thirds of the land is tenant-operated. With the increase in machine size and the ability of a farmer to handle additional crop acreage in a timely fashion, renting is one way to expand the land base without a large capital outlay. Many farmers own part of the land they operate and rent the remainder.

Rental arrangements are based on custom and tradition in many communities. The common types of rental or leasing arrangements used in most parts of the United States include the cash lease, the crop-share lease, the livestock share lease, and the flexible cash lease. Each of these arrangements provides the basis for a business agreement between the land owner and tenant.

To be successful, a lease must combine the resources of land, labor, capital, and management contributed by both the landlord and tenant in an efficient manner. Furthermore, the lease must provide for an equitable sharing of that income between the landlord and tenant. Thus, leasing and rental agreements must be updated periodically as costs, prices, and production efficiency change so that they encourage efficient resource use and equitable sharing of income and expenses.

The Cash Lease

Probably the easiest leasing or rental arrangement to develop and utilize is the cash lease. With a cash lease arrangement, the tenant pays a specified amount of annual cash rent per acre to the landlord. This payment may be made in one lump sum or in installments with the first installment typically due in late winter or spring and the second installment after the crop has been harvested. The landlord may place certain restrictions on the crop rotations that can be used, but other than these restrictions, the tenant has free rein to produce whatever crops he or she chooses. The cash lease is typically utilized for unimproved land, although it may be utilized in a livestock operation as well.

Among the advantages of the cash lease is that of simplicity. Because the arrangement requires no specification or estimate of expenses or income that will be shared, there are few opportunities for misunderstanding. Furthermore, the cash lease may also be the most flexible of any of the lease arrangements, particularly for the tenant. With a cash lease, tenants have almost complete freedom to develop and plan their cropping and livestock programs and enterprises, except for the general restrictions imposed by the landlord. The cash lease also relieves the landlord of any decisions that would be required in planning the operation, purchasing inputs, or marketing the crop since the landlord receives income in cash rather than commodities.

With a cash lease, the landlord receives a fixed income while the tenant takes the full risk of price or yield variability. This fixed cash commitment may result in a high risk for the tenant if crop prices are low or yields are reduced due to drought or disease. Another problem encountered in using the cash lease is maintenance of the productivity of the farm over time. Since the landlord is not participating directly in the sharing of income from the crop, he or she may be less willing to make improvements or investments in such items as tiling, terracing, or even liming. In addition, the appropriate amount of the cash rent may be difficult to determine. Cash rents have a tendency to be too low during periods of rising commodity prices and too high during times of low prices or low yields.

The Crop Share Lease

With a crop share lease, the tenant and the landlord share in both the income and expenses of producing the various crops. Crop share arrangements are used primarily with cash crops such as corn, soybeans, and vegetables. This arrangement may be used with forage crops and pasture as well, although a cash charge based on grazing capacity or yield is more frequently used in these cases.

A typical crop share lease in the Midwest and numerous other regions of the United States is the 50-50 arrangement whereby the landlord receives one-half of the corn or soybean crop as payment for use of the land. The landlord normally pays one-half of such costs as fertilizer, seed, and insecticide expenses as well as furnishing the land in most 50-50 crop share arrangements. The tenant provides all the labor, machinery, and fuel resources, and pays one-half of the other expenses involved in crop production. In some parts of the United States, the crop share arrangement may be on a two-thirds—one-third, or three-fifths—two-fifths tenant-landlord share, depending on the relative importance of the land resource compared to machinery and labor in crop production. Figure 13.1 to be discussed later summarizes a procedure that can be used to determine an equitable sharing arrangement depending on the value of the contributions provided by the landlord and tenant. If the farm does include livestock facilities, the tenant may pay a cash rent for use of the buildings and other improvements.

Because the crop share lease involves the sharing of both expenses and income, it can be more difficult to develop than the cash lease arrangement. Specifically, the landlord and tenant must decide how to share such expenses as commercial fertilizer, seed, lime, chemicals for weed and insect control, and expenses for any custom work. The sharing of the cost of harvesting such as field shelling of corn and drying must also be determined along with the application costs for various types of chemicals. Joint determination of the cropping rotation and participation in government programs as well as who receives government payments must also be negotiated between the landlord and tenant. If the farm includes improvements that can be used for livestock, a decision must be made concerning the appropriate rent to be paid for these facilities and who will pay for improvements in livestock facilities if they should be made. In addition, the appropriate rental rate for pasture land must be determined. For items like fertilizer or lime where a carryover between years might be expected, the appropriate amount of credit that the tenant should receive in case the lease is terminated must be specified. And provision must be made for adjustments in the lease to encourage the adoption of new technology or to accommodate the possibility of storing the owner's and tenant's grain in a common bin and marketing the grain jointly. Hence, the crop share lease requires many joint decisions and detailed negotiations between the landlord and tenant.

The crop share arrangement is nevertheless quite widely used and well understood in many areas; therefore, negotiations between the tenant and landlord may not be particularly difficult. Furthermore, with the crop share arrangement the risk of low prices and yields is shared between both landlord and tenant rather than being shouldered solely by the tenant as occurs with the cash rental arrangement. The landlord is relieved from making many operational decisions in the management of the farm; yet she or he can obtain the benefits of new technology or improved tillage or conservation practices that will increase yields. Furthermore, since the tenant can also expect to benefit from improved practices, both parties will receive a payoff from adopting new technology and maintaining high levels of fertility along with using top-notch management procedures. Finally, if the farm does include livestock facilities, cash renting the improvements does enable the tenant to develop and manage the livestock program without interference by the landlord.

The Livestock Share Lease

The livestock share lease agreement is much like the crop share agreement in that both the landlord and tenant contribute resources to the farming enterprise and share in the costs and returns. With this type of arrangement, however, the livestock is owned jointly by the landlord and tenant, and receipts from both livestock and crop sales are divided. Such livestock expenses as purchased feed, veterinary bills, and other cash costs are shared by the landlord and tenant along with the typical sharing of the crop production expenses. In most cases, the landlord provides the land and buildings, with the tenant furnishing labor and most of the equipment for both crops and livestock. Because of differences in the profitability of livestock enterprises and resources required, the livestock share lease agreement is not as easily standardized as crop share arrangements, and many variations have been developed.

If a farming operation has a substantial set of livestock facilities and improvements, a livestock share arrangement may be the most acceptable way to utilize these facilities. With this arrangement both weather and price risks on crops and livestock are shared by the landlord and tenant, and both parties will benefit from modern, up-to-date livestock facilities and improvements. Consequently, landlords may be more willing to construct new facilities and keep current livestock improvements up-to-date if they share in the income from the livestock enterprise. The livestock share lease also encourages beginning farmers with limited capital to expand their farm operations through livestock intensification rather than attempt to obtain more land. Furthermore, landlords may provide some management and other inputs to improve the potential success of the farm business when they are involved on a share basis in both the crop and livestock program.

As with the crop share arrangement, a livestock share rental agreement requires numerous joint decisions by the tenant and landlord. Determination of the cropping and livestock plan to be used is required, along with joint decisions concerning the time, place, and method of selling both crop and livestock products. Joint decisions must be made concerning the type of livestock to be included and the size of the various livestock enterprises, along with whether homegrown grains or purchased feeds will be used in the livestock feeding program. The issues of purchasing or raising herd replacements and the financial requirements of purchasing feeder livestock can also present problems in a livestock share arrangement, particularly for the tenant, because of the large capital outlays that may be required. Certainly, a good set of records is required with the livestock share arrangement to accurately determine resource contributions and income and expenses that must be shared. In addition, someone must be responsible for making final decisions concerning purchases and sales, and if the landlord and tenant have different goals in terms of the income they want from farming and the risk they are willing to take, the livestock share arrangement may be very difficult to develop. In many dimensions, a livestock share arrangement is much like a partnership in terms of the joint decision-making process. The landlord and tenant must be compatible to make such an arrangement work.

The Flexible Cash Lease

With frequent fluctuations in commodity prices and higher crop yields, many landlords and tenants have expressed interest in variable or flexible cash lease arrangements. Numerous variable or flexible cash lease agreements are in use today. One such flexible arrangement uses a fixed percentage of the production or yield as the variable payment. Typically, with this arrangement the tenant pays all production expenses, and the landlord receives a proportion of the output, say 30 percent. The landlord can then sell her or his proportion of the crop at whatever price is acceptable. With this type of arrangement, both the landlord and tenant share the risk of price and yield variability, but the landlord does not have to share expenses as with a crop share arrangement.

A second type of flexible or variable cash lease arrangement utilizes a base cash rent plus an additional payment based on changes in the price or yield of the crop. For example, the base cash rent would be set below the "going rate." Then, if corn prices went above a specified price, the cash rent would be increased by an agreed upon percentage of the difference between the actual and specified price. Alternatively, the landlord might receive the base cash rent plus the value of a specified number of bushels of corn. For example, the base rent might be specified at $40 per acre plus the price of 20 bushels of corn. As corn prices went up, the cash rent would increase, but decreases in corn prices would result in a decrease in the cash rent.

The basic purpose of the variable or flexible cash lease is to spread the risk of price and yield variability between the tenant and landlord without requiring a share arrangement. With the variable cash lease, the tenant typically has more flexibility concerning what enterprises he or she can produce and what crops to grow. Moreover, the landlord has the opportunity to capitalize on unexpected increases in commodity prices, but must take the risk of declines in commodity prices. In addition, with a flexible cash lease the landlord is not required to provide expense money for producing the crop as is necessary with the crop share arrangement.

The key questions to be answered in the development of a variable cash lease are the base rent (if one is to be paid) and the formula for determining the additional cash payment as a function of prices or yields. The calculations should be well specified in advance, and the returns to each party under different situations should be budgeted to determine the income sharing that will occur under the lease arrangement. Certainly, as with any lease arrangement, the provisions should be put in writing and reexamined each year so that adjustments can be made for changing economic conditions.

Evaluating a Lease Arrangement

There are both legal and economic dimensions to any leasing arrangement. Both parties must evaluate the agreement from their viewpoint and determine whether it is equitable and fair. A properly structured lease arrangement should provide the framework for profitable operation of the farm in the long run. Consequently, it is essential that the arrangement encourage the adoption of new technology and the optimal utilization of both the tenant's and the landlord's capital, labor, and land resources in the farm business.

The lease should be in writing and should specify legal protections for both parties in case of a default or nonperformance. The minimum requirements for a written lease are that it be signed by both parties, specify a definite period of time during which the lease is in effect, contain an accurate description of the property, and specify the kind and amount of rent to be paid and the time and place of payment. Other questions that may require negotiation and should be included in any leasing arrangement are: (1) What is the cost sharing of harvesting, drying, or storing under a crop share lease? Likewise, the sharing of costs incurred in weed control using herbicides that substitute for cultivation practices usually provided by the tenant must be determined. Any new technology that changes the relative contributions of the tenant and landlord must be evaluated and an appropriate cost-sharing arrangement developed. (2) What procedures will be used to add improvements if a cash or crop share lease is being used—particularly improvements that may not have a clear, positive impact on the landlord's net income? And how will the tenant be compensated for improvements made if the lease is terminated? (3) What if the tenant wants to rent additional land? What arrangements are needed to guarantee timely planting and harvesting of the crop to protect the landlord? In addition, adequate records need to be kept to make sure that each landlord pays the appropriate share of the expenses. (4) And what about notice of and actual termination of the farm lease? For example, in Iowa the tenant must be given notice as to the termination of the lease arrangement by September 1, to be effective the following March 1. Both landlord and tenant must be aware of termination deadlines in order to avoid disagreements and legal problems.

Because of the complexity involved in developing the leasing arrangement and the important decisions that must be made, competent legal and management advice should be obtained in developing a leasing agreement. It should be in writing to minimize chances of disagreement or misunderstanding between the parties.

Income Sharing Under the Lease

One of the most difficult decisions that must be made in the development of a cash or livestock share lease arrangement is the sharing of income and expenses between the tenant and landlord. It is crucial that both parties understand their own contribution and income share as well as that of the other participant in the lease arrangement. Only with a complete understanding between the parties will it be possible to develop a lease that is fair to both tenant and landlord. As used here, the concept of "fair" means that the share of income received by each party is in approximate proportion to his/her respective resource contributions to the farm operation. Thus, the first place to start in the development of a "fair" share rental arrangement is the contributions being made by both parties.

Figure 13.1 illustrates the calculation of the value of contributions for an example livestock share rented farm. Once the value of resource contributions provided by each party is determined, the sharing of cash expenses and production from the rented land can be specified. This is accomplished by determining the proportion of the total value of contributions

Figure 13.1 Developing or Testing
Your Rental Arrangement

Item	Contribution			Each Party's Share	
	Cost or Value	Rate (Percent)	Value of Annual Contribution	Tenant	Land Owner
Land and Buildings					
1. Interest (3–6 percent of valuation)	$520,000	4	$20,800		$20,800
2. Real estate tax			2,500		2,500
Buildings, Fences, and Other Permanent Improvements					
3. Depreciation (4–10 percent of replacement value)	Depreciation records		4,800		4,800
4. Repair (2–4 percent of replacement value)	Average annual cost		1,600		1,600
5. Insurance			400		400
Power and Machinery (cost basis of machinery—$70,000; salvage—$10,000)					
6. Interest (8–10 percent of new cost plus salvage value ÷ 2)	40,000	10	4,000	4,000	
7. Depreciation (10–14 percent of new cost less salvage value)	60,000	12	7,200	7,200	
8. Repair (4–6 percent of new cost)	70,000	4	2,800	2,800	
9. Insurance			200	200	
Livestock					
10. Interest (8–10 percent of current value)	35,000	10	3,500	3,500	
11. Depreciation, if any (breeding stock only)			300	300	
12. Insurance			180	180	
13. *Personal Property Tax*			600	600	

(*continued*)

Figure 13.1 (Continued)

Item	Contribution			Each Party's Share	
	Cost or Value	Rate (Percent)	Value of Annual Contribution	Tenant	Land Owner
Labor and Management					
14. Operator <u>12</u> months			$12,000	$12,000	
15. Family help <u>400</u> hrs	400 hrs	3.50	1,400	1,400	
16. Hired labor <u>600</u> hrs	600 hrs	4.50	2,700	2,700	
17. Management (10 percent of gross)			10,000	10,000	
18. *Cash Rent* (paid by land owner to tenant)			—	—	
19. *Subtotal*—major contributions (add lines 1 through 18)			74,980	44,880	30,100
Other Cash Costs					
20. Cash cost of boarding hired labor					
21. Purchased feed for productive livestock					
22. Other livestock expense					
23. Machine work hired					
24. Seed, plants					
25. Twine and baling wire					
26. Fertilizer and chemicals					
27. Tractor fuel			4,000	4,000	
28. Miscellaneous			2,000		2,000
29. *Total Expenses* (add lines 19 through 28)			80,980	48,880	32,100
30. *Percentage of Total Contribution*				60.4	39.6

Source: Myron Bennett, "Livestock Share Rental Arrangements for Your Farm." North Central Regional Extension Publication 107, Columbia: University of Missouri, 1980.

contributed by the tenant and by the landlord and then sharing expenses and production in this same proportion.

The contributions to be included in the calculations include the land and buildings, permanent improvements, power and machinery, livestock, personal property tax, labor and management, and other contributed resources or inputs (Figure 13.1). Land and buildings should be included in the contributions calculation at fair market value based on agricultural use. Interest on the land and building value should be credited to the landlord at a rate that reflects the current cash rate of return on farm real estate investments. In addition, the landlord should be credited for any contribution in the form of real estate taxes.

Most farm businesses include improvements to the farming operation as well. Consequently, the landlord must be compensated for the depreciation, repairs, and insurance that will be required on these improvements. Note that replacement value rather than depreciated value is used in determining the landlord's contribution in the form of improvements. Figure 13.1 and the discussion in Chapter 4 provide guidelines that might be used in the computation of depreciation and repair expenses for improvements as a percentage of replacement value.

In most share rental arrangements, the tenant provides the power and machinery that will be used in the farming operation. Consequently, the tenant must be compensated for the ownership costs associated with the machinery line. As discussed in Chapter 4, these costs include interest, depreciation, repairs, taxes, and insurance. Again, suggestions for the calculation of the value of these fixed contributions by the tenant are provided in Figure 13.1. A similar calculation is required for contributions of livestock to the farming operation if a livestock share arrangement is being developed.

Next, the value of labor and management that will be contributed to the operation by the landlord and tenant must be calculated. A reasonable wage rate must be determined for the tenant that reflects the value of the labor contribution. Moreover, any family or hired help that the tenant plans to contribute to the operation must be included in the computation of the fixed labor contribution to the business. Finally, a management contribution must be included that is based on the relative amounts of management that will be provided by the tenant and landlord. In the example of Figure 13.1, the management return is calculated as 10 percent of the estimated gross return and then allocated between the tenant and landlord based on their respective management contributions to the business. Any adjustments to reflect cash rent paid to the landlord by the tenant for the use of the house or livestock facilities must also be included in the computation. If the landlord or tenant contributes other resources such as fuel or time, these contributions should also be reflected as is the case in the example of Figure 13.1.

Finally, the total value of contributions provided by the tenant and landlord can be determined by summing the value in each category. Then the value of contributions for the entire farming operation can be determined and the relative proportion of contributions provided by the tenant and the landlord calculated. Note that in the example of Figure 13.1 the tenant is contributing approximately 60 percent of the contributed resources and the landlord 40 percent. Thus, cash expenses and production should be shared in these same proportions to reflect a fair and equitable division of expenses and production between the tenant and landlord. If higher productivity and thus value land were being contributed by the landlord, his or her proportion of the contributions would increase and an equitable arrangement would result in the landlord sharing a larger proportion of the cash expenses and production. Likewise, more machinery or labor provided by the tenant would result in the tenant receiving a larger share.

Calculating the value of contributions provides a basis to determine the bargaining position of the tenant and landlord. Certainly, the final determination of an acceptable rental payment and rental agreement depends on the bargaining power of the parties involved. If the number of farms available for rent are few, the landlord is in a strong bargaining position and might be expected to obtain a higher rental income and more favorable leasing terms than if the number of farms available were large. A reverse situation might occur if the number of good tenants was small compared to the number of rental opportunities. However, a good tenant can have a significant impact on the productivity and income of the farm operation. It may be more desirable for the landlord to give up some income share to get a good tenant because he or she will receive more total income over time from the rental arrangement due to increased yields and productivity. Likewise, a tenant may find that some concessions to a good landlord will increase long-run income-generating capacity, and improve chances both to maintain the lease in the future and obtain the benefits of long-run investments to improve productivity and income.

LAND IMPROVEMENT AND CONSERVATION

One of the important tasks in the acquisition and management of farm real estate is that of maintaining and improving the productivity of the land resource. With increased emphasis on food production, more marginal land is being tilled and is thus subject to water and wind erosion. Terraces may be needed in some areas to increase productivity and reduce erosion and stream pollution from soil sediment. Land that has limited productivity because of moisture problems can be tiled to improve yields and timeliness of farming. Timberland can be converted into row cropland by clearing, and gullies can be seeded and streams straightened to increase the amount of tillable or productive acreage and reduce the potential of soil erosion or loss. Some farmers are fortunate to own land where such improvements may not be necessary. However, much of the productive land of the United States is subject to erosion or could benefit from investments in tile, terraces, or other improvements. A basic question facing the farm manager is, what are the benefits of investment in land improvements and will they pay?

Adopting conservation practices and improving the productivity of land can have a payoff to the farmer. Figure 13.2 illustrates hypothetical production relationships, assuming both use of conservation practices and no conservation practices. One possible response from the use of conservation practices is that the fertility of the soil is maintained and improved to such an extent that more output is forthcoming from each level of input of seed, fertilizer, and chemicals. This is illustrated in Figure 13.2 by the higher response function assuming conservation practices compared to the function with no conservation. However, the question of whether or not the increased response from the variable resources will be sufficient to pay for the capital outlay necessary to adopt the conservation practice still must be answered.

The capital budgeting or investment analysis procedures discussed in Chapter 8 can be used to answer this question. Essentially, the analysis involves comparing the present value of the expected increased yield and income for the life of the conservation investment to the capital outlay that

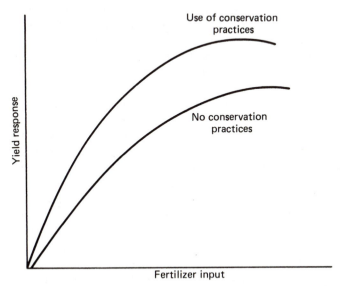

Figure 13.2. Hypothetical impact of use of conservation practices.

is necessary to obtain the improved yield and income. For example, with a tiling project the number of acres that will benefit from tile and the incremental number of bushels expected due to better drainage and more timely planting of the crop would be estimated. A long-run average price would be used to value the increased yield, and the present value of the net income stream would be calculated using the cost of capital to determine the discount factor. This present value of benefits stream would then be compared to the cost of the tiling project to determine if it was, in fact, a profitable investment.

Numerous management considerations must be included in the economic analysis of land conservation and improvement investments. In many cases, the conservation program may result in changes in the rotation that can be used on the land. For example, if terraces are built to reduce water erosion, it may be possible to change from a rotation that includes two or three years of meadow to one that includes fewer years of meadow and more years of higher value row crops such as corn or soybeans. This change in rotation could have a significant impact on the income generated from the land and improve the possibility of the terraces being a profitable investment. Investments that reduce erosion may also reduce the loss of nutrients, thus requiring lower levels of fertilization or better response from current fertilizer and chemical applications. Improvements in stand and more timely planting and harvesting may result in improved yields because of reduced erosion.

Many conservation practices qualify for cost sharing under state and federal programs. Such cost-sharing arrangements have been developed to encourage the adoption of conservation practices by reducing the private cost to the land owner. In some cases, cost sharing up to 50 or even 75 percent of the approved cost of the conservation investment may be available. The farm manager should investigate the availability of cost-sharing programs in his or her area to determine eligibility of various practices and availability of funds.

Land improvement and conservation expenses receive special tax treatment under the federal income tax regulations. A farmer may deduct as a current expenditure outlays for soil and water conservation up to a maximum of 25 percent of the gross income from farming. Outlays in excess of the 25 percent maximum can be carried forward as deductions in succeeding years. However, if the land is sold in subsequent years, these deductions may be recaptured as ordinary income. If the sale occurs within five years after it was acquired (not after the conservation outlay), all soil and water conservation deductions are recaptured. The recapture proportions for land sold in six, seven, eight, or nine years after the date of acquisition are 80 percent, 60 percent, 40 percent, and 20 percent, respectively. No recapture occurs if the improved land is sold ten years after it was acquired. If soil and water conservation expenses are not deducted in the year they are incurred or carried forward, they are added to the basis of the land and recovered at the time of sale.

Expenses for clearing land to make it suitable for farming can also be deducted for tax purposes in the year they are incurred. The land clearing deduction is limited to 25 percent of the taxable income from farming up to a maximum of $5,000. Excess land clearing deductions cannot be carried forward, but are added to the basis of the land and recovered at the time of sale. If the land is subsequently sold, the recapture rules discussed earlier for soil and water conservation expenses also apply to land clearing expenses. These soil and water conservation and land clearing expenses can be valuable deductions in years of abnormally high income.

In many communities, land is not available for purchase or rent, and land expansion is difficult or impossible. However, if a farmer has timberland that can be cleared or more intensively managed, or pasture land that can be placed in row crops if terraced, she or he can increase the intensity of the operation and farm output by making investments in these improvements rather than renting or buying additional land. As with any investment, the benefits and costs of increasing volume by improving the productivity of the current land base compared to adding to that land base must be evaluated. However, land improvements can frequently be a profitable means of expanding the farm business with a smaller capital outlay than is typically required to purchase additional land resources. In addition, such activities as clearing or terracing can be done over a period of years as funds become available, rather than requiring a large one-time financial commitment as is usually the case with the purchase of land.

LAND: A PUBLIC OR PRIVATE RESOURCE

An important issue facing many farmers, particularly those with concerns about the environment and pollution, future food production, and open space, is that of public versus private property rights in farm real estate. Although many farmers feel that their private property rights in real estate should be unconstrained, debates concerning zoning in rural counties as well as soil stewardship and conservation have focused on the important issue of the possible conflict between public and private rights in rural property. Restrictions on land use imposed by zoning and land use planning regulations have in the past been a result of urban sprawl and development. Recent attention on future agricultural productivity has shifted some of the emphasis in land use planning and zoning to protect "prime" farmland from conversion to other uses. Public policy debates concerning the identification and definition of prime farmland as well as restrictions concerning its use along with zoning regulations to control the urbanization process will continue to focus society's attention on the efficient use of land.

A second area where public and private concerns with respect to the land resource often conflict concerns soil erosion and conservation. Although most farmers are very concerned about issues of land stewardship and attempt to control erosion, economic forces and the desire to obtain competitive levels of income sometimes make stewardship of the land resource a costly proposition. In many cases, the benefits accruing directly to the farmer or land owner from conservation practices are not sufficient to offset the cost. However, the social benefits of reducing erosion and the water and air pollution that results as well as maintaining the productivity of the land resource for future food production may still be significant. In fact, soil erosion is the major source of pollution from agriculture and is the key concern of nonpoint source pollution regulations being developed by state and federal environmental protection agencies.

Various techniques can be adopted to reduce soil erosion, including minimum and reduced tillage procedures, terraces, contouring, and strip farming. Some of these techniques, particularly reduced tillage options, may actually result in reduced cost for the individual producer once the new technology has been learned and adopted; in these cases, the public sector need not encourage or force adoption, although additional educational efforts will be necessary to demonstrate the efficiency and productivity impact of the new technology. If the control practices require substantial capital outlays, which is frequently the case with terraces and some reduced tillage options, various incentives may be useful in encouraging and accelerating the rate of adoption. Such incentives include tax write-offs, cost sharing, and even various types of regulations specifying maximum levels of soil erosion. Farmers must, therefore, be aware of the social costs incurred by using tillage and other management practices that increase erosion and reduce productivity, because both state and federal regulations are being implemented to limit soil erosion and encourage land owners to adopt reduced tillage and erosion control practices.

Summary

Land is a key resource in agricultural production. Control of farmland can be accomplished by ownership or leasing arrangements. The concepts of land valuation and appraisal, financial feasibility of land ownership and rental, and alternative methods of transferring real estate must be understood to evaluate alternative ownership and rental opportunities.

Three methods are commonly used in land appraisal: (1) the market approach which attempts to determine what the property would bring if sold, (2) the cost approach which involves estimating the cost that would be incurred to replace the resource, and (3) the income approach which values land according to the present value of its income stream. The income approach to valuation involves use of the income capitalization procedure, which requires calculation of the net earnings per acre and selection of a discount or capitalization rate. Even if land is an economically profitable investment using the income capitalization procedure for determining value, the financial feasibility of that purchase (the ability to service the principal and interest obligation with the income) must be evaluated prior to implementing an actual purchase.

A number of arrangements can be used to transfer the ownership of real estate between two parties. These methods include the cash, mortgage financed, or contract sale, the exchange, the gift, or the transfer of the property through the estate at death. The choice of the appropriate transfer method will depend on such factors as expected rate of inflation, life expectancies of the individuals, cost of the various transfer methods, income and estate tax regulations, cash flow considerations, estate planning objectives, and security interest desired by the seller. For those involved in a real estate transaction, knowledge of the legal considerations including the differences between a warranty and quit claim deed, the importance of the legal description, and the different methods of owning real estate (joint tenancy, tenancy in common, tenancy by the entirety, a life estate, and community property) is essential.

Many farmers rent all or part of the land they farm. Different rental arrangements between the landlord and tenant can be used; popular arrangements in the farmland rental market are the cash lease, crop share lease, and livestock share lease. More recently, various types of flexible cash lease arrangements have been developed to share the risk and return between landlord and tenant. Whatever the lease arrangement, it should be evaluated with respect to equitability to both parties. This evaluation can be done using budgeting procedures first to determine the value of contributions provided by the two parties. These values then provide the base to determine the sharing of income and risk between the tenant and landlord.

Recent concerns about soil erosion have stimulated increased interest on the part of farmers and the public in soil conservation. Frequently, the private benefits received by the land owner from conservation practices are not sufficient to cover their cost. However, the public interest in land as a resource for future food production as well as a potential source of air and water pollution may override private ownership rights. Farmers can be encouraged to adopt soil conservation practices such as reduced tillage techniques, terraces, contouring, and strip farming through public incentives such as tax write-offs, cost sharing, and regulations on maximum levels of soil erosion.

Questions and Problems

1. What are the advantages and disadvantages of owning compared to renting farmland?
2. Discuss the three methods of land appraisal. What are the strengths and weaknesses of each method?

3. Summarize the formula for determining the income or earnings value of farm real estate. What factors must be considered in choosing the capitalization rate used in this formula?
4. List and discuss the steps involved in analyzing the financial feasibility of a land purchase. Obtain the data for a tract of land in your area and complete a financial feasibility analysis.
5. What are the advantages and disadvantages to the seller and buyer of using the contract sale compared to a mortgage-financed sale of land?
6. What are the seven key factors to consider in choosing among different land transfer methods?
7. What are the advantages and disadvantages of using the exchange arrangement for transferring real estate?
8. What are the key differences between a warranty and quit claim deed? Which deed is preferred?
9. What are the differences between owning property as joint tenants and as tenants in common?
10. Describe a common crop share lease arrangement used in your community.
11. Describe the various types of variable or flexible cash lease arrangements that might be developed. Why use a flexible cash lease compared to a fixed cash lease or a crop share lease?
12. How would you proceed to evaluate the equitability or fairness of a lease arrangement?
13. Why are there possibly differences between the public and private benefits and costs of soil conservation practices?
14. Identify specific instances in your community where conflicts exist between the public and private property rights in farmland.

Further Reading

Bennett, Myron, "Livestock Share Rental Arrangements For Your Farm." North Central Regional Extension Publication 107, Columbia: University of Missouri, 1980.

Henderson, Philip A. "Fixed and Flexible Cash Rental Arrangements for Your Farm." North Central Regional Extension Bulletin No. 75, Lease Form, North Central Regional Extension Bulletin No. 76, Lincoln: University of Nebraska.

Henderson, Philip A. "Is Your Lease Fair?" North Central Regional Publication No. 9, EC, 69–814, Lincoln: University of Nebraska, 1969.

Henderson, Philip A. "Your Pasture Lease." EC 77–828, Cooperative Extension Service, Institute of Agriculture and Natural Resources, University of Nebraska-Lincoln, November 1977.

Lins, D. A., Neil E. Harl, and T. L. Frey. *Farmland*. Skokie, Ill.: Century Communications, 1982.

Murray, William, Duane Harris, Gerald Miller, and Neill Thompson. *Farm Appraisal and Valuation*. Ames: Iowa State University Press, 1983.

Nelson, Doug C., and Philip A. Henderson. "Long-term Installment Land Contracts." North Central Regional Extension Publication No. 56, Lincoln: University of Nebraska, 1978.

Pretzer, Don D. "Crop Share or Crop Share-Cash Rental Arrangements for Your Farm." North Central Regional Extension Publication No. 105, Manhattan: Kansas State University, 1980.

Pretzer, Don D. "Irrigation Crop-Share and Cash Rental Arrangements for Your Farm." North Central Regional Extension Publication No. 148, Manhattan: Kansas State University, March 1981.

Rossi, Anthony. "Tax Planning Before the Purchase or Sale." in *Do It Right the First Time*. St. Louis, Mo.: Doane Agricultural Service, 1973, p. 118.

Stoneberg, E. G. "Improving Your Farm Lease Contract." FM 1564 (Rev.) Cooperative Extension Service, Iowa State University, Ames, Iowa, June 1978.

Suter, Robert C. *The Appraisal of Farm Real Estate*. Danville, Ill.: Interstate Printers and Publishers, 1980.

ACQUISITION AND MANAGEMENT OF DEPRECIABLE ASSETS

CHAPTER CONTENTS

Depreciable assets provide a flow of service that is consumed in production over a period of more than one year. These are assets that wear out, decay, become obsolete, or lose value from natural causes. They include buildings, fencing, drain tile, waterwells, machinery, equipment, breeding stock, orchards, and vineyards.

Both the pattern of service flow over time and the value per unit of this flow are important in making acquisition and management decisions on depreciable assets. Physical assets, such as machinery and buildings, provide a relatively constant service flow over the initial portion of their lifetime, followed by a declining flow of service due to wear and decay. Biological assets, such as breeding stock and orchards, typically display increasing physical productivity (the annual yield or production) during earlier years, followed by declining productivity as the asset ages. The value of the service flow also may change, either because of changing market conditions for the products or because a superior asset becomes available. For example, a long-term decline in the market price of fruit will decrease the value of even a high-yielding orchard. Likewise, the development of a machine that reduces harvesting costs per acre may make the current harvesting equipment obsolete before it is worn out.

Two important economic problems related to the acquisition and management of depreciable assets are discussed in this chapter. The first is choosing the appropriate combination of depreciable assets and the method of financing these assets. A depreciable asset that is used over time must eventually be replaced or the production process will cease. This suggests the second problem—selecting the appropriate replacement item (when more than one alternative exists) and the time for replacement. These problems are quite important in both developing and implementing plans.

SELECTING DEPRECIABLE ASSETS AND THE METHOD OF FINANCING

The procedure to select the combination of assets and the method of financing can be divided into three parts. They are:

1. Choose the combination of depreciable assets based on economic profitability.

2. Select the most profitable method of financing.

3. Evaluate the financial feasibility of the asset and the method of financing.

Choosing the Assets Based on Economic Profitability

The first step focuses on the economic profitability of the alternatives being considered. Managers must frequently evaluate alternative types of facilities to store a product or alternative facilities to house a specific type of livestock. They may need to compare obtaining alternative sizes of a machine with hiring a custom operator. In each of these examples, the capital budgeting procedures introduced in Chapter 8 can be used to evaluate the economic profitability of alternative assets that can accomplish a specific task. To review, the procedure described in Chapter 8 is composed of several computational steps.

1. Identify the relevant alternatives for consideration.

2. Choose an appropriate discount rate to adjust future flows of income to their present value. The weighted average cost of capital was recommended as the appropriate discount rate.

3. Calculate the present value of the cash outlay required to purchase the asset.

4. Calculate the annual net cash flow from the investment for each year of its useful life. The annual net cash flow is calculated using equations 8.7 and 8.8 which are repeated here for reference.

$$ANCF_n = CI_n - CE_n - T_n + TSIC_n + S_k \qquad (14.1)$$

$$T_n = [CI_n - (CE_n + D_n)] \, TR_n \qquad (14.2)$$

where

$ANCF_n$ = Net cash flow in year n

CI_n = Cash income in year n

CE_n = Cash expense in year n

T_n = Taxes in year n

$TSIC_n$ = Tax savings due to investment credit available in year n

S_k = Remaining salvage value at the end of the investment period, that is the kth period

D_n = Depreciation in year n

TR_n = Marginal tax rate applicable to income from the investment in year n.

5. Calculate the present value of each annual net cash flow.

6. Sum the present values of the outlay and the annual cash flows to obtain the net present value of the investment. Steps 4 and 5 convert the stream of annual net cash flows over the life of the asset to a single value which is the current or present value of the stream of net income.

7. Accept or reject the investment based on the net present value criterion.

These seven steps were outlined and illustrated in Chapter 8 as a way to compare the economic profitability of alternative investment opportunities, such as comparing the expansion of a livestock enterprise with the purchase of additional land. This chapter applies the procedure to a subset of these alternative investment opportunities—selecting the most profitable method of providing a required service. Such decisions will typically involve a choice among assets that require different combinations of capital outlay, labor, and operating costs, and have different lengths of life. In many cases, the cash income derived year by year will not be affected by the alternative chosen, while in others the income stream will differ from one asset to another. In each case, the profitability can be estimated by comparing the present value of the cash outlay and the present value of the net cash inflow or outflow from the alternative investments for their useful lives. When the appropriate size and type of capital assets have been selected, the economic profitability of the enterprises using these capital assets can be compared with alternative investment opportunities available to the business as described in Chapter 8.

Selecting the Most Profitable Method of Financing

The discussion in Chapter 8 did not include this step because it was assumed that the financing arrangement for the asset was specified and not negotiable. In reality, numerous financing arrangements can be used to acquire depreciable capital assets. Conventional loan terms may involve different combinations of repayment periods and interest rates. Dealer financing arrangements may include an interest-free period before any payments are required as well as favorable interest rates if the dealer is using credit terms to stimulate sales. And in many cases the equipment or facilities can be leased, which in reality is an alternative method of financing the acquisition of services of a capital asset. Different financing arrangements may be more or less profitable and should be evaluated to determine which arrangement is preferred.

The most profitable method of financing can also be evaluated using discounting procedures. The stream of payments to be discounted in this analysis is the downpayment and debt servicing payments. For a loan this involves calculating the present value of the downpayment and annual principal and interest payments. In the case of a lease, the present value of the lease payments is calculated. The financing arrangement with the lowest discounted cost of the after-tax stream of payments and tax savings is the preferred one.

The discount rate used in these calculations is the after-tax cost of debt capital to the firm. Discounting by the cost of debt capital indicates if the lease, dealer credit, or other method of financing is more or less profitable than the general method of acquiring debt capital for the firm. It follows that a financing arrangement with a lower (higher) cost of debt capital than the general method of acquiring debt capital has a lower (higher) discounted cost. Thus, this procedure allows one to compare the profitability of alternative methods of financing an asset and to estimate the value of favorable or unfavorable financing terms.

Evaluating Financial Feasibility

As noted in Chapter 8, the purpose of financial feasibility is to determine if an asset will generate sufficient cash income to make the interest and principal payments on funds borrowed to purchase the asset. This can be done by comparing the annual interest and principal or lease payments with the annual net cash flows for the project. Such a comparison may suggest that, although the asset is profitable and the lowest cost financing terms have been selected, the cash flows will not cover the principal and interest or lease payments. In this case, a cash subsidy may be needed to implement the purchase, or other financing terms may need to be investigated.

These concepts will now be used in evaluating various ways of acquiring machinery and livestock facilitiy services.

SELECTING THE APPROPRIATE MACHINE

Farm operators often have several alternative ways they can acquire required machinery services. Purchasing a new machine, purchasing a used machine, hiring a custom operator, and renting the machine on a short-term basis are four alternatives commonly available. Each method has different initial cash outlays and different annual cash expenses. They may also differ in the timeliness with which the required task can be completed and in the reliability of the service provided. The steps and some of the important factors to consider in analyzing the acquisition of machinery services are discussed below.

Machine Capacity and Timeliness

The initial step in analyzing machine services is to identify the sizes that are expected to perform the required tasks at relatively low costs. A decision to select relatively small machines that require many hours of operation is effective in spreading ownership costs, but it may result in large losses due to untimely operation. Thus, the effect of timeliness should be considered in the analysis.

Two general approaches to timeliness of operations are considered here. In one approach, the operator simply selects an acceptable period of time for completion of the task and only considers machines that can complete the task within the good field days available during the period. This approach requires data on machine capacity and days available for field work within the acceptable period. A second approach calculates the length of time required to complete the task(s) and estimates the field loss associated with each machine size. The second approach requires the data of the first approach and explicit estimates of field losses from untimely operation.

The capacity of machines used for field operations is usually measured in acres per hour. Equation 4.12 can be used to estimate capacity in acres per hour and is repeated here for reference.

$$\text{Acres per hour} = \frac{S \times W \times E}{8.25} \qquad (14.3)$$

where

$$S = \text{Speed in miles per hour,}$$
$$W = \text{Width in feet,}$$
$$E = \text{Efficiency expressed as a decimal.}$$

Data on representative speeds and efficiency for the major field operations are presented in Table 4.6.

Consider the example shown in Table 14.1. A farmer with 400 acres each of corn and soybeans to harvest annually is considering two sizes of combines. The medium-size combine will use a header 15 feet in width for soybeans and should be able to travel 3.15 miles per hour with an efficiency of 70 percent. The machine would be able to harvest 4.0 acres of soybeans per hour [(3.15 × 15 × 0.70) / 8.25]. This combine would require approx-

Table 14.1 Estimated Field Capacity
for Two Combine Sizes

Item	Medium-Size Combine		Large-Size Combine	
	Soybeans	Corn	Soybeans	Corn
Size in feet	15	10	18	15
Speed in mph	3.15	3.0	3.3	3.15
Efficiency	0.70	0.70	0.70	0.70
Acres per hour	4.0	2.5	5.0	4.0
Hours for 400 acres	100	160	80	100
Number of hours per day	6	10	6	10
Number of days required	16⅔	16	13⅓	10

imately 100 hours to harvest the 400 acres of soybeans. The medium-size combine harvests 4–30 inch rows and should be able to harvest the 400 acres of corn in 160 hours. The large combine has an 18 foot header for soybeans and a 6–30 inch row corn head. The large combine will require 80 hours to harvest the 400 acres of soybeans and 100 hours to harvest the 400 acres of corn. If the operator can harvest soybeans an average of 6 hours per day and corn 10 hours per day, the medium-size combine will require 16 2/3 days to harvest soybeans and 16 days for corn. The large combine will require a total of 23 1/3 days to harvest the 800 acres.

The time required for harvesting must be considered in combination with estimates of the field time available to decide if the machines are of acceptable size. The number of field days available have been tabulated for some areas of the country by research workers at the agricultural experiment stations. For example, the number of days suitable for field work in north central Iowa is listed by period of the growing season in Table 14.2. These data indicate that during the two-week period October 4–17 farmers in that area can expect 10.3 or more days suitable for field work 4 years out of 8, 8.0 or more days available 6 years out of 8, and 7.2 or more 7 years out of 8.

The number of calendar days required to complete the harvest can now be estimated with the data on accomplishment rates in Table 14.1 and the data on days suitable for field work in Table 14.2. The major assumptions and calculations are noted in Table 14.3. The operator expects to begin harvesting soybeans on approximately September 20 and to start harvesting corn on approximately October 18, or as soon thereafter as all of the soybeans have been harvested. No excess harvest loss is expected for soybeans harvested by October 17, but a loss of 1 bushel per acre per week is expected after October 17. The operator also expects harvest losses of 2 bushels per acre per week for corn harvested after November 14.

Comparison of the entries on lines 2 and 3 of Table 14.3 suggests that

Table 14.2 Estimated Number of Days Suitable for Field Work in North Central Iowa

Time Period	Expected Number of Years Out of 8-Year Period for Which Field Days Will Equal or Exceed the Number Indicated		
	4	6	7
March 29–April 11	1.7	0.8	0.3
April 12–25	7.5	5.7	4.0
April 26–May 9	10.3	8.7	6.8
May 10–23	10.5	8.5	8.2
May 24–June 6	9.1	7.9	7.7
June 7–July 4	21.8	19.0	17.7
July 5–August 1	22.9	20.4	18.1
August 2–September 5	29.5	28.1	27.1
September 6–19	11.9	8.6	8.5
September 20–October 3	9.6	7.9	6.9
October 4–17	10.3	8.0	7.2
October 18–31	12.4	9.0	8.0
November 1–14	10.6	6.8	4.7
November 15–December 5	16.0	14.3	10.0

Source: David L. Williams and George E. Ayres, *Fieldwork Days in Iowa*, Cooperative Extension Service, PM–695, Iowa State University, Ames, July 1976, p. 2.

the medium-size combine will not be able to complete the soybean harvest by October 17 under the adverse weather conditions that will likely be faced in at least 6 out of 8 and 7 out of 8 years. The number of acres remaining to be harvested after this date are estimated as the number of days of harvesting remaining times the daily accomplishment rate of 24 acres per day. The completion of soybean harvest in the two adverse weather situations (6 out of 8 and 7 out of 8 years) will delay the beginning of corn harvest until October 20 and 22, respectively (line 6), and reduce the days available to harvest corn without incurring excess losses (line 8). The loss of corn due to untimely harvest is calculated in an analogous manner and is recorded on line 10. The total value of the corn and soybean losses is recorded on line 11. The operator assigns a probability of 0.25 to each of these two adverse weather situations (he/she assumes that the harvest can be completed 50 percent of the time without excess losses) and calculates an expected excess harvest loss for the medium-size combine of 0.25 ($288.75) + 0.25 ($1,362.75) = $412.88 per year. The excess harvest loss for the large combine is zero since it can complete both corn and soybean harvest in a timely fashion, even in the adverse weather years.

Economic Profitability Analysis

Once the timeliness and size considerations have been evaluted, the economic profitability of the various alternatives should be determined. The

Table 14.3 Estimated Date of Completing Harvest and the
Associated Harvest Losses

Item	Medium-Size Combine			Large-Size Combine		
1. Expected number of years in an 8-year period	4	6	7	4	6	7
2. Days for soybean harvest (from Table 14.1)	16⅔	16⅔	16⅔	13⅓	13⅓	13⅓
3. Days available September 20–October 17 (from Table 14.2)	19.9	15.9	14.1	19.9	15.9	14.1
4. Estimated acres to harvest after October 17	0	18.5	61.7	0	0	0
5. Estimated loss (bushels of soybeans)	0	18.5	61.7	0	0	0
6. Estimated date to start corn harvest	10/18	10/20	10/22	10/18	10/18	10/18
7. Days for corn harvest (from Table 14.1)	16	16	16	10	10	10
8. Days available October 18–November 14 (from Table 14.2)	23	15.03[a]	10.13[b]	23	15.8	12.7
9. Estimated acres to harvest after November 14	0	25	150	0	0	0
10. Estimated loss (bushels of corn)	0	50	300	0	0	0
11. Value of losses with soybeans @ $7.50 and corn @ $3.00	0	$288.75	$1,362.75	0	0	0

[a]Reduced by 0.77 of a day suitable for field work required to complete the soybean harvest.
[b]Reduced by 2.57 days suitable for field work required to complete the soybean harvest.

first step in analyzing the profitability of the various methods of obtaining combine services is to calculate the weighted average cost of capital based on the long-run capital structure of the firm. The weighted average cost of capital (d) is calculated using equation 8.6, which is repeated below for reference.

$$d = k_e W_e + k_d (1 - t) W_d \qquad (14.4)$$

The weighted average cost of capital is based on the long-run proportion of debt (W_d) and equity capital (W_e) that will be used to operate the firm, the after-tax rate of return on equity capital (k_e), the interest rate on debt (k_d) and the marginal tax rate (t). The operator calculates the proportion of debt and equity capital currently being used in the business from the market value columns of the balance sheet discussed in Chapter 2. Assume the operator's capital structure currently includes 62 percent equity and 38 percent debt (borrowed) capital. The operator selects 60 percent equity and 40 percent debt as the appropriate long-run capital structure because it provides reasonable safety against insolvency and opportunity for relatively rapid growth of the business. The return on equity capital is composed of two parts: a cash return, k_{ce}, and an increase in the market value of assets, k_{ne} (particularly real estate).[1] Defining G as the proportion of the noncash return that is subject to tax, the after-tax cost of equity capital can be calculated as shown in equation 14.5.

$$k_e = k_{ce} (1 - t) + k_{ne} (1 - tG) \qquad (14.5)$$

The operator estimates that k_{ce} is 9 percent, k_{ne} is 8 percent, G is 40 percent, and the marginal tax rate (t) is 30 percent. The resulting after-tax cost of equity capital is 13.3 percent.

$$k_e = 0.09 \, (0.7) + 0.08 \, (1 - 0.3 \cdot 0.4) =$$
$$0.09 \, (0.7) + 0.08 \, (0.88) = 0.063 + 0.070 = 0.133$$

The operator estimates the value of k_d as 14.5 percent. If we substitute the appropriate values into equation 14.4, the resulting weighted average cost of capital is approximately 12 percent.

$$d = (0.133) \, (0.60) + (0.145) \, (0.7) \, (0.40) =$$
$$0.0798 + 0.0406 = 0.1204$$

Remember that it is appropriate to evaluate the present value of costs for alternative combines with a 12 percent discount rate because this represents the average cost of the capital that will be used to finance the farm operation.

The next step in the profitability analysis procedure is to determine the present value of the cash outlay to purchase the asset. With respect to the harvesting service, the farmer has obtained quotes on new medium- and large-size combines, both with corn and soybean heads, of $66,970 and $80,720, respectively. A third alternative is to purchase a used large-size combine that is three years old for $57,800. Since these outlays must be incurred immediately, their present value is equal to the dollar outlay as summarized in Table 14.4.

The annual net cash flow must be calculated for each of the alternative ways to acquire the machine services. If a particular machine service is required or the income attributable to that service will not vary with the method used to obtain the service, the most profitable alternative can be

---

[1]The before-tax cash return can be calculated using the procedure to allocate farm income to unpaid labor and equity capital discussed in Chapter 2. The before-tax cash return was recorded in Figure 2.4 on line 5a, page 49. The before-tax increase in the market value of assets can be estimated from the market value column of the balance sheet (Figure 2.5).

chosen by comparing only the costs and choosing that alternative with the lowest cost. However, the third step in the analysis (financial feasibility) cannot be completed without revenue as well as cost estimates. To illustrate all steps of the analysis procedure, we will assume that the farmer charges the corn and soybean production enterprise a custom rate for the use of combine services. The custom rate is set at $30.00 per acre for corn and $27.50 per acre for soybeans in the first year; the custom rate is assumed to increase 5 percent per year to reflect higher costs in the future.

The cash costs incurred in owning and operating a machine were identified in Chapter 4 as fuel, lubricants, insurance, and repairs. The owner must hire labor during the harvest season which would not be needed if a custom operator was hired. Thus, the cost of the labor to operate the combine is also considered a cash cost. The annual operating costs for both the medium- and large-size combines purchased new are summarized in Table 14.4 along with the costs for a used large-size combine. The procedures to estimate operating costs were explained in Chapter 4. These costs and the timeliness costs noted earlier are used in the computation of the annual net cash flows. Costs are assumed to increase 5 percent per year in this computation. Taxes are computed using equation 14.2.[2] The computations and resulting annual net cash inflows are summarized in Table 14.4.

Next, the present value of each annual net cash inflow is calculated by using the appropriate discount factors as summarized in Table 14.4. Finally, the net present value for each alternative is calculated; the computations of Table 14.4 indicate that the net present value for the new medium-size combine is $4,668. It is $4,756 for the new large-size combine and $313 for the used large-size combine. These net present values refer to an eight-year life for the two new combines and a five-year life for the used combine. They can be compared by converting the net present values to their annual equivalent using the amortization factor for the appropriate length of life and the 12 percent discount rate. The annualized net present value is $940 for the new medium-size machine, $957 for the new large-size machine, and $87 for the used large machine. Thus, the new large-size model is the most profitable combine (of the alternatives considered) to use in providing the harvesting services.

---

[2]The depreciation is calculated using the 1983 regulations. This requires the basis of the asset to be reduced by one-half of the amount of investment credit. It should be noted that the examples assume that the farmer has enough taxable income in year 1 or taxes paid in past years that can be recovered to use all of the investment credit in year 1. In the event only part of the investment credit available could be used in year 1, the remainder would be carried forward and applied to offset taxes owed in the second and succeeding years within the limitations provided by law. Any necessity to delay use of part of the investment credit will reduce the present value of the tax saving from investment credit.

Table 14.4 Economic Profitability Analysis
of the Three Combines

Year	Cash Income[a]	Cash Expense[b]	Taxes[c]	Investment Tax Credit	Annual Net Cash Inflow	Discount Factor (12 Percent)	Present Value of Annual Net Cash Inflow
New Medium Size							
1	$21,000	$ 6,817	$1,392	$6,697	$19,489	0.893	$17,403
2	22,050	8,787	−220	—	13,483	0.797	10,746
3	23,153	10,212	−127	—	13,068	0.712	9,304
4	24,310	11,588	−193	—	12,915	0.636	8,214
5	25,526	12,961	−240	—	12,805	0.567	7,260
6	26,802	14,321	3,744	—	8,737	0.507	4,430
7	28,142	15,540	3,781	—	8,821	0.452	3,987
8	29,549	16,988	3,768	—	8,793	0.404	3,552
Remaining value					$16,687[d]	0.404	6,742
Present value of net cash inflows							$71,638
Cash outlay							66,970
Net present value							$ 4,668
Annual equivalent net present value (12 percent)							$ 940[e]
New Large Size							
1	$21,000	$ 5,561	$1,181	$8,072	$22,330	0.893	$19,941
2	22,050	6,992	−544	—	15,602	0.797	12,435
3	23,153	8,088	−312	—	15,377	0.712	10,948
4	24,310	9,113	−272	—	15,469	0.636	9,838
5	25,526	10,102	−204	—	15,628	0.567	8,861
6	26,802	11,142	4,698	—	10,962	0.507	5,558
7	28,142	12,006	4,841	—	11,295	0.452	5,105
8	29,549	13,113	4,931	—	11,505	0.404	4,648
Remaining value					20,154[d]	0.404	8,142
Present value of cash inflows							$85,476
Cash outlay							80,720
Net present value							$ 4,756
Annual equivalent net present value (12 percent)							$ 957[e]

(*continued*)

The Most Profitable Financing Method

The next question to be answered is that of the most profitable method of financing the combine acquisition. Assume that three options are available: (1) the farmer can pay 25 percent down and borrow the remainder from the major lender at 14.5 percent interest with three equal annual payments on the remaining balance; (2) the farmer can pay 25 percent down and borrow the remainder from the dealer at 15 percent interest with four equal annual payments on the remaining balance; and (3) the farmer can lease the combine on a five-year schedule with annual lease payments at the beginning of each year of $21,794 and an option to buy in year 5. The farmer estimates that the combine can be purchased at the end of the lease

Table 14.4 (*Continued*)

Year	Cash Income[a]	Cash Expense[b]	Taxes[c]	Investment Tax Credit	Annual Net Cash Inflow	Discount Factor (12 Percent)	Present Value of Annual Net Cash Inflow
Used Large Size							
1	$21,000	$ 8,621	$1,243	$5,780	$16,916	0.893	$15,106
2	22,050	9,320	195	—	12,535	0.797	9,990
3	23,153	10,138	445	—	12,570	0.712	8,950
4	24,310	11,085	508	—	12,717	0.636	8,088
5	25,526	11,969	608	—	12,949	0.567	7,342
Remaining value					15,233[d]	0.567	8,637
Present value of cash inflows							$58,113
Cash outlay							57,800
Net present value							$ 313
Annual equivalent net present value (12 percent)							$ 87[e]

[a]Calculated as 400 acres of corn at $30.00 plus 400 acres of soybeans at $27.50 per acre for the first year with a 5 percent increase per year.

[b]The total of fuel, lubricants, insurance, labor, and repair costs was calculated for each year of life using formulas indicated in Chapter 4. The timeliness cost was added and the total converted to nominal dollars with a 5 percent increase per year. To illustrate, for the medium size combine the cash expenses in year 1 were: fuel—$2,074, lubricants—$311, insurance—$252, labor—$1,300, repairs—$2,467, and timeliness cost—$413; for a total of $6,817.

[c]Calculated as the tax rate of 30 percent times cash income minus the sum of cash expenses and depreciation using the ACRS system (assuming a five-year life). The basis is reduced by one-half of the investment tax credit to calculate depreciation, as required by IRS rules of 1983. To illustrate, the depreciation deduction for the medium size combine for year 1 is calculated as the adjusted basis of $63,621 ($66,970 minus $3,349 which is one-half of the investment credit taken) times the 1st year write-off rate of 15 percent which totals $9,543. The taxable income for year 1 is then $4,460 ($21,000 − ($6,817 + $9,543)). The tax liability is calculated as $4,640 times the tax rate of 30 percent which totals $1,392.

[d]Calculated as the remaining value using the formulas of Chapter 4 increased by 5 percent per year for the life of the machine minus taxes that will be incurred on sale of the combine. Since the combine has been fully depreciated, the entire salvage value is ordinary income taxed at the 30 percent marginal tax rate. To illustrate, for the medium size combine the before tax salvage value is calculated as the $66,970 purchase price times the remaining value percentage for eight years from Table 4.2 of 24.1 percent times $(1.05)^8$ which equals $23,838 ($66,970 × .241 × 1.477). The after-tax salvage is then calculated as $23,838 minus the 30 percent tax liability ($7,151) or $16,687.

[e]Calculated as the net present value multiplied by the amortization factor from Appendix Table IV for 12 percent and the appropriate number of years.

for 10 percent of the initial purchase price of $80,720.[3] The cash payments each year for the two loans and the lease are summarized in Table 14.5.

One method of financing is considered more profitable than another if the net present value of the combined financing and tax effects is less for the first than the second. The basis for comparison is the cost of financing by the major lender. The comparison must consider any difference in the after-tax cost of financing the asset, including differences in the after-tax cost of interest and principal payments, the after-tax value of depreciation and investment credit, and the after-tax salvage value. The example in Table 14.5 illustrates that consideration of each of these components may be important.

The discount rate selected for the profitability of financing analysis should compare the advantage or disadvantage of the after-tax financing cost being considered for an asset with the cost of financing from the major lender. The before-tax cost of debt capital from the major lender is 14.5 percent in this example. Assuming a 30 percent marginal tax rate results in an after-tax cost of financing and thus discount rate of 10.15 percent (14.5 percent × 0.7).

The logic of using this rate follows from noting that an asset financed at the same percentage rate as the discount rate will have a present value of the net cash outflow for principal and interest payments (principal payments plus the after-tax interest payments) equal to the amount financed. This is illustrated for the combine example in the first portion of Table 14.5. Financing the combine with a 25 percent downpayment and repayment over three years at 14.5 percent would result in the payment schedule shown in the first three columns. Including the tax savings from interest at the operator's 30 percent marginal tax rate results in the annual after-tax interest and principal payments shown. Discounting by 10.15 percent (14.5 percent × 0.7), the after-tax cost of financing from the major lender, results in a present value of the interest and principal payments for the three years of $60,540 (which is equal to the amount financed). Adding the present value of the downpayment results in a total present value of the outflow of $80,720.

Loan arrangements with a higher (lower) interest rate would have higher (lower) interest costs and, when disounted by 10.15 percent, a higher (lower) present value of the annual net cash outflow. This is illustrated by the second method of financing considered in Table 14.5—dealer financing at 15 percent interest. The net present value of the after-tax loan payments with the higher interest rate is $81,161, $441 higher than financ-

---

[3]It is assumed that the terms of the agreement qualify as a lease rather than a conditional sales contract for tax purposes. The conditions for a financing agreement to be considered a lease rather than a conditional sales contract are discussed in the *Farmers' Tax Guide*, Department of the Treasury, Publication 225 (revised annually).

Table 14.5 Analysis of the Profitability of Alternative Methods of Financing the Combine

Year	Down Payment or Principal	Interest	Total Loan Payment	Tax Savings from Interest[a]	Annual After-Tax Loan Payment	Discount Factor (10.15 Percent)	Present Value of After-Tax Loan Payment	Tax Savings Depreciation[b]	Investment Credit	Total	Present Value of Tax Savings from Depreciation and Investment Credit	Present Value of After-Tax Salvage Value
					Financing with Major Lender at 14.5 Percent Interest							
0	$20,180	—	$20,180	—	$20,180	1.000	$20,180	—	—	—	—	
1	20,180	$8,778	28,958	$2,633	26,325	0.9079	23,901	$3,451	$8,072	$11,523	$10,462	
2	20,180	5,852	26,032	1,756	24,276	0.8242	20,008	5,061	—	5,061	4,171	
3	20,180	2,926	23,106	878	22,228	0.7482	16,631	4,831	—	4,831	3,615	
4	—	—	—	—	—	0.6793	—	4,831	—	4,831	3,282	
5	—	—	—	—	—	0.6167	—	4,831	—	4,831	2,979	
After-tax salvage value in year 8—$20,154[c]						0.4614						$9,299
Total							$80,720				$24,509	$9,299

Net present value of after-tax outflow = $80,720 − $24,509 −$9,299 = $46,912

Year	Down Payment or Principal	Interest	Total Loan Payment	Tax Savings from Interest[a]	Annual After-Tax Loan Payment	Discount Factor (10.15 Percent)	Present Value of After-Tax Loan Payment	Tax Savings Depreciation[b]	Investment Credit	Total	Present Value of Tax Savings from Depreciation and Investment Credit	Present Value of After-Tax Salvage Value
					Dealer Financing at 15 Percent Interest							
0	$20,180	—	$20,180	—	$20,180	1.0000	$20,180	—	—	—	—	
1	15,135	$9,081	24,216	$2,724	21,492	0.9079	19,513	$3,451	$8,072	$11,523	$10,462	
2	15,135	6,811	21,946	2,043	19,903	0.8242	16,404	5,061	—	5,061	4,171	
3	15,135	4,541	19,676	1,362	18,314	0.7482	13,703	4,831	—	4,831	3,615	
4	15,135	2,270	17,405	681	16,724	0.6793	11,361	4,831	—	4,831	3,282	
5	—	—	—	—	—	0.6167	—	4,831	—	4,831	2,979	
After-tax salvage value in year 8—$20,154[c]						0.4614						$9,299
Total							$81,161				$24,509	$9,299

Net present value of after-tax outflow = $81,161 − $24,509 −$9,299 = $47,353

(continued)

Table 14.5 (Continued)

Year	Annual Lease Payment	Tax Savings from Lease Payment[d]	Annual After-Tax Lease Payment	Discount Factor (10.15 Percent)	Present Value of After-Tax Lease Payment	Tax Savings			Present Value of Tax Savings from Depreciation	Present Value of After-Tax Salvage Value
						Depreciation[b]	Investment Credit	Total		
					Leasing Arrangement					
0	$21,794	—	$21,794	1.0000	$21,794	—	—	—	—	—
1	21,794	$6,538	15,256	0.9079	13,851	—	—	—	—	—
2	21,794	6,538	15,256	0.8242	12,574	—	—	—	—	—
3	21,794	6,538	15,256	0.7482	11,415	—	—	—	—	—
4	21,794	6,538	15,256	0.6793	10,363	—	—	—	—	—
5	8,072	6,538	1,534	0.6167	946	—	—	—	—	—
6	—	—	—	0.5599	—	$363	—g	$363	$203	
7	—	—	—	0.5083	—	533	—	533	271	
8	—	—	—	0.4614	—	509	—	509	235	
Salvage value in year 8—$24,901f				0.4614						$11,489
Total					$70,943				$709	$11,489

Net present value of after-tax outflow = $70,943 − $709 − $11,489 = $58,745.

aCalculated as the interest deduction times the marginal tax rate of 30 percent.

bCalculated as the ACRS depreciation allowance times the marginal tax rate of 30 percent.

cCalculated as the salvage value of $28,792 minus taxes on the salvage value of $8,676 ($28,792 × 0.30).

dCalculated as the lease payment times the marginal tax rate of 30 percent.

eCalculated assuming the ACRS depreciation allowance for a five-year life asset purchased at the beginning of year 6 for $8,072 with the ACRS deductions each year multiplied times the marginal tax rate of 30 percent.

fCalculated as the salvage value of $28,792 minus taxes on that salvage value of $3,891. Taxes are computed as the depreciation recapture of $4,682 ($8,072 purchase price minus $3,390 remaining basis) plus 40 percent of the capital gain of $20,720 ($28,792 sale price minus $8,072 purchase price) which totals $12,970 of taxable gain times the 30 percent marginal tax rate.

gThe investment in the used combine is not eligible for investment credit because it was leased previously by the same individual.

ing with the major lender at 14.5 percent interest. If the operator had an opportunity to finance the combine at a rate lower than 14.5 percent, the net present value of the outlay would be less than $80,720.

The tax effects may differ from one financing arrangement to another, particularly when a leasing arrangement is being compared to a loan arrangement. If so, the net effect of this difference must be considered in comparing the profitability of the financing. With a loan option, additional tax savings are available from depreciation and investment credit; the tax savings from the deductibility of interest has already been taken into account. The annual tax savings from depreciation (30 percent of the ACRS depreciation for the year) and investment credit are totaled and discounted by 10.15 percent, the after-tax cost of borrowing from the major lender. The sum of the annual present value of tax savings from depreciation and investment credit is $24,509 for either the loan from the major lender or the dealer financing, as shown in Table 14.5. The tax savings and consequences of leasing will be discussed shortly.

The after-tax salvage value of the asset may also differ from one method of financing to another, making it necessary to consider this difference in evaluating the profitability of financing alternatives. For example, the combine will have a different after-tax salvage value with the loans than with the lease because of differences in the tax basis of the combine at salvage. The before-tax salvage value of the combine in year 8 is $28,792, and the after-tax salvage value (70 percent of the total) is $20,154 for either loan. Discounting by the after-tax cost of borrowing from the major lender (10.15 percent) gives a present value of $9,299; as noted later, this number will be different for the lease.

The present value of each financing alternative is then the sum of the present value of the after-tax loan payments, the present value of tax savings from depreciation and investment credit, and the present value of the after-tax salvage value. This is $46,912 ($80,720 − $24,509 − $9,299) for the 14.5 percent loan and $47,353 for the 15 percent loan. As expected, the higher interest rate for dealer financing results in a present value that is $441 higher than financing with the major lender. All of this difference is in the present value of after-tax loan payments. The present value of tax savings from depreciation and investment credit and the present value of the after-tax salvage value are identical for the two loans considered in this example. However, an analysis of the leasing arrangement in the third section of Table 14.5 illustrates the importance of considering the present value of the tax savings from depreciation and investment credit and the present value of the after-tax salvage value in comparing the profitability of financing alternatives.

The computations for the leasing alternatives are completed in a similar manner, as shown in the third part of Table 14.5. The lease payments are tax deductible in the year after the payment is made (the lease payments are made in advance); thus, the tax savings are delayed one year as shown. The net cash outflow (or the annual after-tax lease payment) is calculated as the annual lease payment minus the tax savings, and the present value of this outflow is calculated using the appropriate discount factor. Note that in year 5 the purchase option is exercised with a payment of $8,072 (10 percent of the initial purchase price). The combine is then placed on a five-year ACRS depreciation schedule which results in tax savings from depreciation in years 6, 7, and 8. The combine is sold at the end of year 8 for the same price as in the borrow and buy option ($28,792), but the cost basis is $3,390 (the remaining ACRS allowances) rather than 0. Since the machine is sold for more than the purchase price, part of the gain is taxed as ordinary income and part as capital gain; thus the after-tax salvage is higher than in the borrow and buy option. The calculations of Table 14.5 show that the net present value of the combined financing and tax effects for the leasing option is $58,745 ($70,943 − $709 − $11,489).

The most profitable financing method is to buy the machine with the loan terms from the major lender. Dealer financing has a present value that is $441 greater, while the present value of the leasing option is $11,833 greater. However, if the farmer could obtain a lower purchase price or other concessions worth more than $441 for agreeing to accept dealer financing, then dealer financing would be the most profitable alternative. Thus, the difference in net present values provides additional information—the amount of change in the purchase price required to offset higher or lower financing charges.[4]

Financial Feasibility

The final step in the analysis is to evaluate the financial feasibility of the combine purchase using the most profitable financing method. The calculations discussed earlier in Chapter 8 are illustrated in Table 14.6 for the new large combine acquired with the dealer financing option. Dealer financing is chosen to illustrate the procedure even though it is not the most profitable financing alternative. This choice was made to illustrate that a project which is financially feasible in the first year may not be feasible in future years. First, the annual net cash inflow from Table 14.4 is listed. This is the after-tax inflow calculated using equation 14.1. It considers the deductibility of cash expenses and depreciation as well as the tax-reducing effects of investment credit. Then the after-tax loan payment schedule

[4]Computing the after-tax cost of lease capital and comparing it to the after-tax cost of borrowing is another method of comparing the profitability of financing. The implicit cost of lease capital considers differences in tax treatment and the cost of purchasing the asset at the end of the lease. The advantage of computing the discounted present values, however, is that it also indicates how much more or less profitable the leasing arrangement is—a significant advantage in negotiating the leasing arrangement. See James C. Van Horne, *Financial Management and Policy,* 5th ed. (Englewood Cliffs, N.J.: Prentice-Hall, 1980), Chapter 19, for a discussion of computing the implicit cost of leasing.

Table 14.6 Financial Feasibility of the Combine Purchase

Dealer Financing Option

Year	Annual Net Cash Inflow (Table 14.4)	Annual After-Tax Loan Payment (Table 14.5)	Surplus (+) Deficit (−)
0	—	$20,180	−$20,180
1	$22,330	21,492	+838
2	15,602	19,903	−4,301
3	15,377	18,314	−2,937
4	15,469	16,724	−1,255
5	15,628	—	+15,628
6	10,962	—	+10,962
7	11,295	—	+11,295
8	11,505	—	+11,505
Salvage value	20,154	—	+20,154

Lease Option

Year	Cash Income (Table 14.4)	Cash Expense (Table 14.4)	Taxes with Lease[a]	Annual Net Cash Inflow with Lease	Annual After-Tax Lease Payment (Table 14.5)	Tax Savings from Depreciation (Table 14.5)	Surplus (+) Deficit (−)
0	—	—	—	—	$21,794	—	−$21,794
1	$21,000	$ 5,561	$4,632	$10,807	15,256	—	−4,449
2	22,050	6,992	4,517	10,541	15,256	—	−4,715
3	23,153	8,088	4,520	10,545	15,256	—	−4,711
4	24,310	9,113	4,559	10,638	15,256	—	−4,618
5	25,526	10,102	4,627	10,797	1,534	—	+9,263
6	26,802	11,142	4,698	10,962	—	363	+11,325
7	28,142	12,006	4,841	11,295	—	533	+11,828
8	29,549	13,113	4,931	11,505	—	509	+12,014
Salvage value				24,901			+24,901

[a]This calculation can and does ignore the tax deductibility of the lease payments because this phenomenon is included by calculating the lease payments on an after-tax basis (i.e., lease payments minus the tax savings from such payments being deductible).

from Table 14.5, which considers the tax deductibility of interest, is summarized. Finally, the surplus or deficit is calculated as the annual net cash inflow minus the after-tax loan payment schedule. Note that a deficit of $20,180 will occur when the machine is purchased because of the downpayment. A surplus cash flow will occur in year 1 because of the investment credit, but deficits will occur in years 2 through 4. These deficits will require a cash subsidy from another part of the farm operation. After the loan has been repaid, sizable cash surpluses will occur for the remainder of the life of the machine. The reader can establish that the financial feasibility would involve larger deficits for years 1 through 3 with the three-year loan from the major lender.

The computations must be altered slightly to complete the financial feasibility analysis for a lease option. The annual net cash inflow of Table 14.4 is calculated based on the assumption that the machine is purchased and investment credit and depreciation allowances are taken. If the machine is leased instead, depreciation is not allowed, and the farmer may or may not be able to take the investment credit depending on the lease terms. In the example, the farmer does not receive the investment credit. Thus, the tax savings from both the depreciation allowance and the investment tax credit must be eliminated from the net cash flow. The computational procedure is illustrated in the second part of Table 14.6 for the combine example. The cash income and cash expenses are obtained from Table 14.4. Taxes with the lease are the difference between cash income and cash expenses multiplied by the marginal tax rate. This is ($21,000 − $5,561) × 0.3 = $4,632 for year 1, and the after-tax net cash inflow for year 1 is $10,807. The annual after-tax schedule of lease payments and the tax savings from the depreciation are taken from Table 14.5. Note that the surplus or deficit column is negative for the leasing arrangement in each of years 0 through 4. The leasing alternative will require larger subsidies from other parts of the business during years 0 through 4 than dealer financing.

SELECTING A DAIRY PARLOR

The choice of alternative ways to obtain the services of buildings and facilities proceeds in much the same fashion as has been discussed for machinery services. The first step in making this decision is to identify the alternative ways of providing the building service. Alternative designs for a new building to be constructed and equipped on site should be considered. In many situations, a manufactured building can be purchased and erected on the farm. It may also be possible to remodel an existing structure for use. In some cases, it is possible to rent an unused facility at another location, particularly if the service is needed for a brief period.

In addition to estimating the original outlay for each alternative, it is also important to estimate the expected life of the facility, its capacity, and the labor requirements. These data provide the basis to estimate the annual cash costs for energy, repairs, maintenance, and hired labor. The capacity for certain types of production or service facilities can most meaningfully be measured as a rate of throughput (such as bushels of grain dried per hour or number of cows milked per hour), while in others the capacity is measured by the number of units of space (such as the number of hogs, laying hens, or bushels of storage). Data on the capacity can usually be estimated from studies of these technologies provided by the land grant college.[5]

A farmer organizing a dairy herd at a new location must decide on the appropriate size and type of milking parlor.[6] Assume the operator narrows the choice to two alternatives, a double-4 herringbone and a double-6 herringbone, based on the labor requirements and the availability of hired labor to perform the milking. The double-4 unit has lower investment costs but higher labor and operating costs since it requires more time to complete each milking. The operator estimates that the repairs, insurance, real estate and property taxes, labor, electricity, and gas are the annual cash expenses that are expected to differ between the two alternative milking systems. Since the cash income generated by the herd, as well as other expenses, is expected to be the same for both of the milking parlors, the analysis will focus on the comparative costs of the various systems.

The first step in the analysis is again to choose a discount rate; assume that the computations discussed earlier in the combine example have been completed and a discount rate of 12 percent has been calculated. The next step in the analysis is to calculate the present value of the cash outlay required to acquire the asset. The capital investment required for each set of milking facilities is listed in Table 14.7. The double-4 herringbone parlor is composed of the building, milking equipment, and other parlor equipment. The double-6 herringbone also includes some mechanization equipment (including automatic detachers and some equipment to facilitate the movement of cows through the parlor). The total initial outlay for the parlor and equipment is listed at the top of Tables 14.8 and 14.9 for the double-4 and double-6 parlors, respectively. These investments must be made at the beginning of the period, making their present value equal to the initial outlay.

The length of useful life is noted for the parlor and each category of equipment in Table 14.7. Although the building is expected to be struc-

---

[5]For example, see Harold R. Jensen, Jeffrey P. Madsen, and Vernon R. Eidman, *Economics of Owning and Operating Corn Drying and Storing Systems with Rising Energy Prices,* St. Paul: University of Minnesota Agricultural Experiment Station, Technical Bulletin 320, 1979.

[6]The data on investment costs and operating requirements were based on Howard Wetzel and B. F. Stanton, *Alternative Milking Systems for Daily Herds of 50 to 500 Cows,* Ithaca, N.Y.: Cornell University Department of Agricultural Economics, AE Res. 80–18, 1980.

Table 14.7 Data for the Two Milking Parlors

Item	Useful Life in Years	Double-4 Herringbone	Double-6 Herringbone
Parlor building	15	$23,500	$28,400
Parlor equipment	10	9,300	11,700
Milking equipment	7.5	11,700	15,100
Mechanization equipment	7.5		12,500
Total Initial outlay		$44,500	$67,700
Annual cash expenses			
Repairs, insurance, and taxes		$ 2,971	$ 4,240
Electricity and gas		782	863
Labor		3586.5 hr @ $6.00 21,519	2984.2 hr @ $6.00 17,905
Total		$25,272	$23,008

594

Table 14.8 Summary of 15-Year Discounted Investment and Operating Costs for a Double-4 Herringbone Parlor Milking 150 Cows

Year		Purchase Cost or Salvage Value	Annual Cash Expense	Income Tax	Investment Tax Credit	Annual Net Cash Flow	Discount Factor (12 Percent)	Present Value of Annual Net Cash Flow	Depreciation
0	Purchase parlor building	$23,500				-$23,500	1.000	-$23,500	
	Purchase parlor equipment	9,300				-9,300	1.000	-9,300	
	Purchase milking equipment	11,700				-11,700	1.000	-11,700	
1			$25,272	-$9,484	$4,450	-11,338	0.893	-10,125	$6,341
2			26,788	-10,826		-15,962	0.797	-12,722	9,300
3			28,406	-11,185		-17,221	0.721	-12,261	8,878
4			30,099	-11,693		-18,406	0.636	-11,706	8,878
5			31,893	-12,231		-19,662	0.567	-11,148	8,878
6			33,814	-10,144		-23,670	0.507	-12,001	
7			35,861	-10,758		-25,103	0.452	-11,347	
8	Replace milking equipment	17,286	38,009	-11,403		-43,892	0.404	-17,732	
9			40,284	-12,824	1,728	-25,732	0.361	-9,289	2,463
10			42,710	-13,897		-28,813	0.322	-9,278	3,613
11	Replace parlor equipment	15,149	45,262	-14,613	1,515	-45,798	0.287	-13,144	3,449
12			47,966	-16,072		-30,379	0.257	-7,807	5,608
13			50,847	-17,239		-33,608	0.229	-7,696	6,615
14			53,905	-17,078		-36,827	0.205	-7,550	3,022
15	Salvage value	12,411ᵃ	57,140	-15,232		-29,497	0.183	-5,398	3,022
Net present value								-$203,704	

ᵃThe undepreciated value of the replacement parlor equipment at the end of year 15 is $3,023. Thus, only $9,388 of this salvage value is taxable as ordinary income.

Table 14.9 Summary of 15-Year Discounted Investment and Operating Costs for a Double-6 Herringbone Parlor Milking 150 Cows

Year		Purchase Cost or Salvage Value	Annual Cash Expense	Income Tax	Investment Tax Credit	Annual Net Cash Flow	Discount Factor (12 Percent)	Present Value of Annual Net Cash Flow	Depreciation
0	Purchase parlor building	$28,400				-$28,400	1.000	-$28,400	
	Purchase parlor equipment	11,700				-11,700	1.000	-11,700	
	Purchase milking equipment	15,100				-15,100	1.000	-15,100	
	Purchase mechanization equipment	12,500				-12,500	1.000	-12,500	
1			$23,008	-$9,796	$6,770	-6,442	0.893	-5,753	$ 9,647
2			24,388	-11,561		-12,827	0.797	-10,223	14,149
3			25,861	-11,810		-14,051	0.712	-10,004	13,506
4			27,403	-12,273		-15,130	0.636	-9,623	13,506
5			29,036	-12,763		-16,273	0.567	-9,227	13,506
6			30,785	-9,236		-21,549	0.507	-10,925	
7			32,648	-9,794		-22,854	0.452	-10,330	
8	Replace milking and mechanization equipment	40,778	34,604	-10,381	4,078	-65,001	0.404	-26,260	
9			36,675	-12,746		-19,851	0.361	-7,166	5,811
10			38,884	-14,222		-24,662	0.322	-7,941	8,523
11	Replace parlor equipment	19,058	41,207	-14,803	1,906	-45,462	0.287	-13,048	8,135
12			43,669	-16,356		-25,407	0.257	-6,530	10,851
13			46,292	-17,523		-28,769	0.229	-6,588	12,118
14			49,076	-15,863		-33,213	0.205	-6,809	3,802
15	Salvage value	17,400[a]	52,021	-12,668		-21,953	0.183	-4,017	3,802
Net present value								-$212,144	

[a] The undepreciated value of the replacement parlor equipment at the end of year 15 is $3,802. Thus, only $13,598 of this salvage value is taxable as ordinary income.

turally sound for more than 15 years, it probably will be obsolete and require extensive modification and repairs. Given the difficulty of predicting the dollar value of the remodeling required after 15 years of use and the relatively small present value of cash costs 16 or more years in the future, a 15-year planning horizon is selected for the analysis. Each type of equipment will have to be replaced one time during the 15-year life of the building as indicated by the outflows in years 8 and 11 in Tables 14.8 and 14.9.

The third step in the analysis is to calculate the annual net cash flow using equation 14.1. As noted earlier, we are assuming that the cash income and all expenses except those associated with the purchase and operation of the milking parlor are identical for the two systems. Under this assumption, the present value of the income and the other expenses is identical and can be ignored in estimating the difference between the parlors in net present value. Because the income is not estimated, the financial feasibility analysis cannot be completed in this case. The dollar values of the outlay and expenses that are different between the two parlor systems are listed by year for the double-4 system in Table 14.8, and comparable data are shown for the double-6 system in Table 14.9.

Consider the entries in Table 14.8 for year 1. The cash expenses of $25,272 include repairs, insurance, real estate taxes, electricity, gas, and labor as listed in Table 14.7. The milking parlor qualifies as a single-purpose livestock structure, making the building and the equipment eligible for the five-year accelerated cost recovery system. The basis is reduced by one-half of the investment tax credit to calculate depreciation, as required by IRS rules for 1983. Fifteen percent of the adjusted basis, or $6,341, can be deducted for depreciation during the first year. The income tax effect of the investment is estimated using equation 14.2. With a 30 percent marginal tax rate, income tax payments will be reduced $9,484 during the first year because of the deductibility of the cash expenses and the depreciation. Investment credit can also be used to reduce income tax payments in year 1. The full $44,500 is five-year ACRS property and is eligible for the 10 percent credit.

The annual net cash flow (ANCF) for each year is calculated with equation 14.1. The operator assumes cash expenses will increase at an annual rate of 6 percent, while cost of replacement equipment increases 5 percent per year. Replacing the milking equipment during the eighth year requires a net outlay of $[(1.05)^8 \times \$11,700]$ or $17,286, and the replacement cost of the parlor equipment after 10 years is $(1.05)^{10} \times \$9,300$ or $15,149. Each of these new investments can be depreciated with the five-year ACRS percentages, and they are eligible for 10 percent investment credit in the year following the investment. The annual cash flow includes the additional outlay for replacement equipment purchased during the year. The salvage value at the end of year 15 includes 10 percent of the purchase price of the building ($2,350), 10 percent of the replacement cost of the milking equipment ($1,729), and 55 percent of the replacement cost of the parlor equipment ($8,332). If the unit is sold at the end of year 15, the salvage value in excess of unrecovered capital is taxed as ordinary income. The resulting net cash flow for year 15 is −$29,497.

The present value of each annual net cash flow is then calculated, and these values are summed to obtain the present value of the cash outlays associated with purchasing the investment, replacing the equipment at the indicated intervals, paying the annual cash expenses, and disposing of the investment at the end of 15 years for the salvage value indicated. The total present value is −$203,704 for the double-4 herringbone parlor.

The comparable analysis for the double-6 herringbone is presented in Table 14.9. The $67,700 investment is composed of the building and three types of equipment. Equipment replacement costs and annual cash expenses are projected to increase at the same annual rate (5 and 6 percent, respectively) used in the analysis for the double-4 facilities. All of the investments and the replacements are again eligible for the five-year accelerated cost recovery system and investment credit. The present value of the annual cash outflows is again computed using a 12 percent discount rate; the present value for the double-4 herringbone is −$212,144.

The analysis indicates the present value of costs is $8,440 higher for the double-6 herringbone. This may seem to be a relatively small difference in costs when the difference in the initial investment outlay is $23,200 ($67,700 − $44,500). However, this higher outlay for the double-6 is partially offset by the tax effects and the higher annual cash expenses associated with the double-4 system. The small difference also suggests that any increase in annual cash costs relative to the initial outlay for buildings and equipment will make the double-6 system more attractive.

Before making a decision, the operator may want to determine how sensitive the present value of the cost is to certain key assumptions. The analysis can be repeated varying the assumptions to estimate the resulting effect on the present value of costs. This may be quite time consuming when the computations are done by hand, but may be an easy task when computing equipment is used to make the calculations.

Labor efficiency and the wage rate may be important in selecting the lowest cost milking parlor. Data on the possible range of labor requirements and operating costs are shown in Table 14.10 for the two milking parlors. The data in the table indicate that a slower (faster) milking rate requires more (less) hours of labor as well as more (less) electricity and fuel since the parlor is used more (less) hours during the year.

The effect of a change in the milking speed can be examined by comparing the present value of costs for a specified wage rate in Table 14.11. For example, with a first-year wage rate of $6 per hour, the present value of costs for the double-4 herringbone ranges from $221,112 for the slow milking rate to $191,204 for the rapid rate. While the range in out-

Table 14.10   Annual Cash Costs for Year 1
by Milking Speed

Item	Double-4 Herringbone			Double-6 Herringbone		
	Slow	Average	Rapid	Slow	Average	Rapid
Repairs, insurance, and taxes	$2,971	$2,971	$2,971	$4,240	$4,240	$4,240
Electricity and fuel	$874	$782	$714	$1,021	$863	$764
Labor hours	4,013.5	3,586.5	3,280.1	3,531.7	2,984.2	2,644.8

comes is rather large, it is probably more important to notice that the present value of costs is greater at each milking speed for the double-6 than for the double-4 herringbone. If we presume that individual(s) will operate at the same relative speed in either facility, the double-4 herringbone is cheaper for any milking speed selected. The range also suggests that it will be advantageous to encourage efficient work patterns in using either of the facilities.

The effect of a change in the beginning wage rate (with the same rate of increase over the 15-year period) can be studied by selecting a relative milking speed (say the average speed) and comparing the present value of costs for the two systems for different wage rates. For example, with the average milking speed, the double-4 herringbone has a significant cost advantage at wage rates of $4 and $6 per hour, but the two milking parlors have very similar present values of costs at a wage rate of $8 per hour ($250,747 compared to $251,287). At the rapid rate of speed and the $8 wage rate, the double-6 herringbone has a lower present value of costs than the double-4 system.

The operator could also include risk considerations in the analysis by identifying important sources of uncertainty and estimating the subjective probability distribution of each variable to be considered. Then the capital budgeting procedure could be used to estimate the present value of costs for each level of the uncertain variable. Given these present values of costs, the procedures in Chapter 11 could be used to select the appropriate milking parlor.

Table 14.11   Sensitivity of Present Value of Costs to Milking
Speed and Wage Rate

Year 1 Wage Rate $/Hour	Double-4 Herringbone			Double-6 Herringbone		
	Slow	Average	Rapid	Slow	Average	Rapid
$4	$178,088	$156,661	$138,561	$188,400	$173,001	$163,446
$6	221,112	203,704	191,204	234,724	212,144	198,137
$8	264,136	250,747	243,847	281,048	251,287	232,828

THE REPLACEMENT DECISION

"Should I buy a new planter this year to replace my current machine, or should I wait until next year?" That is a question farm managers often ask. The replacement decision must be made frequently, and, with the larger outlays for depreciable capital items, it is a decision that can also have a significant impact on the profitability and efficiency of the farm firm. Our discussion will focus on two related but different types of replacement decisions. The first decision is encountered in long-run planning for a new enterprise: the issue here is what is the optimal replacement policy or time for the machinery and equipment that will be purchased. The second issue is faced when trying to answer the question posed earlier: if you currently own a machine, when should the old machine be replaced with a new one?

The replacement decision is influenced by a number of factors: (1) the age, efficiency, and reliability of the present machine; (2) the repair and timeliness costs of the present machine; (3) income tax considerations, including depreciation allowances and investment tax credit as well as tax recapture provisions; (4) new technology which has resulted in increased efficiency and made the present machine obsolete; (5) the pattern of changes in the salvage value of machinery; and (6) size considerations which may suggest that a larger or smaller machine is needed because of expansion or contraction of the enterprise or farm operation. Numerical estimates of the impact of these factors should be included in the calculations that are required to make the replacement decision.

Accurate analysis of the replacement decision again requires use of capital budgeting procedures and discounting.[7] The application of these procedures to the replacement decision is simple in concept but complex in computations. In essence, the analysis proceeds by first calculating the net present value of revenues (or costs) for the machine assuming various replacement periods. For example, the net present value for an orchard or machine would be calculated assuming it was replaced at the end of one year, two years, three years, four years, and so on, until it was completely worn out or until the end of its service life. These calculations for each assumed replacement year must include the tax and tax recapture considerations, the changing repair costs and reliability and timeliness considerations, and the changing salvage value for different lengths of life. The net present values for each replacement period are then converted to

---

[7]The procedure suggested here is an approximation of the theoretically more complex replacement models suggested in the literature. It does not include consideration of the potential income from an infinite number of future replacement machines or the opportunity cost of keeping the current machine compared to continually replacing it with a new machine (R. K. Perrin "Asset Replacement Principles," *American Journal of Agricultural Economics* 54 (1972):60–67). Including this consideration in the computations is impractical because of problems in obtaining reasonable data on an infinite number of future replacement machines. Ignoring this consideration results in a slightly longer replacement period than would occur if the income of future replacement machines were included.

an annual equivalent net present value by using the amortization factor from Appendix Table IV as discussed in Chapter 8. This *annualized* net present value indicates the annualized net revenue or cost for the machine for different lengths of life. The optimal length of life or the optimal replacement policy (assuming a machine is not currently owned) is the number of years with the highest annualized net present value of revenues. If the analysis is based solely on the costs of supplying the machine service (revenue is not considered), the optimal replacement policy is indicated by the number of years having the lowest annualized net present value of cost.

The procedure to follow in deciding whether or not to replace a currently owned machine with a new machine is similar. The computations for the new machine are identical to those noted above. For the old machine, the net present value of continuing to use that machine for an additional one year, an additional two years, an additional three years, and so on, until the end of its service life is calculated using the procedures discussed earlier. These net present values for the old machine assuming different lengths of continued use are again annualized using the amortization factor procedure. If we assume the annualized net present value of net revenues has been calculated, the decision as to whether or not to replace is then made by comparing the largest annualized net present value for the new machine (which reflects the optimal length of life or replacement policy for it) to the largest annualized net present value for the old machine (which again reflects the optimal length of life for the current machine). If the annualized net present value for the new machine exceeds that of the current one, the current machine should be replaced. If the annualized value for the current machine exceeds that of the new machine, the current machine should be kept for another year and the analysis should be repeated again next year.

Determining the Replacement Cycle

To illustrate the applications of these concepts, they will be applied to the decision of replacing a combine. First, let us assume that a farmer whose crops have been custom harvested is contemplating the purchase of a combine. In the process of developing long-run production and financial plans, the farmer wants to determine the optimal replacement cycle for the combine being considered for purchase.

The data presented earlier for the new large-size combine will be used in this analysis. The first step is to determine the annual net cash inflow from the combine which has already been calculated in Table 14.4; these data are summarized in the first column of Table 14.12. Multiplying these annual net cash flows by the appropriate discount rate results in the present value of the annual net cash flow shown in Table 14.12.

The next step is to calculate the after-tax salvage value of the combine if it was sold at the end of each year. This calculation must take into account the change in machinery values over time, the deterioration of the ma-

Table 14.12 Analyses of the Replacement Policy for a New Combine

Year	Annual Net Cash Inflow[a]	Present Value of Annual Net Cash Flow (12 Percent)	Accumulated Present Value of Annual Net Cash Flow	Present Value of After-Tax Salvage Value (12 Percent)[b]	Net Present Value of Sale at End of Each Year	Amortization Factor (12 Percent)	Annualized Net Present Value of Sale at End of Each Year
1	$22,330	$19,941	$19,941	$40,241	–$20,538	—	—
2	15,602	12,435	32,376	31,222	–17,122	—	—
3	15,377	10,948	43,324	24,152	–13,244	—	—
4	15,469	9,838	53,162	18,249	–9,309	—	—
5	15,628	8,861	62,023	13,315	–5,382	—	—
6	10,962	5,558	67,581	12,182	–957	—	—
7	11,295	5,105	72,686	9,795	+1,761	0.2191	$386
8	11,505	4,648	77,334	8,142	+4,756	0.2013	957
9	4,310	1,556	78,890	6,734	+4,904	0.1877	920

[a]From Table 14.4

[b]From Table 14.13.

Table 14.13 Calculation of the Present Value
of the After-Tax Salvage Value

Year	Adjusted Salvage Value[a]	Tax on Salvage Value[b]	Investment Tax Credit Recapture	After-Tax Salvage Value[c]	Present Value of After-Tax Salvage Value (12 Percent)
1	$47,972	−$5,163	$8,072	$45,063	$40,241
2	44,485	−1,148	6,458	39,175	31,222
3	41,574	2,810	4,843	33,921	24,152
4	38,702	6,779	3,229	28,694	18,249
5	35,853	10,756	1,614	23,483	13,315
6	33,315	9,995	—	24,028	12,182
7	30,958	9,287	—	21,671	9,795
8	28,792	8,638	—	20,154	8,142
9	26,650	7,995	—	18,655	6,734

[a]Calculated as the remaining value using the formulas of Chapter 4 increased by 5 percent per year.

[b]Calculated as the remaining value minus the cost basis (using the ACRS system) times the marginal tax bracket of 30 percent. For example, at the end of year 1 the cost basis is calculated as the purchase price of $80,720 reduced by 50 percent of the investment tax credit taken ($4,036) minus the 15 percent ACRS write-off for the first year ($11,503) which amounts to $65,191. With a sale value of $47,972, the tax loss on the sale is $17,209 ($65,181 − $47,972). Assuming a 30 percent tax bracket, this loss results in a tax savings or negative taxes in year 1 of $5,163.

[c]Calculated as the salvage value minus the tax on salvage value minus investment tax credit recapture. For year 1 this would be calculated as $47,972 − (−$5,163) − $8,072 = $45,063.

chine, any obsolescence, and the taxable gain or loss on the difference between the salvage value and depreciated value as well as the recapture of investment credit. The calculations of Table 14.13 assume that the value of the combine at the end of each year declines as indicated by the remaining value formula of Chapter 4, but that used machinery prices increase over time by 5 percent per year. The tax calculations reflect the capital gain or loss (sale price minus the depreciated value using the five-year ACRS system) taxed at the 30 percent rate, plus the appropriate amount of investment credit recapture. This net after-tax salvage value is then discounted to reflect the present value of selling the machine at the end of each year (Table 14.13).

The total net present value assuming a sale at the end of each year is obtained by accumulating the annual net present values to the appropriate year, adding the present value of the salvage, and subtracting the initial outlay of $80,720. For example, the net present value of the sale at the end of year 2 is − $17,122 ($19,941 + $12,435 + $31,222 − $80,720) as shown in Table 14.12. As indicated in Table 14.12, this total is negative for the combine until year 7, which indicates that it must be kept at least seven years before it will generate a profit.

The total net present value is annualized by the appropriate amortization factor (Appendix Table IV) to obtain the optimal year of replacement. The amortization factor for a 12 percent rate of discount is used for the combine. This step in the computations adjusts the net present value for the differences in the assumed ownership period for the asset. It standardizes the net present value by determining the annual contribution to net present value, and thus eliminates the bias that occurs when comparisons are made between alternatives that have different lengths of life. Because the net present value is negative until year 7 (which indicates that it should not even be purchased unless it is kept that long), the annualized net present value computations need not be made for the first seven years.

Once the annualized net present value is computed, the optimal replacement period is determined as the end of the year with the highest annualized value. For the combine, applying this rule indicates that the optimal replacement cycle is eight years; the annualized net present value for an eight-year replacement period is $957. If the combine were replaced in year 7, the annualized generation of net present value over its life and that of its replacements would only be $386 as indicated in Table 14.12. If it is kept until the ninth year, the annualized generation of net present value declines from $957 to $920. Consequently, the combine should be replaced after eight years of service since this will result in the highest annual generation of net present value for this particular machine and its replacements.

Our example assumes that a revenue can be attributed to the machine service, and the objective is to replace the machine when the annualized net present value of revenues is at a maximum and begins to decline. If revenue estimates cannot be obtained and the analysis is completed using only costs, the procedure will be to calculate the annualized present value of after-tax costs and to replace the machine when these costs have reached a minimum and begin to rise.

Should the Current Machine Be Replaced?

The analysis required to make the decision of replacing an old machine with a new one will now be illustrated. Assume that the data in Table 14.12 reflect the efficiency, costs, and revenues of the new machine. The currently owned machine is four years old and is in need of major repairs. Furthermore, the new machine is more fuel efficient, and it is expected to have less harvesting loss than the one currently owned. The new combine is also slightly larger, which will make it more timely in completing the harvest. These differences in costs and efficiency must be reflected in the computation of the net present value of keeping the current machine.

The computations of the net present value of the current machine for various additional years begin with the determination of the outlay. If the current machine is sold, assume it will generate $29,421 of after-tax cash

Table 14.14 Analyses of the Replacement Policy for the Currently Owned Combine

Year	Net Present Value of Sale at End of Each Future Year	Amortization Factor (12 Percent)	Annualized Net Present Value of Sale at End of Each Year
1	$430	1.1200	$482
2	1,007	0.5917	596
3	1,179	0.4163	491
4	1,236	0.3292	407

flow or salvage value. In essence, this is the amount that the operator is "giving up" or paying to continue to own the current machine. Consequently, this salvage value can be viewed as the outlay (in reality, the income forgone) to keep the current machine.[8]

The annual net cash flow of using the current machine is calculated in the traditional way by subtracting operating expenses and taxes from the annual cash income. This computation should reflect any differences in the harvesting losses or timeliness of harvest operations. Given this computation of the outlay and the annual net cash inflow, the computations proceed in the same fashion as discussed earlier with the new combine. Thus, the net present value of owning the current combine one more year and then selling it is computed by adding the present value of the salvage value and the present value of the net cash inflow. Likewise, the present value of owning the current machinery for two years is computed as the accumulated net present value of the net cash inflow for the next two years, plus the present value of the salvage value at the end of the second year. In this fashion, the net present value of continuing to operate the current machine for a number of future years is calculated, and then these net present values are again annualized using the appropriate amortization factors (Table 14.14). The result is the annualized contribution to net present value of the current machine.

As with the new machine, this annualized value can be used to choose the optimal time to replace the current machine. In general, one expects to replace an asset after the annualized net present value begins to decline. But the major question to be answered is, should we keep the current machine for another year or replace it today with a new machine? By comparing the largest annualized net present value of the current machine to the largest value for the replacement (new) machine, the decision as to

---

[8]An alternative approach is to assign an initial outlay of $0 to the old machine and to reduce the purchase cost of the replacement machine by the after-tax salvage value of the current machine. This would have the effect of increasing the net present value of both by $29,421 and would not affect the optimal time of replacement.

whether or not to replace can be made. As indicated by comparing the largest annualized net present values of Tables 14.12 and 14.14, the largest annualized net present value for the new machine is $957, and for the current machine it is $596. By replacing the current machine with a new one now rather than keeping it for two more years which would maximize the annualized net present value of the current combine, $361 ($957 − $596), more annualized net present value will be generated. Consequently, the operator should replace the current machine with a new one and plan to keep the new one for eight years. Once the new machine has been purchased, however, changes in efficiency, costs, prices, etc., may make it advantageous to modify the replacement policy. The reevaluation should be made using the procedure outlined here to determine if, in fact, modifying the replacement policy would be profitable.

Summary

This chapter deals with two important problems in the acquisition and management of depreciable assets. One of these problems is choosing the assets and the method of financing each of these alternative assets. The other is selecting the appropriate replacement item and the time for replacement of depreciable assets.

Managers are concerned not only with the profitability of depreciable assets that they purchase, but also with the best method of financing and the financial feasibility of these assets. The procedure of analysis is divided into three parts to deal with these three issues in a manner that separates the profitability of the investment from the profitability of financing the investment, and that separates both of the profitability analyses from the financial feasibility.

The first step is concerned with identifying the relevant alternative investments for consideration and estimating the economic profitability of each of these alternatives. An important part of this step is to identify or choose an appropriate discount rate. We recommend choosing the weighted average cost of capital as the appropriate discount rate to adjust future flows of income to their present value. The annual net cash flows must be calculated on an after-tax basis, that is, considering the effect of depreciation and investment credit on the cash flow that would be available to the manager if the investment in the asset were made. Then the present value of the outlay and the annual net cash flows after taxes are calculated and summed to obtain the net present value of the investment. For investments considering both revenues and costs, the most profitable alternative is the one with the largest positive net present value. When revenues are not considered, the rule is to select the alternative having the minimum present value of costs.

The second step is to evaluate the profitability of the financing method by discounting the downpayment and annual principal and interest payments or the lease payments to be made under the various financing arrangements. The stream of net cash outflows discounted in this step is an after-tax value that considers the deductibility of interest on the financing, as well as the deductibility of depreciation and investment credit. The appropriate discount rate is the after-tax cost of debt capital from the major lender. The financing arrangement with the lowest discounted cost is the preferred method.

The third step is concerned with financial feasibility, that is, determing if the asset will generate sufficient cash income to make the interest and principal payments on funds borrowed to purchase the asset out of earnings from the asset. The procedure is to estimate the annual cash inflows from the investment and the annual cash outflows for principal and interest payments. The surplus or deficit for each year is calculated considering the income tax savings from depreciation, investment credit, and the deductibility of interest on the financing arrangement. The decision-maker simply considers whether the surplus on a year-by-year basis is positive or negative. Negative surpluses mean that some other part of the business will have to subsidize this investment in those years. The manager can use these data to decide if it is financially feasible to meet the negative surpluses out of cash flow from other parts of the business.

The second major problem considered in this chapter is the replacement decision for depreciable assets. These decisions must be made frequently, and they typically involve large capital outlays which may have a significant impact on the profitability and efficiency of the farm firm. The analysis of replacement decisions again uses capital budgeting procedures. The analysis proceeds by calculating the annualized net revenue or cost for the asset for different lengths of life. The optimal length of life is the number of years with the highest annualized net present value of revenues (or the number of years having the lowest annualized net present value of cost when revenue is not considered).

The rule to use in deciding when to replace a currently owned machine is a logical extension of selecting the optimal replacement life for a new machine. The annualized net present value for the replacement machine is calculated for different lengths of life. The decision as to whether to replace is made by comparing the largest annualized net present value for the new machine to the largest annualized net present value for the machine currently owned. If the annualized net present value for the new machine exceeds that of the current machine, the current machine should be replaced. If it does not, the current machine should be kept and used for another year, and the analysis should be repeated the following year.

This chapter demonstrates that both of these important economic problems can be evaluated using capital budgeting procedures. In each case, the tax considerations were taken into account in the analysis. Thus, the analysis is not only conceptually correct, but also quite useful in making decisions for operating farm businesses.

Questions and Problems

1. Why is it important to distinguish between the profitability of the asset and the profitability of the method of financing?
2. What factors should an operator consider in selecting the appropriate long-run capital structure to use in calculating the weighted average cost of capital?
3. What are the impacts of income taxes and inflation on the costs of debt and equity capital?
4. Identify a depreciable asset that might be used in some farming situation that you specify. What data would be required to estimate the economic profitability of purchasing this asset?
5. How would the analysis of the profitability of the two sizes of combines and the used combine be changed if the analysis was made on the basis of cost alone?
6. Should the sensitivity of the economic profitability of the three combines be evaluated? If so, what factors or variables should be considered for analysis?
7. What conditions would be required to make the used large-size combine a more profitable alternative, perhaps even the most profitable alternative, relative to the two new combines?
8. Some farmers argue that the investment in certain items of equipment and the financing come as a package, and that one cannot separate the profitability of the investment from the financing in those cases. How would you respond?
9. Explain why the after-tax cash flow of the salvage value must be considered as the outlay for the currently owned machine in deciding whether a currently owned machine should be replaced.
10. What is meant by financial feasibility?

Further Reading

Barry, Peter J., John A. Hopkins, and C. B. Baker. *Financial Management in Agriculture.* Danville, Ill: Interstate, 1979, Chapters 11, 12 and 14.

Bierman, Harold, Jr., and Seymour Smidt. *The Capital Budgeting Decision.* 4th ed., New York: Macmillan, 1975.

Chisholm, Anthony H. "Effects of Tax Depreciation Policy and Investment Incentives on Optimal Replacement Decisions." *American Journal of Agricultural Economics* 56 (1974):776–83.

Faris, Edwin J. "Analytical Techniques Used in Determining the Optimum Replacement Pattern." *Journal of Farm Economics* (1960): 41:755–66.

Jose, Douglas, Robert Christensen, and Earl Fuller. "Consideration of Weather Risk in Forage Machinery Selection." *Canadian Journal of Agricultural Economics* 19 (1971):98–109.

Kay, Ronald D., and Edward Rister. "Effects of Tax Depreciation Policy and Investment Incentives on Optimal Equipment Replacement Decisions: Comment." *American Journal of Agricultural Economics* 58 (1976):355–58.

Kletke, Darrel D., and Luther G. Tweeten. *Costs and Replacement Procedures for Farm Tractors.* Stillwater, Okla.: Oklahoma State University Agricultural Experiment Station Bulletin B–687, November 1970.

Menz, Kenneth M. *A Computerized Management Aid for Evaluating the Machinery Lease or Purchase Decision.* St. Lucia, Queensland, Austrialia, Department of Agriculture, Discussion Paper 1/77, February 1977.

Perrin, R. K. "Asset Replacement Principles." *American Journal of Agricultural Economics* 54 (1972):60–67.

CAPITAL ACQUISITION
AND MANAGEMENT

CHAPTER CONTENTS

Any decision to expand or reorganize the farm business must involve an evaluation of the alternative means of obtaining the capital resources. A farm operation requires two types of capital—investment capital and operating capital. Investment capital includes such items as machinery, equipment, land, and other durable inputs, whereas operating capital includes seed, chemicals, fertilizer, and other inventories and supplies.

The funds required to finance the investment and operating capital requirements of the farm business can be obtained from many sources but usually are classified into two basic categories—equity funds and debt funds. Equity funds are supplied by the owner(s) of the farm operation; they provide the backbone of any financing arrangement. Some people refer to equity funds as "risk capital" because, in the event of liquidation of the business, the holder of equity funds has the residual (last) claim on the liquidation proceeds after all other claims have been satisfied. Consequently, the equity capital bears the risk of any financial loss, and it also reaps the benefits of any profits or financial gains. In contrast, debt funds are provided by financial institutions or individuals with no ownership interest in the farm business. Debt funds usually carry a cash cost in the form of interest and have a first claim on net income or proceeds from liquidation.

A third method that can be used to gain control of investment capital items is that of renting or leasing. Renting or leasing a capital item such as machinery or land reduces the investment capital commitment of the farm operator, but it typically increases the cash flow and operating capital requirements. From a capital acquisition or financing viewpoint, rental or leasing agreements require more short-term or operating funds, whereas the purchase of such assets will require investment or long-term funds.

The approach of this chapter will be first to discuss the various sources of debt and equity funds available to farmers and the role that both sources play in the financial structure of the farm business. Then the advantages and disadvantages of leasing capital items will be reviewed. Finally, the role that credit plays in the farm business, and the legal requirements and costs of borrowing money are discussed.

EQUITY SOURCES OF FUNDS

Since equity is the financial backbone of any business, acquiring or accumulating equity funds is essential for the successful farm operator. A farmer can accumulate equity through savings or acquire it through inheritance or marriage and other family arrangements. Alternatively, the farm operator may combine her or his equity capital with that of an outside investor or a family member such as a father or father-in-law in some form of pooling arrangement to obtain a larger equity base, which then can be used to increase the size of the business and improve its efficiency through economies of size. These different sources of equity funds will be briefly discussed, along with the implications for the financial structure of the business and the acquisition of debt funds.

Savings

The most important source of equity funds is savings. Savings is the amount of income that is not consumed and is thus available for reinvestment in the farm business. Thus, savings can be calculated as net farm income plus nonfarm income minus family living expenditures and income taxes. The volume of savings can be increased not only by increasing farm income, but also by reducing family expenditures and taxes. In fact, in many farm businesses, the primary method to increase equity capital accumulation is through reduced consumption, particularly for young farmers. Another method for increasing equity capital accumulation through savings is for the husband or wife to obtain off-farm employment with the earnings being substituted for farm income in meeting consumption requirements.

Savings provides more than just equity funds which can be used to purchase assets. Savings indicates an ability to handle one's finances which will have an impact on the amount of credit or debt that can be obtained. It also indicates a willingness to forgo current consumption for the benefit of a higher level of income and standard of living in the future. Historically, farmers have had a higher savings rate than most people. Analysts have estimated that farmers save almost one-third of their income—that is, approximately one-third of their disposable income is reinvested in their farming operation.

Much of this savings might be described as "forced savings." Forced savings occurs because the farmer does not purchase machinery and equipment with cash, but instead typically makes a 20 to 30 percent cash downpayment and acquires the remainder of the funds from credit sources. Payments on debt take priority over unnecessary consumption, and since the repayment schedule on most items is shorter than the depreciable life, the farmer accumulates equity by making loan payments because the annual principal payments are larger than the annual depreciation. Thus, the farm family is forced to reduce consumption and save so as to make the payments on debt obligations.

In summary, savings is the financial backbone of the farm business, the primary component of the risk portion of the capital structure, and the most important source of equity funds for most farmers. Savings requires reduced consumption, but the expected payoff of savings is a higher level of income and standard of living in future years.

Inheritance

A second important source of equity funds for many businesses is that of inheritances or gifts. For many young farm operators, accumulated savings will not provide an adequate financial base for a viable farm operation with the potential for growth and expansion. One common way to augment savings is through gifts received from relatives and inheritances from the parents. In most cases, accumulating equity funds through gifts and inheritances is part of an overall intergenerational transfer plan that has been developed to transfer the farm business as a "going economic concern" from the parents to the on-farm operating heir. In these situations, the operating heir has typically been active in the business for a number of years, and his or her acquisition of the farm at the death of the parents is a natural step in the transfer plan.

The transfer plan and process may involve various types of legal arrangements and business organizations such as the partnership, the corporation, the life estate, or trust, as well as gifts from the parents to the operating heir during the parent's lifetime. The various business and legal arrangements used in transferring the farm firm between generations were discussed in Chapter 9 and will not be reviewed in depth here. However, it should be noted that these arrangements not only facilitate the transfer of assets between family members and generations, but they also may be essential to provide a sufficient equity base for the beginning farmer to acquire debt funds for expansion and growth of the firm. Most studies have indicated that a beginning farmer cannot accumulate equity funds rapidly enough through savings to acquire the necessary "critical mass" of capital for successful entry into farming. For many beginning farmers, gifts and inheritances play a crucial role in the success of their operation and farming career.

One caution should be noted concerning gifts and inheritances as a source of equity funds for the farm business. Most credit institutions will be somewhat hesitant to loan debt funds to a farm operator who has acquired most of his or her equity through gifts and inheritances. As indicated earlier, equity accumulation through savings provides evidence to credit institutions of the ability to manage finances and sacrifice consumption when necessary to have a better standard of living in the future. An equity base composed primarily of gifts and inheritances does not provide evidence of a willingness to make financial sacrifices if necessary. Consequently, the lender may question whether such a farm operator will be willing to forgo consumption when required to make payments on loan obligations. Hence, gifts and inheritances do not provide the same intangible benefits as a source of equity funds as are provided by savings. Inheritance may be crucial to the initial success of the farm business, but savings are essential to the continued growth and expansion through the use of debt funds.

Investors and Pooling Arrangements

A third source of equity funds for the farm business is that of the investor whether he or she be a doctor, lawyer, or farmer or widow of a farmer. Combining resources with an investor may not directly increase the equity funds of the farm operator, but it does increase the capital base and the size of the business available to manage. This increased size of operation may result in increased efficiency because of economies of size, and thus increase the income-generating capacity of the business and the accumulation of equity over time through increased savings. Thus, the benefits to the farm operator of using someone else's equity funds are primarily those of economies of size and future equity accumulation. In addition, the investor may be the only source of additional equity funds for the beginning farmers who have no family members with sufficient resources to assist them in obtaining the "critical mass" of capital necessary to begin farming. The investor also may play an important role in providing capital to agriculture through the rental market. This contribution occurs through the rental of real estate to operators who may not have adequate resources to purchase a similar tract of land.

One of the disadvantages of the investor as a source of equity funds is that he or she will typically want to exercise some control over the use of funds and the management of the business. A farmer who acquires an investor as a partner must, therefore, be willing to give up some independence and freedom to make investment and operating decisions. Depending on the business arrangement developed, the operator may also have a "partner" to share in the liability and losses that can occur.

The investor (specifically the nonfarm or foreign investor) has been maligned in many discussions of the structure, control, and financing of agriculture, particularly in the Midwest. However, she or he does play an important role in providing access to equity funds and investment capital, particularly farmland through the rental market, to those who do not have the financial base to purchase such capital items. Thus, the investor may be an important source of funds for those operators who do not have a sufficient equity base to purchase a viable farming unit and are not fortunate enough to have family members who can provide assistance and equity funds.

A similar financing arrangement that is becoming more common in agriculture is the equity pooling or joint venture arrangement. Pooling typically involves a number of farm operators and nonfarm investors; common examples in the Midwest include the joint venture livestock operation such as the feeder pig cooperative, the sow corporation, or the cattle feeding cooperative. Those who participate in such joint ventures indicate that they provide a means for farmers to combine their resources and expand with the latest technology. In many cases, such joint ventures enable farm operators to increase their volume of business without any additional labor and management commitment because these resources are purchased by the joint venture. Such arrangements may also appeal to a renter who does not want to construct permanent livestock production facilities on rented land. Therefore, pooling of equity capital funds may provide a financial base for a larger production unit with lower per unit costs, but part of the price that must be paid for such an arrangement is the reduced freedom to make independent decisions concerning production and marketing activities. Any successful joint-venture arrangement involving pooled equity funds will also require joint decision-making.

DEBT SOURCES OF FUNDS

Although equity funds provide the financial backbone of any farm business, most farmers do not generate sufficient equity from savings or other sources to expand as rapidly as they desire. Thus, they are forced to use additional sources of funds in the form of debt or credit to expand their operation. During the past two decades, credit has increased in relative importance as a source of funds for most farmers. Nationally, debt amounted to about 12 percent of total farm assets in 1960. But, by 1980, credit institutions such as banks, insurance companies, and Cooperative Farm Credit System agencies provided the funds and had claims on 17 percent of total farm assets.

The Agricultural Credit Market

Farmers are served by a three-pronged credit market: the private sector, the cooperative sector, and government agencies. The private sector consists of such firms as commercial banks, merchants and dealers, insurance companies, finance companies, and individuals who make personal loans to farmers. For the most part, these financial institutions have been a dependable source of operating and investment capital for farmers and in many cases have developed specific lending programs for agricultural producers.

Although the private sector has historically been an important source of credit for farmers, at times it has had difficulty servicing agriculture because of the higher rates of interest that could be obtained making loans to nonagricultural businesses and because of the limited supply of funds that could be mobilized to loan to farm firms. Consequently, the cooperative credit system was developed to enable farmers, through a cooperative effort, to tap the national money markets. The cooperative credit system includes the Federal Land Banks and Federal Land Bank Associations, the Federal Intermediate Credit Banks and Production Credit Associations, and the Banks for Cooperatives. These banks obtain funds by selling bonds on the national money markets to investors. The proceeds of the bond sales are then loaned to farmers or to grain merchandising or input supply cooperatives. As suggested by the name, the banks of the Cooperative Farm Credit System function as cooperatives and are owned and managed by the users of the system. The Cooperative Farm Credit System has not only increased the availability of funds to farmers through access to national money markets, but it has also provided many new innovations in agricultural lending and has stimulated the private sector to provide more efficient service to farmers.

The third component of the agricultural credit market includes the government agencies. There are three major governmental agencies that provide funds to farmers: the Farmers Home Administration, the Small Business Association, and the Commodity Credit Corporation. In some states, state agencies also make loans to farmers, particularly beginning farmers.

The primary purpose of the Farmers Home Administration (FmHA) is to provide loans to farmers who cannot obtain funds from either the private or cooperative sector. Consequently, the FmHA provides funds for disaster situations and when risks are too high for the private or cooperative credit institutions. The lending authority of the Small Business Administration (SBA) was extended to farmers during the 1970s, but primarily for emergency and disaster purposes. In recent years, FmHA has again taken major responsibilities for emergency and disaster loans to farmers, and SBA lending activity has declined substantially. Commodity Credit Corporation (CCC) loans are part of the income and price support program of the U.S. Department of Agriculture. The CCC provides for grain storage as well as a combined operating loan—income support program to augment farmers' incomes by accepting the commodity as payment in full on the loan if commodity prices are below the "loan value."

The relative market shares of the various lending institutions in the agricultural credit market are summarized in Table 15.1. The characteristics of each of these institutions as to loan policy and operating procedures will be briefly reviewed. The discussion will emphasize the type of funds provided, terms for various loans, and the information or documentation required to qualify for a loan. Additional information on the credit institutions servicing agriculture can be obtained from various sources, including the institutions themselves as well as textbooks on farm finance such as Warren F. Lee, et al., *Agricultural Finance* (Ames: Iowa State University Press, 1980), and Peter J. Barry, John A. Hopkin, and C. B. Baker, *Finan-*

Table 15.1 Market Share of Real Estate and Non-Real Estate Debt by Lender (Percentage Distribution)

	Real Estate Debt (January 1)				
Date	*Federal Land Banks*	*Farmers Home Administration*	*Life Insurance Companies*	*All Operating Banks*	*Individuals and Others*
1950	17	4	21	17	41
1960	19	6	23	13	39
1970	23	8	20	12	38
1980	36	8	13	11	31
1982	43	9	13	8	28

	Non-Real Estate Debt (January 1)					
Date	*All Operating Banks*	*Production Credit Associations*	*Federal Intermediate Credit Banks*	*Farmers Home Administration*	*Individuals and Others*	*CCC*
1950	30	6	1	5	34	25
1960	38	10	1	3	38	9
1970	43	19	1	3	22	11
1980	41	24	1	12	16	6
1982	35	23	1	16	16	9

cial Management in Agriculture (Danville, Ill.: Interstate Printers and Publishers, 1979).

Commercial Banks

The United States is served by approximately 14,000 commercial banks, many of which are located in rural areas and depend on the agricultural sector for their income. The primary sources of funds used by commercial banks to make loans are deposits of their customers—both time and demand deposits. In addition to deposits, rural commercial banks frequently acquire funds from larger city banks through a correspondent arrangement. This arrangement enables the larger bank to assist the smaller, rural bank in making loans that exceed the rural banks' legal lending limit. For a nationally chartered bank, the limit on the size of loan that can be made to any one customer is 25 percent of unimpaired capital and surplus for feeder livestock loans and 10 percent for real estate and other agricultural loans. Individual loan limits for state-chartered banks may be higher or lower than those of nationally chartered banks. In addition to the correspondent arrangement, a rural commercial bank can obtain funds by borrowing (discounting) from the Federal Intermediate Credit Bank of the Cooperative Farm Credit System. Banks also may collaborate with insurance companies or participate with the local Production Credit Association (PCA) or Farmers Home Administration to provide adequate financing for a particular venture.

Since the majority of funds used by a commercial bank to make loans come from deposits which may be volatile from year to year, many commercial banks are reluctant to become heavily involved in long-term lending. The primary type of loan made by commercial banks is the production or operating loan for such purposes as purchasing livestock, feed, or supplies. These loans are typically one year or less in maturity. Intermediate-term loans for the construction of livestock facilities or the purchase of machinery also are made by commercial banks. Some intermediate-term loans are made by banks on a one-year renewal basis, while other banks will make equipment loans with a three- to five-year maturity and facility or building loans on a five- to seven-year payback. In addition to agricultural loans, most commercial banks have alternative places to invest their funds, including installment and business loans as well as government securities and municipal bonds. Consequently, farmers must compete with other borrowers for loan funds and must pay competitive rates of interest.

The interest rates charged by commercial banks will depend on local supply and demand conditions, rates on alternative types of loans such as installment and commercial credit, rates charged by competitors, and the cost of funds, as well as the character, risk-bearing ability, and repayment capacity of the individual borrower. During the late 1970s and early 1980s, the cost of funds increased rapidly for commercial banks, and their interest rates for farm customers rose to unprecedented levels of 17 to 20 percent. Furthermore, interest rates charged by banks have become quite volatile in recent years, with wide fluctuations even within a single year (Table 15.2). This increase in interest rate variability has increased the risk for the farm borrower. Usury laws which impose an upper limit on interest rates may also influence the rate of interest that a bank can charge its customers in some states.

For livestock and crop input supply loans, most banks will use the inventories themselves as security for the note. For machinery loans, the machine being purchased as well as other major items in the farmer's line of equipment will be requested as security. In some cases, a "barnyard" security agreement where all the nonreal estate assets are pledged as collateral may be used.

Many bankers feel that livestock facilities and other permanent improvements are one of the most difficult capital items to finance. Because bankers are using the funds provided by their depositers when they make a loan, they need to protect these deposit funds against the risk of loss. Consequently, the banker will frequently request a mortgage or an assignment of equity in a land contract if possible to secure a livestock facility or improvement loan. In addition, the banker will probably expect a 25 to 40 percent downpayment on improvement or livestock facility loans. Many bankers will encourage a farmer to find longer term financing through an insurance company or the Federal Land Bank if they feel such an arrangement will improve the terms and make it easier for the farmer to repay the loan.

Federal Land Banks

The Federal Land Bank is one of the agencies of the nationwide Cooperative Farm Credit System. Organized in 1916, the Federal Land Bank in each of the twelve districts in the United States and the local Federal Land Bank Associations are now completely capitalized and owned by member-borrowers. To provide the necessary funds for the capital structure and operation of the local association and the district Federal Land Bank, each farmer-owner is required to purchase capital stock in the local association in the amount of 5 to 10 percent of the face value of the loan. This stock purchase requirement increases the effective interest rate on the loan by about 0.5 to 1.0 percent. Because Federal Land Banks are organized as cooperatives, the stock entitles the farmer to one vote regardless of the size of the loan or stock ownership.

Federal Land Banks obtain their funds to make loans to farmers from the proceeds of bond sales in the national money markets. These bonds are sold to private and institutional investors at prevailing rates of interest.

Interest rates charged by the Federal Land Bank are not limited by state usury laws and typically reflect the cost of funds, plus a minimal markup to cover operating costs and accumulate reserves for possible loan losses.

Most Federal Land Bank loans now are made on a variable rate plan whereby the rate is adjusted depending on the cost of money and interest rates in the national money markets. Thus, if interest rates and the cost of bonds sold by the Federal Land Bank decline, the interest rate on Federal Land Bank loans also declines. Conversely, if interest rates in national money markets and the costs of funds increase, Federal Land Bank loan rates will increase. Table 15.2 summarizes recent Federal Land Bank interest rates.

Federal Land Banks make long-term loans to farmers for real estate and permanent improvements to the real estate. Their market share has grown considerably in recent years as noted in Table 15.1. Operating loans to acquire livestock and feed inventories or finance crop expenses are typically not available from the Federal Land Bank. The Federal Land Bank can make loans with 5- to 40-year maturities, but most of their real estate loans have 30 to 35 year maturities. For permanent improvements such as livestock facilities or grain storage and handling, a 7- to 10-year maturity is most common.

Farmers who want to borrow money from the Federal Land Bank may contact the local Federal Land Bank Association and complete a loan application which includes a financial statement. The Federal Land Bank then completes an appraisal of the farm or facility that is to be financed. With respect to real estate, this appraisal is based on its agricultural value or what it can be expected to produce in terms of farm income for an owner-operator. The Federal Land Bank is authorized to loan up to 85 percent of the appraised agricultural value of the land, but the appraised value is typically lower than the market value. Therefore, the actual percentage of the purchase price that is loaned may be closer to 65 to 75 percent. Consequently, a 25 to 35 percent downpayment is usually required to make a land purchase using Federal Land Bank financing. Land Bank repayment terms usually require full amortization of the loan during the 30- to 35-year term, but special arrangements may be possible in unique situations for delayed principal payments or partial amortization programs if they seem necessary and appropriate.

With respect to permanent improvements, Federal Land Bank Associations typically require a larger downpayment than for real estate. As with real estate loans, interest rates are determined by the cost of money on the national money markets plus a margin for operating costs and reserve accumulation. Terms on facility loans usually do not exceed ten years, and in many cases a seven-year repayment schedule is used. Since specific loan policies and terms can vary from association to association, a farm operator

Table 15.2 Average Interest Rates on Business and Farm Borrowings (Percentage)

Date	Business Loans: Effective Rates[a] on Bank Loans, United States, First Week of Second Month of Quarter		Nonreal Estate Farm Loans			Stated Nominal Rates, First Day of Quarter								Farm Credit System[d]	
	Prime Rate[b] Average, Large Banks	All Banks	Large Banks	Other Banks	All Banks	Prime Rate, Large Banks	Average of Most Common Farm Loan Rates at Banks Surveyed in Specified Federal Reserve Districts[c]							Production Credit Associations	Federal Land Banks
							Feeder Cattle Loans			Other Operating Loans					
							Chicago	Kansas City	Dallas	Chicago	Minneapolis	Kansas City	Dallas		
1977–Q1	6.35	7.6	8.3	8.9	8.8	6.25	8.7	8.8	9.3	8.8	9.1	8.9	9.3	8.2	8.5
–Q2	6.35	7.6	8.1	8.9	8.7	6.25	8.7	8.8	9.3	8.8	9.2	9.0	9.2	8.1	8.4
–Q3	6.86	7.9	8.4	8.9	8.7	6.75	8.7	8.8	9.3	8.8	9.2	9.0	9.2	7.9	8.3
–Q4	7.90	8.6	9.1	9.0	9.1	7.25	8.8	8.9	9.3	8.9	9.2	9.0	9.3	8.0	8.3
1978–Q1	8.16	8.9	9.3	9.1	9.2	7.75	8.8	8.9	9.4	8.9	9.2	9.0	9.4	8.4	8.2
–Q2	8.16	9.1	9.6	9.2	9.3	8.00	8.9	8.9	9.4	9.0	9.2	9.1	9.4	8.7	8.3
–Q3	9.20	10.0	10.4	9.3	9.6	9.00	9.1	9.1	9.5	9.2	9.4	9.2	9.5	9.0	8.3
–Q4	10.78	11.4	11.7	10.0	10.4	9.75	9.4	9.3	9.7	9.5	9.5	9.4	9.7	9.2	8.4
1979–Q1	12.09	12.2	12.5	10.4	11.0	11.75	10.1	9.9	10.1	10.2	10.2	9.9	10.1	10.0	8.7
–Q2	12.09	12.3	12.8	10.7	11.2	11.75	10.5	10.2	10.2	10.5	10.4	10.3	10.2	10.6	9.0

—Q3	12.09	12.3	12.9	11.3	10.9	11.50	10.8	10.4	10.3	10.9	10.8	11.1	10.3	10.9	9.3
—Q4	16.39	15.8	16.2	13.6	13.1	13.50	11.7	11.5	11.4	11.7	11.8	11.6	11.3	11.0	9.3
1980—Q1	16.39	15.7	16.0	14.1	13.7	15.25	13.5	13.0	13.1	13.6	13.6	13.1	13.0	12.1	9.8
—Q2	18.81	17.8	18.5	17.4	17.1	19.50	17.1	16.5	16.2	17.1	16.4	16.5	15.8	13.7	10.6
—Q3	11.30	11.6	12.8	13.5	13.7	12.00	14.0	14.0	13.2	14.0	15.3	14.1	13.2	13.3	10.6
—Q4	15.56	15.6	16.3	15.5	15.3	13.50	14.3	14.0	13.3	14.3	14.0	14.1	13.3	12.0	10.3
1981—Q1	20.56	19.8	19.9	17.9	17.5	21.50	17.3	16.9	18.6	17.4	17.6	17.1	18.4	12.9	10.6
—Q2	19.90	19.9	19.5	17.9	17.5	17.50	16.5	16.3	17.6	16.5	17.0	16.3	17.4	14.2	10.9
—Q3	21.55	21.0	20.8	19.6	19.1	20.00	17.7	17.4	19.2	17.8	18.0	17.4	19.0	15.1	11.4
—Q4	18.54	17.4	18.9	18.8	18.7	19.50	18.6	18.1	19.7	18.6	18.9	18.1	19.4	15.8	11.7
1982—Q1	16.64	17.1	18.0	17.7	17.5	15.75	16.9	16.6	17.3	17.0	17.2	16.6	17.5	15.3	12.1
—Q2	17.18	17.1	17.9	17.8	17.7	16.50	17.3	16.9	17.6	17.3	17.4	16.9	17.4	14.9	12.2
—Q3	15.56					16.50	17.2	16.9	17.5	17.2	17.4	16.9	17.6	14.5	12.4

[a] Effective loan rates are a dollar-weighted average of effective rates on loans of $1,000 or more made in the week indicated. Additional data from this quarterly survey of bank lending are published in Statistical Releases E.2 (Survey of Terms of Bank Lending) and E.15 (Agricultural Finance Databook—Quarterly Series), Federal Reserve Board, Washington, D.C. 20551. "Large banks" (survey strata 1–3) correspond roughly to banks with over $500 million in total assets in September 1981.

[b] Effective prime rate is calculated by assuming a loan maturity of six months with all interest paid at maturity.

[c] The type of bank included in quarterly agricultural credit conditions surveys conducted by Federal Reserve Banks varies among Federal Reserve Districts, and so the rates shown are not strictly comparable. See page 58 of the Databook cited above.

[d] Farm Credit System rates are unweighted averages of quoted rates, not adjusted for required stock purchases and loan fees.

Source: Emanuel Melichar and Paul T. Balides, *Agricultural Finance Databook* (Monthly Series), Board of Governors of the Federal Reserve System, Washington, D.C., June 1982, p. 32.

should check with the local Federal Land Bank Association concerning current financing arrangements.

Production Credit Association

As an agency of the Cooperative Farm Credit System, Production Credit Associations (PCAs) provide operating and intermediate-term credit to farmers and livestock producers. PCAs are retail outlets for credit and obtain their funds from the district Federal Intermediate Credit Bank (FICB), which in turn acquires its funds from the national money markets. Like the Federal Land Banks, PCAs were begun initially with federal funds, but all federal monies have since been repaid and local associations and district FICBs are fully owned by the member-patrons. PCA borrowers must purchase stock in a local PCA equivalent to a minimum of 5 percent of the face value of the loan. Currently, many PCAs are requiring their borrowers to purchase stock in the amount of 10 percent of their loan. As with Federal Land Banks, these stock requirements increase the effective interest rate to the borrower. Rates charged by PCAs in recent years are also summarized in Table 15.2. PCAs provide short-term (up to one year) and intermediate-term (up to seven years) credit to farmers and livestock producers.

For a new applicant, PCAs will require a financial statement for the last year and perhaps statements for one or two previous years as well. PCA loan officers may also request the farmer's tax records for the last one or two years and a cash flow for the coming year as well. Once this information has been gathered, a farm visit is typically made to see the farmer's operation. Livestock facility or other permanent improvement loans may be put on a five- to seven-year payback. For large livestock facilities or other enterprises that require more time for repayment, it is possible to set up a special term loan. This arrangement would include a balloon payment at the end of the normal seven years that could be extended. For example, the payment schedule might require 10 percent of the loan to be repaid during the first six years with 40 percent due in year 7. At the end of the seventh year, part of this 40 percent would be refinanced for three more years. This arrangement is available only in special circumstances. Security for the loan may involve the facility itself, but a first or second mortgage on the real estate may also be requested if necessary. PCAs use a variable interest rate plan that adjusts rates with market conditions.

The FICB-PCA organizations also are involved in cooperative ventures with other financial institutions such as commercial banks as a means of supplying credit to farmers. If specific regulations are met, local commercial banks can borrow funds or discount loans they have made to farmers with the district FICB. Alternatively, commercial bank-PCA participation arrangements, whereby the local commercial bank takes a portion of the farmer's loan and the local PCA advances funds for the remainder of the loan request, are also possible. In most cases, a commercial bank will use these options of working jointly with Cooperative Farm Credit System agencies only when it is not possible to handle the total loan request with their own resources or those of their correspondent bank.

Insurance Companies

Life insurance companies are a major source of intermediate- and long-term credit for agriculture, although their market share has declined in recent years. Most insurance company loans have been made for real estate purchase in the past, but insurance companies also are actively involved in financing livestock facilities, particularly joint-venture operations such as integrated poultry, feeder pig or cattle feeding corporations, and limited partnerships. In recent years, insurance companies have also financed drainage and irrigation systems for many farm operators.

Insurance companies obtain their funds from the premiums on insurance policies and the earnings and reserves that are generated by their primary business of selling insurance. These funds are invested in numerous types of investments, including government and corporate securities, corporate stocks, urban, commercial, and industrial real estate loans, and farm mortgages. Thus, unlike the Cooperative Farm Credit System agencies whose primary interest is agriculture, insurance companies will make agricultural loans only if the return is reasonably competitive with alternative uses of the funds in other sectors of the economy. As a means of reducing costs, insurance companies prefer large loans that are well secured and are intermediate to long term in maturity so that annual servicing costs are low. Large livestock facility loans as well as most real estate loans exhibit these characteristics. In periods of high interest rates, demands are made on life insurance companies for policy loans (loans of the cash surrender value of whole life insurance policies) which must be honored. This naturally reduces the volume of funds available for investment in any field, including agriculture, regardless of the rates of interest that could be charged.

As with Federal Land Banks, insurance company loans are usually secured by a first or second mortgage on the real estate or improvement. Insurance company appraisers appraise the real estate or improvement prior to a loan closing. For many permanent improvements as well as real estate loans, the typical loan will amount to 60 to 65 percent of the cost of the real estate or facility. Interest rates are competitive with those charged by other lenders. Because insurance companies are subject to state usury legislation, whereas Federal Land Bank Associations are not, insurance companies may not be particularly active in the farm mortgage market when interest rates are generally high and other industries in other states are willing to pay the higher rates or are not limited by usury legislation.

The repayment terms on life insurance loans have changed dramatically in recent years because of high and volatile interest rates. Many loans are now being made on a 10- to 15-year term with 2 to 3 percent principal reduction per year and a balloon payment at the end of the term. In many cases, the interest rate is fixed for only a set period of years (for example, 3 to 5 years) with the right to negotiate a new rate periodically.

Farmers Home Administration

Farmers Home Administration loans are available to farmers who are unable to obtain adequate credit from other commercial sources on reasonable terms. Typically, this loan source is used by the individual who is attempting to get a start in farming or in disaster or emergency situations. Once FmHA borrowers become established, they are required to "graduate" to conventional commercial lenders.

Among its many other lending activities, FmHA is authorized to make both farm ownership and farm operating loans. The maximum outstanding operating loan that can be made by FmHA is $100,000 ($200,000 if made by a commercial lender guaranteed by FmHA), whereas ownership loans are limited to $200,000 ($300,000 if made by a commercial lender and guaranteed by FmHA). The interest rate on operating loans may be slightly below market rates, and the repayment schedule is tailored to the use of the funds with a normal limit of seven years. Under certain conditions, the repayment on operating loans can be scheduled with a 50 percent balloon at the end of seven years, which can be refinanced and paid off over another five years. On ownership loans, the interest rate is again usually below market rates and the maximum term is 40 years. These lending authorities enable FmHA to assist the farmer in purchasing or improving land; constructing, improving, and repairing livestock and grain storage facilities; and financing the purchase or production of livestock and feed inventories and grain supplies and inputs. In addition to the farm operating and ownership loans available through the Farmers Home Administration, FmHA also provides credit for disaster and emergency situations as well as rural housing, business development, sewer and water projects, and many othe other public services for the rural community. A county loan committee is appointed by the state director of FmHA to determine loan eligibility.

FmHA may also participate with other lenders such as the Federal Land Bank or a commercial bank to assist the farmer in obtaining funds through the guaranteed loan program. The procedure followed in obtaining guaranteed loans is for the producer to request a loan from her or his normal lender. The lender who sees the need for a guarantee to make the loan makes a request to the local Farmers Home Administration for such a guarantee. A county committee of the county FmHA approves or disapproves the request. If the request is approved, the lender makes the loan at conventional interest rates with a 90 percent guarantee of the principal and interest by the federal government. Thus, the availability of the guarantee enables the conventional lender to reduce his or her risk and make loans to producers who otherwise may not qualify for credit.

Commodity Credit Corporation

As indicated earlier, Commodity Credit Corporation (CCC) loans are part of the income and price support program of the U.S. Department of Agriculture. The CCC was not a significant source of loan funds during the early 1970s, but it has been a major supplier of funds to farmers in recent times through the farmer-owned Grain Reserve Program. The CCC loan program provides that a farmer who participates in the government program can obtain a loan for a specified amount per bushel on stored crops. When the loan is due, the borrower can follow two possible courses of action: (1) if the market price for the commodity is lower than the loan value, she/he can deliver the commodity to the CCC as payment in full for the loan obligation; or (2) if the market price for the commodity is higher than the loan value, the farmer can repay the loan and sell her/his commodity. The program is implemented through personnel of the Agricultural Stabilization and Conservation Service. Thus, the CCC loan program provides loan funds that can be used to finance other farming activities as well as income support by accepting the commodity as payment in full on the loan.

Other Sources

Other sources of credit for the farm business include merchants and dealers, individuals, and finance companies. Merchants and dealers supply a significant amount of short-term credit to farmers in the form of open accounts or short-term notes for purchased inputs such as feed and chemicals. In addition, machinery dealers frequently offer financing packages as a sales tool to increase machinery and equipment sales. In many cases, the merchant and dealer who provide longer term credit will actually sell the credit contract to a company credit corporation or to a local commercial bank rather than hold the contract until maturity. Terms on dealer credit depend on the input being purchased. For such items as fertilizer, feed, and chemicals, 30-day open accounts are quite common, with a cash discount given if payment is made within a shorter period of time such as ten days. In some cases, the interest rates on merchant and dealer credit may be higher than that charged by other lending institutions. The farmer should, therefore, carefully compare terms and rates before financing purchases with a local dealer.

Individuals are also an important source of credit in agriculture, not only in terms of the installment land contract which was discussed in Chapter 13, but also in terms of personal loans between family members. The terms and conditions of personal loans are dependent on the individuals involved and are not standardized. Individuals also provide assistance in obtaining credit in the form of a co-signature or guarantee of a loan. In this situation, a father or uncle, for example, may co-sign a note executed by the son or daughter to a commercial bank. This co-signature essentially makes the father or uncle liable to pay the loan should the son or daughter default. Loans between family members to help a child buy the father's share of machinery and inventory in a partnership are also a common occurrence.

As was suggested in Chapter 13, installment land contracts really involve a loan from the seller to the buyer in the amount of the contract. Such arrangements typically involve a low downpayment, lower than market interest rates, and a 15- to 20-year term (or a 10- to 15-year term with a balloon payment at the end of the contract that will be refinanced). Thus, an installment land contract usually will involve a lower downpayment than a real estate purchase financed by an insurance company or the Federal Land Bank, and it may also involve a lower rate of interest. However, it may be necessary for the buyer (particularly in the case of a nonfamily member) to pay a price premium to obtain the privileges of the better terms that are frequently available with the contract arrangement.

Finance companies also provide credit for farm families, although they have not been active in providing agricultural production credit until recently. Generally, finance company credit is used to purchase consumer durables and consumer items or possibly small tools and equipment. In certain areas of the United States, finance companies also will provide funds for operating inputs, but they are not a major source of intermediate- or long-term credit. In many cases, finance companies will charge a higher rate of interest than other lending institutions because of the higher risk and costs associated with making smaller loans on consumer items.

LEASING

Although farmers have traditionally financed the purchase of capital assets using debt and/or equity funds, leasing of capital assets has become more important in recent years. With recent changes in the income tax treatment of leasing arrangements, lease financing is expected to continue to grow as an alternative to debt financing of farm machinery purchases.

Two common types of machinery leasing arrangements are used in many parts of the United States. One is the operating lease, which is a short-term seasonal leasing arrangement whereby the lessee (the farmer) leases the piece of equipment by the hour, day, or on a per acre basis. Many different kinds of operating leases are used, but typically the lessor (the owner of the machine) is responsible for insurance, taxes, and major repairs, while the lessee must pay variable expenses such as fuel, lubricants, and routine maintenance. Custom hiring is a common form of operating lease, and many farmers have their crops fertilized, planted, sprayed, or harvested by custom operators. With custom hire arrangements, the owner of the equipment furnishes the operator as well as the machine and is paid on a per acre or unit of production basis. In some parts of the United States, custom farming arrangements have developed whereby an operator is paid to till, plant, weed, and harvest the crop with payments for each machine operation negotiated on a per acre basis.

Operating leases or custom hiring can be very effective in obtaining the services of specialized machinery which might be underutilized if the farmer were to buy the machine. Such leasing arrangements can also be used to augment the capacity of a farmer's basic machine line in times of critical need. Operating leases may enable a farmer to try out a new machine or piece of equipment prior to the commitment to make a purchase. For example, farmers may lease conservation tillage equipment or other machinery that involves a new or different technology in order to obtain experience in using that machine and technology prior to the commitment of a large capital outlay to a new idea.

A second method of leasing that is becoming increasingly popular is that of financial or capital leases. Such lease arrangements involve longer term commitments than with operating leases, and are best understood as an alternative means of financing the acquisition of a machine. A financial or capital lease for farm machinery is usually a three- to five-year commitment with the lessee responsible for repairs, maintenance, insurance, and other expenses that would be incurred if she or he owned the machine. The lease agreement frequently calls for the lease payments to be made in advance (i.e., at the beginning of the year), and the lessor retains title to the equipment. At the expiration of the lease contract, the lessee usually has an option to buy the equipment at a percentage of fair market value at that time, or with a residual value lease, at a fixed percentage of the original price of the machine. Capital leases are available for buildings and equipment as well as machinery; most buildings and equipment leases are longer term than those for machinery.

There are numerous advantages and disadvantages of leasing capital assets. In many cases, the lease arrangement provides 100 percent financing so that no funds are necessary to make the purchase. However, since most leases require the first payment to be made in advance, a financial commitment is still required, but the amount of funds necessary to make the first least payment may in many cases still be less than the downpayment required with conventional debt financing. Thus, leasing may be a way to stretch the limited capital a farmer may have available to obtain a total machinery line, particularly in the case of beginning farmers. Leasing is also an attractive way to obtain the services of highly specialized equipment which may become obsolete relatively quickly; the lessor assumes the risk of obsolescence in this case but may charge higher lease payments to partially offset this increased risk.

Another potential advantage of leasing is that it is one method of obtaining a longer term financial commitment when acquiring capital assets. Many lenders have been hesitant in recent years to provide term financing for machinery purchases; instead, they have wanted to write the loan agreement for the purchase of machinery and equipment on a one-year renewable basis with the verbal agreement that the loan will be renewed if conditions warrant. In some cases, this has resulted in very large short-term financial commitments, and even if the loan is renewed, usually the interest rate is changed. With a lease arrangement, the farmer can not only obtain a specific commitment to finance the purchase for a period of years, but most lease payment schedules are fixed so that the lessee knows exactly how much must be paid each year on the lease. Thus, the leasing option may not only enable the farmer to obtain term financing, but it may also reduce the financial risk because the lease payment schedule is known, whereas the loan repayment schedule may vary from year to year depending on current rates of interest.

Leasing may be a lower cost method of acquiring machine services compared to conventional debt-financed ownership methods. Historically, leasing has been a much higher cost method of obtaining control of a capital asset, but with recent changes in the tax treatment of leases by the Internal Revenue Service, including the new "finance lease" provisions, some leasing companies are offering very attractive lease terms that may be lower than the cost of ownership.

One of the key considerations in determining the cost of leasing versus buying capital items is the tax treatment of the various expenditures. With ownership, any of the repair, maintenance, insurance, depreciation, and other costs of ownership as well as interest on borrowed funds and any operating expenses are tax deductible. With qualified finance leases, the entire lease payment along with the repair, maintenance, insurance, and operating expense can be treated as deductible expenses in determining income tax liabilities. Depending on the lessee's tax bracket, the after-tax cost of leasing may be lower than that of buying the machine even if the before-tax costs of leasing are higher. Use of the net present value procedure to accurately evaluate the profitability of financing a machine with a lease versus a loan and the financial feasibility of leasing were discussed in Chapter 14. The tax considerations noted earlier are an important component of this analysis; therefore, recent tax rules should be explicitly reviewed in doing the present value computations.

For some farmers, pride of ownership is an important consideration in the decision to lease or buy an asset. This intangible benefit should not be ignored, but it should also be recognized that most lease arrangements have options to buy at the end of the contract period, so the lessee eventually can obtain ownership of the asset. Furthermore, the lease arrangement gives the lessee the flexibility to obtain a different make or model of machine when the contract period is up rather than exercise the purchase option. This may be particularly valuable in situations where the technology has changed so that the leased machine is obsolete, the farming situation has changed so that the equipment does not contribute to overall profitability, or the specific machine has not been as reliable and efficient as the operator would like. Each operator must evaluate the economic costs and benefits of leasing versus buying for his or her own situation, but new leasing arrangements are becoming more cost competitive with ownership than in the past and farmers should give leasing serious consideration when making major capital acquisitions.

USE OF CREDIT IN THE FARM BUSINESS

Credit is an important and necessary resource in nearly all commercial farm businesses. Credit is a somewhat unique resource in that it provides the opportunity to pay for the cost of using additional inputs and capital items *now* from future earnings. Hence, the potential improvement in net farm income should be the determining factor in deciding whether or not to use credit in the farm business.

Credit can contribute to the improvement of net income of a farm operation in several ways.

1. *Create and maintain an adequate size business*. In most farm operations, this means expanding the operation to take advantage of economies of size. For example, studies in the Midwest indicate that per unit costs decline in cattle feeding as feedlot size approaches 1,000 head capacity. Credit can play an important role in acquiring the investment capital to expand the business and take advantage of the economies of size as well as to acquire operating inputs to maintain a high volume of output.

2. *Increase the efficiency of the farm business*. The use of credit may make it possible to substitute one resource for another (such as machinery for labor) as a means of reducing cost, improving timeliness, and increasing the efficiency of the farm business. Credit may also be essential to increase the intensity of production with present resources by using increased quantities of fertilizer and chemicals, better breeding stock, or more efficient machinery to improve the timeliness of crop production.

3. *Adjust the business to changing economic conditions*. New technological developments or changing market conditions can make it essential to make major changes in the farm business. Adopting confinement hog production technology or acquiring conservation tillage, or larger planting, harvesting, or power equipment may be essential to maintain efficiency as

prices decline and costs increase. The change to pipeline milkers and bulk tanks by dairy farmers, or dairying to another enterprise may require major investments to meet new regulations or market requirements. Credit is a major resource that can be used to assist in making these adjustments and changes.

4. *Meet seasonal and annual fluctuations in income and expenditures.* Most farm operations have wide seasonal and annual fluctuations in expenditures and incomes. Cash inflows and outflows typically do not occur at the same time of the year, and cash deficits frequently occur from the planting to harvesting seasons. Using credit to match cash inflows and outflows is essential to efficient operation of the farm business if large cash reserves are not available.

5. *Protect the business against adverse conditions.* Weather, disease, and price are all uncertainties in the farm business. As discussed in Chapter 11, good management can reduce the risk, but it is extremely difficult to eliminate all risks in farming. Credit can play a major role in protecting the farm business from financial failure or liquidation when adverse conditions occur. Maintaining some credit in reserve that can be used in such situations, such as an equity margin in real estate that can be used for refinancing short-term obligations, may be an important method of protecting the farm business from unpredictable risks. Liability management or managing the structure and amount of the liabilities of the farm business may be as important as asset management (diversification, flexible facilities, etc.) in protecting the farm business against the adverse financial consequences associated with risk.

6. *Provide continuity of the farm business.* The transfer of an on-going farm business from one proprietor to another involves large quantities of capital. Without credit, many farm businesses would have to be liquidated during the transfer process because nonfarm heirs frequently want their inheritance in cash and do not want to maintain ownership of farm real estate and other assets. If the transfer is from father to son, credit plays a major role by enabling the son to purchase his father's interest and expand and operate the farm efficiently. In most cases, credit is essential for the successful intergenerational transfer of the business because the tax liability and claims by off-farm heirs erodes the equity capital base, and either assets must be sold or credit used to substitute for the equity that has been lost in the transfer process.

Although there may be other important uses of credit in the farm business, a farm operator should evaluate whether credit will contribute to the improvement of future net income in one of the aforementioned ways. If not, the operator should evaluate whether the credit is being used as part of the business operation, or whether it is, in fact, consumer or personal credit—credit that will be used to satisfy current consumption and will not generate any cash income to use in making principal and interest payments on the debt obligation.

THE THREE R's OF CREDIT

Although numerous factors influence the credit-worthiness of a farmer, lenders typically evaluate farms and farm operators with respect to their risk-bearing ability, the returns that they generate in the business, and their ability to repay loans. These "Three R's of Credit" should be considered by all farmers as they approach their lenders to determine the realism of their loan requests and to provide evidence and documentation to lenders concerning their credit-worthiness.

Risk-Bearing Ability

The basic issue with respect to risk-bearing ability is whether or not the farm operation can withstand financial losses without being forced into liquidation or insolvency. Financial risks are influenced by other business risks, such as production risk, price risk, health risk, risk of obsolescence, and innovation. If production and prices decline, resulting in reduced profits or even losses, these losses must be covered or absorbed out of equity capital or net worth. Financial risks are also influenced by the proportion of debt and equity included in the farm business. Thus, with higher leverage or higher debt in relation to equity or total assets, losses can be magnified and financial risk becomes much greater. The concept of leverage and how it can influence the financial risk of the farm business were discussed in Chapter 8.

The basic document that measures the risk-bearing ability of the farm business is the financial statement. As discussed earlier, the structure of the financial statement includes assets on the left-hand side of the form and liabilities and net worth on the right-hand side. Assets and liabilities are classifed as current, intermediate, and long-term. A key ratio used in the analysis of risk-bearing ability is the relationship of total debts to total assets or debt to net worth. These ratios indicate the proportion of the business financed with debt compared to owner's equity, and thus the claims by others on the assets if liquidation should occur. Consequently, a lower debt to asset or debt to net worth ratio provides an indication of a higher risk-bearing ability since more equity is available to cover potential losses that might occur. For a mature farming operation, a debt to asset ratio of 0.3 or about twice as much equity or net worth as debt is desirable and indicates acceptable risk-bearing ability. Higher ratios may also be acceptable, but the loan terms would be expected to be more restrictive with a higher ratio.

Returns

The basic question with respect to returns is whether or not the use of credit will add to potential profits. Only if business profits will be increased will there be additional income available to use in making principal and interest payments on the borrowed capital.

Two additional questions are of interest in evaluating returns. First is the issue of whether the planned use of the credit is not only profitable, but also the *most* profitable use of credit in this particular farm business. The farm operator may be planning to use the credit to construct new livestock facilities when, in fact, it may be more profitable to update the machinery line and improve the timeliness of the corn planting and harvesting operations. Alternatively, it may be more profitable to borrow operating funds to purchase feeder pigs rather than feeder cattle, or crop production inputs such as fertilizer and chemicals rather than feeder livestock. Thus, even though the planned use of credit funds is profitable, the further question of whether this is the most profitable use of credit must be analyzed.

An additional question with respect to returns is whether the farm business as a whole is generating an adequate income for the farm family as well as for equity accumulation. Although the planned use of credit may be profitable, if the business as a whole is losing money, profits generated through the use of credit will more than likely be used to cover the losses in another part of the business rather than for equity accumulation or to make principal and interest payments on a loan. Hence, the business as a whole must be evaluated to reduce the possibility of income being siphoned off from profitable enterprises to cover losses on unprofitable enterprises rather than being used to repay the loan.

Repayment Capacity

A lender wants to be repaid in cash and has little interest in repossessing the security or collateral as a means of obtaining performance on a debt obligation. Consequently, the ability to repay is the final determination of whether or not credit should be extended. As was discussed in Chapter 2, a cash flow analysis can indicate not only the ability of the firm to repay a particular loan, but also when cash will be available for payment so that the loan terms and payment dates can be tailored to the cash flow of the business.

In evaluating repayment capacity, two different types of loan situations are frequently encountered—the self-liquidating loan and the loan that is not self-liquidating. The self-liquidating loan is made for the purchase of inputs that generate sufficient cash income in use to repay the loan. For example, feeder cattle loans are self-liquidating in most situations; the income generated by the sale of the cattle will be sufficient to repay the loan, with the remaining proceeds being available to cover other cash costs, leaving a residual return to homegrown feed as well as labor, fixed capital, and management. Loans that are not self-liquidating will not generate sufficient cash proceeds to make the principal and interest payments, and thus a cash subsidy is required to make the loan payments. Real estate and most improvement loans are not self-liquidating. The procedures illustrated in Chapters 8 and 14 to analyze the financial feasibility of financing arrangements provide a means of estimating if the loan will be self-liquidating. For non-self-liquidating loans, the payment must come from income after expenditures for operating and family living expenses, taxes, and payments on short-term borrowings. Thus, self-liquidating loans have fewer repayment problems and are easier to structure and set up. In contrast, loans on land and improvements and on breeding stock where replacements are held for expansion rather than sold for cash are more difficult to structure and can present problems in terms of the repayment capacity of the business.

LEGAL CONSIDERATIONS

The legal dimensions of credit use can be confusing and awsome at times. However, some basic instruments or documents and terms will be encountered in almost all credit transactions. The following brief discussion will review some of the basic legal dimensions of credit transactions, but should not substitute for counsel with your attorney or other credit specialists.

For most credit transactions, the promissory note is the first and probably most important document to be encountered. A promissory note essentially is a written promise to pay given by the borrower to the lender. It represents a moral and legal obligation to repay the loan. The note will include not only the amount of principal that must be repaid, but also the interest rate that will be charged and any penalties that will be incurred if performance does not occur. The signature of the promissory note by both borrower and lender makes it a legal and binding obligation; both parties should, therefore, have a full and complete understanding of the terms and conditions of the credit transaction.

The second legal document or set of documents encountered in a credit transaction are those used to provide security for the loan in case the borrower defaults on the promise to pay. This security interest essentially insures the lender that if payments are not made in cash, she or he will have access to specified property that can be used to satisfy the remaining portion of the financial obligation. With respect to credit transactions involving real estate, the typical security instrument is the mortgage (Figure 15.1). The mortgage describes the specific property that is being used as security or collateral for the credit transaction and the provisions whereby that property can be used to satisfy the obligation. In most states, in order to provide evidence to other potential lenders that the real estate has been pledged as collateral for a loan, the mortgage must be recorded or filed at the county courthouse. This filing or recording process establishes the priority of claims among lenders on the same tract of real estate. The mortgage recorded first has first priority if default should occur on the loan for which the mortgage is being used as security, and any other mortgages are "junior" to this first recorded mortgage regardless of when they were signed or delivered. Therefore, the recording process is an important

MORTGAGE

THIS MORTGAGE is made this . . . . day of . . . . ,
19   , between the Mortgagor, ...
. . . . (herein "Borrower"), and the Mortgagee, ...
. . . . , a corporation organized and
existing under the laws of . . . . , whose address is . . . .
. . . . (herein "Lender").

WHEREAS, Borrower is indebted to Lender in the principal sum of . . . .
. . . . Dollars, which indebtedness is evidenced by Borrower's
note dated . . . . (herein "Note"), providing for monthly installments of principal and
interest, with the balance of the indebtedness, if not sooner paid, due and payable on . . . .
. . . . ;

To SECURE to Lender (a) the repayment of the indebtedness evidenced by the Note, with interest thereon, the
payment of all other sums, with interest thereon, advanced in accordance herewith to protect the security of this
Mortgage, and the performance of the covenants and agreements of Borrower herein contained, and (b) the repayment
of any future advances, with interest thereon, made to Borrower by Lender pursuant to paragraph 21 hereof (herein
"Future Advances"), Borrower does hereby mortgage, grant and convey to Lender the following described property
located in the County of . . . . , State of Iowa:

which has the address of . . . . , ... ,
                                                                    [Street]   [City]
. . . . (herein "Property Address");
        [State and Zip Code]

TOGETHER with all the improvements now or hereafter erected on the property, and all easements, rights,
appurtenances, rents, royalties, mineral, oil and gas rights and profits, water, water rights, and water stock, and all
fixtures now or hereafter attached to the property, all of which, including replacements and additions thereto, shall be
deemed to be and remain a part of the property covered by this Mortgage; and all of the foregoing, together with said
property (or the leasehold estate if this Mortgage is on a leasehold) are herein referred to as the "Property".

Borrower covenants that Borrower is lawfully seised of the estate hereby conveyed and has the right to mortgage,
grant and convey the Property, that the Property is unencumbered, and that Borrower will warrant and defend
generally the title to the Property against all claims and demands, subject to any declarations, easements or restrictions
listed in a schedule of exceptions to coverage in any title insurance policy insuring Lender's interest in the Property.

IOWA—1 to 4 Family—6/75—FNMA/FHLMC UNIFORM INSTRUMENT
Maynard Printing, Inc., Des Moines, Iowa

Figure 15.1. Mortgage.

UNIFORM COVENANTS. Borrower and Lender covenant and agree as follows:

1. Payment of Principal and Interest. Borrower shall promptly pay when due the principal of and interest on the indebtedness evidenced by the Note, prepayment and late charges as provided in the Note, and the principal of and interest on any Future Advances secured by this Mortgage.

2. Funds for Taxes and Insurance. Subject to applicable law or to a written waiver by Lender, Borrower shall pay to Lender on the day monthly installments of principal and interest are payable under the Note, until the Note is paid in full, a sum (herein "Funds") equal to one-twelfth of the yearly taxes and assessments which may attain priority over this Mortgage, and ground rents on the Property, if any, plus one-twelfth of yearly premium installments for hazard insurance, plus one-twelfth of yearly premium installments for mortgage insurance, if any, all as reasonably estimated initially and from time to time by Lender on the basis of assessments and bills and reasonable estimates thereof.

The Funds shall be held in an institution the deposits or accounts of which are insured or guaranteed by a Federal or state agency (including Lender if Lender is such an institution). Lender shall apply the Funds to pay said taxes, assessments, insurance premiums and ground rents. Lender may not charge for so holding and applying the Funds, analyzing said account, or verifying and compiling said assessments and bills, unless Lender pays Borrower interest on the Funds and applicable law permits Lender to make such a charge. Borrower and Lender may agree in writing at the time of execution of this Mortgage that interest on the Funds shall be paid to Borrower, and unless such agreement is made or applicable law requires such interest to be paid, Lender shall not be required to pay Borrower any interest or earnings on the Funds. Lender shall give to Borrower, without charge, an annual accounting of the Funds showing credits and debits to the Funds and the purpose for which each debit to the Funds was made. The Funds are pledged as additional security for the sums secured by this Mortgage.

If the amount of the Funds held by Lender, together with the future monthly installments of Funds payable prior to the due dates of taxes, assessments, insurance premiums and ground rents, shall exceed the amount required to pay said taxes, assessments, insurance premiums and ground rents as they fall due, such excess shall be, at Borrower's option, either promptly repaid to Borrower or credited to Borrower on monthly installments of Funds. If the amount of the Funds held by Lender shall not be sufficient to pay taxes, assessments, insurance premiums and ground rents as they fall due, Borrower shall pay to Lender any amount necessary to make up the deficiency within 30 days from the date notice is mailed by Lender to Borrower requesting payment thereof.

Upon payment in full of all sums secured by this Mortgage, Lender shall promptly refund to Borrower any Funds held by Lender. If under paragraph 18 hereof the Property is sold or the Property is otherwise acquired by Lender, Lender shall apply, no later than immediately prior to the sale of the Property or its acquisition by Lender, any Funds held by Lender at the time of application as a credit against the sums secured by this Mortgage.

3. Application of Payments. Unless applicable law provides otherwise, all payments received by Lender under the Note and paragraphs 1 and 2 hereof shall be applied by Lender first in payment of amounts payable to Lender by Borrower under paragraph 2 hereof, then to interest payable on the Note, then to the principal of the Note, and then to interest and principal on any Future Advances.

4. Charges; Liens. Borrower shall pay all taxes, assessments and other charges, fines and impositions attributable to the Property which may attain a priority over this Mortgage, and leasehold payments or ground rents, if any, in the manner provided under paragraph 2 hereof or, if not paid in such manner, by Borrower making payment, when due, directly to the payee thereof. Borrower shall promptly furnish to Lender all notices of amounts due under this paragraph, and in the event Borrower shall make payment directly, Borrower shall promptly furnish to Lender receipts evidencing such payments. Borrower shall promptly discharge any lien which has priority over this Mortgage; provided, that Borrower shall not be required to discharge any such lien so long as Borrower shall agree in writing to the payment of the obligation secured by such lien in a manner acceptable to Lender, or shall in good faith contest such lien by, or defend enforcement of such lien in, legal proceedings which operate to prevent the enforcement of the lien or forfeiture of the Property or any part thereof.

5. Hazard Insurance. Borrower shall keep the improvements now existing or hereafter erected on the Property insured against loss by fire, hazards included within the term "extended coverage", and such other hazards as Lender may require and in such amounts and for such periods as Lender may require; provided, that Lender shall not require that the amount of such coverage exceed that amount of coverage required to pay the sums secured by this Mortgage.

The insurance carrier providing the insurance shall be chosen by Borrower subject to approval by Lender; provided, that such approval shall not be unreasonably withheld. All premiums on insurance policies shall be paid in the manner provided under paragraph 2 hereof or, if not paid in such manner, by Borrower making payment, when due, directly to the insurance carrier.

All insurance policies and renewals shall be in form acceptable to Lender and shall include a standard mortgage clause in favor of and in form acceptable to Lender. Lender shall have the right to hold the policies and renewals thereof, and Borrower shall promptly furnish to Lender all renewal notices and all receipts of paid premiums. In the event of loss, Borrower shall give prompt notice to the insurance carrier and Lender. Lender may make proof of loss if not made promptly by Borrower.

Unless Lender and Borrower otherwise agree in writing, insurance proceeds shall be applied to restoration or repair of the Property damaged, provided such restoration or repair is economically feasible and the security of this Mortgage is not thereby impaired. If such restoration or repair is not economically feasible or if the security of this Mortgage would be impaired, the insurance proceeds shall be applied to the sums secured by this Mortgage, with the excess, if any, paid to Borrower. If the Property is abandoned by Borrower, or if Borrower fails to respond to Lender within 30 days from the date notice is mailed by Lender to Borrower that the insurance carrier offers to settle a claim for insurance benefits, Lender is authorized to collect and apply the insurance proceeds at Lender's option either to restoration or repair of the Property or to the sums secured by this Mortgage.

Unless Lender and Borrower otherwise agree in writing, any such application of proceeds to principal shall not extend or postpone the due date of the monthly installments referred to in paragraphs 1 and 2 hereof or change the amount of such installments. If under paragraph 18 hereof the Property is acquired by Lender, all right, title and interest of Borrower in and to any insurance policies and in and to the proceeds thereof resulting from damage to the Property prior to the sale or acquisition shall pass to Lender to the extent of the sums secured by this Mortgage immediately prior to such sale or acquisition.

6. Preservation and Maintenance of Property; Leaseholds; Condominiums; Planned Unit Developments. Borrower shall keep the Property in good repair and shall not commit waste or permit impairment or deterioration of the Property and shall comply with the provisions of any lease if this Mortgage is on a leasehold. If this Mortgage is on a unit in a condominium or a planned unit development, Borrower shall perform all of Borrower's obligations under the declaration or covenants creating or governing the condominium or planned unit development, the by-laws and regulations of the condominium or planned unit development, and constituent documents. If a condominium or planned unit development rider is executed by Borrower and recorded together with this Mortgage, the covenants and agreements of such rider shall be incorporated into and shall amend and supplement the covenants and agreements of this Mortgage as if the rider were a part hereof.

7. Protection of Lender's Security. If Borrower fails to perform the covenants and agreements contained in this Mortgage, or if any action or proceeding is commenced which materially affects Lender's interest in the Property, including, but not limited to, eminent domain, insolvency, code enforcement, or arrangements or proceedings involving a bankrupt or decedent, then Lender at Lender's option, upon notice to Borrower, may make such appearances, disburse such sums and take such action as is necessary to protect Lender's interest, including, but not limited to, disbursement of reasonable attorney's fees and entry upon the Property to make repairs. If Lender required mortgage insurance as a condition of making the loan secured by this Mortgage, Borrower shall pay the premiums required to maintain such insurance in effect until such time as the requirement for such insurance terminates in accordance with Borrower's and

Figure 15.1. *(Continued)*

Lender's written agreement or applicable law. Borrower shall pay the amount of all mortgage insurance premiums in the manner provided under paragraph 2 hereof.

Any amounts disbursed by Lender pursuant to this paragraph 7, with interest thereon, shall become additional indebtedness of Borrower secured by this Mortgage. Unless Borrower and Lender agree to other terms of payment, such amounts shall be payable upon notice from Lender to Borrower requesting payment thereof, and shall bear interest from the date of disbursement at the rate payable from time to time on outstanding principal under the Note unless payment of interest at such rate would be contrary to applicable law, in which event such amounts shall bear interest at the highest rate permissible under applicable law. Nothing contained in this paragraph 7 shall require Lender to incur any expense or take any action hereunder.

8. Inspection. Lender may make or cause to be made reasonable entries upon and inspections of the Property, provided that Lender shall give Borrower notice prior to any such inspection specifying reasonable cause therefor related to Lender's interest in the Property.

9. Condemnation. The proceeds of any award or claim for damages, direct or consequential, in connection with any condemnation or other taking of the Property, or part thereof, or for conveyance in lieu of condemnation, are hereby assigned and shall be paid to Lender.

In the event of a total taking of the Property, the proceeds shall be applied to the sums secured by this Mortgage, with the excess, if any, paid to Borrower. In the event of a partial taking of the Property, unless Borrower and Lender otherwise agree in writing, there shall be applied to the sums secured by this Mortgage such proportion of the proceeds as is equal to that proportion which the amount of the sums secured by this Mortgage immediately prior to the date of taking bears to the fair market value of the Property immediately prior to the date of taking, with the balance of the proceeds paid to Borrower.

If the Property is abandoned by Borrower, or if, after notice by Lender to Borrower that the condemnor offers to make an award or settle a claim for damages, Borrower fails to respond to Lender within 30 days after the date such notice is mailed, Lender is authorized to collect and apply the proceeds, at Lender's option, either to restoration or repair of the Property or to the sums secured by this Mortgage.

Unless Lender and Borrower otherwise agree in writing, any such application of proceeds to principal shall not extend or postpone the due date of the monthly installments referred to in paragraphs 1 and 2 hereof or change the amount of such installments.

10. Borrower Not Released. Extension of the time for payment or modification of amortization of the sums secured by this Mortgage granted by Lender to any successor in interest of Borrower shall not operate to release, in any manner, the liability of the original Borrower and Borrower's successors in interest. Lender shall not be required to commence proceedings against such successor or refuse to extend time for payment or otherwise modify amortization of the sums secured by this Mortgage by reason of any demand made by the original Borrower and Borrower's successors in interest.

11. Forbearance by Lender Not a Waiver. Any forbearance by Lender in exercising any right or remedy hereunder, or otherwise afforded by applicable law, shall not be a waiver of or preclude the exercise of any such right or remedy. The procurement of insurance or the payment of taxes or other liens or charges by Lender shall not be a waiver of Lender's right to accelerate the maturity of the indebtedness secured by this Mortgage.

12. Remedies Cumulative. All remedies provided in this Mortgage are distinct and cumulative to any other right or remedy under this Mortgage or afforded by law or equity, and may be exercised concurrently, independently or successively.

13. Successors and Assigns Bound; Joint and Several Liability; Captions. The covenants and agreements herein contained shall bind, and the rights hereunder shall inure to, the respective successors and assigns of Lender and Borrower, subject to the provisions of paragraph 17 hereof. All covenants and agreements of Borrower shall be joint and several. The captions and headings of the paragraphs of this Mortgage are for convenience only and are not to be used to interpret or define the provisions hereof.

14. Notice. Except for any notice required under applicable law to be given in another manner, (a) any notice to Borrower provided for in this Mortgage shall be given by mailing such notice by certified mail addressed to Borrower at the Property Address or at such other address as Borrower may designate by notice to Lender as provided herein, and (b) any notice to Lender shall be given by certified mail, return receipt requested, to Lender's address stated herein or to such other address as Lender may designate by notice to Borrower as provided herein. Any notice provided for in this Mortgage shall be deemed to have been given to Borrower or Lender when given in the manner designated herein.

15. Uniform Mortgage; Governing Law; Severability. This form of mortgage combines uniform covenants for national use and non-uniform covenants with limited variations by jurisdiction to constitute a uniform security instrument covering real property. This Mortgage shall be governed by the law of the jurisdiction in which the Property is located. In the event that any provision or clause of this Mortgage or the Note conflicts with applicable law, such conflict shall not affect other provisions of this Mortgage or the Note which can be given effect without the conflicting provision, and to this end the provisions of the Mortgage and the Note are declared to be severable.

16. Borrower's Copy. Borrower shall be furnished a conformed copy of the Note and of this Mortgage at the time of execution or after recordation hereof.

17. Transfer of the Property; Assumption. If all or any part of the Property or an interest therein is sold or transferred by Borrower without Lender's prior written consent, excluding (a) the creation of a lien or encumbrance subordinate to this Mortgage, (b) the creation of a purchase money security interest for household appliances, (c) a transfer by devise, descent or by operation of law upon the death of a joint tenant or (d) the grant of any leasehold interest of three years or less not containing an option to purchase, Lender may, at Lender's option, declare all the sums secured by this Mortgage to be immediately due and payable. Lender shall have waived such option to accelerate if, prior to the sale or transfer, Lender and the person to whom the Property is to be sold or transferred reach agreement in writing that the credit of such person is satisfactory to Lender and that the interest payable on the sums secured by this Mortgage shall be at such rate as Lender shall request. If Lender has waived the option to accelerate provided in this paragraph 17, and if Borrower's successor in interest has executed a written assumption agreement accepted in writing by Lender, Lender shall release Borrower from all obligations under this Mortgage and the Note.

If Lender exercises such option to accelerate, Lender shall mail Borrower notice of acceleration in accordance with paragraph 14 hereof. Such notice shall provide a period of not less than 30 days from the date the notice is mailed within which Borrower may pay the sums declared due. If Borrower fails to pay such sums prior to the expiration of such period, Lender may, without further notice or demand on Borrower, invoke any remedies permitted by paragraph 18 hereof.

NON-UNIFORM COVENANTS. Borrower and Lender further covenant and agree as follows:

18. Acceleration; Remedies. Except as provided in paragraph 17 hereof, upon Borrower's breach of any covenant or agreement of Borrower in this Mortgage, including the covenants to pay when due any sums secured by this Mortgage, Lender prior to acceleration shall mail notice to Borrower as provided in paragraph 14 hereof specifying: (1) the breach; (2) the action required to cure such breach; (3) a date, not less than 30 days from the date the notice is mailed to Borrower, by which such breach must be cured; and (4) that failure to cure such breach on or before the date specified in the notice may result in acceleration of the sums secured by this Mortgage, foreclosure by judicial proceeding and sale of the Property. The notice shall further inform Borrower of the right to reinstate after acceleration and the right to assert in the foreclosure proceeding the non-existence of a default or any other defense of Borrower to acceleration and foreclosure. If the breach is not cured on or before the date specified in the notice, Lender may declare all of the sums secured by this Mortgage to be immediately due and payable without further demand and may foreclose this Mortgage by judicial proceeding. Lender shall be entitled to collect in such proceeding all expenses of foreclosure, including, but not limited to, reasonable attorney's fees, and costs of documentary evidence, abstracts and title reports.

19. Borrower's Right to Reinstate. Notwithstanding Lender's acceleration of the sums secured by this Mortgage, Borrower shall have the right to have any proceedings begun by Lender to enforce this Mortgage discontinued at any time

Figure 15.1. *(Continued)*

prior to entry of a judgment enforcing this Mortgage if: (a) Borrower pays Lender all sums which would be then 'ue under this Mortgage, the Note and notes securing Future Advances, if any, had no acceleration occurred; (b) Borrower cures all breaches of any other covenants or agreements of Borrower contained in this Mortgage; (c) *Borrower pays all reasonable expenses incurred by Lender in enforcing the covenants and agreements of Borrower contained in this Mortgage and in enforcing Lender's remedies as provided in paragraph 18 hereof, including, but not li·.ited to, reasonable attorney's fees; and (d) Borrower takes such action as Lender may reasonably require to assure that the lien of this Mortgage, Lender's interest in the Property and Borrower's obligation to pay the sums secured by this Mortgage shall continue unimpaired. Upon such payment and cure by Borrower, this Mortgage and the obligations secured hereby shall remain in full force and effect as if no acceleration had occurred.

20. Assignment of Rents; Appointment of Receiver. As additional security hereunder, Borrower hereby assigns to Lender the rents of the Property, provided that Borrower shall, prior to acceleration under paragraph 18 hereof or abandonment of the Property, have the right to collect and retain such rents as they become due and payable.

Upon acceleration under paragraph 18 hereof or abandonment of the Property, and at any time prior to the expiration of any period of redemption following judicial sale, Lender shall be entitled to have a receiver appointed by a court to enter upon, take possession of and manage the Property and to collect the rents of the Property including those past due. All rents collected by the receiver shall be applied first to payment of the costs of management of the Property and collection of rents, including, but not limited to, receiver's fees, premiums on receiver's bonds and reasonable attorney's fees, and then to the sums secured by this Mortgage. The receiver shall be liable to account only for those rents actually received.

21. Future Advances. Upon request of Borrower, Lender, at Lender's option prior to release of this Mortgage, may make Future Advances to Borrower. Such Future Advances, with interest thereon, shall be secured by this Mortgage when evidenced by promissory notes stating that said notes are secured hereby. At no time shall the principal amount of the indebtedness secured by this Mortgage, not including sums advanced in accordance herewith to protect the security of this Mortgage, exceed the original amount of the Note.

22. Release. Upon payment of all sums secured by this Mortgage, Lender shall release this Mortgage without charge to Borrower.

23. Waiver of Dower, Homestead and Distributive Share. Borrower hereby relinquishes all right of dower and hereby waives all right of homestead and distributive share in and to the Property. Borrower hereby waives any right of exemption as to the Property.

24. Redemption Period. If the Property is less than ten acres in size and if Lender waives in any foreclosure proceeding any right to a deficiency judgment against Borrower, then the period of redemption from judicial sale shall be reduced to six months. If the court finds that the Property has been abandoned by Borrower and if Lender waives any right to a deficiency judgment against Borrower, then the period of redemption from judicial sale shall be reduced to sixty days. The provisions of this paragraph 24 shall be construed to conform to the provisions of Sections 628.26 and 628.27 of the 1975 Code of Iowa.

IN WITNESS WHEREOF, Borrower has executed this Mortgage.

. . . .
—Borrower

. . . .
—Borrower

STATE OF IOWA, ... County ss:

On this . . . . day of . . . . , 19. . . ., before me, a Notary Public in the State of Iowa, personally appeared . . . . , to me personally known to be the person(s) named in and who executed the foregoing instrument, and acknowledged that . . . . executed the same as . . . . voluntary act and deed.

My Commission expires:

. . . .
Notary Public in and for said County and State

——————————————— (Space Below This Line Reserved For Lender and Recorder) ———————————————

Figure 15.1. (*Continued*)

and essential task in "perfecting" a security interest in a particular parcel of real estate to support a credit transaction.

With the adoption of the Uniform Commercial Code by most states, the method of obtaining a security interest in other types of property (personal property such as tractors, machinery, livestock, and grain inventories) has been somewhat standardized among states. Essentially, a security interest or lien on specific types of property other than real property is obtained by the signing of a security agreement between the borrower and the lender (Figure 15.2), and the filing of a "financing statement" by the lender at either the county courthouse or a centralized filing office in each state (Figure 15.3). The process of filing a "financing statement" perfects the security interest in specific property against claims by other possible lenders or third parties. The "financing statement" should not be confused with the financial statement, which is a document indicating a borrower's assets, liabilities, and net worth or the financial condition of the firm. A "financing statement" indicates the names and addresses of the borrower and lender and describes the specific property being used as security for the credit transaction. It must be signed by both parties and will typically include the property being purchased with the credit proceeds, such as tractor or machine, as well as other property that is deemed to be necessary to provide adequate security for the transaction. Although the lender will take care of the details of completing the security agreement and filing a "financing statement," the borrower should be aware of the implications of signing such documents. Details concerning the legal obligations of each party can be obtained from your lender, legal counsel, or by reviewing Article 9 of the Uniform Commercial Code.

The purchase contract or conditional sales contract is another method that is frequently used to finance and secure the purchase of inputs, including machinery and equipment and even land (Figure 15.4). With respect to machinery and equipment, provisions of the Uniform Commercial Code similar to those noted earlier are applied in a majority of states to conditional sales contracts. Sales contracts utilized in the transfer of land have been discussed previously. However, it should be noted that as a credit and security instrument, the buyer of real estate with a contract arrangement may be in a rather poor bargaining position if he or she defaults on the contract. In many states, a set of "forfeiture procedures" are available that can be invoked by the seller upon the buyer's default on a land contract. For example, forfeiture procedures in Iowa indicate that if the contract has been written properly, the seller can repossess the land within as short a period as 30 days after the buyer defaults. Thus, a land contract arrangement has unique legal features that should be thoroughly investigated before it is signed.

A final set of legal procedures are encountered if the borrower does not make the payments or otherwise perform on the credit arrangement. If

SECURITY AGREEMENT - GENERAL FORM

CONSUMER GOODS, EQUIPMENT, FIXTURES, FARM PRODUCTS OR INVENTORY GOODS
(UNIFORM COMMERCIAL CODE SECTION 554.9109 AND FOLLOWING)

_____, 19_____
(Date)

1. PARTIES—PROPERTY: The undersigned Debtor (jointly and severally) for value received hereby grants to the undersigned Secured Party or Lender, a security interest in the following described property:

all products of, additions to and replacements thereof and all accessories, accessions, parts and equipment now or hereafter affixed thereto or used in connection therewith, and the proceeds of all property secured hereby as set out below.

2. IF FARM PRODUCTS, CROPS OR FIXTURES ARE COLLATERAL: If this instrument includes livestock, then as additional collateral, Debtor assigns, transfers and conveys to Secured Party a security interest in and to all increase and issue thereof and additions, replacements and substitutions therefor, and all feed, both hay and grain, owned by Debtor, all water privileges, and all equipment, used in feeding and handling said livestock and also all of Debtor's right, title and interest in all leases covering lands for pasture and grazing purposes. If crops, this agreement includes annual and perennial crops and products thereof growing or planted on the following described property: either before or after harvest and all additions and substitutions therefor; or if the property covered hereby is livestock, crops or fixtures, it is and will be **located on the following described property in** _____**County, Iowa:**

A. LANDOWNER: If other than Debtor, the record owner of the land above described is_____

3. IF INVENTORY IS COLLATERAL: If this instrument includes inventory then Debtor hereby grants to Lender a security interest in all of his inventory now owned or hereafter acquired and all replacements, substitutions, and additions thereto, and a security interest in all of Debtor's merchandise, raw materials, work in process and finished products.

A. Upon execution of this agreement and upon request of Secured Party at any time while the indebtedness hereby secured remains unpaid, Debtor will furnish to Secured Party a signed statement, in form satisfactory to Secured Party, showing the current status of the inventory herein secured to include for any given period designated by Secured Party the opening inventory, inventory acquired, inventory sold and delivered, inventory sold and held for future delivery, inventory returned or repossessed, inventory used or consumed in Debtor's business, and closing inventory.

B. If at any given time the value of the collateral does not equal or exceed the total amount of indebtedness of Debtor to Secured Party, Debtor shall at once pay the excess of indebtedness to Secured Party or transfer additional collateral to Secured Party to meet Secured Party's satisfaction.

4. OBLIGATIONS SECURED—OPEN END: This security interest is given to secure the performance of the covenants and agreements herein set forth and for the payment of an indebtedness in the face amount of $_____ **as evidenced by a promissory note(s)** or other instrument(s) executed by Debtor payable to the order of said Secured Party as therein provided, and with interest as therein set forth and for all costs and expenses incurred in the collection of same including a reasonable attorney's fee and enforcement of Secured Party's rights thereunder; and for the payment of all extensions and renewals thereof and all changes in form of said indebtedness which may be from time to time effected by agreement between Secured Party and Debtor: and for all advances made by Secured Party for taxes, levies and repairs to or maintenance of said collateral or to protect or preserve the collateral against the claims of others and for all costs and expenses incurred in the collection of same and enforcement of Secured Party's rights thereunder; and all money heretofore and hereafter advanced by Secured Party at his option to or for the account of Debtor and all other present or future direct or contingent liabilities of Debtor to Secured Party of any nature whatsoever and however arising or acquired; and for interest on any money expended by Secured Party for taxes, levies and repairs to or maintenance of said collateral for interest on any money expended for costs and expenses incurred in the collection of said note or instrument and the enforcement of Secured Party's rights hereunder. All sums payable hereunder shall be paid at the place stated in the promissory note or instrument, if any, and if none then at the location of the Secured Party as stated below, and if none, then at the place of residence of the Secured Party.

5. This instrument shall be void upon payment of all obligations secured hereby.

6. INFORMATIONAL (Check one or more).

☐ The address of the Debtor, below, is his residence.　　☐ Such address is the Debtors chief place of business.

☐ Such address is where the Collateral is kept.　　☐ Debtor is a non-resident of Iowa.

7. USE OF PROPERTY: Debtor warrants, covenants and agrees that: The property is or is to be used by Debtor primarily (check 1, 2 or 3):

1. In business　☐ Equipment　☐ Inventory　　2. For personal, family or household purposes:　　3. In farming operations.　☐ Farm Products　☐ Farm Equipment

8. PURPOSE: The security interest herein is given on this collateral　☐ for a purchase money loan;　☐ otherwise.

9. THIS AGREEMENT SPECIFICALLY INCLUDES ALL OF THE ADDITIONAL PROVISIONS SET FORTH ON THE REVERSE SIDE HEREOF, THE SAME BEING INCORPORATED HEREIN BY REFERENCE. DEBTOR ACKNOWLEDGES RECEIPT OF A COPY OF THIS CONTRACT FULLY COMPLETED.

UNIVERSITY BANK & TRUST CO
AMES, IOWA 50012

_____
(Debtor)

★_____
(Debtor)

By_____
(Secured Party)

_____
(Number and Street)

_____
(Number and Street)

_____
(City)

_____
(City)

_____　_____
(County)　　(State)

_____　_____
(County)　　(State)

★Consider the desirability of joinder of spouse. Although Code Section 556.1 has been repealed by Section 554.10102, there remains Code Chapter 627 and the practicalities of determining title to any exempt property.

UNITED STATES **IOWA FORM**

SECURITY AGREEMENT—GENERAL FORM

Figure 15.2. Security agreement—General form.

641

SUCH ADDITIONAL PROVISIONS ARE AS FOLLOWS:

10. PROCEEDS: Proceeds of collateral are also covered by the lien of this instrument; however, such provision shall not be construed to mean that the Secured Party consents to any sale of such collateral, except inventory as described in Section 554.9109 Uniform Commercial Code of Iowa.

11. WARRANTY: If the collateral herein is for a loan, Debtor represents that he is owner of the above described property free and clear of all liens and encumbrances, unless specifically excepted herein, and will not sell, assign or transfer said property or any part thereof without the written consent of the Secured Party.

12. EXAMINATION AND INSPECTION: For right to inspect collateral, see paragraph 20, below: Inspection of Collateral. If the collateral hereunder is inventory or equipment used for business purposes, the Debtor will keep accurate books and records of the collateral and shall allow the Secured Party or representatives of the Secured Party to examine said books and records at any reasonable time.

13. INSURANCE AND TAXES: Debtor promises and agrees to keep said collateral insured from loss or destruction by fire, wind-storm and such other perils as Secured Party requires, in an amount not less than the full insurable value of the collateral, or the amount secured hereby, whichever is lesser, with appropriate endorsement to secure both parties as their interest appears, and to pay any and all taxes or charges which may be assessed against same. In the event the Debtor shall fail to provide adequate insurance or to pay any taxes or charges assessed against said collateral, Secured Party may, without notice, at its option, but without any obligation or liability so to do, procure insurance, pay taxes or other said charges and add said sums to the balance of the debt herein secured. Debtor hereby appoints the Secured Party the agent and attorney for the Debtor in adjusting and cancelling such insurance and endorsing settlement drafts and hereby assigns to the Secured Party all sums including return premiums and dividends, as additional security, specifically agreeing that Secured Party may cancel any said insurance upon any default by Debtor and apply any refund to the balance then due. Insurance policies shall promptly be delivered to Secured Party.

14. CARE OF PROPERTY: Debtor shall take good care of this property; shall shelter it and keep it in good repair; shall keep it free from all other liens, encumbrances, charges and claims, whether contractual or imposed by operation of law; shall not make any material change in said property nor use nor permit the same to be used for any unlawful purpose whatsoever; shall not remove it from the Debtor's residence or place of business without Secured Party's consent; will promptly supply to Secured Party any new residence address and secure permission from him to change the location of the collateral; and shall give Secured Party immediate written notice of any loss of, or damage to, any of said property.

15. SUCCESSORS AND ASSIGNS: The rights and privileges of Secured Party under this agreement shall inure to the benefit of his successors and assigns. All covenants, representations, warranties and agreements of Debtor contained in this agreement are joint and several if Debtor is more than one and shall bind Debtor's personal representatives, heirs, successors and assigns.

16. ASSIGNMENTS AND DEFENSES: That Secured Party shall have the right to negotiate or assign the security interest evidenced by this agreement **and the note which it secures,** and understands that Secured Party may do so without any notice to Debtor. Debtor specifically agrees that if there is any assignment or transfer of the security agreement, debt instrument, or note, the assignee or transferee shall have all of the Secured Party's rights and remedies under this agreement and that Debtor will not assert as a defense, counter-claim, set-off, cross complaint or otherwise, any claim, known or unknown, which he now has or hereafter acquires against the original Secured Party herein in any action commenced by an assignee or transferee of this agreement and the note which it secures, and will pay the indebtedness to the assignee at his place of business as it becomes due.

17. NON-WAIVER, EXTENSIONS, ETC.: That any extension of time for payment of any installment hereunder, or the acceptance of only a part of such installment, or the failure of the Secured Party to enforce the strict performance of any covenant, promise or condition herein contained on the part of the Debtor to be performed, shall not operate as a waiver of the right of the Secured Party thereafter to require that the terms hereof be strictly performed according to the tenor hereof. No party to this agreement shall be discharged from liability to the Secured Party by reason of the Secured Party's extending the time for payment of an installment or installments owing or due upon said loan, or by reason of the Secured Party's waiver or modification of any terms of the note or instrument evidencing such loan, or of any terms of this agreement.

18. LAW APPLICABLE: This agreement shall be deemed to have been made in the State of Iowa and shall be construed according to the laws of said State. If any provision of this agreement shall for any reason be held to be invalid or unenforceable, such invalidity or unenforceability shall not affect any other provision hereof, but this agreement shall be construed as if such invalid or unenforceable provision had never been contained herein.

19. ACCELERATION OF OBLIGATIONS AND DEFAULT: Upon the occurrence of any of the following events, the Secured Party may at his option, orally or in writing, declare the whole unpaid balance of any obligation secured by this agreement, immediately due and payable and if not so paid, then may declare Debtor to be in default under this agreement; said events being as follows:

(a) Debtor fails to make payments to the Secured Party as agreed. (b) Debtor fails to perform the other obligations agreed to be by him performed in any paragraph of this agreement. (c) Debtor or agent has made or furnished a false statement, representation or warranty in a material respect. (d) Debtor fails in his business; or if there occurs the dissolution or termination of its existence; or if there is commenced any proceeding under any bankruptcy or insolvency, laws or against the Debtor or by any guarantor or surety hereon for the Debtor; or if the Debtor shall make any assignment for the benefit of creditors. (e) Occurrence of loss, theft, damage or destruction of the collateral not covered by adequate insurance containing a loss payable clause for the protection of Secured Party.

20. REMEDIES: Upon default as in paragraph 19 above, Secured Party shall have all the rights and remedies of a Secured Party under the Uniform Commercial Code of Iowa (among others see Code Sections 554.9501-554.9507 inclusive) and under any other applicable laws. Debtor will, at Secured Party's request, assemble the collateral and make it available to the Secured Party at such place as is designated by the Secured Party, which shall be reasonably convenient. Debtor agrees that any regular business place in the county where this transaction takes place, as designated by the Secured Party shall be deemed reasonably convenient to both parties. Any requirements of reasonable notice by either party to the other or to the guarantors or sureties of Debtor shall be met if such notice is mailed, postage prepaid to the address of the parties shown on the first page of this agreement [or to such other mailing address as either party in writing later furnishes to the other] at least seven days before the time of the event or contemplated action set forth in said notice. Debtor agrees to pay all expenses of retaking, holding, preparing for sale, selling and reasonable attorney's fees and legal expense as may be allowable by law and incurred by Secured Party in enforcing his rights under this Security Agreement. INSPECTION OF COLLATERAL. Debtor hereby authorizes the Secured Party, agents or assigns to enter upon the premises of the Debtor at any reasonable time, and whether or not in default, to inspect the collateral; and if in default to possess, or attempt to possess said personal property and to assert or attempt to assert the rights of the Secured Party under any of the terms and provisions of this agreement, Debtor waives all rights and claims for trespass or conversion and damages in any manner thereby caused by Secured Party, his agents or assigns. All exemptions in and to any of the collateral are hereby waived. *INSECURITY. If and when, and so long as the Secured Party believes himself insecure, and even though Debtor is not then in default, the Secured Party, at his option, and without liability for trespass, conversion, and damages may repossess and keep possession of any or all of said collateral as provided herein; but without acceleration of maturity, unless and until the Debtor is in default. [Compare Sec. 554.9505 paragraph (1)] RIGHTS AND REMEDIES CUMULATIVE. The rights and remedies herein conferred upon the Secured Party shall be cumulative and not alternative and shall be in addition to and not in substitution of or in derogation of rights and remedies conferred by the Uniform Commercial Code of Iowa, and other applicable law.

21. CONSTRUCTION: Words and phrases herein, including acknowledgment hereof, if any, shall be construed as in the singular or plural number, and as masculine, feminine or neuter gender, according to the context. The paragraph headings of this agreement are for convenience only and shall not limit the terms of this agreement.

O P T I O N A L

STATE OF IOWA _____County, ss:

On this_____day of_____, 19_____, before me, the undersigned, a Notary Public in and for said County, in said State, personally appeared _____

to me known to be the identical persons named in and who executed the within and foregoing Security Agreement, and acknowledged that they executed the same as their voluntary act and deed.

_____
Notary Public in and for said County

ASSIGNMENT

For valuable consideration, receipt of which is hereby acknowledged, the undersigned does hereby sell, assign, and transfer to

_____
this agreement and the debt instruments secured hereby with_____recourse. The undersigned warrants that Debtor had the legal capacity and authority to execute the agreement and notes hereunder; that there has been no default and the security is free and clear of all claims, liens and encumbrances of any nature (except taxes not delinquent). The undersigned waives all demands for payment, notices of default and repossession of the security and agrees that assignee may grant extensions of time or renewals of notes hereunder without notice to the undersigned and without the consent of the undersigned. Upon default of the Debtor or breach of any provision of the Security Agreement or warranty given above, the undersigned agrees to repurchase this agreement from the assignee, for the amount of the unpaid indebtedness and advances, plus accrued interest and interest due and unpaid and all costs and expenses.

Dated this_____day of_____ 19_____.

By_____

*Compare Code Section 554.1208.

Secured Party

Figure 15.2. (Continued)

STATE OF IOWA
UNIFORM COMMERCIAL CODE — FINANCING STATEMENT — FORM UCC-2 (rev. 1-1-75)

UCC-2 (County Only)

Use only for recording Fixture Timber or Mineral security interests in County real estate records,
and cross indexing in County UCC records.

INSTRUCTIONS

1. Please type this form.
2. Remove pages 4 and 5 (Debtor and Secured Party copies) and send the first 3 pages with interleaved carbon paper to the filing officer with $3.00 filing fee. If the space provided for any item(s) is inadequate, the item(s) should be continued on additional sheets. Use of additional sheets or a non-standard form requires an additional fee of $1.00. The filing officer will return the third copy as evidence of filing as soon as it has been filed and assigned to a real estate number.
3. If a security agreement is filed as a financial statement, it must include all information required by UCC 9-402(5); in addition, all signatures must be acknowledged. Also submit a completed set of 3 pages of this form which need not be signed, and an additional fee of $1.00.
4. The real estate description must be sufficient to give constructive notice under the real estate mortgage laws of the State of Iowa (UCC 9-402(5); also the statement must identify the name of a record owner of the described real estate.
5. This instrument must have all of the signatures acknowledged on one of the acknowledgement forms in item No. 9 below before it can be lawfully recorded in county real estate records.
6. This financing statement shall be filed and recorded in county real estate records under the real estate number. This statement shall also be indexed under a real estate number in the UCC files of the County Recorder's office.

This FINANCING STATEMENT is presented to THE FILING OFFICER for filing for record in the real estate records:

1. Debtor(s) (Last Name First) and address(es)	2. Secured Party(ies) and address(es)	3. For Filing Officer (Date, Time, Number, and Filing Office)
	UNIVERSITY BANK & TRUST CO. AMES, IOWA 50010	

4. This Financing Statement covers the following types (or items) of property:

5. Name and Address of Assignee

6. Check appropriate box(s) ☐ The above goods are or are to become fixtures on ☐ The above timber is standing on ☐ The above minerals or the like (including oil and gas) or mineral accounts will be financed at the wellhead of the well or mine located on (Describe real estate below. (See instruction No. 4):

The name of a record owner is

7. ☐ Products of collateral are also covered

8. _____   _____
   Signature of Debtor   Signature of Debtor

   _____   _____
   Type or Print name (Iowa Code 335.2)   Type or Print name (Iowa Code 335.2)

   Secured party or other appropriate signature may be substituted for debtor(s) signature only in cases covered by UCC 9-402(2), 9-408 and 1105(3) (5) (7) and must be identified as such when used.)

9. Acknowledgement (Complete whichever one is applicable).

STATE OF IOWA,_____COUNTY, ss:

On this_____day of_____, A. D. 19_____, before me, the undersigned, a Notary Public in and for the State of Iowa, personally appeared_____

to me known to be the identical persons named in and who executed the within and foregoing instrument, and acknowledged that they executed the same as their voluntary act and deed.

_____Notary Public in and for the State of Iowa

STATE OF IOWA,_____COUNTY, ss:

On this_____day of_____, A. D. 19_____, before me, the undersigned, a Notary Public in and for the State of Iowa, personally appeared_____and

_____, to me personally known, who, being by me duly sworn, did say that they are the

_____and_____

respectively, of said corporation executing the within and foregoing instrument, that (no seal has been procured by the said) (the seal affixed thereto is the seal of said) corporation; that said instrument was signed (and sealed) on behalf of said corporation by authority of its Board of Directors; and that

the said_____and_____

_____as such officers acknowledged the execution of said instrument to be the voluntary act and deed of said corporation, by it and by them voluntarily executed.

_____Notary Public in and for the State of Iowa

Form Approved (1-1-75) By: MELVIN D. SYNHORST, Secretary of State

1. **White—Filing officer copy—alphabetical**

Figure 15.3. Uniform commercial code—Financing statement—Form UCC—2 (rev. 1-1-75).

IOWA STATE BAR ASSOCIATION
Official Form No. 21.2   (Trade-Mark Registered. State of Iowa. 1967)

FOR THE LEGAL EFFECT OF THE USE
OF THIS FORM, CONSULT YOUR LAWYER

REAL ESTATE CONTRACT (SHORT FORM)

It Is Agreed between _____

_____

of _____ County, Iowa, **Sellers,** and, _____

_____

of _____ County, Iowa, **Buyers:**

That Sellers hereby agree to sell and Buyers hereby agree to buy the real estate situated in _____

_____ County, Iowa, described as:

together with all easements and servient estates appurtenant thereto, upon the following terms:

1. **TOTAL PURCHASE PRICE** for said property is the sum of _____

_____ Dollars ($_____)

of which _____

Dollars ($_____) has been paid herewith, receipt of which is hereby acknowledged by Sellers; and Buyers agree to pay the balance to Sellers at residence of Sellers, or as directed by Sellers, as follows:

2. **INTEREST.** Buyers agree to pay interest from_____upon the unpaid balances, at the rate of _____per cent per annum, payable_____annually.

3. **TAXES.** Sellers agree to pay _____

_____, and any unpaid taxes thereon payable in prior years and any and all special assessments for improvements which have been installed at the date of this contract; and Buyers agree to pay, before they become delinquent, all other current and subsequent taxes and assessments against said premises. **Any proration of taxes shall be based upon the taxes for the year currently payable unless the parties state otherwise.***

4. **POSSESSION.** Sellers agree to give Buyers possession of said premises on or before_____, 19_____.

5. **INSURANCE.** Sellers agree to carry existing insurance until date of possession and Buyers agree to accept the insurance recovery instead of replacing or repairing buildings or improvements. Thereafter until final settlement, Buyers agree to keep the improvements upon said premises insured against loss by fire, tornado and extended coverage for a sum not less than $_____ or the balance owing under this contract, whichever is less, with insurance payable to Sellers and Buyers as their interests may appear, and to deliver policies therefor to Sellers.

6. **ABSTRACT.** Sellers agree to forthwith deliver to Buyers for their examination abstract of title to said premises continued to the date of this contract showing merchantable title in accordance with Iowa Title Standards. After examination by Buyers the abstract shall be held by Sellers until delivery of deed. Sellers agree to pay for an additional abstracting which may be required by acts, omissions, death or incompetency of Sellers, or either of them, occurring before delivery of deed.

7. **FIXTURES.** All light fixtures, electric service cable and apparatus, shades, rods, blinds, venetian blinds, awnings, storm and screen doors and windows, attached linoleum, attached carpeting, water heater, water softener, outside TV tower and antenna, attached fencing and gates, pump jacks, trees, shrubs and flowers and any other attached fixtures are a part of the real estate and are included in this sale except

*Decide for yourself if that formula is fair if Buyers are purchasing a lot with newly built improvements.

21.2   **REAL ESTATE CONTRACT (Short Form)**
Current January, 1981

Figure 15.4.   Real estate contract (short form).

8. **CARE OF PROPERTY.** Buyers shall not injure, destroy or remove the improvements or fixtures or make any material alterations thereof without the written consent of Sellers, until final payment is made.

9. **DEED.** Upon payment of all sums owing by Buyers to Sellers by virtue of this contract, Sellers agree to contemporaneously execute and deliver to Buyers a warranty deed upon the form approved by The Iowa State Bar Association and which shall be subject to:

    (a) Liens and encumbrances suffered or permitted by Buyers, and taxes and assessments payable by Buyers.

    (b) Applicable zoning regulations and easements of record for public utilities and established roads and highways.

    (c)

10. **FORFEITURE AND FORECLOSURE.** If Buyers fail to perform this agreement in any respect, time being made the essence of this agreement, then Sellers may forfeit this contract as provided by Chapter 656 of the Iowa Code and all payments made and improvements made on said premises shall be forfeited; or Sellers may declare the full balance owing due and payable and proceed by suit at law or in equity to foreclose this contract, in which event Buyers agree to pay costs and attorney fees and any other expense incurred by Sellers. It is agreed that the periods of redemption after sale on foreclosure may be reduced under the conditions set forth in Sections 628.26 and 628.27, Code of Iowa.

11. **PERSONAL PROPERTY.** If this contract includes personality, then Buyer grants Seller a security interest in such personality. In the case of Buyer's default, Seller may, at his option, proceed in respect to such personality in accordance with the Uniform Commercial Code of Iowa and treat such personality in the same manner as real estate, all as permitted by Section 554.9501(4), Code of Iowa.

12. **JOINT TENANCY IN PROCEEDS AND IN SECURITY RIGHT IN REAL ESTATE.** If, and only if, the Sellers, immediately preceding this sale, hold the title to the above described property in joint tenancy, this sale shall not constitute a destruction of that joint tenancy. In that case, all rights of the Sellers in this contract, in the proceeds thereof, and in any continuing or recaptured rights of Sellers in said real estate, shall be and continue in Sellers as joint tenants with full rights of survivorship and not as tenants in common. Buyers, in the event of the death of one of such joint tenants, agree to pay any balance of the proceeds of this contract to the surviving Seller and to accept deed executed solely by such survivor; but with due regard for the last sentence of paragraph 6, above.

13. **"SELLERS."** Spouse, if not a titleholder immediately preceding this sale, shall be presumed to have executed this instrument only for the purpose of relinquishing all rights of dower, homestead and distributive share and/or in compliance with section 561.13 Code of Iowa; and the use of the word "Sellers" in the printed portion of this contract, without more, shall not rebut such presumption, nor in any way enlarge or extend the previous interest of such spouse in said property, or in the sale proceeds, nor bind such spouse except as aforesaid, to the terms and provisions of this contract.

14. (Here add further terms or provisions)

Words and phrases herein shall be construed as singular or plural and as masculine, feminine or neuter gender according to the context

Dated this_____day of_____ 19_____.

_____   _____

_____   _____
                 BUYERS   **SELLERS**

_____   _____
       Buyers' Address   **Sellers' Address**

STATE OF IOWA, _____COUNTY, ss:
    On this_____day of_____, A. D. 19_____, before me, the undersigned, a Notary Public in and for said State, personally appeared_____

_____

to me known to be the identical persons named in and who executed the foregoing instrument, and acknowledged that they executed the same as their voluntary act and deed

_____,Notary Public in and for State.

Real Estate Contract (Short Form)

TO

Entered for taxation the _____ day of _____ 19___, _____ Auditor By _____ Deputy

Filed for record the _____ day of _____ 19___, o'clock _____ M., and recorded in Book _____ of _____ on page _____ County Records. _____ Recorder By _____ Deputy

WHEN RECORDED RETURN TO

Figure 15.4. *(Continued)*

the terms of a mortgage arrangement are not met, the purchaser is said to have defaulted on the mortgage and he or she may lose title to the land under foreclosure proceedings. The first step in foreclosure after default has occurred is to have court proceedings whereby the lender presents arguments to the court concerning what terms or conditions of the mortgage have not been met. The court then renders a decision concerning whether the borrower has, in fact, defaulted, and if so, a judgment is placed against that person. Following the court proceedings, if a judgment against the borrower results, a foreclosure sale occurs, which typically involves a public auction conducted by the county sheriff. At this foreclosure sale, the lender is usually a bidder and may purchase the land. Following the foreclosure sale, there is a redemption period during which time other creditors or the borrower (previous owner) may purchase the farm by paying the purchase price at the foreclosure sale plus interest and other costs that have been incurred. This redemption period may be as long as two years in many states. If redemption does occur, the purchaser at the foreclosure sales holds clear title and any excess proceeds that might have been obtained at the sale are rebated to the original buyer.

With respect to the transactions covered by the Uniform Commercial Code, the procedures to be followed upon default of the borrower are usually specified in the security agreement signed by both parties. The remedies available to the lender may include repossession of the item or equipment used as collateral for the credit transaction, sale of the item with the lender taking the proceeds from the sale as payment for the debt obligation, or various other judicial procedures as specified by the courts. Again, before a credit agreement is signed a farmer-borrower should have a thorough understanding of the sequence of events that will occur upon her or his default on a credit obligation.

THE COST OF CREDIT

The price that must be paid for borrowed funds is the rate of interest. Interest charges present a special problem for farmers, as well as other borrowers, because they are a fixed cost; they do not vary with the level of production or the amount of output. Thus, interest and, in most cases, principal must be paid whether the farmer had a bumper crop or was hit by drought, whether prices for livestock and crops were high or low, and even whether any crops or livestock were produced at all.

Numerous factors influence the cost of borrowed funds or the interest rate that must be paid by farmers and ranchers. Generally, supply and demand conditions influence the rates of interest in the national money markets. Because many agricultural lending institutions, particularly the Cooperative Farm Credit Banks, obtain funds from these national money markets, rates set in capital markets that are reacting to national and international monetary conditions have a direct impact on the cost of money and interst rates paid by farmers. For example, monetary policy during the late 1970s and early 1980s was quite restrictive in an attempt to reduce the inflation expectations of savers and investors. At the same time, large government deficits placed heavy demands on the capital markets for funds and suggested to participants in the financial markets that fiscal policy would remain stimulative, resulting in more inflation. To be willing to save, savers must receive a real rate of return on their funds (an "inflation premium") or a nominal interest rate that exceeds the rate of inflation; hence, expectations of higher rates of inflation mean that they want higher nominal rates of interest. The combination of tight monetary policy and stimulative fiscal policy resulted in very high market interest rates during the period. These high market rates were reflected in the high rates that farmers had to pay as indicated in Table 15.2. Thus, national montary policies and changes in the economic health of the total U.S. and world economies can have an impact on the cost of credit for farm operators.

A second factor that influences the rate of interest charged on a particular loan is the risk of loss associated with the loan. The risk of loss depends on two important components of the credit agreement and transaction: (1) the credit-worthiness of the customer including the customer's risk-bearing ability, business returns and repayment capacity, and (2) the collateral and security agreement between the borrower and lender. If a farm customer is considered to be a higher risk borrower because of a low equity position or history of unprofitable operations, he or she can expect to pay a higher rate of interest than a neighbor who has a stronger equity position. A farmer who can provide high-quality security such as farm real estate will typically be charged a lower rate of interest than one who only has a set of used machinery to use as collateral or security for the loan.

In many farming communities, lending institutions are careful not to charge significantly different rates of interest to different customers because of potential problems they perceive of "customer good-will." Instead, to adjust for differences in risk among customers, they utilize different standards for such terms as the amount of downpayment required or the amount of operating funds they will advance per acre. Thus, the higher risk borrower either will have to pay a higher rate of interest to obtain credit or will not be able to obtain as much credit for a particular use as a lower risk borrower.

A third factor that influences the cost of credit to farmers is the administration and supervision costs that may be required. In many cases, these costs are associated with the riskiness of the loan—higher risk loans will usually require more supervision and higher administration costs. Since the major costs of administering and supervising a loan are incurred in the actual making, servicing, and collecting process, these costs are related primarily to the number of loans made by a particular financial institution and not to the size of an individual loan. The amount of paperwork and time required to make a loan are about the same for a $10,000 loan as for a $50,000 loan. Consequently, the administration and supervision costs per dollar of credit extended will usually be lower for larger loans than for smaller loans. Thus, one would expect a farmer who is borrowing $100,000 to $200,000 to obtain credit at a lower cost and be charged a lower rate of interest than one who is borrowing $5,000 to $10,000.

The fourth factor that influences the rate of interest charged on various loans is that of competition—or lack thereof. The existence of alternative lenders in the agricultural credit market will certainly influence the rate of interest charged by any one lender, since a rate higher than that of their competition will result in a reduction in credit activity. In addition, competition influences interest rates in numerous other ways. As has been indicated earlier, aggregate demand and supply conditions in the money markets as reflected in the competition for funds and the rates of return that can be obtained in nonagricultural industries influence the interest rates that must be paid by farmers. Likewise, banks and other lenders have alternative uses for their funds, and if nonagricultural businesses and industry are willing to pay higher interest rates than farmers, this will result in fewer funds and less credit available to the farming sector. Thus, farmers must compete with other sectors of the economy for local as well as money market funds.

The competition among various institutions for the funds provided by savers (which are then loaned to the borrower) also influences the cost of money and thus the interest rate. There may be numerous financial institutions in a local area that are willing to acquire the saver's funds, and this competition for savings usually results in higher rates on savings instruments. Consequently, those institutions that use local funds to support their agricultural loans must pay a higher price for these funds. This higher price will in turn result in higher interest rates that must be charged to the farmer-borrower.

A fifth factor that influences interest rates and the cost of money are the legal restrictions set in many states on the maximum interest rates that financial institutions can charge their borrowers. These legal restrictions are commonly called usury laws or usury legislation. Most usury laws specify a maximum interest rate that can be charged on loans. In many states, this rate is sufficiently high as not to restrict the rate set by market competition. However, in some states usury laws do specify maximum rates that are below the equilibrium interest rate that would be determined by the forces of supply and demand interacting in the traditional fashion to set price. Typically the financial institutions do not aggressively service those states or areas with restrictive usury legislation, and they move their funds to other regions where they can receive a competitive rate of return. Historically, this has been one of the reasons why insurance companies have reduced their volume of agricultural real estate loans in the Midwest when interest rates have been high. Insurance companies were able to invest their funds in other states or regions at higher rates than those that could be charged in many of the Midwestern farm belt states where usury legislation was most prevalent. Since each state has different usury laws, it is necessary to evaluate the usury legislation in each state when determining its impact on interest rates and credit use.

CALCULATING THE RATE OF INTEREST

Although interest would seem to be a fairly simple concept and easily calculated, the terms of the loan arrangement may have a significant impact on the actual interest rate that is being charged on a particular loan. Some loans, called discount loans, require that the interest be paid in advance and be subtracted from the loan proceeds. Other loan arrangements may involve calcuation of interest on the beginning balance and adding the amount of interest to the principal payment to determine the amount to be repaid. In addition, the principal may be repaid in annual installments or in quarterly or monthly installments during the year. All of these factors influence the actual rate of interest or annual percentage rate being charged on a particular loan.

The true annual percentage rate charged on a particular loan can be determined using the discounting concepts developed in Chapter 8. The true rate is the rate of interest which, when used to discount the annual principal and interest payments, will equate that stream to the initial loan balance. In essence, the true rate is determined using the internal rate of return formula of Chapter 8, with the loan balance being the initial outlay and the principal and interest payments being annual cash inflows.

Federal Truth-in-Lending legislation requires lenders to disclose the dollar amount of the interest charges and the annual percentage rate being charged on loan transactions, but many farm loans are now exempt from Truth-in-Lending. When Truth-in-Lending does not apply or the trial-and-error procedure of the internal rate of return formula seems cumbersome, the following approximation formula (the direct ratio equation) might be used:

$$r = \frac{6RP}{3D(n + 1) + R(n - 1)} \qquad (15.1)$$

where:

r = Annual percentage rate.
R = Amount of interest.
P = Number of payments per year.

$$D = \text{Initial loan balance.}$$
$$n = \text{Total number of payments.}$$

For example, assume a farmer borrows $1,200 at 7 percent add-on interest to be repaid over one year with 12 monthly payments. The total interest charge would be calculated as: $1,200 × 0.07 = $84, and the monthly payments would be:

$$\frac{\$1,284}{12} = \$107$$

If we use the formula presented earlier, the annual percentage rate on this loan will be calculated as:

$$\frac{6 \cdot 84 \cdot 12}{(3 \cdot 1,200 \cdot 13) + (84 \cdot 11)} = \frac{6,048}{47,724} = 0.1267 \text{ or } 12.67\%$$

As a second example, assume that a tractor is traded with a $6,000 boot payment. It is financed with the dealer on a three-year plan with equal payments yearly of $2,480 at the end of each year. The total payments on the loan would be $2,480 × 3 or $7,440; thus, the interest or finance charge would be $1,440 ($7,440 minus $6,000—the original boot payment). By again using the formula, the annual percentage rate being charged on the finance plan would be calculated as:

$$\frac{6 \cdot 1,440 \cdot 1}{(3 \cdot 6,000 \cdot 4) + (1,440 \cdot 2)} = \frac{8,640}{74,880} = 0.1154 \text{ or } 11.54\%$$

Thus, stated interest rates must be converted to annual percentage rates to get a true picture of the interest rate being charged by the lender.

In addition to the stated rate of interest, other fees and charges must be included in the price of obtaining credit. With some credit transactions, particularly those involving real estate, an appraisal fee or inspection fee may be charged. Some lenders may charge a loan fee to complete the loan transaction such as 1 percent of the face value of the loan. As has been indicated earlier, any stock purchase such as is required to borrow from a PCA or Federal Land Bank must also be included in the cost of credit. These additional stock purchases will increase the effective interest rate on the loan.

Such requirements as compensating balances that require the borrower to maintain a specified percentage of the loan on deposit in the lending institution also will increase the effective interest rate, if this balance is larger than would typically be held by the borrower. Moreover, compulsory insurance such as credit life insurance, crop insurance, and property damage insurance as well as mortgage recording fees and taxes are additional costs that will be incurred in using credit. These additional costs must be added to the interest rate that is paid on borrowed funds to obtain an accurate and complete picture of the cost of a credit transaction.

REPAYMENT PLANS AND CAPACITY

Repayment Plans

A major problem that occurs in using credit in the farm business is matching the repayment terms on loans with the repayment capacity and ability of the farm operation. This problem occurs because (1) repayment capacity changes over time, and thus it is sometimes difficult to measure accurately, (2) farmers frequently need and should use more long-term or intermediate-term credit than they can obtain or their collateral can support, (3) poor planning or repayment schedules result in cash deficits and shorter terms than are acceptable or financially feasible, and (4) credit is obtained from numerous sources with no coordination among them, which makes repayment schedules difficult to estimate and evaluate.

Different repayment plans should be developed for different loans, depending on the use of the funds and the lengths of the loan. With short-term loans used for operating expenses, the payment should coincide with the time of year when cash sales occur and the firm can be expected to have a positive net cash flow.

In most cases, payment due dates on operating loans will be specified as the time when the crop or livestock product is marketed. Monthly payment schedules should be used only if cash income is generated monthly from product sales such as occurs in a dairy operation or hog enterprise that includes monthly marketing of hogs. For a grain operation, payments are more typically made on an annual or semiannual basis on operating loans.

For intermediate- and long-term loans, an amortization plan that requires regular payments should be used. The payments could be quarterly, monthly, or annually. Amortization plans that require periodic payments of principal as well as interest systematically reduce the outstanding loan balance and thus the risk of a financial catastrophe impairing the operator's ability to repay the loan. For intermediate-term assets, particularly machinery and breeding stock, repayment over one-half to two-thirds of the life of the asset is most realistic. Farm mortgages and other real estate financing may include payment terms from 10 to 35 years. In most situations, land contracts have shorter repayment terms than mortgages that are available from such financial institutions as Federal Land Banks and insurance companies. Mortgage terms will typically be 20 to 35 years, whereas many land contracts have 10- to 15-year terms.

For term loans, two kinds of amortization plans can be used—the even total payment plan and the even principal payment plan. The even total

Table 15.3 Even Total Payment Plan—$10,000 Loan, 10 Years, 13 Percent

Year	Payment	Principal	Interest	Balance
1	$1,842.89	$542.89	$1,300.00	$9,457.11
2	1,842.89	613.47	1,229.42	8,843.64
3	1,842.89	693.22	1,149.67	8,150.43
4	1,842.89	783.33	1,059.56	7,367.09
5	1,842.89	885.17	957.72	6,481.93
6	1,842.89	1,000.24	842.65	5,481.69
7	1,842.89	1,130.27	712.62	4,351.42
8	1,842.89	1,277.21	565.68	3,074.21
9	1,842.89	1,443.24	399.65	1,630.97
10	1,842.89	1,630.96	212.03	—
		$10,000.00	$8,429.00	

payment plan results in equal payments over the entire period of the loan, with an increasing proportion of each payment going to principal and a decreasing portion going to interest. Table 15.3 illustrates the total payment and the proportion of each payment attributable to principal and interest for a $10,000, 10-year, 13 percent loan using the even total payment plan.

The even principal plan results in equal principal payments, with decreasing interest payments and thus decreasing total payments as more payments are made. Table 15.4 illustrates the payments on an even principal payment plan for a similar $10,000, 13 percent loan. Note that the total payment in the first year (and future years) is $1,842.89 with the even total payment plan (Table 15.3), where it is $2,300 with the even principal payment plan (Table 15.4). Thus, the even principal payment plan will

Table 15.4 Even Principal Payment Plan—$10,000 Loan, 10 Years, 13 Percent

Year	Payment	Principal	Interest	Balance
1	$2,300	$1,000	$1,300	$9,000
2	2,170	1,000	1,170	8,000
3	2,040	1,000	1,040	7,000
4	1,910	1,000	910	6,000
5	1,780	1,000	780	5,000
6	1,650	1,000	650	4,000
7	1,520	1,000	520	3,000
8	1,390	1,000	390	2,000
9	1,260	1,000	260	1,000
10	1,130	1,000	130	—

require a larger cash flow from the farm business during the first few years of the loan. However, the total interest payments on the even principal payment plan are less than those incurred with the even total payment plan. This occurs because the outstanding balance is reduced faster and the funds are borrowed for a shorter period when equal principal payments are made over the life of the loan.

Repayment Capacity

A basic question asked by many farm operators is, "How far in debt can I safely go?" The answer to this question will be different for each farmer and will depend on the risk, return, and repayment capacity of each farming operation. But essentially, the answer to the question of a safe debt load is answered with a second question, "How much can I repay?" We will illustrate how the concept of repayment capacity can be used to determine how far in debt one can safely go.

If we assume an even total payment amortization schedule, the amount of debt that can be supported with a given amount of repayment capacity can be estimated using a table of annuity factors (Appendix Table III). In essence, an annuity factor indicates the present value of a constant future stream of income. If we borrow money today and plan to repay a specified constant amount each year, we have set up an annuity for the lender, and the annuity factor indicates how much we could borrow today and safely repay with each dollar of annual repayment capacity. For example, at 13 percent interest, $1,000 of annual repayment capacity will pay the principal plus interest on a 10-year loan of $5,426.20 (5.4262 from Appendix Table III × $1,000).

To illustrate the use of the annuity table in estimating repayment capacity, assume that a farmer is considering the replacement of a worn-out livestock facility. The new facility will cost $60,000, and the farmer has about $12,000 that could be used for a downpayment and will need to borrow the remaining $48,000. The producer has provided the following income and expense information on his or her operation.

Cash receipts	$90,000
Cash expenses	66,000
Net cash income	24,000
Living expense	15,000
Cash for debt servicing	9,000

If we assume a 7-year, 13 percent loan, the data in Appendix Table III indicate that each $1 of annual repayment capacity will support a loan with a face value of $4.4226. Consequently, $9,000 of annual repayment capacity could support a loan of only $39,803.40 (4.4226 × $9,000). If the terms of the loan could be expanded to 10 years, $9,000 of annual repayment capacity would support a loan of $48,835.80 (5.4262 × $9,000). Thus, extending the terms of the loan increases the amount of debt that can be supported with a specified amount of annual repayment capacity and in this case would make the purchase financially feasible.

In many cases, it is desirable to know the annual payment required to amortize a loan of a particular size during a specified number of years. This information can be obtained from amortization tables (Appendix Table IV). The formula for determining the amortization factor for a specified year and interest rate was discussed in Chapter 4. To illustrate the use of the amortization table, a 13 percent 10-year loan of $1,000 would require annual payments of $184.30 (0.1843 × $1,000) to pay interest plus repay principal during the 10-year term. If a farmer borrowed $20,000 at 13 percent to purchase a machine and was required to repay the loan in three years, annual payments of $8,470 (0.4235 × $20,000) would be required. Alternatively, the same $20,000 loan amortized over a four-year period would require only $6,724 (0.3362 × $20,000) of cash annually for principal and interest payments, and a six-year loan would require only $5,004 (0.2502 × $20,000) annually.

The data in these tables indicate that one means of increasing the debt-carrying capacity of the farm business is to extend the term of the loan. For example, Appendix Table III indicates that for a 13 percent loan $1,000 of annual repayment capacity would support $3,517.20 of a five-year debt and $4,798.80 of an eight-year debt. Thus, the same amount of cash will support $1,281.60 or about 36 percent more debt if the terms can be lengthened from five to eight years. Alternatively, if a real estate contract in the amount of $1,000 per acre were acquired that had a 10-year term (9 percent), the annual payment would be $155.80 per $1,000 or $155.80 per acre. Extending the terms of the loan to 20 years would reduce the annual cash requirements for principal and interest payments from $155.80 to $109.50 per acre.

Note that for shorter repayment periods the length of the repayment period can have a significantly greater impact on the repayment capacity of the firm or the amount of debt that can be supported than the interest rate. For example, decreasing the interest rate from 13 to 12 percent on a five-year $1,000 loan results in approximately a $7 decrease from $284.30 per year to $277.40 in the annual cash requirement for amortization. If in the process of negotiating the lower interest rate, the lender shortens the term from five to four years, the annual payment does not decline but actually increases from $284.30 to $329.20. Thus, accepting shorter terms to get a lower interest rate may actually impair the financial solvency of the business because the higher cash requirement on the shorter term loan may not be within the repayment capacity of the firm. Adjusting the repayment schedule to the cash flow or cash-generating capacity of the business is a key component of wise and profitable credit management and utilization.

Therefore, to obtain a final answer to the question of how far in debt a farmer can safely go, the cash flow and repayment capacity of the business must be evaluated.

Setting up reasonable repayment schedules may be the single most important factor in keeping the use of credit in the farm business on a sound foundation. For farm operators who want to strengthen the repayment capacity of their business, at least four steps can be taken. First, they can increase their equity in the farm business over time through reduced family expenditures and more reinvestment of earnings which will result in a higher level of income and more cash flow for debt servicing. Second, the farm operator can make the maximum use of self-liquidating loans—loans that will generate sufficient cash to make the loan payments and thus reduce the cash drain required for loan amortization. Moreover, using the total capital structure of the business—particularly land equity if it is available—to lengthen out the repayment schedule will increase the debt-carrying capacity of the farm operation. Longer terms and lower annual payments also reduce the risk of being unable to make a payment if income is reduced because of low prices or unexpected poor weather conditions. Finally, developing a repayment schedule based on a cash flow can not only assist farmers in planning their financial and credit needs, but it can also provide a means of communicating with lenders concerning when and how much they expect to borrow and when they expect to repay. A good set of financial plans that have been developed with realistic assumptions can assist farmers in selling their credit-worthiness to lenders and in increasing their ability to borrow money and expand the farm operation.

Summary

Most commercial farms use a combination of debt and equity funds to finance the investment and operating capital requirements of the business. Equity is the financial backbone of any business and can be acquired in many ways. The most important source of equity funds is savings; savings not only provides funds to finance the purchase of assets, but it also indicates an ability to handle finances and typically provides the financial base to acquire debt funds. Other sources of equity include inheritance and pooling arrangements with family members or investors. Although pooling arrangements may not result in a direct increase in equity funds of the operator, they may provide a larger capital base and increase the efficiency of the operation in the future. However, pooling with others does require the sharing of control and management decision-making.

The agricultural debt or credit market includes three major types of institutions: (1) private sector lenders, (2) cooperative lenders, and (3) government agencies. Commerical banks have historically been a major source of short- and intermediate-term financing for agriculture, although their market share has eroded in recent years. Because of the volatility of funds availability within and between years, commercial banks have become reluctant to be heavily involved in long-term lending.

The Federal Land Bank, one of the agencies of the cooperative credit system, has been an important source of long-term mortgage funds for farmers. Its market share has grown over time, not only because of competitive interest rates, but also because of innovative programs to provide long-term financing for agriculture. Production Credit Associations are the short- and intermediate-term financing institution of the Cooperative Farm Credit System. Like Federal Land Banks, PCAs require the borrower to purchase stock to provide the equity capital base of the local association. PCAs make loans to farmers for purchasing operating inputs as well as machinery, equipment, and various types of improvements.

Life insurance companies are a major source of mortgage funds for farmers interested in buying or improving land. Life insurance companies obtain their funds primarily from the premiums on insurance policies and the earnings and reserves that are generated through the sale of insurance. In recent years, farm lending activity by insurance companies has not kept pace with the growing market, and thus the market share of insurance companies has declined.

The major government agency that lends to farmers is the Farmers Home Administration. This agency is commonly referred to as a "lender of last resort"; this designation reflects the higher risk of the Farmers Home Administration loan portfolio and the fact that many of the borrowers are beginning farmers who are unable to obtain credit from other commercial sources on reasonable terms. The Commodity Credit Corporation also provides loans to farmers as part of the income and price support program for the agricultural sector. In addition, a substantial proportion of the loan funds for farmers comes from individuals, particularly in the form of the installment land contract, as well as from merchants and dealers who finance many of the inputs bought by farmers.

Although debt and equity funds have traditionally been the major sources of capital for U.S. farmers, leasing is becoming more popular in the farm sector. The operating lease is a short-term, seasonal leasing arrangement whereby the lessee leases the equipment for a specified number of hours, days, or on a per acre basis. Custom hiring is one form of an operating lease. In recent years, capital leases which involve a longer time commitment (such as three to five years) have become more popular for some machinery, equipment, and facility purchases. A farmer should analyze the profitability of leasing compared to loan arrangements for capital items before a lease or loan is selected.

Safe use of borrowed money is extremely important in the successful farm business. The credit-worthiness of any farm depends on the risk-bearing ability of the operation, the returns that can be generated in the business, and the repayment capacity of the operation. Furthermore, farmers should be aware of the legal documents involved in borrowing money, including the promissory note, the mortgage or security agreement and financing statement, and the installment contract. In addition, a farmer should be aware of the obligations he or she faces upon default, including foreclosure and bankruptcy procedures.

The most obvious cost incurred in borrowing money is interest. Farmers should calculate the true interest cost (the annual percentage rate or APR) on any loans so that they can compare various types of financing arrangements. Recent changes in Truth-in-Lending rules mean that these interest costs may not always be calculated and disclosed to farm customers. In addition to interest charges, other costs of borrowing include fees for making the loan (frequently called points), appraisal or inspection charges, compensating balances or stock purchases, and credit life insurance premiums.

Structuring the repayment on a loan is one of the most important tasks in safe and successful use of credit. A farmer should estimate the annual repayment capacity and determine how much debt can safely be borrowed before a commitment is made to a loan obligation. Setting up reasonable repayment schedules may be the single most important factor in keeping the use of credit in the farm business on a sound foundation.

Questions and Problems

1. What are the two major sources of funds for most farm operations?
2. Why is savings the most important source of equity funds?
3. What are the three major categories of lending institutions in the agricultural credit market? List the major lenders in each of these categories.
4. Visit with a commercial bank in your community about its lending program for farmers.
5. Describe the Federal Land Bank and the types of loans Federal Land Banks make to farmers.
6. What is the role of the Farmers Home Administration in the U.S. agricultural credit market?
7. What is the difference between an operating lease and a capital lease?
8. Discuss the advantages and disadvantages of leasing compared to purchasing capital assets.
9. List the seven ways that credit can contribute to the improvement of net income in a farm operation.
10. Discuss the "Three R's of Credit".
11. What is a financing statement and why does a lender use such a legal document?
12. Discuss one of the five factors that influence the cost of credit.
13. Contact a lender and find out the terms of the loan for an automobile and for

a machine purchase. Calculate the true annual percentage rate for these loans using the direct ratio equation.

14. Discuss how you might estimate the repayment capacity of a farm business.

Further Reading

American Bankers Association. *Agricultural Credit Analysis Handbook.* Washington, D.C.: American Bankers Association, 1975.

Barry, Peter J., John A. Hopkin, and C. B. Baker. *Financial Management in Agriculture.* 2d ed, Danville, Ill.: Interstate Printers and Publishers, 1979.

Boehlje, Michael D., Everett Stoneberg, and H. B. Howell. "Agricultural Credit and the Farm Business." Cooperative Extension Service FM–1541, Iowa State University, Ames, Iowa, 1979.

Brake, John R., and Emanuel Melichar. "Agricultural Finance and Capital Markets." In Lee R. Martin, ed. *A Survey of Agricultural Economics.* Minneapolis: University of Minnesota Press, 1977.

Bunn, Charles, Harry Snead, Richard Speidel, and Kenneth R. Redden. *An Introduction to the Uniform Commercial Code.* Charlottesville, Va.: Michie, 1964.

Frey, Thomas L., and Danny A. Klinefelter. *Coordinated Financial Statements for Agriculture.* Skokie, Ill.: Agri-Finance, 1978.

James, Syndney, and Everett Stoneberg. *Farm Accounting and Business Analysis.* 2d ed., Ames: Iowa State University Press, 1979.

Lee, Warren F., Michael D. Boehlje, Aaron G. Nelson, and William G. Murray. *Agricultural Finance.* 7th ed., Ames: Iowa State University Press, 1980.

Melichar, Emanuel, and Amy Brooks. *Agricultural Finance Data Book.* Washington, D.C.: Federal Reserve System, Monthly Series.

Penson, John B., Jr., and David A. Lins. *Agricultural Finance,* Englewood Cliffs, N.J.: Prentice-Hall, 1980.

Polakoff, M. E., et al. *Financial Institutions and Markets,* 2d ed. Boston: Houghton Mifflin, 1983.

Smith, Frank, J., and Ken Cooper. "The Financial Management of Agribusiness Firms." Agricultural Extension Service Special Report 26, University of Minnesota, Minneapolis, Minnesota, 1975.

Stelson, Hugh E. *The Mathematics of Finance.* Princeton, N.J.: Van Nostrand, 1963.

Swackhamer, G. L., and R. J. Doll. *Financing Modern Agriculture–Banking Problems and Challenges.* Kansas City: Federal Reserve Bank, 1969.

U.S. Department of Agriculture. *Agricultural Finance Outlook.* Economics, Statistics and Cooperative Service, Annually.

CONTROL

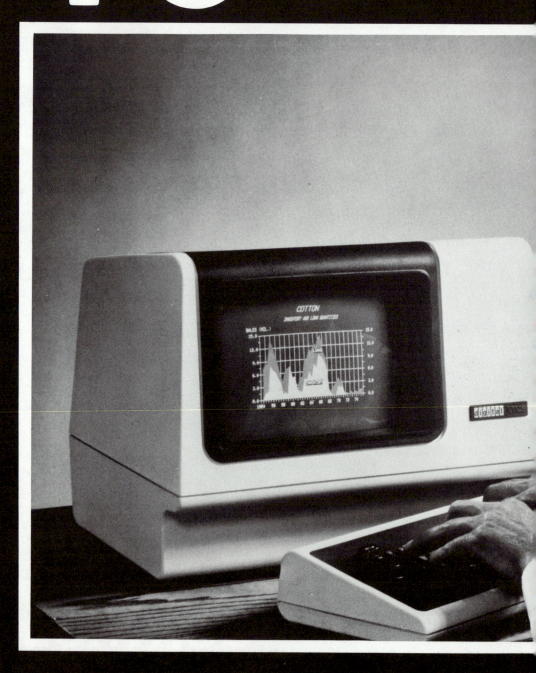

CONTROL CONCEPTS
AND PROCEDURES

CHAPTER CONTENTS

The early chapters emphasize procedures to use in planning the farm business within the expected economic, institutional, social, and climatic environment. Implementation of the plan may result in deviations between actual performance and budgeted expectations. With the passage of time, new information and changes in the environment will make adjustments in the plan desirable, perhaps even imperative. A useful way to deal with the dynamics of managing the farm business is to measure performance of the business (and the plan) through time and to make adjustments as needed to attain the operator's and family's goals. The process of measuring performance, comparing measured performance with the standards established in the plan, and making adjustments to achieve the desired goals is referred to as control.

The control process is such an automatic phenomenon in daily life that it is often overlooked or taken for granted. For example, the human body uses a system of pores and perspiration to regulate body temperature. The thermostat used to control room temperature is another example. The thermostat monitors room temperature, causes the furnace (air conditioner) to be turned on when the temperature drops below (rises above) a prescribed level, and shuts off when the specified temperature has been attained. These procedures are rarely noticed in day-to-day activity when they are working within prescribed limits.

Control implies adaptive behavior. A plan is developed to achieve certain goals, and the status of the operation is monitored over time. The feedback from monitoring is compared with standards specified in the plan to decide when adjustments are needed to achieve the goals. Then through changes or adaptions in the implementation process, adjustments are made to better achieve the specified goals.

This chapter discusses control concepts and the development of a control system for the farm business. The first part of the chapter describes the control process and discusses considerations in carrying out the control function of management. The discussion emphasizes that development of a control system is part of a complete management information system for the business. The final sections outline a strategy that can be followed in developing a control system for the farm business.

THE CONTROL PROCESS

The three parts of the control process were introduced in Chapter 1. They are: (1) establish standards specifying the expected performance; (2) measure the actual performance and compare it to the standards; and (3) correct the deviations from the standards and plans.

Establish Standards

The results of actual performance must be compared to some quantitative standards or they are meaningless statistics. The standards can be expressed in either physical units (such as pounds of feed per pound of milk or meat produced, pigs produced per litter, and yield per acre), or monetary units (such as average cost per hundred pounds produced, average price received, and rate of return on investment). But it is important that the specified standards be written so that a comparison can be made. Producing "as much as possible" or striving for "a high rate of return on the investment" and "a good standard of living" may avoid the difficult task of specifying precise quantitative standards, but they seldom provide an adequate comparison for control.

Establishing standards involves selecting the appropriate characteristics to monitor and the goal or objective for each characteristic. Only a limited number of characteristics can be monitored given the time and cost involved. Thus, it is important to select a limited number of physical and financial characteristics that indicate if the system is operating at acceptable levels of performance. These characteristics must not only be related to overall financial or business performance, but they should also be clearly linked to and be influenced by specific production, marketing, and financial decisions.

Standards must be realistic, yet must provide a challenge. The input-output coefficients used in planning are normally based on past performance in the same business and the level achieved by other farmers operating under comparable conditions. The average price levels based on outlook information and financial performance budgeted during the planning process are the basic source of such standards for control. Farmers frequently use conservative input-output ratios and prices in farm planning to avoid overstating the expected profitability of the business and its ability to meet cash outflows. In these cases, the operator may want to select more challenging levels for the control standards. Thus, the standard selected for control purposes should be based on, but not necessarily be identical to, the value used in planning. To be more specific, it is usually appropriate to compare actual performance (1) to the performance of other operators in the same period of time, (2) to the manager's performance during the past production period, and (3) to the budgeted level for the current period which has been chosen to achieve specific goals of the firm. This three-way comparison should avoid looking at only one inappropriate, unattainable, or poor performance level.

Measure Performance

Measurement of performance and comparison of measured performance with the standards is the indispensable second step in the control process. Three attributes of measurement are important: (1) timeliness; (2) use of appropriate units; and (3) reliability. There is little value in receiving information at the end of the year or even the end of the month if the correction has to be made much earlier to achieve the planned outcome. For example, a system that weighs and reports milk production daily is more timely for the correction of production problems than a system that reports production levels monthly. The measurement should be reported in units relevant to the processes or outcomes which the numbers represent. This may mean reporting not only the number of units but also key measures of quality, such as the number of pigs by weight and grade. For example, if the outcome to be measured is annual volume produced in the hog enterprise, total hundredweight of a particular quality of hogs is more meaningful than just number of head.

Data that are reliable must be accurate and consistent. This may require more objective measurement systems such as scales for weighing livestock and grain or sampling for nutrient composition or moisture content of feedstuffs. And a system of recording such data must be developed so that it can be transferred accurately to the individual responsible for the enterprise or decision.

Take Corrective Action

As mentioned earlier, deviations from the standards can be corrected in three ways: (1) change the plan; (2) adjust the implementation; and (3) change the goals. The appropriate combination of these methods to use in correcting deviations can be determined by considering the following questions. Have the standards adopted been arrived at objectively after considering all of the alternatives? Can adjustments be made in the implementation process? Do the changes being proposed treat the cause of the problem rather than the symptoms? In some cases, corrections for the current period may involve changing the implementation of the plan or changing the standard, while in the longer run the manager can solve the problem by changing the plan. In other cases, the source of the deviation may be outside of the manager's control, unavoidable, and best dealt with by developing a contingency plan to follow when the problem occurs.

As the discussion of the control function suggests, the planning, implementation, and control processes are interrelated. After the annual plan is developed, it is modified, refined, and adapted frequently as information becomes available through the control system on the changing environment and the actual performance of the firm. Information requirements to implement and control the current year's operation are largely dictated by the annual plan and environmental conditions. Furthermore, information developed to monitor this year's plan may also be useful in planning for the following year.

These interrelated information needs suggest that the discussion should be broadened from the development of control systems to the development of an information system that serves all managerial functions. While the development of a complete management information system (MIS) is beyond the scope of this text, introducing the basic concept makes it possible to relate the control system to other parts of an overall MIS for the farm business.

CONTROL AND THE FIRM'S INFORMATION SYSTEM

Information is needed to support decision-making in each of the three functional areas. All farm businesses must process certain data on receipts, expenses, investments, and other financial transactions for tax reporting and the fulfillment of other institutional requirements. Farm operators also develop procedures to collect, record, store, and analyze data on prices, production alternatives, and progress of the business. These data are used to support long-term as well as short-term planning, implementation of the plan, and control of the operation.

Activities conducted within the organization to provide information for managerial decision-making are commonly included under the heading of a management information system (MIS). These activities may be carried out by the manager or by others and are designed to provide information that is of potential value in making management decisions.

The term "management information system" can be defined in several ways. One approach is to describe the physical components of the system. Management information systems may integrate computer hardware and software, manual procedures, decision models, and human time to provide information in support of operations, management, and decision-making functions in the business. Few farmers in earlier years included a computer component in their management information systems.[1] However, the availability of relatively low-cost microcomputer equipment makes consideration of the computer component for data storage and processing realistic for commercial family farms at the current time.[2]

A management information system can also be viewed from the perspective of the managerial responsibilities to be served. The data and tools used in production, market, and financial planning are part of the MIS. It may include computational procedures to prepare enterprise budgets and projected financial statements, linear programming procedures for least cost feed formulation and whole-farm planning, and price prediction procedures. In addition to the tools, the MIS would include data storage (whether a file cabinet or a computer disk) containing the appropriate agronomic, animal husbandry, engineering, and economic data for planning the business. The MIS would also include data on inventory levels of inputs, as well as the relative cost of inputs obtained from alternative sources. The maintenance of production records, financial accounts, and data on market price movements is required for control of the business. Procedures to process these data and report them in a timely manner for control decisions would also be part of the MIS for control.

The MIS is required to produce reports for the manager(s) of the business. It is also used to produce the necessary reports required by others, such as the loan officer, the crop and livestock reporting service, and the Internal Revenue Service. In short, the MIS should facilitate the collection, classification, and storage of appropriate data, and retrieve and process the data as needed to provide information useful in managing the business.

Management information systems are based on several underlying concepts. Perhaps the most funadmental is the distinction between data and information. This distinction provides the basis to develop a conceptual model of the firm's management information system.

Data Versus Information

The two words "data" and "information" are commonly used interchangeably in informal discussions. However, the term "data" is used to refer to

---

[1]For example, see L. M. Eisgruber, "Managerial Information and Decision Systems in the U.S.A.: Historical Developments, Current Status and Major Issues," *American Journal of Agricultural Economics* 55 (1973): 930–37, for a discussion of the relatively small numbers that had used computers as an aid in decision-making as late as the early 1970s.

[2]Considerations in implementing management information systems using microcomputers are discussed by C. L. Pugh, "Microcomputer Facilities for Information Systems," Chapter 6, in *Information Systems for Agriculture,* ed. by M. J. Blackie and J. B. Dent (London: Applied Science Publishers, 1979), pp. 113–29, and Steven T. Sonka, *Computers in Farming: Selection and Use* (New York: McGraw-Hill, 1983).

input collected, while the term "information" is reserved for data that have been processed for use in making decisions. More specifically, data are symbols that represent quantities, actions, prices, and other phenomena that are the raw material for information. Data are not information. Information is data that has been processed into a form that is meaningful to the recipient and is of real or perceived value in making current or prospective decisions.[3] Thus, information influences behavior (decisions). It has "surprise value" or "intelligence value." For example, the numbers from the pig crop report represent data, whereas the analysis of the impact on future hog prices represents information. Likewise, pigs weaned per litter or yield per acre are data, but the analysis of why these physical performance ratios are above or below expected standards will assist in making decisions and thus is information. In this and most other examples, the data collected require processing before they are information useful in making management decisions.

This distinction between data and information is useful in setting up a control system and identifying what data to collect. In essence, if a piece of data cannot be processed into information and thus be useful in improving a decision, it should not be part of the control system. Furthermore, if the cost of collecting and processing data into information exceeds the economic benefit of the improved decision that can be made based on this information, that data also should not be part of the control system. For example, weekly data on the weight of feeder cattle during the feeding period are useful in monitoring the growth process and rate of gain, but if the operator cannot accommodate weekly changes in the ration and other management practices, the weekly weight data will not improve decisions and therefore would not be useful. Monthly data on weight gains would require less time and expense to gather and may be adequate to adjust rations and improve decisions. In a similar vein, scouting programs to monitor pest populations, moisture conditions, and plant growth may provide data but not necessarily information. If the processed data available at no cost improve (the profitability of) decision-making, it is information. However, if the cost of the scouting service is higher than the improved profits that can be generated by the better decisions, the scouting reports should not be purchased.

A System

A system can be defined as an orderly arrangement of interacting parts that operate together to achieve an objective or purpose. For example, a computer system is composed of the equipment that functions together to carry out the prescribed computer processing. A grain storage system might be defined as the equipment required to move grain into storage,

---

[3]Gordon B. Davis, *Management Information Systems: Conceptual Foundations, Structure, and Development* (New York: McGraw-Hill, 1974), p. 32.

Figure 16.1. System with *m* inputs and *n* outputs.

provide storage space, maintain the condition of stored grain, and remove grain from storage as prescribed. An accounting system is composed of certain records, rules, procedures, equipment, and personnel which record data and prepare reports. A management information system is composed of computer hardware and software, manual recording procedures, decision models, and human time. In these examples, as well as others, a system is defined by identifying the various components and the interactions between them.

The general model of a system is illustrated in Figure 16.1. In this simple form, the system is composed of input(s), processor, and output(s). Data represent the major input, and information is the primary output for an MIS.

Each system is composed of subsystems, which are in turn made up of other subsystems. Thus, one can subdivide a complex system, such as the MIS for a farm business, into subsystems of manageable size. Figure 16.2 illustrates a system with two subsystems. The interconnections or interfaces between subsystems (as shown in Figure 16.2) represent the flow of data from one subsystem to another. The MIS for a farm firm can be divided into appropriate subsystems to trace the flow of inputs and products that are useful in planning and controlling the operation.

The basic model of input, process, and output described in Figures 16.1 and 16.2 does not provide for feedback and control of the process—two important elements in developing an MIS. A feedback loop illustrated in Figure 16.3 must be added to introduce control of the system. The control loop involves four decisions and components: (1) certain characteristics to be monitored must be selected; (2) one or more means of monitoring (sensing) the characteristics are developed; (3) a control unit compares the level of the characteristics with the standards established; and (4) the appropriate adjustments are activated to bring or keep the system

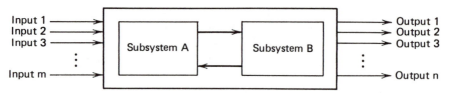

Figure 16.2. System with two subsystems and interconnections.

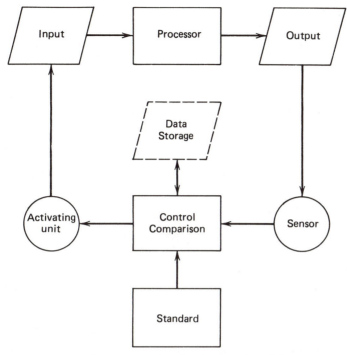

Figure 16.3. System with feedback control and data storage.

under control. This process is typically referred to as *negative* feedback because it seeks to reduce fluctuations around the norm or standard. However, the astute manager may (through either experimentation or accident) discover an alternative combination of inputs that results in more desirable levels of output than the system currently achieves. In this case, the manager will probably want to adjust the plan and the control standards to adopt this *positive* feedback. Thus, managers are interested in developing management information systems that include both positive and negative feedback.

Feedback control loops can be either open or closed. A closed loop is an automated control such as a thermostat or computer-controlled process. While farm managers deal with some examples of closed loop control, such as temperature and humidity control of dryers and certain livestock facilities, we will be dealing primarily with a human decision-maker and open loop control.[4]

The MIS for a farm business follows the basic model described in

---

[4]More technically, an open loop system is one with random disturbances, whether or not the disturbances result from human intervention. However, control loops with human intervention are open loop systems.

Figure 16.3 of input, processor, and output. The system receives inputs of data and instructions. Reports and information are the outputs of MIS. This basic model can be made more useful by broadening the concept in two ways. First, it is important to sense not only the output of the process, but also the progress of the process (such as the growth of crops and livestock) and certain environmental variables (such as weather conditions and prices). Second, the information system may need to collect data over time and then process it when information is needed for decision-making. This can be accommodated by adding data storage to the system as shown by the broken line area (labeled data storage) in Figure 16.3. This basic model of an MIS is useful both to model an overall MIS for the busines and for application to a specific portion of the business.

The Farm as a System

In designing a management information system for the farm business, it is useful to consider the farm firm itself as a system operating within an environment (Figure 16.4). The farm is a complex system comprised of inputs, activities or enterprises, and outputs. A farm receives a large number of inputs, some of which are controllable by the manager and some of which are outside the manager's control. Examples of controllable inputs include purchased inputs, such as feed, seed, fertilizer, borrowed capital, and machinery purchases, while some of the uncontrollable inputs are rainfall and other weather variables, air pollution, input and product prices, and changes in the institutional setting. The uncontrollable inputs are frequently uncertain and largely unpredictable. The outputs include the production of commodities for sale off the farm (to the environment). Other outputs may be unintended and undesirable, such as disease problems and runoff water polluted with silt, fertilizer, and pesticides.

The environment of the farm firm has four major dimensions: (1) the physical environment composed of seasonal weather conditions and the variability of these conditions; (2) the economic environment which determines the relative, as well as the absolute, level of input and product prices; (3) the institutional environment which prescribes (a) rules for the use of debt capital, (b) rules for payment of taxes, (c) limits on water, ground, and air pollution, (d) conditions under which the government will purchase farm products above open market prices, (e) legal rights and obligations, and (f) conditions to transfer wealth and capital assets between generations; and (4) the social environment which provides additional constraints on the business operation.

Farm operators must develop an MIS for the business that monitors both the operation of the business and the environment. The farm operator's MIS should identify relevant data, collect the data, store the data until needed, and process the data into a form that can be used in developing

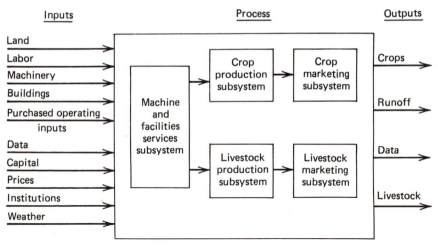

Figure 16.4. The farm system.

plans, implementing plans, and controlling the operation through time to accomplish the firm's goals.

Formal Versus Informal Information Systems

One of the major decisions in developing a management information system is deciding how much of the MIS should be formalized (written). All farm operators have an informal information system composed of discussions with neighbors about varieties, pest control measures, expected prices, and other information needed in decision-making. The use of telephone calls, discussions with extension workers, and material gleaned from newspapers and farm magazines are additional examples of the informal information system. The same farmers may also have formalized informations systems, such as Dairy Herd Improvement (DHI) records, written records of fertilizer application history by field, enterprise accounts, and financial statements.

The replacement of all informal with formal information systems would be costly, overly complicated, and foolish. A management information system should be formalized in those areas where there are opportunities for large returns on the time and effort expended. The payoff may be greater from considering a small deviation in a large cost category than a large deviation in a small cost category.

The equal marginal return principle discussed in Chapter 5 should be applied in allocating the resources (time, effort, and finances) a firm has among alternative uses. Thus, the firm should allocate time, financing, and effort for developing management information systems to the most profitable uses, with the value added by the marginal unit of effort equal to each of its alternative uses. It may be difficult to apply this concept in a very detailed manner, but keeping it in mind should enable managers to avoid formalizing parts of the information system that appear to be of little benefit.

THE STRUCTURE OF A MANAGEMENT INFORMATION SYSTEM FOR CONTROL

Several approaches have been suggested to structure a management information system for a business firm.[5] The discussion here is limited in two major ways. First, the discussion is limited to developing the portion of the management information system used for control. Second, rather than review the alternatives, a strategy is suggested that a farm operator can use to formalize those parts of the control system with the highest payoff. This can be accomplished by identifying the area having the highest ratio of the value of data and information to the cost of obtaining that information, followed by the second highest, etc., with an opportunity to revise the control system in subsequent periods if it becomes apparent that more profitable opportunities for data acquisition and processing exist.

The discussion of whole-farm planning (Chapter 6) started with a determination of the personal/family goals and the business goals of the operation. The personal/family goals included the amount of money required for family living as well as the social and community-oriented goals of the family. The business goals relate to specific production, marketing, and financial goals. The production and marketing goals are tied to measures of efficiency, while the financial goals are tied to measures of profitability, liquidity, and solvency. These goals provide the basis to select the preferred plan for the short, intermediate, and long run. If these goals are to be realized, it is appropriate to relate the control measurements, standards, and procedures to the same goals. This suggests that a control system for a farming operation must monitor production, marketing, and financial performance.

Enterprise or Activity Control

The enterprises represent the major profit and cost centers of the business; an enterprise was defined (Chapter 2) as any portion of the farm business that can be separated from the rest of the operation by accounting procedures that can allocate receipts and expenses to specific activities. Enterprises were classified earlier into three types. *Production enterprises* are those that produce a marketable product, such as the major crop and livestock products. *Service enterprises,* such as machinery, livestock production facili-

---

[5]An overview of alternative ways to structure a management information system for all functions of management including control is presented by G. Davis, *op. cit.,* Chapter 5.

ties, and other buildings, provide a service used by the production enterprises but do not normally produce a marketable product. The third category, *marketing enterprises,* includes the activities of purchasing and storage of inputs required by the farm, and the marketing of farm products.

It is appropriate to note one additional type of activity or "enterprise" that should be considered in developing the control system. The construction of a new livestock facility, coordination of livestock availability with completion of the facility, and learning how to operate the facility can be considered investment or development activities. Likewise, beginning to operate a newly rented or purchased farm may also require learning and developing special skills. The control system must facilitate and monitor the investment and development process to make sure that production schedules are met and that the expected learning and improved performance is in fact taking place.

The manager must select the variables to monitor for each of these enterprises within the business. Some operators have an incomplete control system that provides little information that can be used to improve performance. This may be a particular problem on farms managed by one individual who may also provide the major part of the labor supply and has limited time for formally structured monitoring and control. Other operators fall into the trap of developing a control system that requires data on more aspects of the business than can reasonably be collected throughout the year. Thus, it may be appropriate to formally structure major aspects of the control system for some enterprises, but relatively little development of formalized control systems may be appropriate for others.

Several steps should be followed in developing an enterprise control subsystem. The terminology we will use relates more specifically to production enterprises, although, as we will see later, the same basic steps apply to marketing, service, and developmental activities or enterprises as well. The starting point for development of an enterprise control system is the definition of the specific activity (production, market, service) and its enterprise budget used in the process of planning. The enterprise budget for production systems details the input levels required and the approximate timing of use. The quantity and timing of product expected, and the projected cash outflows associated with the enterprise are also presented in the enterprise budget. Similar estimates of receipts and expenses can be prepared for market and service enterprises as the basis for control. Given the definition of the activity and the enterprise budget, the development of an enterprise control system normally follows six steps.

1. Break the enterprise into subsystems that are meaningful for identification of the important inputs and outputs to monitor. For example, in livestock production it may be important to break the total feeding period as used in the enterprise budget into segments and establish a means of monitoring performance by stage of the biological development of the product.

2. List the inputs and outputs to monitor for each subsystem in approximate order of importance. The tendency is to monitor more items than can be economically considered. Listing the items to monitor in order of importance provides a means of reducing the list for the enterprise to a more manageable length.

3. Specify the monitoring interval for each input and output item selected. A comparison of performance with the standard must be completed in a timely manner if corrective action is to be taken.

4. Identify the appropriate means of monitoring (or sensing) each item selected. The data collected must be reliable if they are to be used for decision-making. In many cases, more reliable data can be collected only at greater cost. Thus, identifying the appropriate means of monitoring price, rate of growth, and environmental variables may not represent an easy and obvious choice. The concepts of the value compared to the cost of information discussed earlier can be useful in determining the level of reliability to choose.

5. Specify the standard and the "in-control" range for each variable being monitored. A standard is usually explicitly used in developing the enterprise budget. The "in-control" range must be specified to indicate whether corrective action is or is not to be taken; deviations outside this range indicate unacceptable performance, and corrective action is necessary. The in-control range is usually based on the judgment and appraisal of the manager. However, it is important to consider the implications of the in-control range for the cash flow and profitability of the enterprise before it is chosen. In some cases a wide in-control range may not have a significant impact on profitability, whereas in other cases a narrow range is necessary to trigger corrective action before the financial consequences become severe.

6. Establish rules of action to apply when the observed variable is outside of the in-control range. These are the actions or contingency plans to follow when the actual performance of the system is outside of the planned performance range.

Financial Control

The system of financial controls is necessary to monitor the overall performance of the farm business. The monthly cash flow has been projected for the coming year as part of the planning process. The income statement for the coming year and the ending balance sheet have been projected as well. The primary component of a financial control system is an accounting and recordkeeping system that permits the business to compare actual periodic (monthly, quarterly, etc.) cash inflows, outflows, and other financial measures of performance with those projected. The comparison of actual to projected cash flows and monitoring of changes in inventory and capital items provide the basis to determine if the firm's overall financial objectives are being achieved.

The financial control system ties the parts of the business together, allowing the operator to observe the financial impact of changes in an enterprise or activity on the entire operation and the achievement of the firm's goals. This information is important to the operator of a farm who provides most of the labor and management as well as to the manager of an operation who has assistants responsible for separate parts of the business. The financial control system also highlights areas of the business where over- and underachievement of the financial projections are occurring. This suggests where the payoff for additional components of the control system might be relatively high.

Revision of the Control System

An additional step should be noted in developing a control system for the farm business. This step requires testing the control system and revising it as required. This testing process involves assessment of the reliability of the data being collected by the control system and evaluation of whether the data are or can be used to make more profitable decisions. The control system may be incomplete in that it can detect deviations between plans and actual performance, but provides no indication of the source of the problem or what corrective action should be taken. Or the system may be unreliable in that the performance characteristics monitored are being measured inaccurately or do not provide a useful indication of actual performance. Maybe the control system is so complex that it is impossible to collect the data and process them into information. The control system must be reviewed and tested for accuracy and usefulness. Inclusion of this step provides a basis to improve on the control system over time. That is, this step provides the mechanism for adaptive control of the adaptive control system!

THREE TYPES OF CONTROLS

Almost all controls can be categorized into one of three types: preliminary controls, concurrent controls, and feedback controls. Preliminary controls focus on the prevention of deviations from the plan. This is accomplished by identifying potential problem areas and specifying inputs that are effective in preventing the occurrence of the deviation. For example, a farmer may use a more expensive pesticide that is effective over the entire potential range of rainfall and temperature conditions to replace a material that is only effective over the typical environmental conditions. Managers can reduce the need for other types of control by correctly anticipating potential problems and developing those preliminary controls that can be economically applied.

Concurrent controls enable adjustments to be made during the event.

They are based on monitoring the system and adjusting the timing, level, and method of using inputs to maintain the quantity and quality at standard levels. For example, irrigators typically monitor the soil water level over the growing season and schedule irrigations week by week based on anticipated crop needs.

Feedback controls are concerned with improving the next attempt. Farm operators frequently observe a deviation from the plan that they may (or may not) be able to solve with concurrent controls but that could be handled more effectively in future periods in another way. Thus, historical data and experience are used to guide development of the plan for future periods, including the development of appropriate preliminary controls, budgets, and standards. Feedback control may also suggest changes that should be made in the monitoring system used and the concurrent control procedure.

Categorizing control procedures into these three types reemphasizes the interrelations of the management functions in the operation of a business over time. Preliminary control emphasizes that planning should be conducted with the control function in mind. More emphasis on the use of preliminary controls may reduce the number of performance characteristics that must be observed and the monitoring frequency, making concurrent control operations less complex. Feedback control emphasizes the use of historical data to improve planning, implementation of the plan, and its control in future periods.

The appropriate combination of the three types of control procedures depends on the cost of the controls, the likelihood of the deviation occurring, the potential loss if the deviation occurs, and the producer's attitude towards risk. Preliminary controls should be emphasized if they are low cost relative to the expected value of the loss, particularly if the deviation cannot be efficiently corrected with concurrent controls. Preliminary controls should also be considered when the loss would be very devastating to the financial success of the business. For example, the purchase of hail insurance or the use of precautions to prevent catastrophic disease in a confinement livestock operation may be justified because of the consequences of a very large loss (even though its probability may be small). Managers can select the combination of preliminary and concurrent controls based on the net benefits for each means of control and the risk considerations.

SOME CONTROL STANDARDS AND TOOLS

Examples of control subsystems for individual production enterprises will be presented in Chapter 17, with examples of marketing and financial control subsystems presented in Chapter 18. However, it is useful to comment briefly here on the development of standards and tools to be used for enterprise and financial control.

The production enterprises in a farm business are based on biological processes. For many of these, control standards should be stated as a rate per unit of time or a quantity over time. Thus, biological growth charts are a source of standard growth rates for crops and livestock. For example, lactation curves that show the amount of milk production per day over the ten-month lactation make it possible to compare the milk yield with the standard on any day of the lactation, even though the standard changes daily.

With respect to control tools, the budgets and planning techniques discussed earlier provide the raw material to set up the specific monitoring devices. The enterprise budgeting technique combined with production records provides the basis for monitoring production efficiency. The enterprise budget contains the production and price standards selected for the enterprise during planning. The budget provides standards on the amount of inputs to use, approximately when the inputs are to be applied, when the output is to be produced, and approximately when it will be available to deliver. It also specifies input and production price standards as well as total cash expenses by month. Thus, the enterprise budget is a key source of standards to use in the control process. The physical performance of the farm operation can also be monitored with the help of production schedules, checklists, time lines that summarize the appropriate sequence of activities, weight sheets, reproduction records, flow charts, and any other mechanism that provides a means of monitoring actual physical activities.

The market and financial control system is based primarily on market records that summarize past and expected prices, quantity and timing of input purchases, and product sales. The financial control system is comprised of a cash flow record and actual income and balance sheet statements structured in a format to make comparisons between budgeted expectations and actual performance. Thus, the tools of a control system are many and varied, but the primary tool is the written summary of physical, financial, or marketing performance structured in such a fashion as to make comparisons between actual performance and budgets or expected performance possible.

In many cases, the control process can be facilitated through the use of computerization. With the availability of microcomputers and the rapid development of software for special application to agriculture, the control process can be accomplished more efficiently and effectively using this tool.

The use of various control tools and techniques for production, marketing, and financial control will be discussed in the next two chapters, with a focus on the concepts and components of the various control systems.

These control systems can be implemented using manual or computer-based data acquisition and information processing procedures.

Summary

The control process involves measuring performance, comparing measured performance with the standards established in the plan, and making adjustments to achieve the desired goals. The time available to monitor business performance is limited, and only the most important input and output variables can be monitored. Performance of these important characteristics must be measured in a timely manner and measured accurately enough to be useful in controlling the operation. Because it is difficult to select the appropriate quantitative standard, performance should be compared (1) to the performance of other operators in the same period of time, (2) to the manager's performance during the past production period, and (3) to the budgeted level for the current period. The manager can take corrective action by changing the plan, adjusting the implementation of the plan, and changing the goals.

Information is needed to support decision-making in each of the three functional areas. A management information system integrates computer hardware and software, manual procedures, decision models, and human time to provide information in support of decision-making for all three functions. A control system designed to gather data and develop information for decision-making is one part of the management information system for the farm business.

Two fundamental concepts are important in developing management information systems. First, the distinction between data and information provides the conceptual basis to identify the types of data that should be collected. Data refer to symbols which represent quantities, actions, prices, and other phenomena that are the raw material for information. Information is data that has been processed into a form that is of real or perceived value in making current or prospective decisions. Obviously, there is little benefit in collecting data that will not be of use in developing information for decision-making.

The definition of a system is a second underlying concept of a management information system. A system is defined as an orderly arrangement of interacting parts that operate together to achieve an objective or purpose. The definition of the management information system is an example. The basic model of the system involves input, processing, and output with a feedback and control loop which includes monitoring the output, providing for storage of data collected, processing the data, comparing the processed data to standards that have been set, and initiating corrective action. This basic model is useful in designing the management information system for a farm business, as well as developing control systems for specific parts of a business.

Control systems for the farm business can be established by developing control systems for the major profit and cost centers in the business—the production, marketing, and service enterprises that make up the total farm business. The enterprise control system is tied to the financial control system of the farm business through the enterprise budget. The enterprise budget provides a way of relating the physical and financial standards at the enterprise level to the cash flow and inventory levels for the total farm business. The primary component of a financial control system for the firm is an accounting and recordkeeping system that permits the business to compare actual periodic (monthly or quarterly) cash inflows and other financial measures of performance with those projected. Comparison of actual to projected cash flows and monitoring the changes in inventory and capital items provide the basis to determine if the firm's overall financial objectives are being achieved.

Almost all controls can be categorized into one of three types. Preliminary controls focus on the prevention of deviations from the plan by identifying potential problem areas and specifying inputs that are effective in preventing the occurrence of the deviation. Concurrent controls are based on monitoring the system and adjusting the timing, level, and method of using inputs to maintain the quantity and quality of output at standard levels. Feedback control recognizes that in some cases the operator may be unable to solve a deviation from the plan with concurrent controls. This third form of control recognizes that the deviation might be handled more effectively in future periods in another way. Thus, historic data and experience are used to guide development of the plan for future periods. Categorizing control procedures into these three types reemphasizes the interrelationships of the management functions in the operation of the business over time.

Questions and Problems

1. Give an example of a mechanical control system and identify the components of control.
2. Would an open loop control system be more sensitive than a closed loop system? Why or why not?
3. Describe an example of closed loop control in the biological sciences.
4. Why is the establishment of goals or objectives important in planning a system?
5. Define the terms "system" and "subsystem." How do the two differ?
6. Give examples of negative and positive feedback for agricultural production, marketing, and service enterprises.

7. Define the subsystems for a farm producing one crop, such as wheat, in a nonirrigated farming area. Define the boundaries between the production, marketing, and service enterprises you identify.
8. List the major inputs and outputs to monitor for each of the subsystems identified in question 7.
9. How would you develop relevant standards for the inputs and outputs selected for monitoring in question 8?
10. Describe the management information system that is used by a farm operator you know. To what extent is it formalized?
11. Give an example of each of the three types of control for a production enterprise, a marketing enterprise, and a service enterprise.

Further Reading

Blackie, Malcolm J. "Management Information Systems for the Individual Farm Firm." *Agricultural Systems* 1(1976):23–36.

Blackie, Malcolm J., and J. B. Dent. *Information Systems for Agriculture.* London: Applied Science Publishers, 1979.

Davis, Gordon B. *Management Information Systems: Conceptual Foundations, Structure and Development.* New York: McGraw-Hill, 1974.

Eisgruber, L. M. "Managerial Information and Decision Systems in the U.S.A.: Historical Developments, Current Status and Major Issues." *American Journal of Agricultural Economics* 55(1973):930–37.

Fuller, Earl I. "Microcomputers: Useful in All of Agricultural Economics and Extension." *American Journal of Agricultural Economics* 64(1982):978–88.

Fuller, Earl I. *Think Systems: Putting a Dairy Farm Management Information and Control System Together.* St. Paul: University of Minnesota Agricultural Experiment Station Unpublished Paper, 1983.

Fulmer, Robert M. "Control Concepts." In *The New Management.* New York: Macmillan Publishing Company, 1974, Chapter 12, pp. 255–76.

Johnson, Richard A., Fremont E. Kast, and James E. Rosenzweig. *The Theory and Management of Systems.* 2d ed., New York: McGraw-Hill, 1967.

Kennedy, John S. "Control Systems in Farm Planning." *European Journal of Agricultural Economics* 1(1974):415–33.

Sonka, Steven T. *Computers in Farming: Selection and Use.* New York: McGraw-Hill, 1983.

Thompson, S. C. "Canfarm: A Farm Management Information System." *Agricultural Administration* 3(1976):181–92.

PRODUCTION
ENTERPRISE CONTROL

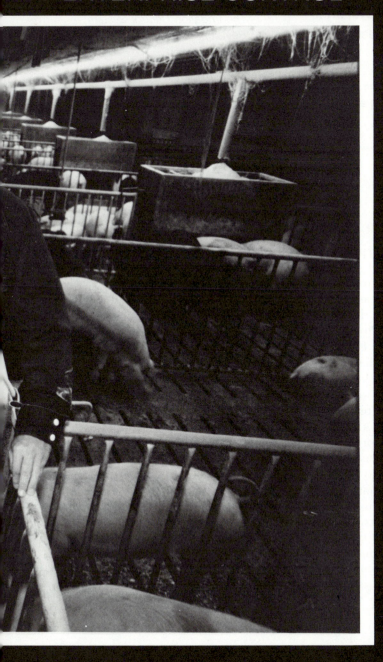

CHAPTER CONTENTS

The purpose of enterprise control is to help the business achieve its goals. The previous chapter outlined the concepts and components of a control system and related it to the firm's objectives. Now we consider the development of production and service enterprise control systems for the farm business. Integrating the enterprise control system into the firm's financial control system is the focus of the next chapter; this integration enables the manager to trace the effect of physical and biological performance on the achievement of the desired financial objectives.

The enterprise budget is the basic tool of production and service enterprise control as well as the vehicle to connect enterprise control to the firm's financial control system. The enterprise budget includes a listing of the expected inputs and outputs, their prices, and the receipts and expenses for the enterprise. This detail on the quantities of inputs and outputs provides the basis to develop a physical and biological control system that can be related to the cash flow, and in turn to the financial performance of the combination of enterprises that comprise the firm.

The discussion in this chapter follows the six steps in the development of an enterprise control system described in Chapter 16. They are:

1. Delineate the production system and break the system into subsystems that are meaningful for identification of the important inputs and outputs to monitor.

2. List the inputs and outputs to monitor for each subsystem in their approximate order of importance.

3. Specify the monitoring time interval for each input and output selected for monitoring.

4. Identify the appropriate means of monitoring (or sensing) each input and output selected.

5. Specify the standard and the "in-control" range for each variable being monitored.

6. Establish rules for action to apply when the observed variable is outside the "in-control" range.

Examples are provided from crop and livestock production for each step. The discussion is intended to illustrate how a manager can develop production enterprise control systems. However, an example of a "complete control system" is not included for two reasons. First, the decision of how much of the control system for a production enterprise should be formalized (with written records) depends on the payoff from developing control systems and the use of similar amounts of time and effort in other parts of the business. Thus, what is formalized as a control system for one operator may be much different from that for another. Obviously, the relative contribution of the enterprise to the financial success of the business is important in making this determination. Second, including enough detail to adequately explain a reasonably complete control system for a production enterprise would add significantly to the length of the presentation.

THE PRODUCTION SYSTEM

The basis for both an effective plan and the control system for a production enterprise is the careful delineation of the system and a thorough understanding of the biological or physical production process. Careful delineation separates the components included within the system from the environment, making it possible to develop a detailed enterprise budget. The division between the system and its environment is commonly referred to as the system boundary. For example, in defining a dairy production system it is important to specify whether raising young stock is to be considered as part of the dairy production enterprise or as a separate enterprise. If dairy production and the raising of young stock are to be separated into two enterprises, at what age is the newborn calf transferred from dairy production to the enterprise responsible for raising young stock? Obviously, the age of transfer affects the inputs, outputs, and control system to be established for the dairy enterprise. In cash grain production, such as corn or small grain, the farmer may decide to separate the production from storage and pricing of the crop. It is important to specify the boundary between grain production and storage rather precisely, in developing both the enterprise budget and the control system. Some farmers may prefer to establish the boundary at the point where undried and unprocessed grain is delivered to the storage site at harvest. With that placement of the boundary, field operations associated with harvest are included in the production system, but drying cleaning, bin filling, aeration, and other tasks to store and preserve grain quality are part of the marketing enterprise. Obviously the boundary can be established at other points, such as the transfer of grain having no more than 15 percent moisture in specified storage bins on December 15 each year. A carefully and clearly delineated boundary is required for construction of the enterprise budget. It also provides the starting point for development of the enterprise control system.

The manager must also have an understanding of the underlying biological or physical production processes to develop a useful enterprise control system. This understanding should include the sequence of operations and the timing of inputs required for uninhibited growth and development of the crop or livestock enterprise. Knowledge of the production process may help the manager define the boundary of the production system. An understanding of how the crop or livestock grows and develops should enable the manager to do a better job of planning and controlling the factors that influence output.

Consider an example from corn production. John Hanway has summarized the stages of growth for a midseason corn hybrid (126 days from emergence to physiological maturity) in central Iowa.[1] He describes the growth of the root system and the aerial parts of the plant under unstressed conditions. The discussion indicates the plant's nutrient uptake by stage of growth, potential insect problems, and considerations in replanting due to hail, frost, and flooding of the crop.

A brief summary of the growth process as described by Hanway is included in Figure 17.1 for illustrative purposes. The figure lists the approximate number of days after emergence that each stage occurs. Plants emerge more rapidly when planted in warm, moist soil, indicating that the date and depth of planting should be selected after monitoring both of these variables. The date of planting should also consider the likelihood of moisture stress approximately 60 to 80 days after emergence because of its detrimental effect on pollination and grain production.

Data on uptake rates indicate that fertilizer should be placed slightly below and to the side of the seed where the roots will normally reach it by the time the plant emerges from the soil, but not with the seed which can

[1]John J. Hanway, *How a Corn Plant Grows,* Ames, Iowa State University, Cooperative Extension Service, Special Report No. 48, 1966, p. 2.

Figure 17.1. Some data on the development of medium season hybrid corn for use in planning and controlling the corn enterprise.

Stage of Growth[a]	Approximate Days after Emergence	Cumulative Percentage of Total Nutrient Uptake[b]			Part of Plant Accumulating Major Portion of Dry Matter	Some Control Considerations
		N	P	K		
Planting	−4 for warm conditions −14 for cool or dry conditions					1. Soil temperature influences the length of time required for emergence. Soil temperatures are cooler at greater depth, and deeper planting tends to increase the period for emergence. Planting depth and date should be based on soil temperature 2. The seed normally supplies the nutrients for emergence. Fertilizer can be placed in a band to the side and slightly below the seed where it will be contacted by the primary roots.
Emergence	0				} Leaves	
4 Leaves	14					1. The growing point is still below the ground surface. Therefore, a light freeze or hail may destroy exposed leaves, but it will reduce the final yield very little. Replanting would probably not be required for such damage. 2. Rootworms may destroy nodal roots and restrict plant growth at this stage.

(*continued*)

Figure 17.1 (*Continued*)

Stage of Growth[a]	Approximate Days after Emergence	Cumulative Percentage of Total Nutrient Uptake[b]			Part of Plant Accumulating Major Portion of Dry Matter	Some Control Considerations
		N	P	K		
8 Leaves	28	6	5	8	Leaves and stalks	1. The growing point is at the surface. Flooding at this or previous stages can kill the corn plants. Flooding at later stages when the growing point remains above the water is not as detrimental. 2. Nutrients can be applied up to this stage providing they are placed in moist soil so that the plant can absorb them.
12 Leaves	42	25	22	35		1. The potential number of kernels on the major ear is determined at about this time. Moisture or nutrient deficiencies, as well as injury by hail or insects, may seriously limit the potential size of the ear harvested.
16 Leaves	56	55	42	72	Stalks and leaves Stalks, Tassel and silks	1. The tip of the tassel has emerged, and the silks are elongating rapidly. Moisture stress or nutrient deficiencies tend to increase from top to bottom of the plant. This delays silking more than pollen shedding. 2. Complete removal (50 percent) of leaves will result in complete (25–30 percent) loss of yield. 1. Moisture stress or nutrient deficiencies may result in poor pollination. This emphasizes the need to

Stage					Description
Silks emerging and pollen shedding	66	67	56	93 ⎫	plant so that this stage will occur when weather conditions are most likely to be favorable. 2. Nutrients in the plant at this stage are highly correlated with final yield. Leaf analysis at this stage may be useful to identify fertility problems that can be corrected in next year's program.
Blister	78	76	69	99 ⎬ Cobs	1. This is the beginning of the rapid increase in grain weight. Moisture stress, nutrient deficiencies, or loss of leaves from hail will result in unfilled kernels.
Dough	90	88	83	⎫	1. The rapid increase in grain weight continues.
Beginning dent	102	96	91	⎬ Grain	1. The rapid increase in grain weight continues.
Full dent	114	99	96		1. Dry matter accumulation is nearing completion.
Physiological maturity	126	100	100	100 ⎭	1. Dry matter accumulation has been completed. The husks are dry, and some leaves are no longer green. The grain will continue to lose moisture.

[a]The stages of growth before silking are defined by the number of leaves that are fully emerged. Stages of growth after silking are defined by the development of the kernels on the ear. An early maturing hybrid may move through the stages more quickly, while a late maturing hybrid may progress more slowly than indicated. The development also is affected by many factors considered constant in this brief illustration. The reader is referred to the source for indications of the effect of temperature, day length, moisture and nutrient deficiencies, and other environmental variables on the development of the plant.

[b]Values interpolated from graphs presented by Hanway, pp. 16–17.

Source: John J. Hanway, *How a Corn Plant Develops*, Ames, Iowa State University, Cooperative Extension Service, Special Report No. 48, 1966.

result in salt injury to the young plant. The rate of nutrient uptake increases rapidly during the fourth week after emergence. The nutrient uptake data emphasize the importance of planning a fertility program that makes the three nutrients available (in moist soil within the root zone) within one month after emergence.

Observation of plant growth from eight leaves through silking and pollen shedding combined with leaf analysis is an effective means of obtaining data or feedback control of next year's crop, even though there may be relatively little that a producer can do in the way of concurrent control. Knowledge of the approximate location of the growing tip is important in making decisions on replanting of the crop after freeze, hail, or flood damage. Information on the effect of moisture deficiencies on ear formation, pollination, and grain filling is important in planning and concurrent control of irrigation applications. Knowledge of the life cycle of harmful insects as they relate to crop production provides the basis to plan for their control, as well as to monitor this population level and apply concurrent control measures.

Similar examples can be presented for livestock production. An understanding of the nutritional and environmental needs of livestock throughout the animal's life is useful in planning the livestock system and developing an enterprise control system. For example, consider the shape of the lactation curve of the dairy cow as presented in Figure 17.2. The amount of milk production commences at calving at a relatively high rate, and the amount produced daily increases for three to six weeks. After the peak is attained, daily milk production gradually declines. Corresponding data on the nutritional requirements to maintain the cow and provide the nutrients for milk production are necessary to plan a feeding program over the lactation and dry period. Furthermore, an understanding of the factors affecting conception rates is important in limiting the time between calving to the desired length so that cows are not milked for longer periods on the low part of the lactation curve. Principles of genetics and nutritional needs of growing calves are important in planning for development of a herd with higher production potential in future years.

Data from the actual lactation curve provide a basis for daily monitoring of milk production. After the curve has been adjusted based on the amount of milk produced during the first few weeks after calving, it can be used to predict production levels for the remainder of the lactation. A daily comparison of actual and predicted milk yields provides a basis to monitor and control production levels. When significant deviations occur, the operator must identify the nutritional, health, or environmental problem causing the deviation and attempt to correct it.

Data on feed requirements over the lactation and dry period combined with decisions on the system of production can be used to develop the animal flow for the enterprise. Data on the system of production would

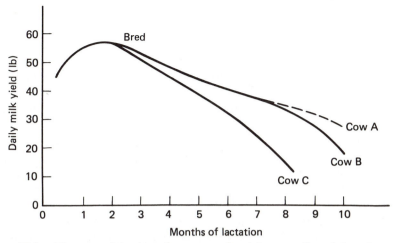

Figure 17.2. Diagram of the lactation curve of a dairy cow. Cow A is not pregnant. Cow B is pregnant. Cow C is not as persistent as cows A or B.

Source: Donald L. Bath, Frank N. Dickenson, H. Allen Tucker, andRobert O. Appleman, *Dairy Cattle: Principles, Practices, Problems, Profits* (Philadelphia: Lea and Febiger, 1978), p. 358. Reprinted by permission.

include whether replacement heifers are to be raised or purchased, the type of feeding program, and the type of facilities. The animal flow is simply a summary of the number of animals by type (lactating cows, dry cows, replacement heifers, and young stock by age) over the production period. The animal flow can be used to estimate detailed feed, labor, and facility requirements month by month throughout the production period. This provides a means to estimate input requirements and product sales for the projected monthly cash flow. The use of enterprise data for production control and as input into the financial control system is illustrated in detail for swine production later in the chapter.

Development of a detailed enterprise budget and cash flow based on the animal flow provides the projected values or standards to use in control of the operation. The challenge is to develop a control system that maintains biological efficiency at the level required to achieve the desired financial objectives.

DEFINE THE SUBSYSTEMS

The purpose of defining subsystems is to break the production system into more manageable parts for control purposes. The boundaries and interfaces of the subsystems must be defined carefully to be sure that the relationships among subsystems are clearly defined and that the sum of the subsystems is the entire system. The process of dividing subsystems into smaller subsystems continues until the resulting subsystems are of manageable size.

Defining a larger number of subsystems reduces the complexity within the subsystem but increases the number of interconnections that must be considered between subsystems. In general, the number of potential interconnections is equal to $n(n - 1)$, where n is the number of subsystems. A larger number of subsystems may lead to more difficulty in relating the subsystems to each other.

Production enterprise control systems should be kept relatively simple until it is clear that more complexity will have a payoff. Experience suggests this can be done by defining a limited number (usually six or less) of subsystems that are meaningful for monitoring and control purposes. To be meaningful for control purposes, the delineation must aid in identifying the important inputs and outputs to monitor. For example, the corn production enterprise (system) with the grain transferred to a marketing enterprise after drying might be divided into six subsystems:

1. Seedbed preparation or tillage.
2. Planting including replanting if necessary.
3. Fertility.
4. Pest control including control of weeds, insects, and diseases.
5. Moisture control.
6. Harvest and drying.

There may be little that is worth monitoring in some subsystems, and the number of subsystems can thus be reduced. For example, there may be little benefit to monitoring soil moisture on a regular basis in nonirrigated corn production; this list of subsystems can therefore be reduced to five by combining tillage and moisture control. However, for irrigated production, moisture control is an important subsystem deserving separate attention.

A system to produce feeder pigs to be sold (weighing approximately 40 pounds at seven to eight weeks of age) for finishing to slaughter weight can be divided into five subsystems:

1. Breeding and gestation.
2. Farrowing.
3. Care of pigs after weaning.
4. Sanitation and disease control.
5. Facilities.

The number of subsystems defined depends in part on the extent to which the control system is to be formalized. Producers with larger operations may want to subdivide some of the subsystems to formalize a larger part of the control system. For example, the production of replacement gilts (including selection of potential replacements before feeder pigs are sold and the care, feeding, and culling of these replacements) may be separated from breeding and gestation, or the sanitation and disease control may be divided into two subsystems. These divisions may be particularly useful if control of the alternative subsystems is the responsibility of different employees. On the other hand, one individual operating the feeder pig production system may feel that a somewhat less formalized control system is desirable. A producer with a small operation may want to reduce the number of subsystems by combining 1 and 2 above into a breeding herd subsystem.

DEVELOPING THE ENTERPRISE CONTROL SYSTEM

Steps 2 through 6 in developing an enterprise control system will be illustrated with an example from confinement livestock production—the production of 40 pound feeder pigs in a continuous farrowing confinement facility. A description of the production system and the plan is provided as a basis for development of the control system. This includes the enterprise budget and the projected cash flow for the enterprise. Then the subsystems are defined, and the enterprise control system is developed. Finally, the biological aspects of control are related to the enterprise cash flow.

The Confinement Feeder Pig Production System

The feeder pig production enterprise (system) includes the breeding herd, the farrowing of pigs, the marketing of eight-week-old feeder pigs weighing approximately 40 pounds (18.2 kgs), and the raising of replacement gilts for the breeding herd. Each farrowing crate is to be filled once every four weeks, resulting in thirteen farrowings per year.

The system uses environmentally controlled confinement buildings with slotted floors. The farrowing house has sixteen farrowing crates. The pigs are weaned at three weeks of age and moved to a fully slotted floor nursery. The breeding herd is housed in a partially slotted floor gestation building with breeding and gestation pens. All of the buildings have manure storage pits that are emptied semiannually.

The production schedule, shown in Figure 17.3, should be completed as part of the planning process that precedes development of the enterprise budget (discussed in Chapter 4). The schedule is designed to utilize facilities fully and requires carefully controlling the animal flow. A major element in controlling the animal flow is to control the breeding program

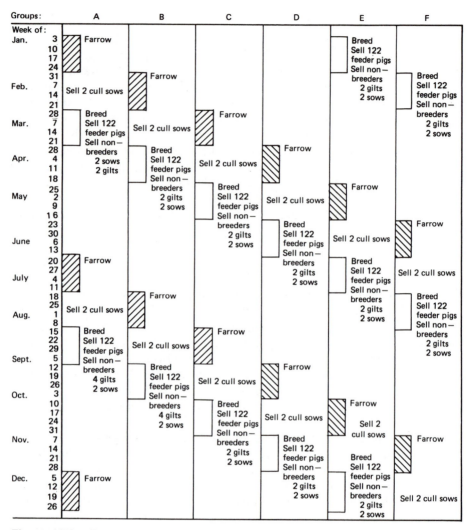

Figure 17.3. Production calendar for average year for continuous farrowing feeder pig system—96 sows.

Source: Duty D. Green and Vernon R. Eidman, *An Economic Analysis of Three Confinement Feeder Pig Production Systems,* Minnesota Agricultural Experimental Station Bulletin 534-1980, p. 20.

so that the number of sows moved into the farrowing house fills, but does not exceed, the number of farrowing crates available. If this is accomplished and the expected growth rates are achieved, the animal flow will utilize the facility space relatively completely. The production schedule lists the number of animals for sale by the approximate date. The complete

animal flow used to estimate feed, building space, and other input requirements is not shown in order to conserve space.

The enterprise budget for this operation is presented in Table 17.1.
The format for the enterprise budget and the computational procedures
used in its development are described in Chapter 4. The budget indicates
that thirteen groups (1,586) of feeder pigs (7.6 per litter) are to be marketed during the year of operation. Receipts from the sale of feeder pigs
are based on an annual average price of $47 per head, which is adjusted to
reflect the seasonal and cyclical fluctuations expected during the coming
year. The quantities of feed and other inputs are listed in the operating
inputs section. The estimated feed requirement is 204.3 tons per year or
1,964 pounds of feed per litter (204.3 tons × 2,000 pounds ÷ 208 litters).
The operator and family plan to provide all of the labor required and a
labor cost has not been included.

The projected cash flow for the enterprise, developed from the enterprise budget as illustrated in Chapters 4 and 6, is shown in Table 17.2. The
projected cash flow lists the total cash receipts and cash expenses for the
enterprise that can be entered directly in the cash flow for the total farm
business. The expected livestock sales and the expected cash operating
expenses are listed by month in the first and second sections of the cash
flow. Feed expense is included at the approximate rate that feed will be
consumed during the year. The operator credits the corn marketing enterprise for the number of bushels of corn consumed during the month at $3
per bushel. The operator has all the feed required processed commercially
and purchases the soybean meal and supplementary feed (vitamins, minerals, and antibiotics) from the dealer as needed. The cash costs listed reflect
the expected monthly consumption of these feeds and the expected
monthly cash expense for the items.[2]

The insurance and tax entries include both the $700 of insurance
premiums noted in the operating inputs section of the budget and the $680
insurance and real estate taxes on the facilities. All cash expenses, including
those in the ownership cost section, should be included in the projected
cash flow. However, an operator who includes these cash overhead expenses in the enterprise cash flow must be sure they are not double counted in completing the cash flow for the total business. The remaining operating expenses in the enterprise budget are listed by month in the
projected cash flow.

The Subsystems   The feeder pig production system is divided into five
subsystems:

---

[2]It should be noted that many operators purchase the protein and supplementary feed in large
quantities and process their feed on the farm. In such a case, the cash flow should reflect the
uneven purchases of feed by month throughout the year.

Table 17.1 Average Annual Costs and Returns for a Continuous Farrowing Feeder Pig Production System

Item	Head Sold	Weight Each	Unit	Price or Cost/Unit	Quantity	Value or Cost	Cost per Litter
1. Gross receipts							
Feeder pigs	122		hd	$44.09	122.00	$5,378.98	
Feeder pigs	122		hd	50.81	122.00	6,198.82	
Feeder pigs	122		hd	54.00	122.00	6,588.00	
Feeder pigs	122		hd	53.77	122.00	6,559.94	
Feeder pigs	183		hd	51.51	183.00	9,426.33	
Feeder pigs	183		hd	45.54	183.00	8,333.82	
Feeder pigs	122		hd	45.68	122.00	5,572.96	
Feeder pigs	122		hd	44.32	122.00	5,407.04	
Feeder pigs	122		hd	44.09	122.00	5,378.98	
Feeder pigs	122		hd	43.24	122.00	5,275.28	
Feeder pigs	122		hd	41.12	122.00	5,016.64	
Feeder pigs	122		hd	43.66	122.00	5,326.52	
Gilts	30	2.9	cwt	47.00	87.00	4,089.00	
Open sows	26	3.6	cwt	45.00	93.60	4,212.00	
Cull sows	24	3.7	cwt	40.00	88.80	3,552.00	
Boars	4	4.5	cwt	30.00	18.00	540.00	
Total						$86,856.31	$417.58
2. Operating costs							
Corn			bu	3.00	5,595.70	$16,787.10 ⎤	
Soybean meal (48.5 percent)			cwt	12.00	773.44	9,281.28 ⎬ 135.28	
Supplement feed			cwt	7.00	178.97	1,252.78	
Grinding and mixing			tons	4.00	204.30	817.20 ⎦	
Insurance			dol			700.00 ⎤	
Veterinarian and medicine			dol			1,586.00	
Electricity and fuel			dol			1,996.58	
Hauling and marketing (pigs)			hd	1.60	1,586.00	2,537.60	
Hauling and marketing (culls)			cwt	1.50	287.40	431.10	
Boars			hd	450.00	4.00	1,800.00 ⎬ 54.86	
Miscellaneous expense			dol			240.00	
Machinery and equipment (fuel, lubrication, repairs)			dol			1,091.10	
Interest on operating capital			dol			1,028.70 ⎦	
Total operating costs						$39,549.44	190.14
3. Income above operating costs						$47,306.87	227.44
4. Ownership costs							
Interest on livestock capital			dol	0.05	33,500.00	$1,675.00 ⎤	
Interest on facilities and machinery			dol	0.05	48,429.54	2,421.48	
Depreciation on facilities and machinery			dol			7,953.47 ⎬ 61.20	
Insurance, taxes on equipment, livestock, and machinery			dol			680.02 ⎦	
Total ownership costs						$12,729.97	

Table 17.1   *(Continued)*

Item	Unit	Price or Cost/Unit	Quantity	Value or Cost	Cost per Litter
5. Total costs shown				$52,279.84	251.34
6. Net returns above costs shown				$34,576.47	166.24

Labor Hours Required by Month

Jan.	Feb.	Mar.	Apr.	May	Jun.	Jul.	Aug.	Sept.	Oct.	Nov.	Dec.	Total
139	138	129	165	137	151	135	130	125	184	132	125	1,690

1. Gestation and breeding

2. Farrowing

3. Weaned pigs

4. Disease control

5. Facilities and sanitation

The gestation and breeding subsystem includes the management of the boars and the sows from the time they leave the farrowing crate until the time they return to the farrowing facilities. It also includes the selection and rearing of replacement gilts for the breeding herd.

The farrowing subsystem includes the management of the sow and litter while they are in the farrowing facilities. The weaned pigs subsystem includes the management of the pigs from the time they are weaned and leave the farrowing house at approximately three weeks of age until they reach market weight (40 pounds).

The disease control subsystem includes observation for disease and efforts to prevent and treat disease in the swine enterprise. The facilities and sanitation subsystem is responsible for maintaining the facilities (buildings and equipment) and the environment, including sanitation within the livestock facilities. Feed is delivered to the farm by the suppliers. However, this subsystem is responsible for feed handling on the farm as well as removal and spreading of manure and transportation of livestock.

Completing the Control System   The remaining steps in developing the control system can be illustrated in tabular form. An example of some of the detail that might be included in a feeder pig production control system is summarized in Figure 17.4. This figure identifies inputs and outputs to monitor, sensor(s) to be used, monitoring time interval, control standards, and actions to bring the system back into control. The following discussion

Table 17.2 Projected Monthly Cash Flow
for the Continuous Farrowing Feeder Pig System

Item	Jan.	Feb.	Mar.	Apr.	May	Jun.	Jul.	Aug.	Sept.	Oct.	Nov.	Dec.	Total
Cash Receipts													
Feeder pigs	$5,379	$6,199	$6,588	$6,560	$9,426	$8,334	$5,573	$5,407	$5,379	$5,275	$5,017	$5,327	$74,464
Gilts	273	273	273	273	273	273	273	273	544	544	273	544	4,089
Open sows	324	324	324	324	324	324	324	324	324	324	324	648	4,212
Cull sows	296	296	296	296	296	296	296	296	296	296	296	296	3,552
Boars	540												540
Total	$6,812	$7,092	$7,481	$7,453	$10,319	$9,227	$6,466	$6,300	$6,543	$6,439	$5,910	$6,815	$86,857
Cash Expenses													
Corn	$1,279	$1,341	$1,446	$1,391	$1,412	$1,376	$1,433	$1,408	$1,411	$1,469	$1,411	$1,410	$16,787
Soybean meal	692	735	805	774	788	761	801	764	779	815	792	775	9,281
Supplementary feed	90	100	110	102	104	107	107	106	104	113	103	107	1,253
Grinding and mixing	61	65	71	67	70	67	70	68	68	72	70	68	817
Insurance and taxes							700			680			1,380
Veterinarian and medicine	132	132	132	132	132	133	132	132	132	132	132	133	1,586
Electricity and fuel	263	239	257	91	101	108	103	103	103	118	252	259	1,997
Hauling and marketing	253	226	226	226	323	323	226	226	234	234	226	246	2,969
Miscellaneous expenses	20	20	20	20	20	20	20	20	20	20	20	20	240
Machinery operating expenses	59	59	59	251	59	59	59	59	59	250	59	59	1,091
Hired labor													—
Capital Expenses													
New facilities and equipment													—
Purchase breeding stock	1,800												1,800
Total	$4,649	$2,917	$3,126	$3,054	$3,009	$2,954	$3,651	$2,886	$2,910	$3,903	$3,065	$3,077	$39,201

Figure 17.4. Some elements of a control system for the feeder pig production enterprise.

Inputs and Outputs Monitored	Sensor Procedure	Monitoring Time Interval	Control Standards	Actions to Bring System Back into Control
		Subsystem: Gestation and Breeding		
Female, individual:				
Flushing	Records	7–10 days prebreeding	Feed 6–8 lbs of the normal gestation ration daily	Keep accurate records.
Breeding: Heat check twice daily	Records, visual and boar in pen	2–10 days postweaning Breed when heat is detected	Double mate the afternoon heat is detected and the following morning. Breed 5 females per week ± changes from breeding adjustment program.	Putting sow into an unfamiliar pen may bring on heat. More observation time. Use "hog holder" if the female will not stand. Use the breeding adjustment program.
Postbreeding feed intake	Records and feed scale	1 day postbreeding to the 86th day of gestation	Reduce the feed intake of the gestation ration to 4 pounds per day.	Maintain accurate records.
Postbreeding heat check	Visual and boar in pen	19–23 days postbreeding	All females should be checked. Not more than 10 percent should come into heat.	If the female does not settle after the second breeding, sell her. Rebreed the female if heat is detected. If more than 40 percent of females that were bred to one boar come back into heat, have veterinarian conduct a semen analysis.
Pregnancy test	Visual and pregnancy detector	30–35 days postbreeding	All the females are to be tested. Those tested not pregnant should be less than 10 percent.	Cull the female if not pregnant after the second breeding. Breed open females that have not been bred twice.

(continued)

Figure 17.4 *(Continued)*

Inputs and Outputs Monitored	Sensor Procedure	Monitoring Time Interval	Control Standards	Actions to Bring System Back into Control
Increase amount of ration	Records and feed scale	86 days postbreeding to the time the female goes to the farrowing facilities	Increase the amount of feed intake to 6–8 lbs of the gestation ration.	Keep accurate records.
Gestation weight gain	Visual and scale	Approximately the 100th day	Gestation weight gain should not exceed 80 lbs; 65 lbs is the standard.	Increase (decrease) feed intake if weight gain is too low (high).
Clean female	Visual	Before moving into farrowing crate	Clean female, especially the underline.	Wash thoroughly with soap and rinse with water.
Move to farrowing	Records	2–4 days prefarrowing date	Move on time (before parturition). Move 4 females to farrowing stalls per week.	Keep accurate records. If the female farrows in the gestation facilities, provide additional heat (heat lamp) and attention to the litter.
Gilts: Weight at breeding	Visual and scale	Prebreeding	Gilts should weigh at least 250 lbs, have sound feet and legs, and have 12 nipples or more that are not inverted.	More observation. Increase the ration. Cull gilts that do not meet the standard.
Boars: Fertility	Records and semen analysis	After boar has mated at least three females	Fertile	Cull boar if infertile.
Subsystem: Farrowing				
Sow: Feed	Visual and feed scale	2–0 days prefarrowing	Reduce feed intake to 1 to 2 lbs per day.	Maintain records. Weigh feed accurately.

700

	Method	Timing	Practice	Recommendation
		Postfarrowing	Feed a 16 percent protein laxative diet for 5–7 days. Feed sow 2 lbs the first day and increase 1–2 lbs per day until on full feed.	Maintain accurate records. Provide careful daily feeding.
Sow: Parturition	Visual and records	110–114 days postbreeding	Assist sow if having difficulty. Produce 9.5 pigs total per litter and 8.75 live pigs per litter.	Maintain accurate records of expected farrowing date and be available during farrowing. Call veterinarian if having difficulty.
Move sow	Records and visual	21 days postfarrowing	Wean 8.25 pigs per litter. Move sow to breeding facilities.	
Litter: Parturition	Visual	110–114 days postbreeding	Disinfect navels with a tincture of iodine. Control excessive bleeding.	Be around during farrowing. Tie off navel cord if excessive bleeding occurs.
Litter: Management practices	Records and visual	0–24 hours	Remove cleanings. Clip needle teeth and tails, leaving ¼ inch of the tail. Prevent chilling. Maintain temperature of 75°–85°F.	Provide extra heat with heat lamp if necessary. Lay a mat down on the slots.
		1–3 days	Equalize litter sizes if necessary.	Mask odor of the baby pig with talcum. Recommend to do before teat order is established.
		5–7 days	Provide a prestarter 20–22 percent protein (dried whey, sucrose).	Provide small amounts to keep it fresh.
		14–20 days	Castrate males. Prevent excessive bleeding and infections.	Sterilize equipment.

(continued)

Figure 17.4 *(Continued)*

Inputs and Outputs Monitored	Sensor Procedure	Monitoring Time Interval	Control Standards	Actions to Bring System Back into Control
		3 weeks	Wean baby pigs. They should weigh 10 pounds or more.	Delay weaning if the farrowing crate is not needed for another female.

Subsystem: Weaned Pigs

Inputs and Outputs Monitored	Sensor Procedure	Monitoring Time Interval	Control Standards	Actions to Bring System Back into Control
Pig: Pen placement	Visual and scale	At weaning	Place in groups of 20–25, depending on the pen size. Group according to size and sex.	
Pig: Temperature	Thermostat and visual	1st week	70°–75°F at pig level.	Adjust the thermostat. If they pile they are too cold, and if they lay over slots they are too warm.
Pig: Nutrition	Visual, reports and scale	Weaning to 20 lbs	20–22 percent protein creep ration.	
Pig: Marketing weight	Visual and scale	At 40 lbs	Sort into even lots.	

Note: Marketing is not included as it is considered to be a separate enterprise.

Subsystem: Facilities and Sanitation

Inputs and Outputs Monitored	Sensor Procedure	Monitoring Time Interval	Control Standards	Actions to Bring System Back into Control
Ventilation and heat	Visual, thermostat, humidostat, and warning system	Continuously Test warning system weekly	Control odors and excess humidity. The temperature referred to in the subsystems should be maintained. Temperature in gestation facilities	Adjust humidostats and thermometers. Repair immediately if not working. Make sure adequate fuel for furnaces is available. Have a backup generator to maintain

702

Item	Method	Frequency		
			should be between 50°F and 75°F.	ventilation. Sprinklers in the gestation facilities should be set to come on if temperature exceeds 75°F.
Water	Visual and water sample	Continuously	Well should provide a constant supply of clean fresh water.	Sample water yearly. Have a backup water supply.
Sanitation: Farrowing house	Visual and records	Continuously	Crates should be cleaned and disinfected after every litter.	
Sanitation: Nursery	Visual and records	Continuously	Pens should be cleaned and disinfected before each group.	
Sanitation: Gestation	Visual	Continuously		Manure that piles up should be scraped into pits daily. Pens should be cleaned and disinfected monthly.
Manure removal	Visual	Check manure holding facilities once a month. Scrape the farrowing building and nursery daily.	Remove manure from holding facilities when it is possible to spread on cropland (every 6 months).	Hire additional labor. Do a better job of scheduling. Work longer hours.
Feeding system nursery	Visual	Check daily	Keep automatic feeding system operating.	Repair immediately. Perform preventive maintenance.

(continued)

Figure 17.4 (Continued)

Subsystem: Disease Control

The sensor procedure used for disease is visual, and tests for disease are based on symptoms observed. The control standard is to prevent the disease and to treat the disease immediately if it occurs. The monitoring time interval is continuous. Only two examples are shown to illustrate the format.

Disease	Symptoms	Prevention	Treatment
Swine Brucellosis	Abortion, lameness, and sterility.	Buy Specific Pathogen Free (SPF) stock. Set up isolation program for new stock (30 days). Test breeding stock annually and re-placement stock.	Cull all infected animals. Dispose of all infected cleanings and aborted materials.
Erysipelas	Acute elevated body temperature. Feed refusal. Reddened areas on skin which blanch out when pressure is applied. Sudden death loss. Large numbers of young die. Chronic arthritis.	Vaccinate sows 3–4 weeks prefarrowing and 10 days prebreeding. Vaccinate pigs 2 weeks postweaning. Use avirulent vaccine.	Cull all infected animals with chronic case. Start prevention program.

elaborates on the records and reports that might be used to formalize some of the control system components detailed in Figure 17.4.

Several records on the biological production process are useful for concurrent control purposes. The basic unit of observation in the system is the individual animal in the breeding herd. An ear tag should be placed in both the right and left ear of all animals brought into the breeding herd for permanent identification. If one tag is lost, the other can be used to identify the animal. The major records and reports required on the biological production processes are the permanent sow record, worksheets to use in detailing daily work schedules, and the breeding adjustment schedule.

The permanent sow record should remain in a central file, regardless of whether the record is computerized or on a card with hand entries. This record should show the sow's ear tag identification numbers and, for each litter, the breeding date(s), location of the animal in the gestation building, the date due, date farrowed, the number of pigs born, the number of pigs born alive, and the number of pigs weaned. An example of such a record is presented in Figure 17.5. Note that the card contains space to record only two breedings per gestation because the control rule is to sell the sow if she remains open after the second breeding.

It is useful to design the record storage in a manner that alerts the operator when pregnancy testing and other tasks are to be performed. This can be accomplished by setting up a 170- to 180-day cycle with date bred as day 1. For the control system outlined in Figure 17.4, the records should provide a means of identifying all females at the following times:

19–23 days after breeding for heat check.

30–35 days after breeding for pregnancy testing.

86th day after breeding to increase the feed.

112th day after breeding to move to the farrowing crates.

141st day after breeding for weaning.

160th day after breeding for flushing and movement into the spaces for heat detection and breeding.

Other dates at which disease control operations or other checks are to be performed.

If the sow records are computerized, the software can be developed to list the sows that must be checked for each phase of the production cycle for the coming week. Numerous hand-kept sow record systems have been developed, some of which have been sold commercially, that include a filing device with a compartment for each day in the cycle. The permanent sow record (typically a card) is placed in the system in compartment 1 on

Figure 17.5.   An example of a permanent sow record.

Identification Number _____   *Right Ear* _____   *Left Ear* _____

Litter no.	Date bred	Date rebred	Location	Date due	Date farrowed	No. born	No. alive	No. weaned

the date the sow is bred. The tray or carousel containing the compartments is advanced one space daily. Markers are placed the appropriate number of days after breeding to indicate each task to be performed. When the group of cards for sows bred reach the nineteenth day, for example, the operator records the sow numbers and location on a worksheet and makes the heat checks as specified by the control system. If the animal does not appear to be in heat on any of the five days, the permanent sow record remains in the space and is advanced through the cycle. If heat is detected, the female either is moved to the breeding area, rebred, and the permanent sow record is returned to day 1 in the cycle; or the animal is moved to the holding area for culling. A worksheet is used by the operator to perform each of the control system checks, with the appropriate entries made on the permanent sow record. Thus, the record system alerts the operator when monitoring is to be performed, and the control rule indicates what action should be taken after the monitoring has been completed.

The breeding adjustment program is another example of a concurrent control procedure that can be used to help achieve the objective of filling each farrowing crate once each four weeks in an effort to maximize profits. A detailed breeding record is required to utilize the breeding adjustment program for a continuous farrowing system of the type described. The number of females to be bred weekly to accomplish the objective of filling the crates every four weeks depends on the conception rate. A conception rate of 80 percent suggests that five females should be bred weekly to fill four crates each week. However, it may not be possible to breed five females per week if an insufficient number come into heat, or if for some other reason females are unavailable for breeding. When this happens, the producer may want to breed more females in subsequent weeks in an effort to "catch up." To do this, bred sows are moved forward and included with those bred earlier. Then more are bred for the week from which they are moved. For example, suppose a producer has the objective of breeding five females per week as shown in Table 17.3, but only three are bred in week 1. The number of matings required to catch up in week 2 is seven, but suppose only four are bred. Then the number of matings needed to catch up in week 3 is eight. In the example, the producer is able to breed six females per week in weeks 3, 4, and 5 to catch up. It is important to recognize that a producer can catch up by combining sows that will farrow over several days into one group and weaning the pigs from later farrowings at a younger age, if necessary, to make room for subsequent farrowings. However, the producer should not work ahead, because breeding more sows than the

Table 17.3 An Example of the Breeding Adjustment Record

Objective: Breed 5 Females per Week

Week	Number Bred	Number of Additional Matings Required the Following Week to Catch Up
1	3	2
2	4	3
3	6	2
4	6	1
5	6	0

objective will result in sows ready to farrow when farrowing crates are not available.

The operator can reduce the need for the catch-up program by maintaining a larger group of females in the breeding area—an example of a preliminary control procedure. Doing so should increase the probability that five or more females will come into heat in any week, reducing the chances of breeding less than the desired number per week. The manager must balance the costs of maintaining the larger group of females in the breeding area—the preliminary control—against the cost of increased use of the breeding adjustment program—the concurrent control procedure. This may be difficult because the cost of increased use of the breeding adjustment program would occur in the form of earlier weaning of some pigs and increased management time. However, it illustrates a case in which the manager must make a choice on the emphasis placed on the two types of control.

A variety of swine herd management record programs have been developed for microcomputers. Some of these programs provide a daily or weekly work schedule listing the tasks to be performed. Figure 17.6 illustrates the type of work schedule data that might be provided by such a system. Although it is not illustrated, the calculation of the number of females to be bred during the week (the objective plus the adjustment to catch up) could be added to the report.

Many additional inputs and outputs can be monitored if the operator feels it will be profitable to do so. For example, the feed consumption and weight of the growing pigs could be monitored per group produced for the total time period or for shorter time intervals. Obviously, some time and effort would be required to collect and analyze the data by group of pigs. Monitoring feed consumption and the rate of gain becomes increasingly important as the hogs increase in size.

The production performance of the enterprise must be summarized periodically and compared to the standards used in planning to determine if the objectives are being achieved. The appropriate monitoring interval depends on the frequency with which the activities being controlled are repeated. In this example, the number of sows bred and the other operations are reported on a weekly basis. Thus, the results could be summarized and reported weekly or any multiple of the week; for this example, the operator may want to summarize the actual results and compare them to the standards every four weeks.

An example production performance summary and comparison is shown in Table 17.4. It is desirable to make three comparisons of production performance. The first is to compare the actual performance to the expected performance that was used in planning the operation and that serves as the standard for the control system. The data in Table 17.4 suggest that the operator is farrowing, weaning, and marketing more pigs

Figure 17.6. An example work schedule for a microcomputer swine herd management program.

Work Schedule for 6/8/83
19–23 Day Heat Check

Sow Identification				No Heat Cycle	Open and Moved to	
Rt. Ear	L. Ear	Pen	Date Bred	Detected	Pen	Bred to Boar No.
G197	G198	2	5/24/83	_____	___	_____
G213	G214	2	5/23/83	_____	___	_____
G215	G216	2	5/19/83	_____	___	_____

33 Day Pregnancy Check

Sow Identification					
Rt. Ear	L. Ear	Pen	Date Bred	Results	Date
G157	G158	3	5/9/83	_____	____
G163	G164	3	5/7/83	_____	____
G183	G184	3	5/9/78	_____	____

86 Day Increase Feed

Sow Identification			
Rt. Ear	L. Ear	Pen	Date Bred
G143	G144	5	3/20/83
G149	G150	5	3/20/83
G153	G154	5	3/21/83

110 Day—Move to Farrowing House

Sow Identification			Predicted	Actual	No.	No.
Rt. Ear	L. Ear	Pen	Farrowing Date	Farrowing Date	Born	Alive
G121	G122	7	6/11/83	_____	___	___
G125	G126	7	6/12/83	_____	___	___
G131	G132	7	6/12/83	_____	___	___

Wean and Move to Nursery

Sow Identification			
Rt. Ear	L. Ear	Date Weaned	No. Weaned
F39	F40	_____	_____
F93	F94	_____	_____
G117	G118	_____	_____

Table 17.4 Example Comparison of a Swine Herd Management System with Standards and Other Operations

Item	Your Herd 2/28 Through 3/27		Your Herd Previous 52 Weeks		Other Farms Previous 52 Weeks		
	Standard	Actual	Standard	Actual	Top 1/3	Middle 1/3	Bottom 1/3
Sows and gilts in herd	96	90	96	93	199	177	154
Died	1	0	16	4			
Culled	6	5	78	62			
Sows and gilts per crate	6.0	5.6	6.0	5.8	6.1	5.8	6.4
Number of sows per boar	24	22.5	24	23.25	15	20	20
Sows and gilts bred	20	18	260	254			
Sows and gilts rebred	4	5	52	59			
Sows and gilts negative on pregnancy check	2	3	26	33			
Sows and gilts farrowed	16	15	208	195			
Percentage of farrowing crates used	100	94	100	94	105	95	87
Total pigs born	152	151	1,976	1,872			
Total pigs born per litter	9.5	10.1	9.5	9.6	10.2	10.1	9.8
Total pigs born alive	140	135	1,820	1,755			
Pigs born alive per litter	8.75	9.0	8.75	9.0	9.5	9.4	9.0
Farrowing house deaths	8	6	104	128			
Farrowing house death/litter	0.5	0.4	0.5	0.66	0.78	1.80	2.95
Pigs weaned	132	129	1,716	1,627			
Pigs weaned per litter	8.25	8.6	8.25	8.34	8.22	7.83	5.83
Nursery deaths	2	4	26	32			
Number of pigs marketed	122	120	1,586	1,524			
Pigs marketed per litter	7.6	8.0	7.6	7.8	7.9	7.48	5.42
Pigs marketed/farrowing crate	7.6	7.5	99.1	95.2	107.8	92.4	61.3
Average weight of feeder pigs marketed (lbs)	40.0	42	40	38	52	48	43
Total lbs feed/feeder pig marketed	258	276	258	268	269	271	317
Price per cwt of feed fed ($)	6.88	7.12	6.88	7.15	7.01	7.04	7.15
Feed cost per pig marketed ($)	17.74	19.65	17.74	19.16	18.86	19.08	22.66
Average price per feeder pig marketed ($)	51.51	48.12	47.00	48.15	58.12	56.14	44.78

per litter than the standards, but that fewer sows are being bred and fewer litters farrowed than the standard. These data suggest that increasing the size of the sow herd may be an appropriate way to achieve the standards.

The second type of comparison is trend analysis or evaluating performance over a longer period. In the example, the comparison is made using 52-week rolling total and average performance levels. The data indicate no difference in the proportion of farrowing crates used, but a larger number of pigs weaned per litter and pigs marketed per litter during the current four weeks than the 52-week rolling average.

Third, it is important to compare the results with what other producers are doing. Comparisons of this type can be obtained by enrolling in a record system for swine producers that provides weekly or monthly summaries. Another source of such comparative data is the annual record summaries published by the farm management associations in many states and made available through the federal-state Cooperative Extension Service. These data are typically published only once per year, making them somewhat less current than data obtained from participation in a record system that provides periodic summaries throughout the year.

Records of the enterprise receipts and expenses should also be kept and summarized monthly by category for comparison with the projected cash flow. Doing so provides a basis to relate the herd management control system to the financial results of the enterprise. A format for comparison of the projected monthly cash flow and the actual cash flow is shown in Table 17.5. The format also includes the percentage deviation by line of the cash flow to indicate where large relative differences are occurring. When deviations occur, the manager must ask if the deviations are occurring because actual prices are different from those used in planning, or because the quantity of input used or product produced is different than planned. If the quantity of hogs sold is below the projected level, the summary of the herd management data (Table 17.4) should indicate which standards have not been met and are resulting in the lower level of output.

Comparisons of the type shown in Table 17.5 should be completed month by month throughout the year. The deviations for any one month may often be easily explained by slight differences in the time of marketing hogs or purchasing inputs. For this reason, it is also useful to compare the budgeted amount for the year to date with the actual amount. When the deviations for both the current month and the year to date are on the wrong side of the standard (that is, negative for receipts and positive for expenses), the manager should be quite concerned about taking corrective actions. For example, the operator of the unit summarized in Table 17.5 finds that receipts are consistently low. The actual production control summaries (Table 17.4) indicate that a substandard number of sows are being farrowed and fewer pigs are produced than planned. Evaluating the production schedules of the herd management control system provides an

Table 17.5 Actual and Projected Monthly Cash Flow for the Confinement Feeder Pig Enterprise

Item	March Budgeted This Period	Actual This Period	Percentage Deviation	Budgeted Year to Date	Actual Year to Date	Percentage Deviation	··· December Budgeted This Period	Actual This Period	Percentage Deviation	Budgeted Year to Date	Actual Year to Date	Percentage Deviation
Cash Receipts												
Feeder pigs	···$6,588	$6,222	-5.6	$18,166	$17,888	-2.5	···$5,327			$74,464		
Gilts	··· 273			819			··· 544			4,089		
Open sows	··· 324			972			··· 648			4,212		
Cull sows	··· 296	810	-9.3	888	3,019	-6.2	··· 296			3,552		
Boars	··· 0			540			···			540		
Total	7,481	7,032	-6.0	21,385	20,907	-2.2	··· 6,815			86,857		
Cash Expenses												
Corn	··· 1,446	1,498	3.6	4,066	4,196	3.2	··· 1,410			16,787		
Soybean meal	··· 805	788	-2.1	2,232	2,462	10.3	··· 775			9,281		
Supplementary feed	··· 110	123	11.8	300	340	13.3	··· 107			1,253		
Grinding and mixing	··· 71	75	5.6	197	212	7.6	··· 68			817		
Insurance and taxes	···						···			1,380		
Veterinarian and medicine	··· 132	143	8.3	396	354	-10.7	··· 133			1,586		
Electricity and fuel	··· 257	278	8.8	759	712	-6.2	··· 259			1,997		
Hauling and marketing	··· 226	218	-3.5	705	688	-2.4	··· 246			2,969		
Miscellaneous expenses	··· 20	84	6.3	60	283	19.4	··· 20			240		
Machine operation expense	··· 59			177			··· 59			1,091		
Hired labor	···						···			0		
Capital Expenditure												
New facilities							···			0		
Purchase breeding stock				1,800	1,740	-3.3	···			1,800		
Total	3,126	3,207	2.6	10,692	10,987	2.8	··· 3,077			39,201		

indication of how the problem can be corrected. The data for March also suggest that more money is being spent for feed than was planned, in spite of the lower number of hogs being sold. The performance standard comparison indicates that both actual feed consumption and feed prices are exceeding the standards. The manager must consider if feed wastage and the price paid can be reduced without reducing performance in other ways.

Revision of the Plan and the Control System   Data of the type obtained in Tables 17.4 and 17.5 suggest that some fine tuning of the plan or control system may be appropriate. The operator may want to increase the number of sows and gilts in the herd or alter the breeding schedules and techniques (such as using pen breeding or artificial insemination) in an effort to achieve the planned number of matings and farrowings per month. It should be clear that the type of herd management record system and cash flow comparison described provides a great deal of data for feedback control—data for use in planning future years.

The Control System for Crop Production

Control systems for crop production are developed in the same manner as those for livestock systems. The development of an effective plan and control system first requires a basic understanding of the biological production process. This is important in specifying the system of production to be used, including the sequence of field operations; the variety or hybrid to plant; the amounts, form, and timing of fertilizer used; the methods of weed, disease, and insect control; irrigation applications; and the harvesting methods. The data in Figure 17.1 illustrate this point for corn production. Furthermore, the enterprise budget is again the basic tool used in summarizing the plan and control standards, including the input and output quantities and related dollar values. The projected cash flow for the enterprise can be developed from the enterprise budget and, as for other enterprises, provides the link between control of the enterprise and financial control at the whole-farm level.

Perhaps the major difference in developing the plan and control system for crops compared to livestock is the difference in emphasis placed on preliminary, concurrent, and feedback control measures. We noted a number of opportunities for concurrent and feedback control in the feeder pig enterprise described earlier. Controlling the number of sows and gilts in the herd, the number of females bred per week, the location of the female in the breeding cycle based on heat and pregnancy checks, and adjusting the amount of feed during gestation based on the amount of weight gain are examples of concurrent control of the operation. Other confinement livestock enterprises also provide opportunities to adjust rations, breeding programs, and other inputs to bring the process under control during the current production period.

There are fewer opportunities for concurrent control in most cropping enterprises. Many decisions such as planting date, seeding rate, plant population, and fertility applications are made with little opportunity for an economically feasible adjustment during the production cycle of the current crop. The limited opportunity for concurrent control places increased emphasis on the use of preliminary and feedback control measures. For example, an operator may apply more fertilizer than is required for the most economic yield because the opportunities to apply the fertilizer after the moisture and other growing conditions are known are very limited. The operator responds by selecting a fertilizer application that will not limit crop production for the conditions experienced in most years. The operator also considers preliminary controls in selecting pesticide treatments, machinery operations, plant population, and other input levels.

Feedback control, that is, adjusting the plan in future production periods, is the second type of control that is widely used in crop production. Frequently, a problem, such as a nutrient deficiency or plant disease, that has not been experienced previously reduces the net returns of the enterprise. The operator may be able to adjust the production system to prevent or at least reduce the impact of this potential pest in future years.

The purpose of this section is to present some examples of control procedures in crop production. The examples are based primarily on the production of corn grain. The subsystems to be considered are described, and some examples of alternative types of control are noted.

Subsystems and the Control System The examples presented are based on the production of corn for grain in a corn-soybean rotation on clay-loam soils in the northern Corn Belt of the United States. The clay-loam soil is high in organic matter and available water-holding capacity. Assume the enterprise budget in Table 4.12 is applicable here. The production subsystems are (1) tillage; (2) fertility; (3) planting and replanting if necessary; (4) pest control; and (5) harvest and drying. Some components of the control system are summarized in Figure 17.7. Again note that the components and schedules of Figure 17.7 also can be used as part of the process of implementing the corn production plan.

The tillage system is selected to enhance moisture retention, to control erosion, insects, and weeds, and to economize on the operator's time. Research evaluating the effect of alternative tillage systems on these and other considerations is typically available from the state agricultural experiment station. The tillage subsystem considered here includes chisel plowing in the fall, disking one time in the spring, and harrowing one time before planting. It is doubtful that the control system should be written or for-

Figure 17.7. Some elements of a control system for corn production.

Inputs and Outputs Monitored	Sensor Procedure	Monitoring Time Interval	Control Standards	Actions to Bring System Back Into Control
			Subsystem: Tillage	
. . .				
Disk 21 ft.	Visual. Check depth with a tape measure	4/15 to 5/15. Start of the season and then weekly	Smooth the seedbed with 10 percent prepared by 4/25 (planting is to start 4/18 to 4/25). Disk at a depth of 5 inches with a range of 3–6 inches. Try to avoid soil compaction.	Work sufficiently long hours and hire help as needed to complete in the designated period. If compaction in the wheel tracks is a problem, it may be necessary to wait a day or two until the soil dries.
			Subsystem: Fertility	
. . .				
. . .				
Phosphate P_2O_5	Visual	Weekly from emergence to 15 inches	Symptoms of deficiency are leaves dark bluish green; more narrow leaves than normal; reddish purple at tips of upper leaves.	Take soil test. It is possible to sidedress liquid phosphoric acid, if it is judged economically feasible.

(continued)

Figure 17.7 (Continued)

Inputs and Outputs Monitored	Sensor Procedure	Monitoring Time Interval	Control Standards	Actions to Bring System Back Into Control
		Maturity	Symptoms of deficiency are small misshapen ears which may appear twisted because a row or parts of rows of kernels are missing on one side.	
	Soil test	Early Sept.	As given in soil test recommendations available from the extension service.	Apply P_2O_5 before chisel plowing in the fall.
	Tissue test on midrib mature leaf	20 leaf stage (early July)	As specified by test.	Apply P_2O_5 before chisel plowing in the fall.

Subsystem: Planting

Inputs and Outputs Monitored	Sensor Procedure	Monitoring Time Interval	Control Standards	Actions to Bring System Back Into Control
· · · · · ·				
Planter 8/38″	Soil thermometer	4/18 to 5/17 (daily)	From 4/18 to 4/25 begin planting if soil temperature at seed depth is above 55°F. On April 25 begin without regard to soil temperature.	If other operations are holding up planting, hire additional labor or a custom operator.

Plant population	Count seeds (visual). Electronic monitor.	First time out or continuously if planter performance is inconsistent	28,000 plants per acre requires a kernel spacing of 5.9 inches within the rows.	Make adjustment to planter. Drive at recommended speeds. Check for worn or broken parts.
Seed depth	Visual and tape measure	4/18 to 5/15 (once/field)	Depth of 2 inches. If dry, 3–4 inches. Range for normal depth 1¾ to 2¼ inches.	Adjust planter to correct depth.
. . .				

Subsystem: Pest Control

Corn rootworms	Visual and records from previous production periods	6 leaf stage to Aug. (weekly)		Crop rotation. Band a rootworm insecticide at planting.
. . .				

(continued)

Figure 17.7 (Continued)

Inputs and Outputs Monitored	Sensor Procedure	Monitoring Time Interval	Control Standards	Actions to Bring System Back Into Control
			Subsystem: Harvest, Drying, and Storage	
. . .				
Moisture content of corn grain to determine start of harvest	Moisture tester and walk fields	10/1 to 11/20	Black layer formed at tip of kernel (indicates maturity). Grain moisture from 20–25 percent during harvest to reduce field loss.	On November 1, begin harvesting any corn with less than 30 percent moisture, harvesting the driest corn first.
. . .				

malized for these operations. However, the operator should monitor each operation to ensure it is carried out in the prescribed time interval and in the prescribed manner. For example, the disking operation should be performed between April 15 and May 15 at a depth of 5 inches (with a range of 3 to 6 inches).

The fertility subsystem includes the application of the appropriate amounts of anhydrous ammonia, phosphate, potash, and limestone. The monitoring procedure is designed to determine that the appropriate amounts of these four inputs are applied. Common monitoring procedures include visual inspection of the crop in previous years, use of tissue tests when nutrient deficiencies are indicated, and soil tests taken in September of the year before the corn is planted. The application rate is typically based on the recommended levels from the soil test, unless visual inspection and tissue tests indicate that nutrient deficiencies persist when soil tests indicate adequate nutrients are present. As the entries for phosphate in Figure 17.7 indicate, the primary use of these monitoring tools is to develop the fertility program for the following year.

The objective of the planting system is to plant approximately 28,000 seeds per acre in an expeditious manner after a suitable soil temperature for rapid emergence has been reached. The data in Figure 17.7 indicate procedures to monitor soil temperature, planting rate, and depth. While these items must be monitored, no formal record is probably required.

The purpose of the pest control system is to identify the major weed, insect, and disease pests and to develop procedures to control them. Herbicides and insecticides are typically applied as preliminary controls on corn before and at planting to reduce populations of broad leaf weeds, grasses, and corn rootworm. A system should be established to monitor the development of weed, insect, and disease infestations through the growing season. This can be done by the operator if the individual is trained to identify the pests accurately and to estimate population levels. Commercial scouting firms and integrated pest management programs developed by the USDA-land grant university system provide scouting services for farmers in many areas of the United States. A copy of a scouting field form used to record pest populations and to report survey findings to potato producers in Minnesota is presented in Figure 17.8.

Producers may want to plant hybrids with a combination of maturity dates to spread the harvest season over a longer period to facilitate better use of machinery and labor available. During the harvest season, one of the important characteristics to monitor is the moisture content of the grain to determine when harvest can begin. Starting harvest at high moisture levels increases grain drying costs, but waiting for more natural drying in the field increases field loses. The last entry in Figure 17.7 indicates a monitoring procedure and a control rule to use to select the beginning harvest date.

There are typically many items to monitor in the production of nonir-

Figure 17.8. Field survey form—Potatoes.

Cooperator I.D. _____
Field I.D. _____
County _____
Growers name _____
Scout name _____
Date _____

Growth Stage

1. Seed piece	**6.** Full bloom
2. Emergence	**7.** Past full bloom
3. Plant erect	**8.** Full grown
4. Flower buds	**9.** Maturity
5. 50% bloom	**10.** Harvested

	NE	NW	SE	SW	CENTER	
AVE						
Stand count	___	___	___	___	___	(# plants/30 feet)
Plant height	___	___	___	___	___	(inches)

Insects

DAMage = % of leaf area
RATing = use 1 to 10 scale (1 = scattered, 10 = severe)

AVE *RAT*	*AVE* *DAM*	*Plants/ Acre*		NE		NW		SE		SW		CENTER
				RAT	*DAM*	*RAT*	*DAM*	*RAT*	*DAM*	*RAT*	*DAM*	*RAT*	*DAM*
___	___	___	2601 Green peach aphid	___	___	___	___	___	___	___	___	___	___
___	___	___	2602 Potato aphid	___	___	___	___	___	___	___	___	___	___

2606	Colorado potato beetle
	Cutworm
2609	Potato flea beetle
	Grasshopper
2519	Potato leafhopper
2526	Cabbage looper
2533	Tarnished plant bug
2626	Potato tuberworm
2150	Wireworm

Diseases

INCidence = # of plants infected/20 plants
SEVerity = % of leaf area infected

		NE		NW		SE		SW		CENTER	
		INC	SEV	INC	SEV	INC	SEV	INC	SEV	INC	SEV
1602	Blackleg										
1620	Late blight										
1622	Early blight										
1603	Rhizoctonia canker										
1632	Verticillium wilt										
1640	Leafroll virus										

AVE INC	AVE SEV	Plants/ Acre

Source: 1983 Minpest Survey Form, Department of Plant Pathology, University of Minnesota.

721

rigated field crops as this example illustrates. However, formal records are usually justified for only a few of the inputs and outputs monitored. Formal crop records are probably justified to summarize the following items by field for most farms: (1) soil tests; (2) fertilizer applied; (3) herbicides applied; (4) crop planted; (5) nutrient deficiencies indicated by observation and tissue tests; (6) weed, insect, and disease infestations; (7) yields; and (8) financial results in terms of receipts from product sales and costs of inputs. These data are useful for feedback control in planning the following year's production.

The Irrigation Control System   Producers of irrigated crops should monitor soil water levels during the growing season and control irrigation applications based on the response of the crop to available soil moisture. Increased interest in controlling irrigation applications has developed recently for several reasons, including: (1) the cost of pumping water has risen as investment costs in irrigation systems and energy costs have increased; (2) water supplies have declined in some areas; (3) the technology to monitor soil water levels more accurately has developed; and (4) applying too much water leaches valuable nutrients from the soil and may pollute the ground water.

Ideally, irrigators should base their control decisions on estimates of the marginal physical productivity of maintaining soil water at alternative levels by stage of crop growth. But many soil, weather, and crop factors interact to determine plant growth and yield. Consequently, estimates of the relationship between soil water and yield require a great deal of experimental data. Until the appropriate marginal physical productivity estimates are available for the crop, soil type, and area, irrigation scheduling is typically based on the consumptive use of water.

The basic approach to controlling irrigation applications is to monitor the amount of available water in storage that is within the root zone of the crop and to maintain this amount through irrigation at a level that does not stress the crop. The available water is the amount of water held in the soil between field capacity (the amount held in the soil after excess water has drained) and wilting point (the amount that remains when no more water is available to the plant). Even though all water between field capacity and wilting point is available to the plant, removing water for plant growth becomes more difficult the closer the soil moisture level is to the wilting point. Therefore, a measure that can be termed "readily available water" is frequently defined as the water that can be withdrawn without stressing the crop. This amount is commonly considered to be one-half the available water.

The amount of available water-holding capacity can be estimated by considering the rooting depth of the crop and the available water-holding capacity of the soils. Irrigation is normally limited to the upper portion of the root zone where most of the roots are located. For example, a common irrigation depth for corn is 3 feet, even though the rooting depth is much greater. The available water-holding capacity of soils varies from as little as 0.5 to 0.7 inches per foot of soil for coarse sand to as much as 2.0 to 2.6 inches per foot for clay loams. Data on the irrigation depths for alternative crops and the available water-holding capacity of soils can usually be obtained from the Soil Conservation Service or the Cooperative Extension Service. The example in Figure 17.9 assumes that the water balance is being prepared for corn production on a loamy sand with 1 inch of *total available* water per foot, or 1.5 inches of *readily available* water in the 3 feet of the irrigation zone.

Rainfall and irrigation result in additions to the soil water inventory. It is important to allow for evaporation losses (particularly for sprinkler irrigation applications) and any runoff to more accurately measure the net addition to the inventory.

Evaporation from the soil, transpiration by the crop, and deep percolation of the water from the root zone are the deletions from the soil water inventory. Various approaches have been devised to estimate the daily withdrawals for these three uses. One of the simplest is illustrated for corn in Table 17.6. Use of this table requires data on only the maximum daily temperature and number of weeks that have passed since the crop emerged. In general, the accuracy of the daily transpiration uses can be improved by considering several other variables. For example, the Jensen-Hayes formula estimates daily evapotranspiration losses for crops based on average daily temperature [(daily maximum + daily minimum)/2], solar radiation, the average minimum temperature and the average maximum temperature for the warmest month of the year, and elevation. The maximum temperature, minimum temperature, and solar radiation must be monitored daily. The remaining items are constant over the growing season for a given location. Including the additional data usually results in more accurate estimates of daily evaporation and transpiration uses.

Thus, an individual who wants to control irrigation applications based on soil water levels must monitor several items. These are (1) the amount of water in storage in the root zone of the crop at the start of the season; (2) the daily additions through rainfall and irrigation applications; (3) the temperature and other weather variables used to estimate daily use; and (4) the amount of water in storage weekly to correct the balance. The equipment required to make these measurements includes rain gauges, thermometers, tensiometers or resistance blocks, and a flow meter to measure the amount of water being applied by the irrigation system.

The controls or rules of action indicate when the irrigation system should be started and the amount of water that should be applied. Irrigation should be initiated at a sufficiently high level of available water so that the entire field can be irrigated before significant stress occurs. The level

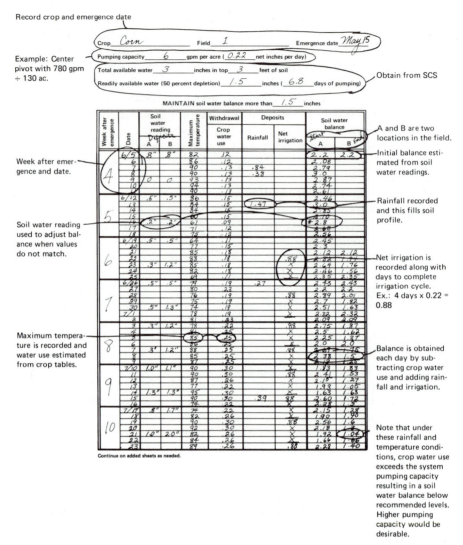

Source: Hal D. Werner, *Irrigation Scheduling: Checkbook Method,* University of Minnesota Agricultural Extension Service, Miscellaneous Publication M–160, 1978, p. 8.

Figure 17.9. Example soil water balance sheet

selected depends on the number of days required to irrigate the entire field and the expected use per day given the crop and season of the year.

An example of the irrigation record that can be used to monitor the process is presented in Figure 17.9. The hand-completed procedure illustrated in this figure records what has happened to date. Soil moisture is measured on the day the record is initiated. In this example, measurements

Table 17.6 Average Water Use per Day for Corn Production

Temperature °F	Week After Emergence																
	1	2	3	4	5	6	7	8	9	10	11	12	13	14	15	16	17
50–59	0.04	0.04	0.04	0.04	0.05	0.07	0.08	0.09	0.09	0.09	0.09	0.09	0.09	0.08	0.08	0.06	0.05
60–69	0.04	0.04	0.05	0.07	0.09	0.11	0.13	0.15	0.15	0.15	0.14	0.14	0.14	0.13	0.13	0.09	0.07
70–79	0.04	0.05	0.07	0.10	0.12	0.15	0.19	0.22	0.22	0.22	0.20	0.20	0.20	0.18	0.17	0.11	0.09
80–89	0.04	0.06	0.08	0.12	0.15	0.18	0.23	0.25	0.26	0.26	0.24	0.24	0.24	0.22	0.21	0.14	0.12
90–99	0.05	0.07	0.10	0.13	0.17	0.21	0.26	0.29	0.30	0.30	0.29	0.29	0.29	0.26	0.25	0.17	0.14
Corn growth stages	3 leaf				12 leaf			1st tassel	silk					hard dent		black layer	

Source: Hal D. Werner, *Irrigation Scheduling: Checkbook Method,* University of Minnesota Agricultural Extension Service, Miscellaneous Publication M–160, 1978, p. 6.

are made at two locations in the field. The available soil water in storage on each subsequent day is estimated by subtracting crop water use and adding the "deposits" from rainfall and irrigation. Notice that soil water readings are used to update or correct the soil water balance periodically, with the frequency of updating being greater when water use is expected to be higher. On June 8 and 13 more deposits were made than the soil could hold, and the excess water was assumed to run off or be lost to increased deep percolation.

Using the monitoring and control system, the operator initiated irrigation when the amount of available water minus expected use by the crop in the next four days was less than 1.5 inches—the estimate of readily available water. This rule maintained available soil water above the 1.5 inch standard at location A throughout the season. Because of the time required to apply one irrigation, the available soil water level dropped below the 1.5 inch level on several days at location B.

Several computerized scheduling services use a similar procedure, but also project water use into the future and recommend the date to start the next irrigation based on expected water use. An obvious advantage of the computer system is that it can use more complex equations considering more weather variables in calculating crop water use, providing the weather data are available. Computerized procedures also eliminate the mathematical errors and reduce the time required to make the calculations when the daily balance must be determined for a large number of locations. Irrigators with several fields and crops to irrigate may not have sufficient pumping capacity or a large enough distribution system to irrigate all crops at the time designated. The decision rules to follow can be included in the computer program so that recommendations on the sequencing of irrigation can be made.

Revisions of the Plan and Control System   Many of the examples given for nonirrigated crop production emphasized the use of data collected for feedback control—revising the plan and control system the following year. Data on fertilizer deficiencies and pest control problems may result in adjusting the inputs for next year's production. For example, suppose observation of plant color and plant tissue tests indicate areas within one or more fields with phosphorus deficiency. The operator may want not only to increase phosphate applications on those areas for next year (a preliminary control), but also to consider more intensive monitoring of those areas next year and to study the cost-effectiveness of within-season applications in the event the problem shows up again (a revision of the control system).

The data in Figure 17.9 indicate that the rule followed to initiate irrigation resulted in available soil water dropping below the standard several days during the year at location B. The operator may want to compare corn yields on those areas of the field where the greatest moisture stress occurred with yields on those areas where the record system indicated no significant stress. If significant yield reductions resulted, the operator might simulate the use of alternative irrigation rules (make hypothetical calculations of the soil water balance for different irrigation initiation rules) with several years of historic weather data and estimate if another rule which initiates irrigation at a higher level of soil water would appear to be economic. This analysis could be used to revise the concurrent control rule.

SERVICE ENTERPRISE CONTROL

Service enterprises were defined in Chapter 2 as those that provide services to each other and to the production enterprises, but do not normally produce a marketable product. Examples are tractors and machinery used to produce crops; buildings, equipment, and facilities used for livestock production; storage facilities for grain and forage; irrigation systems (including the well, pump, power unit, and distribution system); and grain-drying equipment. We have discussed many of the considerations in selecting and planning the use of such equipment and facilities earlier in the text. The concept of animal flow as a means to plan the appropriate size and combination of livestock facilities was discussed earlier in this chapter. The value of timeliness was considered in comparing alternative sizes of combines in Chapter 14. The use of capital budgeting procedures to compare the profitability of alternative sizes and types of investments was illustrated for both machinery and buildings in Chapters 8 and 14.

In some cases, service enterprise control is included as one subsystem of a production or marketing enterprise. For example, the feeder pig production system discussed earlier included facilities and sanitation as one subsystem. In a similar manner, maintenance of storage and grain-handling facilities might be included as a subsystem in a marketing enterprise.

Like production enterprises, the plan and the enterprise budget is the starting point to develop the control system for service enterprises. Then the control system for service enterprises can again be developed following the six steps outlined early in the chapter. Defining the subsystems (components) of the facilities or machine may be helpful in identifying the inputs or outputs to monitor. For example, a center pivot irrigation system might be divided into the well, pump, gearhead, power unit, safety control circuits, water line to the center pivot, and the sprinkler system itself. The division into subsystems should only be made if it is useful in identifying the important inputs and performance indicators to monitor. The standard or in-control range can usually be specified for each subsystem of service enterprises based on previous use (feedback control), design requirements, engineering standards, and manufacturer's specifications. An owner's manual (if it exists) typically outlines the means of monitoring the item's performance and suggests ways to correct deviations.

Controls for most service enterprises can be divided into three types.

Preseason or preliminary controls include repairs and preventive maintenance to be conducted when the (livestock and storage) facilities are empty, and before the season in which machinery is to be used. *In season or concurrent controls* include routine maintenance, as well as monitoring the performance and making adjustments and repairs during the season the facilities or machines are in use. This includes scheduled lubrication, maintenance and adjustments, as well as repairing the system during an unexpected breakdown. Contingency plans which farmers use for concurrent control include maintaining excess capacity (in the form of an older machine, an additional livestock facility, or a storage facility), leasing machinery or equipment, hiring a custom operator, and maintaining a supply of repair parts. The third type of control is *feedback control.* For service enterprises, this involves maintaining a list of items to be repaired or corrected after the season, as well as suggestions for changes in the preventive maintenance and control program for future production periods.

Most farmers formalize some part of their control system for service enterprises. There are three major types of records that a farmer might consider maintaining on one or more of the service enterprises: (1) usage records indicating the number of hours the item is used, the type of work and the production enterprise receiving the service, and the amount of fuel and other operating inputs used; (2) a maintenance record for each item listing routine preventive maintenance (such as changing lubricating oil and checking safety controls) and repairs performed; and (3) a listing of the repairs and modifications needed. A record of the amount of use is frequently maintained on items requiring periodic servicing, but the amount of fuel, repairs, and other operating units are recorded only by machine or facility if it is important to calculate the cost of operating the service enterprise. Such data might be used to decide if the machine or facility should be kept or replaced, or to allocate the expense to the production enterprises. Many farmers also have some formal maintenance records and a somewhat less formal listing of repairs and modifications needed. As the number of service enterprises increases, the difficulty of maintaining control without formal records increases and the payoff for formalizing more of the service enterprise control system tends to increase. Since the development of these records is relatively straightforward, it is not illustrated here.

Summary

The purpose of production and service enterprise control systems is to facilitate control of the individual components of the business, which contribute to control of the firm's total financial performance. The enterprise is linked to the firm's financial performance via an enterprise budget. The enterprise budget (which lists the expected inputs, outputs, receipts, and expenses) is used to develop the enterprise's portion of the firm's cash flow. When large deviations occur in the firm's cash flow, the cause of these changes can be traced to the enterprise and through the monitoring system, to the cause of the deviation within the enterprise.

An understanding of the underlying biological and physical production processes aids in planning the enterprise and in developing the enterprise control system. Such an understanding allows the application of the principle of diminishing marginal returns and the principle of input substitution in determining the most profitable amount and timing of input levels, identifying inputs and outputs to monitor, specifying standards, and selecting the monitoring frequency.

Development of a production or service enterprise control system follows six steps. First, the system is divided into a limited number of subsystems that are meaningful for identification of the important inputs and outputs to monitor. Steps 2 through 4 involve specifying the inputs and outputs to monitor, selecting the monitoring frequency, and choosing an appropriate means of monitoring each item selected. Specifying the standard for comparison and establishing rules for action when the variable is outside the control range are the final two steps.

The examples presented suggest that many inputs and outputs can be monitored for production and service enterprise control. The opportunities to develop concurrent control procedures are greater for those enterprises having inputs "applied" more frequently, such as confinement livestock enterprises. However, even when the opportunity for concurrent control is limited, there is ample challenge to develop preliminary control measures and to use feedback control in adjusting the plan and control system for the following production period.

Questions and Problems

1. Select a crop or livestock enterprise and specify the production system. What attributes of the system must be specified?
2. Develop a detailed plan for the production system chosen in question 1 and summarize the plan with an enterprise budget.
3. Define the subsystems for the production system in question 1. What inputs and outputs should be monitored for each subsystem?
4. Give an example of a preliminary control that might be used with the production system selected in question 1 above.
5. Give an example of a concurrent control that could be included in the control system for the production system described in question 2 above.
6. Give examples of the way feedback control might be used to adjust the enterprise and its control system in later years.
7. What does it mean to formalize the control system? How much of the control system for the production enterprise specified in question 1 above should be

formalized? How would you respond to a farmer who said your control system would be more trouble than it would be worth?

8. Illustrate animal flow for a confinement livestock operation.
9. Illustrate the use of the principle of diminishing marginal returns and the principle of input substitution in developing enterprise control systems.
10. Refute the following statement: "The equal marginal return principle is of little use in developing enterprise control systems."

Further Reading

Allen, David, and Brian Kilkenny. *Planned Beef Production.* London: Granada, 1980.

Bath, Donald L., Frank N. Dickinson, H. Allen Tucker, and Robert D. Appleman. *Dairy Cattle: Principles, Practices, Problems, Profits.* 2d ed., Philadelphia: Lea and Febiger, 1978.

Blackie, M. J., and J. B. Dent, eds. *Information Systems for Agriculture.* London: Applied Science Publishers, 1979.

Bowers, Wendell. *Fundamentals of Machine Operations: Machinery Management.* Moline, Ill.: Deere and Company, 1975.

Confinement Swine Manager's Training Notebook. Rochester, Minn.: Babcock Swine, 1975.

Foth, Henry D. *Fundamentals of Soil Science.* 6th ed., New York: John Wiley and Sons, 1978.

Fuller, Earl I. *Think Systems: Putting a Dairy Farm Management Information and Control System Together.* St. Paul: University of Minnesota Agricultural Experiment Station Unpublished Paper, 1983.

Hanway, John J. *How a Corn Plant Develops.* Ames: Iowa State University, Cooperative Extension Service, Special Report No. 48, 1966.

Horngren, Charles T. *Cost Accounting.* 3d ed., Engelwood Cliffs, N.J.: Prentice-Hall, 1962, Chapters 7, 8, 9.

Hunt, Donald. *Farm Power and Machinery Management.* 7th ed., Ames: Iowa State University Press, 1977.

Neuman, A. L. *Beef Cattle.* 7th ed., New York: John Wiley and Sons, 1977.

O'Mary, Clayton C., and Irwin A. Dyer, eds. *Commercial Beef Cattle Production.* 2d ed., Philadelphia: Lea and Febiger, 1972.

Owen, John B. *Complete Diets for Cattle and Sheep.* Ipswich, England: Farming Press, 1979.

Owen, John B. *Sheep Production.* London: Balliere Tindall, 1976.

Stoskopf, Neal C. *Understanding Crop Production.* Reston, Va.: Reston, Inc., 1981.

18

MARKET AND FINANCIAL CONTROL

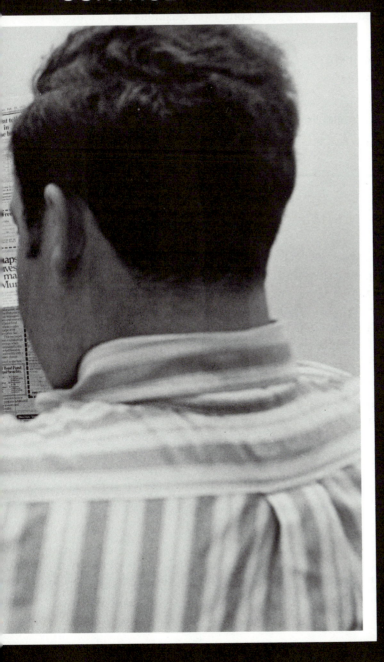

CHAPTER CONTENTS

The previous two chapters have discussed the concept of control and the development of control procedures to monitor the production activities and technical efficiency of the farm business. In this chapter, we will discuss methods of monitoring the marketing and financial performance of the firm—the concepts of market and financial control. As with the production area, specific market and financial plans have been developed as part of the planning process for the farm business, and through the process of implementation, these plans have been put into action. The purpose of the market and financial control system is to assess the results and determine what adjustments are necessary to improve performance.

Our discussion will first focus on evaluating performance with respect to the marketing function. Then we will turn to financial performance and the financial control system, with emphasis on cash flows, income analysis, financial statements, and business analysis ratios. Finally, we will discuss the process of controlling the farm business during the expansion and contraction process.

MARKET CONTROL

As with any control process, the basic structure and components of the market control system come from the specification of the market plan. The market plan includes specific decisions as to when to set the price; where to price; what form, grade, or quality to deliver or purchase; what services to provide or acquire; what method to use in pricing; and when and how to deliver. To monitor market performance, the control system must record information on both prices and quantities of inputs purchased by the firm and products sold.

Product and Input Prices

The first component of the marketing plan that should be monitored and can benefit from concurrent and feedback control mechanisms is the decision of when to price an input or product. A summary of historical prices for the major products sold and inputs acquired by the farm manager can be used to assess past price performance and determine potential price patterns. Price levels and trends can be effectively illustrated in a graphical fashion as shown in Figure 18.1. To obtain the price information that is plotted in the figure for wheat and hogs, daily market quotes as reported in the local newspaper or from local bidders are plotted. By plotting the daily quoted cash prices, a visual impression of cycles, seasonal trends, and other changes in price relationships can be detected. These historical price series can be useful in evaluating past pricing decisions as well as in assessing

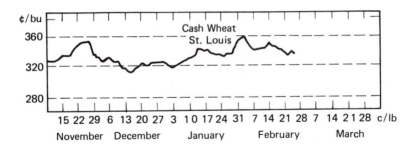

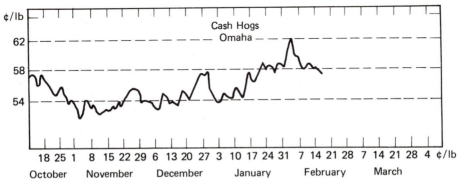

Figure 18.1.   Charts of cash wheat and hog prices for 1982–83.

Futures Prices

Wednesday, September 28, 1983
Open Interest Reflects Previous Trading Day.

	Open	High	Low	Settle	Change	Lifetime High	Low	Open Interest

-GRAINS AND OILSEEDS-

CORN (CBT) 5,000 bu.; cents per bu.
Sept								29
Dec	357¼	358½	348	348½	− 6	376½	253	111,752
Mar84	361	362½	352	352	− 6¼	386¼	278½	54,932
May	362	364½	353	353½	− 6¾	390	285	19,154
July	361½	363½	351	353¼	− 6	388	288¼	25,346
Sept	336¼	336½	326½	327	− 6½	356½	320	2,578
Dec	315	315	307½	308¼	− 3½	330	296	11,155
Est vol 45,266; vol Tues 51,388; open int 224,946, − 4,408.								

CORN (MCE) 1,000 bu.; cents per bu.
Dec	358	360¾	348	348½	− 6	381	253¼	8,160
Mar84	361	362½	352	352	− 6¼	386¼	278½	2,670
May	362	364¾	353	353½	− 6¾	390	285½	1,535
July	362	364	351	353¼	− 6	388½	288¼	2,243
Sept	326	326	320	321	− 6½	355	321	153
Dec	314	314½	307¼	308¼	− 3½	330	296½	1,702
Est vol 4,000; vol Tues 2,260; open int 16,463, + 167.								

OATS (CBT) 5,000 bu.; cents per bu.
Sept								1
Dec	191	193¼	185¼	185¼	− 4½	207½	161¾	6,969
Mar84	202¼	203	196¼	196¾	− 4½	219	172	1,357
May	209½	209½	203	203	− 5	226	177	416
July	214	214	208	208	− 5¼	226	189½	318
Sept	216	218	211½	212	− 4	219	211½	15
Est vol 1,500; vol Tues 1,007; open int 9,076, − 96.								

SOYBEANS (CBT) 5,000 bu.; cents per bu.
Sept								43
Nov	902	910½	865½	868	− 27½	986½	568½	68,928
Jan84	918	922	879	880½	− 28½	982	594	27,614
Mar	929	932	889	889	− 30	993½	616	18,544
May	929	931	889½	889½	− 30	996	630	6,883
July	917	922	880½	880½	− 30	992½	639½	10,742
Aug	883	891	847	847	− 28	956¼	640	2,605
Sept	788	800	766	766	− 19	860	710	2,241
Nov	718	721	694	695½	− 17½	772¼	671	7,416
Est vol 103,400; vol Tues 68,647; open int 145,016, − 1,366.								

SOYBEANS (MCE) 1,000 bu.; cents per bu.
Nov	908	910½	865½	868	− 27½	968½	568	6,709
Jan84	918	922	879	880½	− 28½	982	596	1,497
Mar	930	932	889	889	− 30	993½	620½	1,083
May	928	930	889½	889½	− 30	996	636	827
July	917	924½	880½	847	− 28	992½	645	596
Aug	887	887	845	766	− 19	955	646	222
Sept	793	793	792	766	− 19	851	710	89
Nov	715	721	695	695½	− 17½	722¼	663	269
Est vol 3,800; vol Tues 5,874; open int 11,292, − 61.								

SOYBEAN MEAL (CBT) 100 tons; $ per ton.
Oct	235.50	237.50	227.00	228.50	− 5.80	264.50	163.00	7,971
Dec	240.00	242.50	231.50	233.50	− 5.30	268.50	166.50	28,991
Jan84	242.00	243.50	232.30	235.00	− 5.50	268.50	174.50	10,408
Mar	244.50	245.00	234.00	236.00	− 6.00	268.50	179.50	6,219
May	245.00	246.00	234.00	235.00	− 6.00	267.50	185.00	2,931
July	245.00	246.00	235.00	235.00	− 8.00	267.50	188.00	2,832
Aug	236.00	236.00	224.20	226.50	− 7.80	251.00	192.50	1,479
Sept	222.50	225.00	217.00	217.00	− 10.00	243.00	206.00	1,632
Oct	199.00	201.00	194.00	199.70	− 4.30	240.00	194.00	1,028
Dec	198.00	199.50	194.00	194.70	− 4.50	227.00	194.00	263
Est vol 28m066; vol Tues 19,641; open int 63,754, − 1,328.								

SOYBEAN OIL (CBT) 60,000 lbs.; cents per lb.
Oct	34.00	34.30	32.55	32.55	− 1.00	36.50	17.50	8,683
Dec	34.35	34.48	32.95	32.97	− .98	36.90	17.82	34,025
Jan84	34.50	34.80	32.93	32.93	− 1.00	36.91	18.02	11,035
Mar	34.20	34.50	32.65	32.65	− 1.00	36.65	18.33	10,633
May	33.35	33.60	31.83	31.83	− 1.00	36.85	19.75	4,669
July	32.45	32.70	31.05	31.05	− 1.00	35.25	20.00	2,640
Aug	30.90	31.10	29.50	29.50	− 1.00	33.25	20.30	514
Sept	28.50	28.75	27.20	27.20	− 1.00	31.10	23.45	1,247
Oct	26.30	26.70	25.55	25.55	− 1.00	29.25	24.90	199
Dec	26.25	26.25	24.70	24.70	− 1.00	28.75	24.00	337
Est vol 33,530; vol Tues 22,639; open int 74,562, + 269.								

WHEAT (CBT) 5,000 bu.; cents per bu.
Dec	376½	378¼	371	371¼	− 5¾	428½	356¼	43,772
Mar84	392½	394¼	386¼	386½	− 5¾	437	370	10,978
May	398¾	398¾	392	392	− 6¼	441	377½	4,073
July	390	391¼	385½	386	− 3¾	427	357½	6,539
Sept	398½	398½	393	393	− 4½	432	386¼	720
Dec	410¼	404¼	404½	− 4½	418	403½	316	
Est vol 16,800; vol Tues 13,202; open int 66,398, − 514.								

WHEAT (KC) 5,000 bu.; cents per bu.
Dec	389½	385¾	385¾	− 1¼	423	361½	16,947	
Mar84	397	399	395	395	− 1¾	433	367½	3,697
May	401	397½	397½	− 1½	426	367½	354	
July	389	389	385½	385½	− 1	419	362	843
Sept	392	393	391	391	+ ½	417	390	25
Est vol 2,003; vol Tues 1,837; open int 21,866, + 202.								

WHEAT (MPLS) 5,000 bu.; cents per bu.
Dec	414	414½	411¼	411¼	− 1¼	484½	376¼	4,170
Mar84	412	412¼	409¼	409¼	− 1¼	449	377½	1,746
May	414½	414¾	410½	411	− 1¼	450	380	296
July	414		414		− ¾	414	414	8
Est vol 1,266; vol Tues 1,266; open int 6,220, + 105.								

WHEAT (MCE) 1,000 bu.; cents per bu.
Dec	377½	378½	371½	371¼	− 5¾	428½	357	12,104
Mar84	394	394½	386½	386½	− 5¾	437	371	603
May	399	399	394	394	− 6¼	441	373½	120
July	391	391¾	387½	386	− 3¾	400	380	280
Sept	398	398	394½	393	− 4½	430	389	50
Dec			404½		− 4½	408¼	404	4
Est vol 500; vol Tues 421; open int 13,168, − 9.								

BARLEY (WPG) 20 metric tons; Can. $ per ton
Oct	132.40	132.60	131.00	131.99	+ .90	136.00	96.00	5,361
Dec	133.50	134.40	132.80	133.00	+ .20	137.80	99.60	8,235
Mar84	136.60	137.20	135.00	135.80	+ .30	140.70	103.50	4,437
May	138.50	139.00	136.30	136.90	− .40	142.10	120.70	787
July	140.00	140.00	138.50	138.50		142.50	127.00	75
Est vol 3,400; vol Tues 3,469; open int 18,889, + 536.								

FLAXSEED (WPG) 20 metric tons; Can. $ per ton
Oct	408.00	408.00	396.00	396.00	− 7.50	427.00	294.00	702
Dec	417.00	417.50	404.50	404.50	− 8.50	436.30	299.90	1,117
Mar84	430.00	430.00	419.00	419.00	− 7.00	444.00	313.50	211
May	433.50	434.50	423.00	423.00	− 8.50	454.00	407.00	127
Est vol 720; vol Tues 610; open int 3,181, − 45.								

RAPESEED (WPG) 20 metric tons; Can. $ per ton
Nov	438.00	440.00	428.50	429.50	− 6.10	453.00	299.70	8,536
Jan84	448.00	449.50	437.80	438.90	− 6.10	460.00	306.80	7,657
Mar	452.00	453.20	442.50	443.00	− 5.50	463.00	314.50	3,870
June	454.00	456.50	445.00	445.50	− 5.40	467.40	330.00	1,359
Sept			423.00		+ 1.00	427.00	392.00	44
Nov	407.00	413.50	402.00	402.00	− 3.00	425.50	382.50	707
Est vol 3,000; vol Tues 2,532; open int 20,436, − 238.								

	Open	High	Low	Settle	Change	Lifetime High	Low	Open Interest
Sept	2,110	2,110	2,100	2,105	− 45	2,450	1,987	1,194
Dec	2,100	2,130	2,100	2,140	− 23	2,484	2,040	1,021
Est vol 4,165; vol Tues 3,935; open int 27,677, + 115.								

COFFEE (CSCE) - 37,500 lbs.; cents per lb.
Dec	134.45	136.20	134.30	135.68	+ 3.09	136.20	98.00	6,227
Mar84	131.35	132.42	131.10	131.96	+ 2.31	132.42	110.50	1,658
May	130.25	129.75	128.30	129.29	+ 2.19	129.75	108.50	468
July	126.25	128.00	126.25	127.60	+ 2.27	128.00	106.25	329
Sept	124.01	126.25	124.01	125.75	+ 2.13	126.25	110.50	299
Dec	123.25	124.75	123.00	123.87	+ 1.62	124.75	116.40	98
Est vol 1,909; vol Tues 746; open int 9,079, − 85.								

COTTON (CTN) - 50,000 lbs.; cents per lb.
Oct	75.30	75.65	74.40	74.30	− .05	82.20	65.65	537
Dec	77.25	77.40	75.60	76.16	− .09	83.10	65.50	18,689
Mar84	78.85	79.00	77.55	77.88	−	83.50	67.10	5,964
May	79.85	79.85	78.50	78.70	− .15	83.80	69.38	1,361
July	80.00	80.10	78.65	79.35	− .10	83.90	71.50	1,950
Oct	75.50	75.50	75.45	75.10	− .30	79.00	74.25	674
Dec	74.95	75.00	74.15	74.50	− .10	76.75	73.95	2,555
Mar85				75.15	+ .05	76.10	76.10	2
Est vol 7,000; vol Tues 4,259; open int 31,762, − 125.								

ORANGE JUICE (CTN) - 15,000 lbs.; cents per lb.
Sept								89
Nov	119.00	119.10	118.60	118.80	− .45	131.95	105.70	1,325
Jan84	114.80	114.80	114.05	114.20	− .40	132.50	100.15	1,252
Mar	113.50	113.80	113.05	113.15	− .55	132.50	100.00	2,381
May	112.50	112.50	112.25	112.35	− .75	119.20	100.90	739
July				112.30	− .75	113.30	101.00	353
Sept				112.30	− .70	112.75	103.80	143
Nov				110.75	− .85	111.50	108.00	108
Jan85				102.00	− .20	111.00	109.00	27
Est vol 150; vol Tues 166; open int 6,417, + 14.								

SUGAR - WORLD (CSCE) - 112,000 lbs.; cents per lb.
Oct	9.84	10.23	9.83	10.00	+ .69	13.50	7.05	8,278
Jan84	10.70	10.85	10.60	10.75	+ .63	14.23	7.65	367
Mar	11.15	11.35	11.15	11.26	+ .53	14.48	8.08	48,942
May	11.52	11.74	11.52	11.65	+ .50	14.70	8.35	13,240
July	11.85	12.11	11.83	11.95	+ .50	14.95	8.65	5,264
Sept	12.09	12.30	12.09	12.13	+ .49	14.93	9.65	765
Oct	12.20	12.43	12.20	12.28	+ .48	15.30	10.65	3,454
Jan85	12.75	12.75	12.75	12.76	+ .54	13.10	12.00	11
Est vol 16,250; vol Tues 19,677; open int 80,321, − 2,496.								

SUGAR - DOMESTIC (CSCE) - 112,000 lbs.; cents per lb.
Nov	21.85	21.85	21.80	21.84	+ .03	22.15	20.80	2,237
Jan84				21.88		22.05	21.30	351
Mar	21.90	21.90	21.85	21.85	−	22.15	21.25	3,166
May	22.10	22.10	22.10	22.03	− .04	22.23	21.20	1,775
July	22.25	22.25	22.25	22.18	−	22.35	21.70	2,015
Sept				22.21	+ .01	22.25	21.50	724
Nov				22.21	+	22.15	21.90	443
Jan85				22.21	+	22.22	22.00	203
Est vol 254; vol Tues 696; open int 10,914, + 89.								

-METALS & PETROLEUM-

COPPER (CMX) - 25,000 lbs.; cents per lb.
Sept								338
Oct	68.50	68.50	68.50	68.45	+ .20	75.50	68.50	4
Nov				69.05	+ .10	69.50	69.50	2
Dec	69.30	70.20	69.20	69.70	+ .10	93.00	66.30	63,274
Jan84	70.10	70.30	70.10	70.10	+ .10	89.50	66.95	782
Mar	71.70	72.25	71.30	71.80	+ .10	90.40	68.00	25,053
May	73.10	73.60	72.60	73.10	+ .05	88.40	69.00	5,120
July	74.40	74.80	73.95	74.45	+ .05	89.20	70.35	5,307
Sept	75.60	76.30	75.25	75.80	+ .05	90.40	73.20	4,678
Dec	77.55	78.35	77.35	77.90	+ .15	92.00	77.35	2,782
Jan84	78.45	78.50	78.00	78.55	+ .15	92.00	78.00	245
Mar	79.70	80.10	79.70	79.90	+ .15	93.20	79.60	1,870
May	82.50		82.45	82.40	+ .15	87.95	82.45	1,417
July	82.50	82.75	82.45	82.60	+ .15	87.95	82.45	200
Est vol 14,000; vol Tues 15,665; open int 109,104, − 49.								

GOLD (CMX) - 100 troy oz.; $ per troy oz.
Sept								16
Oct	413.60	415.00	410.00	411.20	− 2.90	548.50	362.00	9,505
Nov				414.30	− 2.90			0
Dec	421.00	421.70	416.00	417.60	− 2.60	554.70	370.00	50,701
Feb84	427.80	428.50	424.00	424.60	− 2.60	550.50	389.00	12,098
Apr	435.00	435.20	431.00	431.80	− 2.40	572.00	385.50	7,400
June	442.00	442.00	440.00	439.10	− 2.30	580.00	409.00	10,108
Aug	448.00	448.00	444.50	446.50	− 2.30	588.00	438.50	12,828
Oct	458.30	458.30	457.00	454.30	− 2.00	597.00	447.00	4,783
Dec	464.50	464.50	461.50	462.10	− 1.90	608.00	455.60	6,527
Feb85	471.50	471.50	471.50	470.00	− 1.80	527.00	468.00	1,941
June	480.00	480.00	479.00	478.00	− 1.70	514.50	484.50	1,171
June				486.10	− 1.50	504.00	484.50	521
Est vol 32,000; vol Tues 30,502; open int 119,769, − 206.								

GOLD (IMM) - 100 troy oz.; $ per troy oz.
Oct	411.00	414.00	411.00	411.20	− 2.60	436.00	408.00	14
Dec	421.00	421.70	416.50	417.60	− 2.60	454.20	407.50	2,344
Mar84	431.40	431.40	427.00	428.00	− 2.70	567.40	418.00	178
Apr	441.00	442.00	438.00	439.10	− 2.50	566.90	427.00	19
June	442.00	442.00	438.00	439.10	− 2.40	574.60	433.00	41
Est vol 3,197; vol Tues 2,932; open int 2,580, − 163.								

PLATINUM (NYM) - 50 troy oz.; $ per troy oz.
Oct	428.50	430.00	422.50	422.90	− 5.50	511.00	298.50	5,823
Jan84	439.50	440.50	432.00	432.90	− 4.60	518.60	325.00	7,339
Apr	445.70	445.70	439.00	438.90	− 4.60	522.50	387.00	2,083
July	453.00	453.00	451.50	446.90	− 4.60	508.20	433.00	751
Oct	460.50	460.50	457.00	455.40	− 4.60	440.00	460.00	4
Est vol 4,554; vol Tues 4,530; open int 16,167, − 4.								

PALLADIUM (NYM) 100 troy oz.; $ per troy oz.
Sept				147.70	− 1.95	153.00	140.00	43
Dec	150.25	150.50	147.50	147.70	− 1.95	157.50	106.00	5,504
Mar84	150.65	150.65	148.50	148.70	− 1.95	157.90	106.10	1,918
June	152.00	152.00	152.00	149.95	− 1.95	158.50	103.00	1,126
Sept				151.20	− 1.95	160.00	140.00	486

	Open	High	Low	Settle	Change	Lifetime High	Low	Open Interest
Apr		.8350	− .0090	.8850	.8020	59		
May	.8500	.8500	.8500	.8450	− .0110	.8850	.7500	18
June				.8450	− .0110	.8850	.8800	2
Est vol 2,802; vol Tues 1,862; open int 12,277, − 702.								

HEATING OIL NO. 2 (NYM) 42,000 gal.; $ per gal.
Oct	.8310	.8320	.8210	.8230	− .0116	.8770	.6950	3,260
Nov	.8460	.8470	.8350	.8358	− .0163	.8775	.7050	8,159
Dec	.8560	.8580	.8445	.8468	− .0169	1.0090	.7100	15,440
Jan84	.8590	.8600	.8470	.8491	− .0163	.9015	.7850	5,467
Feb	.8520	.8520	.8410	.8410	− .0170	.8725	.7910	1,426
Mar	.8240	.8250	.8190	.8190	− .0146	.8710	.8140	2,029
Apr	.8180	.8180	.8100	.8100	− .0100	.8550	.8100	67
May	.8050	.8050	.8025	.8025	− .0095	.8665	.6900	80
June	.8140	.8140	.8140	.8000	− .0130	.8540	.8140	9
Est vol 15,890; vol Tues 10,162; open int 35,937, − 416.								

CRUDE OIL, Light Sweet (NYM) 42,000 gal.; $ per bbl.
Nov	31.06	31.06	30.72	30.76	− .35	32.22	29.40	2,320
Dec	30.99	31.00	30.70	30.78	− .33	32.21	28.60	3,574
Jan84	30.90	30.90	30.65	30.72	− .33	32.15	30.65	1,048
Feb	30.77	30.77	30.55	30.60	− .30	32.00	30.55	340
Mar	30.72	30.72	30.52	30.52	− .33	31.72	30.52	170
Apr				30.50	− .30	30.73	30.50	1
Est vol 1,798; vol Tues 686; open int 7,453, − 27.								

-WOOD-

LUMBER (CME) - 130,000 bd. ft.; $ per 1,000 bd. ft.
Nov	170.00	175.10	169.80	175.10	+ 5.00	229.90	161.10	5,249
Mar	185.40	191.00	185.30	190.80	+ 4.80	234.00	172.30	983
May	198.60	203.50	198.00	203.40	+ 4.80	240.50	184.80	983
July	207.00	211.30	207.00	211.30	+ 4.50	240.50	204.00	257
Sept	213.70	218.70	213.70	218.60	+ 4.60	252.50	204.00	85
Nov	218.70	223.40	218.70	223.40	+ 3.90	236.00	209.80	35
Nov	220.00	223.00	220.00	223.00	+ 3.50	225.10	220.00	1
Est vol 3,358; vol Tues 3,216; open int 8,618, − 189.								

-FINANCIAL-

BRITISH POUND (IMM) - 25,000 pounds; $ per pound
Dec	1.4985	1.5030	1.4980	1.5025	+ .0050	1.6425	1.4460	16,242
Mar	1.5025	1.5060	1.5040	1.5040	+ .0040	1.6010	1.4470	1,742
June	1.5045	1.5060	1.5030	1.5055	+ .0030	1.5400	1.4860	193
Sept				1.5070	+ .0020	1.5240	1.5050	3
Est vol 4,063; vol Tues 5,781; open int 18,180, + 12.								

CANADIAN DOLLAR (IMM) - 100,000 dlrs.; $ per Can $
Dec	.8123	.8123	.8113	.8115	− .0006	.8171	.8005	4,459
Mar84	.8124	.8125	.8120	.8120	− .0004	.8169	.8060	966
June	.8130	.8130	.8125	.8121	− .0009	.8160	.8100	364
Sept	.8128	.8128	.8128	.8125	− .0009	.8147	.8128	1
Est vol 826; vol Tues 1,556; open int 5,790, − 280.								

JAPANESE YEN (IMM) 12.5 million yen; $ per yen (.00)
Dec	.4263	.4266	.4244	.4251	+ .0017	.4416	.4082	37,124
Mar	.4295	.4295	.4277	.4282	+ .0015	.4329	.4125	1,468
June	.4325	.4327	.4313	.4314	+ .0016	.4345	.4180	92
Est vol 15,242; vol Tues 18,044; open int 38,684, + 1,643.								

SWISS FRANC (IMM) - 125,000 francs-$ per franc
Dec	.4755	.4761	.4739	.4747	+ .0007	.5405	.4634	28,635
Mar84	.4813	.4820	.4798	.4806	+ .0011	.5170	.4695	1,268
June	.4866	.4867	.4858	.4858	+ .0010	.5045	.4752	79
Sept	.4945	.4945	.4915	.4915	+ .0010	.4945	.4844	5
Est vol 21,480; vol Tues 21,463; open int 29,949, + 954.								

W. GERMAN MARK (IMM) - 125,000 marks; $ per mark
Dec	.3826	.3828	.3809	.3817	+ .0001	.4400	.3710	24,001
Mar84	.3863	.3863	.3848	.3852	+ .0001	.4100	.3758	1,824
June	.3897	.3899	.3897	.3890	+ .0006	.3960	.3813	71
Sept	.3935	.3935	.3935	.3926	+ .0006	.3885	.3885	13
Est vol 13,330; vol Tues 13,648; open int 25,999, + 2,343.								

EURODOLLAR (IMM) - $1 million; pts of 100%
	Open	High	Low	Settle	Yield Chg Interest			Open
Dec	90.10	90.11	89.96	89.97	− .06	10.03	+ .06	13,405
Mar84	89.77	89.77	89.61	89.61	− .05	10.39	+ .05	11,521
June	89.49	89.50	89.35	89.35	− .05	10.65	+ .05	4,705
Sept	89.20	89.23	89.10	89.11	− .05	10.89	+ .05	242
Dec	89.00	89.00	88.80	88.90	− .05	11.10	+ .05	53
Mar85	88.78	88.78	88.70	88.70	− .05	11.30	+ .05	37
Est vol 5,072; vol Tues 3,770; open int 32,068, + 460.								

GNMA 8% (CBT) - $100,000 prncpl; pts. 32nds. of 100%
Sept								2
Dec	68-00	68-12	67-26	68-25	− .10	13.422	+ .072	25,830
Mar84	68-13	68-13	67-29	67-29	−	9	13.625	+ .066	10,445
June	67-17	67-21	67-05	67-05	−	3	13.801	+ .074	4,133
Sept	67-00	67-01	66-20	66-20	−	10	13.928	+ .074	1,309
Dec	66-18	66-19	66-06	66-06	−	10	14.034	+ .075	282
Mar85	66-06	66-07	65-26	65-26	−	10	14.126	+ .077	184
June			65-13	65-13	−	10	14.202	+ .076	132
Sept	65-07	65-07	65-04	65-04	−	10	14.311	+ .078	24
Est vol 3,000; vol Tues 2,498; open int 47,443, + 43.								

TREASURY BONDS (CBT) - $100,000; pts. 32nds of 100%
								4,944
Dec	72-30	73-04	72-14	72-16	− 7	11.553	+ .037	96,656
Mar84	72-13	72-14	71-29	71-31	− 7	11.648	+ .037	20,953
June	71-31	71-31	71-15	71-16	− 7	11.722	+ .037	12,991
Sept	71-18	71-18	71-03	71-04	− 7	11.787	+ .038	8,729
Dec				71-04	− 7	11.841	+ .038	3,642
Mar85	78-28	78-28	70-18	70-19	− 7	11.884	+ .038	2,328
June	70-10	70-12	70-05	70-05	− 7	11.927	+ .038	1,941
Sept	70-03	70-03	70-00	70-00	− 7	11.956	+ .039	803
Dec				69-28	− 7	11.994	+ .039	184
Mar86				69-28	− 7	12.022	+ .039	2
Est vol 85,000; vol Tues 69,494; open int 155,436, + 2,036.								

TREASURY NOTES (CBT) - $100,000; pts. 32nds of 100%
Dec	81-08	81-09	80-24	80-27	− 5	11.238	+ .031	9,155
Mar84	80-18	80-18	80-07	80-07	− 5	11.380	+ .031	3,173
June	80-02	80-02	79-23	79-23	− 5	11.459	+ .031	1
Sept	79-18	79-18	79-09	79-09	− 5	11.546	+ .031	4
Dec				79-00	− 5	11.622	+ .032	1
Est vol 3,500; vol Tues 2,603; open int 13,264, − 24.								

TREASURY BILLS (IMM) - $1 mil.; pts. of 100%
	Open	High	Low	Settle	Chg	Discount Settle Chg		Open Interest
Dec	91.14	91.14	90.98	91.00	− .08	9.00	+ .08	16,135
Mar84	90.76	90.76	90.64	90.66	− .09	9.37	+ .06	10,347
June	90.42	90.42	90.29	90.29	− .07	9.71	+ .05	3,412
Sept	90.00	90.00	89.89	89.91	− .07	10.09	+ .05	420
Dec				89.72	− .10	10.28	+ .07	419
Mar85				89.72	− .10	10.28	+ .07	162
June				89.31	− .10	10.46	+ .11	132
Est vol 16,101; vol Tues 8,549; open int 45,880, + 432.								

BANK CDs (IMM) - $1 million; pts. of 100%
| Sept | 90.90 | 90.90 | 90.82 | 90.85 | + .03 | 9.15 | − .03 | 111 |

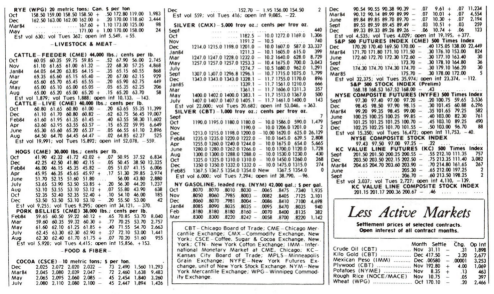

Figure 18.2. Futures price quotations from the *Wall Street Journal* (Reprinted by permission of the *Wall Street Journal*, © Dow Jones & Company, Inc., 1983, all rights reserved).

trends in price relationships as part of the process of developing expectations as to future prices. By noting those days on the chart when commodity sales were made, the manager can have a record of the specific price received and the relationship between that price and what might have been received if the sales were scheduled differently.

Because the futures market can also be used in the pricing of commodities and inputs, information should also be compiled on fluctuations and changes in futures market prices for the major commodities included in the business. Specific commodity charting procedures are commonly used to record price fluctuations on the futures market. These charts can provide information concerning trends and turning points in the market and are used by professional traders and speculators. Because charts and chart formations are used by participants in the futures market, some argue that the basic rules used in the interpretation of charts have a significant impact on short-run price movements in that market. We will not attempt here to discuss the interpretation of chart formations as used by the professional speculator, but to discuss the construction and use of charts as a means of summarizing information on past price movements and obtaining some insight into price fluctuations.

Futures prices are available from any major commodity brokerage firm or advisory service as well as from major metropolitan newspapers and the *Wall Street Journal*. Figure 18.2 illustrates the futures price quotations provided in the *Wall Street Journal* on a daily basis. Note that for each commodity contract, data are summarized concerning the opening price on the exchange, the high and low prices during the trading day, and the closing price for the day. Furthermore, the change in the closing price from the previous day is also recorded, along with the high and low prices for the contract since it has been on the board.

Charting of futures prices enables the market analyst to obtain a quick visual impression of what has happened during any trading day as well as over a period of time. For ease of interpretation, futures prices charts should be kept on graph paper as is illustrated in Figure 18.3, preferably professional chart paper. Figure 18.3 illustrates the chart that has been developed for a July Chicago wheat contract. Note that the horizontal axis refers to trading days and the vertical axis reflects prices. The specific trading days within each month are denoted on the horizontal axis, with the first trading day of the week (Monday) represented by the darker vertical lines; the lighter vertical lines refer to the remaining trading days of the week. The charting process is quite simple: for each day, a vertical line is drawn connecting the high and the low prices on the contract for that day. A cross-hatch is used to denote the closing price for the day. By comparing not only the high and low prices for the day but also the close with price fluctuations that occurred in the past, it is possible to obtain a visual picture of futures market price trends and movements.

For detailed information concerning the methods used by profes-

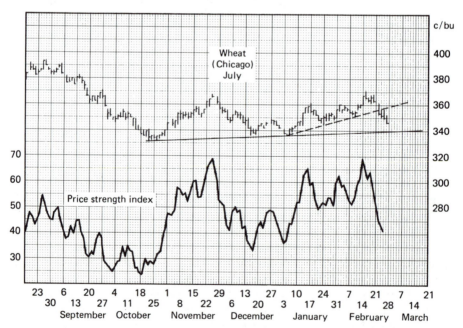

Figure 18.3.   Chart of July Chicago wheat futures contract prices for 1982–83.

sional commodity traders to detect trends, turning points, and other chart formations, contact a commodity broker or a professional commodity charting service. Using charts to speculate in commodity markets is a risky business, and one should not indulge in it without professional assistance, the mental capacity to handle risk, and the financial strength to withstand potentially large losses.

Input Purchase Schedules

A schedule of major input purchases can provide useful marketing as well as financial information. First, such a schedule is useful in making sure that the inputs are available when needed to complete the production process in a timely fashion. Second, as will be discussed shortly, financial records such as the cash flow record that reflects the timing of input purchases and product sales can be useful in evaluating the financial performance of a business. A historical record of monthly cash outlays to purchase inputs can provide useful information in evaluating past and current financial performance and in developing the projected cash flow statement for future years.

A third and extremely important purpose of the record of input purchases is to summarize the terms of the purchase, including the availability of a cash discount and when and if finance charges become due. Many input supply firms such as feed and fertilizer dealers offer discounts if cash is paid within a specified number of days. These dealer credit arrangements can result in significant savings for the farmer. For example, common discount terms utilized in the feed and fertilizer industry are a 2 percent cash discount if a bill is paid within 10 days, with the full balance due in 30 days if the discount is not taken. This type of arrangement means that if a farmer buys $100 worth of seed, he or she need only pay $98 if payment occurs in 10 days but the full $100 if payment is made in 30 days. Thus, the farmer is paying $2 for the privilege of having 20 days of credit if the bill is not paid within the cash discount period. Not taking the cash discount and paying $2 for 20 days of credit is equivalent to a rate of interest in excess of 36 percent. Since most cash discount terms result in high interest rate equivalents, the farmer is usually well advised to take advantage of cash discount arrangements. By keeping a record of when purchases were made and when payment is due to take advantage of the cash discount, significant cost savings can be generated for many farm businesses. In addition, a record of when bills are due may reduce the possibility of incurring costly finance charges or impairing a farmer's credit rating because of forgotten or overdue accounts payable.

One method that can be used to keep a schedule of input purchases, due dates, and cash discounts for various bills and other obligations is illustrated in Table 18.1. In addition, a calendar might be used to record the date when a bill was received, the date when payment is due on that bill,

Table 18.1 Schedule of Input Purchases
and Cash Discounts

Date Purchased	Type of Input	Total Amount	Payment Due Date	Cash Discount			Net Due
				Amount	Due Date		Net Due
1/26	Hog protein	$3,000	2/28	$60	2/4		$2,940
1/30	Vet supplies	1,000	2/28	20	2/10		980
2/10	Tractor repairs	1,500	3/15	45	2/15		1,455
2/15	Seed corn	2,500	3/30	50	2/28		2,450
2/18	Building supplies	2,000	3/18	60	2/28		1,940
3/1	Fertilizer	6,000	4/1	180	3/10		5,820
	.						
	.						
	.						
	.						
	.						
	(other inputs)						

and the cash discount date (possibly two to three days earlier than the due dates if the bill will be paid by mail). This recording of due dates serves as a reminder to pay any outstanding bills before finance charges are imposed and to take advantage of cash discounts if possible. Furthermore, the dated record of receipt and payment of bills is useful in keeping financial records for tax purposes as well as in developing cash flow projections and budgets for future years.

Product Sales Schedules

The third component of the market control system is the product sales record. Table 18.2 illustrates a form to summarize product sales activities which can be used to monitor performance with respect to this dimension of the marketing plan.

As with any control system, the structure of the product sales record facilitates the comparison between planned sales decisions and actual decisions. A separate record should be kept for each major commodity sold by the firm. As noted in Table 18.2, the actual supply and use of the commodity and thus the total amount that is marketable are first documented. The monthly actual and planned sales of this marketable surplus using cash, forward contract, or hedging methods are then summarized. Note that entries are required not only for actual compared to budgeted sales on a monthly basis, but also for year-to-date budgeted and actual sales. The year-to-date comparisons facilitate analysis of whether short-run adjustments in the scheduling of marketings has dramatically altered the longer run market plan; i.e., whether or not plans to market a certain proportion or quantity of the product by a particular date are being carried out,

Table 18.2 Commodity Sales Record

			Unhedged Cash Sales												
			Budgeted Sales			Actual Sales			Budgeted Sales Year to Date			Actual Sales Year to Date		
Crop	Inventory	Month[a]	Quantity	Price/Unit	Gross Sales	Quantity	Price/Unit	Gross Sales	Quantity	Gross Sales	Average Price	Quantity	Gross Sales	Average Price
Corn	Beginning carryover 28,500	January												
	Production 44,700	February	2,000	$2.65	$5,300	2,000	$2.60	$5,200	2,000	$5,300	$2.65	2,000	$5,200	$2.60
	Total 73,200	March												
	Usage on farm 26,200	April												
	Ending carryover 25,000	May												
	Total 51,200	June	6,000	$2.87	$17,220	7,000	$2.70	$18,900	8,000	$22,520	$2.82	9,000	$24,100	$2.68
	Marketable surplus 22,000	July												
		August												
		September												
		October												
		November												
		December												
Soybeans	Beginning carryover	January												
	Production	February												
	Total	March												
	Usage on farm	April												
	Ending carryover	May												
	Total	June												
	Marketable surplus	July												
		August												
		September												
		October												
		November												
		December												

[a]The month for forward contract and hedged sales refers to the month when the commodity will be delivered and cash received.

(continued)

Table 18.2 (Continued)

Crop	Inventory		Month[a]	Forward Contract Sales											
				Budgeted Sales			Actual Sales			Budgeted Sales Year to Date			Actual Sales Year to Date		
				Quantity	Price/Unit	Gross Sales	Quantity	Price/Unit	Gross Sales	Quantity	Gross Sales	Average Price	Quantity	Gross Sales	Average Price
Corn	Beginning carryover	28,500	January												
	Production	44,700	February												
	Total	73,200	March												
	Usage on farm	26,200	April	6,000	$2.94	$17,640	6,000	$2.90	$17,400	6,000	$17,640	$2.94	6,000	$17,400	$2.90
	Ending carryover	25,000	May												
	Total	51,200	June												
	Marketable surplus	22,000	July												
			August												
			September												
			October												
			November												
			December												
Soybeans	Beginning carryover		January												
	Production		February												
	Total		March												
	Usage on farm		April												
	Ending carryover		May												
	Total		June												
	Marketable surplus		July												
			August												
			September												
			October												
			November												
			December												

(continued)

Table 18.2 (Continued)

| | | | | Hedged Sales | | | | | | | |
| | | | | Budgeted Sales | | | Actual Sales | | | | |
Crop	Inventory		Month[a]	Quantity	Net Price/Unit	Gross Sales	Quantity	Cash Price/Unit	Futures Contract Gain or Loss/Unit	Net Price/Unit	Gross Sales
Corn	Beginning carryover	28,500	January								
	Production	44,700	February								
	Total	73,200	March	2,000	$2.63	$5,260	2,000	$2.50	+0.27	$2.77	$5,540
	Usage on farm	26,200	April								
	Ending carryover	25,000	May								
	Total	51,200	June								
	Marketable surplus	22,000	July	5,000	$2.76	$13,800	5,000	$2.40	+0.42	$2.82	$14,100
			August								
			September								
			October								
			November								
			December								
Soybeans	Beginning carryover		January								
	Production		February								
	Total		March								
	Usage on farm		April								
	Ending carryover		May								
	Total		June								
	Marketable surplus		July								
			August								
			September								
			October								
			November								
			December								

(continued)

Table 18.2 *(Continued)*

| | | | Hedged Sales | | | | | |
| | | | Budgeted Sales Year to Date | | | Actual Sales Year to Date | | |
Crop	Inventory		Month[a]	Quantity	Gross Sales	Average Price	Quantity	Gross Sales	Average Price
Corn	Beginning carryover	28,500	January						
	Production	44,700	February						
	Total	73,200	March	2,000	$5,250	$2.63	2,000	$5,540	$2.77
	Usage on farm	26,200	April						
	Ending carryover	25,000	May						
	Total	51,200	June						
	Marketable surplus	22,000	July	7,000	$19,060	$2.72	7,000	$19,640	$2.81
			August						
			September						
			October						
			November						
			December						
Soybeans	Beginning carryover		January						
	Production		February						
	Total		March						
	Usage on farm		April						
	Ending carryover		May						
	Total		June						
	Marketable surplus		July						
			August						
			September						
			October						
			November						
			December						

regardless of short-term adjustments in actual sales between months that might reflect attempts to take advantage of unexpected fluctuations in prices.

The commodity sales record summarizes the actual price received for the product on a monthly basis and the year-to-date average price. This record can again be used to compare the expected prices included in the marketing plan to the actual price received to determine whether prices above or below those expected are being obtained for the commodity. For hedged sales, the futures contract gain or loss per unit is also summarized so that an actual net price—the cash price plus or minus the net gain or loss on futures market transactions—can be determined.

Comparing the market record for Andy's farm operation of Table 18.2 to the market plan of Table 7.6 indicates that the firm actually has an additional 1,000 bushels of corn to sell, and it sold that additional corn for cash in June. The market plan called for an average price for all cash sales of $2.82 per bushel, an average price for forward contract sales of $2.94, and an average price for hedged sales of $2.72. Actual average prices were $2.68, $2.90, and $2.81 per bushel for cash, contract, and hedged sales, respectively. Thus, actual performance was above expectations using the hedging alternative, but below expectations for the cash and contract sale methods. A similar analysis should be used to evaluate Andy's hog marketing plan of Table 7.6.

FINANCIAL CONTROL

The financial control system is probably the most important component of the total control system for the farm business. The purpose of the financial control system is to monitor the solvency, liquidity, profitability, and efficiency of the business and to indicate when performance is not consistent with expectations in these areas. The financial control system may not indicate explicitly what corrective action must be taken to improve performance, *but it should clearly indicate which dimensions of the production, marketing, or financial plan are the cause of unacceptable performance and suggest where further detailed analyses are needed.*

As with the production and marketing control systems, financial control seeks to compare actual performance with specified standards. These standards are reflected in the financial budgets prepared as part of the planning process. These pro forma statements of the expected financial performance of the firm summarize the overall anticipated results of the production, marketing, and financial plan. In addition to the comparison of actual performance to budgeted expectations, the financial performance of the firm can also be evaluated by comparing the firm to other similar farm operations or by comparing performance this year to performance in previous years. These additional forms of comparative analyses (interfirm comparisons and intrafirm trend analysis) may also be useful in evaluating whether the standards of performance are adequate. It is possible that intrafirm and trend analysis may suggest that higher or lower standards should be specified for future years to generate better performance.

The key documents of the financial control system have been discussed in detail in Chapters 2 and 6; they include the cash flow statement, the income or profit and loss statement, the balance sheet, and financial and business performance ratios. We will briefly review these documents here with specific emphasis on how they can be used as part of a financial control system.

The Cash Flow Statement

The cash flow statement is the key component of the financial control system. The income statement and balance sheet to be discussed shortly provide annual summaries of the financial condition and progress of the firm, but these are only annual statements. As noted earlier in the discussion of concurrent and feedback control, a useful control system must monitor performance frequently so that deviations between plans and performance can be detected and adjustments made before disastrous financial consequences occur. The cash flow statement provides the mechanism for continual monitoring of the performance of the firm; it should provide an early warning system concerning the source of potential problems that might result in unacceptable performance.

Construction of the cash flow statement or record follows closely the structure and procedure of the cash flow budget discussed in Chapters 2 and 6. In fact, as indicated in Table 18.3, the headings along the left-hand side of the cash flow record are identical to those in the budget, with cash income summarized at the top of the statement and cash expenses at the bottom of the statement. The headings across the top for the cash flow statement reflect monthly, bimonthly, or quarterly periods of the year as with the cash flow budget. However, additional subheadings are included in the statement compared to the budget form of Chapter 6. As shown in Table 18.3, entries for budgeted cash flow for the period, actual cash flow for the period, percentage deviation between budget and actual, budgeted cash flow year-to-date, actual cash flow year-to-date, and percentage deviation are required for each period of the year.

The numbers for the budgeted entries for each period come from the cash flow budget. These entries indicate what the expected financial consequences will be if the production, marketing, and financial plans are implemented. The "budgeted this period" comes directly from the cash flow budget, whereas the "year-do-date" figure is a simple summation of all of the projected cash flows from the beginning of the year through the current period.

The entries for the "actual" columns in the cash flow record come from monthly, bimonthly, or quarterly income and expense records. One simple way to obtain this type of information is to use the monthly bank statements that summarize the checking account transactions. In addition, a monthly, bimonthly, or quarterly transactions journal will provide such information on income and expenditures by period.

The primary benefit of the cash flow record can now be easily seen. By comparing actual cash flow for the period to projected cash flow, constant monitoring of the income and expense pattern of the business can occur. If actual expenditures for a particular item are higher than expected, the manager immediately knows that some part of the production, financial, or marketing plan is not being properly implemented. It may be that adjustments in the plan were required and the deviations between budgeted and actual cash flows in the cash flow record are no cause for concern. However, the deviations may also indicate that expenses are higher than expected and new cost control measures must be implemented. Likewise, if actual income is lower than expected income, the manager is quickly aware of these deviations so that appropriate adjustments in the implementation of the production or marketing plan can be made.

The "year-to-date" columns for each period in the cash flow record provide two important additional sources of information about the financial condition of the firm. First, a slight deviation in the production and marketing plan may change the timing of cash income or cash expenditures. If these changes in the plans move income and expenses between periods in the cash flow statement, significant differences would occur in the budgeted and actual cash flows for a particular period. These deviations may provide false signals that major problems are occurring in implementing the plans, whereas actually only minor adjustments have been made. The "year-to-date" totals in the cash flow record are more stable and should provide a second check concerning whether major deviations between plans and performance are occurring. For example, the first quarter figures for grain income may show significantly smaller actual income than expected. If the farm manager only analyzed within-period numbers, he or she might suspect that the marketing and financial plan had gone awry. However, if the planned and actual "year-to-date" totals at the end of the second period are quite small, the manager is reassured that, although minor changes may have been made in the timing of sales to take advantage of market opportunities, the overall marketing plan is still being followed and thus there is no major cause for alarm.

The second major use of the "year-to-date" columns of the cash flow record is to provide a concurrent or within-year indication of the actual performance of the business in terms of income generation and expense outlays. Comparing these columns to the "year-to-date" figures for previous years will indicate whether the firm is generating more or less income during the first quarter or during the first half of the year. This informa-

Table 18.3 Actual and Projected Monthly Cash Flow for the First Year of Andy's Operation with Alternative Plan 2

Item	January Budgeted This Period	Actual This Period	Percentage Deviation	Budgeted Year to Date	Actual Year to Date	Percentage Deviation	December Budgeted This Period	Actual This Period	Percentage Deviation	Budgeted Year to Date	Actual Year to Date	Percentage Deviation
Cash Receipts												
1. Corn										36,000	34,150	−5.1
2. Soybeans	5,985	5,540	−7.4	5,985	5,540	−7.4	42,500	40,830	−3.9	48,485	46,370	−4.4
3. Feeder pig enterprise										53,753	55,780	+3.8
4. Farrow-to-finish swine										3,253	2,870	−11.8
5. Miscellaneous income							200	0	−100.0	800	1,000	+25.0
6. Capital sales										0	0	0.0
7. Total cash receipts	5,985	5,540	−7.4	5,985	5,540	−7.4	42,700	40,830	−4.4	142,291	140,170	−1.5
Operating Expense												
8. Corn production										16,754	17,421	+4.0
9. Soybean production										9,242	9,845	+6.5
10. Feeder pig enterprise —Purchased pigs										18,240	17,896	−1.9
11. Other cash expenditures										8,718	8,521	−2.3
12. Farrow-to-finish swine										4,999	5,246	+4.9
13. Utilities	250	230	−8.0	250	230	−8.0	1,321	1,425	+7.9	2,000	2,105	+5.3
14. Other repairs	180	180	0.0	180	180	0.0	250	300	+20.0	2,180	2,148	−1.5
15. Taxes, insurance, and rent										2,900	2,907	+0.2
16. Hired labor										2,495	2,816	+12.9
17. Other										0	0	0.0
Capital Expenditures												
18. New facilities	5,000	4,870	−2.6	5,000	4,870	−2.6	960	850	−11.5	50,000	46,870	−6.3
19. Purchase breeding stock										14,000	15,640	+11.7
Proprietor Withdrawals												
20. Living expenses and taxes	6,500	7,100	+9.2	6,500	7,100	+9.2	1,500	1,400	−6.67	17,500	18,729	+7.0

(*continued*)

Table 18.3 (Continued)

Item	January Budgeted This Period	Actual This Period	Percentage Deviation	Budgeted Year to Date	Actual Year to Date	Percentage Deviation	...	December Budgeted This Period	Actual This Period	Percentage Deviation	Budgeted Year to Date	Actual Year to Date	Percentage Deviation
Payments on Previous Debt													
21. Intermediate—principal											0	0	0.0
—interest											0	0	0.0
22. Long-term—principal											5,355	5,355	0.0
—interest											17,000	17,000	0.0
23. Other											0	0	0.0
24. Total cash outflow	11,930	12,380	+3.8	11,930	12,380	+3.8	...	4,031	3,975	-1.4	171,383	172,499	+0.7
Flow-of-Funds Summary													
25. Beginning cash balance	2,083	2,083	0.0	2,083	2,083	0.0		2,057	1,361	-33.8	2,083	2,083	0.0
26. Cash receipts	5,985	5,540	-7.4	5,985	5,540	-7.4		42,700	40,830	-4.4	142,291	140,170	-1.5
27. Cash outflow	11,930	12,380	+3.8	11,930	12,380	+3.8		4,031	3,975	-1.4	171,383	172,499	+0.7
28. Cash difference	(3,862)	(4,757)	-23.2	(3,862)	(4,757)	-23.2	...	40,762	38,216	-6.2	(27,009)	(30,246)	-12.0
29. Borrowing this month													
—Intermediate and long	5,000	5,000	0.0	5,000	5,000	0.0		960	850	-11.5	64,000	64,000	0.0
—Operating	900	900	0.0	900	900	0.0	...				55,700	58,937	+5.8
30. Payment on operating loan													
—principal								36,100	34,443	-4.6	85,700	86,409	+0.8
31. Interest								1,412	1,862	+31.9	2,817	3,521	+25.0
32. Ending cash balance	2,038	1,143	-43.9	2,038	1,143	-43.9	...	4,174	2,761	-33.9	4,174	2,761	-33.9
Debt Outstanding													
33. Long-term	170,000	170,000	0.0	170,000	170,000	0.0		164,645	164,645	0.0	164,645	164,645	0.0
34. Intermediate loan	5,000	5,000	0.0	5,000	5,000	0.0		64,000	64,000	0.0	64,000	64,000	0.0
35. Operating loan	30,900	30,900	0.0	30,900	30,900	0.0	...	0	2,528	a	0	2,528	a
36. Accrued interest on operating loan	309	309	0.0	309	309	0.0	...	0	0	0	0	0	0

[a] A percentage deviation from budgeted cannot be calculated because the budgeted amount was zero (0).

tion again can be useful in evaluating the performance of the firm during the year and monitoring whether changes in plans or their implementation are required. For example, if accumulated cash income from the first of the year to the end of the second quarter is significantly less than in previous years, a farm manager may want to evaluate the production or marketing plan to determine whether adjustments should be made. Likewise, if accumulated expenditures during the first half of the year are running higher than in previous years, the manager should evaluate whether additional cost control measures should be implemented.

The data in Table 18.3 for Andy's operation illustrate the usefulness of the cash flow record as part of the financial control system. Andy expected to sell $5,985 of soybeans in January, but actual sales totaled $5,540 or 7.4 percent less than expected. Cash outlays in January were budgeted at $11,930, but actual outflows were $12,380, about 3.8 percent higher than expected. The higher outlays are primarily attributable to a 9.2 percent larger outlay for taxes and living expenses. The final result in January is a 43.9 percent smaller actual ending cash balance compared to budget. Thus, the data for January suggest that the farm manager should monitor closely the soybean marketing plan and family living expenses to avert future financial problems.

The data for December indicate the performance of the firm for both that period and the entire year. Andy's actual receipts were 4.4 percent below expectations in December, but outflows were 1.4 percent below budget. The actual cash difference for December was 6.2 percent below budget, primarily because of the lower cash receipts and the smaller actual beginning cash balance. The year-to-date figures for December indicate that actual cash receipts for the year are 1.5 percent lower than expected (higher feeder pig sales are more than offset by lower corn, soybean, and market hog sales), whereas cash outlays are 0.7 percent above budget (higher corn, soybean, market hog, labor, and living expenses are not completely offset by lower feeder pig expenses and capital expenditures). At the year's end, the actual cash deficit is 12.0 percent larger than expected, the actual interest paid on the operating loan is 25.0 percent higher than budget, and the actual ending cash balance is 33.9 percent smaller than expected. Note also that the actual long-term and intermediate-term debt are no different than expected by the end of the year. However, it was expected that the operating debt would be completely repaid by the end of the year, but in actuality an operating loan of $2,528 is still outstanding at year's end.

Thus, with the cash flow record or statement, potential problems can be detected during the year so that corrections can be made. If a farm manager relies solely on the annual income statement and net worth statement as the major components of her or his financial control system, problems that can have serious financial consequences are not detected until the end of the year. By year's end, losses may already be significant and control measures difficult to implement. The cash flow statement provides a much more timely system for monitoring the financial condition of the firm, detecting potential problems, and suggesting alternative procedures that could be implemented to correct such problems. This is the essence of a financial control system.

The Income Statement

The second component of the financial control system is the income statement. As indicated in Table 18.4, the format of the income statement is similar to that noted earlier, with the addition of blanks for summarizing budgeted as well as actual numbers for major entries and the percentage deviation between budgeted and actual results. The data of Table 18.4 for Andy's operation indicate that actual operating receipts were 1.3 percent below expectations and actual operating expenses were 2.4 percent above expectatons. Consequently, actual net cash operating income was $3,819 or 7.4 percent below budget. Note how only relatively small percentage changes in opposite directions in receipts and expenses can have a significant impact on net income.

Although the actual adjustment for inventory changes indicates that Andy had 0.9 percent more inventory at the end of the year than he expected, the net capital adjustment which reflects new purchases, sales, and depreciation is a larger deficit than expected because of fewer purchases and more depreciation. If we include these adjustments, actual farm profit is $5,042 or 10.3 percent less than expected. With the 7.0 percent larger proprietor withdrawal for family living and taxes as noted earlier, the amount of funds available for reinvestment is reduced by 20.0 percent compared to expectation. This sizable reduction is a function of both lower income and higher withdrawals. It suggests that care must be exercised in controlling business and family expenses and increasing income in future years if the long-run planned expansion of the firm is to be financially successful.

The Balance Sheet

The third component of the financial control system is the balance sheet. As suggested by Table 18.5, the format of the balance sheet is again similar to that noted earlier, with the additional blanks for summarizing budgeted and actual numbers for major entries as well as the percentage deviation between budgeted and actual results. The data in Table 18.5 for Andy's operation indicate that actual current assets are 1.3 percent lower than expected (primarily because of less cash and a lower inventory of crops), and actual intermediate assets are 0.6 percent lower than expected (the lower value of machinery and equipment is not completely offset by the higher value of breeding stock). Actual long-term assets are 0.3 percent lower than expected because of lower valued buildings and improvements. Thus, actual year-end assets are $3,743 or 0.5 percent lower than expected.

Table 18.4 Actual and Projected Income Statement for Andy's Operation

Period Covered: ___1/1___ Year 1 to ___12/31___ Year 1

Farm Operating Receipts	Units				Farm Operating Expenses	Actual	Projected	Percentage Deviation
Livestock and livestock products					Corn enterprise	$ 17,421		
Slaughter hogs	( )	$ 55,780						
	( )				Soybean enterprise	9,845		
	( )							
	( )				Finishing feeder pigs	26,417		
	Subtotal	$ 55,780	(1)					
Crop sales					Farrow-to-finish	5,246		
Corn	( )	$ 34,150						
	( )							
Soybeans	( )	46,370						
	( )							
	Subtotal	$ 70,520	(2)		Utilities	2,105		
Other operating receipts					Other repairs	2,148		
Miscellaneous		$ 1,000			Taxes, real estate, sales	2,100		
					Insurance	807		
					Rents			
	Subtotal	$ 1,000	(3)		Trucking and market			
Gross farm operating receipts					Hired labor	2,816		
(1) + (2) + (3)—Actual	(4)	$137,300			Farm interest paid			
					($3,521 + $17,000)	20,521		
Projected		$139,038			Other			
Percentage deviation		−1.3			Gross farm operating expense (5)	$ 89,426	$87,345	+2.4
					Net cash operating income (6)	$ 47,874	$51,693	−7.4
					(4) − (5)			

Value of farm products produced on the farm and consumed by the household

Adjustment for Changes in Inventory

	Feed and Grain	Market Livestock	Accounts Receivable	Supplies and Prepaid Expenses	Accounts Payable		Actual		Projected	Percentage Deviation
					$12,930	Begin				
					17,633	End				
Ending inventory	$64,870	$22,855	$0	$2,100	(4,703)					
Beginning inventory	63,000	0	0	913						
Net adjustment	1,870	22,855	0	1,187			$ 21,209	(8)	$21,024	+0.9

Adjustments for Changes in Capital Items

	Breeding Livestock	Machinery and Equipment	Buildings and Improvements		Actual		Projected	Percentage Deviation
Ending inventory	$ 9,621	$102,870	$101,500					
Plus sales	2,870	0	0					
Subtotal	12,491	102,870	101,500	(9)				
Beginning inventory	0	108,191	71,500					
Plus purchases	15,640	8,870	38,000					
Subtotal	15,640	117,061	109,500	(10)				
Net capital adjustment (9) − (10)	(3,149)	(14,191)	(8,000)					
Farm profit or loss (6) + (7) + (8) + (11)					($25,340)	(11)	($23,932)	−5.9
(returns to unpaid labor, operator's labor, equity capital, and management)								
Off-farm income					$43,743	(12)	$48,785	−10.3
					0	(13)	0	0.0
Total net income (12) + (13)					$43,743	(14)	$48,785	−10.3
Proprietor withdrawal					$18,729	(15)	$17,500	+7.0
Addition to retained earnings (14) − (15)					$25,014	(16)	$31,285	−20.0

Table 18.5 Actual and Projected Balance Sheets for Andy's Operation

Assets	Projected	Actual	Percentage Deviation
Current Business			
1. Cash and checking account	$4,174	$2,761	
2. Farm notes and accounts receivable	0	0	
3. Livestock held for sale	22,440	22,855	
4. Crops held for sale	65,200	64,870	
5. Value in growing crops	0	0	
6. Farm supplies	2,000	2,100	
7. Prepaid expenses	0	0	
8. Other	0	0	
9. Total current assets	93,814	92,586	−1.3
Intermediate Business			
10. Machinery, equipment, and vehicles	104,742	102,870	
11. Breeding livestock	8,464	9,621	
12. Movable farm buildings	0	0	
13. Securities not readily marketed	0	0	
14. Other	0	0	
15. Total intermediate assets	113,206	112,491	−0.6
16. Total current and intermediate assets (9 and 15)	207,020	205,077	−0.9
Long-Term Business			
17. Farmland	480,000	480,000	0.0
18. Permanent buildings and improvements	103,300	101,500	−1.7
19. Other	0	0	
20. Total long-term assets	583,300	581,500	−0.3
21. Total business assets (16 and 20)	790,320	786,577	−0.5

Liabilities	Projected	Actual	Percentage Deviation
Current Business			
27. Accounts payable	$0	$2,528	
28. Notes payable within 12 months	0	0	
29. Principal payments on longer term debt due within 12 months	12,233	12,233	
30. Estimated accrued interest	17,633	17,633	
31. Estimate accrued tax	0	0	
32. Accrued rent	0	0	
33. Total current liabilities	29,866	32,394	+8.5
Intermediate Business			
34. Deferred principal owed	57,658	57,658	
35. Deferred accounts payable	0	0	
36. Deferred notes payable	0	0	
37. Contingent income tax liabilities	0	0	
38. Total intermediate	57,658	57,658	0.0
39. Total current and intermediate (33 + 38)	87,524	90,052	+2.9

Table 18.5 (*Continued*)

Liabilities	Projected	Actual	Percentage Deviation
Long-Term Business			
40. Deferred principal on-farm real estate	158,754	158,754	
41. Other	0	0	
42. Contingent capital gains tax liability on real estate	52,700	52,700	
43. Total long-term liabilities	211,454	211,454	
44. Total business liabilities (39 + 43)	298,978	301,506	+0.8
45. Net worth (21 − 44)	491,342	485,071	−1.3
46. Total business liabilities and net worth	790,320	786,577	−0.5

The liability entries indicate that actual current liabilities are 8.5 percent higher than expected because the operating loan was not completely repaid, and total liabilities are 0.8 percent higher than expected. End of the year net worth was projected to total $491,342, but actual year-end net worth is 1.3 percent lower at $485,071. Although these deviations appear to be within tolerable limits, Andy will want to monitor future performance carefully to make sure that it is not consistently lower than planned expectations.

Liquidity, Solvency, Profitability

The fourth component of the financial control system consists of the financial ratios that indicate the liquidity, solvency, and profitability of the business. These ratios are summarized in Table 18.6 for Andy's operation. The liquidity analysis suggests that the actual current capital and current ratios are not as high as planned (5.9 percent and 8.9 percent lower than expectations, respectively), but are still reflective of a strong liquidity position. The net capital ratio is not much different than expected (1.1 percent lower) and is also reflective of a strong solvency position. And with the larger appreciation than was anticipated, the net worth with appreciation is slightly higher than expected. As to profitability, the 10.3 percent lower profit than budgeted results in a 24.0 percent lower return to operator's labor and management and an approximate 15 percent lower return to equity capital. Thus, although the solvency and liquidity position are relatively strong, the profitability is significantly below expectations and suggests that Andy should monitor future income and expenses very carefully.

BUSINESS EXPANSION AND CONTRACTION

Most of the discussion thus far has concentrated on implementing control and monitoring procedures for the short (and possibly intermediate) run.

Table 18.6 Actual and Projected Liquidity, Solvency, and Profitability for Andy's Operation

Item		Projected	Actual	Percentage Deviation
1.	Current working capital	$63,948	$60,192	−5.9
2.	Current ratio	3.14	2.86	−8.9
3.	Net worth without appreciation	491,342	485,071	−1.3
4.	Net capital ratio	2.64	2.61	−1.1
5.	Appreciation of capital assets	48,000	57,000	+18.8
6.	Total assets with appreciation	838,320	843,577	+0.6
7.	Increase in contingent tax liability	9,600	11,400	+18.8
8.	Total liabilities with appreciation	308,578	312,906	+1.4
9.	Net worth with appreciation	529,742	530,671	+0.2
10.	Net capital ratio with appreciation	2.72	2.70	−0.7
11.	Profit or loss	48,785	43,743	−10.3
12.	Interest on net worth in business at 6 percent	28,542	28,354	−0.7
13.	Opportunity cost of unpaid family labor	0	0	0.0
14.	Opportunity cost of operator's labor	15,000	15,000	0.0
15.	Residual return to operator's labor and management	20,243	15,389	−24.0
16.	Operator's labor and management earnings per month	1,687	1,282	−24.0
17.	Residual return to equity capital	33,785	28,743	−14.9
18.	Rate of return on equity capital (%)	7.10	6.08	−14.4

But most farm firms are not in a stable state in the long run; they are in the process of either expanding and increasing the size of the business, or contracting and phasing out various enterprises, activities, and responsibilities. Managing and controlling the long-run expansion and contraction process is difficult because it requires the projection of possible events further into the future. Our discussion here will focus on the key factors to consider in the expansion or contraction process and the methods that might be used to monitor the firm's longer run performance.

As with any control or monitoring techniques, the first crucial step in monitoring the long-run expansion or contraction process is to have a well-developed set of plans that can provide the standards of performance. Note that these plans may call for contraction or dissolution of the firm in the long run as well as expansion. Most of the focus of farm management is on improving the efficiency of a growing or expanding firm in the long run. Much of the earlier discussion of financial planning, for example, has focused on the acquisition of additional resources, and the control techniques discussed earlier in this chapter have emphasized the efficient utilization of those additional resources. But the focus of the control function need not always be on a growing firm or one that is increasing in size over time. At some stage in the life cycle of many entrepreneurs, their objective becomes one of reducing the size of the business and the managerial and entrepreneurial responsibilities associated with the farming operation. If there are no heirs or others to take over the business, a systematic, controlled contraction is appropriate. It is just as important to specify the long-run plans and set up performance standards for this contraction phase as for the expansion phase. Just as efficiency is necessary to enhance the growth prospects of the firm, efficiency is also essential to minimize the losses or costs that can be incurred in the contraction process. The point to remember is that a control process and performance standards are necessary during all phases of the life cycle of the business, and that efficient performance is important during contraction as well as expansion.

Planning Considerations

A farm manager must consider a number of dimensions of the firm in planning the expansion or contraction process. These dimensions will be briefly noted here since the control function requires monitoring of performance in these same dimensions. Clearly, one of the first considerations in any expansion plan is the expected profitability of that expansion. In the long run, will the commitment of additional resources to a particular enterprise increase the profit potential of the firm? Is there a more profitable enterprise or activity that the firm should consider? Similarly, if part of the resource base of the firm is being liquidated or a particular enterprise is being phased out, how does this change in organization influence the longer run profits of the firm?

A second consideration in planning the expansion or contraction process is that of risk. Does the change in enterprise organization and business activity increase or decrease the business and financial risk faced by the manager? If the risk is increased, can the firm withstand this higher risk and does it increase the chances of financial failure? Are there methods and techniques that might be used to reduce risk?

A third consideration in the expansion or contraction process is the change in resource requirements that will result from a larger or smaller firm. Resource utilization issues relate not only to the availability of labor and physical facilities, but also to management and financial resources. For example, expansion in a particular enterprise may require the acquisition of additional physical facilities such as buildings and equipment as well as additional labor with specialized skills to carry out the production activities. Furthermore, additional funds (debt or equity) will typically be required to purchase the resources needed for the expansion, suggesting the importance of a long-run cash flow budget. And different management expertise may be essential to successfully operate the expanded business. Similarly, contraction will typically result in fewer resources being committed to the productive activities of the firm. The issues of which resources to cut back or leave idle, and how to efficiently redeploy excess resources can make a significant difference in the efficiency of the contraction process. For example, if physical resources sit idle when they could possibly be rented to other operators, the contraction process is not being implemented efficiently. Likewise, the redeployment or reallocation of management and labor resources to alternative uses (which in some cases may include increased leisure time) is necessary to implement the contraction process in an efficient fashion.

Tax considerations are an increasingly important dimension of the expansion and contraction process. As has been noted earlier, depreciation rules, investment tax credit, capital gains tax provisions, and managing tax liabilities with the use of the cash accounting system are important determinants of the financial consequences of the expansion process. Tax provisions may not only affect the type of facilities or equipment acquired and the technology adopted during expansion, but they may also significantly influence the timing and schedule of expansion. Tax rules may even have a significant impact on the choice of enterprise to emphasize during expansion. Likewise, during contraction the tax recapture provisions with respect to depreciation and investment tax credit combined with the differential treatment of ordinary income and capital gains and the progressive nature of the income tax rates may affect the timing and sequence of liquidation of assets or dissolution of the firm. Again recall that after-tax results are the focus of management decisions, regardless of whether the firm is in the expansion or contraction process.

Monitoring and Control

How does the manager monitor and control the expansion or contraction process? Many of the budgeting and scheduling procedures noted earlier can be applied; the major modification must be in the time frame of the analysis since the focus of contraction or expansion planning is long run in nature. If the expansion process involves the acquisition of new machinery and equipment or construction of facilities, a time schedule of the process can be very useful in monitoring progress. For example, construction schedules might be used to make sure that various phases of the work are completed on time. This is particularly important when various subcontractors are used in the construction activities and monitoring the completion of these activities is necessary to eliminate undue delays in completing the project.

Similarly, it is important to have all of the related phases of the project completed within a consistent time frame. For example, if the expansion involves a breeding barn for hogs or other livestock, the breeding animals must be ready to enter the facility when it is completed or there will be a delay in getting the facility into productive use, and labor, feed, and other supplies must be purchased and available when needed. Thus, the financing, production, and marketing schedules must be coordinated so that the funds are available to pay the contractor and subcontractors when construction is finished, the production inputs have been purchased, and the production process is ready to begin. Contingency plans that can be implemented if the construction process is delayed should also be developed.

Time schedules can also be used to monitor the acquisition of other resources such as chemicals, fertilizer, and other supplies and services. For example, the expansion may require changes in the auxiliary services such as the water supply or the crop product storage, processing, and/or handling facilities and equipment. Schedules and checklists are essential to monitor the physical process of expansion or contraction.

Monitoring and controlling the financial results of expansion and contraction are equally important. A long-run set of financial budgets (cash flow, net worth, and income statements) should be part of the expansion planning process. Comparisons between these budgets and actual long-run performance, much like the comparisons noted earlier on a seasonal or annual basis between actual results and budgeted expectations, provide the mechanism to make sure that expansion or contraction is having the expected result and to facilitate corrective action if that is not the case. However, because of the longer run nature of expansion and contraction planning, one does not expect to have the same amount of detail included in the monitoring and control process.

The Learning Curve

In monitoring the performance of the operation during the expansion process, the manager must be aware of the "start-up year" problem. Most managers will find that it takes some time to get accustomed to an expanded farming operation. If the expansion involves new technology, it may take some time to adapt to this new technology and learn how to utilize it efficiently. Even if the technology has not changed, increased size may require the commitment of more time and the changing of priorities on the part of management. The learning curve or the speed with which the manager can adapt to and efficiently operate the larger farm is important in obtaining a high payoff from the expansion. The control process should recognize this potential difference between the start-up phase and the "steady state" phase. The budgets and schedules should be developed based on the expected differences in performances between these two phases, and the control system should be used to monitor the improved efficiency as the manager and employees become familiar with the new technology to make sure that the learning process is, in fact, taking place.

Finally, the expansion and contraction process may require changes in the monitoring system itself. The increased size of a particular enterprise may mean that more detailed data should be collected to monitor the physical and financial performance in that enterprise. More detailed analysis may be required to detect potential problems and to take corrective action. With a larger operation, the manager may have less time to be physically involved in actually carrying out the work activities; consequently, he or she is not as readily able to observe the physical performance on a first-hand basis. It may be necessary to record additional physical and financial data and to make other changes in the control system so that the manager can accurately monitor performance of the operation. Thus, changes in the size of the farming operation may require changes not only in the planning and implementation process, but also in the control and monitoring activities of management.

Long-Run Business and Financial Performance

The long-run business and financial performance of the farm operation can best be monitored and controlled using the four key criteria of (1) solvency, (2) liquidity, (3) profitability, and (4) efficiency. To review our discussion of Chapter 2, solvency measures the long-run ability of the firm to pay all obligations and debts without any losses if the firm were "sold out" or liquidated. It is a measure of the firm's long-run risk-bearing ability in terms of the use of debt funds. In contrast, liquidity refers to the short-run ability of the firm to pay current obligations when due without impairing the profitability of the business or liquidating part of the productive plant. Thus, liquidity measures the ability of the firm to repay current obligations by converting assets into cash "in the normal course of business operations."

Profitability indicates the firm's capacity to generate profits or income rather than losses over a period of time. This income or return to unpaid labor, equity capital, and management is the source of funds for family living as well as reinvestment and expansion of the farm business. Without profits, solvency problems will eventually result. Efficiency ratios can provide additional insight into the profitability (or lack thereof) of the farm business. These efficiency measures indicate whether or not the physical resources of the farm business are being utilized and combined in a profitable fashion. And if losses are being incurred, the efficiency measures can provide some indication of the source of the problem and what enterprises and resources should be reorganized and reallocated to improve the profitability of the business.

Whereas the earlier discussion of the financial control system focused on comparisons between actual performance and planned expectations in the short run, the focus here will be on trends in actual performance over time and comparisons with other farm firms that have similar characteristics. These comparative analyses can provide more useful measurements of long-run business and financial performance. Table 18.7 illustrates the trends over time in these key financial and business performance ratios using the data for Andy's farm operation.

Table 18.7   Actual Financial and Business Performance Trend
Analysis for Andy's Operation

Item	Actual Last Year	Actual This Year	Actual Next Year
Solvency (December 31)			
Assets (December 31)			
Current	$65,996	$92,586	
Intermediate	108,191	112,491	
Long-term	551,500	581,500	
Total assets	725,687	786,577	
Liabilities (debt) (December 31)			
Current	48,285	32,394	
Intermediate	0	57,658	
Long-term	217,345	211,454	
Total liabilities	265,630	301,506	
Net worth	460,057	485,071	
Annual change in net worth	17,826	25,014	
Debt to total asset ratio	0.37	0.38	
Debt to net worth ratio	0.58	0.62	
Liquidity (December 31)			
Current ratio			
(current assets ÷ current liabilities)	1.37	2.86	
Current liability to total liability ratio	0.18	0.11	
Current or working capital			
(current assets − current liabilities)	$17,711	$60,192	
Profitability			
Gross farm operating receipts	$118,610	$137,300	
Gross farm operating expenses	76,784	89,426	
Net cash operating income	41,826	47,874	
Farm profit or loss	37,645	43,743	
Operator and family labor charge	14,000	15,000	
Residual return to equity capital	23,645	28,743	
Capital turnover ratio			
(gross product ÷ total assets)	0.15	0.18	
Rate of return on equity			
(return to equity capital ÷ net worth) (%)	5.27	6.08	
Efficiency			
Crop value per acre	$289	$274	
Livestock increase per $100 feed fed	137	140	
Gross product per person	83,425	92,523	
Gross product per $100 of expense	163	188	
Machinery and power costs per rotated acre	72	78	

Solvency Documentation of the solvency of the farm business is obtained from the financial statement. As indicated in Chapter 2, the absolute size of the net worth of equity that the farm operator has accumulated provides some indication of the solvency and risk-bearing ability of the business. An increase over time in the equity in the business which can be attributable to earnings provides an indication of financial progress and increased solvency of the operation. However, the best measures of solvency require a comparison between the equity funds in the business and the claims on assets that are held by the creditors of the firm. One such measure is the ratio of debt to total assets which indicates the relative dependence of the firm on debt and the ability to utilize additional credit without impairing the risk-bearing ability of the business. Although each firm is different and general guidelines for this ratio or percentage are difficult to develop, most lenders are hesitant to loan additional funds to a firm that has a debt to total asset ratio that exceeds 0.5 since the lender would then have more at stake in the business than the owner.

A second ratio which is frequently used to measure the solvency of the farm business is the ratio of total liabilities or debt to net worth, or the leverage ratio as discussed in Chapter 8. As with the debt to total asset ratio, the leverage ratio measures the combination of equity and debt funds used in the business. A ratio of 1:1 for the leverage ratio is typically the maximum acceptable. Such a ratio would indicate that for each dollar of equity the farmer has contributed to the business, he or she is borrowing a dollar. Note that because of the relationship between assets, liabilities, and equity as indicated in Chapter 2, the liabilities to net worth or leverage ratio and the debt to total asset ratio really measure the same phenomena. The choice of which ratio to use in evaluating the solvency of the business depends on the preference of the analyst.

Liquidity The liquidity of the farm business indicates the ability to pay current obligations and can be measured in many ways. One of the most accurate measures of liquidity is obtained from the projected cash flow statement as discussed earlier. This statement clearly indicates the cash flow requirements for debt repayment and whether or not sufficient cash income has been generated to make the required debt servicing payments.

An additional measure of the liquidity of the business can be obtained from the net worth statement. As indicated earlier, current liabilities include the amount of debt obligations that must be repaid during the coming year, whereas current assets are the inventory items that will be liquidated and converted into cash during the coming year. Thus, a common measure of the liquidity of the business is the ratio of current assets to current liabilities. A modern standard for the current ratio is 1.5:1 which indicates that there are $1.50 of current assets for every $1.00 of current liabilities in the firm. Such a ratio indicates that even if commodity prices declined or losses occurred which destroyed one-third of the inventory or current assets of the business, sufficient inventory would still be left to pay off current obligations.

The ratio for current liabilities to total liabilities also is useful in evaluating the liquidity of the business. This ratio indicates the debt structure of the business or the proportion of total debt that must be repaid within the year. A low current ratio combined with a high current liability to total liability ratio would indicate that repayment problems are imminent and that some restructuring of debt to move a higher proportion into the intermediate- or long-term category would be desirable. Such a combination of ratios may indicate that the total capital structure of the farm business is not being completely utilized, particularly the debt-carrying capacity of the real estate. It may be possible to acquire additional real estate debt in this case with a long-term repayment plan to reduce the liquidity problems of the business. Such action would improve the ability to meet debt obligations in a reasonable fashion and have sufficient funds for the purchase of additional operating inputs and inventories to expand the business.

A further measure of the liquidity of the farm business is the current working capital. Current capital is defined as current assets minus current liabilities and indicates the amount of funds that will be available to purchase inventory items, expand the business, or satisfy family living needs. Although it is again difficult to develop simple rules concerning the amount of current or working capital that is necessary for a farm business, many lenders prefer to have current capital in the amount of at least 30 to 40 percent of the current liabilities. Furthermore, trends in current capital should be evaluated carefully—a consistent decline in current capital over time provides a strong indication that the liquidity position is deteriorating and problems in terms of debt servicing and credit availability are imminent.

Profitability   The basic measure of profitability of the farm business is net farm income as documented by the income statement. However, as indicated earlier, net farm income must be allocated to the unpaid labor, equity capital, and management contributions to determine if these resources are generating a competitive return. Once such allocations have been made, a comparison of income to that of other similar-sized farm businesses can provide some indication of the profitability of a specific farm operation.

A second measure of the profitability of the farm business is the relationship between gross product and total assets. This ratio provides an indication of the volume of business or the effectiveness of the use of assets in the farm operation. Gross product is measured as gross cash farm receipts minus feed and livestock purchases plus or minus inventory adjustments. Also known as the capital turnover ratio, this profitability measure indicates whether farm resources are being fully employed or whether excess capacity exists. In addition, the trend of this ratio over time as expansion occurs should provide an indication of whether the additional resources are being efficiently managed. For example, a declining capital turnover ratio during expansion indicates that the volume of business per dollar of assets is actually going down rather than remaining constant or going up as a result of economies of size. This declining ratio may indicate that the expansion has been too rapid and the operator has lost control of the business. If this is the case, it may be desirable to postpone future expansion and gain control over the current resource base so that ineffi-ciencies can be eliminated and the profitability of the operation improved.

An additional measure of profitability is the ratio of total return to equity capital to net worth; this ratio indicates the rate of return that is being earned on equity capital. The rate of return on equity capital pro-vides a useful comparison between farm and nonfarm businesses or other investments as the earnings or profitability for most nonfarm investments are quoted as a rate of return. Again, evaluating the trends over time with the use of a comparative analysis form such as that shown in Table 18.7 will provide the most useful information concerning trends in profitability and potential problems in terms of financial performance.

Efficiency Although the performance criteria discussed thus far provide an indication of the financial progress being made by the business over time, they provide only limited diagnostic information that could be useful in determining the source of poor financial performance. The ratios to be discussed now provide additional diagnostic information that can be useful in evaluating the efficiency of enterprise organization and resource use as well as the potential sources of inefficiency. Specifically, the following five ratios provide basic information concerning efficiency in the cropping and livestock enterprises, the efficient utilization of labor and machinery, and the effectiveness of the cost control program. Table 18.8 illustrates the 1981 values for these five efficiency ratios for different types and profit levels for Iowa farm businesses and can be used as guidelines for compar-ison.

Crop Value per Acre This efficiency factor is calculated as the gross value of crop production divided by either total acres or rotated acres. The value of crop production is obtained as total crop sales adjusted for changes in inventory from the annual income statement, and the number of acres is obtained from the financial statement or resource inventory. This efficien-cy ratio reflects both the yield level and the crop selection dimensions of the farm business that result from implementation of the production plan, and the price received for crop products which measures the effectiveness of the marketing plan.

Table 18.8 Efficiency Ratios by Size, Profit Level,
and Enterprise Type, Iowa, 1981

Item	Crop Value per Rotated Acre	Livestock Increase per $100 Feed Fed	Gross Product per Person	Gross Product per $100 Expense	Machine and Power Cost per Rotated Acre
Size (Acres)					
100–179	$270	$138	$64,181	$125	$113
180–259	280	136	71,035	123	100
260–359	274	130	80,887	120	93
360–499	271	131	92,662	127	81
500+	270	128	117,388	125	73
Profit Level					
Average	273	134	77,325	114	93
Low-profit farms	270	116	70,385	96	108
High-profit farms	273	159	98,690	165	81
Enterprise Type					
Hog raising	—	139	76,820	122	104
Beef feeding	—	115	89,956	108	76
Dairy	—	169	59,078	136	125
Cash grain	—	146	99,140	140	69
Beef raising	—	122	97,020	119	73
Hog-beef raising	—	117	58,301	111	94
Hog-beef feeding	—	110	67,650	109	106
Hog-dairy	—	169	75,879	135	143

Livestock Increase per $100 Feed Fed This efficiency ratio provides an
indication of livestock production efficiency, including reproductive efficiency and feed conversion as well as marketing effectiveness in the livestock enterprises. "Livestock increase" is calculated as livestock sales plus or
minus inventory changes minus the purchase price of feeder livestock.
"Feed fed" is calculated as the value of purchased feeds plus the opportunity cost of homegrown grain. A consistent downward trend in this ratio
over time would provide an indication of decreased production efficiency
or an inadequate marketing plan in the livestock enterprises. High-profit
farms had a livestock income per $100 feed fed that was 37 percent higher
than low-profit farms in Iowa in 1981.

Gross Product per Person This efficiency factor measures the productivity of the labor input in the farm business. Gross product is a value
added concept and is calculated as gross receipts minus feed purchases and
the purchase price of livestock such as feeder cattle or feeder pigs plus or
minus inventory adjustments. The manpower figures include operator and
family members as well as hired labor. All labor should be converted to a

full-time equivalent; for example, seasonal labor hired for three months each year would be counted as 0.25 of a person. Labor costs have increased very rapidly for many farm businesses, and record analyses show a direct correlation between net farm income and gross product per person. Note from Table 18.8 that the more efficient firms had gross product per person that was 40 percent higher than the low-profit farms in Iowa in 1981.

Gross Product per $100 of Expense   Expanding the farm business without adequate cost control has caused many farm operations to fail. This efficiency measure indicates the overall cost control being exercised in the farm business. Gross product is measured as defined earlier, and expense figures are obtained from the income statement and include only cash expenditures less purchases of livestock and feed. Table 18.8 indicates that higher profit Iowa farms had a 72 percent higher gross product per $100 of expense compared to low-profit farms in 1981.

Machinery and Power Costs per Rotated Acre   This efficiency ratio measures the costs associated with machinery and equipment and is again a major cost control ratio. Cost items include repairs, depreciation, fuel, oil, and custom hire associated with the cropping enterprises. Again, typical figures for various farming situations are show in Table 18.8.

As with many of the ratios discussed earlier, industry standards for these efficiency ratios are difficult to specify because of differences in enterprise type, size, and other characteristics of farm businesses. An important issue is the trend in these efficiency ratios for a specific business over time. For example, if production per person is declining over time, declining farm income over this period is probably due to reductions in labor efficiency rather than in inefficient utilization of the capital resources. Moreover, interrelationships between the trends in various efficiency factors can provide an indication of problems in the farm business. For example, if machinery and power costs per acre are increasing (indicating that additional machinery inputs are being used in the operation) but production per person is decreasing, machinery inputs apparently are not being substituted for labor and labor efficiency is not improving. Thus, the farm manager may want to evaulate whether or not machinery and power costs are excessive or whether the labor force is being productively employed. Trends in these ratios over time provide the best indication of the performance and efficiency of the farm business.

Summary

In addition to a control system to monitor the technical efficiency of the farm business, the successful farmer must also monitor the marketing and financial performance of the firm. The market control system should include three key components: (1) a summary of historical prices of inputs and products, (2) an input purchase record, and (3) a commodity sales record. Input and commodity prices can be charted to obtain a visual impression of historical price movements which can be very useful in determining expected future trends in prices. The input purchase record is useful in monitoring the buying schedule for inputs and the cash expenses of the farm. It can also be used to summarize the availability of cash discounts and other special arrangements that can reduce the cost of purchased inputs.

The commodity or product sales record can be used to evaluate the manager's performance in planning and implementing the product component of the marketing program. The record should summarize forward contracting and hedging as well as cash sales activities. A comparison between expected or budgeted sales and actual sales of various products as well as the expected compared to the actual price of those products should indicate whether the marketing program is being implemented so as to accomplish the goals and objectives delineated in the market plan.

The financial control system is probably the most important component of the total control system for the farm business. The financial control system should include four key components: (1) the cash flow record or statement, (2) the income or profit and loss statement, (3) the financial statement, and (4) financial and business performance ratios. The cash flow statement provides a mechanism for continual monitoring of the financial performance of the business; it provides concurrent feedback to indicate whether or not the firm is on target with respect to the financial plan. The structure of the cash flow record is to compare periodic (monthly, bimonthly, etc.) actual cash income and expenses to budgeted expectations. Absolute or percentage deviations are then calculated; they provide the basis to determine whether income is being generated and expenses are being incurred as anticipated in the financial plan. In addition to comparisons on a periodic basis, the cash flow record should involve year-to-date comparisons as well so that overall financial performance can be monitored both during the year and at the end of the year.

The income statement and balance sheet components of the financial control system have a similar function to that of the cash flow component, but they provide only annual comparisons between budgeted expectations and actual performance. However, these statements do indicate whether or not the firm generated the net income that was expected during the year, and they provide an indication of the actual value of assets, liabilities, and net worth compared to planned values developed at the beginning of the year.

In addition to monitoring the seasonal cash flow, annual income, and annual financial structure of the farm business, the financial control system should also provide an indication of the long-run financial performance of the business. Monitoring the long-run expansion or contraction process requires information on the liquidity, solvency, profitability, and efficiency of the farm operation. Comparison of the key financial and business performance ratios over time provides an indication of whether the financial condition of the firm is improving or deteriorating in the longer run. In addition, many of the financial and business performance ratios can be compared to firms with similar characteristics so that the farm operator can determine whether she or he has a competitive operation, or whether improvements are needed compared to other similar farms. This component of the control system can be particularly important in monitoring changes in performance in the farm operation during the expansion or contraction process.

Questions and Problems

1. What are the major components of a market control system?
2. Why should a farmer keep a schedule of input purchases?
3. Obtain information from the local newspaper or *Wall Street Journal* and develop a price chart for a major agricultural commodity produced in your area.
4. How can the product sales schedule be used to improve performance in implementing the marketing plan?
5. What are the four components of the financial control system?
6. How can a cash flow record be used to monitor short-run financial performance?
7. Why compare actual cash income and expenses to planned income and expenses on both a periodic and year-to-date basis in a cash flow record?
8. What are the major considerations in planning the expansion or contraction of the farm business?
9. What financial measures can be used to monitor the long-run business and financial performance of the farm operation?
10. Distinguish between the concepts of liquidity and solvency. How might you measure the liquidity and solvency of the farm business?
11. Identify some of the key efficiency ratios that might be used in evaluating the performance of a farm operation?
12. What is the concept of the "learning curve"? Why is it important in specifying performance standards and monitoring business performance?

Further Reading

Frey, Thomas L., and Danny A. Klinefelter. *Coordinated Financial Statements for Agriculture*. Skokie, Ill.: Agri-Finance, 1978.
Helmkamp, John G., LeRoy F. Imdieke, and Ralph E. Smith. *Principles of Accounting*. New York: John Wiley and Sons, 1983, Chapters 24 and 25.

Horngren, Charles T. *Cost Accounting: A Managerial Emphasis.* 3d ed., Englewood Cliffs, N.J.: Prentice-Hall, 1972, Chapters 5–11.

James, Sydney, and Everett Stoneberg. *Farm Accounting and Business Analysis.* 2d ed., Ames: Iowa State University Press, 1979.

Pyle, William W., and Merkit D. Larson. *Fundamental Accounting Principles.* 9th ed., Homewood, Ill.: Richard D. Irwin, 1981.

Smith, Frank J., Jr., and Ken Cooper. "Financial Management of Agri-Business Firms." University of Minnesota Agricultural Extension Service, Special Report No. 26, 4th printing, 1975.

APPENDIX

APPENDIX TABLE I: Amount of 1 at compound interest

$$V_n^f = (1 + i)^n$$

n	1%	2%	3%	4%	5%	6%	n
1	1.0100	1.0200	1.0300	1.0400	1.0500	1.0600	1
2	1.0201	1.0404	1.0609	1.0816	1.1025	1.1236	2
3	1.0303	1.0612	1.0927	1.1249	1.1576	1.1910	3
4	1.0406	1.0824	1.1255	1.1699	1.2155	1.2625	4
5	1.0510	1.1041	1.1593	1.2167	1.2763	1.3382	5
6	1.0615	1.1262	1.1941	1.2653	1.3401	1.4185	6
7	1.0721	1.1487	1.2299	1.3159	1.4071	1.5036	7
8	1.0829	1.1717	1.2668	1.3686	1.4775	1.5938	8
9	1.0937	1.1951	1.3048	1.4233	1.5513	1.6895	9
10	1.1046	1.2190	1.3439	1.4802	1.6289	1.7908	10
11	1.1157	1.2434	1.3842	1.5395	1.7103	1.8983	11
12	1.1268	1.2682	1.4258	1.6010	1.7959	2.0122	12
13	1.1381	1.2936	1.4685	1.6651	1.8856	2.1329	13
14	1.1495	1.3195	1.5126	1.7317	1.9799	2.2609	14
15	1.1610	1.3459	1.5580	1.8009	2.0789	2.3966	15
16	1.1726	1.3728	1.6047	1.8730	2.1829	2.5404	16
17	1.1843	1.4002	1.6528	1.9479	2.2920	2.6928	17
18	1.1961	1.4282	1.7024	2.0258	2.4066	2.8543	18
19	1.2081	1.4568	1.7535	2.1068	2.5269	3.0256	19
20	1.2202	1.4859	1.8061	2.1911	2.6533	3.2071	20
21	1.2324	1.5157	1.8603	2.2788	2.7860	3.3996	21
22	1.2447	1.5460	1.9161	2.3699	2.9253	3.6035	22
23	1.2572	1.5769	1.9736	2.4647	3.0715	3.8197	23
24	1.2697	1.6084	2.0328	2.5633	3.2251	4.0489	24
25	1.2824	1.6406	2.0938	2.6658	3.3864	4.2919	25
26	1.2953	1.6734	2.1566	2.7725	3.5557	4.5494	26
27	1.3082	1.7069	2.2213	2.8834	3.7335	4.8223	27
28	1.3213	1.7410	2.2879	2.9987	3.9201	5.1117	28
29	1.3345	1.7758	2.3566	3.1187	4.1161	5.4184	29
30	1.3478	1.8114	2.4273	3.2434	4.3219	5.7435	30
31	1.3613	1.8476	2.5001	3.3731	4.5380	6.0881	31
32	1.3749	1.8845	2.5751	3.5081	4.7649	6.4534	32
33	1.3887	1.9222	2.6523	3.6484	5.0032	6.8406	33
34	1.4026	1.9607	2.7319	3.7943	5.2533	7.2510	34
35	1.4166	1.9999	2.8139	3.9461	5.5160	7.6861	35
40	1.4889	2.2080	3.2620	4.8010	7.0400	10.2857	40
45	1.5648	2.4379	3.7816	5.8412	8.9850	13.7646	45
50	1.6446	2.6916	4.3839	7.1067	11.4674	18.4201	50
55	1.7285	2.9717	5.0821	8.6464	14.6356	24.6503	55
60	1.8167	3.2810	5.8916	10.5196	18.6792	32.9877	60

APPENDIX TABLE I (*continued*): Amount of 1 at compound interest

$$V_n^f = (1 + i)^n$$

n	7%	8%	9%	10%	11%	12%	n
1	1.0700	1.0800	1.0900	1.1000	1.1100	1.1200	1
2	1.1449	1.1664	1.1881	1.2100	1.2321	1.2544	2
3	1.2250	1.2597	1.2950	1.3310	1.3676	1.4049	3
4	1.3108	1.3605	1.4116	1.4641	1.5181	1.5735	4
5	1.4026	1.4693	1.5386	1.6105	1.6851	1.7623	5
6	1.5007	1.5869	1.6771	1.7716	1.8704	1.9738	6
7	1.6058	1.7138	1.8280	1.9487	2.0762	2.2107	7
8	1.7182	1.8509	1.9926	2.1436	2.3045	2.4760	8
9	1.8385	1.9990	2.1719	2.3579	2.5580	2.7731	9
10	1.9672	2.1589	2.3674	2.5937	2.8394	3.1058	10
11	2.1049	2.3316	2.5804	2.8531	3.1518	3.4785	11
12	2.2522	2.5182	2.8127	3.1384	3.4984	3.8960	12
13	2.4098	2.7196	3.0658	3.4523	3.8833	4.3635	13
14	2.5785	2.9372	3.3417	3.7975	4.3104	4.8871	14
15	2.7590	3.1722	3.6425	4.1772	4.7846	5.4736	15
16	2.9522	3.4259	3.9703	4.5950	5.3109	6.1304	16
17	3.1588	3.7000	4.3276	5.0545	5.8951	6.8660	17
18	3.3799	3.9960	4.7171	5.5599	6.5435	7.6900	18
19	3.6165	4.3157	5.1417	6.1159	7.2633	8.6128	19
20	3.8697	4.6610	5.6044	6.7275	8.0623	9.6463	20
21	4.1406	5.0338	6.1088	7.4002	8.9492	10.8038	21
22	4.4304	5.4365	6.6586	8.1403	9.9336	12.1003	22
23	4.7405	5.8715	7.2579	8.9543	11.0263	13.5523	23
24	5.0724	6.3412	7.9111	9.8497	12.2391	15.1786	24
25	5.4274	6.8485	8.6231	10.8347	13.5855	17.0000	25
26	5.8074	7.3964	9.3992	11.9182	15.0799	19.0401	26
27	6.2139	7.9881	10.2451	13.1100	16.7386	21.3249	27
28	6.6488	8.6271	11.1671	14.4210	18.5799	23.8838	28
29	7.1143	9.3173	12.1722	15.8631	20.6237	26.7499	29
30	7.6123	10.0627	13.2677	17.4494	22.8923	29.9599	30
31	8.1451	10.8677	14.4618	19.1943	25.4104	33.5551	31
32	8.7153	11.7371	15.7633	21.1138	28.2056	37.5817	32
33	9.3253	12.6760	17.1820	23.2251	31.3082	42.0915	33
34	9.9781	13.6901	18.7284	25.5477	34.7521	47.1425	34
35	10.6766	14.7853	20.4140	28.1024	38.5748	52.7995	35
40	14.9745	21.7245	31.4094	45.2592	65.0008	93.0508	40
45	21.0024	31.9204	48.3273	72.8904	109.5301	163.9873	45
50	29.4570	46.9016	74.3575	117.3908	184.5645	289.0015	50
55	41.3150	68.9138	114.4082	189.0590	311.0017	509.3196	55
60	57.9464	101.2570	176.0312	304.4812	524.0562	897.5950	60

APPENDIX TABLE II: Present value of 1 at compound interest

$$V_n^P = \frac{1}{(1 + i)^n}$$

n	1%	2%	3%	4%	5%	6%	n
1	0.9901	0.9804	0.9709	0.9615	0.9524	0.9434	1
2	0.9803	0.9612	0.9426	0.9246	0.9070	0.8900	2
3	0.9706	0.9423	0.9151	0.8890	0.8638	0.8396	3
4	0.9610	0.9238	0.8885	0.8548	0.8227	0.7921	4
5	0.9515	0.9057	0.8626	0.8219	0.7835	0.7473	5
6	0.9420	0.8880	0.8375	0.7903	0.7462	0.7050	6
7	0.9327	0.8706	0.8131	0.7599	0.7107	0.6651	7
8	0.9235	0.8535	0.7894	0.7307	0.6768	0.6274	8
9	0.9143	0.8368	0.7664	0.7026	0.6446	0.5919	9
10	0.9053	0.8203	0.7441	0.6756	0.6139	0.5584	10
11	0.8963	0.8043	0.7224	0.6496	0.5847	0.5268	11
12	0.8874	0.7885	0.7014	0.6246	0.5568	0.4970	12
13	0.8787	0.7730	0.6810	0.6006	0.5303	0.4688	13
14	0.8700	0.7579	0.6611	0.5775	0.5051	0.4423	14
15	0.8613	0.7430	0.6419	0.5553	0.4810	0.4173	15
16	0.8528	0.7284	0.6232	0.5339	0.4581	0.3936	16
17	0.8444	0.7142	0.6050	0.5134	0.4363	0.3714	17
18	0.8360	0.7002	0.5874	0.4936	0.4155	0.3503	18
19	0.8277	0.6864	0.5703	0.4746	0.3957	0.3305	19
20	0.8195	0.6730	0.5537	0.4564	0.3769	0.3118	20
21	0.8114	0.6598	0.5375	0.4388	0.3589	0.2942	21
22	0.8034	0.6468	0.5219	0.4220	0.3418	0.2775	22
23	0.7954	0.6342	0.5067	0.4057	0.3256	0.2618	23
24	0.7876	0.6217	0.4919	0.3901	0.3101	0.2470	24
25	0.7798	0.6095	0.4776	0.3751	0.2953	0.2330	25
26	0.7720	0.5976	0.4637	0.3607	0.2812	0.2198	26
27	0.7644	0.5859	0.4502	0.3468	0.2678	0.2074	27
28	0.7568	0.5744	0.4371	0.3335	0.2551	0.1956	28
29	0.7493	0.5631	0.4243	0.3207	0.2429	0.1846	29
30	0.7419	0.5521	0.4120	0.3083	0.2314	0.1741	30
31	0.7346	0.5412	0.4000	0.2965	0.2204	0.1643	31
32	0.7273	0.5306	0.3883	0.2851	0.2099	0.1550	32
33	0.7201	0.5202	0.3770	0.2741	0.1999	0.1462	33
34	0.7130	0.5100	0.3660	0.2636	0.1904	0.1379	34
35	0.7059	0.5000	0.3554	0.2534	0.1813	0.1301	35
40	0.6717	0.4529	0.3066	0.2083	0.1420	0.0972	40
45	0.6391	0.4102	0.2644	0.1712	0.1113	0.0727	45
50	0.6080	0.3715	0.2281	0.1407	0.0872	0.0543	50
55	0.5785	0.3365	0.1968	0.1157	0.0683	0.0406	55
60	0.5504	0.3048	0.1697	0.0951	0.0535	0.0303	60

APPENDIX TABLE II (*continued*): Present value of 1 at compound interest

$$V_n^P = \frac{1}{(1 + i)^n}$$

n	7%	8%	9%	10%	11%	12%	n
1	0.9346	0.9259	0.9174	0.9091	0.9009	0.8929	1
2	0.8734	0.8573	0.8417	0.8264	0.8116	0.7972	2
3	0.8163	0.7938	0.7722	0.7513	0.7312	0.7118	3
4	0.7629	0.7350	0.7084	0.6830	0.6587	0.6355	4
5	0.7130	0.6806	0.6499	0.6209	0.5935	0.5674	5
6	0.6663	0.6302	0.5963	0.5645	0.5346	0.5066	6
7	0.6227	0.5835	0.5470	0.5132	0.4817	0.4523	7
8	0.5820	0.5403	0.5019	0.4665	0.4339	0.4039	8
9	0.5439	0.5002	0.4604	0.4241	0.3909	0.3606	9
10	0.5083	0.4632	0.4224	0.3855	0.3522	0.3220	10
11	0.4751	0.4289	0.3875	0.3505	0.3173	0.2875	11
12	0.4440	0.3971	0.3555	0.3186	0.2858	0.2567	12
13	0.4150	0.3677	0.3262	0.2897	0.2575	0.2292	13
14	0.3878	0.3405	0.2992	0.2633	0.2320	0.2046	14
15	0.3624	0.3152	0.2745	0.2394	0.2090	0.1827	15
16	0.3387	0.2919	0.2519	0.2176	0.1883	0.1631	16
17	0.3166	0.2703	0.2311	0.1978	0.1696	0.1456	17
18	0.2959	0.2502	0.2120	0.1799	0.1528	0.1300	18
19	0.2765	0.2317	0.1945	0.1635	0.1377	0.1161	19
20	0.2584	0.2145	0.1784	0.1486	0.1240	0.1037	20
21	0.2415	0.1987	0.1637	0.1351	0.1117	0.0926	21
22	0.2257	0.1839	0.1502	0.1228	0.1007	0.0826	22
23	0.2109	0.1703	0.1378	0.1117	0.0907	0.0738	23
24	0.1971	0.1577	0.1264	0.1015	0.0817	0.0659	24
25	0.1842	0.1460	0.1160	0.0923	0.0736	0.0588	25
26	0.1722	0.1352	0.1064	0.0839	0.0663	0.0525	26
27	0.1609	0.1252	0.0976	0.0763	0.0597	0.0469	27
28	0.1504	0.1159	0.0895	0.0693	0.0538	0.0419	28
29	0.1406	0.1073	0.0822	0.0630	0.0485	0.0374	29
30	0.1314	0.0994	0.0754	0.0573	0.0437	0.0334	30
31	0.1228	0.0920	0.0691	0.0521	0.0394	0.0298	31
32	0.1147	0.0852	0.0634	0.0474	0.0355	0.0266	32
33	0.1072	0.0789	0.0582	0.0431	0.0319	0.0238	33
34	0.1002	0.0730	0.0534	0.0391	0.0288	0.0212	34
35	0.0937	0.0676	0.0490	0.0356	0.0259	0.0189	35
40	0.0668	0.0460	0.0318	0.0221	0.0154	0.0107	40
45	0.0476	0.0313	0.0207	0.0137	0.0091	0.0061	45
50	0.0339	0.0213	0.0134	0.0085	0.0054	0.0035	50
55	0.0242	0.0145	0.0087	0.0053	0.0032	0.0020	55
60	0.0173	0.0099	0.0057	0.0033	0.0019	0.0011	60

APPENDIX TABLE II (*continued*): Present value of 1 at compound interest

$$V_n^P = \frac{1}{(1 + i)^n}$$

n	13%	14%	15%	16%	18%	20%	n
1	0.8850	0.8772	0.8696	0.8621	0.8475	0.8333	1
2	0.7831	0.7695	0.7561	0.7432	0.7182	0.6944	2
3	0.6931	0.6750	0.6575	0.6407	0.6086	0.5787	3
4	0.6133	0.5921	0.5718	0.5523	0.5158	0.4823	4
5	0.5428	0.5194	0.4972	0.4761	0.4371	0.4019	5
6	0.4803	0.4556	0.4323	0.4104	0.3704	0.3349	6
7	0.4251	0.3996	0.3759	0.3538	0.3139	0.2791	7
8	0.3762	0.3506	0.3269	0.3050	0.2660	0.2326	8
9	0.3329	0.3075	0.2843	0.2630	0.2255	0.1938	9
10	0.2946	0.2697	0.2472	0.2267	0.1911	0.1615	10
11	0.2607	0.2366	0.2149	0.1954	0.1619	0.1346	11
12	0.2307	0.2076	0.1869	0.1685	0.1372	0.1122	12
13	0.2042	0.1821	0.1625	0.1452	0.1163	0.0935	13
14	0.1807	0.1597	0.1413	0.1252	0.0985	0.0779	14
15	0.1599	0.1401	0.1229	0.1079	0.0835	0.0649	15
16	0.1415	0.1229	0.1069	0.0930	0.0708	0.0541	16
17	0.1252	0.1078	0.0929	0.0802	0.0600	0.0451	17
18	0.1108	0.0946	0.0808	0.0691	0.0508	0.0376	18
19	0.0981	0.0829	0.0703	0.0596	0.0431	0.0313	19
20	0.0868	0.0728	0.0611	0.0514	0.0365	0.0261	20
21	0.0768	0.0638	0.0531	0.0443	0.0309	0.0217	21
22	0.0680	0.0560	0.0462	0.0382	0.0262	0.0181	22
23	0.0601	0.0491	0.0402	0.0329	0.0222	0.0151	23
24	0.0532	0.0431	0.0349	0.0284	0.0188	0.0126	24
25	0.0471	0.0378	0.0304	0.0245	0.0160	0.0105	25
26	0.0417	0.0331	0.0264	0.0211	0.0135	0.0087	26
27	0.0369	0.0291	0.0230	0.0182	0.0115	0.0073	27
28	0.0326	0.0255	0.0200	0.0157	0.0097	0.0061	28
29	0.0289	0.0224	0.0174	0.0135	0.0082	0.0051	29
30	0.0256	0.0196	0.0151	0.0116	0.0070	0.0042	30
31	0.0226	0.0172	0.0131	0.0100	0.0059	0.0035	31
32	0.0200	0.0151	0.0114	0.0087	0.0050	0.0029	32
33	0.0177	0.0132	0.0099	0.0075	0.0042	0.0024	33
34	0.0157	0.0116	0.0086	0.0064	0.0036	0.0020	34
35	0.0139	0.0102	0.0075	0.0055	0.0030	0.0017	35
40	0.0075	0.0053	0.0037	0.0026	0.0013	0.0007	40
45	0.0041	0.0027	0.0019	0.0013	0.0006	0.0003	45
50	0.0022	0.0014	0.0009	0.0006	0.0003	0.0001	50
55	0.0012	0.0007	0.0005	0.0003	0.0001	0.0000	55
60	0.0007	0.0004	0.0002	0.0001	0.0000	0.0000	60

APPENDIX TABLE II (*continued*): Present value of 1 at compound interest

$$V_n^P = \frac{1}{(1 + i)^n}$$

n	25%	30%	35%	40%	45%	50%	n
1	0.8000	0.7692	0.7407	0.7143	0.6897	0.6667	1
2	0.6400	0.5917	0.5487	0.5102	0.4756	0.4444	2
3	0.5120	0.4552	0.4064	0.3644	0.3280	0.2963	3
4	0.4096	0.3501	0.3011	0.2603	0.2262	0.1975	4
5	0.3277	0.2693	0.2230	0.1859	0.1560	0.1317	5
6	0.2621	0.2072	0.1652	0.1328	0.1076	0.0878	6
7	0.2097	0.1594	0.1224	0.0949	0.0742	0.0585	7
8	0.1678	0.1226	0.0906	0.0678	0.0512	0.0390	8
9	0.1342	0.0943	0.0671	0.0484	0.0353	0.0260	9
10	0.1074	0.0725	0.0497	0.0346	0.0243	0.0173	10
11	0.0859	0.0558	0.0368	0.0247	0.0168	0.0116	11
12	0.0687	0.0429	0.0273	0.0176	0.0116	0.0077	12
13	0.0550	0.0330	0.0202	0.0126	0.0080	0.0051	13
14	0.0440	0.0254	0.0150	0.0090	0.0055	0.0034	14
15	0.0352	0.0195	0.0111	0.0064	0.0038	0.0023	15
16	0.0281	0.0150	0.0082	0.0046	0.0026	0.0015	16
17	0.0225	0.0116	0.0061	0.0033	0.0018	0.0010	17
18	0.0180	0.0089	0.0045	0.0023	0.0012	0.0007	18
19	0.0144	0.0068	0.0033	0.0017	0.0009	0.0005	19
20	0.0115	0.0053	0.0025	0.0012	0.0006	0.0003	20
21	0.0092	0.0040	0.0018	0.0009	0.0004	0.0002	21
22	0.0074	0.0031	0.0014	0.0006	0.0003	0.0001	22
23	0.0059	0.0024	0.0010	0.0004	0.0002	0.0001	23
24	0.0047	0.0018	0.0007	0.0003	0.0001	0.0001	24
25	0.0038	0.0014	0.0006	0.0002	0.0001	0.0000	25
26	0.0030	0.0011	0.0004	0.0002	0.0001		26
27	0.0024	0.0008	0.0003	0.0001	0.0000		27
28	0.0019	0.0006	0.0002	0.0001			28
29	0.0015	0.0005	0.0002	0.0001			29
30	0.0012	0.0004	0.0001	0.0000			30
31	0.0010	0.0003	0.0001				31
32	0.0008	0.0002	0.0001				32
33	0.0006	0.0002	0.0001				33
34	0.0005	0.0001	0.0000				34
35	0.0004	0.0001					35
40	0.0001	0.0000					40
45	0.0000						45

APPENDIX TABLE III   Annuity Factors
(Present value of 1 per annum at compound interest)

$$a = \frac{1 - (1 + i)^{-n}}{i}$$

n	1%	2%	3%	4%	5%	6%	n
1	0.9901	0.9804	0.9709	0.9615	0.9524	0.9434	1
2	1.9704	1.9416	1.9135	1.8861	1.8594	1.8334	2
3	2.9410	2.8839	2.8286	2.7751	2.7232	2.6730	3
4	3.9020	3.8077	3.7171	3.6299	3.5460	3.4651	4
5	4.8534	4.7135	4.5797	4.4518	4.3295	4.2124	5
6	5.7955	5.6014	5.4172	5.2421	5.0757	4.9173	6
7	6.7282	6.4720	6.2303	6.0021	5.7864	5.5824	7
8	7.6517	7.3255	7.0197	6.7327	6.4632	6.2098	8
9	8.5660	8.1622	7.7861	7.4353	7.1078	6.8017	9
10	9.4713	8.9826	8.5302	8.1109	7.7217	7.3601	10
11	10.3676	9.7868	9.2526	8.7605	8.3064	7.8869	11
12	11.2551	10.5753	9.9540	9.3851	8.8633	8.3838	12
13	12.1337	11.3484	10.6350	9.9856	9.3936	8.8527	13
14	13.0037	12.1062	11.2961	10.5631	9.8986	9.2950	14
15	13.8651	12.8493	11.9379	11.1184	10.3797	9.7122	15
16	14.7179	13.5777	12.5611	11.6523	10.8378	10.1059	16
17	15.5623	14.2919	13.1661	12.1657	11.2741	10.4773	17
18	16.3983	14.9920	13.7535	12.6593	11.6896	10.8276	18
19	17.2260	15.6785	14.3238	13.1339	12.0853	11.1581	19
20	18.0456	16.3514	14.8775	13.5903	12.4622	11.4699	20
21	18.8570	17.0112	15.4150	14.0292	12.8212	11.7641	21
22	19.6604	17.6580	15.9369	14.4511	13.1630	12.0416	22
23	20.4558	18.2922	16.4436	14.8568	13.4886	12.3034	23
24	21.2434	18.9139	16.9355	15.2470	13.7986	12.5504	24
25	22.0232	19.5235	17.4131	15.6221	14.0939	12.7834	25
26	22.7952	20.1210	17.8768	15.9828	14.3752	13.0032	26
27	23.5596	20.7069	18.3270	16.3296	14.6430	13.2105	27
28	24.3164	21.2813	18.7641	16.6631	14.8981	13.4062	28
29	25.0658	21.8444	19.1885	16.9837	15.1411	13.5907	29
30	25.8077	22.3965	19.6004	17.2920	15.3725	13.7648	30
31	26.5423	22.9377	20.0004	17.5885	15.5928	13.9291	31
32	27.2696	23.4683	20.3888	17.8736	15.8027	14.0840	32
33	27.9897	23.9886	20.7658	18.1476	16.0025	14.2302	33
34	28.7027	24.4986	21.1318	18.4112	16.1929	14.3681	34
35	29.4086	24.9986	21.4872	18.6646	16.3742	14.4982	35
40	32.8347	27.3555	23.1148	19.7928	17.1591	15.0463	40
45	36.0945	29.4902	24.5187	20.7200	17.7741	15.4558	45
50	39.1961	31.4236	25.7298	21.4822	18.2559	15.7619	50
55	42.1472	33.1748	26.7744	22.1086	18.6335	15.9905	55
60	44.9550	34.7609	27.6756	22.6235	18.9293	16.1614	60

APPENDIX TABLE III   (*continued*):   Annuity Factors
(Present value of 1 per annum at compound interest)

$$a = \frac{1 - (1 + i)^{-n}}{i}$$

n	7%	8%	9%	10%	11%	12%	n
1	0.9346	0.9259	0.9174	0.9091	0.9009	0.8929	1
2	1.8080	1.7833	1.7591	1.7355	1.7125	1.6901	2
3	2.6243	2.5771	2.5313	2.4869	2.4437	2.4018	3
4	3.3872	3.3121	3.2397	3.1699	3.1024	3.0373	4
5	4.1002	3.9927	3.8897	3.7908	3.6959	3.6048	5
6	4.7665	4.6229	4.4859	4.3553	4.2305	4.1114	6
7	5.3893	5.2064	5.0330	4.8684	4.7122	4.5638	7
8	5.9713	5.7466	5.5348	5.3349	5.1461	4.9676	8
9	6.5152	6.2469	5.9952	5.7590	5.5370	5.3282	9
10	7.0236	6.7101	6.4177	6.1446	5.8892	5.6502	10
11	7.4987	7.1390	6.8052	6.4951	6.2065	5.9377	11
12	7.9427	7.5361	7.1607	6.8137	6.4924	6.1944	12
13	8.3577	7.9038	7.4869	7.1034	6.7499	6.4235	13
14	8.7455	8.2442	7.7862	7.3667	6.9819	6.6282	14
15	9.1079	8.5595	8.0607	7.6061	7.1909	6.8109	15
16	9.4466	8.8514	8.3126	7.8237	7.3792	6.9740	16
17	9.7632	9.1216	8.5436	8.0216	7.5488	7.1196	17
18	10.0591	9.3719	8.7556	8.2014	7.7016	7.2497	18
19	10.3356	9.6036	8.9501	8.3649	7.8393	7.3658	19
20	10.5940	9.8181	9.1285	8.5136	7.9633	7.4694	20
21	10.8355	10.0168	9.2922	8.6487	8.0751	7.5620	21
22	11.0612	10.2007	9.4424	8.7715	8.1757	7.6446	22
23	11.2722	10.3711	9.5802	8.8832	8.2664	7.7184	23
24	11.4693	10.5288	9.7066	8.9847	8.3481	7.7843	24
25	11.6536	10.6748	9.8226	9.0770	8.4217	7.8431	25
26	11.8258	10.8100	9.9290	9.1609	8.4881	7.8957	26
27	11.9867	10.9352	10.0266	9.2372	8.5478	7.9426	27
28	12.1371	11.0511	10.1161	9.3066	8.6016	7.9844	28
29	12.2777	11.1584	10.1983	9.3696	8.6501	8.0218	29
30	12.4090	11.2578	10.2737	9.4269	8.6938	8.0552	30
31	12.5318	11.3498	10.3428	9.4790	8.7331	8.0850	31
32	12.6466	11.4350	10.4062	9.5264	8.7686	8.1116	32
33	12.7538	11.5139	10.4644	9.5694	8.8005	8.1354	33
34	12.8540	11.5869	10.5178	9.6086	8.8293	8.1566	34
35	12.9477	11.6546	10.5668	9.6442	8.8552	8.1755	35
40	13.3317	11.9246	10.7574	9.7791	8.9511	8.2438	40
45	13.6055	12.1084	10.8812	9.8628	9.0079	8.2825	45
50	13.8007	12.2335	10.9617	9.9148	9.0417	8.3045	50
55	13.9399	12.3186	11.0140	9.9471	9.0617	8.3170	55
60	14.0392	12.3766	11.0480	9.9672	9.0736	8.3240	60

APPENDIX TABLE III (*continued*): Annuity Factors
(Present value of 1 per annum at compound interest)

$$a = \frac{1 - (1 + i)^{-n}}{i}$$

n	13%	14%	15%	16%	18%	20%	n
1	0.8850	0.8772	0.8696	0.8621	0.8475	0.8333	1
2	1.6681	1.6467	1.6257	1.6052	1.5656	1.5278	2
3	2.3612	2.3216	2.2832	2.2459	2.1743	2.1065	3
4	2.9745	2.9137	2.8550	2.7982	2.6901	2.5887	4
5	3.5172	3.4331	3.3522	3.2743	3.1272	2.9906	5
6	3.9975	3.8887	3.7845	3.6847	3.4976	3.3255	6
7	4.4226	4.2883	4.1604	4.0386	3.8115	3.6046	7
8	4.7988	4.6389	4.4873	4.3436	4.0776	3.8372	8
9	5.1317	4.9464	4.7716	4.6065	4.3030	4.0310	9
10	5.4262	5.2161	5.0188	4.8332	4.4941	4.1925	10
11	5.6869	5.4527	5.2337	5.0286	4.6560	4.3271	11
12	5.9176	5.6603	5.4206	5.1971	4.7932	4.4392	12
13	6.1218	5.8424	5.5831	5.3423	4.9095	4.5327	13
14	6.3025	6.0021	5.7245	5.4675	5.0081	4.6106	14
15	6.4624	6.1422	5.8474	5.5755	5.0916	4.6755	15
16	6.6039	6.2651	5.9542	5.6685	5.1624	4.7296	16
17	6.7291	6.3729	6.0472	5.7487	5.2223	4.7746	17
18	6.8399	6.4674	6.1280	5.8178	5.2732	4.8122	18
19	6.9380	6.5504	6.1982	5.8775	5.3162	4.8435	19
20	7.0248	6.6231	6.2593	5.9288	5.3527	4.8696	20
21	7.1016	6.6870	6.3125	5.9731	5.3837	4.8913	21
22	7.1695	6.7429	6.3587	6.0113	5.4099	4.9094	22
23	7.2297	6.7921	6.3988	6.0442	5.4321	4.9245	23
24	7.2829	6.8351	6.4338	6.0726	5.4509	4.9371	24
25	7.3300	6.8729	6.4641	6.0971	5.4669	4.9476	25
26	7.3717	6.9061	6.4906	6.1182	5.4804	4.9563	26
27	7.4086	6.9352	6.5135	6.1364	5.4919	4.9636	27
28	7.4412	6.9607	6.5335	6.1520	5.5016	4.9697	28
29	7.4701	6.9830	6.5509	6.1656	5.5098	4.9747	29
30	7.4957	7.0027	6.5660	6.1772	6.6168	4.9789	30
31	7.5183	7.0199	6.5791	6.1872	5.5227	4.9824	31
32	7.5383	7.0350	6.5905	6.1959	5.5277	4.9854	32
33	7.5560	7.0482	6.6005	6.2034	5.5320	4.9878	33
34	7.5717	7.0599	6.6091	6.2098	5.5356	4.9898	34
35	7.5856	7.0700	6.6166	6.2153	5.5386	4.9915	35
40	7.6344	7.1050	6.6418	6.2335	5.5482	4.9966	40
45	7.6609	7.1232	6.6543	6.2421	5.5523	4.9986	45
50	7.6752	7.1327	6.6605	6.2463	5.5541	4.9995	50
55	7.6830	7.1376	6.6636	6.2482	5.5549	4.9998	55
60	7.6873	7.1401	6.6651	6.2492	5.5553	4.9999	60

APPENDIX TABLE III (*continued*): Annuity Factors (Present value of 1 per annum at compound interest)

$$a = \frac{1 - (1 + i)^{-n}}{i}$$

n	25%	30%	35%	40%	45%	50%	n
1	0.8000	0.7692	0.7407	0.7143	0.6897	0.6667	1
2	1.4400	1.3609	1.2894	1.2245	1.1653	1.1111	2
3	1.9520	1.8161	1.6959	1.5889	1.4933	1.4074	3
4	2.3616	2.1662	1.9969	1.8492	1.7195	1.6049	4
5	2.6893	2.4356	2.2200	2.0352	1.8755	1.7366	5
6	2.9514	2.6427	2.3852	2.1680	1.9831	1.8244	6
7	3.1611	2.8021	2.5075	2.2628	2.0573	1.8829	7
8	3.3289	2.9247	2.5982	2.3306	2.1085	1.9220	8
9	3.4631	3.0190	2.6653	2.3790	2.1438	1.9480	9
10	3.5705	3.0915	2.7150	2.4136	2.1681	1.9653	10
11	3.6564	3.1473	2.7519	2.4383	2.1849	1.9769	11
12	3.7251	3.1903	2.7792	2.4559	2.1965	1.9846	12
13	3.7801	3.2233	2.7994	2.4685	2.2045	1.9897	13
14	3.8241	3.2487	2.8144	2.4775	2.2100	1.9931	14
15	3.8593	3.2682	2.8255	2.4839	2.2138	1.9954	15
16	3.8874	3.2832	2.8337	2.4885	2.2164	1.9970	16
17	3.9099	3.2948	2.8398	2.4918	2.2182	1.9980	17
18	3.9279	3.3037	2.8443	2.4941	2.2195	1.9986	18
19	3.9424	3.3105	2.8476	2.4958	2.2203	1.9991	19
20	3.9539	3.3158	2.8501	2.4970	2.2209	1.9994	20
21	3.9631	3.3198	2.8519	2.4979	2.2213	1.9996	21
22	3.9705	3.3230	2.8533	2.4985	2.2216	1.9997	22
23	3.9764	3.3254	2.8543	2.4989	2.2218	1.9998	23
24	3.9811	3.3272	2.8550	2.4992	2.2219	1.9999	24
25	3.9849	3.3286	2.8556	2.4994	2.2220	1.9999	25
26	3.9879	3.3297	2.8560	2.4996	2.2221	1.9999	26
27	3.9903	3.3305	2.8563	2.4997	2.2221	2.0000	27
28	3.9923	3.3312	2.8565	2.4998	2.2222	2.0000	28
29	3.9938	3.3317	2.8567	2.4999	2.2222	2.0000	29
30	3.9950	3.3321	2.8568	2.4999	2.2222	2.0000	30
31	3.9960	3.3324	2.8569	2.4999	2.2222	2.0000	31
32	3.9968	3.3326	2.8569	2.4999	2.2222	2.0000	32
33	3.9975	3.3328	2.8570	2.5000	2.2222	2.0000	33
34	3.9980	3.3329	2.8570	2.5000	2.2222	2.0000	34
35	3.9984	3.3330	2.8571	2.5000	2.2222	2.0000	35
40	3.9995	3.3332	2.8571	2.5000	2.2222	2.0000	40
45	3.9998	3.3333	2.8571	2.5000	2.2222	2.0000	45
50	3.9999	3.3333	2.8571	2.5000	2.2222	2.0000	50
55	4.0000	3.3333	2.8571	2.5000	2.2222	2.0000	55
60	4.0000	3.3333	2.8571	2.5000	2.2222	2.0000	60

APPENDIX TABLE IV: Capital Recovery or Amortization Factors (Annuity whose present value at compound interest is 1)

$$\frac{1}{a} = \frac{i}{1 - (1 + i)^{-n}}$$

n	1%	2%	3%	4%	5%	6%	n
1	1.0100	1.0200	1.0300	1.0400	1.0500	1.0600	1
2	0.5075	0.5150	0.5226	0.5302	0.5378	0.5454	2
3	0.3400	0.3468	0.3535	0.3603	0.3672	0.3741	3
4	0.2563	0.2626	0.2690	0.2755	0.2820	0.2886	4
5	0.2060	0.2122	0.2184	0.2246	0.2310	0.2374	5
6	0.1725	0.1785	0.1846	0.1908	0.1970	0.2034	6
7	0.1486	0.1545	0.1605	0.1666	0.1728	0.1791	7
8	0.1307	0.1365	0.1425	0.1485	0.1547	0.1610	8
9	0.1167	0.1225	0.1284	0.1345	0.1407	0.1470	9
10	0.1056	0.1113	0.1172	0.1233	0.1295	0.1359	10
11	0.0965	0.1022	0.1081	0.1141	0.1204	0.1268	11
12	0.0888	0.0946	0.1005	0.1066	0.1128	0.1193	12
13	0.0824	0.0881	0.0940	0.1001	0.1065	0.1130	13
14	0.0769	0.0826	0.0885	0.0947	0.1010	0.1076	14
15	0.0721	0.0778	0.0838	0.0899	0.0963	0.1030	15
16	0.0679	0.0737	0.0796	0.0858	0.0923	0.0990	16
17	0.0643	0.0700	0.0760	0.0822	0.0887	0.0954	17
18	0.0610	0.0667	0.0727	0.0790	0.0855	0.0924	18
19	0.0581	0.0638	0.0698	0.0761	0.0827	0.0896	19
20	0.0554	0.0612	0.0672	0.0736	0.0802	0.0872	20
21	0.0530	0.0588	0.0649	0.0713	0.0780	0.0850	21
22	0.0509	0.0566	0.0627	0.0692	0.0760	0.0830	22
23	0.0489	0.0547	0.0608	0.0673	0.0741	0.0813	23
24	0.0471	0.0529	0.0590	0.0656	0.0725	0.0797	24
25	0.0454	0.0512	0.0574	0.0640	0.0710	0.0782	25
26	0.0439	0.0497	0.0559	0.0626	0.0696	0.0769	26
27	0.0424	0.0483	0.0546	0.0612	0.0683	0.0757	27
28	0.0411	0.0470	0.0533	0.0600	0.0671	0.0746	28
29	0.0399	0.0458	0.0521	0.0589	0.0660	0.0736	29
30	0.0387	0.0446	0.0510	0.0578	0.0651	0.0726	30
31	0.0377	0.0436	0.0500	0.0569	0.0641	0.0718	31
32	0.0367	0.0426	0.0490	0.0559	0.0633	0.0710	32
33	0.0357	0.0417	0.0482	0.0551	0.0625	0.0703	33
34	0.0348	0.0408	0.0473	0.0543	0.0618	0.0696	34
35	0.0340	0.0400	0.0465	0.0536	0.0611	0.0690	35
40	0.0305	0.0366	0.0433	0.0505	0.0583	0.0665	40
45	0.0277	0.0339	0.0408	0.0483	0.0563	0.0647	45
50	0.0255	0.0318	0.0389	0.0466	0.0548	0.0634	50
55	0.0237	0.0301	0.0373	0.0452	0.0537	0.0625	55
60	0.0222	0.0288	0.0361	0.0442	0.0528	0.0619	60

APPENDIX TABLE IV (*continued*): Capital Recovery or Amortization Factors (Annuity whose present value at compound interest is 1)

$$\frac{1}{a} = \frac{i}{1 - (1 + i)^{-n}}$$

n	7%	8%	9%	10%	11%	12%	n
1	1.0700	1.0800	1.0900	1.1000	1.1100	1.1200	1
2	0.5531	0.5608	0.5685	0.5762	0.5839	0.5917	2
3	0.3811	0.3880	0.3951	0.4021	0.4092	0.4163	3
4	0.2952	0.3019	0.3087	0.3155	0.3223	0.3292	4
5	0.2439	0.2505	0.2571	0.2638	0.2706	0.2774	5
6	0.2098	0.2163	0.2229	0.2296	0.2364	0.2432	6
7	0.1856	0.1921	0.1987	0.2054	0.2122	0.2191	7
8	0.1675	0.1740	0.1807	0.1874	0.1943	0.2013	8
9	0.1535	0.1601	0.1668	0.1736	0.1806	0.1877	9
10	0.1424	0.1490	0.1558	0.1627	0.1698	0.1770	10
11	0.1334	0.1401	0.1469	0.1540	0.1611	0.1684	11
12	0.1259	0.1327	0.1397	0.1468	0.1540	0.1614	12
13	0.1197	0.1265	0.1336	0.1408	0.1482	0.1557	13
14	0.1143	0.1213	0.1284	0.1357	0.1432	0.1509	14
15	0.1098	0.1168	0.1241	0.1315	0.1391	0.1468	15
16	0.1059	0.1130	0.1203	0.1278	0.1355	0.1434	16
17	0.1024	0.1096	0.1170	0.1247	0.1325	0.1405	17
18	0.0994	0.1067	0.1142	0.1219	0.1298	0.1379	18
19	0.0968	0.1041	0.1117	0.1195	0.1276	0.1358	19
20	0.0944	0.1019	0.1095	0.1175	0.1256	0.1339	20
21	0.0923	0.0998	0.1076	0.1156	0.1238	0.1322	21
22	0.0904	0.0980	0.1059	0.1140	0.1223	0.1308	22
23	0.0887	0.0964	0.1044	0.1126	0.1210	0.1296	23
24	0.0872	0.0950	0.1030	0.1113	0.1198	0.1285	24
25	0.0858	0.0937	0.1018	0.1102	0.1187	0.1275	25
26	0.0846	0.0925	0.1007	0.1092	0.1178	0.1267	26
27	0.0834	0.0914	0.0997	0.1083	0.1170	0.1259	27
28	0.0824	0.0905	0.0989	0.1075	0.1163	0.1252	28
29	0.0814	0.0896	0.0981	0.1067	0.1156	0.1247	29
30	0.0806	0.0888	0.0973	0.1061	0.1150	0.1241	30
31	0.0798	0.0881	0.0967	0.1055	0.1145	0.1237	31
32	0.0791	0.0875	0.0961	0.1050	0.1140	0.1233	32
33	0.0784	0.0869	0.0956	0.1045	0.1136	0.1229	33
34	0.0778	0.0863	0.0951	0.1041	0.1133	0.1226	34
35	0.0772	0.0858	0.0946	0.1037	0.1129	0.1223	35
40	0.0750	0.0839	0.0930	0.1023	0.1117	0.1213	40
45	0.0735	0.0826	0.0919	0.1014	0.1110	0.1207	45
50	0.0725	0.0817	0.0912	0.1009	0.1106	0.1204	50
55	0.0717	0.0812	0.0908	0.1005	0.1104	0.1202	55
60	0.0712	0.0808	0.0905	0.1003	0.1102	0.1201	60

APPENDIX TABLE IV (*continued*): Capital Recovery or Amortization Factors (Annuity whose present value at compound interest is 1)

$$\frac{1}{a} = \frac{i}{1 - (1 + i)^{-n}}$$

n	13%	14%	15%	16%	17%	18%	n
1	1.1300	1.1400	1.1500	1.1600	1.1700	1.1800	1
2	0.5995	0.6073	0.6151	0.6230	0.6308	0.6387	2
3	0.4235	0.4307	0.4380	0.4453	0.4526	0.4599	3
4	0.3362	0.3432	0.3503	0.3574	0.3645	0.3717	4
5	0.2843	0.2913	0.2983	0.3054	0.3126	0.3198	5
6	0.2502	0.2572	0.2642	0.2714	0.2786	0.2859	6
7	0.2261	0.2332	0.2404	0.2476	0.2549	0.2624	7
8	0.2084	0.2156	0.2229	0.2302	0.2377	0.2452	8
9	0.1949	0.2022	0.2096	0.2171	0.2247	0.2324	9
10	0.1843	0.1917	0.1993	0.2069	0.2147	0.2225	10
11	0.1758	0.1834	0.1911	0.1989	0.2068	0.2148	11
12	0.1690	0.1767	0.1845	0.1924	0.2005	0.2086	12
13	0.1634	0.1712	0.1791	0.1872	0.1954	0.2037	13
14	0.1587	0.1666	0.1747	0.1829	0.1912	0.1997	14
15	0.1547	0.1628	0.1710	0.1794	0.1878	0.1964	15
16	0.1514	0.1596	0.1679	0.1764	0.1850	0.1937	16
17	0.1486	0.1569	0.1654	0.1740	0.1827	0.1915	17
18	0.1462	0.1546	0.1632	0.1719	0.1807	0.1896	18
19	0.1441	0.1527	0.1613	0.1701	0.1791	0.1881	19
20	0.1424	0.1510	0.1598	0.1687	0.1777	0.1868	20
21	0.1408	0.1495	0.1584	0.1674	0.1765	0.1857	21
22	0.1395	0.1483	0.1573	0.1664	0.1756	0.1848	22
23	0.1383	0.1472	0.1563	0.1654	0.1747	0.1841	23
24	0.1373	0.1463	0.1554	0.1647	0.1740	0.1835	24
25	0.1364	0.1455	0.1547	0.1640	0.1734	0.1829	25
26	0.1357	0.1448	0.1541	0.1634	0.1729	0.1825	26
27	0.1350	0.1442	0.1535	0.1630	0.1725	0.1821	27
28	0.1344	0.1437	0.1531	0.1625	0.1721	0.1818	28
29	0.1339	0.1432	0.1527	0.1622	0.1718	0.1815	29
30	0.1334	0.1428	0.1523	0.1619	0.1715	0.1813	30
31	0.1330	0.1425	0.1520	0.1616	0.1713	0.1811	31
32	0.1327	0.1421	0.1517	0.1614	0.1711	0.1809	32
33	0.1323	0.1419	0.1515	0.1612	0.1710	0.1808	33
34	0.1321	0.1416	0.1513	0.1610	0.1708	0.1806	34
35	0.1318	0.1414	0.1511	0.1609	0.1707	0.1806	35
40	0.1310	0.1407	0.1506	0.1604	0.1703	0.1802	40
45	0.1305	0.1404	0.1503	0.1602	0.1701	0.1801	45
50	0.1303	0.1402	0.1501	0.1601	0.1701	0.1800	50
55	0.1302	0.1401	0.1501	0.1600	0.1700	0.1800	55
60	0.1301	0.1401	0.1500	0.1600	0.1700	0.1800	60

APPENDIX TABLE IV (*continued*): Capital Recovery or Amortization Factors (Annuity whose present value at compound interest is 1)

$$\frac{1}{a} = \frac{i}{1 - (1 + i)^{-n}}$$

n	19%	20%	21%	22%	23%	24%	n
1	1.1900	1.2000	1.2100	1.2200	1.2300	1.2400	1
2	0.6466	0.6545	0.6625	0.6705	0.6784	0.6864	2
3	0.4673	0.4747	0.4822	0.4897	0.4972	0.5047	3
4	0.3790	0.3863	0.3936	0.4010	0.4085	0.4159	4
5	0.3271	0.3344	0.3418	0.3492	0.3567	0.3642	5
6	0.2933	0.3007	0.3082	0.3158	0.3234	0.3311	6
7	0.2699	0.2774	0.2851	0.2928	0.3006	0.3084	7
8	0.2529	0.2606	0.2684	0.2763	0.2843	0.2923	8
9	0.2402	0.2481	0.2561	0.2641	0.2722	0.2805	9
10	0.2305	0.2385	0.2467	0.2549	0.2632	0.2716	10
11	0.2229	0.2311	0.2394	0.2478	0.2563	0.2649	11
12	0.2169	0.2253	0.2337	0.2423	0.2509	0.2596	12
13	0.2121	0.2206	0.2292	0.2379	0.2467	0.2556	13
14	0.2082	0.2169	0.2256	0.2345	0.2434	0.2524	14
15	0.2051	0.2139	0.2228	0.2317	0.2408	0.2499	15
16	0.2025	0.2114	0.2204	0.2295	0.2387	0.2479	16
17	0.2004	0.2094	0.2186	0.2278	0.2370	0.2464	17
18	0.1987	0.2078	0.2170	0.2263	0.2357	0.2451	18
19	0.1972	0.2065	0.2158	0.2251	0.2346	0.2441	19
20	0.1960	0.2054	0.2147	0.2242	0.2337	0.2433	20
21	0.1951	0.2044	0.2139	0.2234	0.2330	0.2426	21
22	0.1942	0.2037	0.2132	0.2228	0.2324	0.2421	22
23	0.1935	0.2031	0.2127	0.2223	0.2320	0.2417	23
24	0.1930	0.2025	0.2122	0.2219	0.2316	0.2414	24
25	0.1925	0.2021	0.2118	0.2215	0.2313	0.2411	25
26	0.1921	0.2018	0.2115	0.2213	0.2311	0.2409	26
27	0.1917	0.2015	0.2112	0.2210	0.2309	0.2407	27
28	0.1915	0.2012	0.2110	0.2208	0.2307	0.2406	28
29	0.1912	0.2010	0.2108	0.2207	0.2306	0.2405	29
30	0.1910	0.2008	0.2107	0.2206	0.2305	0.2404	30
31	0.1909	0.2007	0.2106	0.2205	0.2304	0.2403	31
32	0.1907	0.2006	0.2105	0.2204	0.2303	0.2402	32
33	0.1906	0.2005	0.2104	0.2203	0.2302	0.2402	33
34	0.1905	0.2004	0.2103	0.2203	0.2302	0.2402	34
35	0.1904	0.2003	0.2103	0.2202	0.2302	0.2401	35
40	0.1902	0.2001	0.2101	0.2201	0.2301	0.2400	40
45	0.1901	0.2001	0.2100	0.2200	0.2300	0.2400	45
50	0.1900	0.2000	0.2100	0.2200	0.2300	0.2400	50
55	0.1900	0.2000	0.2100	0.2200	0.2300	0.2400	55
60	0.1900	0.2000	0.2100	0.2200	0.2300	0.2400	60

INDEX